ROBERT B. BURCKEL

AN INTRODUCTION TO CLASSICAL COMPLEX ANALYSIS

Vol. 1

MATHEMATISCHE REIHE
BAND 64

LEHRBÜCHER UND MONOGRAPHIEN
AUS DEM GEBIETE DER EXAKTEN WISSENSCHAFTEN

An Introduction to Classical Complex Analysis

Vol. 1

by Robert B. Burckel
Kansas State University

1979
BIRKHÄUSER VERLAG BASEL
UND STUTTGART

CIP-Kurztitelaufnahme der Deutschen Bibliothek

Burckel, Robert B.:
An introduction to classical complex analysis /
by Robert B. Burckel. - Basel, Stuttgart:
Birkhäuser.

Vol. 1. - 1979.
(Lehrbücher und Monographien aus dem Gebiete der
exakten Wissenschaften: Math. Reihe; Bd. 64)

North and South America Edition published by
ACADEMIC PRESS, INC.
111 Fifth Avenue, New York, New York 10003
(Pure and Applied Mathematics, A Series of Monographs and Textbooks, Volume 82)
ISBN-13: 978-3-0348-9376-3 e-ISBN-13: 978-3-0348-9374-9
DOI: 10.1007/978-3-0348-9374-9
Library of Congress Catalog Card Number 78-67403

Contents

Volume I

PREFACE 9

Chapter 0
PREREQUISITES AND PRELIMINARIES 13
§ 1 Set Theory 13
§ 2 Algebra 14
§ 3 The Battlefield 14
§ 4 Metric Spaces 15
§ 5 Limsup and All That 18
§ 6 Continuous Functions 20
§ 7 Calculus 21

Chapter I
CURVES, CONNECTEDNESS AND CONVEXITY 22
§ 1 Elementary Results on Connectedness 22
§ 2 Connectedness of Intervals, Curves and Convex Sets 23
§ 3 The Basic Connectedness Lemma 28
§ 4 Components and Compact Exhaustions 29
§ 5 Connectivity of a Set 33
§ 6 Extension Theorems 37
Notes to Chapter I 39

Chapter II
(COMPLEX) DERIVATIVE AND (CURVILINEAR) INTEGRALS 41
§ 1 Holomorphic and Harmonic Functions 41
§ 2 Integrals along Curves 44
§ 3 Differentiating under the Integral 47
§ 4 A Useful Sufficient Condition for Differentiability 49
Notes to Chapter II 50

Chapter III
POWER SERIES AND THE EXPONENTIAL FUNCTION 53
§ 1 Introduction 53
§ 2 Power Series 53
§ 3 The Complex Exponential Function 60
§ 4 Bernoulli Polynomials, Numbers and Functions 73
§ 5 Cauchy's Theorem Adumbrated 77
§ 6 Holomorphic Logarithms Previewed 78
Notes to Chapter III 80

Chapter IV
THE INDEX AND SOME PLANE TOPOLOGY 83
§ 1 Introduction 83

§ 2 Curves Winding around Points 83
§ 3 Homotopy and the Index 90
§ 4 Existence of Continuous Logarithms 92
§ 5 The Jordan Curve Theorem 102
§ 6 Applications of the Foregoing Technology 106
§ 7 Continuous and Holomorphic Logarithms in Open Sets 111
§ 8 Simple Connectivity for Open Sets 113
Notes to Chapter IV 115

Chapter V
CONSEQUENCES OF THE CAUCHY–GOURSAT THEOREM—MAXIMUM PRINCIPLES AND THE LOCAL THEORY 120
§ 1 Goursat's Lemma and Cauchy's Theorem for Starlike Regions 120
§ 2 Maximum Principles 127
§ 3 The Dirichlet Problem for Disks 134
§ 4 Existence of Power Series Expansions 144
§ 5 Harmonic Majorization 151
§ 6 Uniqueness Theorems 165
§ 7 Local Theory 172
Notes to Chapter V 183

Chapter VI
SCHWARZ' LEMMA AND ITS MANY APPLICATIONS 191
§ 1 Schwarz' Lemma and the Conformal Automorphisms of Disks 191
§ 2 Many-to-one Maps of Disks onto Disks 197
§ 3 Applications to Half-planes, Strips and Annuli 198
§ 4 The Theorem of Carathéodory, Julia, Wolff, *et al.* 203
§ 5 Subordination 207
Notes to Chapter VI 215

Chapter VII
CONVERGENT SEQUENCES OF HOLOMORPHIC FUNCTIONS 218
§ 1 Convergence in $H(U)$ 218
§ 2 Applications of the Convergence Theorems; Boundedness Criteria 228
§ 3 Prescribing Zeros 237
§ 4 Elementary Iteration Theory 242
Notes to Chapter VII 251

Chapter VIII
POLYNOMIAL AND RATIONAL APPROXIMATION—RUNGE THEORY 256
§ 1 The Basic Integral Representation Theorem 256
§ 2 Applications to Approximation 260
§ 3 Other Applications of the Integral Representation 265
§ 4 Some Special Kinds of Approximation 268
§ 5 Carleman's Approximation Theorem 273
§ 6 Harmonic Functions in a Half-plane 276
Notes to Chapter VIII 289

Chapter IX
THE RIEMANN MAPPING THEOREM 293
§ 1 Introduction 293
§ 2 The Proof of Carathéodory and Koebe 298
§ 3 Fejér and Riesz' Proof 303
§ 4 Boundary Behavior for Jordan Regions 303
§ 5 A Few Applications of the Osgood–Taylor–Carathéodory Theorem 310
§ 6 More on Jordan Regions and Boundary Behavior 315
§ 7 Harmonic Functions and the General Dirichlet Problem 322
§ 8 The Dirichlet Problem and the Riemann Mapping Theorem 333
Notes to Chapter IX 337

Chapter X
SIMPLE AND DOUBLE CONNECTIVITY 344
§ 1 Simple Connectivity 344
§ 2 Double Connectivity 348
Notes to Chapter X 355

Chapter XI
ISOLATED SINGULARITIES 359
§ 1 Laurent Series and Classification of Singularities 359
§ 2 Rational Functions 366
§ 3 Isolated Singularities on the Circle of Convergence 375
§ 4 The Residue Theorem and Some Applications 377
§ 5 Specifying Principal Parts—Mittag-Leffler's Theorem 390
§ 6 Meromorphic Functions 395
§ 7 Poisson's Formula in an Annulus and Isolated Singularities of Harmonic Functions 398
Notes to Chapter XI 406

Chapter XII
OMITTED VALUES AND NORMAL FAMILIES 411
§ 1 Logarithmic Means and Jensen's Inequality 411
§ 2 Miranda's Theorem 417
§ 3 Immediate Applications of Miranda 432
§ 4 Normal Families and Julia's Extension of Picard's Great Theorem 436
§ 5 Sectorial Limit Theorems 441
§ 6 Applications to Iteration Theory 450
§ 7 Ostrowski's Proof of Schottky's Theorem 451
Notes to Chapter XII 456

BIBLIOGRAPHY 462
NAME INDEX 544
SUBJECT INDEX 554
SYMBOL INDEX 568
SERIES SUMMED 569
INTEGRALS EVALUATED 570

Contents
Volume II

Chapter XIII
WERMER'S MAXIMALITY THEOREM AND ITS COROLLARIES

Chapter XIV
PHRAGMÉN–LINDELÖF THEORY

Chapter XV
TERATOLOGY

Chapter XVI
OVERCONVERGENCE AND NATURAL BOUNDARIES

Chapter XVII
THE GAP THEOREMS

Chapter XVIII
UNIVALENT FUNCTIONS—GEOMETRIC METHODS

Chapter XIX
APPLICATIONS TO BANACH ALGEBRA

BIBLIOGRAPHY

Preface

This book evolved from lectures at the University of Oregon and at Kansas State University. The subject matter is analytic functions of a single complex variable, a truly glorious area of mathematics, abounding in deep and beautiful theorems. It is written, for better or for worse, in a modified Landau "Satz-Beweis" style. Thus the book is almost wholly self-contained logically, but the role of the instructor in providing heuristics and motivation has not been usurped. I undertook the task partly, to quote Leo Rosten, in order that I might have the book; certainly there are many who could have done a far better job but who, deeply committed to research, have not found the time and energy to do so. On the other hand, a certain evangelical zeal motivates every author and I wanted to communicate some of these beautiful results to as large an audience as possible. To this end, the prerequisites are rather modest; they are detailed in Chapter 0 but consist mainly of an $\varepsilon - \delta$ familiarity with (and attitude toward) analysis, at about the level of Walter Rudin's classic *Principles of Mathematical Analysis*. (However, complete mastery of that book is by no means necessary for the prospective reader of this one.)

The material is divided into two volumes. Volume I, consisting of twelve chapters, covers basics, while Volume II is topical and its seven chapters are largely independent of one another. There is a large bibliography and chapter notes which are keyed to it. It is hoped that this will bring the novice into contact with the journal literature.

Cauchy's integral formula and the power series that flows from it are of course the basic first step in the study of complex analysis. The reader will not encounter them, however, until Chapter V. There are few geodesics in this book and subordinate themes are usually followed up where they are broached, which is as soon as they are relevant and the machinery needed to treat them is available. Often, however, special cases of major later themes are anticipated in exercises. But the main reason Cauchy's theorem is deferred is to allow an examination of the topology of the plane. In Chapters I and IV most of the deeper results like the Invariance of Domain and the Jordan Curve Theorem are proved. Thus, in particular, all the purely topological support material for the most general Cauchy theorem is disposed of at this point. I believe the reader will find Chapter IV gratifyingly elegant. Following Borsuk, Eilenberg and Čech, everything is based on the existence of continuous logarithms (i.e., for *cognoscenti*, on cohomology) and Jordan's theorem emerges surprisingly painlessly. No knowledge of algebraic topology is supposed nor is a long machinery-building prologue needed. In fact, the efficiency and elegance of the techniques here might even serve as a motivation to the student to explore algebraic topology *per se*. In Chapter IX Jordan's theorem is completed with Schönflies'. This is a by-product of an analysis of the boundary behavior of conformal maps on Jordan regions;

this deduction of Schönflies' theorem, even though it yields so much more, is still the most economical. In Chapter X a climax is reached when all the topological and analytic features of simple connectivity are shown to be equivalent for plane regions. There too the fundamental form of the most general doubly connected region is established.

In the meantime, harmonic and subharmonic functions, harmonic majorization, various maximum principles and the theory of convergence (including the results on induced convergence and the elementary theory of iteration) are developed. The polynomial approximation theorems of Weierstrass, Runge and Carleman are presented in Chapter VIII. Chapter IX also contains the Wiener–Perron–Kellogg solution of the Dirichlet problem for (bounded) simply-connected regions. The treatment of the mapping theorem there follows the beautiful "snuggling up" (to the boundary) method of Koebe and Carathéodory and the boundary treatment borrows ideas from several sources, mainly Lindelöf and Carathéodory. Chapter XI is devoted to isolated singularities of holomorphic and harmonic functions, including Mittag-Leffler's theorems, the Residue Theorem and its applications (e.g., Gauss sums and a Theta formula), rationality criteria, singularities on the circle of convergence, and the ideal theory of the ring of holomorphic functions. In Chapter XII Ostrowski's beautiful proof, with explicit constants, of Schottky's remarkable theorem is presented. In fact, I go all out and give Miranda's extension of Schottky's theorem, following Valiron; the reader thereby experiences an introduction to the methods of Nevanlinna theory. From this height we rappel through the Pólya–Saxer–Csillag theorem, normal families, Julia's extension of Picard's Great Theorem and the sectorial limit theorems of Montel, Lindelöf, Hardy, Ingham and Pólya.

In Volume II gap theorems, including several proofs of Fabry's, Carlson's theorem on power series with integer coefficients and Duffin and Schaeffer's generalization of Szegö's theorem on power series with only finitely many different coefficients are presented. Chapter XV is devoted to Bohr's construction of a non-zero entire function convergent to 0 along every algebraic curve to infinity, a power series absolutely convergent in the unit disk which maps $\{|z| = 1\}$ onto $[0, 1] \times [0, 1]$, and other exotica. Included are some startling Baire Category results on overconvergence and natural boundaries. In Chapter XVIII a little more background is expected of the reader: Double integrals (but only over open or compact sets with continuous integrands), Fubini's theorem and the Jacobian change-of-variable formula are used as the fruitful area method of Bieberbach is exploited. Among the other highlights there are the various coefficient estimates of Bieberbach, Littlewood, Nevanlinna and Privalov, a careful treatment of the equivalence of the geometric and analytic notions of convex and starlike functions and the extension of the Riemann mapping theorem by de Possel and Grötzsch affirming that any region is conformal to a plane with parallel slits removed. Chapter XIX is addressed to the reader who has some knowledge of Banach algebra and functional analysis (albeit the actual knowledge presupposed is minimal). There myriad applications of complex analysis to those subjects are explored, including a careful treatment of the holomorphic functional calculus, criteria for commutativity (of an algebra) and multiplicativity (of a linear functional), the Müntz–Szász closure theorem and the metric definition of hermiticity, as well as several applications to operators on Hilbert space. Because a prerequisite knowledge of measure theory is being eschewed, one rich and beautiful analytic topic is regrettably absent: the theory of the Hardy spaces on the unit disk or the upper half-plane. However, this matter is already extremely well exposed in the extant literature (cf. HOFFMAN [1962], chapter 17 of RUDIN [1974] and chapter 19 of HILLE [1962]).

I have borrowed freely from many of the great treatises and texts, but four especially have influenced me: my debt, in content, spirit, and method, to the books of SAKS and ZYGMUND [1971], RUDIN [1974] and LANDAU [1929a] is enormous (and obvious to the knowledgeable reader) and I have made frequent forays into what is perhaps the greatest classic of all, PÓLYA and SZEGÖ [1964].

The Bibliography is fairly extensive, certainly not exhaustive. Not every item in it is mentioned in the notes and comments, but most are related to material discussed in the text and could be profitably looked at, even by the beginner; especially this applies to papers from the *American Mathematical Monthly*, the *Mathematics Magazine*, *Elemente der Mathematik* and *l'Enseignement Mathématique*. On the other hand, not every paper I looked at and perhaps not every paper I borrowed an idea from appears in the bibliography—*ars longa, vita brevis.*

The instructor using this book as a text or the reader pursuing it on his own should be alerted that most of its "exercises" are really just continuations of the text, with slightly less fulsome proofs. Sometimes calculations are relegated there, as are special results which will be needed as mortar later in the structure; in any case, most are teleological and there are no *Fingerübungen*. The serious reader will therefore want to supplement his study here with one of the many available problem collections listed in the bibliography, e.g., PÓLYA and SZEGÖ [1972], [1976], KNOPP [1948], [1953], KRZYŻ [1971], VOLKOVYSKII *et al.* [1965], FEYEL and PRADELLE [1973] or EVGRAFOV *et al.* [1974].

In researching the literature for this book, I was of course constantly encountering the giants (Landau, Pólya, Nevanlinna, Lindelöf, Bieberbach, Ostrowski, Carathéodory, to mention only a few at random) and was left with a permanent sense of awe of them. I hope that this exposition of their work conveys that fact and even induces a similar awe in the reader.

It is a pleasure to thank the several ladies who worked on typing this monstrosity, in various stages of its evolution: Judy Bernhart, Lynn Caldwell, Phyllis

Pickel, Judy Toburen, Marie Davis, Lorraine Douglas, Gail Buckner, Marlyn Logan and Elisabeth Quitsch. I thank my wife for help with some of the foreign language literature and the staff of William Clowes and Sons for their careful type-setting of a difficult manuscript. Special thanks go to Mrs. Ellyn Taylor and her staff at the Kansas State University library who secured for me a steady flow of interlibrary loans, principally from the non-pareil Linda Hall Science Library in Kansas City; and I express deep gratitude to that institution for its generous lending policy. The hospitality and good working conditions extended to me during the last phase of writing by the mathematics department of the University of the Saarland, particularly by my host Prof. Dr. Heinz König, are also gratefully and happily acknowledged. The extraordinary patience and accommodations of Mr. C. Einsele and his staff at Birkhäuser in the face of my numerous late corrections cannot go unacknowledged. Finally I want to thank Prof. Dr. Alexander Ostrowski for valuable criticisms of the text and for his kind help in placing my book with Birkhäuser Verlag and Prof. Edwin Hewitt for encouragement early in the project.

The author welcomes correspondence from readers with criticisms and suggestions, especially in regard to errors (mathematical, historical or typographical) and obscurities, but also concerning better proofs, overlooked references, gems that could be profitably inserted here or there, and even a better ordering or presentation of the material (consistent with the obvious overall philosophy of the work). In the event of a second edition I will incorporate and acknowledge as many of these improvements as possible.

R. B. B.

Chapter 0
Prerequisites and Preliminaries

As indicated in the Preface, only very modest specific knowledge is required of the prospective reader. Those (salient) *facts* that will be used in the text without proof I will attempt to state fully below with references. What is really expected of the reader is a level of mathematical maturity comparable to that fostered by books like STROMBERG [1980], RUDIN [1976] and LANDAU [1950], but by no means a mastery of the contents of these books. A junior or senior mathematics major at any good college or university nowadays should therefore be adequately equipped.

Caveat: No one should read this chapter linearly (*who could survive it?*), *but only refer to it as the need arises.*

§ 1 Set Theory

We'll be using the usual language and notation of set theory: $\in$, $\forall$, $\exists$, $\cup$, $\cap$, $\subset$, etc. When $P(x)$ is a proposition about x and S is a set, the set of all x in S for which $P(x)$ is true is denoted by $\{x \in S: P(x)\}$. When $P(x)$ is the statement "x does not belong to A," then $\{x \in S: P(x)\}$ is denoted by $S \backslash A$. As usual, the empty set is denoted by $\varnothing$. The reader is assumed to have a modest acquaintance with naive (i.e., couched in a language and logic which have not been fully formalized), axiomatic set theory. Primarily he should know the bare rudiments of cardinal arithmetic. For over a decade the standard testament at this level has been HALMOS [1960]. This pretty little basic book is warmly recommended to the reader, even though in the present work only a small fraction of its contents will be utilized. Consult that work in particular for the standard terminology like map, function, injective, bijective, surjective, one-to-one, onto, countable, etc.

The rubric $f: S \to T$ or $S \xrightarrow{f} T$ means that f is a function with domain S and range a part of T. If $A \subset S$, then $f(A) = \{f(x): x \in A\}$ is called the *image of A under f* and $f|A$ or $f|_A$ traditionally denotes the *restriction of f to A*. If $B \subset T$ (and not necessarily $B \subset f(S)$ or even $B \cap f(S) \neq \varnothing$), then $f^{-1}(B) = \{x \in S: f(x) \in B\}$. This is called the *inverse image of B under f*; it may be void. When B is a singleton, $B = \{t\}$, we write simply $f^{-1}(t)$. When f is one-to-one, we use f^{-1} to denote the inverse function of f (and the above notations are consistent). In case of zero-free complex-valued functions f, we write $\frac{1}{f}$ or $1/f$ for the function whose value at each x is $1/f(x)$. We will never use f^{-1} for the latter. The composition of two functions f and g whose domains and images are properly related is denoted by $f \circ g$. The cartesian product of the sets A and B is denoted by $A \times B$. The cross is never used to signify multiplication, which is usually indicated by juxtaposition. Occasionally in long formulas dots are used to indicate multiplication and dots are used with the inner product in euclidean spaces. (See below.)

If $\mathscr{S}$ is a family of sets, we write $\bigcup \mathscr{S}$ (or $\bigcap \mathscr{S}$) for the union (or intersection) of all the sets in $\mathscr{S}$. If $\mathscr{S} = \{S_\alpha : \alpha \in I\}$, we write $\bigcup_{\alpha \in I} S_\alpha$ (or $\bigcap_{\alpha \in I} S_\alpha$). When I is the integers or the natural numbers, we write $\bigcup_{n=-\infty}^{\infty} S_n$ or $\bigcup_{n=1}^{\infty} S_n$, etc. Other obvious variations on this notation are also employed. E.g., $\bigcup_{P(\alpha)} S_\alpha = \bigcup \{S_\alpha : P(\alpha)\}$ for the union of all the S_α such that the statement $P(\alpha)$ is true. We follow the convention that the union of a void family of subsets of a set S is $\varnothing$ and the intersection of a void family of subsets of S is all of S.

§ 2 Algebra

I use freely the basic language of algebra, like semigroup, group, ring, algebra, field, ideal, integral domain (for which consult any contemporary algebra text) but almost no facts from algebra. The third isomorphism theorem for (commutative) groups is used implicitly once (Chapter IV), as is the fact that a permutation is a product of disjoint cycles (Chapter VIII). I assume that the reader has a passing acquaintance with vector spaces and their linear transformations (matrices). Once (Chapter XI) I invoke the row-rank equals column-rank theorem and from time to time a dimension argument intervenes, often implicitly. A few elementary properties of determinants are used in § 3 of Chapter XI.

§ 3 The Battlefield

$\mathbb{N}$ shall denote the natural numbers: $1, 2, 3, \ldots$, $\mathbb{Z}$ (for *Zahl*) the integers: $\mathbb{N} \cup \{0\} \cup -\mathbb{N}$, $\mathbb{Q}$ (for quotient) the rational numbers: $\{m/n : m, n \in \mathbb{Z}, n \neq 0\}$. We use $\mathbb{R}$ to denote the real numbers and $\mathbb{C}$ to denote the complex numbers. The latter is just the cartesian product $\mathbb{R} \times \mathbb{R}$ with coordinate-wise addition and the multiplication $(a, b)(c, d) = (ac - bd, ad + bc)$. That is, the multiplication is the one determined by the decree that $(0, 1)$ have square equal to $(-1, 0)$, that the subset $\{(x, 0) : x \in \mathbb{R}\}$ of $\mathbb{C}$ multiply in the usual way, and that the multiplication throughout $\mathbb{C}$ obey the field axioms (commutativity, associativity, distributivity). $\mathbb{C}$ is thus constructed from $\mathbb{R}$, which in turn is constructed from $\mathbb{Q}$ via the Dedekind cut (see RUDIN [1976], chapter 1). Of course, $\mathbb{Q}$ is manufactured from $\mathbb{Z}$, which comes from $\mathbb{N}$ and the latter can be brought into being from set theory and a "successor" function. For the (consummate) details of the whole *Aufbau* see LANDAU [1951].

The order relation in $\mathbb{N}$ is transmitted up through $\mathbb{Z}$, $\mathbb{Q}$ and into $\mathbb{R}$ (LANDAU, *loc. cit.*) We write $\mathbb{Z}^+$, $\mathbb{Q}^+$, $\mathbb{R}^+$ for the non-negative elements of $\mathbb{Z}$, $\mathbb{Q}$, $\mathbb{R}$, respectively. For $a, b \in \mathbb{R}$ we write $a \leq b$ and $b \geq a$ if $b - a \in \mathbb{R}^+$, $a < b$ and $b > a$ if $b - a \in \mathbb{R}^+ \backslash \{0\}$, i.e., if $b - a$ is *positive*. When $a \leq b$ we write

$[a, b] = \{x \in \mathbb{R} : a \leq x \leq b\}$, called a closed interval,

$(a, b) = \{x \in \mathbb{R} : a < x < b\}$, called an open interval,

$[a, b) = \{x \in \mathbb{R}: a \leq x < b\}$, called a half-open (on the right) interval,
$(a, b] = \{x \in \mathbb{R}: a < x \leq b\}$, called a half-open (on the left) interval.

In each case a is called the *left endpoint*, b the *right endpoint* of the interval. There is no order in $\mathbb{C}$ but an interval notation is handy there too and will be introduced in Chapter I. We rely on context to resolve the ambiguity in the notation (a, b)—is it an interval or an ordered pair of real numbers? The problem seldom (never?) arises because we don't often use ordered pair notation for complex numbers and then not in the presence of intervals.

For any $a, b \in \mathbb{R}$ we also write

$$[a, \infty) = \{x \in \mathbb{R}: a \leq x\}$$
$$(a, \infty) = \{x \in \mathbb{R}: a < x\}$$
$$(-\infty, b] = \{x \in \mathbb{R}: x \leq b\}$$
$$(-\infty, b) = \{x \in \mathbb{R}: x < b\},$$

with the obvious verbalizations. [No ontological significance attaches to the symbols ∞, $-\infty$ here (as yet).]

For any $z = (x, y) \in \mathbb{C}$ we write $\bar{z}$ for $(x, -y)$ and call it the *complex conjugate* of z. Also $|z|$ is $\sqrt{z\bar{z}}$, the *non-negative* square root of $z\bar{z} = x^2 + y^2$. (See RUDIN *loc. cit.* for existence.) This number is called the *modulus* or the *absolute value* of z. x is called the *real part* or *abscissa* of z, y the *imaginary part* or *ordinate* of z and we write $x = \operatorname{Re} z$, $y = \operatorname{Im} z$. The number $(0, 1)$ is called the *imaginary* unit and is universally denoted (following Euler) by i. I have striven to avoid ever using i as an index of summation, but once or twice it was ineluctable; no confusion should arise there however. [The terms "real" and "imaginary" as well as "complex" and "rational" are relics from the days when these concepts had not yet been formalized, were imperfectly understood, especially their ontological status, and were generally rather mysterious. They have been quietly de-mythologized over the years so there is now neither any need nor any zeal to eradicate these so characteristically unmathematical terms from the official lexicon. Of course, no scintilla of their common parlance meaning still attaches to them.]

§ 4 Metric Spaces

Recall that a *metric space* is a (non-empty) set X together with a *distance function* or *metric* $d: X \times X \to [0, \infty)$ which satisfies for all $x_1, x_2, x_3 \in X$

(1) $d(x_1, x_2) = d(x_2, x_1)$ (symmetry)

(2) $d(x_1, x_2) = 0 \Leftrightarrow x_1 = x_2$ (reflexivity)

(3) $d(x_1, x_3) \leq d(x_1, x_2) + d(x_2, x_3)$ (triangle inequality).

Formally, a metric space is a pair (X, d) as above but one speaks loosely of "the metric space X." If $x_0 \in X$ and $r \in [0, \infty)$, the set $\{x \in X: d(x, x_0) < r\}$ is called the *(open) ball of radius r and center* x_0. It is denoted by $B(x_0, r)$. A set in a metric space is *bounded* if it lies wholly in some ball, *open* if it is a union of open balls and *closed* if its complement is open. The *closure* (or *adherence*) of a set S is the intersection of all the closed supersets. It is evidently closed. We denote it by $\bar{S}$. The union of all the open subsets of S is called the *interior of* S and is denoted $\mathring{S}$; it is obviously an open set but may well be empty (see the convention at the end of § 1). Notice that $\mathring{S} = X\backslash(\overline{X\backslash S})$. A *neighborhood* of a point $x \in X$ is any set which contains some $B(x, r)$, $r > 0$. Generally N (neighborhood), V (*voisinage*) or U (*Umgebung*) are used to designate such sets. Most often they are open but this is not part of the definition here (though it is in some books).

The function $d(z, w) = |z - w|$, $z, w \in \mathbb{C}$ satisfies properties (1)–(3) above and is the standard (euclidean) metric in $\mathbb{C}$. Here balls are called *disks* (to be more suggestive) and we use $D(z_0, r)$ instead of $B(z_0, r)$, calling this set the *(open) disk of radius r and center* z_0. Notice that $\bar{D}(z_0, r) = \{z \in \mathbb{C}: |z - z_0| \le r\}$ (called the *closed disk of radius r and center* z_0). In a general metric space however $\bar{B}(x_0, r)$ lies in, but need not coincide with, $\{x \in X: d(x, x_0) \le r\}$. (E.g., $X = \mathbb{N}$, r a positive integer.) The set $C(z_0, r) = \{z \in \mathbb{C}: |z - z_0| = r\}$ is called the *circle of center* z_0, *radius r*. For $C(0, 1)$ the notation $\mathbb{T}$ (for torus) is also occasionally used. The set $\{z \in \mathbb{C}: r < |z - c| < R\}$ is called the *annulus of center c, inner radius r and outer radius R*. It is defined for every $c \in \mathbb{C}$ and $0 \le r < R$. The notation $A(c; r, R)$ is used for this set, or simply $A(r, R)$ if $c = 0$.

For each positive integer n we let $\mathbb{C}^n = \mathbb{C} \times \cdots \times \mathbb{C}$, the cartesian product of n copies of $\mathbb{C}$. This is a vector space over $\mathbb{C}$ (coordinate-wise operations). A conjugate bilinear form is defined in $\mathbb{C}^n$ by

$$(z_1, \ldots, z_n)\cdot(w_1, \ldots, w_n) = \sum_{j=1}^{n} z_j\bar{w}_j.$$

This is called the *scalar* or *inner* or *dot product* in $\mathbb{C}^n$ and together with its obvious linearity properties it satisfies the so-called *Cauchy–Schwarz Inequality*

$$(4) \qquad |(z_1, \ldots, z_n)\cdot(w_1, \ldots, w_n)|^2 \le \sum_{j=1}^{n} |z_j|^2 \sum_{k=1}^{n} |w_k|^2.$$

This product defines a *norm* or *length* function $\| \ \|: \mathbb{C}^n \to [0, \infty)$ by means of

$$\|z\| = \sqrt{z\cdot z}.$$

In this language (4) reads simply

$$(5) \qquad |z\cdot w| \le \|z\| \, \|w\| \quad \forall z, w \in \mathbb{C}^n.$$

[There is another potential notational conflict here when $n = 1$: $z\cdot z$ means $z\bar{z}$ according to the above, yet dot sometimes denotes ordinary field multiplication in $\mathbb{C}$. The problem really never arises because the inner product is used so seldom in this book.] The usual *euclidean metric* in $\mathbb{C}^n$ is that defined by the above norm:

$$d(z, w) = \|z - w\| \quad \forall z, w \in \mathbb{C}^n.$$

It is, of course, via (5) that we verify property (3) [(1) and (2) being obviously true for this d].

A *sequence* in a metric space X is (formally) a function x from $\mathbb{N}$ or $\mathbb{Z}^+$ into X, with $x(n)$ usually written as x_n. [Functions from $\mathbb{Z}$ to X are called *bilateral* sequences.] We denote the sequence x variously by $\{x_n\}_{n=1}^{\infty}$, $\{x_n\}_{n=0}^{\infty}$ (as the case may be) or simply $\{x_n\}$. A sequence $\{x_n\}$ is *Cauchy* or *fundamental* if for each $\varepsilon > 0$ there exists an $N(\varepsilon) \in \mathbb{N}$ such that $d(x_n, x_m) < \varepsilon$ whenever $n, m \geq N(\varepsilon)$. The sequence $\{x_n\}$ *converges* to $x \in X$ if for each $\varepsilon > 0$ there exists an $N(\varepsilon)$ such that $d(x_n, x) < \varepsilon$ for all $n \geq N(\varepsilon)$. We write $\lim_{n\to\infty} x_n = x$ and sometimes $x_n \to x$ *as* $n \to \infty$, or simply $x_n \to x$. The sequence $\{x_n\}$ is called *convergent* if such an x exists, *divergent* otherwise. (Notice that a convergent sequence is necessarily Cauchy and a Cauchy sequence is necessarily bounded.) A metric space (X, d) is called *complete* if every Cauchy sequence in it is convergent.

If $k\colon \mathbb{N} \to \mathbb{N}$ (or $\mathbb{Z}^+ \to \mathbb{Z}^+$) is any unbounded, order preserving (see § 5) function and x is a sequence, then the composite $x \circ k$ is called a *subsequence* of x. With the above conventions we write k_n for $k(n)$ and x_{k_n} for $x(k_n) = x(k(n))$.

An *isolated point* of a set S in a metric space X is a point $x \in X$ such that $B(x, r) \cap S = \{x\}$ for some $r > 0$. A *cluster point* of S is a point x such that every ball centered at x meets S. Synonyms for cluster point are *limit point* and *accumulation point*. (Sometimes, though not in this work, it is necessary to distinguish those cluster points of S which are not isolated points of S. One of the above synonyms is then charged with this assignment.) The cluster points of S are just the limits of sequences in S; the set of all of them coincides with $\bar{S}$. The *boundary* or *frontier* of S, denoted ∂S or *bdry*(S), is the set $\bar{S}\backslash\mathring{S} = \bar{S} \cap \overline{(X\backslash S)}$. It consists of those $x \in X$ every neighborhood of which meets both S and $X\backslash S$.

A metric space X is *compact* if whenever a class $\mathcal{O}$ of open subsets of X satisfies $X \subset \bigcup \mathcal{O}$ (we say $\mathcal{O}$ *covers* X), then $X \subset \bigcup \mathfrak{F}$ for some finite $\mathfrak{F} \subset \mathcal{O}$ (we say X *admits a finite subcover* from $\mathcal{O}$). This is equivalent to the condition that every sequence in X possess a convergent subsequence. (See RUDIN [1976], pp. 36–40.) If we take complements of open sets, the definition of compactness is that the space have the *finite intersection property*: Whenever $\mathscr{C}$ is a set of closed subsets of X such that $\bigcap \mathfrak{F} \neq \varnothing$ for each finite $\mathfrak{F} \subset \mathscr{C}$, then $\bigcap \mathscr{C} \neq \varnothing$.

If (X, d) is a metric space and $Y \subset X$, then $(Y, d|Y \times Y)$ is a metric space. Notice that for $y_0 \in Y$ the ball of center y_0 and radius r *in* Y is the intersection with Y of the corresponding ball in X. $S \subset Y$ is called *relatively* open (or closed) if it is open (or closed) in the metric space $(Y, d|Y \times Y)$. This is equivalent to $S = T \cap Y$ where T is open (or closed) in (X, d). Y is called a *subspace* of X. It follows that the metric space $(Y, d|Y \times Y)$ is compact if and only if every cover of Y by open subsets of X admits a finite subcover and from this it follows that Y must be closed if it is compact. A subset Y of X is called *pre-compact* or *conditionally compact* or *relatively compact* if $\overline{Y}$ is compact. The two most important features of $\mathbb{C}^n$ are that it is a complete metric space and a subset of it is compact if and only if closed and bounded (*Heine–Borel* theorem).

Though the terminology is paradoxical, there is nothing logically wrong with a subset Y of a metric space being both closed and open. (*Every* subset of the metric space $\mathbb{Z}$ has this property, for example.) For such a set the neologism *clopen* is employed, though many linguistic purists prefer the less colorful term closed-and-open (or open-and-closed).

A metric space (X, d) is called *connected* if the only clopen subsets of X are $\varnothing$ and X. A *disconnection* of a metric space X is a pair of disjoint sets X_1, X_2 which cover X and are each non-empty and open (equivalently, by looking at complements, each closed). Then, evidently, X is connected if and only if it possesses no disconnection. This important concept is explored in some depth in Chapter I and together with compactness is of fundamental importance for all that follows.

The most important *fact* about metric spaces which we will use in this book is the *Baire Category Theorem*: a complete metric space cannot be a countable union of closed subsets which each have empty interior. The profundity of this theorem lies not in the proof (which is exceedingly simple: see, e.g., RUDIN [1976], p. 82) but in the viewpoint that it engenders. It is a powerful tool for proving existence theorems, plenty of examples of which will occur in the sequel (see esp. Chapters XV and XVI). The name derives from the fact that a countable union of closed subsets of a metric space X, each having empty interior, is called a set of the *first* (*Baire*) *category* (in X). The complement of such a set is called *residual*, while a set whose closure has empty interior is called *nowhere dense* and a set whose closure is all of X is called *dense*. These notions are all *relative to* X. In this jargon the theorem says that a complete metric space is not of the first category in itself. As a corollary to the theorem itself or as a scholium to the canonical proof, we have the version which we shall use most often: In a complete metric space a countable intersection of open, dense subsets is a dense subset.

§ 5 Limsup and All That

In the Dedekind construction of $\mathbb{R}$ (i.e., that in LANDAU [1951]) the order relation is absolutely primal and one of the first things proved about $\mathbb{R}$ so constructed is its *order-completeness*: for every non-empty subset of $\mathbb{R}$ which is *bounded above* (i.e., lies in $(-\infty, b]$ for some $b \in \mathbb{R}$) there is a *least upper bound* (i.e., there is a least such b). When $\mathbb{R}$ is not manufactured but only axiomatically summoned, this property must be postulated: $\mathbb{R}$ is a (the unique, in some sense) field with an order relation that behaves in the expected way with respect to addition and multiplication and is complete in the above sense (so-called *Least Upper Bound Axiom*). In any case, it is from this fundamental property that the metric completeness of $\mathbb{R}$ and the compactness of closed-and-bounded sets (as well as the connectedness of intervals, to be proved in Chapter I) is derived. [The extension of these properties to $\mathbb{C}$, thence to $\mathbb{C}^n$, is then routine.]

The order relation in $\mathbb{R}$ can be rendered unconditionally complete by adjoining a least element $-\infty$ and a greatest element $+\infty$ (often written just ∞). These are two (distinct) points not in the set $\mathbb{R}$. The order relation in $\mathbb{R}$ is extended to $\overline{\mathbb{R}} = \mathbb{R} \cup \{\pm\infty\}$ by the decree

$$-\infty < x < +\infty \quad \text{for all } x \in \mathbb{R}.$$

The algebraic structure of $\mathbb{R}$ is not extended (or sometimes only partially extended) to $\overline{\mathbb{R}}$. In $\overline{\mathbb{R}}$ the above axiom reads: every subset of $\overline{\mathbb{R}}$ has a least upper bound. The abbreviation l.u.b.(S) for the least upper bound of a set S is common. The usual convention about vacuous hypotheses leads to l.u.b. $\varnothing = -\infty$. Another synonym for least upper bound is *supremum* and the supremum of S is noted $\sup(S)$. *Greatest lower bound*, g.l.b.(S), *infimum* and $\inf(S)$ are defined in the obvious way. A few self-explanatory notational variants prove quite convenient. E.g., $\sup_{\alpha\in I}\{x_\alpha\}$ for $\sup\{x_\alpha : \alpha \in I\}$ or $\sup_{n\geq k}\{x_n\}$ for $\sup\{x_n : n \geq k\}$.

For a subset S of a metric space (X, d) the element $\sup\{d(x, y): x, y \in S\}$ of $[0, +\infty]$ is called the *diameter of* S, sometimes noted diam(S). For any point $x \in X$ the *distance from* x *to* S is denoted $d(x, S)$ and is defined as

$$\inf\{d(x, y): y \in S\}.$$

(See 1.32.)

For a sequence $\{x_n\}_{n=1}^{\infty} \subset \overline{\mathbb{R}}$ we define $\overline{\lim}_{n\to\infty} x_n = \limsup_{n\to\infty} x_n$ (called the *limit superior* or *upper limit* of the sequence $\{x_n\}$) as $\inf_{n\geq 1}[\sup_{k\geq n}\{x_k\}]$. Analogously $\underline{\lim}_{n\to\infty} x_n = \liminf_{n\to\infty} x_n$ (called the *limit inferior* or *lower limit*) is defined as $\sup_{n\geq 1}[\inf_{k\geq n}\{x_k\}]$. If all $x_n \in \mathbb{R}$, we have

$$(*) \qquad \underline{\lim}_{n\to\infty} x_n \leq x \leq \overline{\lim}_{n\to\infty} x_n$$

for every subsequential limit x of the sequence $\{x_n\}$ and the sequence $\{x_n\}$ itself converges to a number $x \in \mathbb{R}$ if and only if equality holds in (*). With the obvious (based on order rather than distance) notions of sequential convergence in $\overline{\mathbb{R}}$, it follows easily from the definition that $\underline{\lim}_{n\to\infty} x_n$ is the smallest subsequential limit of $\{x_n\}$ in $\overline{\mathbb{R}}$ and $\overline{\lim}_{n\to\infty} x_n$ is the largest.

For a sequence $\{x_n\} \subset \mathbb{C}^N$ we write $\|x_n\| \to \infty$ if the sequence $\{\|x_n\|\}$ converges to $+\infty$ in $\overline{\mathbb{R}}$.

If $S \subset \overline{\mathbb{R}}$, $f: S \to \overline{\mathbb{R}}$ is called *non-decreasing* or *order-preserving* [*increasing*] if $f(x) \leq f(y)[f(x) < f(y)]$ whenever $x, y \in S$ and $x < y$. Sometimes *strictly in-*

creasing is used for increasing. *Non-increasing, decreasing,* etc. have then the obvious meanings. The term *monotonic* is generic for all these properties.

§ 6 Continuous Functions

If (X_1, d_1), (X_2, d_2) are metric spaces, $S \subset X_1, f: S \to X_2$, $a \in \bar{S}$, and $b \in X_2$, we say that $f(x)$ *converges to b as x converges to a* or that the *limit of f at a is b,* written $\lim_{x\to a} f(x) = b$, if for every $\varepsilon > 0$ there exists a $\delta = \delta(f, a, \varepsilon) > 0$ such that $d_2(f(x), b) < \varepsilon$ whenever $x \in S$ and $d_1(x, a) < \delta$. We say that *f is continuous at a* if $\lim_{x\to a} f(x) = f(a)$. Notice that $\lim_{x\to a} f(x) = b$ is equivalent to the implication

$$\{x_n\} \subset S \ \&\ \lim_{n\to\infty} x_n = a \Rightarrow \lim_{n\to\infty} f(x_n) = b.$$

For a subset T of S, we say *f is continuous on T* if f is continuous at each point of T. That f be continuous on S is equivalent to the requirement that $f^{-1}(U) \cap S$ be a relatively open subset of S whenever U is an open subset of X_2. If f is continuous on S and for each $\varepsilon > 0$ the $\delta(f, a, \varepsilon)$ can be chosen the same (say, $\delta(f, \varepsilon)$) for each $a \in S$, then f is called *uniformly continuous on S.* A family $\mathfrak{F}$ of continuous functions on S is called *uniformly equicontinuous on S* if in addition, for each $\varepsilon > 0$ and all $f \in \mathfrak{F}$ a common value $\delta(\varepsilon)$ can be selected for the $\delta(f, \varepsilon)$.

If (X, d) is a metric space, $S \subset X, f: S \to \bar{\mathbb{R}}$ and $a \in \bar{S}$, we define $\overline{\lim}_{x\to a} f(x) = \limsup_{x\to a} f(x)$ as $\inf_{r>0}[\sup f(B(a, r) \cap S)]$ with a similar definition for $\underline{\lim}_{x\to a} f(x) = \liminf_{x\to a} f(x)$. Evidently $\overline{\lim}_{x\to a} f(x)$ is the maximum of the elements $\overline{\lim}_{n\to\infty} f(x_n) \in \bar{\mathbb{R}}$ over all possible sequences $\{x_n\} \subset S$ which converge to a. When X is $\mathbb{R}$, $\overline{\lim}_{x\to\infty}$, $\overline{\lim}_{x\to-\infty}$, etc. have the obvious meanings.

Well-known elementary facts about continuous functions which we shall use: the composite of two continuous functions is continuous. The functions $x \to \|x\|$, $x \to \lambda x$ $(\lambda \in \mathbb{C})$ are continuous from $\mathbb{C}^n$ to $\mathbb{R}$ and $\mathbb{C}^n$, respectively; the functions $(x, y) \to x \cdot y$ and $(x, y) \to x + y$ are continuous from $\mathbb{C}^n \times \mathbb{C}^n$ to $\mathbb{C}$ and $\mathbb{C}^n$, respectively. The function $(x, y) \to xy$ is continuous from $\mathbb{C} \times \mathbb{C}$ to $\mathbb{C}$ and $z \to 1/z = z^{-1}$ is continuous from $\mathbb{C}\backslash\{0\}$ into $\mathbb{C}\backslash\{0\}$. With the composition law the continuity of all the elementary real- and complex-valued functions is thus assured.

The inverse of a continuous, one-to-one function need not be continuous (those functions for which the inverse *is* continuous are said to be *bicontinuous* and are called *homeomorphisms*), but when the domain and range are each open subsets of $\mathbb{R}$, or of $\mathbb{C}$, continuity of the inverse follows. (These non-trivial facts are proved in Chapter IV.)

The restriction of a continuous function to a subspace is continuous. However, if Y is a subspace of X and f is defined on X, then $f|_Y$ may be continuous at a point y of Y without f being continuous at y. On the other hand, if A and B are two closed subsets of X and $f|A, f|B$ are each continuous functions, then

$f|A \cup B$ is a continuous function. This elementary fact is used repeatedly in the sequel to build up a continuous function from continuous pieces.

On the line $\mathbb{R}$ there are the notions of *left-* and *right-hand continuity*. We use the notation $x \uparrow a$ or $x \to a^-$ to indicate x approaching a in $(-\infty, a)$, i.e., from the left. Similarly for $x \downarrow a$ or $x \to a^+$. Generally to indicate x approaching a from inside some set S (in any metric space) we write $\lim_{x \to a, x \in S}$.

The continuous image of a compact space is compact and the continuous image of a connected space is connected. (For a proof of the latter, see Chapter I.) When $X = [a, b]$ and $f(X) \subset \mathbb{R}$, this latter result amounts to the *Intermediate Value Theorem* for continuous functions: if c is a number between $f(a)$ and $f(b)$, then $c = f(x)$ for some x between a and b. A function which is continuous on a metric space is uniformly continuous on any compact subset.

If (X, d) is a metric space, we denote by $C(X)$ the set of all bounded continuous complex-valued functions on X, by $C_{\mathbb{R}}(X)$ the real-valued functions in $C(X)$. $C(X)$ is an algebra over $\mathbb{C}$ and $C_{\mathbb{R}}(X)$ is an algebra over $\mathbb{R}$, both with "pointwise" operations, e.g., fg is the function $(fg)(x) = f(x)g(x)$ $(\forall x \in X)$. The sets $C(X)$, $C_{\mathbb{R}}(X)$ are usually made into metric spaces by introducing the *supremum* or *uniform norm*

$$\|f\|_\infty = \|f\|_{C(X)} \stackrel{\text{def.}}{=} \sup|f(X)|$$

and the accompanying *uniform metric*

$$d_\infty(f, g) = \|f - g\|_\infty.$$

It is a fundamental and elementary fact that $C(X)$ is then complete and $C_{\mathbb{R}}(X)$ is a closed subset (hence also complete) of $C(X)$.

§ 7 Calculus

The elements of differential and integral calculus of real- and complex-valued functions on $\mathbb{R}$ as presented in chapters 5 and 6 of Rudin [1976] will be taken for granted. However, only (piecewise) continuous functions are ever integrated and neither Lebesgue nor Riemann–Stieltjes integrals are employed. We often cite the *Mean Value Theorem* for differentiable *real-valued* functions on a subinterval of $\mathbb{R}$, and the corollary thereof which says that a function with a positive derivative is increasing, as well as the *Chain Rule* for differentiating composites. (See 2.3 for the canonical proof of the latter.) *Integration-by-parts*, the integral analog of the Chain Rule, is used, as are the *Cauchy–Schwarz* inequality for integrals and the most elementary facts about the relation of uniform convergence to integration. Partial differentiation with respect to the kth coordinate variable is denoted by D_k and the ∂ notation is never used for differentiation. Ordinary differentiation on the line is usually denoted by $'$, rarely is d/dx used.

Chapter I
Curves, Connectedness and Convexity

We want to develop here as expeditiously as possible some routine facts about the topology of the plane centering on connectedness; these are essential in all that follows. A deeper exploration of the topology of the plane occurs in Chapter IV, after the exponential function has been developed (in Chapter III) and the useful concept of the index of a curve is available.

§ 1 Elementary Results on Connectedness

Theorem 1.1 *If* $\mathfrak{F}$ *is a collection of connected subsets of* $\mathbb{C}$ *and* $\bigcap \mathfrak{F}$ *is not void, then* $\bigcup \mathfrak{F}$ *is connected.*

Proof: Let $x \in \bigcap \mathfrak{F}$. It suffices to show that the only relatively clopen subset of $\bigcup \mathfrak{F}$ which contains x is $\bigcup \mathfrak{F}$. Let S be such a set. Then for each $F \in \mathfrak{F}$, $S \cap F$ is a relatively clopen subset of F and contains at least x, consequently is non-void and so must be all of the connected set F. It follows that $S \supset F$, and this for each $F \in \mathfrak{F}$. Therefore $S \supset \bigcup \mathfrak{F}$.

Theorem 1.2 *If S is a connected subset of* $\mathbb{C}$, *then any set T satisfying* $S \subset T \subset \bar{S}$ *is also connected.*

Proof: If V is any non-void relatively clopen subset of T, then $V \cap S$ is a relatively clopen subset of S. Moreover $V \cap S \neq \varnothing$, for the closure of S in T is $\bar{S} \cap T = T$; that is, S is dense in T, and V is open in T. It follows from the connectedness of S that $V \cap S = S$. But then the relatively closed subset V of T contains also the closure in T of S, viz., T itself.

Exercise 1.3 *Prove the following useful extension of* 1.1. *If* $A, B \subset \mathbb{C}$ *are each connected and* $\bar{A} \cap B \neq \varnothing$, *then* $A \cup B$ *is connected.*

Hints: Let $p \in \bar{A} \cap B$. By 1.2 the set $A \cup \{p\}$ is connected and by 1.1 the union $A \cup \{p\} \cup B = A \cup B$ is also, since $p \in B$.

Theorem 1.4 *If X, Y are subsets of* $\mathbb{C}$ *and* $f\colon X \to Y$ *is continuous, then* $f(X)$ *is connected if X is.*

Proof: If V is a proper, non-void, relatively clopen subset of $f(X)$, then by continuity of f the set $f^{-1}(V)$ is a proper, non-void, clopen subset of X.

Exercise 1.5 *Show that if* $X_1 \supset X_2 \supset \cdots$ *are closed, connected subsets of* $\mathbb{C}$ *at least one of which is compact, then* $X = \bigcap_{n=1}^{\infty} X_n$ *is connected.*

Hints: If not, there exist disjoint open U, V which each meet X such that $X \subset U \cup V$. Then $X_n \subset U \cup V$ for some n.

Exercise 1.6 *Show that if $X_1 \supset X_2 \supset \cdots$ are closed, connected subsets of $\mathbb{C}$ and $X = \bigcap_{n=1}^{\infty} X_n$ is compact and non-void, then some X_n is compact and X is connected.* (*Cf.* W. M. WHYBURN [1935].)

Hint: Take open V with $X \subset V \subset \bar{V}$ compact. Then $V \supset X = X \cap \bar{V} = \bigcap_{n=1}^{\infty} (X_n \cap \bar{V})$. These sets being compact, it follows $X_n \cap \bar{V} \subset V$ for some n. Then $X_n \cap \bar{V}$ is clopen in X_n, hence not a proper subset.

Exercise 1.7 *If $F_1, F_2\colon \Omega \to \mathbb{C}\backslash\{0\}$ are continuous, Ω is connected and $F_1^2 \equiv F_2^2$, prove that either $F_1 \equiv F_2$ or $F_1 \equiv -F_2$. What happens if the F_j are allowed to take the value zero?*

Hint: Ω is the disjoint union of the two relatively closed subsets $(F_1 + F_2)^{-1}(0)$ and $(F_1 - F_2)^{-1}(0)$.

§ 2 Connectedness of Intervals, Curves and Convex Sets

Theorem 1.8 *If $a, b \in \mathbb{R}$, then $[a, b]$ is connected.*

Proof: We are to show that there is no proper, non-void, relatively clopen subset A of $[a, b]$. With such an A the set $B = [a, b]\backslash A$ is another such and one of A and B contains a. Let, say, $a \in A$. Since A is relatively open in $[a, b]$, there is some $a < x \le b$ such that $[a, x] \subset A$. Let c denote the supremum of such x. Then by definition of c we have $[a, x] \subset A$ for every $x < c$, and so since A is closed, $[a, c] \subset A$. If $c < b$, then since $c \in A$ and A is relatively open in $[a, b]$, there exists $c < d < b$ such that $[c, d] \subset A$; and then $[a, d] = [a, c] \cup [c, d] \subset A$, contrary to the maximality of c. It follows that $c = b$, so $[a, b] = [a, c] \subset A$, contrary to the propriety of A.

Corollary 1.9 *Any (non-void) interval in $\mathbb{R}$ is connected.*

Proof: If I is an interval and $x \in I$ is arbitrary, then

$$I = \bigcup_{\substack{a,\, b \in I \\ a \le x \le b}} [a, b]$$

An appeal to the last result and 1.1 finishes the proof.

Exercise 1.10 *Establish the converse of the last corollary, namely that every connected subset of $\mathbb{R}$ is an interval.*

Definition 1.11 (i) A *curve* is a continuous function γ from a compact interval $[a, b]$ in $\mathbb{R}$ into $\mathbb{C}$. $\gamma(a)$ is called the *initial* point and $\gamma(b)$ the *terminal* point of γ and γ is called *closed* if $\gamma(a) = \gamma(b)$. When there is no possibility of confusion, we sometimes permit the linguistic abuse of writing simply γ for the set $\gamma[a, b]$. This set is called the *range* of γ. A closed curve is also called a *loop*. A one-to-one curve is called an *arc*. A loop $\gamma: [a, b] \to \mathbb{C}$ is called *simple* if γ is one-to-one on $[a, b)$.

(ii) A curve $\gamma: [a, b] \to \mathbb{C}$ is *continuously differentiable* or *smooth* if it is differentiable in (a, b), has left- and right-hand derivatives at b and a respectively and the derivative function γ' is continuous on $[a, b]$.

(iii) A curve $\gamma: [a, b] \to \mathbb{C}$ is *piecewise continuously differentiable* or *piecewise smooth* if there exist $a = t_0 < t_1 < \cdots < t_n = b$ such that γ is continuously differentiable on each $[t_{j-1}, t_j]$.

(iv) For a closed curve $\gamma: [a, b] \to \mathbb{C}$ define $0 \bullet \gamma: [a, b] \to \mathbb{C}$ to be the constant curve $\gamma(a)$, $(-1) \bullet \gamma: [a, b] \to \mathbb{C}$ by $(-1) \bullet \gamma(t) = \gamma(b + a - t)$, $n \bullet \gamma: [a, b] \to \mathbb{C}$ for each positive integer n by

$$n \bullet \gamma(t) = \gamma(a + n(t - t_{j-1})) \quad \text{for } t \in [t_{j-1}, t_j] \quad \text{and} \quad t_j = a + (b - a)j/n, \quad j = 1, 2, \ldots, n$$

and for every negative integer k define $k \bullet \gamma$ to be $(-k) \bullet (-1) \bullet \gamma$.

(v) A *Jordan-curve* is a homeomorph in $\mathbb{C}$ of a non-degenerate circle.

Definition 1.12 (i) If $a_0, \ldots, a_n \in \mathbb{C}$, then $[a_0, a_1, \ldots, a_n]$ is used to denote both the piecewise smooth curve $\gamma: [0, n] \to \mathbb{C}$ defined by

$$\gamma(t) = a_{k-1} + (t - k + 1)(a_k - a_{k-1}), \quad k - 1 \le t \le k, 1 \le k \le n$$

and the range of this curve. Any such curve is called a *polygon*, more precisely, an *n-gon*. When $n = 1$ the term *interval* is also used. Furthermore we let $(a_0, \ldots, a_n]$, $[a_0, \ldots, a_n)$ and $(a_0, \ldots, a_n)$ denote the restriction of the above γ to $(0, n]$, $[0, n)$ and $(0, n)$ respectively, as well as the ranges of these restriction functions.

(ii) If $a, b, c, d \in \mathbb{R}$, $a < b$, $c < d$ and $S = (a, b) \times (c, d)$, then

$$\partial S = \partial \bar{S} = [a + ci, b + ci, b + di, a + di, a + ci].$$

(Note that here the convention in 1.11(i) is consistent with the usual definition of topological boundary. See § 4, Chapter 0.)

Definition 1.13 A non-void subset S of $\mathbb{C}$ is *polygonally connected* if each pair of points of S lie on some polygon wholly contained in S. (Compare 1.21.)

When this happens a one-to-one polygon can always be selected, a technical fact which will be quite useful in Chapter IV. This follows from

Exercise 1.14 *Let S be a connected set which is a finite union of (closed) intervals. Then any two points of S lie on a polygonal arc in S.*

Hints: Go by induction on the (minimal) number of intervals in S. The result is clearly true if S is comprised of a single interval. If the result is true whenever

S is comprised of fewer than n intervals, consider a connected set S' which is the union of n intervals $I_1, \ldots, I_n$. Consider any two points $a', b' \in S'$, say $a' \in I_n$. We want to find a polygonal arc in S' joining a' to b'. There is nothing to prove if $b' \in I_n$, so we suppose $b' \notin I_n$. Let S denote the union of all the connected subsets of $I_1 \cup \cdots \cup I_{n-1}$ which contain b'. By 1.1, S is connected and if I_j meets S for some $j \leq n - 1$, then by 1.1 and 1.8, $S \cup I_j$ is connected. It follows that S is a union of some of the intervals $I_1, \ldots, I_{n-1}$ and is disjoint from each of the others. Let S'' denote the union of the latter. If I_n were disjoint from S, then S and $I_n \cup S''$ would constitute a disconnection of S'. We infer that I_n meets S. Letting $I_n = [a_n, b_n]$, it follows that one of $[a', a_n]$ and $[a', b_n]$ meets S. We take a to be the first point of this interval which lies in S. By the induction hypothesis there is a polygonal arc p in S joining b' to a. Then $p \cup [a, a']$ is a polygonal arc in S' joining b' to a'.

Definition 1.15 (i) If $S \subset \mathbb{C}$, an arc with one endpoint on ∂S and all other points in S is called an *end-cut in S*. An arc with both endpoints on ∂S but all other points in S or a Jordan-curve with one point on ∂S but all other points in S is called a *cross-cut in S*.

(ii) A point $z \in \partial S$ is *accessible from* S if it is an endpoint of an end-cut in S.

Exercise 1.16 *For any open subset U of $\mathbb{C}$ the accessible points are dense in ∂U.*

Hints: Let $z_0 \in \partial U$ and $\varepsilon > 0$. Since z_0 is a boundary point, there exists $w \in U \cap D(z_0, \varepsilon/2)$. There obviously exists (elementary compactness argument) at least one point $z_1 \in \partial U$ which is at a minimum distance δ from w: $|z_1 - w| = \inf\{|z - w| : z \in \partial U\} = \delta > 0$. Then $|z_1 - w| \leq |z_0 - w| < \varepsilon/2$, whence $|z_1 - z_0| \leq |z_1 - w| + |w - z_0| < \varepsilon$. By minimality of δ, $[w, z_1)$ does not meet ∂U. Since it is connected and meets U, it must lie wholly in U. Therefore $[w, z_1]$ is an end-cut, and z_1 is an accessible point of ∂U lying in $D(z_0, \varepsilon)$.

Definition 1.17 (i) A subset S of $\mathbb{C}$ is *starlike with respect to the point z* if $[z, w] \subset S$ for every $w \in S$. We say simply that S is *starlike* if it is starlike with respect to some point.

(ii) A subset S of $\mathbb{C}$ is *convex* if it is starlike with respect to each of its points, that is, if $[z, w] \subset S$ for every $z, w \in S$.

(iii) If S is a subset of $\mathbb{C}$, the *convex hull of S*, noted co S, is the intersection of all convex subsets of $\mathbb{C}$ which contain S.

Exercise 1.18 *A* convex combination *of elements of a subset S of $\mathbb{C}$ is any sum $\sum_{j=1}^{n} \lambda_j x_j$ where $x_j \in S$, $0 \leq \lambda_j \leq 1$ and $\sum_{j=1}^{n} \lambda_j = 1$.*

(i) *Show that S is convex if and only if it contains all convex combinations of its elements.*

Hint: Use induction on n.

(ii) *Show that co S is convex and consists of all convex combinations of elements of S.*

Theorem 1.19 (i) *If S is a bounded subset of $\mathbb{C}$, then* co S *is bounded and has the same diameter as S.*
(ii) *If K is a compact subset of $\mathbb{C}$, then* co K *is compact.*

Proof: (i) Let d be the diameter of S. It evidently suffices to show that $|x - y| \le d$ for $x, y \in \text{co } S$. For each $w \in S$ we have $|w - z| \le d$ for all $z \in S$, that is, S lies in the convex set $\bar{D}(w, d)$. Therefore $\text{co } S \subset \bar{D}(w, d)$, that is, $|w - y| \le d$ for every $y \in \text{co } S$ and every $w \in S$. This says that S lies in the convex set $\bar{D}(y, d)$, whence $\text{co } S \subset \bar{D}(y, d)$, for every $y \in \text{co } S$; that is, $|x - y| \le d$ for every $x \in \text{co } S$.

(ii) Suppose $n > 3$, $z_1, \ldots, z_n \in K$, $t_1, \ldots, t_n$ are positive and $t_1 + \cdots + t_n = 1$. The $n - 1$ vectors $z_j - z_n$ $(1 \le j \le n - 1)$ in $\mathbb{R}^2$ are linearly dependent over $\mathbb{R}$ (since $n - 1 > 2$), so there exist $a_1, \ldots, a_{n-1} \in \mathbb{R}$ not all 0 such that

$$\sum_{j=1}^{n-1} a_j(z_j - z_n) = 0.$$

Set $a_n = -\sum_{j=1}^{n-1} a_j$ and choose k so that

$$|a_j|/t_j \le |a_k|/t_k, \quad 1 \le j \le n.$$

This inequality shows that $a_k \neq 0$ (since not all a_j are 0). Furthermore, replacing all the a_j by their negatives if necessary, we can suppose that $a_k > 0$. Set then

$$c_j = t_j - t_k a_j/a_k, \quad 1 \le j \le n$$

and note that

$$c_j \ge 0, \sum_{j=1}^{n} c_j = 1 \quad \text{and} \quad \sum_{j=1}^{n} t_j z_j = \sum_{\substack{j=1 \\ j \neq k}}^{n} c_j z_j.$$

An obvious inductive argument then shows that

$$t_1 z_1 + \cdots + t_n z_n \in \{ax + by + cz : a, b, c \in [0, 1], x, y, z \in K, a + b + c = 1\}.$$

The latter set, call it C, is a continuous image of the compact set $\{(a, b, c, x, y, z) \in [0, 1]^3 \times K^3 : a + b + c = 1\}$ and so is compact, while the sum on the left is the generic point of co K by 1.18(ii). Since evidently $C \subset \text{co } K$, we have finally $\text{co } K = C$.

Theorem 1.20 *Every curve is connected.*

Proof: 1.4 and 1.8.

Corollary 1.21 *Every polygonally connected set is connected.*

Proof: If S is polygonally connected, fix $x \in S$ and for every $z \in S$ let p_z be a polygon in S which contains x and z. p_z is connected by the last theorem and evidently $S = \bigcup_{z \in S} p_z$, so by 1.1, S is connected.

Corollary 1.22 *Every starlike set and (hence) every convex set is polygonally connected and (hence) connected.*

Corollary 1.23 *Center-punctured disks, annuli and circles in $\mathbb{C}$ are connected.*

Proof: We treat the case of open disks and annuli centered at the origin. The reader can easily extend this to all other cases. Let $D = D(0, 1)$. Then

$$D\backslash\{0\} = ((D\backslash(-1, 0]) \cup (D\backslash[0, 1)).$$

The two sets on the right are starlike (with respect to any point in $(0, 1)$ or $(-1, 0)$ respectively) and so connected. Also they have points in common (for example, $\frac{1}{2}i$). Therefore their union is connected.

Now if $0 \leq a < b < \infty$, then $\psi(t) = tb + (1 - t)a$ maps $(0, 1)$ continuously onto (a, b), so $z \to \psi(|z|)\cdot z/|z|$ maps $D\backslash\{0\}$ continuously onto the annulus $A(a, b)$. Now cite 1.4.

Take $a = b$ in (an argument similar to) the above to see that the circle $\{z \in \mathbb{C}: |z| = a\}$ is connected.

Theorem 1.24 *If Ω is an open connected subset of $\mathbb{C}$ and A is a subset of Ω which has no limit point in Ω, then $\Omega\backslash A$ is open and connected.*

Proof: Since A has no limit point in Ω, each point of $\Omega\backslash A$ is the center of a disk which does not meet A. Therefore $\Omega\backslash A$ is open. Suppose U_1, U_2 are disjoint, non-void, (relatively) open subsets of $\Omega\backslash A$ such that $\Omega\backslash A = U_1 \cup U_2$. For each $a \in A$ there is an $r_a > 0$ such that $D_a = D(a, r_a) \subset \Omega$ and $D_a \cap A = \{a\}$. Each subset $D_a\backslash\{a\}$ of $\Omega\backslash A$ is connected (by 1.23) and so lies entirely in one of the U_j. Let $A_j = \{a \in A: D_a\backslash\{a\} \subset U_j\}$ $(j = 1, 2)$. Then $V_j = U_j \cup \bigcup\{D_a: a \in A_j\}$ $(j = 1, 2)$ are disjoint, non-void, open subsets of Ω, that is, a disconnection of Ω.

Exercise 1.25 (i) *If C is a proper subset of a circle, then $\mathbb{C}\backslash C$ is connected.*

Hints: Let the circle be $C(0, 1)$, set $D = D(0, 1)$ and select $p \in C(0, 1)\backslash C$. Note that

(1) $D \subset \bar{D}\backslash C \subset \bar{D}$,

(2) $\mathbb{C}\backslash\bar{D} \subset (\mathbb{C}\backslash\bar{D}) \cup \{p\} \subset \mathbb{C}\backslash D = \overline{\mathbb{C}\backslash\bar{D}}$.

Now D is connected (1.22) and so is $\mathbb{C}\backslash\bar{D}$, being the image of the connected set $D\backslash\{0\}$ under the continuous map $z \to 1/z$. Thus (1), (2) and 1.2 yield

(3) $\bar{D}\backslash C$ and $(\mathbb{C}\backslash\bar{D}) \cup \{p\}$ are each connected.

But then, since p lies in both sets, their union $\mathbb{C}\backslash C$ is connected, by 1.1.

(ii) *If Ω is an open, connected subset of $\mathbb{C}$ and D is an open disk such that $\bar{D} \subset \Omega$, then $\Omega\backslash\bar{D}$ is connected.*

Hints: Let D_1 be an open disk concentric with D such that

$$\bar{D} \subset D_1 \subset \Omega.$$

Suppose U and V are a disconnection of $\Omega\backslash\bar{D}$: U and V are non-empty, open subsets of $\Omega\backslash\bar{D}$ and $U = (\Omega\backslash\bar{D})\backslash V$. The subset $D_1\backslash\bar{D}$ of $\Omega\backslash\bar{D}$ is connected (1.23) and so lies wholly in one of U or V, say

$$D_1\backslash\bar{D} \subset U.$$

Then

$$U \cup \bar{D} = U \cup D_1, \quad \text{an open subset of } \Omega.$$

We have

$$V = (\Omega\backslash\bar{D})\backslash U = \Omega\backslash(U \cup \bar{D}),$$

so V and $U \cup \bar{D}$ are disjoint, non-empty, open subsets of Ω which cover Ω, contradiction.

Open connected subsets of $\mathbb{C}$ occur so frequently in all that follows that it will be extremely convenient to have a generic name for this species of set. The traditional one is

Definition 1.26 A *region* is an open, connected, non-empty subset of $\mathbb{C}$.

§ 3 The Basic Connectedness Lemma

We codify here a useful and simple device for exploiting connectedness. It is useful because we often encounter in function theory situations where it is easy to show that a certain subset A of a set B contains a fixed percentage of each disk which is centered in A and lies in B.

Theorem 1.27 (Basic Connectedness Lemma) *Let U be an open subset of $\mathbb{C}$ and $\varnothing \neq A \subset U$ have the property that for some $\theta > 0$ there holds:*

$$a \in A \quad \textit{and} \quad \bar{D}(a, r) \subset U \Rightarrow D(a, \theta r) \subset A.$$

Then A is relatively clopen in U. In particular, if in addition U is connected, then $A = U$.

Proof: Evidently A is open in U, since U is open. To see that $U\backslash A$ is also open in U, let $b \in U\backslash A$ and pick $R > 0$ so that $\bar{D}(b, R) \subset U$. It follows that if $D(b, \theta R/(1 + \theta))$ meets A, say in a, then we have

$$(*) \qquad |a - b| < \frac{\theta R}{1 + \theta}.$$

We have for $r = R - |a - b|$

$$\bar{D}(a, r) \subset \bar{D}(b, R) \subset U.$$

But then $D(a, \theta r) \subset A$ and from (*) we get

$$|a - b| < \theta R - \theta|a - b| = \theta r,$$

putting $b \in D(a, \theta r) \subset A$, contradiction.

Corollary 1.28 *Every region is polygonally connected.*

Proof: Let Ω be a region. It suffices to show that for each fixed $a \in \Omega$ the set A of z in Ω for which there is a polygon in Ω with initial point a and terminal point z is all of Ω. This set is non-void ($a \in A$) and if $z \in A$ and $r > 0$ is such that $\bar{D}(z, r) \subset \Omega$, then $D(z, r) \subset A$, because if $[a, \ldots, z]$ is a polygon in Ω and $w \in D(z, r)$, then $[a, \ldots, z, w]$ is also a polygon in Ω. Thus A meets the hypothesis of the Basic Connectedness Lemma with $\theta = 1$.

Exercise 1.29 *Any two distinct accessible points on the boundary of a region Ω lie on a cross-cut in Ω.*

Hints: Let a, b be distinct accessible boundary points of Ω. Let A, B be end-cuts in Ω with a the initial point of A, b the terminal point of B. By ignoring an appropriate initial segment of B, we can suppose A and B disjoint. By 1.28 and 1.14 there is an arc P: $[0, 1] \to \Omega$ with initial point on A and terminal point on B. Let t_a be the largest $t \in [0, 1]$ such that $P(t) \in A$. We have $t_a < 1$ since $A \cap B = \varnothing$ and $P(1) \in B$. Let t_b be the smallest $t \in [t_a, 1]$ such that $P(t) \in B$. Then A, $P(t_a, t_b)$, B are mutually disjoint and, since each of the functions involved here is one-to-one, this implies that their union is an arc. It is evidently the desired cross-cut from a to b.

§ 4 Components and Compact Exhaustions

Recall that a *component* of a set is a maximal connected subset.

Theorem 1.30 *Let X be a non-empty subset of $\mathbb{C}$. Then*

(i) *Each component of X is relatively closed.*
(ii) *Distinct components are disjoint.*
(iii) *Each connected subset of X lies in a unique component.*
(iv) *X is the (disjoint) union of its components.*

If X is open, then:

(i)′ *Each component of X is open and there are at most countably many of them.*

Proof: (i) follows from the fact that a component is a maximal connected set while its relative closure is still connected (by 1.2).

(ii) If C_1 and C_2 are components with a point in common, then $C_1 \cup C_2$ is connected by 1.1 and so by the maximality of C_1 and C_2 we see that $C_1 = C_1 \cup C_2 = C_2$.

(iii) If C is a connected subset of X, then the union of all the connected subsets of X which contain C is connected by 1.1 and evidently is maximal among connected subsets of X, i.e., is a component.

(iv) Since single points are connected sets, (iv) follows from (iii).

(i)′ If C is a component and $x \in C$ and D is an open disk centered at x and contained in X, then D is connected (being convex) and so $C \cup D$ is connected by 1.1 and consequently by maximality $C \cup D \subset C$, that is, $D \subset C$. Thus C contains a neighborhood of each of its points and so is open. Being open, C contains a point with rational coordinates. As there are only countably many such points and components are disjoint, there can be only countably many components.

Theorem 1.31 *Let U be an open subset of $\mathbb{C}$. Then there exist compact sets K_n such that*

(1) $K_n \subset \mathring{K}_{n+1}$.

(2) $\displaystyle\bigcup_{n=1}^{\infty} K_n = U$.

(3) *Every bounded component of $\mathbb{C}\backslash K_n$ contains a component of $\mathbb{C}\backslash U$. In particular, if $\mathbb{C}\backslash U$ has no bounded components, then each set $\mathbb{C}\backslash K_n$ is connected.*

Proof: For each positive integer n form

(4) $C_n = \{z \in \mathbb{C}: d(z, \mathbb{C}\backslash U) \geq 1/n\} \subset U$,

where $d(z, \mathbb{C}\backslash U) = \inf\{|z - w|: w \in \mathbb{C}\backslash U\}$. Notice that

(5) $$\mathbb{C}\backslash C_n = \{z \in \mathbb{C}: d(z, \mathbb{C}\backslash U) < 1/n\} = \bigcup_{a \in \mathbb{C}\backslash U} D(a, 1/n)$$

is obviously open, so C_n is closed. Form

(6) $K_n = C_n \cap \bar{D}(0, n)$.

Then K_n is compact and lies in U. Moreover it is obvious that (2) holds: given $z \in U$, pick positive integer $n \geq |z|$ so that $D(z, 1/n) \subset U$. Then $z \in K_n$.

To prove (1), we set $r_n = 1/n - 1/(n + 1)$ and show that $D(z, r_n) \subset K_{n+1}$ for every $z \in K_n$. Indeed, given such a z and $w \in D(z, r_n)$, we have

$$\begin{aligned} |w| \leq |z| + r_n < n + r_n, &\quad \text{since } z \in K_n \subset \bar{D}(0, n),\\ \leq n + 1, & \end{aligned}$$

so $w \in \bar{D}(0, n + 1)$. But also for any $a \in \mathbb{C}\backslash U$ we have

$$\begin{aligned} |w - a| \geq |z - a| - |w - z| &\geq \frac{1}{n} - |w - z|, \quad \text{since } z \in C_n\\ &> \frac{1}{n} - r_n = \frac{1}{n + 1}, \quad \text{since } w \in D(z, r_n). \end{aligned}$$

It follows that $w \in C_{n+1}$ and so finally $w \in \bar{D}(0, n+1) \cap C_{n+1} = K_{n+1}$.

It remains to look after (3). Notice that

$$\begin{aligned} \mathbb{C}\backslash K_n &= [\mathbb{C}\backslash \bar{D}(0, n)] \cup [\mathbb{C}\backslash C_n] \quad \text{by (6)} \\ &= [\mathbb{C}\backslash \bar{D}(0, n)] \cup \bigcup_{a \in \mathbb{C}\backslash U} D(a, 1/n) \quad \text{by (5).} \end{aligned} \tag{7}$$

The set $\mathbb{C}\backslash \bar{D}(0, n)$ is the image of the center-punctured disk $D(0, 1/n)\backslash\{0\}$ under the continuous function $z \to 1/z$ and so is connected. Therefore all the sets in the union (7) are connected. Now let C be a bounded component of $\mathbb{C}\backslash K_n$. Since C is the union of all the connected subsets of $\mathbb{C}\backslash K_n$ which meet it, we see from (7) that on the one hand C does not meet $\mathbb{C}\backslash \bar{D}(0, n)$ and on the other hand

$$\bigcup_{a \in \mathbb{C}\backslash U} D(a, 1/n) \cap C = \bigcup_{a \in C \cap (\mathbb{C}\backslash U)} D(a, 1/n).$$

In particular, $C \cap (\mathbb{C}\backslash U) \neq \varnothing$. Being non-void, this set meets some component K of $\mathbb{C}\backslash U$. But then K is a connected subset of $\mathbb{C}\backslash U \subset \mathbb{C}\backslash K_n$ (by (2)) which meets C and hence lies in C.

Exercise 1.32 *Let S be a non-empty subset of $\mathbb{R}^n$, $g: S \to [0, \infty)$ a bounded, continuous function. For each $x \in \mathbb{R}^n$ define*

$$d(x) = d_g(x) = \inf\{\|x - y\| g(y) : y \in S\},$$

where $\| \ \|$ is the euclidean norm in $\mathbb{R}^n$.

(i) *Show that d is a continuous function on $\mathbb{R}^n$.*

Hints: Let M be a bound for g. Then for any $x_1, x_2 \in \mathbb{R}^n$, $y \in S$ $d(x_1) \leq \|x_1 - y\| g(y) \leq [\|x_1 - x_2\| + \|x_2 - y\|] g(y) \leq M\|x_1 - x_2\| + \|x_2 - y\| g(y)$, that is, $d(x_1) - M\|x_1 - x_2\|$ is a lower bound for the set $\{\|x_2 - y\| g(y) : y \in S\}$. Consequently

$$d(x_1) - M\|x_1 - x_2\| \leq d(x_2).$$

Reversing the roles of x_1 and x_2, we are therefore led to

$$|d(x_1) - d(x_2)| \leq M\|x_1 - x_2\|.$$

(ii) *For the function $g = 1$ we have $d_1^{-1}(\{0\}) = \bar{S}$.*

Theorem 1.33 *Let X be a compact subset of $\mathbb{C}$ and C a component of X. Then C is the intersection of all the relatively clopen subsets of X which contain it.*

Proof: If $\mathscr{C}$ is the class of all relatively clopen subsets of X which contain C, then $\mathscr{C}$ is not void ($X \in \mathscr{C}$) and the closed set $K = \bigcap \mathscr{C}$ contains C. Because C is a maximal connected set, it suffices to show that K is connected in order to conclude that $K = C$. This we do by *reductio ad absurdum*. If K is not connected, there exist disjoint non-void closed sets K_1, K_2 such that $K = K_1 \cup K_2$. Now there are disjoint open $V_j \supset K_j$. In fact, if $d_j(z)$ denotes the distance from z to the closed set K_j, i.e., $d_j(z) = \inf\{|z - w| : w \in K_j\}$, then by 1.32 d_j is continuous and so the function $f = d_1 - d_2$ is continuous on $\mathbb{C}$. Evidently $K_1 \subset f^{-1}(-\infty, 0)$ and $K_2 \subset f^{-1}(0, \infty)$. Now $\bigcap \mathscr{C}$ is compact and lies in $K = K_1 \cup K_2 \subset V_1 \cup V_2$ and so some finite intersection of elements of $\mathscr{C}$, i.e., another element of $\mathscr{C}$, lies in $V_1 \cup V_2$. Say, $C_0 \in \mathscr{C}$, $C_0 \subset V_1 \cup V_2$. Then

$$\begin{aligned} C_0 \cap \overline{V}_1 &= [C_0 \cap (V_1 \cup V_2)] \cap \overline{V}_1 \\ &= (C_0 \cap V_1 \cap \overline{V}_1) \cup (C_0 \cap V_2 \cap \overline{V}_1) \\ &= C_0 \cap V_1 \end{aligned}$$

since V_2, being open, cannot meet $\overline{V}_1$ without meeting V_1. Therefore $C_0 \cap V_1$ is relatively clopen. Similarly for $C_0 \cap V_2$. Now not both these sets, whose union C_0 contains $K \supset C$, can meet C, because C is connected. Let, say, $C \subset C_0 \cap V_1$. Then $C_0 \cap V_1 \in \mathscr{C}$ and so $C_0 \cap V_1 \supset K$, by definition of K. Therefore $V_1 \supset K$, so $V_2 \cap K = \varnothing$, a contradiction, since $V_2 \cap K \supset K_2$.

Corollary 1.34 (Šura-Bura [1941]) *Let X be a closed subset of $\mathbb{C}$ and C a compact component of X. Then C is the intersection of all the compact, relatively open subsets of X which contain it.*

Proof: It suffices to produce for each open subset U of $\mathbb{C}$ with $C \subset U$ a compact relatively open subset of X which contains C and lies in U. Cover C with finitely many open disks whose closures lie in U and let V be their union. Thus $C \subset V \cap X \subset \overline{V}$ compact $\subset U$. If $\mathscr{C}$ is the family of all compact, relatively open subsets of $\overline{V} \cap X$ which contain C, then $C = \bigcap \mathscr{C}$ by the last theorem. By the finite intersection property of compacta, since $\bigcap \mathscr{C} = C \subset V$, an intersection of finitely many members of $\mathscr{C}$, which is again a member K of $\mathscr{C}$, must already fall into the open set V. Thus $C \subset K \subset V \cap X \subset U$. K is compact and K is relatively open in $\overline{V} \cap X$, whence in $V \cap X$, whence in X, since V is open in $\mathbb{C}$.

Theorem 1.35 *The boundary of an open subset Ω of $\mathbb{C}$ homeomorphic to an open disk is connected if Ω is bounded and otherwise has no bounded components.*

Proof: Let $D = D(0, 1)$ and $\phi: D \to \Omega$ be an onto homeomorphism. Set $D_n = D(0, 1 - 1/n)$, $A_n = D \backslash \overline{D}_n$, $X_n = \overline{\phi(A_n)}$. Thus each X_n is closed and connected (by 1.4).

If all X_n are unbounded, then $X = \bigcap_{n=1}^{\infty} X_n$ has no bounded component. For if C is a bounded component of X, then C lies in a compact subset K of X which is relatively open in X, that is, $X \backslash K$ is closed. There exists then an open set V such that

$$K \subset V \subset \overline{V} \text{ compact, disjoint from } X \backslash K.$$

We have

$$\bigcap_{n=1}^{\infty} (X_n \cap \overline{V}) = X \cap \overline{V} = K \cap \overline{V} = K \subset V,$$

since $X \backslash K$ is disjoint from $\bar{V}$. Therefore, since the sets $X_n \cap \bar{V}$ are each compact, the finite intersection property of compacta implies that not all of them can meet the closed set $\mathbb{C} \backslash V$. So

$$X_n \cap \bar{V} \subset V \quad \text{for some } n.$$

Then the set

$$W = X_n \cap \bar{V} = X_n \cap V$$

is relatively clopen in X_n, contains $X \cap V \supset K$ and so is non-void. Since X_n is connected, it follows that $X_n = W$. But then X_n is compact, a contradiction.

On the other hand, if some X_{n_0} is compact, then $\bigcap X_n$ is connected by 1.5. In this case $\Omega \subset \phi(\bar{D}_{n_0}) \cup X_{n_0}$, a union of compact sets, so Ω is bounded.

Finally notice that $\partial\Omega = \bigcap X_n$ (exercisette).

§ 5 Connectivity of a Set

The theorems of this section are technical results which will be needed later (in Chapter X). They require only material presented so far and it is a good idea to dispose of them now. This will mean one less distraction when the proofs of these later results, often lengthy enough in their own right, are carried out. The impatient reader may skip this material and return to it later as needed.

Definition 1.36 A subset A of $\mathbb{C}$ is called *simply-connected* if its complement $\mathbb{C} \backslash A$ has no bounded component, *doubly-connected* if $\mathbb{C} \backslash A$ has exactly one bounded component. In general we will say A has *connectivity* $n \in \mathbb{N} \cup \{\infty\}$ if $\mathbb{C} \backslash A$ has exactly $n - 1$ bounded components. [Roughly speaking, connectivity counts the holes in A.]

Exercise 1.37 (i) *Show that an open subset of* $\mathbb{C}$ *is simply-connected if either* (a) *its boundary has no bounded component or* (b) *it is bounded and has a connected boundary.*

Hint: Every component of the complement meets the boundary.

(ii) *Show that if the subset A of $\mathbb{C}$ is simply-connected, then A contains every bounded subset S of $\mathbb{C}$ whose boundary lies in A.*

Hints: $\partial S = \bar{S} \backslash \mathring{S} \subset A$ implies $\bar{S} \cap (\mathbb{C} \backslash A) = \mathring{S} \cap (\mathbb{C} \backslash A)$, so that this set is relatively clopen in $\mathbb{C} \backslash A$. It therefore contains every component which it meets. As it is bounded and $\mathbb{C} \backslash A$ has no bounded components, $\bar{S} \cap (\mathbb{C} \backslash A)$ is empty.

Theorem 1.38 *Let U be an open subset of $\mathbb{C}$ such that $\mathbb{C} \backslash U$ has exactly one bounded component C. Then $U \cup C$ is open and simply-connected. If in addition U is connected, then so is $U \cup C$.*

Proof: The components of $\mathbb{C} \backslash (U \cup C)$ are just the unbounded components (if any) of $\mathbb{C} \backslash U$. Thus there are no bounded ones, so $U \cup C$ is simply-connected.

According to 1.34 there is a compact, relatively open subset V of $\mathbb{C}\backslash U$ which contains C. Since V is compact and all the other components of $\mathbb{C}\backslash U$ are unbounded, V contains none of them. On the other hand, the intersection of V with any of these components is a relatively clopen subset of the component. This intersection is therefore void. I.e., $V = C$, so C is relatively open in $\mathbb{C}\backslash U$. It follows easily that $U \cup C$ is open. Now $C \cap \bar{U} \neq \varnothing$; in fact $\partial C \subset \bar{U}$. For if $x \in \partial C\backslash\bar{U}$ and $r > 0$ is such that $D(x, r) \cap \bar{U} = \varnothing$, then $C \cup D(x, r)$ is a connected (by 1.22 and 1.1) subset of $\mathbb{C}\backslash U$ which properly contains the component C, a contradiction. From $C \cap \bar{U} \neq \varnothing$ and 1.3 it follows that $U \cup C$ is connected if U is.

Theorem 1.39 *Let Ω be a region, C_1, C_2 two distinct bounded components of $\mathbb{C}\backslash\Omega$. Then there is an unbounded, connected, closed subset of $\mathbb{C}\backslash C_1$ which contains C_2.*

Proof: Since C_1, C_2 are disjoint and compact, there exists an open neighborhood V of C_2 disjoint from C_1. Now C_2 meets $\bar{\Omega}$. Pick any point $c_2 \in C_2 \cap \bar{\Omega}$ and take a convex neighborhood N_2 of c_2 which is disjoint from C_1 and pick $b_2 \in N_2 \cap \Omega$. Consider then two cases:

Case I: Ω is bounded. Then $\mathbb{C}\backslash\Omega$ has (exactly) one unbounded component C_3. Pick $c_3 \in C_3 \cap \bar{\Omega}$ and a convex neighborhood N_3 of c_3 disjoint from C_1. Pick $b_3 \in N_3 \cap \Omega$ and let (1.28) Γ be a curve in Ω from b_2 to b_3. Consider then

$$C_2 \cup [c_2, b_2] \cup \Gamma \cup [b_3, c_3] \cup C_3.$$

This is an unbounded, connected, closed set which contains C_2 and is disjoint from C_1.

Case II: Ω is unbounded. Then use 1.28 to construct a curve Γ in Ω joining b_2 to some $\omega \in \Omega$ with $|\omega| > \sup|C_1|$.

Then

$$C_2 \cup [c_2, b_2] \cup \Gamma \cup [1, \infty)\omega$$

is an unbounded, connected, closed set which contains C_2 and is disjoint from C_1.

Theorem 1.40 *Let Ω be a region such that $\mathbb{C}\backslash\Omega$ has (only) finitely many bounded components $C_1, C_2, \ldots, C_n$. Then there exists an open subset Ω_1 of Ω such that C_1 is the unique bounded component of $\mathbb{C}\backslash\Omega_1$.*

Proof: Let $\mathscr{C}$ be the set of unbounded components of $\mathbb{C}\backslash\Omega$. For each j the last theorem provides unbounded, closed sets $K_2, \ldots, K_n$ with $C_j \subset K_j \subset \mathbb{C}\backslash C_1$ $(j = 2, \ldots, n)$. Set $\Omega_1 = \Omega\backslash(K_2 \cup \cdots \cup K_n)$. Then

$$(1) \qquad \mathbb{C}\backslash\Omega_1 = (\mathbb{C}\backslash\Omega) \cup K_2 \cup \cdots \cup K_n.$$

Every bounded component of $\mathbb{C}\backslash\Omega$ different from C_1 lies in some K_j, which is connected and unbounded. Therefore every point of $\mathbb{C}\backslash\Omega_1$ not in C_1 lies in an unbounded component of $\mathbb{C}\backslash\Omega_1$. But, as in the proof of 1.38, $\Omega \cup C_1 \cup \cdots \cup C_n$ is open, so its complement $\bigcup \mathscr{C}$ is closed. Evidently from (1)

$$\text{(2)} \qquad \mathbb{C}\backslash\Omega_1 = C_1 \cup K_2 \cup \cdots \cup K_n \cup \bigcup \mathscr{C},$$

which shows that C_1 is relatively open in $\mathbb{C}\backslash\Omega_1$. Being also compact and connected, C_1 is therefore a component.

Theorem 1.41 *Let K be a compact, connected subset of $\mathbb{C}$, $x \in K$. Then x is a limit point of each component of $K\backslash\{x\}$.*

Proof: Suppose contrariwise that C is a component of $K\backslash\{x\}$ and $x \notin \bar{C}$. Then C, being closed in $K\backslash\{x\}$, is closed in K, hence compact. Since $x \notin C$, there is then an $r > 0$ sufficiently small that $\bar{D}(x, r) \cap C = \varnothing$. Evidently C is then a component of the compact set $K\backslash D(x, r)$. By 1.33 there is a compact set U which is relatively open in $K\backslash D(x, r)$, which contains C and which lies in the open set $\mathbb{C}\backslash\bar{D}(x, r)$. Then $K\backslash D(x, r)\backslash U$ is relatively closed in $K\backslash D(x, r)$ and so is compact, while

$$K\backslash U = [K \cap \bar{D}(x, r)] \cup [K\backslash D(x, r)\backslash U].$$

This decomposition shows that $K\backslash U$ is compact. Therefore U disconnects K.

Theorem 1.42 *There is a countable set $\mathscr{K}$ of simply-connected, compact subsets of $\mathbb{C}\backslash\bar{D}(0, 1)$ with the property that every simply-connected, compact subset of $\mathbb{C}\backslash\bar{D}(0, 1)$ is a subset of some element of $\mathscr{K}$.*

Proof: In the space of continuous complex-valued functions on $[0, 1]$ which are 1 at 0 there is a countable set $\{f_1, f_2, \ldots\}$ which is uniformly dense. Indeed, by an easy uniform continuity argument one sees that the piecewise linear functions with rational complex "vertices" are such a class.

For each pair of positive integers n and m define

$$\text{(1)} \qquad U(n, m) = D(0, 1 + 1/m) \cup \bigcup_{t\in[0,1]} D(f_n(t), 1/m) \cup [\mathbb{C}\backslash\bar{D}(0, |f_n(1)| - 1/m)].$$

This is an open, connected set. For

$$\text{(2)} \qquad f_n[0, 1] \cup D(f_n(t), 1/m), \quad t \in [0, 1]$$

is connected, being the union of connected sets with a point in common. Similarly, as $f_n[0, 1]$ meets both the disk $D(0, 1 + 1/m)$ (in $f_n(0) = 1$) and $\mathbb{C}\backslash\bar{D}(0, |f_n(1)| - 1/m)$ (in $f_n(1)$), the set

$$\text{(3)} \qquad D(0, 1 + 1/m) \cup f_n[0, 1] \cup [\mathbb{C}\backslash\bar{D}(0, |f_n(1)| - 1/m)]$$

is connected. Then the union of the set (3) and all the sets in (2), viz., (1), is connected. Define

$$(4)\qquad K(n, m) = \mathbb{C}\backslash U(n, m).$$

Thus $K(n, m)$ is a compact set which is disjoint from $\bar{D}(0, 1)$ and has connected complement.

Now let K be any simply-connected, compact subset of $\mathbb{C}\backslash\bar{D}(0, 1)$. We will show that $K \subset K(n, m)$ for some n, m. Choose $r > 1 + \sup|K|$. Then 1 and r are two points in the open connected set $\mathbb{C}\backslash K$. So by 1.28 there is a continuous γ: $[0, 1] \to \mathbb{C}\backslash K$ with $\gamma(0) = 1$, $\gamma(1) = r$. Choose a positive integer m so large that

$$(5)\qquad D(0, 1 + 1/m) \subset \mathbb{C}\backslash K,$$

and

$$(6)\qquad \frac{2}{m} < \operatorname{dist}(\gamma[0, 1], K).$$

Then choose n so that

$$(7)\qquad \sup_{t\in[0,1]} |f_n(t) - \gamma(t)| < 1/m.$$

It follows from (7) and (6) that

$$(8)\qquad \bigcup_{t\in[0,1]} D(f_n(t), 1/m) \subset \mathbb{C}\backslash K$$

and also that

$$\begin{aligned}\sup|K| < r - 1 = \gamma(1) - 1 &\leq |f_n(1)| + |\gamma(1) - f_n(1)| - 1\\ &< |f_n(1)| + 1/m - 1\\ &\leq |f_n(1)| - 1/m \quad (\text{if } m \geq 2),\end{aligned}$$

that is,

$$(9)\qquad K \subset D(0, |f_n(1)| - 1/m).$$

It follows from (5), (8) and (9) that $K \subset K(n, m)$.

Exercise 1.43 *Let U be an open subset of $\mathbb{C}$ and $C_1, \ldots, C_M$ distinct bounded components of $\mathbb{C}\backslash U$. Show that there exists compact $K \subset U$ such that $C_1, \ldots, C_M$ lie in distinct bounded components of $\mathbb{C}\backslash K$.*

Hints: Choose bounded, open $V_j \supset C_j$ such that $V_1, \ldots, V_M$ are pairwise disjoint. Apply 1.34 to select compact $K_j \subset \mathbb{C}\backslash U$, relatively open in $\mathbb{C}\backslash U$ and satisfying $C_j \subset K_j \subset V_j$. $\mathbb{C}\backslash(U \cup K_1 \cup\cdots\cup K_M) = (\mathbb{C}\backslash U)\backslash(K_1 \cup\cdots\cup K_M)$ is relatively closed in $\mathbb{C}\backslash U$ (since the K_j are relatively open there) and so is closed in $\mathbb{C}$. Therefore $U \cup K_1 \cup \cdots \cup K_M$ is open in $\mathbb{C}$. Consequently there exist U_j open in $\mathbb{C}$ such that

(1) $K_j \subset U_j$ open $\subset \bar{U}_j$ compact $\subset U \cup K_1 \cup \cdots \cup K_M$

and

(2) the $\bar{U}_j$ are mutually disjoint.

Define

$$K = \partial U_1 \cup \cdots \cup \partial U_M.$$

This set is compact and is disjoint from $K_1 \cup \cdots \cup K_M$ by (1) and (2), but lies in $U \cup K_1 \cup \cdots \cup K_M$ by (1). Therefore $K \subset U$. (1) and (2) also show that $U_j \subset \mathbb{C}\backslash K$. Since $\bar{U}_j\backslash U_j = \partial U_j \subset K$, we have $U_j = U_j \cap (\mathbb{C}\backslash K) = \bar{U}_j \cap (\mathbb{C}\backslash K)$, so U_j is open and relatively closed in $\mathbb{C}\backslash K$. Hence U_j contains any component of $\mathbb{C}\backslash K$ which it meets. As $C_j \subset U_j$ and the U_j are bounded and disjoint, it follows that $C_1, \ldots, C_M$ lie in different bounded components of $\mathbb{C}\backslash K$.

§ 6 Extension Theorems

These results, pretty enough in themselves, will be of the utmost importance in Chapter IV when we want to produce logarithms.

Theorem 1.44 (Lebesgue [1907]) *Let C be a closed subset of $\mathbb{R}^n$ and $f: C \to \mathbb{R}$ a continuous function. Then f has a continuous extension $F: \mathbb{R}^n \to \mathbb{R}$. If in addition f is bounded, then such an F exists which satisfies*

$$\inf F(\mathbb{R}^n) = \inf f(C), \qquad \sup F(\mathbb{R}^n) = \sup f(C).$$

Proof: (H. Bohr) The reader may check the routine details involved in the following outline. For each $r \geq 0$ and $x \in \mathbb{R}^n$ define

$$d(x) = \inf\{\|x - y\| : y \in C\}.$$

This is a continuous function by 1.32. Set

$$\bar{B}(x, r) = \{y \in \mathbb{R}^n: \|x - y\| \leq r\}$$
$$S = \{(x, r) \in \mathbb{R}^n \times [0, \infty): d(x) \leq r\}$$
$$\psi(x, r) = \sup f(C \cap \bar{B}(x, r)), \quad (x, r) \in S.$$

Since f is uniformly continuous on each compact subset of C, it is easy to check that ψ is continuous on the closed set S. Form

$$\phi(x) = \frac{1}{d(x)} \int_{d(x)}^{2d(x)} \psi(x, r)dr, \quad x \in \mathbb{R}^n\backslash C.$$

Prove that ϕ is continuous and satisfies in addition $\phi(x_k) \to f(x)$ whenever $x_k \in \mathbb{R}^n \backslash C$ and $x_k \to x \in C$. Finally define the desired extension by

$$F = \begin{cases} f & \text{in } C \\ \phi & \text{in } \mathbb{R}^n \backslash C. \end{cases}$$

Theorem 1.45 (Borsuk [1937]) *Let A be a compact subset of a closed subset B of $\mathbb{R}^n$. Let $f\colon A \to \mathbb{C}\backslash\{0\}$ and $g\colon B \to \mathbb{C}\backslash\{0\}$ be continuous and suppose there exists a continuous function $h\colon A \times [0, 1] \to \mathbb{C}\backslash\{0\}$ so that*

$$\left.\begin{aligned} h(x, 0) &= f(x) \\ h(x, 1) &= g(x) \end{aligned}\right\} \quad x \in A.$$

Then f has a continuous extension $F\colon B \to \mathbb{C}\backslash\{0\}$.

Proof: Let $C = A \times [0, 1] \cup B \times \{1\}$ and extend h to C by the decree $h(x, 1) = g(x)$ for all $x \in B$. Apply the last theorem to Re h and Im h separately to come up with a continuous $H\colon \mathbb{R}^{n+1} \to \mathbb{C}$ which extends h. Since $H(C) = h(C)$ lies in $\mathbb{C}\backslash\{0\}$, $H^{-1}(\mathbb{C}\backslash\{0\})$ is an open neighborhood of C. The compact sets $A_k = \{x \in \mathbb{R}^n\colon d(x, A) \leq 1/k\}$ satisfy

$$\bigcap_{k=1}^{\infty} A_k \times [0, 1] = A \times [0, 1] \subset C \subset H^{-1}(\mathbb{C}\backslash\{0\}).$$

Therefore by compactness there is some k such that

$$A_k \times [0, 1] \subset H^{-1}(\mathbb{C}\backslash\{0\}).$$

Then for every $x \in B$ the point $(x, \min\{1, kd(x, A)\})$ lies in $A_k \times [0, 1] \cup B \times \{1\} \subset H^{-1}(\mathbb{C}\backslash\{0\})$ and so $F(x) = H(x, \min\{1, kd(x, A)\})$ is the desired continuous extension of f.

Remark: This is called Borsuk's homotopy extension theorem. The latter notion will be defined and explored in Chapter IV.

Exercise 1.46 *Let $a, b, A, B \in \mathbb{R}$ with $a < b$, $A < B$ and ϕ be a homeomorphism of (a, b) onto (A, B). Show that ϕ extends to a homeomorphism of $[a, b]$ onto $[A, B]$ with endpoints mapping to endpoints.*

Hints: After changes of scale and translations we can assume $a = A = 0$, $b = B = 1$. For each integer $n \geq 2$ the set $\phi^{-1}[1/n, (n - 1)/n]$ is a compact connected subset of $(0, 1)$ and so is an interval $[a_n, b_n]$, say. Since ϕ is one-to-one, $\phi(0, a_2)$ is a connected subset of $(0, \frac{1}{2}) \cup (\frac{1}{2}, 1)$ and so $\phi(0, a_2)$ lies wholly in one of these intervals. Replacing ϕ by $1 - \phi$ if necessary, we can assume that

(1) $\quad \phi(0, a_2) \subset (0, \frac{1}{2})$.

Since $\phi(0, a_2) \cup \phi(b_2, 1) = (0, \frac{1}{2}) \cup (\frac{1}{2}, 1)$, it follows from (1) that

(2) $\quad \phi(b_2, 1) \subset (\frac{1}{2}, 1)$.

Again the univalence of ϕ and the definition of a_n and b_n give

$$\phi(0, a_n) \cup \phi(b_n, 1) \subset \left(0, \frac{1}{n}\right) \cup \left(\frac{n-1}{n}, 1\right) \quad \forall n \geq 2$$

and from (1), (2) and connectedness then follow

$$(3) \qquad \phi(0, a_n) \subset \left(0, \frac{1}{n}\right) \quad \text{and} \quad \phi(b_n, 1) \subset \left(\frac{n-1}{n}, 1\right) \quad \forall n \geq 2.$$

The conclusions

$$\lim_{t \downarrow 0} \phi(t) \text{ exists and equals } 0$$

and

$$\lim_{t \uparrow 1} \phi(t) \text{ exists and equals } 1$$

are immediate from (3).

Exercise 1.47 *Replace the hypothesis on ϕ in* 1.46 *by the formally weaker one that it be (continuous and) bijective,* i.e., *one-to-one and onto (with no* a priori *continuity claim for the inverse). Prove that then it is either strictly increasing or strictly decreasing and (so) the conclusion of* 1.46 *still follows.*

Notes to Chapter I

The results of this chapter belong to the realm of general point-set topology and all of them are valid in the appropriately general topological space (most in any [Hausdorff] topological space; some require local compactness or connectedness). For some of the history of some of the concepts here consult the notes at the end of HAUSDORFF's books [1914], [1957]. On the important concept of connectedness see especially FREUDENTHAL's notes on pp. 486–488 of BROUWER [1976].

In almost any function theory book (e.g., CONWAY [1973], pp. 8–10) the reader will find a discussion of the stereographic projection of the plane onto the sphere $\mathbb{C}_\infty = \{(x_1, x_2, x_3) \in \mathbb{R}^3 : x_1^2 + x_2^2 + x_3^2 = 1\}$ minus its north pole $(0, 0, 1)$. One regards $\mathbb{C}$ as a subset of $\mathbb{C}_\infty$ and does all of function theory on the sphere. This has several advantages (not the least being esthetical), since now the point "at infinity" does not enjoy a privileged role (topologically, at least). The statement and proofs of many results thus acquire greater cohesion and uniformity. This point of view is essential in the theory of Riemann surfaces and differentiable manifolds and in the study of meromorphic functions. A penalty is, however, extracted, which we have elected not to pay; namely, the consideration of cases returns when one engages in algebraic operations. $\mathbb{C}_\infty$ is called the Riemann sphere—see § 8 of OSGOOD [1901].

Simple connectivity for subsets S of $\mathbb{C}$ is usually defined as $\mathbb{C}_\infty \backslash S$ being connected. It is not hard to show that our definition agrees with this one when S is bounded or open, essentially the only kinds of sets with which we will be working. For arbitrary S, however, the two definitions do not agree. To see this, consider a connected subset C of $\mathbb{C}_\infty$ with an "explosion point" x, i.e., a point x such that $C\backslash\{x\}$ is totally disconnected (meaning that each component is a single point). Such exotica indeed exist; see page 145 of STEEN and SEEBACH [1970]. We may situate C in $\mathbb{C}_\infty$ so that $x = (0, 0, 1)$. Then let $S = \mathbb{C}_\infty\backslash C$, a subset of $\mathbb{C}_\infty\backslash\{(0,0,1)\} = \mathbb{C}$. Then of course this set has a connected complement in $\mathbb{C}_\infty$, but its complement in $\mathbb{C}$ is the totally disconnected set $C\backslash\{(0, 0, 1)\}$ and so has infinitely many bounded components.

A much more general version of 1.14 is true: any two points on any curve lie on an arc which lies on the curve. For a proof of this surprisingly non-trivial fact (due to R. L. MOORE [1916/17]) see TIETZE [1919], MAZURKIEWICZ [1920], KAMIYA [1931] or G. T. WHYBURN [1932]. This fact would be quite useful in Chapter IX (in the proof of 9.14) but appeal to it there will be circumvented by using 1.16 and 1.29. Contained in the proof of 1.19(ii) is a famous result of CARATHÉODORY [1911] (p. 200): if $S \subset \mathbb{R}^k$, then every element of co S is a convex combination of $k + 1$ or fewer points of S. The proof in the text is due to STEINITZ [1913] (p. 153).

The idea of the proof of 1.35 comes from HEINS [1962] (p. 150) and that for 1.39 is from SAKS and ZYGMUND [1971] (p. 210). See also DE VITO [1957]. Definition 1.36 is due to HAUSDORFF (p. 351 of his book [1914]) and supersedes a less tractable one which Carathéodory had made earlier.

There is another proof of 1.44 (due to F. Riesz) in DIEUDONNÉ [1969] (pp. 89, 90). It is quite similar to the second of two proofs which HAUSDORFF [1919] offers. Both of Hausdorff's proofs are elegant and interesting; the first involves inserting a continuous function between an upper semicontinuous function and a lower semicontinuous function. (See also pp. 276–277 and 281–283 of HAUSDORFF [1957].) For other proofs see R. RADO [1931] and VIZZINI [1932]. Theorem 1.44 was extended to metric spaces by TIETZE [1914] and is often called the "Tietze extension theorem"; this designation has even come to be applied to the euclidean space version of the theorem. The plane version of 1.44 with C compact can already be found in PHRAGMÉN [1900].

Chapter II
(Complex) Derivative and (Curvilinear) Integrals

This short chapter is comprised of very basic and very easy material, probably familiar to most readers in one form or other. Even the reader with only modest experience can probably supply his own proofs for most of these results as quickly as he can read mine! We record them here for later reference and to fix notation and terminology.

§ 1 Holomorphic and Harmonic Functions

Definition 2.1 If f is a complex-valued function defined in a neighborhood of the point z_0, we say f is *differentiable at* z_0 if the quotient of complex numbers $(f(z) - f(z_0))/(z - z_0)$ has a limit in $\mathbb{C}$ as z approaches z_0. When it exists, this limit is called the *derivative of* f *at* z_0 and is noted $f'(z_0)$. If f is differentiable at each point of an open subset U of $\mathbb{C}$, we say f is *differentiable in* U or *holomorphic in* U and by $H(U)$ denote the set of all such functions. We denote by $H^\infty(U)$ the subset of all functions in $H(U)$ which are bounded on U. A one-to-one holomorphic function is called *conformal*; other synonyms are *injective*, *univalent* and *schlicht*. A function which is holomorphic in $\mathbb{C}$ is called *entire*.

Remarks: Often the term conformal is bestowed on holomorphic functions which are only locally univalent, i.e., univalent in some neighborhood of each point. We shall see later (5.89) that this just means that the derivative is never zero. The rationale for this term may be found in the nice discussion on pp. 73–75 of AHLFORS [1966a]. It should also be noted that British usage for entire is *integral*. Sometimes *analytic* is also used for holomorphic, but equally often that term has a more technical use. (See chapter VI of SAKS and ZYGMUND [1971].) The terms *monogenic* and *regular*, now largely archaic, are other synonyms for holomorphic. But by now perhaps we are teetering on the edge of Ockham's Razor.

Definition 2.2 A function f defined in an open set U is called *harmonic in* U if U is a union of open sets in each of which f is the real part of some holomorphic function; that is, for each $a \in U$ there is an $r > 0$ and a function F_a holomorphic in $D(a, r)$ such that $f = \text{Re}\, F_a$ in $U \cap D(a, r)$.

Remarks: This is not the customary definition. Later the myriad equivalences of this important concept will be established. One of them is the traditional definition of harmonic: f satisfies Laplace's partial differential equation. (See 9.29, 9.31.) A complex-valued function F is defined to be harmonic if each of $\text{Re}\, F$ and $\text{Im}\, F$ is harmonic.

The expected elementary results prevail (and thus, for example, $H(U)$ and $H^\infty(U)$ are algebras):

Theorem 2.3 *Let f, g, h be complex-valued functions with f and g defined in a neighborhood of z_0 and differentiable at z_0 and h defined in a neighborhood of $f(z_0)$ and differentiable at $f(z_0)$. Then*

(i) *f is continuous at z_0.*

(ii) *$f + g$ and $f \cdot g$ are each differentiable at z_0 and* $(f + g)'(z_0) = f'(z_0) + g'(z_0)$, $(f \cdot g)'(z_0) = f'(z_0)g(z_0) + f(z_0)g'(z_0)$.

(iii) *The composite $h \circ f$ is defined in a neighborhood of z_0, is differentiable at z_0 and $(h \circ f)'(z_0) = h'(f(z_0)) \cdot f'(z_0)$. The latter formula is called the* Chain Rule.

Proof: For (i) write

$$|f(z) - f(z_0)| = |z - z_0|\left|\frac{f(z) - f(z_0)}{z - z_0}\right|$$

and let z approach z_0. For (ii) write

$$\frac{(f \cdot g)(z) - (f \cdot g)(z_0)}{z - z_0} = \frac{f(z)g(z) - f(z_0)g(z_0)}{z - z_0}$$

$$= \frac{f(z) - f(z_0)}{z - z_0} \cdot g(z_0) + f(z) \cdot \frac{g(z) - g(z_0)}{z - z_0},$$

let z approach z_0 and cite (i). For (iii) define two functions δ and ε near z_0 and $f(z_0)$, respectively, by

$$\text{(1)} \qquad \delta(z) = \begin{cases} \dfrac{f(z) - f(z_0)}{z - z_0} - f'(z_0) & z \neq z_0 \\ 0 & z = z_0, \end{cases}$$

$$\text{(2)} \qquad \varepsilon(w) = \begin{cases} \dfrac{h(w) - h(f(z_0))}{w - f(z_0)} - h'(f(z_0)), & w \neq f(z_0) \\ 0 & w = f(z_0). \end{cases}$$

The definitions of $f'(z_0)$ and $h'(f(z_0))$ are just such as to make δ and ε continuous at z_0 and $f(z_0)$, respectively. Then for all z sufficiently near z_0 we get from (1)

$$\text{(3)} \qquad f(z) - f(z_0) = (z - z_0)[f'(z_0) + \delta(z)]$$

and for all w sufficiently near $f(z_0)$ we get from (2)

$$\text{(4)} \qquad h(w) - h(f(z_0)) = (w - f(z_0))[h'(f(z_0)) + \varepsilon(w)].$$

Now by (i), $f(z)$ is an eligible w for z sufficiently close to z_0. For such z insertion of (3) into (4) gives

$$h(f(z)) - h(f(z_0)) = (z - z_0)[f'(z_0) + \delta(z)][h'(f(z_0)) + \varepsilon(f(z))],$$

that is,

$$(5)\qquad \frac{(h\circ f)(z)-(h\circ f)(z_0)}{z-z_0}=[f'(z_0)+\delta(z)][h'(f(z_0))+\varepsilon(f(z))].$$

Let z approach z_0. Then $\delta(z)$ and $f(z)$ approach $\delta(z_0)=0$ and $f(z_0)$ respectively and so $\varepsilon(f(z))$ approaches $\varepsilon(f(z_0))=0$ and the assertion (iii) follows from (5).

Exercise 2.4 *Let f be holomorphic in an open set U. Define $U^*=\{\bar{z}: z\in U\}$ and f^* on U^* by $f^*(z)=\bar{f}(\bar{z})$. Show that f^* is holomorphic in U^*.*

Hint: For $z, z_0\in U^*$ with $z\neq z_0$ we have $\bar{z}, \bar{z}_0\in U$ and

$$\frac{f^*(z)-f^*(z_0)}{z-z_0}=\left[\frac{f(\bar{z})-f(\bar{z}_0)}{\bar{z}-\bar{z}_0}\right]^{-}.$$

Then f^* is differentiable at z_0 with derivative $\overline{f'(\bar{z}_0)}$.

Exercise 2.5 *Consider the function*

$$\phi(z)=\frac{z+1}{z-1}=1+\frac{2}{z-1},\quad z\in\mathbb{C}\backslash\{1\}.$$

(i) *Show that ϕ is univalent, its range is $\mathbb{C}\backslash\{1\}$ and $\phi^{-1}=\phi$.*
(ii) *Show that ϕ maps $D(0,1)$ onto $(-\infty,0)\times\mathbb{R}$ and maps $\mathbb{C}\backslash\bar{D}(0,1)$ onto $(0,\infty)\times\mathbb{R}\backslash\{1\}$.*

Hints: If $w=\phi(z)$, then compute

$$|w|^2-1=\frac{z+1}{z-1}\cdot\frac{\bar{z}+1}{\bar{z}-1}-1=\frac{2(z+\bar{z})}{|z-1|^2}=\frac{4}{|z-1|^2}\operatorname{Re} z.$$

(iii) *Show that ϕ maps $[-1,1)$ onto $(-\infty,0]$ and (consequently) maps $\mathbb{C}\backslash[-1,1]$ onto $\mathbb{C}\backslash(-\infty,0]\backslash\{1\}$.*
(iv) *Show that $\psi(z)=z+1/z$ maps $\mathbb{C}\backslash\bar{D}(0,1)$ conformally onto $\mathbb{C}\backslash[-2,2]$.*

Hints: Set $s(z)=z^2$ and note that in

$$\mathbb{C}\backslash\bar{D}(0,1)\xrightarrow{\phi}(0,\infty)\times\mathbb{R}\backslash\{1\}\xrightarrow{s}\mathbb{C}\backslash(-\infty,0]\backslash\{1\}\xrightarrow{\phi}\mathbb{C}\backslash[-1,1]$$

all maps are one-to-one and onto. Then check that $\psi(z)=2\phi(\phi^2(z))$.

Exercise 2.6 *Let $D=D(0,1)$ and for each $a\in D$ define f_a by*

$$f_a(z)=\frac{z-a}{1-\bar{a}z}.$$

Show that f_a is one-to-one and continuous on $\bar{D}$, holomorphic in a neighborhood of $\bar{D}$, that $f_a(D)=D$, $f_a(\bar{D})=\bar{D}$ and that the inverse of f_a on $\bar{D}$ is f_{-a}. (See 6.2(ii) for the converse of this.)

Hints: Evidently f_a is holomorphic in $\mathbb{C}\backslash\{1/\bar{a}\} \supset \bar{D}$ in case $a \neq 0$ and entire if $a = 0$. Direct computation leads, for $z \in D$, to

$$\left|\frac{z-a}{1-\bar{a}z}\right|^2 = 1 - \frac{(1-|z|^2)(1-|a|^2)}{|1-\bar{a}z|^2} < 1.$$

It follows that $f_a(D) \subset D$. Another simple calculation confirms that for all $z \in \bar{D}$, $f_{-a}(f_a(z)) = z$. Thus f_a is one-to-one in $\bar{D}$. Applying all the conclusions so far secured to $-a$ in the role of a, we learn that for all $z \in D$, $f_{-a}(z) \in D$ and $f_a(f_{-a}(z)) = z$. Thus $f_a(D) \supset D$. We now have $f_a(D) = D$. By continuity then $f_a(\bar{D}) = \bar{D}$ and all claims about f_a and f_{-a} are established.

§ 2 Integrals along Curves

Exercise 2.7 *Let $\gamma\colon [a, b] \to \mathbb{C}$ be a piecewise smooth curve. Let, say, $a = t_0 < t_1 < \cdots < t_n = b$, $a = s_0 < s_1 < \cdots < s_m = b$ and γ be continuously differentiable on each $[t_{j-1}, t_j]$ ($j = 1, 2, \ldots, n$) and on each $[s_{k-1}, s_k]$ ($k = 1, 2, \ldots, m$). (Recall definition* 1.11(ii).)

(i) *Verify that*

$$\sum_{j=1}^{n}\int_{t_{j-1}}^{t_j}|\gamma'| = \sum_{k=1}^{m}\int_{s_{k-1}}^{s_k}|\gamma'|.$$

(ii) *Verify that for each complex-valued function f which is continuous on the range of γ*

$$\sum_{j=1}^{n}\int_{t_{j-1}}^{t_j}(f\circ\gamma)\gamma' = \sum_{k=1}^{m}\int_{s_{k-1}}^{s_k}(f\circ\gamma)\gamma'.$$

Hint: Take a "common refinement" of the two partitions of $[a, b]$.

Definition 2.8 The sums in (i) and (ii) above are noted $l(\gamma)$ and $\int_\gamma f$ (or sometimes $\int_\gamma f(\xi)d\xi$) and called *the length of γ* and *the integral of f along γ*, respectively.

Remarks: My attitude toward curves and integrals is frankly utilitarian. There is no need to integrate over curves more general than piecewise smooth ones, so we will not. There is, however, some esthetic loss in *defining* the length of (a smooth) γ to be $\int|\gamma'|$. One ought to call a curve $\gamma\colon [a, b] \to \mathbb{C}$ *rectifiable* if there is a finite M such that the length $\sum_{j=1}^{n}|\gamma(t_j) - \gamma(t_{j-1})|$ of the polygon $[\gamma(t_0), \ldots, \gamma(t_n)]$ inscribed in γ is not greater than M whatever be the points $a \leq t_0 \leq t_1 \leq \cdots \leq t_n \leq b$. In such cases call the least upper bound of the lengths of these inscribed polygons the *length* of γ. It is then a theorem (and an easy one, see RUDIN [1976], p. 137) that if γ is smooth, then it is rectifiable (see 2.13(ii) below) and its length in the above sense equals $\int|\gamma'|$. Only this latter integral intervenes in our subsequent considerations and I chose for it its traditional name, the length of γ; the above theorem is the rationale for this.

Again esthetic considerations might lead one to want to have $\int_\gamma f$ defined for any curve γ (lying in the domain of the holomorphic function f). After Cauchy's theorem (in a disk) is proved in Chapter V, this can easily be done by approximating γ uniformly with piecewise smooth curves (whether γ is rectifiable in the above sense or not). See PERRY AND YOUNGS [1947] or REDHEFFER [1969] for details. (Cf. also *Mathematical Reviews* vol. 53 #3270.)

Exercise 2.9 (i) *If γ is a piecewise smooth curve and f is continuous on the range of γ and M is a bound for $|f|$ there, then*

$$\left|\int_\gamma f\right| \leq M \cdot l(\gamma).$$

(ii) *If $a_0, \ldots, a_n \in \mathbb{C}$ and f is continuous on $[a_0, \ldots, a_n]$, then*

$$l([a_0, \ldots, a_n]) = \sum_{j=1}^{n} |a_j - a_{j-1}|,$$

$$\int_{[a_0,\ldots,a_n]} f = \sum_{j=1}^{n} \int_{[a_{j-1},a_j]} f,$$

$$\int_{[a_0,a_1]} f = -\int_{[a_1,a_0]} f,$$

and if $a_1 \in [a_0, a_2]$, then

$$\int_{[a_0,a_2]} f = \int_{[a_0,a_1]} f + \int_{[a_1,a_2]} f.$$

Theorem 2.10 *Let U be an open subset of $\mathbb{C}$, $F \in H(U)$ and suppose that F' is continuous on U. [This last hypothesis will later be shown to be redundant.] Then for any piecewise smooth curve γ in U with initial point z_0 and terminal point z_1*

(i) $$\int_\gamma F' = F(z_1) - F(z_0).$$

If in addition γ is closed, then

(ii) $$\int_\gamma F' = 0.$$

If U is connected, there is for each z_0 and z in U a polygon γ_z in U with initial point z_0 and terminal point z (by 1.28). For any such polygons

(iii) $$F(z) = F(z_0) + \int_{\gamma_z} F' \quad \forall z \in U.$$

(When U is starlike with respect to z_0, the interval $[z_0, z]$ is an appropriate γ_z.)

(iv) *If U is connected and $F' = 0$ throughout U, then F is constant in U.*

Proof: If the domain of γ is $[a, b]$ and $a = t_0 < t_1 < \cdots < t_n = b$ are such that γ is continuously differentiable on each $[t_{j-1}, t_j]$, then by definition

$$\begin{aligned}
\int_\gamma F' &= \sum_{j=1}^{n} \int_{t_{j-1}}^{t_j} (F' \circ \gamma)\gamma' \\
&= \sum_{j=1}^{n} \int_{t_{j-1}}^{t_j} (F \circ \gamma)' && \text{(Chain Rule)} \\
&= \sum_{j=1}^{n} [(F \circ \gamma)(t_j) - (F \circ \gamma)(t_{j-1})] && \text{(Fundamental Theorem of Calculus)} \\
&= (F \circ \gamma)(t_n) - (F \circ \gamma)(t_0) = F(\gamma(b)) - F(\gamma(a)) \\
&= F(z_1) - F(z_0) && \text{(by definition of } z_1 \text{ and } z_0).
\end{aligned}$$

This establishes (i), and (ii), (iii) and (iv) are trivial corollaries of (i).

There is a converse to part (ii):

Theorem 2.11 *Let Ω be a region and f a continuous function in Ω with the property that $\int_\gamma f = 0$ for every piecewise smooth loop γ in Ω. Then there is a (continuously) differentiable function F in Ω such that $F' = f$* (cf. 5.39).

Proof: According to 1.28 Ω is polygonally connected. Fix $z_0 \in \Omega$ and for each $z \in \Omega$ let γ_z be a polygon in Ω with initial point z_0 and terminal point z and define

$$F(z) = \int_{\gamma_z} f.$$

We will show that at each $z_1 \in \Omega$, $F'(z_1)$ exists and equals $f(z_1)$. Pick $r > 0$ so that $D(z_1, r) \subset \Omega$. If $z \in D(z_1, r)$, we may consider the piecewise smooth loop $\gamma\colon [0, n + m + 1] \to \Omega$ defined by

$$\gamma(t) = \begin{cases} \gamma_{z_1}(t) & 0 \le t \le n \\ z_1 + (t - n)(z - z_1) & n \le t \le n + 1 \\ \gamma_z(n + m + 1 - t) & n + 1 \le t \le n + m + 1, \end{cases}$$

where $[0, n]$, $[0, m]$ are the parameter intervals of γ_{z_1} and γ_z respectively. After some trivial changes of variable we see that

$$\int_\gamma f = \int_{\gamma_{z_1}} f + \int_{[z_1, z]} f - \int_{\gamma_z} f = F(z_1) - F(z) + \int_{[z_1, z]} f.$$

By hypothesis this integral is zero, that is,

$$F(z) - F(z_1) = \int_{[z_1, z]} f = \int_0^1 f(z_1 + t(z - z_1))(z - z_1)dt,$$

whence

$$\left|\frac{F(z) - F(z_1)}{z - z_1} - f(z_1)\right| = \left|\int_0^1 [f(z_1 + t(z - z_1)) - f(z_1)]dt\right|$$
$$\leq \sup_{0 \leq t \leq 1} |f(z_1 + t(z - z_1)) - f(z_1)|.$$

From the continuity of f at z_1 the desired conclusion is now immediate.

Definition 2.12 For a subset S of $\mathbb{C}$ and a bounded function $f: S \to \mathbb{C}$ let

$$\omega_f(\delta) = \sup\{|f(z) - f(w)| : z, w \in S, |z - w| \leq \delta\}$$

for each $\delta \geq 0$. The function $\omega_f: [0, \infty) \to [0, \infty)$ so defined is called the *oscillation of* f or, in case f is continuous on S, the *modulus of (uniform) continuity of* f.

Exercise 2.13 (i) *Let S be a subset of $\mathbb{C}$ and $f: S \to \mathbb{C}$ a bounded function. Show that ω_f is continuous at 0 if and only if f is uniformly continuous on S.*

(ii) *Let $\gamma: [0, 1] \to \mathbb{C}$ be a piecewise smooth curve. Show that*

$$\sum_{j=1}^{n} |\gamma(t_j) - \gamma(t_{j-1})| \leq l(\gamma)$$

whenever $0 = t_0 < t_1 < \cdots < t_n = 1$.

Hint: By adding more points t_j the left side is not decreased, so we can suppose that all the points of discontinuity of γ' are among the t_j. Then $\gamma(t_j) - \gamma(t_{j-1}) = \int_{t_{j-1}}^{t_j} \gamma'$ and $l(\gamma) = \sum_{j=1}^{n} \int_{t_{j-1}}^{t_j} |\gamma'|$.

(iii) *With $\gamma, t_0, \ldots, t_n$ as above, set $\delta = \max_{1 \leq j \leq n}(t_j - t_{j-1})$. Show that if f is a continuous function on the range of γ and all the points of discontinuity of γ' are among the t_j, then*

$$\left|\int_\gamma f - \sum_{j=1}^{n} f \circ \gamma(t_j)[\gamma(t_j) - \gamma(t_{j-1})]\right| \leq \omega_{f\circ\gamma}(\delta) l(\gamma).$$

Hint: Use the hint to (ii) and the fact that

$$\int_\gamma f = \sum_{j=1}^{n} \int_{t_{j-1}}^{t_j} (f \circ \gamma)\gamma'.$$

§ 3 Differentiating under the Integral

We will use (and prove) as we progress, several successively stronger theorems on differentiating behind the integral sign. (See e.g., 7.18.) The easiest and most useful of these is codified in

Theorem 2.14 *(Cf. with 3.8.) Let γ be a piecewise smooth curve, g a function continuous on the range of γ. Then the functions*

$$G_n(z) = \int_\gamma \frac{g(\xi)}{(\xi - z)^n}\, d\xi, \quad n = 1, 2, 3, \ldots$$

are (infinitely) differentiable in the complement of the range of γ and satisfy

$$(*) \qquad G_n' = nG_{n+1}.$$

Proof: We base the proof on a simple algebraic identity which we derive first. Consider two unequal, non-zero complex numbers a and b.

$$\frac{1}{b^n} - \frac{1}{a^n} = \frac{1}{b^n a^n}(a^n - b^n) = \frac{a-b}{b^n a^n}\sum_{j=0}^{n-1} b^{n-1-j}a^j$$

$$= \frac{a-b}{a^{n+1}}\sum_{j=0}^{n-1}\left(\frac{a}{b}\right)^{j+1} = \frac{a-b}{a^{n+1}}\sum_{j=1}^{n}\left(1 + \frac{a-b}{b}\right)^j$$

$$= \frac{a-b}{a^{n+1}}\sum_{j=1}^{n}\sum_{k=0}^{j}\binom{j}{k}\left(\frac{a-b}{b}\right)^k.$$

$$(1) \qquad \frac{1}{a-b}\left(\frac{1}{b^n} - \frac{1}{a^n}\right) - \frac{n}{a^{n+1}} = \frac{1}{a^{n+1}}\sum_{j=1}^{n}\left[\sum_{k=0}^{j}\binom{j}{k}\left(\frac{a-b}{b}\right)^k - 1\right]$$

$$= \frac{1}{a^{n+1}}\sum_{j=1}^{n}\sum_{k=1}^{j}\binom{j}{k}\left(\frac{a-b}{b}\right)^k$$

$$= \frac{1}{a^{n+1}}\sum_{j=1}^{n}\sum_{l=0}^{j-1}\binom{j}{l+1}\left(\frac{a-b}{b}\right)^{l+1}$$

$$= (a-b)\cdot\frac{1}{a^{n+1}b}\cdot\sum_{j=1}^{n}\sum_{l=0}^{j-1}\binom{j}{l+1}\left(\frac{a-b}{b}\right)^{l}.$$

If $|a - b| \le r$ while $|a| \ge r$ and $|b| \ge r$, then this gives

$$\left|\frac{1}{a-b}\left(\frac{1}{b^n} - \frac{1}{a^n}\right) - \frac{n}{a^{n+1}}\right| \le \frac{|a-b|}{r^{n+2}}\sum_{j=1}^{n}\sum_{l=0}^{j-1}\binom{j}{l+1}$$

$$(2) \qquad \le \frac{|a-b|}{r^{n+2}}\sum_{j=1}^{n}(1+1)^j \le \frac{2^{n+1}}{r^{n+2}}|a-b|.$$

Consider now any z, z_0 not on γ. We have

$$(3) \qquad \frac{G_n(z) - G_n(z_0)}{z - z_0} - nG_{n+1}(z_0)$$

$$= \int_\gamma g(\xi)\left[\frac{1}{z - z_0}\left(\frac{1}{(\xi - z)^n} - \frac{1}{(\xi - z_0)^n}\right) - \frac{n}{(\xi - z_0)^{n+1}}\right]d\xi.$$

Now choose $r > 0$ so small that $D(z_0, 2r)$ is disjoint from γ and consider only $z \in D(z_0, r)$ with $z \neq z_0$. Then $|\xi - z| \geq r$ and $|\xi - z_0| \geq r$ for any $\xi \in \gamma$ and so for the last integrand above, (2) provides the upper bound

$$|g(\xi)|\,|z - z_0| \cdot \left(\frac{2}{r}\right)^{n+2}.$$

From this and (3) follows (recalling 2.9(i))

$$\left|\frac{G_n(z) - G_n(z_0)}{z - z_0} - nG_{n+1}(z_0)\right| \leq M|z - z_0|\left(\frac{2}{r}\right)^{n+2} l(\gamma),$$

where M is the maximum of $|g|$ on γ. Now let z approach z_0 to see that $G_n'(z_0)$ exists and equals $nG_{n+1}(z_0)$.

Exercise 2.15 *If γ is a piecewise smooth loop, show that for every integer $n \neq 1$*

$$\int_\gamma \frac{dz}{(z - w)^n} = 0 \quad \forall w \in \mathbb{C}\backslash\gamma.$$

Hint: For such n the integrand is a derivative, so 2.10(ii) may be cited.

§ 4 A Useful Sufficient Condition for Differentiability

If U is an open subset of $\mathbb{C}$ and $f: U \to \mathbb{C}$ a function, there are various conditions on f which insure that it is holomorphic in U. (See, for example, Morera's theorem 5.39.) One such condition is that f have continuous first partial derivatives which satisfy the so-called *Cauchy–Riemann equation* in U (see p. 13 of OSGOOD [1901]):

(C.R.) $D_1 f = -iD_2 f.$

There is a slick proof in HOFFMAN [1975], p. 202. Here we prove a stronger pointwise result by more pedestrian means; 2.18 below gives a slight extension.

Theorem 2.16 *Let U be an open subset of $\mathbb{C}$, $z_0 \in U$, $f: U \to \mathbb{C}$ a function such that $D_1 f$ and $D_2 f$ exist in U and are continuous at z_0. Suppose that $D_1 f(z_0) = -iD_2 f(z_0)$. Then f is differentiable at z_0 and $f'(z_0) = D_1 f(z_0)$.*

Proof: Write $z_0 = (x_0, y_0)$ and consider $r > 0$ such that $Q_r = [x_0 - r, x_0 + r] \times [y_0 - r, y_0 + r] \subset U$. Consider in what follows real h, k with $|h|, |k| \leq r$. The hypothesis $D_1 f(z_0) = -iD_2 f(z_0)$ gives

$$\begin{aligned}
&|f(x_0 + h, y_0 + k) - f(x_0, y_0) - (h, k)D_1 f(x_0, y_0)| \\
&\quad \leq |f(x_0 + h, y_0 + k) - f(x_0, y_0 + k) \\
&\qquad\qquad - h[\mathrm{Re}\, D_1 f(x_0, y_0) + i\,\mathrm{Im}\, D_1 f(x_0, y_0)]| \\
&\qquad + |f(x_0, y_0 + k) - f(x_0, y_0) \\
&\qquad\qquad - k[\mathrm{Re}\, D_2 f(x_0, y_0) + i\,\mathrm{Im}\, D_2 f(x_0, y_0)]|.
\end{aligned}$$

Apply the Mean Value Theorem to the real and imaginary parts of each expression on the right to come up with real numbers a, b, c, d (dependent on x_0, y_0, h, k) satisfying $|a|, |b| \le |h|$; $|c|, |d| \le |k|$ and such that

$$\begin{aligned}
&|f(x_0 + h, y_0 + k) - f(x_0, y_0) - (h, k)D_1 f(x_0, y_0)| \\
&\le |hD_1 \operatorname{Re} f(x_0 + a, y_0 + k) - h \operatorname{Re} D_1 f(x_0, y_0) \\
&\qquad\qquad + ihD_1 \operatorname{Im} f(x_0 + b, y_0 + k) - ih \operatorname{Im} D_1 f(x_0, y_0)| \\
&\quad + |kD_2 \operatorname{Re} f(x_0, y_0 + c) - k \operatorname{Re} D_2 f(x_0, y_0) \\
&\qquad\qquad + ikD_2 \operatorname{Im} f(x_0, y_0 + d) - ik \operatorname{Im} D_2 f(x_0, y_0)| \\
&\le 2(|h| + |k|)\Big(\max_{z \in Q_r} |D_1 f(z) - D_1 f(z_0)| + \max_{z \in Q_r} |D_2 f(z) - D_2 f(z_0)|\Big).
\end{aligned}$$

Divide both sides by $|(h, k)| = \sqrt{h^2 + k^2} \ge \dfrac{|h| + |k|}{2}$ to see that

$$\begin{aligned}
&\left|\frac{f(x_0 + h, y_0 + k) - f(x_0, y_0)}{(h, k)} - D_1 f(x_0, y_0)\right| \\
&\qquad \le 4 \max_{z \in Q_r} |D_1 f(z) - D_1 f(z_0)| + 4 \max_{z \in Q_r} |D_2 f(z) - D_2 f(z_0)|
\end{aligned}$$

Now let $r \downarrow 0$ and use the continuity of $D_1 f$ and $D_2 f$ at z_0.

Exercise 2.17 (i) *If f is differentiable at z_0, then $D_1 f(z_0)$, $D_2 f(z_0)$ exist and satisfy* (C.R.).

Hint: The hypothesis is that the quotient

$$\frac{f(z_0 + h) - f(z_0)}{h}$$

has a limit as the complex number h approaches 0. Consider first $h \in \mathbb{R}$, then $h \in i \cdot \mathbb{R}$.

(ii) *If f is holomorphic in a region Ω and* Re *f is constant, then f is constant.*

Hint: Equating real and imaginary parts in (C.R.) shows that $D_1(\operatorname{Im} f) = -D_2(\operatorname{Re} f)$. Therefore, $f' = D_1 f = D_1(\operatorname{Re} f) - iD_2(\operatorname{Re} f)$. The latter is 0 if $\operatorname{Re} f$ is constant. Cite 2.10(iv).

Exercise 2.18 (D. R. Curtiss [1901/02]) *Modify the mean value argument above to prove the differentiability of f at z_0 when the second hypothesis in* 2.16 *is replaced by the weaker one that both* Re *$D_1 f$ and* Im *$D_2 f$ [or, alternatively, both* Im *$D_1 f$ and* Re *$D_2 f$] are continuous at z_0.*

Notes to Chapter II

For more on the two geometric features ("streckentreue" and "winkeltreue") which earn locally univalent holomorphic functions the name "conformal," see pp. 42–46 of Bieberbach [1934], Remak [1914], Bohr [1918], Rademacher [1919a], R. K. Williams [1973].

Sometimes it is convenient to speak of a function $f: A \to \mathbb{C}$ being holomorphic in A without A being open. There would be two possible definitions, a local one and a global one: (i) For each $a \in A$ there is an open neighborhood V_a of a and an $f_a \in H(V_a)$ such that f coincides with f_a in $A \cap V_a$. (ii) There is an open neighborhood V of A and $F \in H(V)$ such that f coincides with F in A. We will not use this language, but the two definitions are equivalent, a not entirely trivial fact: see MINDA [1973].

Though adumbrated by Gauss and Poisson, the idea of the complex contour integral evolved at the hands of Cauchy, beginning with his famous 1825 memoir. (See STÄCKEL [1900], ETTLINGER [1921–22] and YUŠKEVIČ [1947].)

Much work has gone into efforts to weaken the conditions in 2.16 and these conditions can indeed be considerably relaxed. Here is a sample result:

Theorem (LOOMAN [1923], RIDDER [1930a], WILKOSZ [1933], MENCHOFF [1936]) *If f is continuous on U, $D_1 f$ and $D_2 f$ each exist at all but countably many points of U and the equation* (C.R.) *holds almost everywhere in U* (*in the sense of plane Lebesgue measure*), *then f is holomorphic in U.*

The requirement of differentiability at every point in the definition 2.1 of holomorphy can also be relaxed. 5.41 and 9.28 are examples; here is another:

Theorem (POMPEIU [1905a], LOOMAN [1925], RIDDER [1930a]) *If f is continuous in U, $\overline{\lim}_{h \to 0} |(f(z+h) - f(z))/h|$ is finite for all but countably many z in U and f' exists almost everywhere in U, then f is holomorphic in U.*

For details of these results, which are heavily real-variable in nature, the reader may consult MENCHOFF [1936], pp. 195–201 of SAKS [1937] or the exhaustive (but terse) treatise of TROKHIMCHUK [1964]. For a somewhat more elementary account of somewhat weaker results in this direction, see MEIER [1951], [1960] and ARSOVE [1955]. For a survey of results and history (together with a proof of the first theorem above) see GRAY and MORRIS [1978]; in TROKHIMCHUK [1962] the reader will find a longer survey and some proofs. See also TAR [1969], [1971], [1973], MEIER [1970], PONOMAREV [1970], and BERENŠTEĬN [1971] for recent results. Readers knowledgeable of Bulgarian may also want to consult the paper reviewed in *Mathematical Reviews* vol. 36 #5311 and *Zentralblatt* vol. 311 #30001. For one-to-one functions even weaker conditions suffice to ensure analyticity. See pages 5, 6 of MONTEL [1933b] and the references there, as well as MENCHOFF [1936], [1937] and BRODOVICH [1970], [1974]. Chapter 1 of MARKUSCHEWITSCH [1976] is also devoted to these questions and contains a proof of the first theorem above.

With definition 2.1 a commitment is made that has profound historical significance. The theory to be erected here did not spring fully-armed from the head of Zeus, but condensed gradually out of the primordial vapors. There are at least three distinct lines of development: the theory of Augustin Cauchy, based on 2.1 and on integrals of the type 2.14, has a geometric character; that of Bernhard Riemann, based on the equation (C.R.) and closely connected with potential theory, is analytic; and that of Karl Weierstrass (and Charles Méray), based on power series, is purely arithmetic. Each of these has its strengths and weaknesses. For example, the (unadulterated) Weierstrass theory did not deal so well with mapping problems and the beautiful theorem 9.7 eluded it. On the other hand, it was the most logically rigorous and philosophically satisfactory of the three because it started with only the arithmetic of complex numbers (i.e., rational functions) and added just one more ("natural") ingredient—uniform convergence. As is customary nowadays, the account to be presented here imperiously conscripts the best features from each of these. The strategy consists in exploiting first one (until it encounters a chasm it cannot bridge) and then another of these points of view. And, as with any anabasis, extensive logistical preparation is necessary. This is the work of the next two chapters.

In Fouët [1904] the reader will find a detailed comparison of the three theories with many references to the original works and in Pringsheim [1925], [1932] a complete and careful (though sometimes pedantic) presentation of the Weierstrass theory. (Incidentally, the latter tome is a good place to observe the price which methodological purity can exact: results which will be secured very expeditiously in the course of Chapter V often require many pages there.) For the historical evolution of Weierstrass' theory see Manning [1975] and for the master's own comparison of the three approaches, Weierstrass [1925].

Chapter III
Power Series and the Exponential Function

§ 1 Introduction

As with integrals, power series (though fascinating) are a tool here and are not pursued extensively. (However, some special kinds of power series, those with many zero coefficients or with integer coefficients, are examined in some detail in Chapters XVI and XVII.) For in-depth treatises on power series the reader should consult KNOPP [1951] or BROMWICH [1926]. To make our definitions and get started only one simple result is needed:

Theorem 3.1 $\lim_{n\to\infty} n^{1/n} = 1$ *and for any* $a > 0$, $\lim_{n\to\infty} a^{1/n} = 1$.

Proof: Set $x_n = n^{1/n} - 1$. Since this is non-negative for $n \geq 2$

$$n = (1 + x_n)^n = \sum_{j=0}^{n} \binom{n}{j} x_n^j \geq \binom{n}{2} x_n^2 = \frac{n(n-1)}{2} x_n^2.$$

Therefore

$$0 \leq x_n \leq \sqrt{\frac{2}{n-1}}$$

and so $x_n \to 0$. If $a \geq 1$, then $1 \leq a^{1/n} \leq n^{1/n}$ for large n, so $a^{1/n} \to 1$. If $0 < a < 1$, apply this conclusion to $1/a$.

Remarks: We are, of course, assuming known the theory of rational roots and powers on $[0, \infty)$. This is an easy consequence of the completeness of $\mathbb{R}$. (See, e.g., chapter 1 of RUDIN [1976].) Irrational powers on $[0, \infty)$ can be dealt with at the same level of elementariness but will not be needed. Everything is finally subsumed under theorem 3.14. (For an alternative elementary derivation of 3.1 see PRACHAR [1975].)

§ 2 Power Series

Definition 3.2 An infinite series of the form $\sum_{n=0}^{\infty} c_n(z - z_0)^n$, where $c_n, z, z_0 \in \mathbb{C}$, is called a *power series*, more precisely a *power series about* z_0, or a *power series centered* at z_0. (No *a priori* convergence claim is being made.)

Theorem 3.3 *Let* $c_n, z_0 \in \mathbb{C}$ *and consider the power series*

$$(1) \qquad \sum_{n=0}^{\infty} c_n(z - z_0)^n.$$

Then there exists an $R \in [0, \infty]$ *such that:*

(i) *The series* (1) *converges absolutely for all* $z \in \mathbb{C}$ *with* $|z - z_0| < R$ *and uniformly in* $D(z_0, \rho)$ *for any* $\rho < R$.

(ii) *The terms of the series* (1) *are unbounded (and hence the series diverges) for every* $z \in \mathbb{C}$ *with* $|z - z_0| > R$.

(iii) *The extended real number* R *is the reciprocal of the extended real number* $\overline{\lim}_{n\to\infty} \sqrt[n]{|c_n|}$ *and is called the* radius of convergence *of the series* (1).

Proof: Define R to be the least upper bound in $[0, \infty]$ of the $r \in [0, \infty)$ for which $\{|c_n|r^n\}$ is a bounded sequence. If $|z - z_0| > R$, then the sequence $\{c_n(z - z_0)^n\}$ is unbounded, while if $\rho < R$ and we pick r satisfying $\rho < r < R$, then by definition of R we shall have that $\{|c_n|r^n\}$ is a bounded sequence, say $|c_n|r^n \le M < \infty$ for all n. Then for all z in $D(z_0, \rho)$

$$|c_n(z - z_0)^n| = |c_n|r^n \cdot \left[\frac{|z - z_0|}{r}\right]^n \le M \cdot \left(\frac{\rho}{r}\right)^n$$

and since $\rho/r < 1$, the assertion (i) follows.

Now if $0 < r < R$, then $\{|c_n|r^n\}$ is bounded, say by $M(r) \in (0, \infty)$:

$$|c_n|r^n \le M(r)$$

$$|c_n|^{1/n} \le \frac{1}{r}[M(r)]^{1/n} \quad \forall n.$$

Let $n \to \infty$ and cite 3.1 to conclude that

$$\overline{\lim_{n\to\infty}} |c_n|^{1/n} \le 1/r$$

and since r is arbitrary in $(0, R)$,

(2) $$\overline{\lim_{n\to\infty}} |c_n|^{1/n} \le 1/R.$$

Equality necessarily holds in (2) if $R = \infty$. If $R < \infty$ and $R < r < \infty$, then by definition of R the sequence $\{|c_n|r^n\}$ is not bounded and so in particular infinitely many of its terms exceed 1 and therefore $|c_n|^{1/n}r > 1$ for infinitely many n. It follows that

$$\overline{\lim_{n\to\infty}} |c_n|^{1/n} \ge 1/r.$$

Again as $r \in (R, \infty)$ is arbitrary, it follows that

(3) $$\overline{\lim_{n\to\infty}} |c_n|^{1/n} \ge 1/R,$$

which with (2) establishes the claim (iii).

Example 3.4 The radius of convergence of the power series $\sum_{n=0}^{\infty} (1/n!)z^n$ is infinite. For if m is any positive integer, then for $n > m$ we have

$$n! \ge n(n-1)\cdots(m+1)m > m^{n-m+1}$$

$$\frac{m^n}{n!} \le m^{m-1}.$$

Recalling how R is defined in the first line of the above proof, we conclude that $R \geq m$ for all m.

Theorem 3.5 *If $c_n, z_0 \in \mathbb{C}$, then the two series $\sum_{n=0}^{\infty} c_n(z-z_0)^n$ and*

$$\sum_{n=1}^{\infty} nc_n(z-z_0)^{n-1} = \sum_{n=0}^{\infty} (n+1)c_{n+1}(z-z_0)^n$$

have the same radius of convergence.

Proof: Since $\sum_{n=0}^{\infty} nc_n(z-z_0)^n = (z-z_0)\cdot\sum_{n=1}^{\infty} nc_n(z-z_0)^{n-1}$, it suffices to show that the two series $\sum_{n=0}^{\infty} nc_n(z-z_0)^n$ and $\sum_{n=0}^{\infty} c_n(z-z_0)^n$ have the same radius of convergence. And by part (iii) of the last theorem, for this it suffices to show that

$$\overline{\lim_{n\to\infty}} |nc_n|^{1/n} = \overline{\lim_{n\to\infty}} |c_n|^{1/n}.$$

This in turn follows from 3.1:

$$\overline{\lim_{n\to\infty}} |c_n|^{1/n} \leq \overline{\lim_{n\to\infty}} |nc_n|^{1/n} \leq \overline{\lim_{n\to\infty}} n^{1/n}\cdot \overline{\lim_{n\to\infty}} |c_n|^{1/n} = \overline{\lim_{n\to\infty}} |c_n|^{1/n}.$$

Theorem 3.6 *Let $c_n, z_0 \in \mathbb{C}$ and let the radius of convergence R of the series $\sum_{n=0}^{\infty} c_n(z-z_0)^n$ be positive. Then the function f defined in the open disk $D(z_0, R)$ by*

$$(1) \qquad f(z) = \sum_{n=0}^{\infty} c_n(z-z_0)^n$$

is infinitely differentiable and for every positive integer k satisfies

$$(2.k) \qquad f^{(k)}(z) = \sum_{n=k}^{\infty} n(n-1)\cdots(n-k+1)c_n(z-z_0)^{n-k} \quad \forall z \in D(z_0, R).$$

Proof: It suffices to prove that f is differentiable in $D(z_0, R)$ and satisfies equation (2.1). For then we may, via 3.5, induce on k. Also, by considering the function $F(z) = f(z+z_0)$ in $D(0, R)$, we may assume without loss of generality that $z_0 = 0$. Let g denote the function defined (by 3.5) in the disk $D(0, R)$ by the right side of equation (2.1). Fix $w \in D(0, R)$ and choose $|w| < r < R$. Then for $z \in D(0, r)\backslash\{w\}$

$$(3) \qquad \frac{f(z)-f(w)}{z-w} - g(w) = \sum_{n=1}^{\infty} c_n\left[\frac{z^n - w^n}{z-w} - nw^{n-1}\right].$$

The expression in square brackets is 0 if $n = 1$ and for $n \geq 2$ direct multiplication confirms that it equals

$$(4) \qquad (z-w)\sum_{k=1}^{n-1} kw^{k-1}z^{n-k-1}.$$

Since $w, z \in D(0, r)$, the absolute value of the sum in (4) is less than

$$\sum_{k=1}^{n-1} kr^{k-1}r^{n-k-1} = r^{n-2}\sum_{k=1}^{n-1} k = \frac{n(n-1)}{2}r^{n-2}$$

and so finally

$$(5) \qquad \left|\frac{f(z)-f(w)}{z-w} - g(w)\right| \le \frac{|z-w|}{2}\sum_{n=2}^{\infty} n(n-1)|c_n|r^{n-2} \quad \forall z \in D(0,r)\backslash\{w\}.$$

Two successive applications of the last theorem show that the series

$$\sum_{n=0}^{\infty} n(n-1)|c_n|z^{n-2}$$

has the same radius of convergence as the series $\sum_{n=0}^{\infty} |c_n|z^n$, viz., R and so, since $r < R$, the sum in (5) is finite. Therefore, since $D(0, r)$ is a neighborhood of w, we have only to let $z \to w$ in (5) to conclude the $f'(w)$ exists and equals $g(w)$, as claimed.

Exercise 3.7 *If the series $f(z) = \sum_{n=0}^{\infty} c_n z^n$ has a radius of convergence $R > 0$ and $f'(0) = c_1 \neq 0$, then for some $0 < r \le R$, f is one-to-one in $D(0, r)$.*

Hints: If $0 < r < R$ and $z, w \in D(0, r)$ then, as in the proof of 3.6,

$$f(z) - f(w) = c_1(z-w) + (z-w)\sum_{n=2}^{\infty} c_n \sum_{k=1}^{n} w^{k-1}z^{n-k},$$

$$|f(z)-f(w)| \ge |c_1|\,|z-w| - |z-w|\sum_{n=2}^{\infty}|c_n|\sum_{k=1}^{n}|w|^{k-1}|z|^{n-k}$$

$$\ge \left[|c_1| - \sum_{n=2}^{\infty}|c_n|\sum_{k=1}^{n} r^{n-1}\right]|z-w|$$

$$= \left[|c_1| - \sum_{n=2}^{\infty} n|c_n|r^{n-1}\right]|z-w|.$$

By 3.5 the series $\sum_{n=1}^{\infty} n|c_n|z^{n-1}$ has radius of convergence R, so for sufficiently small $0 < r < R$, $\sum_{n=2}^{\infty} n|c_n|r^{n-1} < |c_1|/2$ and it follows that

$$|f(z) - f(w)| > (|c_1|/2)|z-w|.$$

Remark: These same techniques show, with a little more work (see KRAFFT [1932]), that $f(D(0, r))$ is open and f^{-1} is given by a power series near c_0, facts which we will secure in Chapter V by other means.

Exercise 3.8 *In the notation of* 2.14 *show that each G_n is of power series type near each point of $\mathbb{C}\backslash\gamma$, i.e., for each $a \in \mathbb{C}\backslash\gamma$ there is a power series about a which agrees with G_n in some neighborhood of a. Deduce along the way a new proof of* 2.14.

Hints: (after Cauchy) From $1/(1-w) = \sum_{k=0}^{\infty} w^k$ for all $w \in D(0, 1)$ and 3.6 follows by differentiation that

$$(1) \qquad \frac{1}{(1-w)^{n+1}} = \sum_{k=n}^{\infty} \binom{k}{n} w^{k-n} \quad \forall w \in D(0, 1).$$

Given $a \in \mathbb{C}\backslash\gamma$, choose $R > 0$ so that $D(a, R) \subset \mathbb{C}\backslash\gamma$. Then for any $0 < r < R$, $z \in D(a, r)$ and $\xi \in \gamma$ we have $\left|\frac{z-a}{\xi-a}\right| < \frac{r}{R} < 1$, and so

$$\frac{1}{(\xi - z)^{n+1}} = \frac{1}{(\xi - a)^{n+1}} \frac{1}{\left(1 - \frac{z-a}{\xi - a}\right)^{n+1}}$$

$$\stackrel{(1)}{=} \frac{1}{(\xi - a)^{n+1}} \sum_{k=n}^{\infty} \binom{k}{n} \left(\frac{z-a}{\xi-a}\right)^{k-n},$$

with convergence uniform in such z and ξ. Consequently, for $z \in D(a, r)$

$$nG_{n+1}(z) = \int_\gamma \frac{ng(\xi)}{(\xi - z)^{n+1}}\, d\xi = \sum_{k=n}^{\infty} \int_\gamma \frac{g(\xi)}{(\xi - a)^{n+1}}\, n \binom{k}{n} \left(\frac{z-a}{\xi-a}\right)^{k-n} d\xi$$

$$(2.n) \qquad = \sum_{k=n}^{\infty} \left(\int_\gamma \frac{g(\xi)d\xi}{(\xi - a)^{k+1}}\right) n \binom{k}{n} (z-a)^{k-n}$$

$$= \frac{d}{dz}\left[\sum_{k=n-1}^{\infty} \left(\int_\gamma \frac{g(\xi)d\xi}{(\xi - a)^{k+1}}\right) \frac{n}{k-n+1} \binom{k}{n} (z-a)^{k-n+1}\right],$$

by 3.6,

$$= \frac{d}{dz}\left[G_n(z)\right],$$

by equation $(2.n-1)$ and the fact that

$$\frac{n}{k-n+1}\binom{k}{n} = \binom{k}{n-1}.$$

Exercise 3.9 (Cf. Cauchy [1821], p. 59.) *Show that if a_n is positive for all (sufficiently large) n, then*

$$\varliminf_{n\to\infty} \frac{a_{n+1}}{a_n} \le \varliminf_{n\to\infty} \sqrt[n]{a_n} \le \varlimsup_{n\to\infty} \sqrt[n]{a_n} \le \varlimsup_{n\to\infty} \frac{a_{n+1}}{a_n}.$$

Hints: Call the last limit superior $a \in [0, \infty]$. It suffices to prove the last inequality. For if $a = 0$, this finishes our work, while if $a > 0$, this result can be applied to the sequence $\{1/a_n\}$.

In proving this last inequality we may assume $a < \infty$. For an $\varepsilon > 0$ there is an $N = N(\varepsilon)$ such that

$$\frac{a_{n+1}}{a_n} < a + \varepsilon \quad \forall n \ge N.$$

Multiply these inequalities for $n = N, \ldots, m-1$ to obtain

$$\frac{a_m}{a_N} < (a+\varepsilon)^{m-N} \quad \forall m > N$$

$$\sqrt[m]{a_m} < (a+\varepsilon)\left[\frac{a_N}{(a+\varepsilon)^N}\right]^{1/m} \quad \forall m > N.$$

It follows from this and 3.1 that $\overline{\lim}_{m\to\infty} \sqrt[m]{a_m} \leq a + \varepsilon$.

Exercise 3.10 *Show that for each positive integer* k

$$\lim_{n\to\infty} \sqrt[n]{\binom{kn}{n}} = \frac{k^k}{(k-1)^{k-1}}.$$

Hint: The corresponding limit for ratios is trivial, so use the last exercise.

Exercise 3.11 *Show that if* $a, a_0, a_1, \ldots, \in \mathbf{C}$ *and* $\lim_{n\to\infty} a_n = a$, *then*

$$\lim_{n\to\infty} 2^{-n} \sum_{k=0}^{n} \binom{n}{k} a_k = a.$$

Hints: Since

(1) $$\sum_{k=0}^{n} \binom{n}{k} = (1+1)^n = 2^n,$$

we have

(2) $$a - 2^{-n} \sum_{k=0}^{n} \binom{n}{k} a_k = 2^{-n} \sum_{k=0}^{n} \binom{n}{k} (a - a_k).$$

Now divide and conquer: given $\varepsilon > 0$ choose $k_0 = k_0(\varepsilon)$ so that

(3) $|a - a_k| < \varepsilon/2 \quad \forall k \geq k_0;$

then use the fact that $\lim_{n\to\infty} n^{k_0}/2^n = 0$ to pick $n_0 = n_0(\varepsilon) \geq k_0$ so that

$$\frac{n^{k_0}}{2^n} < \frac{\varepsilon}{4Mk_0} \quad \forall n \geq n_0$$

for some $M > \sup\{|a|, |a_k| : k = 0, 1, 2, \ldots\}$. Then for $k = 0, 1, \ldots, k_0$

$$2^{-n}\binom{n}{k} = \frac{n(n-1)\cdots(n-k+1)}{2^n k!} \leq \frac{n^k}{2^n} \leq \frac{n^{k_0}}{2^n} < \frac{\varepsilon}{4Mk_0} \quad \forall n \geq n_0,$$

and so

(4) $$2^{-n} \sum_{k=0}^{k_0-1} \binom{n}{k} |a - a_k| \leq 2M \sum_{k=0}^{k_0-1} 2^{-n}\binom{n}{k} < 2M \sum_{k=0}^{k_0-1} \frac{\varepsilon}{4Mk_0} = \frac{\varepsilon}{2} \quad \forall n \geqslant n_0.$$

From (2), (3) and (4)

$$\left|a - 2^{-n}\sum_{k=0}^{n}\binom{n}{k}a_k\right| \le 2^{-n}\sum_{k=0}^{k_0-1}\binom{n}{k}|a - a_k| + 2^{-n}\sum_{k=k_0}^{n}\binom{n}{k}\frac{\varepsilon}{2}$$

$$< \frac{\varepsilon}{2} + 2^{-n}\sum_{k=0}^{n}\binom{n}{k}\frac{\varepsilon}{2} \overset{(1)}{=} \frac{\varepsilon}{2} + \frac{\varepsilon}{2} = \varepsilon \quad \forall n \ge n_0.$$

Exercise 3.12 *Let $c_{n,m} \in \mathbb{C}$ be such that $\sum_{n=1}^{\infty}[\sum_{m=1}^{\infty}|c_{n,m}|] < \infty$. Show that then*

(i) $$\sum_{m=1}^{\infty}\left[\sum_{n=1}^{\infty}|c_{n,m}|\right] = \sum_{n=1}^{\infty}\left[\sum_{m=1}^{\infty}|c_{n,m}|\right] = \sup_{M,N}\left\{\sum_{m=1}^{M}\left[\sum_{n=1}^{N}|c_{n,m}|\right]\right\}.$$

Consequently,

(ii) *for each n the series $\sum_{m=1}^{\infty} c_{n,m}$ converges (say to s_n), for each m the series $\sum_{n=1}^{\infty} c_{n,m}$ converges (say to t_m) and the series $\sum_{n=1}^{\infty} s_n$, $\sum_{m=1}^{\infty} t_m$ are convergent and equal.*

Hints: Let L denote the (finite) number in (i) and set

$$\delta_{N,M} = L - \sum_{n=1}^{N}\left[\sum_{m=1}^{M}|c_{n,m}|\right].$$

Show that $\sum_{n=N+1}^{N'}|s_n| \le \delta_{N,M}$ and $\sum_{m=M+1}^{M'}|t_m| \le \delta_{N,M}$ for all N and M and all $N' > N$, $M' > M$. In the inequality

$$\left|\sum_{n=1}^{N}\left[\sum_{m=1}^{M'}c_{n,m}\right] - \sum_{m=1}^{M}\left[\sum_{n=1}^{N'}c_{n,m}\right]\right| \le 2\delta_{N,M},$$

valid whenever $N' \ge N$ and $M' \ge M$, first let $M' \to \infty$ and then let $N' \to \infty$ to see that $|\sum_{n=1}^{N} s_n - \sum_{m=1}^{M} t_m| \le 2\delta_{N,M}$.

Exercise 3.13 (i) *Let $X \subset \mathbb{C}$, $a_k\colon X \to \mathbb{C}$ and $b_k \in \mathbb{R}$ with $b_k \downarrow 0$. Set $A_n = \sum_{k=0}^{n} a_k$ and suppose that $\{A_n\}$ is uniformly bounded on X. Show that $\sum_{k=0}^{\infty} b_k a_k$ is uniformly convergent on X.*

Hints: Since $a_k = A_k - A_{k-1}$, a little juggling with indexes of summation leads to (*Abel's partial summation formula*):

$$\sum_{k=p+1}^{q} b_k a_k = \sum_{k=p+1}^{q} A_k(b_k - b_{k+1}) + A_q b_{q+1} - A_p b_{p+1}, \quad 0 \le p < q.$$

If M is a bound for $\{A_n\}$ on X, we then get

$$\left|\sum_{k=p+1}^{q} b_k a_k\right| \le M\sum_{k=p+1}^{q}(b_k - b_{k+1}) + Mb_{q+1} + Mb_{p+1}$$

$$= M(b_{p+1} - b_{q+1}) + Mb_{q+1} + Mb_{p+1}$$

$$= 2Mb_{p+1},$$

showing that the sequence $\{\sum_{k=0}^{n} b_k a_k\}$ is uniformly Cauchy on X.

(ii) *If $b_k \in \mathbb{R}$ and $b_k \downarrow 0$, then the series $\sum_{k=0}^{\infty} b_k z^k$ converges throughout $\bar{D}(0, 1)\backslash\{1\}$ and uniformly on each compact subset.*

Hints: For any $0 < r < 1$ and all $n \geq 0$ we have

$$\left|\sum_{k=0}^{n} z^k\right| = \left|\frac{z^{n+1} - 1}{z - 1}\right| \leq \frac{2}{|z - 1|} \leq \frac{2}{r} \quad \forall z \in \bar{D}(0, 1)\backslash D(1, r);$$

now cite (i).

§ 3 The Complex Exponential Function

The most important power series in analysis will now be introduced and studied. Besides the interesting numerical results which we will derive from it in this chapter, it will serve as the basis for a rigorous logical treatment of the idea "the number of times a curve winds around a point." This will be essential to the whole Cauchy theory of holomorphic functions and also will be one of the main tools in the rather complete topological analysis of the plane offered in Chapter IV. The presentation here, while maximally concise, has a *deus ex machina* character. Since the function $y \to e^{iy}$ is, up to a constant, the unique (continuous) homomorphism of the topological group $\mathbb{R}$ onto the topological group $\mathbb{T}$ (see 3.25), it can be introduced "naturally." One has only to show somehow, without the exponential, that these two groups are locally isomorphic. Similarly with the real exponential $x \to e^x$. This elegant, but arduous, program is carried out in BOURBAKI [1966], chapter V, §§ 3 and 4, and chapter VIII, § 2. Cf. also DZEWAS [1968].

Theorem 3.14 *The function E defined for all $z \in \mathbb{C}$ (cf. 3.4) by*

$$E(z) = \sum_{n=0}^{\infty} \frac{1}{n!} z^n$$

has the following properties:

(i) $E' \equiv E$.

(ii) *For all $z_1, z_2 \in \mathbb{C}$, $E(z_1 + z_2) = E(z_1)E(z_2)$.*

(iii) *There is a positive number π such that $E\left(\frac{\pi}{2}i\right) = i$ and*

$$\forall z_1, z_2 \in \mathbb{C}, \qquad E(z_1) = E(z_2) \Leftrightarrow \frac{z_1 - z_2}{2\pi i} \text{ is an integer.}$$

(iv) *$E|\mathbb{R}$ is strictly increasing and $E(\mathbb{R}) = (0, \infty)$.*

(v) *The mapping $x \to E(ix)$ maps $\mathbb{R}$ onto the unit circle $\{z \in \mathbb{C}: |z| = 1\}$.*

(vi) *$E(\mathbb{C}) = \mathbb{C}\backslash\{0\}$.*

Note: (ii), (iii) and (vi) say that E is a homomorphism of the additive group $\mathbb{C}$ onto the multiplicative group $\mathbb{C}\backslash\{0\}$ with kernel $2\pi i\mathbb{Z}$.

Proof: That E is differentiable and that

$$E'(z) = \sum_{n=1}^{\infty} n \cdot \frac{1}{n!} z^{n-1} = E(z)$$

for all z follows from 3.6.

Given $c \in \mathbb{C}$, define a function $f: \mathbb{C} \to \mathbb{C}$ by

$$f(z) = E(c - z).$$

Then by (i) (and the Chain Rule), the function f is differentiable throughout $\mathbb{C}$ and satisfies $f' \equiv -f$. Therefore the function $F = f \cdot E$ satisfies $F' \equiv 0$ whence (by 2.10) $F \equiv F(0)$, that is, for all $z \in \mathbb{C}$

$$E(c - z)E(z) = F(z) = F(0) = E(c)E(0) = E(c).$$

Taking $c = z_1 + z_2$ and $z = z_1$, we get (ii).

For positive real x we have from the series definition of E

$$E(x) > 1 + x,$$

so

(1) $\quad \lim_{x \to \infty} E(x) = \infty;$

while from (ii) follows that for all $z \in \mathbb{C}$

(2) $\quad 1 = E(0) = E(z)E(-z)$

and then (1) yields

(3) $\quad \lim_{t \to -\infty} E(t) = \lim_{-t \to \infty} \frac{1}{E(-t)} = \lim_{x \to \infty} \frac{1}{E(x)} = 0.$

Now the coefficients in the series for E are real, so for any $z \in \mathbb{C}$ we see that

(4) $\quad \overline{E(z)} = E(\bar{z}).$

In particular $E(\mathbb{R}) \subset \mathbb{R}$ and in fact, bearing in mind (2) that E is never zero, for any $x \in \mathbb{R}$

$$E(x) = E(x/2 + x/2) = [E(x/2)]^2 > 0.$$

Therefore $E(\mathbb{R})$ is a connected subset of $(0, \infty)$, which by (1) and (3) must be all of $(0, \infty)$.

Define functions $C, S: \mathbb{R} \to \mathbb{R}$ by

$$C(x) = \operatorname{Re} E(ix), \qquad S(x) = \operatorname{Im} E(ix).$$

Now the derivative of the function $x \to E(ix) = C(x) + iS(x)$ (exists and) is $iE'(ix) = iE(ix) = iC(x) - S(x)$, so

$$C'(x) + iS'(x) = iC(x) - S(x),$$

(5) $\quad -C' = S, \qquad S' = C.$

Also from (4) and (2) we have for all $x \in \mathbb{R}$

$$C^2(x) + S^2(x) = |E(ix)|^2 = E(ix)\overline{E(ix)} = E(ix)E(-ix) = 1,$$

$$C^2 + S^2 \equiv 1. \tag{6}$$

The definition of C gives

$$\begin{aligned}
C(x) &= \frac{E(ix) + \overline{E(ix)}}{2} = \frac{E(ix) + E(-ix)}{2} \\
&= \sum_{n=0}^{\infty} \left[\frac{1 + (-1)^n}{2}\right] \frac{i^n x^n}{n!} = \sum_{k=0}^{\infty} \frac{i^{2k} x^{2k}}{(2k)!} \\
&= \sum_{k=0}^{\infty} \frac{(-1)^k x^{2k}}{(2k)!} = 1 + \lim_{N\to\infty} \sum_{k=1}^{2N} \frac{(-1)^k x^{2k}}{(2k)!} \\
&= 1 + \lim_{N\to\infty} \sum_{n=1}^{N} \left[\frac{(-1)^{2n-1} x^{4n-2}}{(4n-2)!} + \frac{(-1)^{2n} x^{4n}}{(4n)!}\right] \\
&= 1 + \sum_{n=1}^{\infty} \left[\frac{x^2}{(4n)!} - \frac{1}{(4n-2)!}\right] x^{4n-2}.
\end{aligned}$$

Observe that for $0 \le x \le 2$ and $n \ge 1$

$$\frac{x^2}{(4n)!} \le \frac{4}{(4n)(4n-1)(4n-2)!} < \frac{1}{(4n-2)!}$$

and so

$$\begin{aligned}
C(x) &= 1 + \left[\frac{x^2}{4!} - \frac{1}{2!}\right] x^{4-2} + \sum_{n=2}^{\infty} \left[\frac{x^2}{(4n)!} - \frac{1}{(4n-2)!}\right] x^{4n-2} \\
&< 1 + \left[\frac{x^2}{4!} - \frac{1}{2!}\right] x^{4-2}.
\end{aligned}$$

Therefore $C(\sqrt{3}) < -\frac{1}{8}$. Since $C(0) = \operatorname{Re} E(0) = 1$, we have that $C[0, \sqrt{3}]$ is a connected subset of $\mathbb{R}$ which contains positives and negatives and hence zero; that is, the compact set $[0, \sqrt{3}] \cap C^{-1}(0)$ is not void. Let $\pi/2$ denote its least element:

$$\pi/2 = \min[0, \sqrt{3}] \cap C^{-1}(0). \tag{7}$$

Thus

$$0 < \pi/2 < \sqrt{3} \quad \text{and} \quad C(\pi/2) = 0. \tag{8}$$

From (8) and (6) we get

$$S(\pi/2) = \pm 1. \tag{9}$$

However by (5) and the definition of π

$$S' = C > 0 \quad \text{in } [0, \pi/2), \tag{10}$$

so that S is strictly increasing in $[0, \pi/2]$. Therefore $S(\pi/2) > S(0) = 0$ and from (9) we get

(11) $S(\pi/2) = 1.$

Putting (8) and (11) together:

(12) $E\left(\frac{\pi}{2} i\right) = C(\pi/2) + iS(\pi/2) = i,$

whence

$$E(2\pi i) \stackrel{\text{(ii)}}{=} \left[E\left(\frac{\pi}{2} i\right)\right]^4 = 1$$

(13) $E(z + 2\pi i) = E(z)E(2\pi i) = E(z) \quad \forall z \in \mathbb{C}.$

Now suppose $z = x + iy$ $(x, y \in \mathbb{R})$ is such that $E(z) = 1$. Then $1 = |E(z)| = |E(x)E(iy)| = E(x)|E(iy)| = E(x)$, since $E(\mathbb{R}) \subset (0, \infty)$ and $|E(iy)| = 1$ by (6). Since $E(0) = 1$ and $E|\mathbb{R}$ is strictly increasing, it follows that $x = 0$. To prove that $y/2\pi$ is an integer it is enough, in view of (13), to show that $E(it) \neq 1$ if $0 < t < 2\pi$. Given such a t, we have $0 < t/4 < \pi/2$ and, since C and S are both positive in $(0, \pi/2)$ (by (10)), we have

(14) $C(t/4) > 0, \qquad S(t/4) > 0.$

Writing $z^4 - 1 = (z + 1)(z - 1)(z + i)(z - i)$, shows that ± 1, $\pm i$ are the only fourth roots of 1. Equation (14) shows that $E(it/4) = C(t/4) + iS(t/4)$ is none of these numbers. Therefore $E(it) = [E(it/4)]^4$ is not 1.

Now if $u^2 + v^2 = 1$ and $u \geq 0$, $v \geq 0$, then $C(\pi/2) = 0 \leq u \leq 1 = C(0)$ and so by continuity of C there is a $t \in [0, \pi/2]$ such that $C(t) = u$. But S is non-negative in $[0, \pi/2]$ (by (10)), so (6) gives

$$S(t) = \sqrt{1 - C^2(t)} = \sqrt{1 - u^2} = v,$$

$$E(it) = u + iv.$$

If $u^2 + v^2 = 1$ and $u < 0$, $v \geq 0$, pick t according to the previous recipe so that $E(it) = v - iu$ and then have by (12)

$$E\left(i\left(t + \frac{\pi}{2}\right)\right) = E(it)E\left(i\frac{\pi}{2}\right) = (v - iu)\cdot i = u + iv.$$

If $u^2 + v^2 = 1$ and $v < 0$, pick t according to one of the previous two recipes so that $E(it) = -u - iv$ and then have

$$E(i(t + \pi)) = E(it)E(i\pi) = (-u - iv)(-1) = u + iv.$$

The assertion (v) is thus proven.

If $w \in \mathbb{C}\backslash\{0\}$, then $|w| \in (0, \infty) = E(\mathbb{R})$, so $|w| = E(x)$ for some $x \in \mathbb{R}$; while $|w/|w|| = 1$, so $w/|w| = E(iy)$ for some $y \in \mathbb{R}$ by the result (v). Thus

$$E(x + iy) = E(x)E(iy) = w$$

and we have $\mathbb{C}\backslash\{0\} \subset E(\mathbb{C})$, completing the proof of (vi).

Remarks: If n is a positive integer, then $E(1/n)$ is a positive real number whose nth power is $E(1)$, that is, $E(1/n)$ is the unique positive nth root $[E(1)]^{1/n}$ of $E(1)$. If m is any integer, then

$$E(m/n) = [E(1/n)]^m = [E(1)]^{m/n}.$$

Consequently, if we follow the universal custom and, in honor of Euler, let e denote the number $E(1)$, then for every rational r

$$E(r) = e^r.$$

This equality and the homomorphism property of E (the so-called "law of exponents") explain another universal notation, namely e^z, as a name for $E(z)$. Thus

Definition 3.15 $e = \sum_{n=0}^{\infty} 1/n!$ and $e^z = \sum_{n=0}^{\infty} z^n/n!$ for every $z \in \mathbb{C}$. This function is called the (*complex*) *exponential* function and is sometimes also noted exp. Its restriction to $\mathbb{R}$ is called the *real* exponential function.

Definition 3.16 For any $a \in \mathbb{C}$ and $r > 0$ the *circle of center a and radius r* is the curve γ: $[0, 1] \to \mathbb{C}$ defined by $\gamma(t) = a + re^{2\pi it}$. We denote this curve by $C(a, r)$; this is consistent with the definition in Chapter 0 and the convention of 1.11(i), thanks to 3.14(v).

According to 3.14 the real exponential function is strictly increasing and has range exactly $(0, \infty)$. Consequently there is a function from $(0, \infty)$ onto $\mathbb{R}$ which is inverse to the real exponential. The notation and terminology are again traditional:

Definition 3.17 The (*natural*) *logarithm* function, noted log, is the function with domain $(0, \infty)$ and range $\mathbb{R}$ which is inverse to the real exponential.

Most of the properties of this function are immediate consequences of 3.14. For example,

Scholium 3.18 Log is strictly increasing and, since its range is $\mathbb{R}$, we have

(1) $$\lim_{x \downarrow 0} \log x = -\infty, \quad \lim_{x \uparrow \infty} \log x = \infty.$$

Log is continuous and (therefore) differentiable:

(2) $$\log'(x) = \lim_{y \to x} \frac{\log y - \log x}{y - x} = \lim_{y \to x} \frac{1}{\left(\dfrac{e^{\log y} - e^{\log x}}{\log y - \log x}\right)} = \frac{1}{e^{\log x}}$$
$$= \frac{1}{x},$$

since the exponential function is its own derivative. Since the real exponential is an isomorphism of the additive group $\mathbb{R}$ onto the multiplicative group

$(0, \infty)$, its inverse log is an isomorphism of the multiplicative group $(0, \infty)$ onto the additive group $\mathbb{R}$, so

(3) $\quad \log(xy) = \log x + \log y \quad$ for all $x, y \in (0, \infty)$.

We noted in the proof of 3.14 that for positive x we have $e^x > 1 + x$. Therefore $x > \log(1 + x)$ and so

$$\log y < y - 1 \quad \text{for all } y > 1.$$

Since also $\log 1/x = -\log x$ from (3), it follows that, in spite of (1), we have for any positive integer n

$$|x^{1/n} \log x| = |2nx^{1/n} \log x^{1/2n}| = 2nx^{1/n} \log \frac{1}{x^{1/2n}}$$

$$< 2nx^{1/n} \cdot \frac{1}{x^{1/2n}} = 2nx^{1/2n}, \quad \text{if } x < 1;$$

(4) $\displaystyle \lim_{x \downarrow 0} x^{1/n} \log x = 0.$

Now the exponential function is not one-to-one in $\mathbb{C}$ but it is locally so, and we have the following important theorem on the (local) existence of holomorphic logarithms:

Theorem 3.19 *If $c \in \mathbb{C}\backslash\{0\}$, then there is a function L $(= L_c)$ defined in $D(c, |c|)$ by a convergent power series such that $e^{L(z)} = z$ for all $z \in D(c, |c|)$.*

Proof: Evidently the series $\sum_{k=1}^{\infty} z^k/k$ converges for all $|z| < 1$ and therefore defines a holomorphic function l in $D(0, 1)$. By 3.6 we have

$$l'(z) = \sum_{k=1}^{\infty} z^{k-1} = \sum_{n=0}^{\infty} z^n.$$

Since

$$\sum_{n=0}^{\infty} z^n = \lim_{N \to \infty} \sum_{n=0}^{N} z^n = \lim_{N \to \infty} \frac{1 - z^{N+1}}{1 - z} = \frac{1}{1 - z} \quad \text{for } z \in D(0, 1),$$

it follows that

$$l'(z) = \frac{1}{1 - z}$$

(1) $\quad (1 - z)l'(z) = 1 \quad \forall z \in D(0, 1)$,

whence, for the function

$$F(z) = (1 - z)e^{l(z)}, \quad z \in D(0, 1),$$

the Chain Rules gives

$$F'(z) = (1 - z)l'(z)e^{l(z)} - e^{l(z)} \overset{(1)}{=} 0 \quad \forall z \in D(0, 1)$$

It follows from 2.10 that

$$F(z) \equiv F(0) = e^{l(0)} = e^0 = 1 \quad \forall z \in D(0, 1);$$

that is, upon recalling the definition of F,

(2) $$1 - z \equiv \frac{1}{e^{l(z)}} = e^{-l(z)} \quad \forall z \in D(0, 1).$$

Now given $c \in \mathbb{C}\backslash\{0\}$, for each $w \in D(c, |c|)$ apply (2) to

$$z = \frac{c - w}{c} \in \frac{-1}{c}[D(c, |c|) - c] = \frac{-1}{c} D(0, |c|) = D(0, 1)$$

and get

$$\frac{w}{c} = 1 - \frac{c - w}{c} = 1 - z = e^{-l(z)} = e^{-l((c-w)/c)}$$

(3) $$w = ce^{-l((c-w)/c)} \quad \forall w \in D(c, |c|).$$

Now by 3.14, $c = e^b$ for some $b \in \mathbb{C}$, so if we define L in $D(c, |c|)$ by

$$L(w) = b - l\left(\frac{c - w}{c}\right), \quad w \in D(c, |c|),$$

then L is a power series centered at c and by (3) satisfies

$$w = e^{L(w)} \quad \forall w \in D(c, |c|),$$

as desired.

Corollary 3.20 *The function defined by the series* $\sum_{k=1}^{\infty} z^{2k-1}/(2k - 1)$ *maps the unit disk* $D(0, 1)$ *univalently onto the strip* $\mathbb{R} \times (-\pi/4, \pi/4)$.

Proof: Recall (proof of 3.19) that

$$l(z) = \sum_{k=1}^{\infty} \frac{1}{k} z^k$$

converges for all $z \in D = D(0, 1)$ and satisfies

(1) $$e^{-l(z)} = 1 - z.$$

Replacing z with $-z$ and dividing the two versions of (1), gives

$$e^{l(z) - l(-z)} = \frac{1 + z}{1 - z} \stackrel{\text{def.}}{=} \psi(z).$$

Therefore

$$L(z) = l(z) - l(-z) = 2 \sum_{k=1}^{\infty} \frac{z^{2k-1}}{2k - 1}, \quad z \in D,$$

satisfies

(2) $$e^L = \psi.$$

According to 2.5 ψ maps D univalently onto $H = (0, \infty) \times \mathbb{R}$. The univalence of L therefore follows from (2). Now exponential maps $\mathbb{R} \times \{m\pi/2 : m \text{ odd integer}\}$ into $i \cdot \mathbb{R} \subset \mathbb{C} \backslash H$. Therefore (2) shows that we have

$$L(D) \subset \bigcup_{n \in \mathbb{Z}} \mathbb{R} \times ((2n - 1)\pi/2, (2n + 1)\pi/2).$$

As $L(D)$ is connected [1.4 and 1.22], it follows from the last inclusion that $L(D) \subset \mathbb{R} \times ((2n - 1)\pi/2, (2n + 1)\pi/2)$ for a single n. Then $L(0) = 0$ shows that $n = 0$. Moreover, if $w \in \mathbb{R} \times (-\pi/2, \pi/2)$, then $e^w \in H = \psi(D) = e^{L(D)}$, that is, $e^w = e^{L(z)}$ for some $z \in D$. By periodicity, $L(z) - w = 2\pi k i$. But $L(z) - w \in \mathbb{R} \times (-\pi, \pi)$. Therefore $k = 0$, $w = L(z)$ and we have shown that $L(D) \supset \mathbb{R} \times (-\pi/2, \pi/2)$.

There is an important limit expression for e^z which we will derive now. It can serve as the basis for the whole theory, but not quite so smoothly as the power series definition. (See VAN YZEREN [1970] and A. J. MACINTYRE [1949].)

Theorem 3.21 $\lim_{n \to \infty}(1 + z/n)^n$ *exists and equals* e^z *for each* $z \in \mathbb{C}$ *and the convergence is uniform on any compact set. In fact, for all* $z \in \mathbb{C}$ *and all positive integers* m

$$\left| e^z - \left(1 + \frac{z}{m}\right)^m \right| \le e^{|z|} - \left(1 + \frac{|z|}{m}\right)^m \le \frac{|z|^2 e^{|z|}}{2m}.$$

Proof: (Cf. BIEHLER [1887].) The binomial expansion gives

$$\left(1 + \frac{z}{m}\right)^m = \sum_{k=0}^{m} \binom{m}{k} \left(\frac{z}{m}\right)^k = \sum_{k=0}^{m} \left(1 - \frac{1}{m}\right) \cdots \left(1 - \frac{k-1}{m}\right) \frac{z^k}{k!}$$

and so

$$\text{(1)} \quad e^z - \left(1 + \frac{z}{m}\right)^m = \sum_{k=2}^{m} \left[1 - \left(1 - \frac{1}{m}\right) \cdots \left(1 - \frac{k-1}{m}\right)\right] \frac{z^k}{k!} + \sum_{k=m+1}^{\infty} \frac{z^k}{k!}.$$

Two applications of (1) show that

$$\text{(2)} \quad \left| e^z - \left(1 + \frac{z}{m}\right)^m \right|$$

$$\le \sum_{k=2}^{m} \left[1 - \left(1 - \frac{1}{m}\right) \cdots \left(1 - \frac{k-1}{m}\right)\right] \frac{|z|^k}{k!} + \sum_{k=m+1}^{\infty} \frac{|z|^k}{k!}$$

$$= e^{|z|} - \left(1 + \frac{|z|}{m}\right)^m.$$

Notice that for $2 \le k \le m$

$$\left(1 - \frac{1}{m}\right) \cdots \left(1 - \frac{k-1}{m}\right) \ge 1 - \sum_{j=1}^{k-1} \frac{j}{m} = 1 - \frac{k(k-1)}{2m}.$$

Therefore (2) yields

$$(3)\qquad e^{|z|} - \left(1 + \frac{|z|}{m}\right)^m \le \sum_{k=2}^{m} \frac{k(k-1)}{2m}\cdot\frac{|z|^k}{k!} + \sum_{k=m+1}^{\infty} \frac{|z|^k}{k!}.$$

Now $m \ge 1$ means that $m + 1 \ge 2, (m + 1)m \ge 2m$. Consequently $k(k - 1) \ge 2m$ for $k \ge m + 1$ and we can estimate the second sum in (3) thus:

$$\begin{aligned} e^{|z|} - \left(1 + \frac{|z|}{m}\right)^m &\le \sum_{k=2}^{m} \frac{k(k-1)}{2m}\frac{|z|^k}{k!} + \sum_{k=m+1}^{\infty} \frac{|z|^k}{2m(k-2)!} \\ &= \frac{1}{2m}\sum_{k=2}^{\infty} \frac{|z|^k}{(k-2)!} = \frac{1}{2m}\sum_{n=0}^{\infty}\frac{|z|^{n+2}}{n!} \qquad (\text{set } n = k-2) \\ &= \frac{|z|^2}{2m} e^{|z|}. \end{aligned}$$

About the constant π we will need very little beyond the inequality (8) of the proof of 3.14. In fact only

Theorem 3.22 $\int_0^1 1/(1 + x^2)dx = \pi/4$.

Proof: We have, in the notation of the proof of 3.14,

$$\begin{aligned} 0 = C(\pi/2) &= \operatorname{Re}[e^{i\pi/4}]^2 = [\operatorname{Re} e^{i\pi/4}]^2 - [\operatorname{Im} e^{i\pi/4}]^2 \\ &= C^2(\pi/4) - S^2(\pi/4), \\ C^2(\pi/4) &= S^2(\pi/4). \end{aligned}$$

But S and C are both positive in $(0, \pi/2)$, so this last equality means that

$$(1)\qquad C(\pi/4) = S(\pi/4) > 0.$$

Since C is never zero in $[0, \pi/2)$, by definition of π, we may form the function $T = S/C$ on $[0, \pi/4]$ and have from equation (5) of the proof of 3.14 that

$$(2)\qquad T' = \frac{S'C - SC'}{C^2} = \frac{C^2 + S^2}{C^2} = 1 + (S/C)^2 = 1 + T^2 > 0.$$

Therefore T is strictly increasing, $T(0) = 0$ and by (1) $T(\pi/4) = 1$. It follows from (2) then that

$$\int_0^1 \frac{1}{1 + x^2}dx = \int_{T^{-1}(0)}^{T^{-1}(1)} \frac{T'}{1 + T^2} = \int_{T^{-1}(0)}^{T^{-1}(1)} 1 = T^{-1}(1) - T^{-1}(0) = \frac{\pi}{4}.$$

Exercise 3.23 *With the convention* $0^0 = 1$, *the function* x^{-x} *is continuous on* $[0, 1]$. (*Cf.* 3.18(4) *and recall the customary definition* $x^{-x} = e^{-x \log x}$ *for* $x > 0$.) *Prove the satisfyingly symmetric formula*

$$\int_0^1 x^{-x}dx = \sum_{n=1}^{\infty} n^{-n}.$$

Hints: $x^{-x} = \sum_{n=0}^{\infty} (-1)^n (x \log x)^n / n!$. Interchange summation and integration and evaluate $\int_0^1 x^n \log^n (x) dx$ by repeated integration-by-parts.

Exercise 3.24 *Show that if f is differentiable in an open, connected subset Ω of $\mathbb{C}$ or $\mathbb{R}$ and $f' \equiv cf$ for some constant c, then*

$$f(z) = f(a)e^{-ca}e^{cz} \quad \forall a, z \in \Omega.$$

Hint: Let $g(z) = e^{-cz}$ and note that $(f \cdot g)' = 0$ throughout Ω.

Exercise 3.25 *Find all continuous homomorphisms f of the additive group $\mathbb{R}$ with the multiplicative group $\mathbb{T}$.*

Hints: Since f is a homomorphism, $f(0) = 1$. By continuity then there exists $a > 0$ such that $\int_0^a f = b \neq 0$. The homomorphic property $f(x + t) = f(x)f(t)$ (all $x, t \in \mathbb{R}$) then implies that

$$bf(x) = f(x) \int_0^a f(t)dt = \int_0^a f(x + t)dt = \int_x^{x+a} f.$$

Since f is continuous, the last expression is a differentiable function of x. It follows that bf, hence f, is differentiable. We can therefore differentiate the functional equation with respect to t and set $t = 0$ to learn that $f' = cf$, where $c = f'(0)$. This leads via 3.24 to $f(x) = f(0)e^{cx} = e^{cx}$. Since $|f| \equiv 1$, we must have $\operatorname{Re} c = 0$, that is, $c = iy$ for some $y \in \mathbb{R}$.

Exercise 3.26 *Let ϕ: $[0, 1] \to \mathbb{C}$ be continuous and satisfy $\phi(1) = \phi(0)$. Show that there exists a continuous Φ: $C(0, 1) \to \mathbb{C}$ satisfying*

$$(*) \qquad \Phi(e^{2\pi it}) = \phi(t) \quad \forall t \in [0, 1].$$

Hints: By 3.14(v) and the $2\pi i$-periodicity of the exponential function, the map $t \to e^{2\pi it}$ sends $[0, 1)$ one-to-one onto the circle $C(0, 1)$. Hence we can define Φ on $C(0, 1)$ by the decree

$$(1) \qquad \Phi(e^{2\pi it}) = \phi(t) \quad \forall t \in [0, 1).$$

The condition $\phi(1) = \phi(0)$ ensures that in fact (*) holds. Suppose now that Φ is not continuous at some $z_0 \in C(0, 1)$. This means that there exists an $\varepsilon > 0$ and a sequence $\{z_n\} \subset C(0, 1)$ such that $z_n \to z_0$ and

$$(2) \qquad |\Phi(z_n) - \Phi(z_0)| \geq \varepsilon, \quad n = 1, 2, 3, \ldots$$

For each $n \geq 1$ there is a unique $t_n \in [0, 1)$ such that $z_n = e^{2\pi it_n}$ and there is a subsequence of $\{t_n\}$ which converges to some $t_0 \in [0, 1]$. Passing to that subsequence, we can simply suppose that $\{t_n\}$ itself converges to t_0. Then $z_n = e^{2\pi it_n} \to e^{2\pi it_0}$, and so $z_0 = e^{2\pi it_0}$. But then

$$\Phi(z_n) = \phi(t_n) \to \phi(t_0) \overset{(*)}{=} \Phi(e^{2\pi it_0}) = \Phi(z_0),$$

in violation of (2).

It is customary to introduce functions cos, sin, tan, and cot by

Definition 3.27 The functions cos and sin are defined for all $z \in \mathbb{C}$ by

$$\cos(z) = \frac{e^{iz} + e^{-iz}}{2} = \sum_{n=0}^{\infty} \frac{(-1)^n z^{2n}}{(2n)!}$$

$$\sin(z) = \frac{e^{iz} - e^{-iz}}{2i} = \sum_{n=0}^{\infty} \frac{(-1)^n z^{2n+1}}{(2n+1)!}.$$

The function tan is defined as sin/cos. Since the zeros of cos are evidently the odd integer multiples of $\pi i/2$ the domain of tan is $\mathbb{C}\backslash\pi i(\mathbb{Z} + 1/2)$. The function cot is defined as cos/sin, with the appropriate domain, viz., $\mathbb{C}\backslash\pi i\mathbb{Z}$.

Remarks: The functions C and S in the proof of 3.14 were just cos and sin restricted to $\mathbb{R}$. The functions cos and sin are infinitely differentiable throughout $\mathbb{C}$ and $\cos' = -\sin$, $\sin' = \cos$.

Exercise 3.28 $|\sin x| \le |x|$ *for all real* x.

Hint: For each $x \neq 0$ the Mean Value Theorem provides an x' such that

$$\left|\frac{\sin x}{x}\right| = \left|\frac{\sin(x) - \sin(0)}{x - 0}\right| = |\sin'(x')| = |\cos(x')| \le 1.$$

Exercise 3.29 $\cos x \ge 1 - 2x/\pi$ *for all* x *in* $[0, \pi/2]$.

Hints: The function $g(x) = \cos x + 2x/\pi$ satisfies $g(0) = g(\pi/2) = 1$. Also $g' = 2/\pi - \sin$. From equation (10) in the proof of 3.14, sin increases on $[0, \pi/2]$ from $\sin(0) = 0$ to $\sin(\pi/2) = 1 > 2/\pi$. Therefore g' has a unique zero $x_0 \in [0, \pi/2]$ and $g' > 0$ on $[0, x_0)$, $g' < 0$ on $(x_0, \pi/2]$. It follows that g is increasing on $[0, x_0]$, decreasing on $[x_0, \pi/2]$ and so $\min g[0, \pi/2] = g(0) = g(\pi/2) = 1$.

Exercise 3.30 $\sin x \ge 2x/\pi$ *for all* x *in* $[0, \pi/2]$.

Hints: The function $h(x) = \sin x - 2x/\pi$ satisfies $h(0) = h(\pi/2) = 0$, $h' = \cos - 2/\pi$. $\cos' = -\sin$ is negative on $(0, \pi/2]$, so h' decreases from $h'(0) = 1 - 2/\pi > 0$ to $h'(\pi/2) = -2/\pi < 0$. As in the hints of 3.29, it follows that $\min h[0, \pi/2] = h(0) = h(\pi/2) = 0$.

Exercise 3.31 $\int_0^{\pi/4} e^{-R\cos 2x}dx \le \pi/4R$ *for all* $R > 0$.

Hint: By 3.29, $e^{-R\cos 2x} \le e^{-R(1-4x/\pi)} = e^{-R}e^{(4R/\pi)x}$, so

$$\int_0^{\pi/4} e^{-R\cos 2x}dx \le e^{-R}\frac{\pi}{4R}e^{(4R/\pi)x}\Big]_0^{\pi/4} < e^{-R}\cdot\frac{\pi}{4R}\cdot e^{4(R/\pi)(\pi/4)}.$$

Exercise 3.32 (Parseval's Formula) *If the power series $f(z) = \sum_{n=0}^{\infty} c_n z^n$ has radius of convergence R, then for every $0 \le r < R$ we have*

$$\frac{1}{2\pi}\int_0^{2\pi} |f(re^{it})|^2 dt = \sum_{n=0}^{\infty} |c_n|^2 r^{2n}.$$

Hints: The functions $S_N(z) = \sum_{n=0}^{N} c_n z^n$ satisfy

$$|S_N(re^{it})|^2 = \left[\sum_{n=0}^{N} c_n r^n e^{int}\right]\left[\sum_{k=0}^{N} \bar{c}_k r^k e^{-ikt}\right] = \sum_{n=0}^{N}\sum_{k=0}^{N} c_n \bar{c}_k r^{n+k} e^{i(n-k)t},$$

so

$$\int_0^{2\pi} |S_N(re^{it})|^2 dt = \sum_{n=0}^{N}\sum_{k=0}^{N} c_n \bar{c}_k r^{n+k} \int_0^{2\pi} e^{i(n-k)t} dt = \sum_{n=0}^{N} 2\pi |c_n|^2 r^{2n}.$$

By 3.3(i), $|S_N(re^{it})|$ is close to $|f(re^{it})|$ uniformly in $t \in \mathbb{R}$ for large N.

Exercise 3.33 (Liouville) *Deduce from the last exercise that if $R = \infty$ and f is bounded in $\mathbb{C}$, then f is constant.*

Exercise 3.34 *Deduce from* 3.32 *the so-called* Cauchy Estimates: *With the notation and hypotheses of that exercise, if we set $M(r) = \sup_{|z|=r}|f(z)|$, then for every non-negative integer k*

$$\frac{|f^{(k)}(0)|}{k!} = |c_k| \le \frac{M(r)}{r^k}.$$

Exercise 3.35 *Suppose that $\sum_{k=0}^{\infty} c_k z^k$ converges for all $z \in \mathbb{C}$. Call it $f(z)$ and for each $r \ge 0$ set $M(r) = \sup\{|f(z)| : |z| = r\}$.*

(i) *Show that*

$$\sum_{k=0}^{\infty} |c_k| r^k \le 2M(2r) \quad \forall r \ge 0.$$

Hints: For all $r > 0$ the Cauchy estimates (3.34) give

$$|c_k| \le \frac{M(2r)}{(2r)^k}, \quad k = 0, 1, 2, \ldots,$$

whence

$$\sum_{k=0}^{\infty} |c_k| r^k \le \sum_{k=0}^{\infty} \frac{M(2r)}{(2r)^k} r^k = M(2r) \sum_{k=0}^{\infty} \frac{1}{2^k}.$$

(ii) *Let c be a non-negative real number. Show that the following boundedness condition on the coefficients of f:*

(1) $\overline{\lim_{n\to\infty}}\, n|c_n|^{1/n} \le ce$

is equivalent to the following global boundedness condition on f itself:

(2) $M(r)e^{-(c+\varepsilon)r}$ *is a bounded function of* $r \in [0, \infty)$, *for each* $\varepsilon > 0$.

Hints: (1) ⇒ (2). Consider any $\varepsilon > 0$ and define $\delta = \delta(\varepsilon) > 1$ by

(3) $$\left(c + \frac{\varepsilon}{2}\right)\delta = c + \varepsilon.$$

By (1) there exists an $N = N(\varepsilon)$ such that

(4) $$n|c_n|^{1/n} < \left(c + \frac{\varepsilon}{2}\right)e \quad \forall n > N.$$

Moreover by 3.9 we have

$$\lim_{n\to\infty}\left[\frac{n!}{n^n}\right]^{1/n} = \lim_{n\to\infty}\left[\frac{(n+1)!}{(n+1)^{n+1}}\Big/\frac{n!}{n^n}\right] = \lim_{n\to\infty}\left(1 + \frac{1}{n}\right)^{-n} \overset{(3.21)}{=} e^{-1}.$$

Consequently we can also suppose the N above so large that

(5) $$\left[\frac{n!}{n^n}\right]^{1/n} < \delta e^{-1} \quad \forall n > N.$$

Multiplying (4) and (5), we get from (3)

$$(n!|c_n|)^{1/n} < \left(c + \frac{\varepsilon}{2}\right)\delta = c + \varepsilon \quad \forall n > N$$

and so

$$|f(z)| \le \left|\sum_{n=0}^{N} c_n z^n\right| + \sum_{n=N+1}^{\infty} \frac{(c+\varepsilon)^n}{n!}|z|^n$$

(6) $$< \sum_{n=0}^{N} |c_n|\,|z|^n + e^{(c+\varepsilon)|z|} \quad \forall z \in \mathbb{C}.$$

For any $n > 0$, however, the power series for the exponential function yields

$$\frac{e^{\varepsilon|z|}}{|z|^n} > \frac{(\varepsilon|z|)^{n+1}/(n+1)!}{|z|^n} = \frac{\varepsilon^{n+1}|z|}{(n+1)!} \to \infty \quad \text{as } |z| \to \infty$$

and so there exists $M = M(\varepsilon)$ such that

(7) $$\sum_{n=0}^{N} |c_n|\,|z^n| < e^{\varepsilon|z|} \quad \forall |z| \ge M.$$

From (6) and (7)

$$|f(z)| \le 2e^{(c+\varepsilon)|z|} \quad \forall |z| \ge M$$

and this proves that $f(z)e^{-(c+\varepsilon)|z|}$ is bounded in $\mathbb{C}$. As $\varepsilon > 0$ is arbitrary, (2) is established.

(2) ⇒ (1). If the limit superior in (1) exceeds ce, then we can select $\varepsilon > 0$ sufficiently small that

(8) $\quad n|c_n|^{1/n} > (c + 2\varepsilon)e$ for infinitely many n.

According to 3.34 we have

$$|c_n| \le \frac{1}{r^n} M(r),$$

(9) $\quad M(r) \ge r^n|c_n| \quad \forall r > 0, n = 1, 2, 3, \ldots$

If in (9) we choose $r = n/(c + 2\varepsilon)$ for an n as in (8), we get by (8)

$$M(r) \ge \left[\frac{n}{c + 2\varepsilon}\right]^n \left[\frac{(c + 2\varepsilon)e}{n}\right]^n = e^n = e^{(c+2\varepsilon)r},$$

so

$$M(r)e^{-(c+\varepsilon)r} \ge e^{\varepsilon r}.$$

This holds for a set of arbitrarily large r and so shows that $M(r)e^{-(c+\varepsilon)r}$ is not a bounded function of $r \in [0, \infty)$.

(iii) *Show that* $\overline{\lim}_{n\to\infty}(n/e)|c_n|^{1/n}$ *is equal to the least element* c *of* $[0, \infty]$ *for which* (2) *holds.*

§ 4 Bernoulli Polynomials, Numbers and Functions

In this section we use the theory of the exponential and logarithm functions to compute $\sum_{k=1}^{\infty} k^{-2n}$. None of this material is used later.

Exercise 3.36 *The function L of 3.19 has a continuous extension (call it L also) to* $\bar{D}(c, |c|)\backslash\{0\}$ *which satisfies* $e^{L(z)} = z$ *for all such z. In fact, the series for L about c converges uniformly on each compact subset of* $\bar{D}(c, |c|)\backslash\{0\}$.

Hints: It suffices to deal with $c = 1$. In proving 3.19 we defined

(1) $\quad l(z) = \sum_{k=1}^{\infty} \frac{1}{k} z^k, \quad z \in D(0, 1)$

and saw that

(2) $\quad e^{-l(z)} = 1 - z \quad \forall z \in D(0, 1).$

By 3.13(ii) this series converges uniformly on each compact subset of $\bar{D}(0, 1)\backslash\{1\}$ and so defines a continuous function l on $\bar{D}(0, 1)\backslash\{1\}$. By continuity (2) holds for all z in this latter set.

Exercise 3.37 (i) $\sum_{n=1}^{\infty} \sin(n\theta)/n = (\pi - \theta)/2$ *for all* $\theta \in (0, 2\pi)$ *and convergence is uniform on each compact subset of* $(0, 2\pi)$.

Hints: Consider the logarithm l of the previous exercise:

$$e^{-l(z)} = 1 - z$$

$$\text{(1)} \qquad e^{-i\,\mathrm{Im}\,l(z)} = \frac{e^{-l(z)}}{|e^{-l(z)}|} = \frac{1-z}{|1-z|} \quad \forall z \in \bar{D}(0, 1)\backslash\{1\}.$$

For $z \in \bar{D}(0, 1)\backslash\{1\}$ we have $1 - z \in \bar{D}(1, 1)\backslash\{0\} \subset \mathbb{C}\backslash(-\infty, 0]$. It follows then from (1) that $-\mathrm{Im}\, l(z)$ is not an odd integer multiple of π, that is, $\mathrm{Im}\, l$ maps $\bar{D}(0, 1)\backslash\{1\}$ into $\bigcup_{k=-\infty}^{\infty} ((2k-1)\pi, (2k+1)\pi)$. As $\bar{D}(0, 1)\backslash\{1\}$ is connected [use 1.2 or 1.22], the range of $\mathrm{Im}\, l$ must lie in only one of the intervals $((2k-1)\pi, (2k+1)\pi)$. Since $l(0) = 0$, $k = 0$. For $\theta \in (0, 2\pi)$ we have $\sin \theta/2 > 0$ and so from

$$\begin{aligned} 1 - e^{i\theta} &= [e^{-i\theta/2} - e^{i\theta/2}]e^{i\theta/2} = -i(2 \sin \theta/2)e^{i\theta/2} \\ &= (2 \sin \theta/2)e^{i(\theta-\pi)/2}, \end{aligned}$$

follows

$$\text{(2)} \qquad \frac{1 - e^{i\theta}}{|1 - e^{i\theta}|} = e^{i(\theta-\pi)/2}.$$

Since $(\theta - \pi)/2 \in (-\pi, \pi)$, we get from (1), (2) and the above observation about $\mathrm{Im}\, l$,

$$-\mathrm{Im}\, l(e^{i\theta}) = (\theta - \pi)/2.$$

Since $l(e^{i\theta}) = \sum_{k=1}^{\infty} e^{ik\theta}/k$, the assertion follows.

$$\text{(ii)} \qquad \sum_{n=1}^{\infty} (-1)^{n+1} \frac{\sin(nx)}{n} = \frac{x}{2} \quad \forall x \in (-\pi, \pi)$$

with convergence uniform on compacta.

Hint: Replace θ by $\pi - x$ in (i).

$$\text{(iii)} \qquad \sum_{n=1}^{\infty} (-1)^{n+1} \frac{1 - \cos(n\theta)}{n^2} = \frac{\theta^2}{4} \quad \forall \theta \in [-\pi, \pi].$$

Hint: Both sides of the alleged equation are continuous functions of θ (since the series is uniformly convergent), so it suffices to confirm the equality for $\theta \in (-\pi, \pi)$. For such θ, integrate (ii) from $x = 0$ to $x = \theta$ and use the uniform convergence on $[-|\theta|, |\theta|]$ to interchange sum and integral.

$$\text{(iv)} \qquad \sum_{n=1}^{\infty} \frac{(-1)^{n+1}}{n^2} = \frac{\pi^2}{12}.$$

Hint: The series (iii), being uniformly convergent, can be integrated term by term over $[-\pi, \pi]$.

$$\text{(v)} \qquad \sum_{n=1}^{\infty} \frac{\cos(n\theta)}{n^2} = \frac{3\theta^2 - 6\pi\theta + 2\pi^2}{12}, \quad 0 \le \theta \le 2\pi.$$

Hint: Subtract (iv) from (iii) and change x to $\pi - \theta$.

Exercise 3.38 (i) *There exists a unique sequence* $\{P_n\}$ $(n = 0, 1, 2, \ldots)$ *of polynomials (called* Bernoulli polynomials) *such that*

(1) $P_0 = 1$

(2) $P'_{n+1} = P_n \quad \forall n \geq 0$

(3) $\int_0^1 P_n = 0 \quad \forall n \geq 1.$

(ii) *The polynomials* P_n *satisfy in addition*

(4) $P_1(z) \equiv z - \frac{1}{2}$

(5) $P_2(z) \equiv z^2/2 - z/2 + \frac{1}{12}$

(6) $P_n(z) = (-1)^n P_n(1 - z)$ for all $z \in \mathbb{C}$, $n \geq 0$.

Hint: Show that the polynomials $p_n(z) = (-1)^n P_n(1 - z)$ have properties (1), (2), (3).

(7) $P_n(0) = P_n(1/2) = P_n(1) = 0$ for all odd $n > 1$.

(8) $P_n(0) = P_n(1) \quad \forall n \geq 1.$

(9) $P_{n+1}(z + 1) - P_{n+1}(z) = \int_{[0,z]} P_n(\xi + 1)d\xi - \int_{[0,z]} P_n(\xi)d\xi$

for all $z \in \mathbb{C}$ and all $n \geq 1$. (Use induction.)

(10) $P_n(z + 1) = P_n(z) + \dfrac{z^{n-1}}{(n - 1)!}$ for all $z \in \mathbb{C}$ and all $n \geq 1$.

(iii) *Use* (4) *and* 3.37(i) *to deduce that*

(11) $P_1(t) = -\dfrac{1}{\pi} \sum_{k=1}^{\infty} \dfrac{\sin(2\pi kt)}{k^2} \quad \forall t \in (0, 1)$

with convergence uniform on compacta. Similarly use (5) *and* 3.37(v) *to deduce that*

(12) $P_2(t) = \dfrac{1}{2\pi^2} \sum_{k=1}^{\infty} \dfrac{\cos(2\pi kt)}{k^2} \quad \forall t \in [0, 1].$

Evidently the convergence is uniform on [0, 1] *and so successive integrations lead to*

(13) $P_{2m}(t) = (-1)^{m-1} \dfrac{2}{(2\pi)^{2m}} \sum_{k=1}^{\infty} \dfrac{\cos(2\pi kt)}{k^{2m}} \quad \forall t \in [0, 1]$

(14) $P_{2m+1}(t) = (-1)^{m-1} \dfrac{2}{(2\pi)^{2m+1}} \sum_{k=1}^{\infty} \dfrac{\sin(2\pi kt)}{k^{2m+1}} \quad \forall t \in [0, 1].$

(The periodic functions on $\mathbb{R}$ *defined by these series are called the* Bernoulli functions.)

(iv) *Define* $B_n = n!\,P_n(0)$ *(the so-called* Bernoulli numbers*) for each* $n \geq 0$ *and show via* (2) *that*

$$(15)\quad P_n(z) = \frac{1}{n!}\sum_{k=0}^{n}\binom{n}{k}B_k z^{n-k} \quad \text{for all } z \in \mathbb{C},\ n \geq 0;$$

then use (8) *to conclude that*

$$(16)\quad B_n = -\frac{1}{n+1}\sum_{k=0}^{n-1}\binom{n+1}{k}B_k \quad \forall n \geq 1.$$

Deduce from (13) *that*

$$(17)\quad \sum_{k=1}^{\infty}\frac{1}{k^{2m}} = \frac{(-1)^{m-1}(2\pi)^{2m}B_{2m}}{2(2m)!} \quad \forall m \geq 1.$$

Using this and (16), *deduce the formulas*

$$(18)\quad \sum_{k=1}^{\infty}\frac{1}{k^2} = \frac{\pi^2}{6},\quad \sum_{k=1}^{\infty}\frac{1}{k^4} = \frac{\pi^4}{90},\quad \sum_{k=1}^{\infty}\frac{1}{k^6} = \frac{\pi^6}{945},\quad \sum_{k=1}^{\infty}\frac{1}{k^8} = \frac{\pi^8}{9450}.$$

By breaking (17) *into sums over even and odd k, deduce that*

$$(19)\quad \sum_{k=1}^{\infty}\frac{1}{(2k-1)^{2m}} = (-1)^{m-1}\pi^{2m}\frac{2^{2m}-1}{2(2m)!}B_{2m} \quad \forall m \geq 1$$

$$(20)\quad \sum_{k=1}^{\infty}\frac{(-1)^{k-1}}{k^{2m}} = (-1)^{m-1}\pi^{2m}\frac{2^{2m-1}-1}{(2m)!}B_{2m} \quad \forall m \geq 1.$$

Use (10), (15) *and* (16) *to deduce that*

$$(21)\quad \sum_{k=1}^{N}k^n = \frac{1}{n+1}\sum_{k=0}^{n}\binom{n+1}{k}B_k(N+1)^{n+1-k} \quad \text{for all } n, N \geq 1.$$

Confirm then the following identities

$$(22)\quad \begin{cases} \displaystyle\sum_{k=1}^{N}k^2 = \frac{N^3}{3} + \frac{N^2}{2} + \frac{N}{6}, & \displaystyle\sum_{k=1}^{N}k^3 = \frac{N^4}{4} + \frac{N^3}{2} + \frac{N^2}{4} = \left[\sum_{k=1}^{N}k\right]^2, \\ \displaystyle\sum_{k=1}^{N}k^4 = \frac{N^5}{5} + \frac{N^4}{2} + \frac{N^3}{3} - \frac{N}{30}, & \displaystyle\sum_{k=1}^{N}k^5 = \frac{N^6}{6} + \frac{N^5}{2} + \frac{5N^4}{12} - \frac{N^2}{12}. \end{cases}$$

(v) *For* $t \in \mathbb{R}$, $z \in \mathbb{C}$ *with* $0 \leq t \leq 1$ *and* $0 < |z| < 2\pi$ *we have*

$$(23)\quad \frac{ze^{tz}}{e^z - 1} = \sum_{n=0}^{\infty}P_n(t)z^n.$$

Hints: Fix $z \in \mathbb{C}$ with $|z| < 2\pi$. It follows easily from (11)–(14) that for any $n \geq 0$ and $t \in [0, 1]$

$$(24)\quad |P_n(t)| \leq \frac{2}{(2\pi)^n}\sum_{k=1}^{\infty}\frac{1}{k^2} \overset{(18)}{=} \frac{\pi^2}{3}\frac{1}{(2\pi)^n}.$$

Consequently the series in (23) converges uniformly for $t \in [0, 1]$. Call the limit $f(t)$. Using (2),

$$\frac{f(t) - f(t_0)}{t - t_0} - zf(t_0) = \sum_{n=1}^{\infty} \left[\frac{P_n(t) - P_n(t_0)}{t - t_0} - P_{n+1}(t_0)\right] z^n$$

(25)
$$= \sum_{n=1}^{\infty} [P_{n+1}(t_n) - P_{n+1}(t_0)] z^n,$$

for some t_n between t and t_0 (Mean Value Theorem). It follows from (24) and (25) that f' exists in $[0, 1]$ and satisfies $f'(t) = zf(t)$. Consequently $e^{-tz}f(t)$ has derivative 0 with respect to t and so

(26) $f(t) = c(z)e^{tz} \quad \forall t \in [0, 1],$

where $c(z) = f(0)$. Integrate (26) with respect to t. Computing $\int_0^1 f(t)dt$ termwise through the series for f and recalling (1) and (3), gives for this integral the value z, whence (23).

(vi) Take $t = 0$ in (23) and recall (7) to get

$$\frac{z}{e^z - 1} = 1 - \tfrac{1}{2}z + \sum_{m=1}^{\infty} \frac{B_{2m}}{(2m)!} z^{2m}, \quad z \in \mathbb{C}, 0 < |z| < 2\pi.$$

Transpose $-\frac{1}{2}z$, multiply and divide by $e^{-z/2}$ to get

$$\frac{z}{2} \cdot \frac{e^{z/2} + e^{-z/2}}{e^{z/2} - e^{-z/2}} = \sum_{m=0}^{\infty} \frac{B_{2m}}{(2m)!} z^{2m}$$

and conclude by a change of variable that

(27)
$$z \cot z = \sum_{m=0}^{\infty} (-1)^m \frac{2^{2m} B_{2m}}{(2m)!} z^{2m}, \quad z \in \mathbb{C}, 0 < |z| < \pi.$$

Use the identity $\tan z = \cot z - 2 \cot 2z$ to deduce from (27) the series

(28)
$$\tan z = \sum_{m=1}^{\infty} (-1)^{m-1} \frac{2^{2m}(2^{2m} - 1)B_{2m}}{(2m)!} z^{2m-1}, \quad z \in \mathbb{C}, |z| < \frac{\pi}{2}.$$

§ 5 Cauchy's Theorem Adumbrated

Exercise 3.39 (i) *For any* $a \in \mathbb{C}$, $r > 0$

$$\int_{C(a,r)} \frac{d\xi}{\xi - z} = 2\pi i \quad \forall z \in D(a, r).$$

Hint: Take $g = 1$, $\gamma = C(a, r)$ in the series appearing in 3.8 (and recall 2.15).

(ii) *Let f be differentiable and f' be continuous in a neighborhood of* $\bar{D}(a, r)$. *Show that*

(*)
$$f(z) = \frac{1}{2\pi i} \int_{C(a,r)} \frac{f(\xi)}{\xi - z} d\xi \quad \forall z \in D(a, r),$$

whence, via 3.8, *f is locally of power series type in* $D(a, r)$.

Hints: Fix such a z and form

$$F(t) = \int_{C(a,r)} \frac{f(z + t(\xi - z))}{\xi - z}\,d\xi,$$

$$G(t) = \int_{C(a,r)} f'(z + t(\xi - z))d\xi \quad (0 \le t \le 1).$$

Show that F is differentiable and that $F' = G$, thus:

$$\frac{F(t) - F(t_0)}{t - t_0} - G(t_0)$$

$$= \int_{C(a,r)} \left[\frac{f(z + t(\xi - z)) - f(z + t_0(\xi - z))}{t - t_0} - f'(z + t_0(\xi - z))(\xi - z)\right]\frac{d\xi}{\xi - z}$$

$$= \int_{C(a,r)} \frac{1}{t - t_0}\int_{t_0}^{t} [f'(z + \tau(\xi - z)) - f'(z + t_0(\xi - z))]d\tau d\xi,$$

since (Chain Rule) $\frac{d}{d\tau}f(z + \tau(\xi - z)) = f'(z + \tau(\xi - z))(\xi - z)$. Now make the obvious appraisals of the last integrand. Note also that $\frac{d}{d\xi}f(z + t(\xi - z)) = tf'(z + t(\xi - z))$, so by 2.10(ii), $G = 0$. It follows that $F(1) = F(0)$. Use (i) to evaluate $F(0)$.

Remarks: The proof in (ii) works in any convex set to show that

$$(**) \quad f(z)\int_\gamma \frac{d\xi}{\xi - z} = \int_\gamma \frac{f(\xi)d\xi}{\xi - z} \quad \forall z \in U\backslash\gamma$$

whenever U is open and convex, $f \in H(U)$, f' is continuous in U, and γ is a piecewise smooth loop in U. [Convexity is needed so that the points $z + t(\xi - z)$ $(0 \le t \le 1)$ remain in U.] Notice that according to 2.14 and 2.15 the integral on the left in (**) has derivative 0 as a function of z and so by 2.10(iv) is constant in each component of $\mathbb{C}\backslash\gamma$. In the next chapter we will prove that it is always an integer multiple of $2\pi i$. Formula (**) then looks like (*). It is in fact valid for a much larger class of open sets than the convex ones and can be proved without the *a priori* assumption of the continuity of f'. [That continuity and more of course follow from (**) via 3.8.] This is the theme of Chapters X and V respectively. New techniques, not at all as straightforward and simple as those in 3.39, will be required. Their creation was a major advance in the foundations of the subject, which prior to 1900 had always hypothesized continuity of f', in addition to its existence.

§ 6 Holomorphic Logarithms Previewed

Following 3.17 and tradition we shall call any function L which satisfies $e^{L(z)} = z$ for all z in its domain, a *logarithm function*. Thanks to 3.14(vi), such functions exist throughout $\mathbb{C}\backslash\{0\}$, but what is wanted and turns out to be of paramount importance and utility are continuous logarithms. Their existence on a given set is intimately tied to the geometry of the set, i.e., to its disposition in $\mathbb{C}$. This will all be thoroughly explored in the next chapter, but here are a few elementary cases.

Exercise 3.40 *Let U_1, U_2 be open subsets of $\mathbb{C}\backslash\{0\}$ such that $U_1 \cap U_2$ is connected. Show that if there exists a continuous logarithm L_j on U_j for each j, then there exists a continuous logarithm on $U_1 \cup U_2$.*

Hint: Show that $L_1 - L_2$ is constant in $U_1 \cap U_2$. (Cf. the proof of 4.67(ii).)

Exercise 3.41 (Cf. 4.60) *Prove that a continuous logarithm is necessarily holomorphic. More precisely, prove that if L is a continuous logarithm in an open subset U of $\mathbb{C}\backslash\{0\}$, then L is holomorphic and $L'(z) = 1/z$ for every $z \in U$.*

Hints: Imitate the argument in 3.18. Alternatively, use 3.19 to get local holomorphic logarithms and use continuity to show that locally L differs only by constants from such functions.

Exercise 3.42 (i) *Prove that in every open half-plane which does not contain 0 there is a holomorphic logarithm.*

Hints: It suffices to consider $H = \{z \in \mathbb{C}: \operatorname{Re} z > 0\}$. For each positive integer n there is a holomorphic logarithm L_n in $D(n, n)$, by 3.19. Then $[L_{n+1} - L_n]/2\pi i$ is continuous and integer-valued in $D(n + 1, n + 1) \cap D(n, n) = D(n, n)$. Therefore, by adding an appropriate (constant) integer multiple of $2\pi i$ to L_{n+1}, we can assume that L_{n+1} extends L_n. Do this *successively* for each $n = 1, 2, 3, \ldots$ It is then possible to define the desired logarithm L in H by the simple decree $L = L_n$ in $D(n, n)$, for every n, bearing in mind that $\bigcup_{n=1}^{\infty} D(n, n) = H$.

(ii) *Prove that there is a holomorphic logarithm in any slit-plane $S = \mathbb{C}\backslash\{tu: t \leq 0\}$, $u \in \mathbb{C}\backslash\{0\}$. Show moreover that any such function differs only by an additive constant from that defined by the integral $\int_{[u,z]} d\xi/\xi$, $z \in S$.*

Hints: $S = H_1 \cup H_2 \cup H_3$ where H_j are open half-planes such that $H_1 \cap H_2$ and $H_2 \cap H_3$ are connected while $H_1 \cap H_3$ is void. Part (i) provides holomorphic logarithms L_j in each H_j. Add appropriate constants to these to synthesize a single holomorphic logarithm L in $H_1 \cup H_2 \cup H_3$, as in 3.40. (Cf. with the proof of 4.67(iii).) From $e^{L(\xi)} = \xi$ and the Chain Rule follows $L'(\xi) = 1/\xi$. Observing that S is starlike with respect to u, it follows then from 2.10 that $L(z) - L(u) = \int_{[u,z]} d\xi/\xi$, for all $z \in S$. [It is easy to see—from the proof of 5.2, e.g.—that the *existence* of L can also be secured by means of this integral.]

(iii) *Show that in $\mathbb{C}\backslash[0, \infty)$ there is a unique holomorphic logarithm L such that* $\operatorname{Im} L(z) \in (0, 2\pi)$ *for all $z \in \mathbb{C}\backslash[0, \infty)$. Thus for any $r>0$ and $\theta \in (0, 2\pi)$,* $L(re^{i\theta}) = \log r + i\theta$.

Hints: If l is any holomorphic logarithm in $\mathbb{C}\backslash[0, \infty)$ (such exist by (ii)), then

$e^{l(z)} = z \notin [0, \infty)$ implies that $\operatorname{Im} l(z) \neq 2\pi n$ for any integer n. Since $\operatorname{Im} l$ is a continuous real-valued function and $\mathbb{C}\setminus[0, \infty)$ is connected (starlike), the set $\operatorname{Im} l(\mathbb{C}\setminus[0, \infty))$ is a connected subset of $\bigcup_{n=-\infty}^{\infty} (2\pi n, 2\pi(n + 1))$ and so lies wholly in $(2\pi n, 2\pi(n + 1))$ for some one integer n. Set $L = l - 2\pi i n$. Alternatively, notice that $[l(re^{i\theta}) - \log r - i\theta]/2\pi i$ is a continuous, integer-valued function of (r, θ) in the connected set $(0, \infty) \times (0, 2\pi)$.

Remark 3.43: The function L above is customarily called the *Principal Branch of the Logarithm in* $\mathbb{C}\setminus[0, \infty)$. There is similarly a unique holomorphic logarithm in $\mathbb{C}\setminus(-\infty, 0]$ which assigns to $re^{i\theta}$ the value $\log r + i\theta$ whenever $r > 0$ and $-\pi < \theta < \pi$. It is called the *Principal Branch of the Logarithm in* $\mathbb{C}\setminus(-\infty, 0]$.

Exercise 3.44 *Let $a, b \in \mathbb{R}$ with $a < b$ and consider $\phi(z) = (z - ia)/(z - ib)$ for $z \in \mathbb{C}\setminus\{ib\}$ and the Principal Branch l of the Logarithm in $\mathbb{C}\setminus(-\infty, 0]$.*

(i) *Describe the mapping properties of ϕ. Show, in particular, that ϕ maps the open right half-plane onto itself. (Cf. 2.5.)*

(ii) *Show that for all z in the open right half-plane*

$$(*) \qquad l(z - ia) - l(z - ib) - l\left(\frac{z - ia}{z - ib}\right) = 0.$$

Hints: All the parenthetic terms in (*) lie in the open right half-plane, so each of the l values is well-defined and falls into $\mathbb{R} \times (-\pi/2, \pi/2)$. Therefore the left-hand side of (*) lies in $\mathbb{R} \times (-3\pi/2, 3\pi/2)$. Since this number exponentiates to 1, it is an integer multiple of $2\pi i$. The only such in $\mathbb{R} \times (-3\pi/2, 3\pi/2)$ is 0.

(iii) *If* $\operatorname{Re} z > 0$, *then*

$$\operatorname{Im} l(\phi(z)) = \int_a^b \frac{\operatorname{Re} z}{|z - it|^2}\, dt.$$

Hint: $e^{l(z)} = z$ and the Chain Rule show that $l'(z) = 1/z$ and so (see 2.10)

$$l(z - ia) - l(z - ib) = \int_{[z-ia, z-ib]} \xi^{-1} d\xi = \int_a^b \frac{i\,dt}{z - it}.$$

Consequently, using (ii)

$$\operatorname{Im} l\left(\frac{z - ia}{z - ib}\right) = \int_a^b \operatorname{Im}\left(\frac{i}{z - it}\right) dt.$$

Notes to Chapter III

The history and development of the ideas in this chapter are well detailed in the encyclopedia article of PRINGSHEIM, FABER and MOLK [1911].

3.3(iii) is due to CAUCHY [1821], p. 286. It was rediscovered by HADAMARD [1888], p. 259 and [1892b], pp. 107–108. It is therefore customarily called the Cauchy–Hadamard radius of convergence formula.

All sorts of behavior of a power series $\sum_{n=0}^{\infty} c_n z^n$ are possible *on* its circle of convergence: for example, the series may converge uniformly on this circle but not absolutely; an example due to FEJÉR [1917] is found on p. 122 of HILLE [1959] and one due to Hardy in LANDAU [1929a] or NOVINGER [1975b]. See also GAIER [1953]. The series may converge at every point of this circle but represent a discontinuous function there (SIERPINSKI [1916]). We may have divergence at every point of this circle and yet have $c_n z^n \to 0$ for each z on the circle and it is also possible that the series converge at exactly one point on this circle. Examples of the latter two phenomena, due to Lusin and Sierpinski respectively, can also be found in LANDAU [1929a]. See also NEDER [1920], MAZURKIEWICZ [1922b], HERZOG [1953–54], DVORETSKY and ERDÖS [1955], and VIJAYARAGHAVAN [1947].

Two elementary and especially important facts about power series are conspicuously absent from the text. These are the theorem on multiplication of power series and the existence of a power series expansion centered at *any* point within the (open) disk of convergence of a given power series. The reader can easily formulate and prove these as exercises now or find them in BROMWICH [1926] or KNOPP [1951]. However, they will not be needed before Chapter V and there they will appear as effortless spin-off from other results. Similar remarks apply to the theorems affirming that the composition of two or the uniform limit of a sequence of power series is again a power series.

My account of 3.5 and 3.6 was influenced by APOSTOL [1952].

In connection with the definition of radius of convergence contained in the proof of 3.3, there is the following curious result: If $\theta, \psi \in [0, 2\pi)$, $(\theta - \psi)/2\pi$ is irrational but algebraic and the two sets $\{\mathrm{Re}(c_n e^{in\theta})\}$, $\{\mathrm{Re}(c_n e^{in\psi})\}$ are bounded, then the series $\sum c_n z^n$ has radius of convergence at least 1. See TEISSIER DU CROS [1944] and VALIRON [1944].

I have followed PRINGSHEIM [1925] and RUDIN [1974] in the concise treatment of the exponential function. Other interesting approaches occur in WIGNER [1952], EBERLEIN [1954], [1960], ROBISON [1968] and chapter II of WEIL [1976].

A very nice alternative proof of 3.25 can be found on page 35 of GUICHARDET [1968]. Yet another proof runs thus: use an analog of 4.1 to find a continuous function $\phi: \mathbb{R} \to \mathbb{C}$ such that $f = e^{\phi}$ and $\phi(0) = 0$. Deduce that ϕ is additive, whence $\phi(r) = r\phi(1)$ for all rational r, whence by continuity $\phi(x) = cx$ for all $x \in \mathbb{R}$ $(c = \phi(1))$.

Though proofs of 3.32 appear in GUTZMER [1887], [1888], [1889], the result predates this; he acknowledges the priority of A. Harnack, but in fact the formula goes back to an 1806 memoir of M. A. PARSEVAL (see § 23 of BURKHARDT [1914]) and is customarily referred to in the literature as Parseval's formula. It is valid (with the same proof as given in the text) for bilateral series as well. See Chapter XI. The Cauchy estimates, valid also for bilateral series, date from 1831 and were discovered independently in 1841 by WEIERSTRASS (*Werke* I, pp. 67–74 and

II, pp. 224–226), who gave an integral-free treatment. On the latter see also pp. 278, 280, 606 of PRINGSHEIM [1925] and MONTEL [1936].

3.33 is due, curiously, to CAUCHY [1844]—see BURKHARDT [1914], p. 1014; its ultimate generalization is 12.17, Picard's Little Theorem. For an improvement in 3.34 see TAMMI [1957] and for material related to 3.35(i) see SIDON [1927], LANDAU [1929a], § 4 and TOMIĆ [1962]. Embryonic versions of 3.35(ii),(iii) were discovered by HADAMARD [1893] and further elucidated by LINDELÖF [1903] and PRINGSHEIM [1902], [1904].

The numbers e and π introduced in this chapter have many remarkable properties. For example, each is transcendental, that is, satisfies no polynomial equation with rational coefficients. A. BAKER [1975], pp. 4, 5 gives nice short proofs of both these facts, as well as the relevant historical references. The appendix of MAHLER [1976] gives a unified treatment of many of the beautiful classical proofs, together with more history, and HESSENBERG [1912] is a completely self-contained account of the classical proofs.

For other elementary and *ab initio* treatments of the sums 3.38(17) see K. S. WILLIAMS [1971], STARK [1972], [1974] and BERNDT [1975]. The latter two papers contain extensive bibliographies, as does ELY [1882]. See also GOULD [1972] and pp. 350–364, vol. I of LEVEQUE [1974]. SAALSCHÜTZ [1893] and NIELSEN [1923] are exhaustive treatises. These formulas were first deduced by Euler. (See AYOUB [1974].)

Odd powers will be treated in Chapter XIV. The sum $\sum_{n=1}^{\infty} n^{-k}$ $(k > 1)$ is the value at k of an important function which is holomorphic in the open half-plane $H = \{z \in \mathbb{C}: \operatorname{Re} z > 1\}$. This is Riemann's Zeta function, about which there is an enormous literature. Any of the basic texts in the bibliography discuss the elementary theory of this function and in addition several (e.g., VEECH [1967]) discuss its role in the proof of the Prime Number Theorem. One easily checks that the series $\zeta(z) = \sum_{n=1}^{\infty} n^{-z} = \sum_{n=1}^{\infty} e^{-z \log n}$ converges absolutely in H and uniformly on each compact subset. From the results of Chapter VII (or from 5.44) it follows that ζ is indeed holomorphic in H. (This series is a typical example of what are called Dirichlet series, another extensive chapter of function theory with which this book will not deal at all.)

For recent literature on the sums 3.38(21) see pp. 335–339, vol. I of LEVEQUE [1974]. The material of 3.38 can also be treated with the aid of Cauchy's Residue Theorem (Chapter XI). For such a treatment see § 5, chapter III of BEHNKE and SOMMER [1965] or pp. 32–37 of LINDELÖF [1905]. For a complete history of who did what and when, more relations and alternative developments, of the material of § 4 see BURKHARDT [1914], § 7.

Chapter IV
The Index and Some Plane Topology

§ 1 Introduction

In this chapter we complete the preliminary investigations of Chapter I on the topology of the plane. The chief tools are the index and deformation and extension techniques. The former rests squarely on the theory of the exponential function developed in the last chapter.

§ 2 Curves Winding around Points

The exponential function makes clear how to formalize the notion of "the number of times the loop γ in $\mathbb{C}\backslash\{0\}$ winds counterclockwise around 0": we have only to write $\gamma = |\gamma|e^{i\theta}$, where θ is continuous and real-valued. Evidently the function θ tells how much "angle γ sweeps out." Our first theorem along these lines is a bit more general.

Theorem 4.1 *Let $\gamma\colon [a, b] \to \mathbb{C}$ be continuous, z_0 a complex number not on γ. Then there is a disk D_0 about z_0 disjoint from γ and a continuous function $\phi\colon [a, b] \times D_0 \to \mathbb{C}$ such that $\gamma(t) - z = e^{\phi(t,z)}$ for all $(t, z) \in [a, b] \times D_0$. Moreover, for each $z \in D_0$ and any such function ϕ, the function $t \to \phi(t, z)$ enjoys at each point of $[a, b]$ whatever smoothness properties γ does.*

Proof: Select $r > 0$ so that $D(z_0, 2r)$ is disjoint from γ and let $D_0 = D(z_0, r)$. Use uniform continuity of γ to select $a = t_0 < t_1 < \cdots < t_n = b$ so that $\gamma[t_{j-1}, t_j]$ has diameter less than r. Define $\Gamma\colon [a, b] \times D_0 \to \mathbb{C}\backslash\{0\}$ by $\Gamma(t, z) = \gamma(t) - z$ and consider the disks $D_j = D(\Gamma(t_{j-1}, z_0), 2r)$, $(j = 1, 2, \ldots, n)$. Notice that $|\Gamma(t, z_0)| = |\gamma(t) - z_0| \geq 2r$ for every t, by definition of r, so $0 \notin D_j$. Moreover Γ maps $[t_{j-1}, t_j] \times D_0$ into D_j; for if $(t, z) \in [t_{j-1}, t_j] \times D_0$, then

$$\begin{aligned} |\Gamma(t, z) - \Gamma(t_{j-1}, z_0)| &\leq |\gamma(t) - \gamma(t_{j-1})| + |z - z_0| \\ &\leq \text{diameter } \gamma[t_{j-1}, t_j] + |z - z_0| \\ &< r + r = 2r. \end{aligned}$$

By 3.19 there exists a continuous logarithm L_j in D_j and we may form $\phi_j = L_j \circ \Gamma$ on $[t_{j-1}, t_j] \times D_0$. This function is continuous on $[t_{j-1}, t_j] \times D_0$. Notice that for every $z \in D_0$, $\Gamma(t_j, z) \in D_j \cap D_{j+1}$ and so

$$e^{L_j(\Gamma(t_j,z))} = \Gamma(t_j, z) = e^{L_{j+1}(\Gamma(t_j,z))}$$

whence by 3.14(iii)

$$L_j(\Gamma(t_j, z)) - L_{j+1}(\Gamma(t_j, z)) = 2\pi i N_j(z)$$

for some integer $N_j(z)$; that is,

$$(*)\qquad \phi_j(t_j, z) - \phi_{j+1}(t_j, z) = 2\pi i N_j(z).$$

The continuous integer-valued function N_j maps the connected set D_0 onto a connected subset of integers. As the only such sets of integers are single points, N_j is constant on D_0. We can then finally define ϕ on

$$[a, b] \times D_0 = \bigcup_{j=1}^{n} [t_{j-1}, t_j] \times D_0$$

by

$$\phi = \begin{cases} \phi_1 & \text{on } [t_0, t_1] \times D_0 \\ \phi_j + 2\pi i \sum_{k=1}^{j-1} N_k & \text{on } [t_{j-1}, t_j] \times D_0, \quad j = 2, 3, \ldots, n. \end{cases}$$

Then ϕ is well-defined and continuous (by (*)) and in each $[t_{j-1}, t_j] \times D_0$, satisfies

$$e^{\phi(t,z)} = e^{\phi_j(t,z)} = e^{L_j(\Gamma(t,z))} = \Gamma(t, z) = \gamma(t) - z.$$

For $(\tau, z) \in [a, b] \times D_0$, a logarithm L can be defined by a power series in a neighborhood of $\gamma(\tau) - z$ (3.19). As above, it follows from 3.14(iii) that in a connected neighborhood of τ, $\phi(t, z)$ differs only by a constant from $L(\gamma(t) - z)$. The latter function of t has (thanks to the Chain Rule and the power series form of L) any differentiability properties near τ which γ does; hence so does ϕ.

Definition 4.2 If $\gamma: [a, b] \to \mathbb{C}$ is any loop and z any point not on γ, then we define the (circulation) *index* or *order* or *winding number* of γ with respect to z, noted $\text{Ind}_\gamma(z)$, as $[\phi(b) - \phi(a)]/2\pi i$ for *any* continuous logarithm ϕ on $[a, b]$ of the function $\gamma - z$. (Notice that such ϕ exist by the last theorem.)

Corollary 4.3 *If* $\gamma: [a, b] \to \mathbb{C}$ *is a loop, then* Ind_γ *is a well-defined, continuous, integer-valued function on* $\mathbb{C}\backslash\gamma[a, b]$ *which is (therefore) constant on the components of* $\mathbb{C}\backslash\gamma[a, b]$. *Moreover,* Ind_γ *is* 0 *on the unbounded component of* $\mathbb{C}\backslash\gamma[a, b]$.

Proof: To see that $\text{Ind}_\gamma(z_0)$ is well-defined for each $z_0 \in \mathbb{C}\backslash\gamma[a, b]$, let ϕ, ψ: $[a, b] \to \mathbb{C}$ be two continuous logarithms of the function $\gamma - z_0$: $e^{\phi} = \gamma - z_0 = e^{\psi}$. Then $[\phi - \psi]/2\pi i$ is a continuous integer-valued (by 3.14(iii)) function on the connected set $[a, b]$, and so is constant. Thus $\phi(b) - \phi(a) = \psi(b) - \psi(a)$. To see that $\text{Ind}_\gamma(z_0)$ is an integer, note that

$$e^{\phi(b)-\phi(a)} = e^{\phi(b)}/e^{\phi(a)} = (\gamma(b) - z_0)/(\gamma(a) - z_0) = 1$$

(since γ is closed), so $[\phi(b) - \phi(a)]/2\pi i$ is an integer. Next, given $z_0 \in \mathbb{C}\backslash\gamma[a, b]$, the last theorem provides a disk D_0 about z_0 and a continuous $\phi: [a, b] \times D_0 \to \mathbb{C}$ which satisfies $e^{\phi(t,z)} = \gamma(t) - z$ for all $(t, z) \in [a, b] \times D_0$. Then for all $z \in D_0$ we have $\text{Ind}_\gamma(z) = [\phi(b, z) - \phi(a, z)]/2\pi i$, showing that Ind_γ is continuous at

z_0. As noted before, it follows that the continuous integer-valued function Ind_γ is constant on connected subsets of its domain.

Also, if we consider a fixed $|z| > \sup|\gamma[a, b]|$ and (3.19) a continuous logarithm L in the open disk $D_1 = D(-1, 1)$, then we may form $\psi(t) = L(z^{-1}\gamma(t) - 1)$. If we write $z = e^w$ for some complex number w (see 3.14(vi)) and set $\phi = \psi + w$, then $e^{\phi(t)} = e^w e^{\psi(t)} = e^w[z^{-1}\gamma(t) - 1] = \gamma(t) - z$, so by definition of the index, $\mathrm{Ind}_\gamma(z) = [\phi(b) - \phi(a)]/2\pi i = [\psi(b) - \psi(a)]/2\pi i = 0$ (since $\gamma(b) = \gamma(a)$). Thus $\mathrm{Ind}_\gamma(z)$ vanishes for all sufficiently large $|z|$. Since $\{z \in \mathbb{C}: |z| > \sup|\gamma[a, b]|\}$ is a connected subset of $\mathbb{C}\backslash\gamma[a, b]$ (for example, it is the image of $D(0, 1)\backslash\{0\}$ under $z \to c/z$ for appropriate constant c), it lies in some component of $\mathbb{C}\backslash\gamma[a, b]$. All other components, being disjoint from this one, lie in the disk of radius $\sup|\gamma[a, b]|$ and so are bounded. Thus $\mathbb{C}\backslash\gamma[a, b]$ has exactly one unbounded component and Ind_γ vanishes on it.

Corollary 4.4 *If $\gamma: [a, b] \to \mathbb{C}$ is a piecewise smooth loop, then*

$$\mathrm{Ind}_\gamma(z) = \frac{1}{2\pi i}\int_\gamma \frac{1}{\xi - z}\, d\xi \quad \forall z \in \mathbb{C}\backslash\gamma[a, b].$$

Proof: Let $z_0 \in \mathbb{C}\backslash\gamma[a, b]$. By 4.1 there is a continuous $\phi: [a, b] \to \mathbb{C}$ such that $\gamma - z_0 = e^\phi$ and a partition $a = t_0 < t_1 < \cdots < t_n = b$ such that both γ and ϕ are smooth on each $[t_{j-1}, t_j]$. Differentiating the relation $\gamma - z_0 = e^\phi$, gives

$$\gamma' = (\gamma - z_0)' = \phi' e^\phi = \phi'[\gamma - z_0] \quad \text{in } [t_{j-1}, t_j],\ j = 1, 2, \ldots, n.$$

Therefore

$$\begin{aligned}\frac{1}{2\pi i}\int_\gamma \frac{1}{\xi - z_0}\, d\xi &= \sum_{j=1}^n \frac{1}{2\pi i}\int_{t_{j-1}}^{t_j} \frac{\gamma'}{\gamma - z_0}\\ &= \frac{1}{2\pi i}\sum_{j=1}^n \int_{t_{j-1}}^{t_j} \phi' = \frac{1}{2\pi i}\sum_{j=1}^n [\phi(t_j) - \phi(t_{j-1})]\\ &= \frac{1}{2\pi i}[\phi(t_n) - \phi(t_0)] = \frac{1}{2\pi i}[\phi(b) - \phi(a)]\\ &= \mathrm{Ind}_\gamma(z_0).\end{aligned}$$

Example 4.5 *Let $a \in \mathbb{C}$, $s > 0$ and set $b = \mathrm{Re}\, a$, $c = \mathrm{Im}\, a$, $Q = (b, b + s) \times (c, c + s)$. Then*

$$\mathrm{Ind}_{\partial Q}(z) = 1 \quad \forall z \in Q.$$

Proof: (Pringsheim [1896a]) The very definition of index shows that for any loop γ, any $w \in \mathbb{C}$ and any z not on γ we have

$$\mathrm{Ind}_{\gamma - w}(z - w) = \mathrm{Ind}_\gamma(z).$$

(In either case logarithms of $\gamma - z$ are involved.) Moreover, $\partial(Q - w) = \partial Q - w$ and so without loss of generality we may replace Q by any $Q - w$. Taking $w = a + \frac{1}{2}s(1 + i)$, it therefore suffices to consider the case where $a = -\frac{1}{2}s(1 + i)$, that is, with $r = \frac{1}{2}s$

$$Q = (-r, r) \times (-r, r).$$

By 4.3 it then suffices to establish that

$$\mathrm{Ind}_{\partial Q}(0) = 1.$$

Let f be the function in $\mathbb{C}\backslash\{0\}$ defined by $f(z) = 1/z$. Then by 4.4 and 2.9

$$\begin{aligned}
2\pi i \,\mathrm{Ind}_{\partial Q}(0) &= \int_{\partial Q} f = \int_{[r(-1-i), r(1-i)]} f + \int_{[r(1-i), r(1+i)]} f \\
&\quad + \int_{[r(1+i), r(-1+i)]} f + \int_{[r(-1+i), r(-1-i)]} f \\
&= \int_0^1 f((1-t)r(-1-i) + tr(1-i))2r\,dt \\
&\quad + \int_0^1 f((1-t)r(1-i) + tr(1+i))2ri\,dt \\
&\quad + \int_0^1 f((1-t)r(1+i) + tr(-1+i))(-2r)\,dt \\
&\quad + \int_0^1 f((1-t)r(-1+i) + tr(-1-i))(-2ri)\,dt \\
&= 2r\int_0^1 f(r(-1-i) + 2rt)\,dt \\
&\quad + 2ri\int_0^1 f(r(1-i) + 2rit)\,dt \\
&\quad - 2r\int_0^1 f(r(1+i) - 2rt)\,dt \\
&\quad - 2ri\int_0^1 f(r(-1+i) - 2rit)\,dt \\
&= 2r\int_0^1 \frac{dt}{r(-1-i) + 2rt} + 2ri\int_0^1 \frac{dt}{r(1-i) + 2rit} \\
&\quad - 2r\int_0^1 \frac{dt}{r(1+i) - 2rt} - 2ri\int_0^1 \frac{dt}{r(-1+i) - 2rit}.
\end{aligned}$$

Combining the first and second and the third and fourth terms,

$$\begin{aligned}2\pi i \operatorname{Ind}_{\partial Q}(0) &= 8\int_0^1 \frac{dt}{2t-1-i} = 4\int_{-1}^1 \frac{du}{u-i} = 4\int_{-1}^1 \frac{u+i}{u^2+1}\,du\\ &= 4\int_{-1}^1 \frac{u}{u^2+1}\,du + 4i\int_{-1}^1 \frac{1}{u^2+1}\,du\\ &= 8i\int_0^1 \frac{1}{u^2+1}\,du,\end{aligned}$$

since $u/(u^2+1)$ is an odd function and $1/(u^2+1)$ is an even function. The assertion now follows from 3.22.

According to 4.3, $\operatorname{Ind}_\gamma(z)$ is a continuous function of z. We next show that it is also a continuous function of γ.

Exercise 4.6 (i) *Let* $\gamma_0\colon [a, b] \to \mathbb{C}$ *be a loop and* $z \in \mathbb{C}\backslash\gamma_0$. *Then there exists a positive* $\delta = \delta(z, \gamma_0)$ *with this property: for any loop* $\gamma_1\colon [a, b] \to \mathbb{C}$ *with* $|\gamma_0 - \gamma_1| < \delta$ *on* $[a, b]$, *we have* $z \in \mathbb{C}\backslash\gamma_1$ *and*

(*) $\operatorname{Ind}_{\gamma_1}(z) = \operatorname{Ind}_{\gamma_0}(z)$.

Hints: As in the proof of 4.5, replacing γ_0 by $\gamma_0 - z$ and γ_1 by $\gamma_1 - z$, enables us to assume that $z = 0$. Then $\inf|\gamma_0[a, b]|$ is positive. This number has the properties sought for δ. For if $\gamma_1\colon [a, b] \to \mathbb{C}$ is a loop satisfying $|\gamma_0 - \gamma_1| < \delta$, then $|\gamma_0 - \gamma_1| < |\gamma_0|$ on $[a, b]$. In particular, γ_1 never takes the value 0: $0 \in \mathbb{C}\backslash\gamma_1$. Also the quotient $\gamma = \gamma_1/\gamma_0$ is a well-defined continuous function, in fact a loop, which maps $[a, b]$ into $D(1, 1)$. Since $\mathbb{C}\backslash D(1, 1)$ is connected and contains 0, it follows from 4.3 that

(1) $\operatorname{Ind}_\gamma(0) = 0$.

But it is clear from the definition of index in terms of logarithms that

(2) $\operatorname{Ind}_{\gamma_0\gamma}(0) = \operatorname{Ind}_{\gamma_0}(0) + \operatorname{Ind}_\gamma(0)$.

Since $\gamma_0\gamma = \gamma_1$, the assertion (*) follows from (1) and (2).

(ii) *Formulate and prove a theorem on the joint continuity of* $\operatorname{Ind}_\gamma(z)$ *as a function of the pair* (γ, z).

Exercise 4.7 (i) *If* $\Gamma\colon [a, b] \to \mathbb{C}$ *is a simple loop, then* Γ *is bicontinuous on* (a, b).

Hints: We have to show that if $\Gamma(t_n) \to \Gamma(t)$ for some $t_n, t \in (a, b)$, then $t_n \to t$. It suffices to show that t is the only cluster point of $\{t_n\}$ in $[a, b]$. If $t_0 \in [a, b]$ and $t_{n_j} \to t_0$, then by continuity $\Gamma(t_0) = \Gamma(t)$. Since $\{t, t_0\} \subset [a, b)$ or $\{t, t_0\} \subset (a, b]$ and since Γ is one-to-one on $[a, b)$ and on $(a, b]$, it follows that $t_0 = t$.

(ii) *If* Γ_1, Γ_2 *are two simple loops with the same range* J *and* f *is continuous and zero-free on* J, *then*

$$|\operatorname{Ind}_{f\circ\Gamma_1}(0)| = |\operatorname{Ind}_{f\circ\Gamma_2}(0)|.$$

Hints: Let the parameter interval of Γ_j be $[a_j, b_j]$. Let ϕ_1 be a continuous logarithm for $f \circ \Gamma_1$. Then the definition

(1) $\phi(t + n(b_1 - a_1)) = \phi_1(t) + n[\phi_1(b_1) - \phi_1(a_1)], \quad t \in [a_1, b_1], n \in \mathbb{Z}$

provides a continuous extension of ϕ_1 to $\mathbb{R}$. Select $t_1 \in [a_1, b_1]$ such that

(2) $\Gamma_1(t_1) = \Gamma_2(a_2)$.

Extend Γ_1 by periodicity to $\mathbb{R}$ and set

$$\Gamma(t) = \Gamma_1(t), \quad t_1 \leq t \leq t_1 + (b_1 - a_1).$$

We have

$$f \circ \Gamma(t) = e^{\phi(t)}, \quad t_1 \leq t \leq t_1 + (b_1 - a_1)$$

and so

$$\begin{aligned} 2\pi i \operatorname{Ind}_{f \circ \Gamma}(0) &= \phi(t_1 + (b_1 - a_1)) - \phi(t_1) \\ &\overset{(1)}{=} \phi_1(b_1) - \phi_1(a_1) = 2\pi i \operatorname{Ind}_{f \circ \Gamma_1}(0). \end{aligned}$$

Therefore it suffices to prove the assertion with Γ_1 replaced by Γ, the upshot of which is that we can assume

(3) $\Gamma_1(a_1) = \Gamma_2(a_2)$.

Now by (i), Γ_j is a homeomorphism of (a_j, b_j) onto $J \backslash \Gamma_j(a_j)$. By (3) we may therefore form $\Gamma_2^{-1} \circ \Gamma_1$ and have a homeomorphism of (a_1, b_1) onto (a_2, b_2). This extends by 1.46 to a homeomorphism ψ of $[a_1, b_1]$ onto $[a_2, b_2]$. Thus $\Gamma_1 = \Gamma_2 \circ \psi$ and if ϕ_2 is a continuous logarithm for $f \circ \Gamma_2$ then $f \circ \Gamma_2 = e^{\phi_2}$, $f \circ \Gamma_1 = e^{\phi_2 \circ \psi}$, whence

$$\begin{aligned} 2\pi i \operatorname{Ind}_{f \circ \Gamma_1}(0) &= \phi_2 \circ \psi(b_1) - \phi_2 \circ \psi(a_1) \\ &= \pm [\phi_2(b_2) - \phi_2(a_2)], \\ &\quad \text{since } \psi \text{ maps endpoints to endpoints,} \\ &= \pm 2\pi i \operatorname{Ind}_{f \circ \Gamma_2}(0). \end{aligned}$$

Definition 4.8 (i) If $\gamma_0, \gamma_1: [a, b] \to A \subset \mathbb{C}$ are loops, we say γ_0 and γ_1 are *A-homologous* or *homologous in A* if $\operatorname{Ind}_{\gamma_0}(z) = \operatorname{Ind}_{\gamma_1}(z)$ for every $z \in \mathbb{C} \backslash A$.

(ii) If γ_1 (as above) is constant, we say γ_0 is *A-nullhomologous* or *nullhomologous in A*, i.e., if $\operatorname{Ind}_{\gamma_0}(z) = 0$ for every $z \in \mathbb{C} \backslash A$.

(iii) If every loop in A is nullhomologous in A, we call A *homologically-connected.*

We close this section with a technical result which we will need later:

Exercise 4.9 *Let* $0 < r < R < \infty$, $S = \{z \in \mathbb{C}: \operatorname{Re} z > 0, r < |z| < R\}$, $\gamma: [0, 1] \to \{z \in \mathbb{C}: \operatorname{Re} z > 0\}$ *a curve such that* $|\gamma(0)| = r$, $|\gamma(1)| = R$. *If C is a component of* $S \backslash \gamma$, *show that* $\bar{C}$ *does not meet both the negative and the positive y-axis.*

Hints: Since γ is a compact subset of the open right half-plane, if $\bar{C}$ meets both the positive and negative y-axis, then $\bar{C}$ is a relative neighborhood in $\bar{S}$ of both $[ir, iR]$ and $[-ir, -iR]$. It follows easily, via 1.28, that there exists a polygon Γ_1 which joins $-ir$ to ir and lies, except for $\pm ir$, in C. We may suppose the domain of Γ_1 to be $[0, 1]$ and then define

$$\Gamma_2(t) = re^{i\pi t/2} \qquad t \in [1, 3],$$

$$\Gamma_3(t) = \begin{cases} \Gamma_1(t) & t \in [0, 1] \\ \Gamma_2(t) & t \in [1, 3], \end{cases}$$

$$\Gamma_4(t) = \begin{cases} (2t - 1)ir & t \in [0, 1] \\ \Gamma_2(t) & t \in [1, 3], \end{cases}$$

$$\Gamma_5(t) = \begin{cases} (1 - t)ir & t \in [0, 2] \\ \Gamma_1(t - 2) & t \in [2, 3]. \end{cases}$$

Since Γ_1 lies in $C \cup \{\pm ir\} \subset \bar{S}\backslash\gamma$, we see that $\Gamma_3 \subset \bar{A}(r, R)\backslash\gamma$. The set

$$\mathbb{C}\backslash[\bar{A}(r, R)\backslash\gamma] = [\mathbb{C}\backslash\bar{A}(r, R)] \cup \gamma = D(0, r) \cup \gamma \cup [\mathbb{C}\backslash\bar{D}(0, R)]$$

is connected (by 1.3), unbounded and lies in $\mathbb{C}\backslash\Gamma_3$. Consequently

$$\operatorname{Ind}_{\Gamma_3}(c) = 0 \quad \forall c \in D(0, r).$$

For $c \in (-r, 0)$ we compute

$$\begin{aligned} 0 = 2\pi i \operatorname{Ind}_{\Gamma_3}(c) &= \int_{\Gamma_3} \frac{dz}{z - c} \\ &= \int_{\Gamma_1} \frac{dz}{z - c} + \int_{[-ir, ir]} \frac{dz}{z - c} + \int_{[ir, -ir]} \frac{dz}{z - c} + \int_{\Gamma_2} \frac{dz}{z - c} \\ &= \int_{\Gamma_4} \frac{dz}{z - c} + \int_{\Gamma_5} \frac{dz}{z - c}. \end{aligned} \tag{1}$$

However Γ_5 and $-\Gamma_4$ are each closed curves and c lies in an open connected unbounded subset of $\mathbb{C}$ disjoint from each, viz., the open left half-plane. Consequently

$$\int_{\Gamma_5} \frac{dz}{z - c} = 0 = \int_{-\Gamma_4} \frac{dz}{z - c}. \tag{2}$$

Using this in (1) gives

$$\begin{aligned} 0 &= \int_{\Gamma_4} \frac{dz}{z - c} + \int_{-\Gamma_4} \frac{dz}{z - c} = \int_{C(0,r)} \frac{dz}{z - c} = 2\pi i \operatorname{Ind}_{C(0,r)}(c), \quad \text{by 4.4} \\ &= 2\pi i \operatorname{Ind}_{C(0,r)}(0), \quad \text{by 4.3}. \end{aligned}$$

This is an obvious contradiction.

§ 3 Homotopy and the Index

Definition 4.10 (i) Let $S \subset \mathbb{R}^n$, $f_0, f_1: S \to A \subset \mathbb{C}$ continuous functions. An *A-homotopy* or a *homotopy in A* of f_0 with f_1 is a continuous function $h: [0, 1] \times S \to A$ such that

$$\left.\begin{array}{l} h(0, x) = f_0(x) \\ h(1, x) = f_1(x) \end{array}\right\} \quad \forall x \in S.$$

We sometimes write $f_t(x)$ for $h(t, x)$. We say f_0 is *A-homotopic* to f_1 or *homotopic in A* to f_1 if such an h exists.

(ii) If f_1 is constant, we say f_0 is *A-nullhomotopic* or *nullhomotopic in A*.

Definition 4.11 (i) Let $[a, b] \subset \mathbb{R}$, $\gamma_0, \gamma_1: [a, b] \to A \subset \mathbb{C}$ be loops. An *A-loophomotopy* or a *loophomotopy in A* of γ_0 with γ_1 is an A-homotopy h such that each curve $\gamma_t(x) = h(t, x)$ is a loop, i.e.,

$$h(t, a) = h(t, b) \quad \forall t \in [0, 1].$$

We say γ_0 is *A-loophomotopic* to γ_1 or *loophomotopic in A* to γ_1 if such an h exists.

(ii) If γ_1 is constant, we say γ_0 is *A-loopnullhomotopic* or *loopnullhomotopic in A*.

(iii) Call A *loophomotopically-connected* if every loop in A is loopnullhomotopic in A.

Remarks: It is the work of a moment to confirm that both these notions define equivalence relations. Thus in particular the roles of f_0 and f_1 (resp., γ_0 and γ_1) can be interchanged above without affecting the definitions and so symmetric terminology like "f_0, f_1 are A-homotopic," etc., is employed.

Theorem 4.12 *Let S be a subset of $\mathbb{C}$, $\gamma_0, \gamma_1: [a, b] \to S$ loops which are loophomotopic in S. Then γ_0, γ_1 are S-homologous.*

Proof: Let H be a loophomotopy in S of γ_0 with γ_1. Then for each $\tau \in [0, 1]$ the function $\gamma_\tau: [a, b] \to S$ defined by $\gamma_\tau(t) = H(\tau, t)$ is a loop and $\tau \to \gamma_\tau$ is a continuous map of $[0, 1]$ into $C[a, b]$. It follows easily then from 4.6 that, for each $z \in \mathbb{C}\backslash S$, $\tau \to \operatorname{Ind}_{\gamma_\tau}(z)$ is a continuous function. Since it is integer-valued and its domain is the connected set $[0, 1]$, it is constant. Thus $\operatorname{Ind}_{\gamma_0}(z) = \operatorname{Ind}_{\gamma_1}(z)$. This holds for every $z \in \mathbb{C}\backslash S$ and therefore says that γ_0 and γ_1 are S-homologous.

Exercise 4.13 *Prove the converse of* 4.12 *in the special case $S = \mathbb{C}\backslash\{0\}$. That is, show that if $\gamma_0, \gamma_1: [a, b] \to \mathbb{C}\backslash\{0\}$ are homologous in $\mathbb{C}\backslash\{0\}$, i.e., have the same index with respect to* 0, *then γ_0 and γ_1 are loophomotopic in $\mathbb{C}\backslash\{0\}$.*

Hints: If $\gamma_j(t) = e^{h_j(t)}$ $(j = 0, 1)$, define $H: [0, 1] \times [a, b] \to \mathbb{C}\backslash\{0\}$ by $H(\tau, t) = e^{\tau h_1(t) + (1-\tau)h_0(t)}$. Then $H(0, t) = \gamma_0(t)$, $H(1, t) = \gamma_1(t)$ for all $t \in [a, b]$, and for each $\tau \in [0, 1]$

$$\begin{aligned}[\tau h_1(1) + (1 - \tau)h_0(1)] &- [\tau h_1(0) + (1 - \tau)h_0(0)] \\ &= \tau[h_1(1) - h_1(0)] + (1 - \tau)[h_0(1) - h_0(0)] \\ &= 2\pi i\tau \operatorname{Ind}_{\gamma_1}(0) + 2\pi i(1 - \tau) \operatorname{Ind}_{\gamma_0}(0) = 2\pi i \operatorname{Ind}_{\gamma_0}(0),\end{aligned}$$

an integer multiple of $2\pi i$, so $H(\tau, t)$ is a *closed* curve.

Exercise 4.14 *If* γ *is a loop and* $z \in \mathbb{C}\backslash\gamma$, *then for every integer* n, $\operatorname{Ind}_{n\cdot\gamma}(z) = n \operatorname{Ind}_\gamma(z)$.

Exercise 4.15 *Let* U *be* $A(r, R)$ *for some* $0 \leq r \leq R < \infty$.

(i) *Show that if* γ_0, γ_1 *are two loops in* U *and* $\operatorname{Ind}_{\gamma_0}(0) = \operatorname{Ind}_{\gamma_1}(0)$, *then* γ_0 *and* γ_1 *are U-loophomotopic.*

Hints: The proof offered for 4.13 works here too. Just note that h_0, h_1 satisfy $\log r < \operatorname{Re} h_j < \log R$ and so the same is true for each $\tau h_1 + (1 - \tau)h_0$, showing that the range of H is indeed in U.

(ii) *Show that if* γ_0, γ_1 *are two loops in* U, *then there exist integers* n_0, n_1, *not both* 0, *such that* $n_0 \bullet \gamma_0$ *is U-loophomotopic to* $n_1 \bullet \gamma_1$.

Hints: If $\operatorname{Ind}_{\gamma_0}(0)$ equals $\operatorname{Ind}_{\gamma_1}(0)$, then take $n_0 = n_1 = 1$ and cite (i). Otherwise, not both of the integers $n_0 = \operatorname{Ind}_{\gamma_1}(0)$ and $n_1 = \operatorname{Ind}_{\gamma_0}(0)$ are 0 and according to 4.14 we have $\operatorname{Ind}_{n_0 \bullet \gamma_0}(0) = \operatorname{Ind}_{n_1 \bullet \gamma_1}(0)$. So again (i) may be applied.

Exercise 4.16 *Every starlike subset of* $\mathbb{C}$ *is loophomotopically-connected.*

Hint: If S is starlike with respect to z_0 and $\gamma: [a, b] \to S$ is a loop, then

$$H(t, \tau) = (1 - t)\gamma(\tau) + tz_0, \quad (t, \tau) \in [0, 1] \times [a, b]$$

is obviously a loophomotopy of γ with the constant z_0. By starlikeness of S the range of H is in S.

We close this section with the computation of another specific index which we will need later.

Exercise 4.17 *Let* $c = 1 + i$, $r > 0$ *and*

$$\gamma = [-rc - \tfrac{1}{2}, -rc + \tfrac{1}{2}, rc + \tfrac{1}{2}, rc - \tfrac{1}{2}, -rc - \tfrac{1}{2}].$$

Show that $0 \in \mathbb{C}\backslash\gamma$ *and* $\operatorname{Ind}_\gamma(0) = 1$.

Hints: Define $\Gamma = [-c/2, \bar{c}/2, c/2, -\bar{c}/2, -c/2]$ and $h: [0, 1] \times \mathbb{C} \to \mathbb{C}$ by $h(s, z) = \operatorname{Re} z - s \operatorname{Im} z + i[s/2r + 1 - s] \operatorname{Im} z$. Show that $h(0, \gamma) = \gamma$, $h(1, \gamma) = \Gamma$ and $0 \notin h([0, 1] \times \gamma)$. It follows from 4.12 that $\operatorname{Ind}_\gamma(0) = \operatorname{Ind}_\Gamma(0)$, and the latter is 1 by 4.5.

§ 4 Existence of Continuous Logarithms

Theorem 4.18 (Abstract Monodromy Theorem) *Let U be an open subset of $\mathbb{C}$ and $f: U \to \mathbb{C}\backslash\{0\}$ a continuous function. Then the following are equivalent:*

(i) $\text{Ind}_{f\circ\gamma}(0) = 0$ *for every loop γ in U.*

(ii) *f has a continuous logarithm in U.*

Proof: (ii) $\Rightarrow$ (i). Trivial.

(i) $\Rightarrow$ (ii). The components of U inherit property (i) and are open. Since we have only to manufacture the logarithm in each component, let us assume that U is connected. Fix $z_0 \in U$ and let $w_0 \in \mathbb{C}$ satisfy $e^{w_0} = f(z_0)$. For each $z \in U$ let $\gamma_z: [0, 1] \to U$ be a curve joining z_0 to z. (See 1.28.) Let ϕ_z be a continuous logarithm for $f \circ \gamma_z$:

(1) $\quad f \circ \gamma_z = e^{\phi_z}, \quad z \in U.$

(See 4.1.) Evidently we may require that

(2) $\quad \phi_z(0) = w_0 \quad \forall z \in U.$

We define $\Phi(z)$ to be $\phi_z(1)$ and have

$$e^{\Phi(z)} = e^{\phi_z(1)} \stackrel{(1)}{=} f(\gamma_z(1)) = f(z), \quad \text{by definition of } \gamma_z.$$

So only continuity of Φ has to be established. To this end, fix $z_1 \in U$ and pick $r > 0$ so small that $\bar{D}(z_1, r) \subset U$ and $f(\bar{D}(z_1, r)) \subset D(f(z_1), |f(z_1)|)$. Let L be a continuous logarithm in the latter disk. (See 3.19.) We may require that

(3) $\quad L(f(z_1)) = \phi_{z_1}(1),$

since both these numbers are logarithms of $f(z_1)$. Now, given $z \in D(z_1, r)$, let $\gamma = [z_1, z]$ and consider the loop Γ defined by

$$(4) \qquad \Gamma(t) = \begin{cases} \gamma_z(1 - t) & 0 \le t \le 1 \\ \gamma_{z_1}(t - 1) & 1 \le t \le 2 \\ \gamma(t - 2) & 2 \le t \le 3 \end{cases}$$

and the function ϕ defined by

$$(5) \qquad \phi(t) = \begin{cases} \phi_z(1 - t) & 0 \le t \le 1 \\ \phi_{z_1}(t - 1) & 1 \le t \le 2 \\ L(f(\gamma(t - 2))) & 2 \le t \le 3. \end{cases}$$

Conditions (2) and (3) ensure that ϕ is consistently defined at $t = 1$ and $t = 2$. Evidently ϕ is continuous on each of the intervals $[0, 1]$, $[1, 2]$, $[2, 3]$. Therefore ϕ is continuous. It is evidently a logarithm for $f \circ \Gamma$. By hypothesis therefore

$$0 = 2\pi i \, \text{Ind}_{f\circ\Gamma}(0) = \phi(3) - \phi(0) = L(f(z)) - \phi_z(1),$$

$$\Phi(z) = \phi_z(1) = L(f(z)).$$

This shows that Φ coincides with $L \circ f$ in $D(z_1, r)$ and hence is continuous in that disk.

Exercise 4.19 *If U is an open subset of $\mathbb{C}$ and $f: U \to \mathbb{C}\backslash\{0\}$ has a continuous nth root ψ_n for every positive integer n (or even for just an infinite set of n), then f has a continuous logarithm.*

Hints: Let γ be any loop in U and ϕ_n a continuous logarithm for $\psi_n \circ \gamma$. Then $n\phi_n$ is a continuous logarithm for $f \circ \gamma$ and so

$$\operatorname{Ind}_{f\circ\gamma}(0) = n \operatorname{Ind}_{\psi_n\circ\gamma}(0).$$

Since both indexes are integers and n does not feature on the left side, the index on the right cannot be non-zero for large n. It follows that $\operatorname{Ind}_{f\circ\gamma}(0) = 0$. Now appeal to 4.18.

Corollary 4.20 *Let U be an open subset of $\mathbb{C}$. If U is loophomotopically-connected, in particular (see 4.16) if U is starlike, then every continuous, zero-free function on U has a continuous logarithm.*

Proof: Let $f: U \to \mathbb{C}\backslash\{0\}$ be continuous and let $\gamma: [a, b] \to \mathbb{C}$ be any loop in U. By hypothesis there is a homotopy $h: [0, 1] \times [a, b] \to U$ of γ with a constant. Then $H = f \circ h$ is a homotopy in $f(U) \subset \mathbb{C}\backslash\{0\}$ of $f \circ \gamma$ with a constant. Consequently by 4.12, $\operatorname{Ind}_{f\circ\gamma}(0) = 0$. Now cite 4.18.

Example 4.21 *Let K be either $[-1, 1] \times [-1, 1]$ or $\bar{D}(0, 1)$. Then any continuous, zero-free function on K has a continuous logarithm.*

Proof: Let $U = (-1, 1) \times (-1, 1)$ or $D(0, 1)$ according as K is $[-1, 1] \times [-1, 1]$ or $\bar{D}(0, 1)$ and let continuous $f: K \to \mathbb{C}\backslash\{0\}$ be given. 1.44 furnishes an extension $F: \mathbb{C} \to \mathbb{C}$ of f. Then $F^{-1}(\mathbb{C}\backslash\{0\})$ is an open neighborhood of K and so contains rU for some $r > 1$. Apply 4.20 to get a continuous logarithm for F in rU.

Exercise 4.22 (i) *Prove that if $h: C(0, 1) \to \mathbb{C}\backslash\{0\}$ is continuous and satisfies $h(-z) = -h(z)$ for all z, then h does not have a continuous square root (hence no continuous logarithm either). The conclusion applies in particular to the identity function.*

Hints: If g is a continuous square root of h, then g is never zero, so $\phi(z) = g(-z)/g(z)$ is a continuous map of $C(0, 1)$ into $\mathbb{C}\backslash\{0\}$ and $\phi^2 \equiv -1$. Thus by 1.7 and the connectedness of $C(0, 1)$ we infer that $\phi \equiv k$ where k is either i or $-i$. It follows that

$$-1 = k^2 = \phi(z)\phi(-z) = 1 \rightarrow\leftarrow.$$

(ii) *If $f: C(0, 1) \to \mathbb{C}\backslash\{0\}$ is continuous, then there exists an integer n and a continuous $\phi: C(0, 1) \to \mathbb{C}$ such that $f(z) \equiv z^n e^{\phi(z)}$.*

Hints: Let $n = \text{Ind}_\gamma(0)$, where γ is the loop $\gamma(t) = f(e^{2\pi i t})$. Thus $\gamma(t) = e^{2\pi i \psi(t)}$ where ψ is continuous and $\psi(1) - \psi(0) = n$. Then

(*) $\quad e^{-2\pi i n t} f(e^{2\pi i t}) = e^{2\pi i[\psi(t) - nt]} \quad \forall t \in [0, 1].$

The function $2\pi i[\psi(t) - nt]$ has the value $2\pi i \psi(0)$ at $t = 0$ and also the value $2\pi i \psi(0)$ at $t = 1$. According to 3.26 there is then a continuous $\phi: C(0, 1) \to \mathbb{C}$ such that

$$\phi(e^{2\pi i t}) = 2\pi i[\psi(t) - nt] \quad \forall t \in [0, 1].$$

We get then from (*)

$$e^{-2\pi i n t} f(e^{2\pi i t}) = e^{\phi(e^{2\pi i t})} \quad \forall t \in [0, 1],$$

that is,

$$z^{-n} f(z) = e^{\phi(z)} \quad \forall z \in C(0, 1).$$

(iii) *The set $\mathscr{G}$ of all continuous, zero-free, complex-valued functions on $C(0, 1)$ is a commutative group under multiplication and the set $\mathscr{E}$ of all exponentials of continuous, complex-valued functions on $C(0, 1)$ constitutes a subgroup of $\mathscr{G}$. What do* (i) *and* (ii) *say about the quotient group $\mathscr{G}/\mathscr{E}$?*

(iv) *Let $\mathscr{G}_1$ be the subgroup of functions in $\mathscr{G}$ of modulus* 1, *$\mathscr{E}_1 = \mathscr{E} \cap \mathscr{G}_1$. Show that $\mathscr{G}_1$ is complete in the uniform metric and that $\mathscr{E}_1$ is both open and closed in $\mathscr{G}_1$ and connected. Conclude that $\mathscr{E}_1$ is the component of the metric space $\mathscr{G}_1$ which contains* 1. *Show also that $\mathscr{G}_1/\mathscr{E}_1 = \mathscr{G}/\mathscr{E}$.*

(v) *Show that this quotient group is torsion-free, that is, if $f: C(0, 1) \to \mathbb{C}\backslash\{0\}$ is continuous and $f^n = e^\phi$ for some non-zero integer n and some continuous ϕ, then $f = e^\psi$ for some continuous ψ.*

Hint: The sets $\{z \in C(0, 1): f(z) = e^{[\phi(z) + 2\pi i k]/n}\}$ $(k = 1, 2, \ldots, n)$ are disjoint, closed and their union is $C(0, 1)$.

Remark: Evidently the nature of the domain of f is irrelevant to the truth of (v), as long as it is connected. Cf. 1.7.

Exercise 4.23 *Prove the analog of* 4.22(ii) *for squares: If $Q = (-1, 1) \times (-1, 1)$ and $f: \partial Q \to \mathbb{C}\backslash\{0\}$ is continuous, then there exists an integer n and continuous $\Phi: \partial Q \to \mathbb{C}$ such that $f(w) = w^n e^{\Phi(w)}$ for all $w \in \partial Q$.*

Hints: Define $\psi: \partial Q \to C(0, 1)$ by $\psi(w) = w/|w|$. Then ψ is a surjective homeomorphism. Apply 4.22(ii) to $f \circ \psi^{-1}$ to come up with an integer n and a continuous ϕ on $C(0, 1)$ such that

$$f(\psi^{-1}(z)) = z^n e^{\phi(z)} \quad \forall z \in C(0, 1),$$

whence

$$\begin{aligned} f(w) &= \psi(w)^n e^{\phi(\psi(w))} \\ &= w^n e^{\phi(\psi(w)) - n \log|w|} \quad \forall w \in \partial Q. \end{aligned}$$

4.21 and 4.22(i) show that the domain of a continuous, zero-free function affects whether it has a continuous logarithm or not. In particular, if continuous, zero-free extension to a disk or to $\mathbb{C}$ is possible, continuous logarithms are assured. Thus the extension theorems 1.44 and 1.45 (and, with the latter, homotopy properties) come to play a big role.

Theorem 4.24 (BORSUK [1932]) *For a compact subset C of $\mathbb{C}$ and a continuous function $f: C \to \mathbb{C}\backslash\{0\}$ the following three conditions are equivalent:*

(i) *f is homotopic in $\mathbb{C}\backslash\{0\}$ to a constant,*

(ii) *f has an extension to a continuous function $F: \mathbb{C} \to \mathbb{C}\backslash\{0\}$,*

(iii) *f has a continuous logarithm.*

Proof: (i) $\Rightarrow$ (ii). Immediate from 1.45.
(ii) $\Rightarrow$ (iii). F will have a continuous logarithm, courtesy of 4.20.
(iii) $\Rightarrow$ (i). If h is a continuous logarithm for f on C, then $H(t, z) = e^{th(z)}$ ($z \in C, t \in [0, 1]$) is evidently a homotopy in $\mathbb{C}\backslash\{0\}$ of the constant function 1 with the function $e^h = f$.

Corollary 4.25 *Let C be a compact subset of $\mathbb{C}$, $f_0, f_1: C \to \mathbb{C}\backslash\{0\}$ continuous functions which are homotopic in $\mathbb{C}\backslash\{0\}$. Then f_0 has a continuous logarithm if and only if f_1 does.*

Proof: If f_0 has a continuous logarithm, then f_0 is homotopic in $\mathbb{C}\backslash\{0\}$ to a constant, by 4.24. But then f_1 is also homotopic in $\mathbb{C}\backslash\{0\}$ to a constant, homotopy being a transitive relation. Finally, by 4.24 again, it follows that f_1 has a continuous logarithm.

Corollary 4.26 *Let C be a compact subset of $\mathbb{C}$, $f_0, f_1: C \to \mathbb{C}$ continuous functions which satisfy $|f_0 - f_1| < |f_1|$ throughout C. Then both functions are zero-free and one has a continuous logarithm if and only if the other does.*

Proof: The strict inequality evidently prevents either function from having a zero. Moreover, it also implies that for any $t \in [0, 1]$,

$$|tf_0 + (1 - t)f_1| = |f_1 + t(f_0 - f_1)| \geq |f_1| - t|f_0 - f_1| > 0$$

throughout C. Therefore $H(t, z) = tf_0(z) + (1 - t)f_1(z)$ defines a homotopy in $\mathbb{C}\backslash\{0\}$ of f_1 with f_0 and the last corollary applies.

Exercise 4.27 (i) *Give a direct proof of* 4.26 *based on the fact that $f = f_0/f_1$ maps into $D(1, 1)$ and on $D(1, 1)$ there is a continuous logarithm.*

(ii) *Show that the conclusion in* 4.26 *still follows if the hypothesis is weakened to $|f_0 - f_1| < |f_0| + |f_1|$, and no hypothesis is imposed on C.*

Hints: Each function is still zero-free and the quotient f_0/f_1 maps into $\mathbb{C}\backslash(-\infty, 0]$. The latter set is starlike with respect to any point on the positive real axis, so by 4.20 the identity function has a continuous logarithm L there.

(See also 3.43.) Then $L \circ (f_0/f_1)$ is a continuous logarithm in C for the quotient f_0/f_1.

Exercise 4.28 *Let K be a compact subset of $\mathbb{C}$ and $f: K \to \mathbb{C}\backslash\{0\}$ continuous. Show that if 0 belongs to the unbounded component of $\mathbb{C}\backslash f(K)$, then f is homotopic in $\mathbb{C}\backslash\{0\}$ to a constant (and then by 4.25 f has a continuous logarithm).*

Hints: Choose $r > \sup|f(K)|$ and use 1.28 and 1.30 to find a continuous $\gamma: [0, 1] \to \mathbb{C}\backslash f(K)$ *such that* $\gamma(0) = 0$, $\gamma(1) = r$. Define h by

$$h(t, z) = \begin{cases} f(z) - \gamma(t) & (t, z) \in [0, 1] \times K \\ (2 - t)f(z) - r, & (t, z) \in [1, 2] \times K. \end{cases}$$

Then h is a homotopy in $\mathbb{C}\backslash\{0\}$ of f with the constant $-r$.

Alternatively, extend γ to $\Gamma: [0, \infty) \to \mathbb{C}$ by setting $\Gamma(t) = rt$ for $t \geq 1$. Then apply 4.18 to $U = \mathbb{C}\backslash\Gamma$ to get a continuous logarithm L for the identity function on U. The composite $L \circ f$ is a continuous logarithm for f.

Theorem 4.29 *Let K be a compact subset of $\mathbb{C}$, $f: K \to \mathbb{C}\backslash\{0\}$ continuous. Then there exist finitely many points $p_1, \ldots, p_N \in \mathbb{C}\backslash K$ and integers $n_1, \ldots, n_N$ such that the function $F(z) = f(z) \prod_{j=1}^{N} (z - p_j)^{n_j}$ has a continuous logarithm on K.*

Proof: As a matter of convenience subject $\mathbb{C}$ to a preliminary transformation $z \to az + b$ in order to have

(1) $\quad K \subset (0, 1) \times (0, 1)$.

Apply 1.44 to get a continuous extension $f_0: Q = [0, 1] \times [0, 1] \to \mathbb{C}$. Let $L = f_0^{-1}(\{0\})$. This is a closed subset of Q disjoint from K, so by compactness there exists an $r > 0$ such that

(2) $\quad |z - w| \geq r \quad \forall z \in K, w \in L$.

Let m be a positive integer greater than $\sqrt{2}/r$ and consider the squares

$$Q_{jk} = \left[\frac{j-1}{m}, \frac{j}{m}\right] \times \left[\frac{k-1}{m}, \frac{k}{m}\right], \quad \text{center } p_{jk} = \frac{j-\frac{1}{2}}{m} + i\frac{k-\frac{1}{2}}{m}$$

for $j, k = 1, 2, \ldots, m$. Set

$$\mathscr{K} = \{(j, k): 1 \leq j, k \leq m \quad \text{and} \quad Q_{jk} \cap K \neq \varnothing\}$$

$$\mathscr{L} = \{(j, k): 1 \leq j, k \leq m \quad \text{and} \quad Q_{jk} \cap K = \varnothing\}.$$

It follows from the choice of m and r that

(3) $\quad K \subset \bigcup\{Q_{jk}: (j, k) \in \mathscr{K}\} = K_1 \text{ closed} \subset Q\backslash L$.

Let f_1 be the restriction of f_0 to K_1. Since K_1 is a union of squares Q_{jk}, each interval $\{j/m\} \times [(k - 1)/m, k/m]$ and each interval $[(j - 1)/m, j/m] \times \{k/m\}$ either lies wholly in K_1 or meets K_1 only at endpoints. At each endpoint where f_1

is not already defined give it the value 1. Then for any interval $I = [a, b]$ of the above kind which does not lie wholly in K_1, $f_1(a)$ and $f_1(b)$ are non-zero complex numbers and f_1 is not defined on (a, b). Since $\mathbb{C}\backslash\{0\}$ is open and connected, 1.28 furnishes a continuous extension of f_1 to a map of I into $\mathbb{C}\backslash\{0\}$. We now have a continuous, zero-free extension f_2 of f to the closed subset

$$(4) \qquad K_2 = K_1 \cup \bigcup_{j,k=1}^{m} \partial Q_{jk}$$

of Q. The definitions of $\mathscr{L}$ and K_1 then show that

$$(5) \qquad K_2 \cap Q_{jk} = \partial Q_{jk} \quad \forall (j, k) \in \mathscr{L}.$$

For each such (j, k) there exists an integer n_{jk} such that $(z - p_{jk})^{n_{jk}} f_2(z)$ has a zero-free, continuous extension F_{jk} to Q_{jk}. This follows from 4.23 and 4.24. We can then consistently define F_0 on Q by

$$F_0(z) = \begin{cases} f_2(z) \prod_{(j,k)\in\mathscr{L}} (z - p_{jk})^{n_{jk}} & z \in K_2 \\ F_{j'k'}(z) \prod_{(j,k)\in\mathscr{L}\backslash(j',k')} (z - p_{jk})^{n_{jk}}, & z \in Q_{j'k'}, (j', k') \in \mathscr{L}. \end{cases}$$

This function is continuous and zero-free on Q, hence has a continuous logarithm on Q by 4.21. The restriction of F_0 to K has the product form stated in the theorem.

Corollary 4.30 *Let K be a compact subset of $\mathbb{C}$, $\mathscr{C}$ the set of bounded components of $\mathbb{C}\backslash K$ and for each $C \in \mathscr{C}$ let $p_C \in C$. Then given continuous $f: K \to \mathbb{C}\backslash\{0\}$, there exist $C_1, \ldots, C_M \in \mathscr{C}$ and integers $n_1, \ldots, n_M$ such that the function $F(z) = f(z) \prod_{j=1}^{M} (z - p_{C_j})^{n_j}$ has a continuous logarithm on K. The case $\mathscr{C} = \varnothing$ is allowed, void products being interpreted as* 1.

Proof: Let $p_1, \ldots, p_N$ and $n_1, \ldots, n_N$ be as provided for f by the last theorem. It suffices to show that if $p_j \in C_j \in \mathscr{C}$, then $g(z) = (z - p_{C_j})/(z - p_j)$ has a continuous logarithm on K and if p_j belongs to the unbounded component of $\mathbb{C}\backslash K$ then $z - p_j$ itself has such a logarithm. For then we just multiply the representation 4.29 by the appropriate powers of these factors. In the second case the existence of the desired logarithm is guaranteed by 4.28. In the first case, recalling that C_j is open (1.30(i)′), we use 1.28 to select a curve $\gamma: [0, 1] \to C_j$ joining p_{C_j} to p_j. Then

$$H(t, z) = \frac{z - \gamma(t)}{z - p_j}, \quad (t, z) \in [0, 1] \times K$$

is a homotopy in $\mathbb{C}\backslash\{0\}$ of g with 1 and the existence of the desired logarithm follows from 4.25.

Theorem 4.31 *Let K be a compact subset of $\mathbb{C}$, C a bounded component of $\mathbb{C}\backslash K$ and p a point of C. Then the function f defined on K by $f(z) = z - p$ has no continuous, zero-free extension to $K \cup C$; neither does any non-zero power of f.*

Proof: We may suppose $p = 0$. Suppose, contrariwise, that for some integer $n \neq 0$ the function f^n has a continuous extension $F: K \cup C \to \mathbb{C}\backslash\{0\}$. Thus $F(z) = z^n$ for all $z \in K$. Choose $r > \sup|C|$ and define

$$g(z) = \begin{cases} z^n & z \in \bar{D}(0, r)\backslash C \\ F(z), & z \in \bar{C}. \end{cases}$$

Notice that $\bar{C} \cap [\bar{D}(0, r)\backslash C] = \bar{C}\backslash C \subset K$ (C being open and relatively closed in $\mathbb{C}\backslash K$). This shows that the definition of g is consistent and that in fact g is continuous. Moreover, we see that g maps into $\mathbb{C}\backslash\{0\}$ (remember, $0 \in C$ and F is zero-free). Then 4.21 furnishes a continuous logarithm ϕ for g. In particular,

$$(*) \qquad z^n = g(z) = e^{\phi(z)} \quad \forall z \in C(0, r).$$

This contradicts 4.22(i) and (v). Alternatively, we can compute indexes at 0 on both sides of (*) and reach the contradiction $n = 0$.

Exercise 4.32 *Let K be a compact subset of $\mathbb{C}$. Form the groups $\mathscr{G} = \mathscr{G}(K)$ and $\mathscr{E} = \mathscr{E}(K)$ of all continuous, zero-free functions on K and all exponentials of continuous functions on K, respectively, as in 4.22.*

(i) *Let $C_1, \ldots, C_n$ be distinct bounded components of $\mathbb{C}\backslash K$ and $p_j \in C_j$. Let $f_j(z) = z - p_j$, $z \in K$. Then $f_1, \ldots, f_n \in \mathscr{G}$. Show that their cosets are independent in the quotient group $\mathscr{G}/\mathscr{E}$. That is, show that $f_1^{k_1} \cdots f_n^{k_n} \in \mathscr{E}$ for integers $k_1, \ldots, k_n$ only if $k_1 = \cdots = k_n = 0$.*

Hints: Use 4.31: If $f_1^{k_1} \cdots f_n^{k_n} \in \mathscr{E}$ and some k_j is not zero, say $k_1 \neq 0$, we write

$$f_1^{k_1} = f_2^{-k_2} \cdots f_n^{-k_n} e^{\phi}$$

for some continuous ϕ on K. That is,

$$f_1^{k_1}(z) = \frac{e^{\phi(z)}}{(z - p_2)^{k_2} \cdots (z - p_n)^{k_n}} \quad \forall z \in K.$$

The denominator on the right naturally extends to a continuous, zero-free function on C_1, since $p_2, \ldots, p_n \notin C_1$, and ϕ has by 1.44 a continuous extension to $\mathbb{C}$. Thus $f_1^{k_1}$ admits a continuous, zero-free extension to $K \cup C_1$, contradicting 4.31.

(ii) *Use* (i) *and* 4.30 *to show that the rank of the group $\mathscr{G}/\mathscr{E}$ equals n if $\mathbb{C}\backslash K$ has exactly n bounded components and this rank is infinite if $\mathbb{C}\backslash K$ has infinitely many bounded components.*

(iii) *Noting that the groups $\mathscr{G}(K_j)$, as well as the groups $\mathscr{E}(K_j)$ ($j = 1, 2$), are naturally isomorphic whenever the compact sets K_j are homeomorphic, deduce that the number of components of the complement of a compactum is a positional invariant.*

Corollary 4.33 (BORSUK [1932]). *For a compact subset K of $\mathbb{C}$ the following are equivalent:*

(i) *$\mathbb{C}\backslash K$ is connected.*

(ii) *Every continuous, zero-free function on K has a continuous logarithm.*

Proof: (i) ⇒ (ii). Use 4.30.
(ii) ⇒ (i). Use 4.31 and 4.24.

Corollary 4.34 *The property of having a connected complement is a positional invariant for compact subsets of $\mathbb{C}$.*

Exercise 4.35 *A compact subset of $\mathbb{C}$ is simply-connected if and only if each of its components is. The sequence of steps below (mainly from* EILENBERG [1936]) *will carry the reader through to this conclusion.*

(i) *Let C closed $\subset K$ closed $\subset \mathbb{C}$, $f: K \to \mathbb{C}\backslash\{0\}$ continuous and suppose that f has a continuous logarithm on C. Show that there is a relatively open subset U of K containing C such that f has a continuous logarithm on U.*

Hints: Let $\phi: K \to \mathbb{C}$ be a continuous extension to K (after 1.44) of a continuous logarithm of f on C and consider $U = \{z \in K: |f(z)e^{-\phi(z)} - 1| < 1\}$. This set is relatively open in K and contains C. It is mapped by $fe^{-\phi}$ into $D(1, 1)$ and there is (by 3.19) a continuous logarithm L in $D(1, 1)$. Therefore we may form $\psi = L \circ (fe^{-\phi})$ on U and have $f = e^{\phi}e^{\psi} = e^{\phi+\psi}$ on U.

(ii) *Let K be a compact subset of $\mathbb{C}$, $f: K \to \mathbb{C}\backslash\{0\}$ continuous. Suppose that f has a continuous logarithm on each component of K. Show that f has a continuous logarithm on K.*

Hints: For each component C of K let U_C be a neighborhood of C as provided by (i). Use 1.33 to select a compact, relatively open (in K) neighborhood V_C of C lying in U_C. Cover compact K with finitely many of the sets V_C, say with $V_{C_1}, \ldots, V_{C_n}$. Then the sets $A_1 = V_{C_1}$, $A_k = V_{C_k}\backslash\bigcup_{j=1}^{k-1} V_{C_j}$ $(1 < k \leq n)$ are disjoint, compact, they cover K and on each of them f has a continuous logarithm.

(iii) *If K is a compact subset of $\mathbb{C}$ and each component of K is simply-connected, then K is simply-connected.*

Hint: Use (ii) and 4.33.

(iv) *If C_0 is a component of a compact, simply-connected subset K of $\mathbb{C}$, then C_0 is simply-connected.*

Hints: It suffices to show that for each (other) component C of K, $\overline{\mathbb{C}\backslash K}$ meets C. For then $(\mathbb{C}\backslash K) \cup C$ is connected (by 1.3) and $\mathbb{C}\backslash C_0$ is a union of such sets. If $\overline{\mathbb{C}\backslash K} \cap C = \varnothing$, then for each $z \in C$ there exists an $r_z > 0$ such that $(\mathbb{C}\backslash K) \cap D(z, r_z) = \varnothing$. Then $U = \bigcup_{z\in C} D(z, r_z) = \bigcup_{z\in C} C \cup D(z, r_z)$ is connected, contains C and lies in K. Hence by maximality of components, $U = C$. But then C is clopen in $\mathbb{C}$.

Exercise 4.36 *Say that the set* $S \subset \mathbb{C}$ separates *two points* a, b *if* a, b *lie in different components of* $\mathbb{C}\backslash S$.

(i) (EILENBERG [1936]) *Let* K *be a compact subset of* $\mathbb{C}$, $a, b \in \mathbb{C}\backslash K$. *Let* $f\colon K \to \mathbb{C}$ *be defined by* $f(z) = (z - a)/(z - b)$. *Show that* K *does not separate* a *and* b *if and only if* f *has a continuous logarithm on* K.

Hints: ($\Rightarrow$) Examine the proof of 4.30.
($\Leftarrow$) According to 4.24, f has a continuous zero-free extension F to $\mathbb{C}$. If a, b are not in the same component of $\mathbb{C}\backslash K$, then one of them, say a, belongs to a bounded component C of $\mathbb{C}\backslash K$ which does not contain the other. Then the function $z - a = (z - b)f(z)$ $(z \in K)$ has the zero-free continuous extension $(z - b)F(z)$ $(z \in K \cup C)$, contradicting 4.31.

(ii) (JANISZEWSKI [1915]) *Suppose that* K_1, K_2 *are two compact subsets of* $\mathbb{C}$ *neither of which separates* a, b. *Show that if* $K_1 \cap K_2$ *is connected then* $K_1 \cup K_2$ *does not separate* a, b *either.*

Hint: Form $f(z) = (z - a)/(z - b)$ $(z \in K_1 \cup K_2)$. According to (i) this function has a continuous logarithm on K_1 and a continuous logarithm on K_2. Use connectedness of $K_1 \cap K_2$ to manufacture a single continuous logarithm on $K_1 \cup K_2$. Then cite the other direction of (i).

Exercise 4.37 *This exercise is designed to culminate in a strengthened version of the non-trivial implication in* 4.18. *Let* U *be an open subset of* $\mathbb{C}$ *and* $f\colon U \to \mathbb{C}\backslash\{0\}$ *a continuous function satisfying*

(*) $\operatorname{Ind}_{f\circ\Gamma}(0) = 0$ *for every simple loop* Γ *in* U.

(i) *Consider two* arcs $\gamma_1, \gamma_2\colon [0, 1] \to U$ *such that*

(1.0) $\gamma_1(0) = \gamma_2(0)$

(1.1) $\gamma_1(1) = \gamma_2(1)$

and let $\phi_1, \phi_2\colon [0, 1] \to \mathbb{C}$ *satisfy*

(2) $f \circ \gamma_j = e^{\phi_j}$ $(j = 1, 2)$

(3.0) $\phi_1(0) = \phi_2(0)$.

Show that $\phi_1(1) = \phi_2(1)$.

Hints: Argue by contradiction and suppose that

(3.1) $\phi_1(1) \neq \phi_2(1)$.

Notice that $h = \gamma_2^{-1} \circ \gamma_1$ is a homeomorphism of the compact subset $\gamma_1^{-1}(\gamma_2[0, 1])$ of $[0, 1]$ onto the compact subset $\gamma_2^{-1}(\gamma_1[0, 1])$ of $[0, 1]$ which maps 0 to 0 and 1 to 1. Consider the set

$$T = \{t \in [0, 1]\colon \gamma_1(t) \in \gamma_2[0, 1] \ \&\ \phi_1(t) = \phi_2(h(t))\}.$$

T is closed, $0 \in T$ [by (1.0) and (3.0)] and $1 \notin T$ [by (1.1) and (3.1)]. Consequently there is a largest element $t_1 \in T$. Setting $s_1 = h(t_1)$, we have

(4) $\quad 0 \leq t_1, s_1 < 1 \;\&\; \gamma_1(t_1) = \gamma_2(s_1) \;\&\; \phi_1(t_1) = \phi_2(s_1)$

and, by the maximality of t_1,

(5) $\quad t_1 < t \leq 1 \;\&\; \gamma_1(t) \in \gamma_2[0, 1] \Rightarrow \phi_1(t) \neq \phi_2(h(t))$.

Next form

$$S = \{s \in [s_1, 1]: \gamma_2(s) \in \gamma_1[t_1, 1] \;\&\; \phi_2(s) \neq \phi_1(h^{-1}(s))\}.$$

By (4), $s_1 \notin S$ and by (1.1) and (3.1), $1 \in S$. If $\gamma_2(s) \in \gamma_1[t_1, 1]$ and we set $t = h^{-1}(s)$, then $\gamma_1(t) = \gamma_2(s)$. Consequently

$$e^{\phi_1(t)} = f(\gamma_1(t)) = f(\gamma_2(s)) = e^{\phi_2(s)}$$

and therefore $\phi_2(s) - \phi_1(t)$ is an integer multiple of $2\pi i$. This shows that the set S can also be described as

$$S = \{s \in [s_1, 1]: \gamma_2(s) \in \gamma_1[t_1, 1] \;\&\; |\phi_2(s) - \phi_1(h^{-1}(s))| \geq 2\pi\},$$

which makes it clear that S is closed. Therefore S has a smallest element s_2. Setting $t_2 = h^{-1}(s_2)$, we have

(6) $\quad s_1 < s_2 \leq 1, \qquad t_1 < t_2 \leq 1, \qquad \gamma_2(s_2) = \gamma_1(t_2) \;\&\; \phi_2(s_2) \neq \phi_1(t_2)$

and, by the minimality of s_2,

(7) $\quad s_1 \leq s < s_2 \;\&\; \gamma_2(s) \in \gamma_1[t_1, 1] \Rightarrow \phi_2(s) = \phi_1(h^{-1}(s))$.

It follows easily from (5) and (7) that

(8) $\quad \gamma_1(t_1, t_2) \cap \gamma_2[s_1, s_2] = \varnothing$.

Indeed, suppose

(9) $\quad t_1 < t < t_2$ and $\gamma_1(t) \in \gamma_2[s_1, s_2]$, say,

(10) $\quad \gamma_1(t) = \gamma_2(s)$, with $s \in [s_1, s_2]$.

Now $t \neq t_1, t_2$ and h one-to-one show that $s_1 = h(t_1) \neq h(t) \neq h(t_2) = s_2$, that is, $s = h(t) \neq s_1, s_2$ and so

(11) $\quad s_1 < s < s_2$.

It follows on the one hand from (9), (10) and (5) that $\phi_1(t) \neq \phi_2(s)$ and on the other hand from (11), (10) and (7) that $\phi_2(s) = \phi_1(t)$. With (8) confirmed and (4) and (6) in mind, the equations

$$\Gamma(\tau) = \begin{cases} \gamma_1(t_2 + \tau(t_1 - t_2)) & 0 \leq \tau \leq 1 \\ \gamma_2(s_1 + (1 - \tau)(s_2 - s_1)) & 1 \leq \tau \leq 2 \end{cases}$$

well-define a simple loop in U. Moreover, by (4) the equations

$$\Phi(\tau) = \begin{cases} \phi_1(t_2 + \tau(t_1 - t_2)) & 0 \leq \tau \leq 1 \\ \phi_2(s_1 + (1 - \tau)(s_2 - s_1)) & 1 \leq \tau \leq 2 \end{cases}$$

well-define a continuous function which satisfies $f \circ \Gamma = e^{\Phi}$. Consequently

$$2\pi i \operatorname{Ind}_{f\circ\Gamma}(0) = \Phi(2) - \Phi(0) = \phi_2(s_2) - \phi_1(t_2) \neq 0, \quad \text{by (6).}$$

This violates the hypothesis on f.

(ii) *Prove that under hypothesis* (*), *f has a continuous logarithm in U.*

Hints: As in 4.18 we reduce at once to connected U. Fix a point $z_0 \in U$ and a $w_0 \in \mathbb{C}$ such that $e^{w_0} = f(z_0)$. Set $\gamma_{z_0} = z_0$ and for each $z \in U$ different from z_0 select [by 1.28 and 1.14] an *arc* $\gamma_z: [0, 1] \to U$ joining z_0 to z. Let ϕ_z and Φ be defined as in 4.18. To prove Φ continuous at a point z_1, select r as in 4.18 and let t_1 be the smallest $t \in [0, 1]$ such that $\gamma_{z_1}(t) \in \bar{D}(z_1, r)$. Then choose L as in 4.18 but with the normalization (3) replaced by

(3)′ $\quad L(f(\gamma_{z_1}(t_1))) = \phi_{z_1}(t_1).$

For each $z \in D(z_1, r)$ define

$$\Gamma_z(t) = \begin{cases} \gamma_{z_1}(t) & 0 \le t \le t_1 \\ \gamma_{z_1}(t_1) + \dfrac{t - t_1}{1 - t_1}(z - \gamma_{z_1}(t_1)), & t_1 \le t \le 1, \end{cases}$$

$$\Phi_z(t) = \begin{cases} \phi_{z_1}(t) & 0 \le t \le t_1 \\ L(f(\Gamma_z(t))) & t_1 \le t \le 1. \end{cases}$$

We note that Φ_z is well-defined at t_1, by (3)′ and so is continuous on [0, 1]. Also clearly $\Gamma_z = e^{\Phi_z}$. If $z_1 \neq z_0$, then γ_{z_1} is one-to-one and if $z_1 = z_0$, then $t_1 = 0$. In either case therefore γ_{z_1} is one-to-one on $[0, t_1]$. Moreover, by minimality of t_1, $\Gamma_z[0, t_1)$ is disjoint from $\bar{D}(z_1, r) \supset \Gamma_z[t_1, 1]$. Hence Γ_z is one-to-one. Finally, since $\phi_z(0) = w_0 = \phi_{z_1}(0) = \Phi_z(0)$, the conclusion $\phi_z(1) = \Phi_z(1)$ follows from (i). This says that

$$\Phi(z) = L(f(\Gamma_z(1))) = L(f(z)) \quad \forall z \in D(z_1, r).$$

§ 5 The Jordan Curve Theorem

Corollary 4.38 *For any Jordan-curve J the complement $\mathbb{C}\backslash J$ has exactly one bounded component.*

Proof: $J = \phi(C)$ for some homeomorphism ϕ of $C = C(0, 1)$. Since $\mathbb{C}\backslash C$ has exactly one bounded component, so does $\mathbb{C}\backslash J$ by 4.32(iii). Alternatively, if one wishes to avoid the algebra, we can argue thus: If $\mathbb{C}\backslash J$ had no bounded component, then by 4.33 $\mathscr{G}(J) = \mathscr{E}(J)$, whence, via ϕ, $\mathscr{G}(C) = \mathscr{E}(C)$. This contradicts 4.22(i). If $\mathbb{C}\backslash J$ has two distinct bounded components C_1, C_2, we select points $p_j \in C_j$ and form functions $f_j(z) = z - p_j$ $(j = 1, 2)$ on J. By 4.22(ii) there exist integers n_j and continuous functions ψ_j on C such that

(*) $\quad \phi(w) - p_j = w^{n_j} e^{\psi_j(w)} \quad \forall w \in C, j = 1, 2.$

Then

$$(\phi(w) - p_1)^{n_2}(\phi(w) - p_2)^{-n_1} = e^{n_2\psi_1(w) - n_1\psi_2(w)} \quad \forall w \in C,$$

whence

$$(z - p_1)^{n_2}(z - p_2)^{-n_1} = e^{n_2\psi_1(\phi^{-1}(z)) - n_1\psi_2(\phi^{-1}(z))} \quad \forall z \in J.$$

Then 4.32(i) implies that $n_1 = n_2 = 0$. Returning to (*), we have then

$$\phi(w) - p_1 = e^{\psi_1(w)} \qquad \forall w \in C,$$
$$f_1(z) = z - p_1 = e^{\psi_1(\phi^{-1}(z))} \quad \forall z \in J.$$

Thus f_1 extends to a continuous, zero-free function on $\mathbb{C}$ (since $\psi_1 \circ \phi^{-1}$ is extendable), violating 4.31.

Exercise 4.39 *If K is a proper, closed subset of a Jordan-curve J, then $\mathbb{C}\backslash K$ is connected.*

Hints: By 4.34 it suffices to demonstrate this for the special case $J = C(0, 1)$. However, that case was handled in 1.25(i).

Exercise 4.40 *If K is a compact subset of $\mathbb{C}$ homeomorphic to a subset of $\mathbb{R}$, then $\mathbb{C}\backslash K$ is connected.*

Hint: Treat the case $K \subset \mathbb{R}$ directly (cf. the proof of 1.25(i)), then cite 4.34.

Theorem 4.41 *If J is a Jordan-curve and C a component of $\mathbb{C}\backslash J$, then $\partial C = J$.*

Proof: By 1.30, C is open and relatively closed in $\mathbb{C}\backslash J$. Therefore $\partial C = \bar{C}\backslash C = [\bar{C} \cap J \cup \bar{C} \cap (\mathbb{C}\backslash J)]\backslash C = [\bar{C} \cap J \cup C]\backslash C = \bar{C} \cap J\backslash C = \bar{C} \cap J$. In particular,

(*) $\quad \partial C \subset J.$

By 4.38, $\mathbb{C}\backslash J$ is not connected, so there is another component K of $\mathbb{C}\backslash J$ disjoint from C. Thus K is disjoint from $\bar{C}$, since $\bar{C} \subset C \cup J$ (as was shown above). It follows that $\mathbb{C}\backslash\bar{C}$ is not void. Therefore the equality

$$\mathbb{C}\backslash\partial C = C \cup (\mathbb{C}\backslash\bar{C})$$

displays $\mathbb{C}\backslash\partial C$ as a disjoint union of non-void, open sets, thus showing it to be disconnected. Consequently 4.39 tells us that the inclusion (*) cannot be proper.

Theorem 4.42 *Let J be a Jordan-curve, Γ any homeomorphism of $C(0, 1)$ onto J. Then for the loop γ defined by $\gamma(t) = \Gamma(e^{2\pi it})$, $t \in [0, 1]$, we have $\mathrm{Ind}_\gamma = \pm 1$ throughout the bounded component of $\mathbb{C}\backslash J$.*

Proof: We have $f = \Gamma^{-1}: J \to \mathbb{C}\backslash\{0\}$ and so by 4.30 there exists p in the (unique!) bounded component of $\mathbb{C}\backslash J$, an integer n and a continuous function ϕ on J such that

$$(z - p)^{-n}f(z) = e^{\phi(z)} \quad \forall z \in J.$$

Consequently

$$w = (\Gamma(w) - p)^n e^{\phi(\Gamma(w))} \quad \forall w \in C(0, 1).$$

In particular,

$$C(0, 1) = (\gamma - p)^n e^{\phi \circ \gamma}.$$

Taking indexes at 0 on both sides gives

$$\begin{aligned} 1 &= n \operatorname{Ind}_{\gamma - p}(0) + \operatorname{Ind}_{e^{\phi \circ \gamma}}(0) \\ &= n \operatorname{Ind}_\gamma(p) + 0. \end{aligned}$$

It follows that $\operatorname{Ind}_\gamma(p) = \pm 1$.

Exercise 4.43 *Any homeomorphism of one circle onto another can be extended to a homeomorphism of* $\mathbb{C}$ *onto* $\mathbb{C}$.

Hint: After translations and dilations in the domain and range it suffices to treat the case of a homeomorphism ϕ of the unit circle $C(0, 1)$ onto itself. Show that

$$\Phi(z) = \begin{cases} 0 & z = 0 \\ |z|\phi\left(\dfrac{z}{|z|}\right), & z \neq 0 \end{cases}$$

defines an extension of the desired kind.

Remarks 4.44 More generally, any homeomorphism of one Jordan-curve in $\mathbb{C}$ onto another extends to a homeomorphism of $\mathbb{C}$ onto $\mathbb{C}$. (Schönflies [1906], pp. 319–324 and [1908], pp. 209–213, modulo some errors; see also Tietze [1913], Kline [1920], Antoine [1921], Nöbeling [1950] and p. 198 of Guggenheimer [1977].) This will be deduced in Chapter IX from a deeper extension theorem for conformal maps. A corresponding extension is possible for homeomorphisms of intervals in $\mathbb{C}$, i.e., for arcs. One first extends the arc to a simple loop (§ 7 of Feigl [1928] or p. 164 of M. H. A. Newman [1951]).

Definition 4.45 (i) For a Jordan-curve J the unique bounded component of $\mathbb{C}\backslash J$ is called the *inside* of J and is denoted $\mathscr{I}(J)$. The unique unbounded component of $\mathbb{C}\backslash J$ is called the *outside* of J and is denoted $\mathscr{O}(J)$.

(ii) A *Jordan region* is any set $\mathscr{I}(J)$.

(iii) A homeomorphism Γ of $C(0, 1)$ onto J is called *positively* (*negatively*) *oriented* if $\operatorname{Ind}_\gamma$ equals 1 (-1) throughout the inside of J, where $\gamma(t) = \Gamma(e^{2\pi i t})$.

Remark 4.46: The fundamental facts 4.38 and 4.41 can be summarized in the above language as follows:

(i) $\mathscr{I}(J)$, $\mathscr{O}(J)$ are disjoint, open, connected sets
(ii) $\mathscr{I}(J)$ is bounded, $\mathscr{O}(J)$ is unbounded
(iii) $\mathbb{C}\backslash J = \mathscr{I}(J) \cup \mathscr{O}(J)$
(iv) $\overline{\mathscr{I}(J)} = J \cup \mathscr{I}(J)$, $\overline{\mathscr{O}(J)} = J \cup \mathscr{O}(J)$.

Exercise 4.47 *Let J be a Jordan-curve, a, b two distinct points on J and C a cross-cut in $\mathscr{I}(J)$ with endpoints a and b.*

(i) *Show that there are two arcs A_1, A_2 whose union is J and whose endpoints are each a and b.*
(ii) *Show that $J_1 = A_1 \cup C$ and $J_2 = A_2 \cup C$ are Jordan-curves.*
(iii) *Show that $\mathscr{I}(J)\backslash C = \mathscr{I}(J_1) \cup \mathscr{I}(J_2)$.*

Hints: The set $A_1' = A_1\backslash\{a, b\}$ lies in J and J lies in the closure of the unbounded connected set $\mathscr{O}(J)$ (See 4.46.) Therefore $A_1' \cup \mathscr{O}(J)$ is unbounded and connected [by 1.2]. It is also disjoint from J_2. Hence it lies in $\mathscr{O}(J_2)$, whence

(1) $\mathscr{I}(J_2) \subset \mathbb{C}\backslash(A_1' \cup \mathscr{O}(J)) = \overline{\mathscr{I}(J)}\backslash A_1'$.

Of course $\mathscr{I}(J_2)$ is disjoint from $J_2 = (J\backslash A_1') \cup C$ and so the inclusion (1) implies

$$\mathscr{I}(J_2) \subset \overline{\mathscr{I}(J)}\backslash J\backslash C = \mathscr{I}(J)\backslash C.$$

Similarly for $\mathscr{I}(J_1)$. Thus

(2) $\mathscr{I}(J_1) \cup \mathscr{I}(J_2) \subset \mathscr{I}(J)\backslash C$.

To reverse the inclusion, consider any point $z \in \mathscr{I}(J)\backslash C$ and select

(3) $w \in \mathscr{O}(J)$.

If z belongs to neither $\mathscr{I}(J_1)$ nor $\mathscr{I}(J_2)$, then (since $z \notin J \cup C = J_1 \cup J_2$) z belongs to each of $\mathscr{O}(J_1)$, $\mathscr{O}(J_2)$, to which also the point w belongs by (2) and (3). Thus neither J_1 nor J_2 separate z and w. Since $J_1 \cap J_2 = C$ is connected, it follows from Janiszewski's theorem that $J_1 \cup J_2$ does not separate z and w. *A fortiori* then J, a subset of $J_1 \cup J_2$, does not separate z and w. That is false however, since $z \in \mathscr{I}(J)$ and $w \in \mathscr{O}(J)$.

(iv) *Enunciate and prove the analog of* (i)–(iii) *in case $a = b$.*

For use later we record here a technical result.

Lemma 4.48 *Let Ω be a Jordan region, $z \in \partial\Omega$. Then for each $\varepsilon > 0$ there exists $\delta > 0$ such that any two points of $\Omega \cap D(z, \delta)$ lie on a polygon in $\Omega \cap D(z, \varepsilon)$.*

Proof: Let Ω be the inside of the Jordan-curve J. There is a homeomorphism ϕ of the unit circle onto J such that $\phi(1) = z$. Let $\gamma(t) = \phi(e^{it})$ ($t \in \mathbb{R}$). Choose $0 < t_0 < \pi$ sufficiently small that

(1) $J_1 = \gamma[-t_0, t_0] \subset D(z, \varepsilon)$.

Now z is disjoint from the compact set $J_2 = J\backslash\gamma(-t_0, t_0) = \gamma[t_0, -t_0 + 2\pi]$, so for some $0 < \delta < \varepsilon$, $D(z, \delta)$ is also disjoint from this set. Consider any two points $p, q \in \Omega \cap D(z, \delta)$. Setting $C = \partial D(z, \varepsilon)$, it is obvious that

(2) p, q are not separated by $C \cup J_2$,

since the latter set lies in the complement of $D(z, \delta)$. Of course

(3) p, q are not separated by J either,

since they both belong to the component Ω of $\mathbb{C}\backslash J$. Moreover, since J_1 is disjoint from C (see (1)) and $J_1 \cup J_2 = J$, it follows that

$$C \cap J \subset J_2$$

and so

(4) $(C \cup J_2) \cap J = (C \cap J) \cup (J_2 \cap J) = J_2.$

From (2), (3), (4) and Janiszewski's theorem we have that

(5) p, q are not separated by $(C \cup J_2) \cup J = C \cup J$,

that is, p, q belong to the same component U of $\mathbb{C}\backslash(C \cup J) = (\mathbb{C}\backslash C) \cap (\mathbb{C}\backslash J)$. Since we know which of the two components of $\mathbb{C}\backslash C$ contains p and q (viz., $D(z, \varepsilon)$) and which of the two components of $\mathbb{C}\backslash J$ contains p and q (viz., Ω), it follows that $U \subset \Omega \cap D(z, \varepsilon)$. Since U is open there is, thanks to 1.28, a polygon in U joining p to q.

§ 6 Applications of the Foregoing Technology

Theorem 4.49 *If $f: \mathbb{C} \to \mathbb{C}$ is continuous and for some positive integer n satisfies $\lim_{|z| \to \infty} z^{-n}f(z) = c \in \mathbb{C}\backslash\{0\}$, then f has a zero. In particular, every non-constant polynomial has a zero.*

Proof: Suppose contrariwise that $0 \notin f(\mathbb{C})$. We may divide f by the hypothesized non-zero limit and so suppose from the start that it is 1. Choose then positive R sufficiently large that

$$|z^{-n}f(z) - 1| < 1 \quad \forall |z| \geq R$$

(1) $|f(z) - z^n| < |z^n| \quad \forall |z| \geq R.$

If we define

(2) $\Gamma(t) = [Re^{2\pi it}]^n = e^{2\pi int + n \log R}, \quad t \in [0, 1]$

and

(3) $\gamma_r(t) = h(r, t) = f(re^{2\pi it}), \quad (r, t) \in [0, R] \times [0, 1],$

then (1) implies that

$$|\gamma_R(t) - \Gamma(t)| < |\Gamma(t)| \quad \forall t \in [0, 1].$$

It follows from (the proof of) 4.6 that

$$\mathrm{Ind}_{\gamma_R}(0) = \mathrm{Ind}_{\Gamma}(0).$$

But from the form (2) of Γ and the definition of index we have at once that $\mathrm{Ind}_{\Gamma}(0) = n$ and so

(4) $\quad \mathrm{Ind}_{\gamma_R}(0) = n.$

However, by hypothesis $h([0, R] \times [0, 1]) \subset f(\mathbb{C}) \subset \mathbb{C}\backslash\{0\}$, and so by 4.12 we have

(5) $\quad \mathrm{Ind}_{\gamma_R}(0) = \mathrm{Ind}_{\gamma_0}(0).$

As γ_0 is constant, we have $\mathrm{Ind}_{\gamma_0}(0) = 0$ and (5) contradicts (4).

Finally, note that if p is a non-constant polynomial and $n \geq 1$ is its degree, then $z^{-n}p(z)$ approaches the leading coefficient of p as $|z| \to \infty$, so that p is an eligible f.

Exercise 4.50 (i) *For any polynomial p over $\mathbb{C}$, $p(z) - p(z_0)$ is divisible by $z - z_0$, that is, $p(z) - p(z_0) = (z - z_0)q(z)$ for some polynomial q.*

Hint: For any positive integer k, $z^k - z_0^k = (z - z_0)\sum_{j=0}^{k-1} z_0^{k-1-j}z^j$.

(ii) *Use* (i), *induction on n, and* 4.49 *to show that if p is any nth degree polynomial over $\mathbb{C}$ then there exist $c, z_1, \ldots, z_n \in \mathbb{C}$ such that $p(z) \equiv c\prod_{j=1}^{n}(z - z_j)$.*

(iii) *If p is a non-constant polynomial, then all zeros of p' lie in* co $p^{-1}(0)$.

Hint: Let $z_1, \ldots, z_N$ be the distinct zeros of p and n_j the multiplicity of z_j. Then from (ii)

$$p(z) = c\prod_{j=1}^{N}(z - z_j)^{n_j} \quad (\text{some } c \in \mathbb{C}\backslash\{0\}).$$

Then for $z \notin p^{-1}(0)$

$$\frac{p'(z)}{p(z)} = \sum_{j=1}^{N} \frac{n_j}{z - z_j} = \sum_{j=1}^{N} \frac{n_j}{|z - z_j|^2}(\bar{z} - \bar{z}_j).$$

If in addition $p'(z) = 0$, it follows that

$$\bar{z}\sum_{j=1}^{N} \frac{n_j}{|z - z_j|^2} = \sum_{j=1}^{N} \frac{n_j\bar{z}_j}{|z - z_j|^2}.$$

Take conjugates to see that

$$z = \sum_{j=1}^{N} \lambda_j z_j,$$

where

$$\lambda_j = \frac{n_j}{|z - z_j|^2}\left[\sum_{l=1}^{N} \frac{n_l}{|z - z_l|^2}\right]^{-1} \in (0, 1] \quad \text{and} \quad \sum_{j=1}^{N} \lambda_j = 1.$$

Exercise 4.51 (i) *For each positive integer n there are multinomials $s_{n,0}, \ldots, s_{n,n-1}$ with integer coefficients such that the coefficient of z^j in the product $(z - \alpha_1)\cdots(z - \alpha_n)$ is $s_{n,j}(\alpha_1, \ldots, \alpha_n)$ for each $j = 0, 1, \ldots, n - 1$ and all choices of complex numbers $\alpha_1, \ldots, \alpha_n$. In particular, the coefficients of a monic polynomial are continuous functions of its zeros.*

Hint: Induction on n.

(ii) *Let n be a positive integer and $\alpha_1, \ldots, \alpha_n \in \mathbb{C}$. Suppose that the polynomial $p(z) = (\alpha_1 - z)\cdots(\alpha_n - z)$ has integer coefficients. Show that then the polynomial $P(z) = (\alpha_1^2 - z)\cdots(\alpha_n^2 - z)$ also has integer coefficients.*

Hint: (Ostrowski) The polynomial $p(z)p(-z)$ has integer coefficients and equals $(\alpha_1 - z)(\alpha_1 + z)\cdots(\alpha_n - z)(\alpha_n + z) = (\alpha_1^2 - z^2)\cdots(\alpha_n^2 - z^2) = P(z^2)$.

(iii) *Let P be a monic polynomial with integer coefficients all of whose zeros lie on the unit circle. Show that each of these zeros is in fact a root of unity.*

Hints: Fix positive integer n and let $\mathscr{P}(n)$ denote the set of nth degree monic polynomials with integer coefficients whose zeros lie on the unit circle. It follows from part (i) that the coefficients of the polynomials in $\mathscr{P}(n)$ are bounded. As these coefficients are integers, it follows that $\mathscr{P}(n)$ is finite. Consequently the set $Z(n)$ of all the zeros of all the polynomials in $\mathscr{P}(n)$ is also finite. But if $\alpha \in Z(n)$, then so does α^2 by part (ii). Inductively, $\{\alpha^{2^k}: k \text{ positive integer}\} \subset Z(n)$. Then for some $k_1 < k_2$ we must have $\alpha^{2^{k_2}} = \alpha^{2^{k_1}}$, $\alpha^{(2^{k_2} - 2^{k_1})} = 1$.

Exercise 4.52 (cf. 11.42) *Let $\mathscr{R}$ denote the set of all polynomials with rational coefficients. Let P, Q be (non-zero) elements of $\mathscr{R}$ which have no common non-constant factor in $\mathscr{R}$. Show that there exist $A, B \in \mathscr{R}$ such that $AP + BQ = 1$.*

Hint: All is trivial if P or Q is constant. For example, if Q is constant take $A = Q$ and $B = -P + 1/Q$. So suppose $\deg P > 0$ and $\deg Q > 0$. Use induction on the degree of PQ. If, say, $\deg Q \le \deg P$, let B_0 be the quotient of the leading term of P by that of Q. Thus $B_0 \in \mathscr{R}$ and

$$P - B_0 Q = P_1$$

has degree less than that of P. Therefore $\deg P_1 Q < \deg PQ$. Moreover P_1 and Q have no common factor in $\mathscr{R}$ (any such would be one for P and Q too). Therefore the induction hypothesis provides $A, B_1 \in \mathscr{R}$ such that $AP_1 + B_1 Q = 1$, that is,

$$A(P - B_0 Q) + B_1 Q = 1$$

$$AP + (B_1 - AB_0)Q = 1.$$

In 4.53 through 4.57 we abbreviate $D(0, 1)$ to D and $C(0, 1)$ to C.

Theorem 4.53 (Wavre and Bruttin [1926]). *Let $f: \bar{D} \to \mathbb{C}$ be continuous and satisfy $f(C) \subset \bar{D}$. Then there exists a point $z \in \bar{D}$ such that $f(z) = z$. In particular, every continuous map of $\bar{D}$ into $\bar{D}$ fixes some point.*

Proof: Define $r: \mathbb{C}\backslash\{0\} \to C$ by $r(z) = z/|z|$. Suppose, contrariwise, that f fixes no point in $\bar{D}$. Then

$$(2 - 2t)z - f((2 - 2t)z) \neq 0 \quad \forall t \in [\tfrac{1}{2}, 1], z \in C,$$

since for such t and z the point $(2 - 2t)z \in \bar{D}$. Moreover,

$$z - 2tf(z) \neq 0 \quad \forall t \in [0, \tfrac{1}{2}), z \in C,$$

since for such t and z, $|2tf(z)| \leq 2t$ (due to $f(C) \subset \bar{D}$) $< 1 = |z|$. Therefore we can define

$$H(t, z) = \begin{cases} r(z - 2tf(z)) & (t, z) \in [0, \tfrac{1}{2}] \times C \\ r((2 - 2t)z - f((2 - 2t)z)), & (t, z) \in [\tfrac{1}{2}, 1] \times C \end{cases}$$

and will get thereby a (well-defined and continuous) homotopy in C of the identity function on C with the constant function $-f(0)/|f(0)|$. Since the latter has a continuous logarithm while (by 4.22(i)) the former does not, a contradiction to 4.25 is reached.

Exercise 4.54 (i) *Let $F: \bar{D} \to C$ be continuous. Show that for each $\lambda \in C$ there exists some $z = z_\lambda \in C$ such that $F(z) = \lambda z$. In particular, there is no continuous retraction of $\bar{D}$ onto C fixing all the points of C.*

Hints: Consider $f = \lambda^{-1}F$. If no such z exists, then f fixes no point of C. No point of D is fixed either, since f maps D into the disjoint set C. Therefore f violates 4.53.

(ii) (BROUWER [1912b]) *Let K be a non-empty, convex and compact subset of $\mathbb{C}$, $F: K \to K$ a continuous function. Show that F has a fixed point.*

Hints: We can assume that $K \subset \bar{D}$. Show first that to each point z of $\bar{D}$ there corresponds a unique nearest point $g(z)$ in K and that g is a continuous function: Existence is an easy compactness argument; uniqueness is the strict convexity of the norm in $\mathbb{R}^2$ (the average of two candidates for $g(z)$ still lies in K and is another candidate). Continuity follows easily from uniqueness. (Cf. with the hints to 1.32.) Now apply 4.53 to $f = F \circ g$. Since $f(\bar{D}) = F(K) \subset K$, any point of $\bar{D}$ fixed by f lies in K and is (therefore) fixed by g.

Exercise 4.55 *Let $f: \bar{D} \to \mathbb{C}$ be continuous, $x_0 \in D$, $f(x_0) = y_0 \notin f(C)$. Suppose that the map $f_0: C \to \mathbb{C}\backslash\{0\}$ defined by*

$$f_0(z) = f(z) - y_0$$

does not have a continuous logarithm. Show that then y_0 belongs to the interior of $f(\bar{D})$.

Hints: $f_0(C)$ is compact and does not contain 0. Let $r = \inf|f_0(C)|$. Then $r > 0$ and we can show that $D(y_0, r) \subset f(\bar{D})$. To this end, consider any $y_1 \in D(y_0, r)$ and form $f_1: C \to \mathbb{C}$ by

$$f_1(z) = f(z) - y_1.$$

We have

$$(*) \qquad |f_1(z) - f_0(z)| = |y_1 - y_0| < r \le |f_0(z)| \quad \forall z \in C.$$

Since f_0 does not have a continuous logarithm, it follows from (*) and 4.26 that f_1 does not have a continuous logarithm. *A fortiori* then the map

$$F_1(z) = f(z) - y_1, \quad z \in \bar{D}$$

does not have a continuous logarithm. Consequently by 4.21 F_1 must have a zero.

Exercise 4.56 *Let $f: \bar{D} \to \mathbb{C}$ be continuous and one-to-one. Show that $f(0)$ is an interior point of $f(\bar{D})$.*

Hints: Define

$$f_0(z) = f(z) - f(0) \qquad z \in C$$

$$f_1(z) = f(\tfrac{1}{2}z) - f(-\tfrac{1}{2}z) \qquad z \in C$$

$$h(z, t) = f\left(\frac{z}{1+t}\right) - f\left(\frac{-tz}{1+t}\right), \quad z \in C, t \in [0, 1].$$

Since f is one-to-one, the function h never vanishes, i.e., maps into $\mathbb{C}\backslash\{0\}$. Evidently f_1 satisfies $f_1(-z) = -f_1(z)$ for all $z \in C$, and so by 4.22(i) f_1 does not have a continuous logarithm. Therefore by 4.25 f_0 does not have a continuous logarithm, and so the last exercise may be applied.

Exercise 4.57 (Open Map Theorem) *If U is an open subset of $\mathbb{C}$ and $f: U \to \mathbb{C}$ is continuous and one-to-one, then $f(U)$ is an open subset of $\mathbb{C}$.*

Hints: Given $a \in U$, choose $r > 0$ so that $\bar{D}(a, r) \subset U$ and apply the last exercise.

Exercise 4.58 *Prove the* Borsuk Antipoden Satz *in dimension two: If $S = \{(x_1, x_2, x_3) \in \mathbb{R}^3 : x_1^2 + x_2^2 + x_3^2 = 1\}$ is the unit sphere in $\mathbb{R}^3$ and $f: S \to \mathbb{C}$ is continuous, then $f(p) = f(-p)$ for some $p \in S$. What does this result say about the existence at any instant of antipodes on the earth which have the same temperature and the same barometric pressure?*

Hint: Consider the function $h: \bar{D}(0, 1) \to \mathbb{C}$ defined by

$$h(x, y) = f(x, y, \sqrt{1 - x^2 - y^2}) - f(-x, -y, -\sqrt{1 - x^2 - y^2}).$$

If no such point p exists, then h maps into $\mathbb{C}\backslash\{0\}$ and so has a continuous logarithm by 4.21. Since $h(-x, -y) = -h(x, y)$ for all $(x, y) \in C(0, 1)$, a contradiction of 4.22(i) ensues.

Remark: It follows from 4.58 that no continuous complex-valued function on S can be one-to-one and so $\mathbb{R}^2$ contains no subset homeomorphic to S. In particular, $\mathbb{R}^2$ is not homeomorphic to $\mathbb{R}^3$.

§ 7 Continuous and Holomorphic Logarithms in Open Sets

The following variations on 4.30 are the first phase of a program to be completed in Chapter X of extending to open sets results of previous sections on existence of continuous logarithms and connectivity of the set.

Theorem 4.59 *Let U be an open subset of $\mathbb{C}$ such that $\mathbb{C}\backslash U$ has only finitely many bounded components $C_1, \ldots, C_M$. Let $p_j \in C_j$. Then for any continuous, zero-free f on U there exist integers $k_1, \ldots, k_M$ and continuous function ϕ such that*

$$f(z) = \prod_{j=1}^{M} (z - p_j)^{k_j} e^{\phi(z)} \quad \forall z \in U.$$

If $\mathbb{C}\backslash U$ has no bounded components, then $f = e^{\phi}$ for such a ϕ.

Proof: Let $\{K_n\}$ be a compact exhaustion of U as provided by 1.31. Let K be a compact subset of U as furnished by 1.43. Since the sets $\mathring{K}_n$ increase and cover U, they eventually all contain the compact set K. Ignoring early terms as necessary, we can suppose that $K \subset K_n$ for all n. Now according to 1.43, $C_1 \ldots, C_M$ lie in distinct bounded components $C_1(K), \ldots, C_M(K)$ of $\mathbb{C}\backslash K$. We have $C_j \subset \mathbb{C}\backslash U \subset \mathbb{C}\backslash K_n \subset \mathbb{C}\backslash K$. Therefore C_j lies in a component $C_j(K_n)$ of $\mathbb{C}\backslash K_n$ and in turn $C_j(K_n)$ lies in a (unique) component of $\mathbb{C}\backslash K$. The latter is perforce $C_j(K)$. Thus $C_1(K_n), \ldots, C_M(K_n)$ are distinct. On the other hand, by 1.31 the number of bounded components of $\mathbb{C}\backslash K_n$ does not exceed M. It follows that $C_1(K_n), \ldots, C_M(K_n)$ account for them all and so $p_1, \ldots, p_M$ represent points from each of the distinct bounded components of $\mathbb{C}\backslash K_n$. Consequently, we may appeal to 4.30 to come up with integers $k_1(n), \ldots, k_M(n)$ and continuous functions ϕ_n on K_n such that

$$(1) \qquad f(z) = \prod_{j=1}^{M} (z - p_j)^{k_j(n)} e^{\phi_n(z)} \quad \forall z \in K_n.$$

The uniqueness result 4.32(i) ensures that we have

$$k_j(1) = k_j(2) = k_j(3) = \cdots$$

for each $j = 1, 2, \ldots, M$. Call the common value k_j. Then (1) says that the function

$$F(z) = f(z) \prod_{j=1}^{M} (z - p_j)^{-k_j}, \quad z \in U$$

has a continuous logarithm on each K_n. Since any loop γ in U lies in some K_n, it follows that $F \circ \gamma$ has index 0 at 0. By 4.18 therefore F has a continuous logarithm on U. (For holomorphic f a different proof is offered in 10.12.)

Exercise 4.60 *Let U be an open subset of $\mathbb{C}$, $f\colon U \to \mathbb{C}\backslash\{0\}$ holomorphic. Show that any continuous logarithm (resp., square root) of f is holomorphic. Cf. 5.65.*

Hints: Let $z \in U$ be given. Choose $r > 0$ so small that $D(z, r) \subset U$ and $f(D(z, r)) \subset D(f(z), |f(z)|)$. Let L be a holomorphic logarithm in the latter disk (as furnished by 3.19). Then we can form $L \circ f$ in $D(z, r)$ and have a holomorphic function ϕ such that $e^{\phi} = f$ in $D(z, r)$. If $g: U \to \mathbb{C}$ is continuous and satisfies $e^{g} = f$ (resp., $g^2 = f$) throughout U, then $e^{g-\phi} = 1$ (resp., $g^2 - (e^{\phi/2})^2 = 0$) in $D(z, r)$. It follows from 3.14(iii) (resp., 1.7) that $(g - \phi)/2\pi i$ is a continuous integer-valued function on the connected set $D(z, r)$ (resp., that $g = e^{\phi/2}$ or $g = -e^{\phi/2}$ throughout $D(z, r)$). Thus $(g - \phi)/2\pi i$ is constant, say N, in $D(z, r)$ and $g = \phi + 2\pi i N$ (resp., $g = e^{\phi/2}$ or $g = -e^{\phi/2}$) exhibits g as a holomorphic function in $D(z, r)$.

Corollary 4.61 *If f in 4.59 is holomorphic, then so is ϕ.*

Proof: ϕ is a continuous logarithm for the holomorphic function $F(z) = f(z) \prod_{j=1}^{M} (z - p_j)^{-k_j}$ in U.

Exercise 4.62 *Let G be a torsion-free abelian group. Suppose that $\{f_1, \ldots, f_N\}$ is an independent set in G and for some $g_1, \ldots, g_n \in G$ we have $\{f_1, \ldots, f_N\} \subset \operatorname{grp}\{g_1, \ldots, g_n\}$, the group generated by $g_1, \ldots, g_n$. Show that $n \geq N$.*

Hints: Writing G multiplicatively, let

(1) $$f_j = \prod_{k=1}^{n} g_k^{m_{kj}}, \quad m_{kj} \in \mathbb{Z}.$$

If $n < N$, the system of equations

(2) $$\sum_{j=1}^{N} c_j m_{kj} = 0, \quad k = 1, 2, \ldots, n$$

has a non-trivial rational solution $(c_1, \ldots, c_N)$ because the N vectors $(m_{1j}, \ldots, m_{nj})$ $(j = 1, 2, \ldots, N)$ are linearly dependent in the n-dimensional space $\mathbb{Q}^n$. Upon clearing denominators, there is then a non-trivial integer solution (c_j) to (2). From (1) and (2) together and the commutativity of the group we get

$$\prod_{j=1}^{N} f_j^{c_j} = \prod_{k=1}^{n} \prod_{j=1}^{N} g_k^{c_j m_{kj}} = \prod_{k=1}^{n} g_k^{\sum_{j=1}^{N} c_j m_{kj}} = \prod_{k=1}^{n} g_k^0 = 1.$$

Since not all c_j are 0, this violates the independence of $\{f_1, \ldots, f_N\}$.

Exercise 4.63 *Let U be an open subset of $\mathbb{C}$. Show that a necessary and sufficient condition that $\mathbb{C} \backslash U$ have at most n bounded components is this: there exist zero-free holomorphic functions $\Phi_1, \ldots, \Phi_n$ such that every zero-free $f \in H(U)$ determines integers $k_1, \ldots, k_n$ and $\phi \in H(U)$ for which $f = e^{\phi} \prod_{j=1}^{n} \Phi_j^{k_j}$. [In fact, sufficiency requires only the continuity of the Φ_j and the ϕ.]*

Hints: Necessity follows from 4.59 and 4.61. For sufficiency, suppose $C_1, \ldots, C_N$ are distinct bounded components of $\mathbb{C} \backslash U$ and $N > n$. Pick $p_j \in C_j$ and define $f_j(z) = z - p_j$ $(z \in U)$. These are zero-free holomorphic functions, hence they have a representation of the advertised kind. But then by the last exercise the cosets $f_j\mathscr{E}(U)$ are dependent in the group $\mathscr{G}(U)/\mathscr{E}(U)$. That is, there exist integers $k_1, \ldots, k_N$, not all zero, and continuous function ϕ on U such that

$$(*) \qquad f_1^{k_1} \cdots f_N^{k_N} = e^{\phi} \quad \text{on } U.$$

But by 1.43 there is a compact subset K of U such that $p_1, \ldots, p_N$ are representatives of distinct bounded components of $\mathbb{C}\backslash K$. Then the validity of the equality (*) on K plus $|k_1| + \cdots + |k_N| > 0$ contradicts 4.32(i).

§ 8 Simple Connectivity for Open Sets

Definition 4.64 Let U be an open subset of $\mathbb{C}$. Say that U has
property (CL) if every zero-free continuous function in U has a continuous logarithm;
property (HL) if every zero-free holomorphic function in U has a holomorphic logarithm;
property (CS) if every zero-free continuous function in U has a continuous square root;
property (HS) [resp., (HS_1)] if every zero-free holomorphic [and one-to-one] function in U has a holomorphic square root.

These definitions and previous results enable us to state succinctly some important topological facts about $\mathbb{C}$:

Theorem 4.65 *For an open subset U of $\mathbb{C}$ the following eight conditions are equivalent:*
(i) *U is simply-connected*
(ii) *U has property (CL)*
(iii) *U has property (CS)*
(iv) *U has property (HS)*
(v) *U has property (HL)*
(vi) *U has property (HS_1)*
(vii) *U is homologically-connected*
(viii) *U contains the inside of each Jordan-curve in it.*

Proof:
(i) ⇒ (ii) from 4.59.
(ii) ⇒ (iii) trivial.
(iii) ⇒ (iv) 4.60.
(iv) ⇒ (v). If $f\colon U \to \mathbb{C}\backslash\{0\}$ is holomorphic, repeated application of the hypothesis produces (zero-free) $f_n \in H(U)$ such that $f_n^{2^n} = f$. It follows from 4.19 that f has a continuous logarithm in U, which by 4.60 must in fact be holomorphic.
(v) ⇒ (vi) trivial.
(vi) ⇒ (vii). For each $p \in \mathbb{C}\backslash U$ the function $f(z) = z - p$, being one-to-one, holomorphic and zero-free on U, has a holomorphic logarithm. The argument

is the same as that in the proof (iv) $\Rightarrow$ (v). It follows that $\text{Ind}_\gamma(p) = \text{Ind}_{f\circ\gamma}(0) = 0$ for all loops γ in U.
(vii) $\Rightarrow$ (viii). If J is a Jordan-curve in U, let ϕ be a homeomorphism of the unit circle onto J and let $\gamma(t) = \phi(e^{it})$. For any $z \in \mathcal{I}(J)$ we have $\text{Ind}_\gamma(z) \neq 0$ by 4.42. Consequently by (vii), $z \notin \mathbb{C}\backslash U$.
(viii) $\Rightarrow$ (i). Let C be a bounded component of $\mathbb{C}\backslash U$, $p \in C$ and form $f(z) = z - p$ as before. If γ is a simple loop in U, then $\mathcal{I}(\gamma)$ lies wholly in U and so Ind_γ vanishes in $\mathbb{C}\backslash U$. Therefore $0 = \text{Ind}_\gamma(p) = \text{Ind}_{f\circ\gamma}(0)$, so by 4.37 this leads (via 1.43) to a contradiction of 4.32(i).

Remarks: In Chapter X these equivalences will be augmented by several more, the most dramatic of which is that each component of U is homeomorphic to a disk.

Corollary 4.66 *An open subset U of $\mathbb{C}$ is simply-connected if and only if each of its components is.*

Proof: The components are open and disjoint, so a function on U is continuous if and only if it is continuous on each component. Therefore the corollary follows from 4.65(ii).

Exercise 4.67 *Let U and V be open, simply-connected subsets of $\mathbb{C}$.*
(i) *Show that $U \cap V$ is simply-connected.*

Hint: Use 4.65(vii).
(ii) *Show that if $U \cap V$ is connected then $U \cup V$ is simply-connected. Supply a counterexample to the conclusion when this intersection is not connected.* (*Cf.* 3.40.)

Hints: If $f: U \cup V \to \mathbb{C}\backslash\{0\}$ is continuous, 4.65(ii) provides continuous logarithms ϕ_U, ϕ_V for f on U, V respectively. We have $e^{\phi_U - \phi_V} = 1$ on the connected set $U \cap V$, hence the integer-valued function $(\phi_U - \phi_V)/2\pi i$ is constant there, say N. Then the function ϕ defined by

$$\phi = \begin{cases} \phi_U - 2\pi i N & \text{on } U \\ \phi_V & \text{on } V \end{cases}$$

is a (well-defined) continuous logarithm for f on $U \cup V$.
(iii) *Let U_j ($j = 1, 2, 3$) be simply-connected regions such that $U_1 \cap U_2$ and $U_2 \cap U_3$ are each connected while $U_1 \cap U_3 = \varnothing$. Show that then $U_1 \cup U_2 \cup U_3$ is simply-connected.* (*Cf.* 3.42(ii).)

Hints: The sets $U_1 \cup U_2$ and $U_2 \cup U_3$ are each simply-connected by (ii). Their intersection is the connected set U_2, due to $U_1 \cap U_3 = \varnothing$. Now use (ii) again.

Exercise 4.68 *If U, V are open subsets of $\mathbb{C}$ which are homeomorphic, then one is simply-connected if and only if the other is. In fact, if $\mathbb{C}\backslash V$ has exactly n (finite) bounded components, then so does $\mathbb{C}\backslash U$.*

Hints: If $p_1, \ldots p_n$ are points from each of the bounded components of $\mathbb{C}\backslash V$, $f_j(z) = z - p_j$ ($j = 1, 2, \ldots, n$; $z \in V$) and $h: U \to V$ a homeomorphism, then consider $\Phi_j = f_j \circ h$. Using these functions, it follows from 4.59 and 4.63 that $\mathbb{C}\backslash U$ has at most n bounded components. If the actual number is $m < n$, make the same argument all over with the roles of U, V reversed, to get a contradiction.

Exercise 4.69 *If U is an open, loophomotopically-connected subset of $\mathbb{C}$, then U is simply-connected. In particular, every open starlike set is simply-connected.*

Hints: 4.20 + 4.65. One can argue directly for starlike sets thus: If U is starlike with respect to 0, then for each $z \in \mathbb{C}\backslash U$ all the points tz, $t \geq 1$, also lie in $\mathbb{C}\backslash U$. Hence $\mathbb{C}\backslash U$ has no bounded component.

Remark: The converse of this is proved in 10.3.

Notes to Chapter IV

The ideas in this approach to the index find their appropriate culmination in the theory of covering spaces in topology. There is an important higher dimensional analog of index, called degree (due to Brouwer). See, e.g., HEINZ [1959], chapters 4, 5 of EISENACK and FENSKE [1978] or LLOYD [1978].

For elementary accounts of aspects of plane topology similar in spirit and method to this one see PÁL [1922/23], TUCKER [1945], EILENBERG [1948], FORT [1952], [1967], the last chapter of KURATOWSKI [1972], chapters VI and VII of ČECH [1969] and R. F. BROWN [1974]. The classic KERÉKJÁRTÓ [1923] should also be consulted. (And the reader equipped to do so may also want to look at CSÁSZÁR [1962] and ČERNÝ [1967].) The definitive treatment of the topology of $\mathbb{R}^n$ is probably KURATOWSKI [1968]. His account (especially chapter X) clearly influenced the presentation here. A very readable, less encyclopedic account, with excellent historical notes, which also introduces the reader to algebraic methods, is AGOSTON [1976]. In chapter 7 of his book, among other things relevant to the material here, the reader will find a discussion of the degree of a map mentioned above. Another elegant introduction to algebraic topology is the little book of WALL [1972].

The Greek constituents of "monodromy" show what an apt description of 4.18 this word is. There is also an important holomorphic monodromy theorem, which relates simple-connectivity and analytic continuation. See the Chapter X notes.

For 4.22(i) see p. 162 of EILENBERG [1935].

For a beautiful direct proof of the implication (i) $\Rightarrow$ (iii) in 4.24 (without the intervention of Lebesgue's extension theorem) see pp. 326–327 of KURATOWSKI [1945].

A Banach algebra version of 4.28 will be proved in Chapter XIX.

The important theorems 4.29, 4.30 are from EILENBERG [1936]. There too (p. 90) the independence result 4.32(i) appears (with an inductive proof). See also KURATOWSKI [1945] and [1958]. These results (indeed most material of this chapter, with the notable exception of the Schönflies extension theorem mentioned in 4.44) are valid, appropriately phrased, in $\mathbb{R}^n$ as well as $\mathbb{C} = \mathbb{R}^2$. The important result 4.32(iii), that the number of components of the complement of a compactum in $\mathbb{R}^2$ is a positional invariant, goes back to BROUWER [1912c]. It has many extensions. EILENBERG [1941] showed that if A, B are homeomorphic subsets of the sphere S^n (in $\mathbb{R}^{n+1}$) and $S^n \backslash A$ has finitely many components, then this number equals the number of components of $S^n \backslash B$. (If both numbers are infinite cardinals, however, they need not coincide—he adduces an example.) BORSUK [1950] supplies an elementary proof (no algebraic topology) of this when A is compact (see also p. 47 of LLOYD [1978]) and KURATOWSKI [1950] (see also [1957], [1959]) modifies Borsuk's proof to cover all A. FORT [1961] shows that even in case the number of components of $\mathbb{R}^n \backslash A$ is infinite, this cardinal equals that of the set of components of $\mathbb{R}^n \backslash B$ whenever A, B are bounded, connected open subsets of $\mathbb{R}^n$ which are homeomorphic to each other. (Cf. 4.68.) Finally I mention that if A, B are open, connected subsets of S^2 whose complements have the same finite number N of components, then A and B are homeomorphic. (See ČECH [1969], p. 264.)

The proof of 4.37 is adapted from ČECH [1969]. For more on 4.36(ii) see KAMPKE [1928].

The union of 4.38 and 4.41 is called the Jordan Curve Theorem, even though Jordan's original proof (1887) was incomplete; he at least recognized that something here needed proof in spite of how obvious it all seems. The intuitions according to which everything here was obvious to the 19th century mind were severely jolted in 1903 when Osgood produced a Jordan-curve having non-zero area. (See, e.g., KNOPP [1917].) For the history of the many proofs of the Jordan Curve Theorem which have appeared in the literature see § 13 of BOREL and ROSENTHAL [1924], KLINE [1942] and GUGGENHEIMER [1977]. Most of these proofs are, not to put too fine a point on it, quite unesthetic and equally unenlightening. (The indirect proofs like that in KURATOWSKI [1926] are exceptions to this indictment.) Remnants of the kind of geometric toil with which they abound can be found in a few of the proofs in Chapter IX in which we discuss boundary behavior of conformal maps. The proofs of 4.38–4.42 in the text are essentially from ČECH [1969]. There is a complete converse of the Jordan theorem due to SCHÖNFLIES [1904] (and subsequently elaborated by many others) which affirms that if C is a compact, connected subset of $\mathbb{C}$ and $\mathbb{C} \backslash C$ has exactly two components, for each of which every point of C is an accessible boundary point, then C is a Jordan-curve. See, e.g., M. H. A. NEWMAN [1951] for a complete proof.

Exercise 4.50(iii) is called the Gauss–Lucas theorem. See in connection with it the papers of HAYASHI [1914–15], SPECHT [1959], RUBEL [1966], MARDEN [1966], [1968], [1976] and LABELLE [1971].

The special case of a polynomial in 4.49 is called the Fundamental Theorem of Algebra and indeed it looks like an algebraic result. But, as the conclusion is false if the coefficients and zeros sought are restricted to $\mathbb{Q} + i\mathbb{Q}$, the completeness (order, or metric) of $\mathbb{R}$ plays an essential role. In fact, this is the irreducible kernel of topology that is needed: given that every odd degree polynomial with real coefficients has a real zero [a trivial consequence of the connectedness of the topological space $\mathbb{R}$], the Fundamental Theorem of Algebra can be proved by several purely algebraic arguments. See, e.g., the standard algebra treatises of van der Waerden and Jacobson or J. KÖNIG [1879], NETTO [1880], WALECKI [1883], BRISSE [1886], A. KNESER [1887], VAHLEN [1897], GIUDICE [1907], MERTENS [1911], HOROWITZ [1966] $\subset$ ARTIN and SCHREIER [1926] (exceedingly elegant), VAN DER CORPUT [1950], VITALI [1928], DÖRGE [1928], PUCCIO [1933], H. KNESER [1939], THOMAS [1940], MIGNOSI [1941], EATON [1960], and ZASSENHAUS [1967]. Interesting, more or less constructive proofs occur in MANSION [1880] (which contains a historical appendix), BLUTEL [1911], MONTEL [1913], ROSENBLOOM [1945] and WOLFENSTEIN [1967]. Unusual proofs occur in WEIERSTRASS [1891] (see also *Werke* I, pp. 247–256 and III, pp. 251–269), MÉRAY [1891], JAMET [1894], H. KNESER [1940], IGLISCH [1940], MOHR [1942], [1953a] and SCARF [1952]. A popular proof which recurs repeatedly in the literature uses compactness to select z_0 at which $|P|$ has its minimum, P being a given polynomial of degree $n > 0$. If $P(z_0) \neq 0$, an easy argument then shows that for an appropriate kth root u of 1 ($k \leq n$), the contradiction $|P(z_0 + ru)| < |P(z_0)|$ ensues for sufficiently small $r > 0$. This is supposed to be a totally elementary argument, but very often the writer falls back on the theory of the exponential function to produce such a u. The beautiful proofs in LITTLEWOOD [1941] are exceptions. (Cf. also MANSION, *op. cit.*) Recognizing that the existence of roots of unity is a special case of the theorem to be proved, he uses (in the second of two proofs) a modification of the argument above in conjunction with induction on the degree of P, so arranged that the requisite root of unity is provided by the induction hypothesis. This point is also recognized and circumvented nicely in POMEY [1905], in PRINGSHEIM [1925], pp. 184 ff., in SPIVAK [1967], p. 460 (this is Littlewood's first proof—cf. MÉRAY [1885] and § 3 of MERTENS [1892]), in ESTERMANN [1956] and in REDHEFFER [1964]. The proof in the latter and that in REDHEFFER [1957] uses elementary ideas about harmonic functions and maximum principles (see next chapter); both these papers are recommended reading. SCHMIDT [1936] and WIGERT [1936] are other interesting proofs using harmonic functions. Of course, using results about holomorphic functions, many proofs are possible: the Fundamental Theorem of Algebra is an immediate consequence of (1) Liouville's theorem (3.33) applied to the reciprocal of a zero-free polynomial, (2) the

Maximum Modulus Principle (see 5.15), (3) the Open Map Theorem (5.77), (4) Rouché's theorem (5.87), or (5) the mean value theorem 12.3. See NETTO and LEVAVASSEUR [1907], §§ 80–88 for history of this theorem and references to over 100 proofs and "proofs" of it which have occurred in the literature (up to that time!). See also LORIA [1891], AGOSTINI [1924], GASAPINA [1957], BAŠMAKOVA [1957], [1961], § 5 of SPECHT [1958] and PETROVA [1971], [1973], [1974]. The first proof which is acceptable by modern standards was given by C. F. Gauss, who altogether gave four more-or-less correct proofs, that in the text being a modern variant of one of them. See GAUSS [1890]. Prior to Gauss, J. d'Alembert had given a not entirely rigorous (but "rigorizable") geometric proof and (so) many French writers refer to this result as "le théorème de d'Alembert." (See PETROVA, *op. cit.*)

The result 4.51(iii) is due to KRONECKER [1857]. The suggested proof is from BIEBERBACH [1953] and is quite close to the original. For a generalization see KLOOSTERMAN [1927]. Part (ii) is a trivial consequence of the Fundamental Theorem on Symmetric Polynomials, but the suggested proof is self-contained and avoids appeal to this basic result from algebra (for a proof of which the interested reader may see, e.g., § 25 of PRINGSHEIM [1925].)

The n-dimensional version of 4.53 was secured by KNASTER, KURATOWSKI and MAZURKIEWICZ [1929]. An infinite dimensional version dates from 1937. (See, e.g., p. 51 of GRANAS [1962].) The first part of 4.53 was rediscovered by ABIAN [1961] and other proofs appear in ABIAN and BROWN [1962] and A. B. BROWN [1962].These authors get the conclusion of 4.53 with D replaced by the inside of any Jordan curve, a corollary of 4.53 and 9.19 taken together.

4.57 first occurs in JÜRGENS [1879] but is often attributed to SCHÖNFLIES [1899]. Proofs were also supplied by OSGOOD [1900], F. BERNSTEIN [1900], SCHÖNFLIES [1908] (pp. 162–164), PÁL [1922/23], WINTERNITZ [1927], and NÖBELING [1950]. The n-dimensional version was secured by BROUWER [1912a] and another proof occurs in EILENBERG [1936]. (See also OSTROWSKI [1946].) The proof in the text is from GRANAS [1962], chapter VII. This result is called the "invariance of domain" and is closely related to the problem of the (topological) invariance of the dimension number n in $\mathbb{R}^n$. (Cf. the Remark following 4.58.) The latter problem was treated by J. Lüroth, G. Cantor and others in the late 19th century (see DAUBEN [1975]) but only definitively resolved by BROUWER in [1911], amid an interesting controversy. In a note following Brouwer's in the 1911 *Mathematische Annalen* Henri Lebesgue claimed to give a simple proof of the invariance of the dimension number. This proof was in error at several points and totally inadequate from the point of view of logical completeness. Though Lebesgue never recanted (as Brouwer intensely hoped and campaigned that he would), in subsequent patch-up attempts (none completely successful) he gave birth to the important concept that lies at the basis of modern (covering) dimension theory.

(See SPERNER [1928] and HUREWICZ [1929]. For the fascinating details of the controversy, including the names of other protagonists, the reader should see FREUDENTHAL [1975] and his notes on pp. 435–452 of BROUWER [1976].

Exercise 4.58 was conjectured by Ulam and proved by BORSUK [1933a]. Its meteorological interpretation seems to be due to H. Steinhaus. For history and extensions see § 59.V of KURATOWSKI [1968], as well as HADWIGER [1959], [1960]. E.g., in 1944 H. Hopf proved the following: with f, S as in 4.58, for every $d \in [0, 2]$ there exist points $x, y \in S$ with $\|x - y\| = d$ and $f(x) = f(y)$. See WILLE [1970] and NOWINSKI [1974] for extensions of Hopf's result.

HEILBRONN [1958] gives the following extension of 4.59: for *any* open $U \subset \mathbb{C}$ and continuous $f: U \to \mathbb{C}\backslash\{0\}$ there exist holomorphic $h: U \to \mathbb{C}$ and continuous $\phi: U \to \mathbb{C}$ such that $f = he^{\phi}$. On the theme of 4.63 see RUDIN [1955b].

For extensions of many of the results of § 6 to the infinite dimensional (i.e., Banach space) setting see the beautiful little monograph of GRANAS [1962] or the book of EISENACK and FENSKE [1978].

Chapter V

Consequences of the Cauchy–Goursat Theorem—Maximum Principles and the Local Theory

§ 1 Goursat's Lemma and Cauchy's Theorem for Starlike Regions

Of great utility in the differential calculus of functions on $\mathbb{R}$ are primitives or antiderivatives. The Fundamental Theorem here affirms that integration always produces such primitives. In particular, every continuous function on $\mathbb{R}$ has a primitive. [We will see that this is not so for continuous functions on regions in $\mathbb{C}$.] Naturally we look to integration to produce primitives in the plane too. But now we must face the fact that integration can be carried out over many paths from z_0 to z, whereas if the integrand f has a primitive F, then the integral over any such path must equal $F(z) - F(z_0)$ (see 2.10) and be consequently independent of the path chosen. This independence of path is equivalent to the integral over any *closed* path being 0. Conversely, when the integral is path independent, a primitive can be manufactured by using an "indefinite" integral (see the proof of 2.11). Thus the corresponding fundamental theorem(s) in the plane affirm that certain integrals over closed paths are 0. Everything flows from

Theorem 5.1 (Cauchy–Goursat) *Let U be an open subset of $\mathbb{C}$, f holomorphic in U, $a, b, c \in \mathbb{C}$ and* $\mathrm{co}\{a, b, c\} \subset U$. *Then* $\int_{[a,b,c,a]} f = 0$.

Proof: Write Δ for $\mathrm{co}\{a, b, c\}$ and d for the diameter of Δ. If $d = 0$, then $a = b = c$ and the assertion is trivial. So we suppose $d > 0$. Consider

$$a' = \frac{b+c}{2}, \qquad b' = \frac{a+c}{2}, \qquad c' = \frac{a+b}{2}.$$

By 2.9

$$\int_{[a,b,c,a]} f = \int_{[a,b]} f + \int_{[b,c]} f + \int_{[c,a]} f$$

$$= \int_{[a,c']} f + \int_{[c',b]} f + \int_{[b,a']} f + \int_{[a',c]} f + \int_{[c,b']} f + \int_{[b',a]} f$$

$$= \left(\int_{[a,c']} f + \int_{[c',b']} f + \int_{[b',a]} f\right) + \left(\int_{[c',b]} f + \int_{[b,a']} f + \int_{[a',c']} f\right)$$

$$+ \left(\int_{[a',c]} f + \int_{[c,b']} f + \int_{[b',a']} f\right) + \left(\int_{[c',a']} f + \int_{[a',b']} f + \int_{[b',c']} f\right)$$

(1) $$= \int_{[a,c',b',a]} f + \int_{[c',b,a',c']} f + \int_{[a',c,b',a']} f + \int_{[a',b',c',a']} f.$$

Notice that the sets co$\{a, c', b'\}$, co$\{c', b, a'\}$, co$\{a', c, b'\}$ and co$\{a', b', c'\}$ each lie in co$\{a, b, c\}$ and each has diameter not greater than $\frac{1}{2}$ the diameter of co$\{a, b, c\}$, by 1.19. It follows from (1) that at least one of the integrals on its right side has modulus not less than $\frac{1}{4}|\int_{[a,b,c,a]} f|$; that is, there are points a_1, b_1, c_1 with co$\{a_1, b_1, c_1\} = \Delta_1$ such that

$$(1.1)\quad \begin{cases} \Delta_1 \subset \Delta, \operatorname{diam} \Delta_1 \le 2^{-1} \operatorname{diam} \Delta \quad \text{and} \\ \left| \int_{[a_1,b_1,c_1,a_1]} f \right| \ge 4^{-1} \left| \int_{[a,b,c,a]} f \right|. \end{cases}$$

Proceed inductively to construct, for every positive integer n, points a_n, b_n, c_n with $\Delta_n = $ co$\{a_n, b_n, c_n\}$ satisfying

$$(1.n)\quad \begin{cases} \Delta_n \subset \Delta_{n-1}, \operatorname{diam} \Delta_n \le 2^{-1} \operatorname{diam} \Delta_{n-1} \quad \text{and} \\ \left| \int_{[a_n,b_n,c_n,a_n]} f \right| \ge 4^{-1} \left| \int_{[a_{n-1},b_{n-1},c_{n-1},a_{n-1}]} f \right|. \end{cases}$$

Now by induction on this we have for all positive integers n

$$(2.n)\quad \Delta_n \subset \Delta_{n-1} \subset \cdots \subset \Delta.$$

$$(3.n)\quad \operatorname{diam} \Delta_n \le 2^{-n} d.$$

$$(4.n)\quad \left| \int_{[a_n,b_n,c_n,a_n]} f \right| \ge 4^{-n} \left| \int_{[a,b,c,a]} f \right|.$$

According to 1.19(ii) Δ_n is compact. From (2.n) therefore it follows that there exists $z_0 \in \bigcap_{n=1}^{\infty} \Delta_n \subset U$. Therefore given $\varepsilon > 0$, there exists $r = r(\varepsilon) > 0$ so small that

$$0 < |z - z_0| < r \Rightarrow z \in U \quad \text{and} \quad \left| \frac{f(z) - f(z_0)}{z - z_0} - f'(z_0) \right| \le \varepsilon,$$

that is,

$$(5)\qquad |z - z_0| < r \Rightarrow z \in U \quad \text{and} \quad |f(z) - f(z_0) - f'(z_0)(z - z_0)| \le \varepsilon |z - z_0|.$$

Now it follows from (3.n) that for $n > (\log d/r)/\log 2$ all points of Δ_n lie in $D(z_0, r)$ and so (5) and (3.n) show that

$$(6)\qquad n > \frac{\log d/r}{\log 2} \quad \text{and} \quad z \in \Delta_n \Rightarrow |f(z) - f(z_0) - f'(z_0)(z - z_0)| \le 2^{-n} d\varepsilon.$$

By 2.9(ii) the length of the curve $[a_n, b_n, c_n, a_n]$ is not greater than 3 diam $\Delta_n \le 3 \cdot 2^{-n} d$. It follows then from (6) and 2.9(i) that

$$(7)\qquad n > \frac{\log d/r}{\log 2} \Rightarrow \left| \int_{[a_n,b_n,c_n,a_n]} [f(z) - f(z_0) - f'(z_0)(z - z_0)] dz \right| \le 3 \cdot 4^{-n} d^2 \varepsilon.$$

But $\int_{[a_n,b_n,c_n,a_n]} [f(z_0) + f'(z_0)(z - z_0)]dz = 0$ by 2.10(ii), because any polynomial is a derivative. Thus (7) says that

$$(8) \qquad n > \frac{\log d/r}{\log 2} \Rightarrow \left| \int_{[a_n,b_n,c_n,a_n]} f \right| \le 3 \cdot 4^{-n} d^2 \varepsilon.$$

It follows from this and (4.n) that

$$(9) \qquad \left| \int_{[a,b,c,a]} f \right| \le 3d^2\varepsilon.$$

As $\varepsilon > 0$ is arbitrary in (9), the theorem follows.

Corollary 5.2 (Cauchy's Theorem for Starlike Regions) *If Ω is an open starlike subset of $\mathbb{C}$ and f is holomorphic in Ω, then there is a holomorphic function F in Ω such that $F' = f$. Therefore $\int_\gamma f = 0$ for every piecewise smooth loop γ in Ω.*

Proof: Let Ω be starlike with respect to a. Then $[a, z] \subset \Omega$ for every $z \in \Omega$ and it is possible to form the integrals

$$F(z) = \int_{[a,z]} f \quad \forall z \in \Omega.$$

Let $z_1 \in \Omega$ be given. Take $r > 0$ so that $D(z_1, r) \subset \Omega$ and consider in what follows only $z \in D(z_1, r)$. For such z we have

$$[z_1, z] \subset D(z_1, r) \subset \Omega$$

and this implies that

$$\mathrm{co}\{a, z_1, z\} \subset \Omega.$$

For if $w \in \mathrm{co}\{a, z_1, z\}$, then $w = \lambda_0 a + \lambda_1 z_1 + \lambda_2 z$ for $\lambda_j \ge 0$ with $\lambda_0 + \lambda_1 + \lambda_2 = 1$. If $\lambda_1 + \lambda_2 > 0$, then

$$w = \lambda_0 a + (\lambda_1 + \lambda_2)\left[\frac{\lambda_1}{\lambda_1 + \lambda_2} z_1 + \frac{\lambda_2}{\lambda_1 + \lambda_2} z\right]$$

and, calling the square-bracketed term z', we have that $z' \in [z_1, z] \subset \Omega$, and so $w \in [a, z'] \subset \Omega$. The last theorem therefore applies and gives

$$F(z) - F(z_1) = \int_{[a,z]} f - \int_{[a,z_1]} f = \int_{[z_1,z]} f.$$

It follows just as in the proof of 2.11 that $F'(z_1)$ exists and equals $f(z_1)$. Then $\int_\gamma f = 0$ follows from 2.10(ii).

Exercise 5.3 (Sneak preview) *Deduce from 5.2 and 3.39 that if f is holomorphic in an open set U, then in fact f is infinitely differentiable and even locally of power series type in U.*

Hints: The problem is a local one, so assume U is a disk. Then 5.2 furnishes a primitive F for f. Since $F' = f$, we know F' is continuous. Consequently by 3.39(ii) F has a power series expansion valid throughout U. By 3.6 so does f.

As an immediate and not atypical application of 5.2 we have

Exercise 5.4 (i) *Prove that*

$$\left|\int_{-r}^{r} \frac{\sin x}{x}\,dx - \pi\right| < \frac{\pi}{r} \quad \textit{for all } r > 0.$$

Hints: The function

$$h(z) = \frac{e^{iz} - 1}{z} = \sum_{n=1}^{\infty} \frac{i^n}{n!} z^{n-1}$$

is entire. Therefore for the piecewise smooth loop

$$\Gamma_r(t) = \begin{cases} re^{\pi i t} & 0 \le t \le 1 \\ (2t - 3)r, & 1 \le t \le 2 \end{cases}$$

Cauchy's Theorem for Convex Regions gives $\int_{\Gamma_r} h = 0$. It follows that

$$\int_{-r}^{r} h(x)dx - \pi i \quad = -i\int_0^{\pi} \exp(ire^{ix})dx$$

$$\left|\int_{-r}^{r} h(x)dx - \pi i\right| \le \int_0^{\pi} |\exp(ire^{ix})|dx = \int_0^{\pi} \exp(\mathrm{Re}(ire^{ix}))dx$$

$$= \int_0^{\pi} e^{-r\sin x}dx = 2\int_0^{\pi/2} e^{-r\sin x}dx$$

$$\le 2\int_0^{\pi/2} e^{-2rx/\pi}dx \quad \text{by 3.30}$$

$$= -\frac{\pi}{r} e^{-2rx/\pi}\Big]_0^{\pi/2} < \frac{\pi}{r}.$$

Since

$$\int_{-r}^{r} \frac{\sin x}{x}\,dx - \pi = \mathrm{Im}\left[\int_{-r}^{r} h(x)dx - \pi i\right],$$

the result follows.

(ii) *Show that*

$$\left|\int_{-r}^{r} \left(\frac{\sin x}{x}\right)^2 dx - \pi\right| < \frac{4}{r} \quad \textit{for all } r > 0.$$

Hints: Integrating by parts, we see that

$$\int_{-2r}^{2r} \frac{\sin y}{y}\,dy = \int_{-r}^{r} \frac{\sin 2x}{x}\,dx = 2\int_{-r}^{r} \frac{\sin x \cos x}{x}\,dx$$

$$= \left[\frac{\sin^2 x}{x}\right]_{-r}^{r} + \int_{-r}^{r} \frac{\sin^2 x}{x^2}\,dx = \frac{2\sin^2 r}{r} + \int_{-r}^{r} \left(\frac{\sin x}{x}\right)^2 dx.$$

Therefore from part (i) follows

$$\left|\int_{-r}^{r} \left(\frac{\sin x}{x}\right)^2 dx - \pi\right| = \left|-\frac{2\sin^2 r}{r} + \int_{-2r}^{2r} \frac{\sin y}{y}\,dy - \pi\right|$$

$$\le \frac{2\sin^2 r}{r} + \frac{\pi}{2r} \le \frac{2}{r} + \frac{\pi}{2r} < \frac{4}{r}.$$

(iii) *Show that*

$$\left|\int_{-r}^{r} \frac{\sin^4 x}{x^2}\,dx - \frac{\pi}{2}\right| < \frac{5}{r} \quad \textit{for all } r > 0.$$

Hints:

$$\int_{-r}^{r} \left(\frac{\sin x}{x}\right)^2 dx = \int_{-r}^{r} \frac{\sin^2 x(\sin^2 x + \cos^2 x)}{x^2}\,dx$$

$$= \int_{-r}^{r} \frac{\sin^4 x}{x^2}\,dx + \int_{-r}^{r} \left(\frac{\sin x \cos x}{x}\right)^2 dx\,.$$

$$= \int_{-r}^{r} \frac{\sin^4 x}{x^2}\,dx + \int_{-r}^{r} \left(\frac{\sin 2x}{2x}\right)^2 dx$$

$$= \int_{-r}^{r} \frac{\sin^4 x}{x^2}\,dx + \frac{1}{2}\int_{-2r}^{2r} \left(\frac{\sin y}{y}\right)^2 dy.$$

Therefore

$$\left|\int_{-r}^{r} \frac{\sin^4 x}{x^2}\,dx - \frac{\pi}{2}\right| \le \left|\int_{-r}^{r} \left(\frac{\sin x}{x}\right)^2 dx - \pi\right| + \left|\frac{\pi}{2} - \frac{1}{2}\int_{-2r}^{2r} \left(\frac{\sin y}{y}\right)^2 dy\right|$$

$$< \frac{4}{r} + \frac{1}{r} = \frac{5}{r} \quad \text{by part (ii).}$$

(iv) *Show that*

$$\left|\int_{-r}^{r} \left(\frac{\sin x}{x}\right)^4 dx - \frac{2\pi}{3}\right| < \frac{2}{3r^3} \quad \textit{for all } r > 0.$$

Hints:

$$\int_{-2r}^{2r} \left(\frac{\sin y}{y}\right)^2 dy = 2\int_{-r}^{r} \left(\frac{\sin 2x}{2x}\right)^2 dx = 2\int_{-r}^{r} \left(\frac{\sin x \cos x}{x}\right)^2 dx$$

$$= 2\int_{-r}^{r} [\sin^2 x \cos x]\left[\frac{\cos x}{x^2}\right] dx.$$

Integrate by parts twice to get

$$\int_{-2r}^{2r}\left(\frac{\sin y}{y}\right)^2 dy = \left[2\frac{\sin^3 x\cos x}{3}\frac{}{x^2}\right]_{-r}^{r} - 2\int_{-r}^{r}\frac{\sin^3 x}{3}\left[-\frac{\sin x}{x^2} - \frac{2\cos x}{x^3}\right]dx$$

$$= \frac{4}{3}\frac{\sin^3 r\cos r}{r^2} + \frac{2}{3}\int_{-r}^{r}\frac{\sin^4 x}{x^2}dx + \frac{4}{3}\int_{-r}^{r}\frac{\sin^3 x\cos x}{x^3}dx$$

$$= \frac{4}{3}\frac{\sin^3 r\cos r}{r^2} + \frac{2}{3}\int_{-r}^{r}\frac{\sin^4 x}{x^2}dx$$

$$+ \frac{4}{3}\left[\frac{\sin^4 x}{4x^3}\right]_{-r}^{r} + \int_{-r}^{r}\frac{\sin^4 x}{x^4}dx.$$

It follows then from parts (ii) and (iii) that

$$\lim_{r\to\infty}\int_{-r}^{r}\frac{\sin^4 x}{x^4}dx \quad \text{exists and equals } \frac{2\pi}{3}.$$

Therefore

$$\frac{2\pi}{3} - \int_{-r}^{r}\left(\frac{\sin x}{x}\right)^4 dx = 2\lim_{R\to\infty}\int_0^R\frac{\sin^4 x}{x^4}dx - 2\int_0^r\frac{\sin^4 x}{x^4}dx$$

$$= 2\lim_{R\to\infty}\int_r^R\frac{\sin^4 x}{x^4}dx < 2\lim_{R\to\infty}\int_r^R\frac{1}{x^4}dx$$

$$= 2\lim_{R\to\infty}\left[\frac{1}{3r^3} - \frac{1}{3R^3}\right] = \frac{2}{3r^3}.$$

(v) *Show that*

$$\left|\int_{-r}^{r}\frac{\sin(x^2)}{x}dx - \frac{\pi}{2}\right| < \frac{\pi}{2r^2} \quad \textit{for all } r > 0.$$

Hint: In $\int_{-r^2}^{r^2}(\sin y/y)dy$ set $y = x^2$ and recall part (i).

Lemma 5.5 *Let $c \in \mathbb{C}$, $0 < a < b < \infty$ and g be holomorphic in the annulus $A = \{z \in \mathbb{C}: a < |z - c| < b\}$. Then*
(i) $\int_0^1 g(c + re^{2\pi it})dt$ *is a constant function of $r \in (a, b)$.*
If g is continuous in $D(c, b)$ and holomorphic in $D(c, b)\backslash\{c\}$, then
(ii) $g(c) = \int_0^1 g(c + re^{2\pi it})dt \quad \forall 0 < r < b.$

Proof: We may assume $c = 0$. Define $E(z) = e^z$ for $z \in \Omega = (\log a, \log b) \times \mathbb{R}$. It is easy to see that Ω is open and convex and E maps Ω into A. Let $r, R \in (a, b)$ be given and set

$$\gamma = [\log r, \log R, \log R + 2\pi i, \log r + 2\pi i, \log r].$$

Evidently γ maps into Ω, is piecewise smooth and is closed. Apply 5.2 to the function $f = g \circ E$, which is holomorphic in Ω. We get

$$\begin{aligned} 0 = \int_\gamma f = &\int_0^1 f(t\log R + (1-t)\log r)(\log R - \log r)dt \\ &+ \int_1^2 f(\log R + 2\pi i(t-1))2\pi i dt \\ &+ \int_2^3 f((t-2)\log r + (3-t)\log R + 2\pi i)(\log r - \log R)dt \\ &- \int_3^4 f(\log r + 2\pi i(4-t))2\pi i dt \\ = (\log R - \log r)&\int_0^1 g \circ E(t\log R + (1-t)\log r)dt \\ &+ 2\pi i\int_0^1 g \circ E(\log R + 2\pi i t)dt \\ &+ (\log r - \log R)\int_0^1 g \circ E(t\log r + (1-t)\log R)dt \\ &- 2\pi i\int_0^1 g \circ E(\log r + 2\pi i t)dt, \end{aligned}$$

by changing variables and using the $2\pi i$-periodicity of E. By the change of variable $\tau = 1 - t$ the first integral is found to be equal to the third. Thus we get

$$\int_0^1 g \circ E(\log r + 2\pi i t)dt = \int_0^1 g \circ E(\log R + 2\pi i t)dt,$$

that is,

$$\int_0^1 g(re^{2\pi i t})dt = \int_0^1 g(Re^{2\pi i t})dt.$$

Since r, R are arbitrary in (a, b), part (i) is established.

In the last equation let $R \downarrow 0$ and cite the continuity of g at $c = 0$ to see that

$$\left| \int_0^1 g(Re^{2\pi i t})dt - g(0) \right| = \left| \int_0^1 [g(Re^{2\pi i t}) - g(0)]dt \right| \leq \sup_{|z| \leq R} |g(z) - g(0)|$$

tends to 0, thus proving (ii).

§ 2 Maximum Principles

With a view toward later applications it is convenient to atomize the theory and proceed a bit more generally here than our immediate goals dictate. For the moment the next definition serves merely as a convenient abbreviation; the rationale for the terminology will emerge later.

Definition 5.6 Let U be an open subset of $\mathbb{C}$, $f\colon U \to \mathbb{R}$ a continuous function. Say that f is *subharmonic in U* if

$$(*) \qquad f(a) \leq \frac{1}{2\pi}\int_0^{2\pi} f(a + re^{i\theta})d\theta$$

whenever $\bar{D}(a, r) \subset U$. Say that f is *superharmonic in U* if $-f$ is subharmonic in U; i.e., if the reverse inequality always obtains in (*). The integrals in (*) are called *circumferential means* of f.

Exercise 5.7 *Let U be an open subset of $\mathbb{C}$, $F\colon U \to \mathbb{C}$ holomorphic. Show that $f = \operatorname{Re} F$ is both subharmonic and superharmonic in U and that $|F|$ is subharmonic in U.*

Hint: Whenever $\bar{D}(a, r) \subset U$ use 5.5(ii) to show that equality holds in (*) above. As to $|F|$, pass $|\ |$ through the integral in 5.5(ii).

Remark: In 5.21(ii) we will see that for harmonic functions equality holds in (*). The proof suggested for 5.7 then shows that the modulus of a harmonic function is subharmonic.

Theorem 5.8 (Maximum Principle) *Let Ω be a region, f a subharmonic function in Ω. If f attains a maximum in Ω, then f is constant.*

Proof: Suppose f attains a maximum over Ω at some point $a \in \Omega$. Then whenever $R > 0$ is such that $D(a, R) \subset \Omega$, we have from (*) above that

$$0 \leq \int_0^{2\pi} [f(a + re^{i\theta}) - f(a)]d\theta, \quad 0 \leq r < R.$$

On the other hand, $f - f(a) \leq 0$ throughout Ω, by definition of a, so the integrand above is non-positive. Being continuous, its integral can only be non-negative if it is identically 0. We infer that

$$f(a + re^{i\theta}) - f(a) = 0 \quad \forall \theta \in [0, 2\pi].$$

This holds for all $r \in [0, R)$, so we conclude that

$$f(z) = f(a) \quad \forall z \in D(a, R).$$

Let Ω_0 be the set of all points at which f attains its maximum value, M say, over Ω. Thus $\Omega_0 = f^{-1}(M)$ and by hypothesis Ω_0 is not empty. Since f is continuous, Ω_0 is relatively closed in Ω. The result of the first paragraph says that Ω_0 is open. Since Ω is connected, it follows that $\Omega_0 = \Omega$.

Scholium This proof does not use the full force of the definition 5.6. In fact an examination of the proof shows that the desired conclusion follows if f is continuous and for each $a \in \Omega$ there is *some* $r_a > 0$ such that inequality (*) of 5.6 holds for all $r < r_a$. (Cf. 5.23 for a further weakening.)

Corollary 5.9 *If U is a bounded open subset of $\mathbb{C}$, $f\colon \overline{U} \to \mathbb{R}$ is continuous and f is subharmonic in U, then*

$$\sup f(U) = \sup f(\partial U).$$

Proof: Since the boundary of each component of U lies wholly in ∂U, it suffices to treat connected U. $\overline{U}$ is compact, so f attains a maximum in $\overline{U}$. If this occurs in U, then f is constant in U (hence in $\overline{U}$) by 5.8. Otherwise this maximum occurs on ∂U. In either case the asserted equality is obvious.

Corollary 5.10 (Maximum Modulus Principle) *Let Ω be a region, F a holomorphic function in Ω. If* Re F, $|\text{Re } F|$ *or* $|F|$ *attains a maximum in Ω, then F is constant.*

Proof: If $|\text{Re } F|$ attains a maximum in Ω, then this is either the maximum of Re F or of $-\text{Re } F = \text{Re}(-F)$. In either case Re F is constant by 5.7 and 5.8. Then F is constant by 2.17(ii).

If $|F|$ attains a maximum in Ω at z_0, write $F(z_0) = e^{i\theta}|F(z_0)|$ for some real θ and observe that the maximum of the function $\text{Re}(\theta^{-i\theta}F)$ in Ω is attained at z_0 and is $|F(z_0)|$. From the result of the first paragraph it follows that $e^{-i\theta}F$, whence F itself, is constant.

Corollary 5.11 *Let Ω be a region, $f\colon \Omega \to \mathbb{R}$ harmonic. If either f or $|f|$ attains a maximum in Ω, then f is constant.*

Proof: As in the proof of 5.10, it suffices to treat the case where f attains a maximum in Ω. In this case cite 5.5(ii) and the Scholium above.

Theorem 5.12 *Let U be a bounded open subset of $\mathbb{C}$, h subharmonic in U and suppose that $\overline{\lim}_{z \to z_0} h(z) \le M$ for every $z_0 \in \partial U$. Then h is bounded above by M in U. In particular, if in addition h is continuous on $\overline{U}$, then* $\max h(\overline{U}) = \max h(\partial U)$.

Proof: Since, as noted several times before, the boundary of each component of U lies in that of U, the hypothesis is also satisfied in each component of U. Therefore it is enough to treat the case of connected U. Let $M_0 = \sup\{h(z) : z \in U\}$. This is non-negative and possibly infinite. In any case there exists a sequence $\{z_n\} \subset U$ such that $h(z_n) \to M_0$. As $\overline{U}$ is compact, a subsequence of $\{z_n\}$, say without loss of generality $\{z_n\}$ itself, converges to an element $z_0 \in \overline{U}$. If $z_0 \in U$, then M_0 is finite and $h(z_0) = M_0$ by continuity of h at z_0. Then by the Maximum Principle h is constant in U. Of course that constant is not greater than M. (Why?) If $z_0 \in \overline{U} \backslash U = \partial U$, then by hypothesis $M_0 = \lim_{n \to \infty} h(z_n) \le M$.

Remark: The conclusion $h \leq M$ throughout U still obtains if the hypothesis that U be bounded is dropped, provided we interpret the other hypothesis to include the condition that

$$\overline{\lim_{\substack{|z|\to\infty \\ z\in U}}} h(z) \leq M.$$

The above proof needs only trifling changes. See 7.15 for another extension (which subsumes the next theorem as well).

Theorem 5.13 *Let U be an open proper subset of $\mathbb{C}$, f a bounded holomorphic function in U and suppose that $\overline{\lim}_{z\to\xi}|f(z)| \leq B$ for every $\xi \in \partial U$. Then f is bounded by B throughout U.*

Proof: Replacing B by $B + \varepsilon$, $\varepsilon > 0$, it is enough to suppose that

(1) $\quad \overline{\lim_{z\to\xi}}|f(z)| < B \quad \forall \xi \in \partial U$

and to show that this implies $|f| \leq B$ throughout U. (Then let $\varepsilon \downarrow 0$.) Now U is open but not closed (since $\mathbb{C}$ is connected), so $\partial U = \bar{U}\backslash U \neq \varnothing$. Let $b \in \partial U$. Then by (1) there is an open neighborhood V of b such that

(2) $\quad |f| < B \quad \text{in } U \cap V.$

Since V is a neighborhood of a boundary point, this intersection is not empty and so it contains the closure of some non-empty open disk D. Now

$$\begin{aligned}\partial(U\backslash\bar{D}) &= \overline{U\backslash\bar{D}}\backslash(U\backslash\bar{D}) \subset (\bar{U}\backslash D)\backslash(U\backslash\bar{D}) \subset (\bar{U}\backslash U) \cup \bar{D} \\ &= \partial U \cup \bar{D} \subset (\partial U) \cup (V \cap U).\end{aligned}$$

Therefore by (1) and (2) we have

(1)′ $\quad \overline{\lim_{z\to\xi}}|f(z)| \leq B \quad \forall \xi \in \partial(U\backslash\bar{D}).$

Since we already have $|f| \leq B$ in $\bar{D}$ and (1)′ says that our original hypothesis is fulfilled by $U\backslash\bar{D}$ in the role of U, it suffices to prove the whole theorem for $U\backslash\bar{D}$, i.e., to suppose that there is a non-empty open disk D whose closure is disjoint from U. This we now assume, as well as the strict inequality (1). Consideration of $F(z) = f(c + Rz)$ for appropriate c and $R > 0$ shows that we may further assume D to be $D(0, 1)$.

Let $D_r = D(0, r)$ and consider a fixed $z_1 \in U$. For $r > |z_1|$, D_r contains z_1. So by 5.12

(3) $\quad |f^n(z_1)/z_1| \leq M_n$

where we may take for M_n any number satisfying

(4) $\quad \overline{\lim_{z\to\xi}}\left|\dfrac{f^n(z)}{z}\right| \leq M_n \quad \forall \xi \in \partial(U \cap D_r).$

(Notice that $0 \notin U$, so $f^n(z)/z$ is holomorphic in $U \cap D_r$.) Now we have

$$(5) \qquad \partial(U \cap D_r) = \overline{U \cap D_r} \backslash U \cap D_r = (\overline{U \cap D_r} \backslash U) \cup (\overline{U \cap D_r} \backslash D_r)$$
$$\subset (\overline{U} \backslash U) \cup (\overline{D}_r \backslash D_r) = \partial U \cup \partial D_r.$$

Consider any $\xi \in \partial(U \cap D_r)$ and any $z \in U \cap D_r$ convergent to ξ. We have $|z| > 1$, since U is disjoint from the closed unit disk. Glancing back at (1) we see therefore that if $\xi \in \partial U$,

$$\overline{\lim_{z \to \xi}} \left| \frac{f^n(z)}{z} \right| \leq B^n.$$

But if ξ is not in ∂U, then by (5) it is in ∂D_r, $|\xi| = r$, and we get

$$\overline{\lim_{z \to \xi}} \left| \frac{f^n(z)}{z} \right| \leq \frac{M^n}{r},$$

where M is some bound for $|f|$ throughout U. It follows that $\max\{B^n, M^n/r\}$ is an eligible M_n in (4). We have then from (3)

$$|f(z_1)| \leq [|z_1| \max\{B^n, M^n/r\}]^{1/n} = |z_1|^{1/n} \max\{B, Mr^{-1/n}\},$$

valid for every positive integer n and all $r > |z_1|$. Let $r \to \infty$ and get

$$|f(z_1)| \leq B|z_1|^{1/n}$$

and then let $n \to \infty$ and get (3.1)

$$|f(z_1)| \leq B.$$

Theorem 5.14 (LINDELÖF [1915]) *Let U be an open subset of $\mathbb{C}$, $z_0 \in U$, $r > 0$ such that $(\mathbb{C} \backslash \overline{U}) \cap C(z_0, r)$ contains an arc of length $2\pi r/k$ for some positive integer k. Let f be holomorphic and bounded by M in U and for some $\varepsilon \geq 0$ satisfy*

$$(*) \qquad \overline{\lim_{z \to \xi}} |f(z)| \leq \varepsilon \quad \forall \xi \in D(z_0, r) \cap \partial U.$$

Then

$$|f(z_0)| \leq (\varepsilon M^{k-1})^{1/k}.$$

Proof: We may suppose that $z_0 = 0$ and that

$$(1) \qquad \mathbb{C} \backslash \overline{U} \supset \left\{ re^{i\theta} : 0 \leq \theta \leq \frac{2\pi}{k} \right\} = A.$$

For each $j = 1, 2, \ldots, k$ let

$$U_j = e^{2\pi i j/k} U, \qquad A_j = e^{2\pi i j/k} A,$$

$$U_0 = \bigcap_{j=1}^{k} U_j.$$

The arc A_j lies outside $\overline{U}_j$ and so

$$C(0, r) = \bigcup_{j=1}^{k} A_j \text{ lies outside } \bigcap_{j=1}^{k} \overline{U}_j,$$

that is,

$$\overline{U}_0 \subset \mathbb{C}\backslash C(0, r).$$

Consequently, if Ω_0 is the component of U_0 which contains 0, we have

$$(2) \qquad \overline{\Omega}_0 \subset D(0, r).$$

Now

$$(3) \qquad \partial\Omega_0 \subset \bigcup_{j=1}^{k} \partial U_j,$$

since if every neighborhood N of a point z meets both Ω_0 and

$$\mathbb{C}\backslash\Omega_0 \subset \bigcup_{j=1}^{k} (\mathbb{C}\backslash U_j),$$

then every such N meets some one $\mathbb{C}\backslash U_j$. From (2) and (3) we have

$$(4) \qquad \partial\Omega_0 \subset \bigcup_{j=1}^{k} D(0, r) \cap \partial U_j = \bigcup_{j=1}^{k} e^{2\pi ij/k}[D(0, r) \cap \partial U].$$

We form

$$(5) \qquad F(z) = \prod_{j=1}^{k} f(ze^{-2\pi ij/k}), \quad z \in \Omega_0.$$

Then any $\xi \in \partial\Omega_0$ satisfies $\xi e^{-2\pi ij/k} \in D(0, r) \cap \partial U$ for some j. Consequently, since $e^{-2\pi ij/k}\Omega_0 \subset U$,

$$\begin{aligned}\overline{\lim_{\substack{z\to\xi\\ z\in\Omega_0}}} |f(ze^{-2\pi ij/k})| &= \overline{\lim_{\substack{ze^{-2\pi ij/k}\to\xi e^{-2\pi ij/k}\\ z\in\Omega_0}}} |f(ze^{-2\pi ij/k})| \\ &\le \overline{\lim_{\substack{w\to\xi e^{-2\pi ij/k}\\ w\in U}}} |f(w)| \overset{(*)}{\le} \varepsilon.\end{aligned}$$

Therefore

$$(6) \qquad \overline{\lim_{\substack{z\to\xi\\ z\in\Omega_0}}} |F(z)| \le \varepsilon M^{k-1} \quad \forall \xi \in \partial\Omega_0.$$

Since Ω_0 is bounded (see (2)), it follows from (6) and 5.12 that

$$|F| \le \varepsilon M^{k-1} \text{ throughout } \Omega_0.$$

In particular,

$$|f(0)|^k = |F(0)| \le \varepsilon M^{k-1}.$$

Exercise 5.15 *Deduce the Fundamental Theorem of Algebra from the Maximum Modulus Principle.*

Hint: If the non-constant polynomial p has no zero, then the holomorphic function $1/p$ is small on the boundaries of large disks.

Exercise 5.16 (LINDELÖF [1909] *and* [1915]) *Let* $\theta_1 < \theta_2$ *with* $\theta_2 - \theta_1 \le \pi$, $S = \{re^{i\theta}: r > 0, \theta_1 < \theta < \theta_2\}$, $f: \bar{S} \to \mathbb{C}$ *bounded and continuous, holomorphic in* S.

(i) *Suppose that for some* m *we have*

$$\overline{\lim_{r\to\infty}}|f(re^{i\theta_1})| \le m \quad \text{and} \quad \overline{\lim_{r\to\infty}}|f(re^{i\theta_2})| \le m.$$

Show that then $\overline{\lim_{\substack{|z|\to\infty \\ z\in S}}} |f(z)| \le m.$

Hints: After a rotation we may assume that $\theta_2 = -\theta_1 > 0$. Let M be a bound for f in S. Given $0 < \varepsilon < M/(m+1)$, there exists by hypothesis $r_1 = r_1(\varepsilon)$ such that

(1) $\quad |f(re^{\pm i\theta_1})| \le m + \varepsilon \quad \forall r \ge r_1.$

Set $\lambda = \lambda(\varepsilon) = r_1 M/(m + \varepsilon)$ and form

(2) $\quad F_\varepsilon(z) = \dfrac{z}{z+\lambda} f(z), \quad z \in \bar{S}.$

Since $\theta_2 - \theta_1 = 2\theta_2 \le \pi$, we have $(-\infty, 0) \cap \bar{S} = \varnothing$, so F_ε is well-defined and (so) continuous on $\bar{S}$, holomorphic in S. We have

(3) $\quad |F_\varepsilon(re^{i\theta})| = \dfrac{r}{\sqrt{r^2 + 2r\lambda\cos\theta + \lambda^2}}|f(re^{i\theta})| \le \dfrac{r}{\sqrt{r^2+\lambda^2}}|f(re^{i\theta})|$

for $r \ge 0$ and $|\theta| \le \theta_2 \le \pi/2$. In particular,

(4) $\quad |F_\varepsilon| \le M \quad \text{in } \bar{S}$

and from (1) and (3)

(5) $\quad |F_\varepsilon(re^{\pm i\theta_1})| \le m + \varepsilon \quad \forall r \ge r_1.$

Furthermore, from the definition of λ

$$\frac{r_1}{\sqrt{r_1^2+\lambda^2}} \le \frac{r_1}{\lambda} = \frac{m+\varepsilon}{M},$$

so (3) and the definition of M give

(6) $\quad |F_\varepsilon(r_1 e^{i\theta})| \le m + \varepsilon \quad \forall |\theta| \le \theta_2.$

With (4), (5) and (6) in hand we cite 5.13 to conclude that

$$|F_\varepsilon| \le m + \varepsilon \quad \text{throughout } S\backslash\bar{D}(0, r_1),$$

and so

$$|f(z)| = \left|1 + \frac{\lambda}{z}\right| |F_\varepsilon(z)| \le \left(1 + \frac{\lambda}{|z|}\right)(m + \varepsilon) < (1 + \varepsilon)(m + \varepsilon)$$

for all $z \in S$ with $|z| > \lambda/\varepsilon = r_1(\varepsilon)M/\varepsilon(m + \varepsilon)$.

(ii) *Suppose that for some $a, b \in \mathbb{C}$ we have*

$$(*) \qquad \lim_{r \to \infty} f(re^{i\theta_1}) = a, \qquad \lim_{r \to \infty} f(re^{i\theta_2}) = b.$$

Show that $a = b$ and that $\lim_{|z| \to \infty, z \in S} f(z) = a$.

Hints: Apply (i) with the function $[f - a][f - b]$ in the role of the f there to learn that

$$(7) \qquad \lim_{\substack{|z| \to \infty \\ z \in S}} |f(z) - a|\,|f(z) - b| = 0.$$

Given $\varepsilon > 0$, there exists by (*) and (7) an $R > 0$ such that

$$(8) \qquad |f(re^{i\theta}) - a|\,|f(re^{i\theta}) - b| < \varepsilon^2 \quad \forall r \ge R,\ \theta \in [\theta_1, \theta_2], \text{ and}$$

$$(9) \qquad |f(re^{i\theta_1}) - a| < \varepsilon, \quad |f(re^{i\theta_2}) - b| < \varepsilon \quad \forall r \ge R.$$

Therefore the sets

$$A = \{\theta \in [\theta_1, \theta_2] : |f(Re^{i\theta}) - a| \le \varepsilon\}$$
$$B = \{\theta \in [\theta_1, \theta_2] : |f(Re^{i\theta}) - b| \le \varepsilon\}$$

are each closed (continuity of f) and non-void (by (9)) and by (8)

$$A \cup B = [\theta_1, \theta_2].$$

Since the interval is connected, we cannot therefore have $A \cap B = \varnothing$. So let $\theta \in A \cap B$. Then

$$|a - b| \le |f(Re^{i\theta}) - a| + |b - f(Re^{i\theta})|$$
$$\le \varepsilon + \varepsilon, \quad \text{since } \theta \in A \cap B.$$

As $\varepsilon > 0$ is arbitrary, we conclude that $a = b$.

Remarks: The restriction $\theta_2 - \theta_1 \le \pi$ is only a convenience. For a holomorphic square root transformation (see 3.43) can be employed to map S into a half-plane if $\theta_2 - \theta_1 \le 2\pi$. Moreover, the result (ii) follows with no restrictions from two citations of 12.30. See 9.20 for a more drastic permissible change in the geometry of the problem and 5.56, 12.31 for related results. Even without hypothesis (*) one has $f^{(k)}(re^{i\theta}) \to 0$ as $r \to \infty$ for each $\theta \in [\theta_1, \theta_2]$, $k \ge 1$. (See p. 18 of PAINLEVÉ [1888], ROGOSINSKI [1945], EGGLESTON [1951] and 6.23(iii) below.)

There will be more maximum modulus results but they must await further developments.

§ 3 The Dirichlet Problem for Disks

The utility of the mean value formula 5.5(ii) was well confirmed by the results of the previous section. Such an integral representation formula is likely to be even more useful if it can be made to yield the value of the function at any point a inside the disk as an integral around the boundary circle. Such a representation is not hard to find. In effect all we do is move 0 to a by a map T of the disk and then apply 5.5(ii) to the composite. Here are the details.

Theorem 5.17 (Poisson Formula) *Let $D = D(0, 1)$, f be holomorphic in D and continuous on $\bar{D}$. Then*

$$f(a) = \frac{1}{2\pi}\int_0^{2\pi} \operatorname{Re}\left[\frac{e^{i\theta} + a}{e^{i\theta} - a}\right] f(e^{i\theta})d\theta \quad \forall a \in D.$$

Proof: Let such an f and such an a be given and consider

$$T(z) = \frac{z + a}{1 + \bar{a}z}, \quad z \in \bar{D}.$$

As noted in 2.6, T maps D conformally onto D and $\bar{D}$ homeomorphically onto $\bar{D}$. Therefore 5.5(ii) is applicable to $g = f \circ T$ and it yields

$$g(0) = \frac{1}{2\pi}\int_0^{2\pi} g(re^{it})dt \quad \forall 0 < r < 1.$$

Use uniform continuity of g on $\bar{D}$ and let $r \uparrow 1$ to get

$$(1) \quad f(a) = g(0) = \frac{1}{2\pi}\int_0^{2\pi} g(e^{it})dt = \frac{1}{2\pi}\int_0^{2\pi} f(T(e^{it}))dt.$$

Now 4.1 supplies us with a smooth $\phi: [0, 2\pi] \to \mathbf{C}$ such that

$$(2) \quad e^{i\phi(t)} = T(e^{it}), \quad 0 \le t \le 2\pi.$$

We compute a little with ϕ and T. From (2) and the Chain Rule

$$i\phi'(t)e^{i\phi(t)} = T'(e^{it})ie^{it}$$

$$(3) \quad \phi'(t) = \frac{T'(e^{it})}{e^{i\phi(t)}}e^{it} = \frac{T'(e^{it})}{T(e^{it})}e^{it} \quad \forall t \in [0, 2\pi].$$

On the other hand,

$$T'(z) = \frac{1 - |a|^2}{(1 + \bar{a}z)^2} \quad \forall z \in \bar{D},$$

$$(4) \quad \frac{T'(z)}{T(z)} = \frac{1 - |a|^2}{(1 + \bar{a}z)(z + a)} = \frac{1 - |a|^2}{z(1/z + \bar{a})(z + a)} = \frac{1 - |a|^2}{z|z + a|^2} \quad \forall z \in C(0, 1).$$

From (3) and (4) we get

$$(5) \quad \phi'(t) = \frac{1 - |a|^2}{|e^{it} + a|^2} \quad \forall t \in [0, 2\pi].$$

In particular ϕ is strictly increasing, so

$$\frac{\phi(2\pi) - \phi(0)}{2\pi} > 0.$$

On the other hand, by 4.7(ii) [with $\Gamma_1 = C(0, 1)$, $\Gamma_2 = T \circ \Gamma_1$, and f the identity function there], this quotient has modulus 1. We infer that

(6) $\quad \phi(2\pi) - \phi(0) = 2\pi.$

Since ϕ is strictly increasing, we can form ϕ^{-1}. We have from (2) and a simple calculation

$$e^{i\phi^{-1}(s)} = T^{-1}(e^{is}) = \frac{e^{is} - a}{1 - \bar{a}e^{is}},$$

(7) $$e^{i\phi^{-1}(s)} + a = \frac{1 - |a|^2}{e^{-is} - \bar{a}} \quad \forall s \in [\phi(0), \phi(2\pi)].$$

We are now in a position to make a change of variable in the integral in (1):

$$f(a) = \frac{1}{2\pi}\int_0^{2\pi} f(e^{i\phi(t)})dt = \frac{1}{2\pi}\int_{\phi(0)}^{\phi(2\pi)} f(e^{is})[\phi^{-1}]'(s)ds$$

$$= \frac{1}{2\pi}\int_{\phi(0)}^{\phi(2\pi)} f(e^{is})\frac{ds}{\phi'(\phi^{-1}(s))} \overset{(5)}{=} \frac{1}{2\pi}\int_{\phi(0)}^{\phi(2\pi)} f(e^{is})\frac{|e^{i\phi^{-1}(s)} + a|^2}{1 - |a|^2}ds$$

$$\overset{(7)}{=} \frac{1}{2\pi}\int_{\phi(0)}^{\phi(2\pi)} f(e^{is})\frac{1 - |a|^2}{|e^{-is} - \bar{a}|^2}ds$$

(8) $$= \frac{1}{2\pi}\int_{\phi(0)}^{\phi(2\pi)} f(e^{is})\,\mathrm{Re}\left[\frac{e^{is} + a}{e^{is} - a}\right]ds.$$

In the last step a direct calculation confirms that for $z \in C(0, 1)$

$$2\,\mathrm{Re}\left[\frac{z + a}{z - a}\right] = \frac{z + a}{z - a} + \frac{\bar{z} + \bar{a}}{\bar{z} - \bar{a}} = \frac{2(1 - |a|^2)}{|\bar{z} - \bar{a}|^2}, \quad \text{since } z\bar{z} = 1.$$

Finally, because all the functions in the integrand of (8) are 2π-periodic, we may replace the interval of integration by

$$[0, \phi(2\pi) - \phi(0)] \overset{(6)}{=} [0, 2\pi]$$

and the desired integral formula for $f(a)$ is at hand.

Corollary 5.18 (Cauchy–Schwarz Formula) *If f is holomorphic in $D = D(0, 1)$ and continuous on $\bar{D}$, then*

$$f(z) = i\,\mathrm{Im}\,f(0) + \frac{1}{2\pi}\int_0^{2\pi}\left[\frac{e^{it} + z}{e^{it} - z}\right]\mathrm{Re}\,f(e^{it})dt \quad \forall z \in D.$$

Proof: According to 5.17 we have

$$(1)\qquad f(z) = \frac{1}{2\pi}\int_0^{2\pi} \mathrm{Re}\left[\frac{e^{it}+z}{e^{it}-z}\right] f(e^{it})\,dt \quad \forall z \in D.$$

Upon taking real parts of both sides

$$(2)\qquad \mathrm{Re}\, f(z) = \mathrm{Re}\left[\frac{1}{2\pi}\int_0^{2\pi} \frac{e^{it}+z}{e^{it}-z}\, \mathrm{Re}\, f(e^{it})\,dt\right].$$

Now according to 2.14 the expression

$$F(z) = \frac{1}{2\pi}\int_0^{2\pi} \frac{e^{it}+z}{e^{it}-z}\, \mathrm{Re}\, f(e^{it})\,dt$$

$$(3)\qquad = \frac{1}{2\pi}\int_0^{2\pi} \frac{e^{it}\, \mathrm{Re}\, f(e^{it})\,dt}{e^{it}-z} + \frac{z}{2\pi}\int_0^{2\pi} \frac{\mathrm{Re}\, f(e^{it})\,dt}{e^{it}-z}$$

is a holomorphic function of $z \in D$. By (2) we have $\mathrm{Re}\, F = \mathrm{Re}\, f$, so from 2.17(ii) we infer that

$$(4)\qquad f - F \equiv f(0) - F(0) \quad \text{throughout } D.$$

But by (3)

$$F(0) = \frac{1}{2\pi}\int_0^{2\pi} \mathrm{Re}\, f(e^{it})\,dt = \mathrm{Re}\left[\frac{1}{2\pi}\int_0^{2\pi} f(e^{it})\,dt\right] \stackrel{(2)}{=} \mathrm{Re}\, f(0).$$

Therefore $f(0) - F(0) = i\, \mathrm{Im}\, f(0)$ and the conclusion follows from (4).

Scholium It follows from (3) and (the proof of) 3.8 that f equals a power series about 0 in D.

Definition 5.19 Let U be an open subset of $\mathbb{C}$. The *Dirichlet problem for U* (also called the *first boundary value problem of potential theory*) is the problem of finding for any continuous $h: \partial U \to \mathbb{R}$ a continuous extension $H: \overline{U} \to \mathbb{R}$ of h which is harmonic in U.

Solving the Dirichlet problem for a disk is easy. For 5.17 tells us that if h is harmonic in $D = D(0, 1)$ and is given by $h = \mathrm{Re}\, f$ on $\overline{D}$, where f is holomorphic in D and continuous on $\overline{D}$, then

$$h(z) = \frac{1}{2\pi}\int_0^{2\pi} \mathrm{Re}\left[\frac{e^{it}+z}{e^{it}-z}\right] h(e^{it})\,dt \quad \forall z \in D.$$

Naturally, we will use this integral to create a harmonic function $h(z)$ when all that is given is a continuous function h on ∂D.

Theorem 5.20 (Dirichlet Problem for the Disk) *Let $D = D(0, 1)$, $h: \partial D \to \mathbb{R}$ be continuous and define*

$$f(z) = \begin{cases} \dfrac{1}{2\pi}\displaystyle\int_0^{2\pi} \frac{e^{it} + z}{e^{it} - z} h(e^{it})dt, & z \in D \\ h(z) & z \in \partial D. \end{cases}$$

Then f equals a power series about 0 in D and $\operatorname{Re} f$ is continuous on $\bar{D}$.

Proof: For fixed $z \in D$ the series $\sum_{k=0}^{\infty} (ze^{-it})^k$ converges uniformly in $t \in [0, 2\pi]$ to $1/(1 - ze^{-it})$. Therefore the series

(*) $\quad 1 + \sum_{k=1}^{\infty} 2(ze^{-it})^k$ converges uniformly in $t \in [0, 2\pi]$ to $\dfrac{e^{it} + z}{e^{it} - z}$.

It follows that

$$(1) \quad f(z) = \frac{1}{2\pi}\int_0^{2\pi} h(e^{it})dt + \sum_{k=1}^{\infty} \frac{1}{2\pi}\int_0^{2\pi} 2(ze^{-it})^k h(e^{it})dt = \sum_{k=0}^{\infty} c_k z^k,$$

where

$$c_0 = \frac{1}{2\pi}\int_0^{2\pi} h(e^{it})dt \quad \text{and} \quad c_k = \frac{1}{\pi}\int_0^{2\pi} e^{-ikt} h(e^{it})dt \quad \text{for } k > 1.$$

Now

$$(2) \quad 2\operatorname{Re}\left[\frac{e^{it} + z}{e^{it} - z}\right] = \frac{e^{it} + z}{e^{it} - z} + \frac{e^{-it} + \bar{z}}{e^{-it} - \bar{z}} = 2\frac{1 - |z|^2}{|e^{it} - z|^2}$$

and so, recalling that h is real-valued,

$$(3) \quad \operatorname{Re} f(z) = \frac{1}{2\pi}\int_0^{2\pi} \operatorname{Re}\left[\frac{e^{it} + z}{e^{it} - z}\right] h(e^{it})dt = \frac{1}{2\pi}\int_0^{2\pi} \frac{1 - |z|^2}{|e^{it} - z|^2} h(e^{it})dt.$$

It remains only to show that for each $u \in D(0, 1)$,

$$\lim_{\substack{z \to u \\ z \in \bar{D}}} \operatorname{Re} f(z) \text{ exists and equals } h(u).$$

Moreover, because of the definition of f on the boundary of $\bar{D}$ and the continuity of h, it is enough to establish this limit relation for approach from within D. If we write $u = e^{is}$, then for each $z \in D$

$$f(zu) = \frac{1}{2\pi}\int_0^{2\pi} \frac{e^{it} + ze^{is}}{e^{it} - ze^{is}} h(e^{it})dt = \frac{1}{2\pi}\int_0^{2\pi} \frac{e^{i(t-s)} + z}{e^{i(t-s)} - z} h(e^{it})dt$$

$$= \frac{1}{2\pi}\int_0^{2\pi} \frac{e^{it} + z}{e^{it} - z} h(e^{i(t+s)})dt = \frac{1}{2\pi}\int_0^{2\pi} \frac{e^{it} + z}{e^{it} - z} h(ue^{it})dt.$$

We see therefore, by replacing h with the function $h_u(\xi) = h(u\xi)$, that it is enough to deal with the case $u = 1$. We return to (*) and notice via (2) that for $z \in D$

$$\int_0^{2\pi} \frac{1-|z|^2}{|e^{it}-z|^2}\,dt = \operatorname{Re}\int_0^{2\pi} \frac{e^{it}+z}{e^{it}-z}\,dt$$

$$(4) \qquad = \operatorname{Re}\left[\int_0^{2\pi} 1\,dt + \sum_{k=1}^{\infty}\int_0^{2\pi} 2(ze^{-it})^k\,dt\right] = 2\pi.$$

From (3) and (4) we have for $z \in D$.

$$\operatorname{Re} f(z) - h(1) = \frac{1}{2\pi}\int_0^{2\pi} \frac{1-|z|^2}{|e^{it}-z|^2}[h(e^{it}) - h(1)]\,dt$$

$$(5) \qquad |\operatorname{Re} f(z) - h(1)| \le \frac{1}{2\pi}\int_0^{2\pi} \frac{1-|z|^2}{|e^{it}-z|^2}|h(e^{it}) - h(1)|\,dt.$$

We are to show that this expression is small for $z \in D$ close to 1. What follows is a standard "approximate identity" argument. Let $M > 0$ be a bound for the continuous function $|h|$. If $\varepsilon > 0$ is given, pick $\pi > \delta > 0$ so that

$$(6) \qquad |h(e^{it}) - h(1)| \le \varepsilon/2 \quad \text{whenever } |t| \le \delta.$$

Then consider $z \in D$ with

$$(7) \qquad |z-1| < \min\left\{\sin\delta/2, \quad \frac{\varepsilon}{8M}\sin^2\delta/2\right\}.$$

Because of 2π-periodicity of its integrand, the integral in (5) may be taken over $[-\delta, 2\pi-\delta]$ instead of $[0, 2\pi]$ and we get

$$\begin{aligned}
|\operatorname{Re} f(z) - h(1)| &\le \frac{1}{2\pi}\int_{-\delta}^{\delta} \frac{1-|z|^2}{|e^{it}-z|^2}|h(e^{it}) - h(1)|\,dt \\
&\quad + \frac{1}{2\pi}\int_{\delta}^{2\pi-\delta} \frac{1-|z|^2}{|e^{it}-z|^2}|h(e^{it}) - h(1)|\,dt \\
&\overset{(6)}{\le} \frac{1}{2\pi}\int_{-\delta}^{\delta} \frac{1-|z|^2}{|e^{it}-z|^2}\cdot\frac{\varepsilon}{2}\,dt + \frac{1}{2\pi}\int_{\delta}^{2\pi-\delta} \frac{1-|z|^2}{|e^{it}-z|^2}\cdot 2M\,dt \\
&\le \frac{\varepsilon}{4\pi}\int_{-\delta}^{2\pi-\delta} \frac{1-|z|^2}{|e^{it}-z|^2}\,dt + \frac{M}{\pi}\int_{\delta}^{2\pi-\delta} \frac{1-|z|^2}{|e^{it}-z|^2}\,dt
\end{aligned}$$

$$(8) \qquad \overset{(4)}{=} \frac{\varepsilon}{2} + \frac{M}{\pi}\int_{\delta}^{2\pi-\delta} \frac{1-|z|^2}{|e^{it}-z|^2}\,dt.$$

Now

$$\begin{aligned}
|e^{it}-z| &\ge |1-e^{it}| - |1-z| = \sqrt{2-2\cos t} - |1-z| \\
&= 2\sin t/2 - |1-z| \ge 2\sin\delta/2 - |1-z| \quad \text{for } t \in [\delta, 2\pi-\delta] \\
&\overset{(7)}{>} \sin\delta/2.
\end{aligned}$$

It follows from this and (8) that for z satisfying (7)

$$|\mathrm{Re}\, f(z) - h(1)| \leq \frac{\varepsilon}{2} + \frac{M}{\pi}\int_{\delta}^{2\pi-\delta} \frac{1 - |z|^2}{\sin^2 \delta/2}\, dt$$

$$\leq \frac{\varepsilon}{2} + \frac{M}{\pi}(1 - |z|)(1 + |z|)\frac{2(\pi - \delta)}{\sin^2 \delta/2}$$

$$< \frac{\varepsilon}{2} + \frac{4M(1 - |z|)}{\sin^2 \delta/2} \leq \frac{\varepsilon}{2} + \frac{4M}{\sin^2 \delta/2}|1 - z| \overset{(7)}{<} \varepsilon.$$

Exercise 5.21 (i) *Let D be an open disk, f a (real) harmonic function in D. Show that $f =$ Re F for some $F \in H(D)$.*

Hints: Let $D = D(0, 1)$. First assume f is defined and continuous on $\bar{D}$. Define F on D by

$$F(z) = \frac{1}{2\pi}\int_0^{2\pi} \frac{e^{it} + z}{e^{it} - z} f(e^{it})dt.$$

Then $|f - \mathrm{Re}\, F|$ is continuous in $\bar{D}$ and equal to 0 on $C(0, 1)$, by 5.20. Therefore its maximum over $\bar{D}$ occurs (at least once) in D and hence is 0 by 5.11.

For the general f there are, by what was just shown, holomorphic functions F_n in $D(0, 1 - 1/n)$ such that $f = \mathrm{Re}\, F_n$ in $D(0, 1 - 1/n)$. 2.17(ii) insures that by adding appropriate imaginary constants to successive F_n we may suppose that $F_{n+1} = F_n$ in $D(0, 1 - 1/n)$. The desired F is then well-defined by the decree $F = F_n$ in $D(0, 1 - 1/n)$, $n = 2, 3, \ldots$. See 10.7 for a different proof.
(ii) *Show that every harmonic function is both subharmonic and superharmonic.*

Hint: Use (i) and take real parts in the equation of 5.5(ii).
(iii) *Though* 5.17 *served to motivate* 5.20, *the proof of the latter is independent of the former. Show that in fact* 5.17 *is a consequence of* 5.20.

Hints: Consider the function F defined by the integral formula in 5.17. By 5.20, Re F is continuous on $\bar{D}$ and Re $F =$ Re f on ∂D, so by the Maximum Principle Re$(F - f)$ is 0; hence by 2.17(ii), $F - f$ is constant. 5.5(ii) shows that this constant is 0.

Exercise 5.22 (i) *Let U be an open subset of $\mathbb{C}$, $f\colon U \to \mathbb{R}$ a continuous function. Say that f has the* harmonic majorant *property* (HM) *if*

$$f \leq F \quad \text{in } V$$

whenever V is a bounded open set with $\bar{V} \subset U$ and F is a continuous function on $\bar{V}$ which is harmonic in V and coincides with f on ∂V. Show that this condition is equivalent to the subharmonicity of f.

Hints: If f is subharmonic and F is as above, then $f - F$ is subharmonic in V and assumes a maximum in the compact set $\bar{V}$. Now cite 5.9 and remember that $f - F = 0$ on ∂V.

Suppose f has property (HM). If $\bar{D}(a, r) \subset U$, let F be the solution of the Dirichlet problem in $D(a, r)$ with boundary data f. Then by (HM)

$$f(a) \leq F(a) \overset{5.21(\text{ii})}{\leq} \frac{1}{2\pi}\int_0^{2\pi} F(a + re^{i\theta})d\theta = \frac{1}{2\pi}\int_0^{2\pi} f(a + re^{i\theta})d\theta.$$

(ii) *Use the characterization in* (i) *to show that if U and W are subsets of $\mathbb{C}$, $u: U \to \mathbb{R}$ is subharmonic and $\phi: W \to U$ is conformal, then $u \circ \phi$ is subharmonic in W.*

Exercise 5.23 (LITTLEWOOD [1927]) (i) *Suppose that U is an open subset of $\mathbb{C}$ and $f: U \to \mathbb{R}$ is a continuous function with this property: for each $a \in U$ there is a sequence $\{r_n(a)\}_{n=1}^{\infty}$ of positive numbers such that $\bar{D}(a, r_n(a)) \subset U$, $r_n(a) \to 0$ and*

$$(**) \quad f(a) \leq \frac{1}{2\pi}\int_0^{2\pi} f(a + r_n(a)e^{i\theta})d\theta, \quad n = 1, 2, \ldots.$$

Show that f is subharmonic in U.

Hints: Use the harmonic majorant property: let V be a bounded open set with $\bar{V} \subset U$ and let F be continuous on $\bar{V}$, harmonic in V and equal to f on ∂V. We have to show that $f \leq F$ throughout V. Suppose this inequality fails. Then the maximum of the continuous function $f - F$ on the compact set $\bar{V}$ is positive, say M, and occurs (only) in V. The set K where it occurs is compact and so contains a point a nearest to the disjoint closed set $\mathbb{C}\backslash V$. Choose $R > 0$ so that $\bar{D}(a, R) \subset V$. Each circle $C(a, r)$ with $0 < r < R$ then contains a (non-degenerate) arc outside K on which consequently $f - F < M$. This follows easily from the fact that some point of this circle is closer to $\mathbb{C}\backslash V$ than is a and so, by the extremal property of a, necessarily lies outside K. Therefore

$$\frac{1}{2\pi}\int_0^{2\pi} (f - F)(a + re^{i\theta})d\theta < \frac{1}{2\pi}\int_0^{2\pi} M = M$$

and so (applying 5.21(ii) to F)

$$\frac{1}{2\pi}\int_0^{2\pi} f(a + re^{i\theta})d\theta < M + \frac{1}{2\pi}\int_0^{2\pi} F(a + re^{i\theta})d\theta \leq M + F(a)$$

$$= f(a) \quad \text{(definition of } M \text{ and } a\text{)}.$$

This holds for all $0 < r < R$ and contradicts (**).

(ii) (*Extension of* (i) *due to* SAKS [1930–32].) *Suppose that U is an open subset of $\mathbb{C}$ and $f: U \to \mathbb{R}$ is a continuous function with this property:*

$$\overline{\lim_{r \to 0}}\, r^{-2}\left[\frac{1}{2\pi}\int_0^{2\pi} f(a + re^{i\theta})d\theta - f(a)\right] \geq 0$$

for each $a \in U$. Show that f is subharmonic in U.

Hint: For every $\varepsilon > 0$ the function $f_\varepsilon(z) = f(z) + \varepsilon|z^2|$ fulfills the hypothesis of (i), and $f_\varepsilon \to f$ locally uniformly in U as $\varepsilon \downarrow 0$.

Exercise 5.24 *Show that for continuous real-valued functions f on open subsets U of $\mathbb{C}$ the* circumferential mean value *property,*

$$\text{(CM)}\quad f(a) = \frac{1}{2\pi}\int_0^{2\pi} f(a + re^{i\theta})d\theta \quad \textit{whenever } \bar{D}(a, r) \subset U,$$

characterizes harmonic functions. That is, show that a function which is both subharmonic and superharmonic is harmonic. (*Converse of* 5.21(ii).) *Formulate the appropriate harmonic version of* 5.23.

Hints: If $\bar{D}(a, r) \subset U$ then, since both f and $-f$ are subharmonic, we have by 5.22(i) that $\pm f \leq \pm F_a$ in $\bar{D}(a, r)$, where F_a (respectively, $-F_a$) is the solution of the Dirichlet problem in $D(a, r)$ with boundary data f (respectively, $-f$). Thus f coincides in the disk $D(a, r)$ with the harmonic function F_a.

Exercise 5.25 (i) *Let U be an open subset of $\mathbb{C}$, $f\colon U \to \mathbb{R}$ superharmonic. Suppose V is a bounded open set with $\bar{V} \subset U$, $g\colon \bar{V} \to \mathbb{R}$ is continuous, harmonic in V and equal to f on ∂V. Show that the function*

$$F = \begin{cases} g & \text{in } \bar{V} \\ f & \text{in } U\backslash\bar{V} \end{cases}$$

is superharmonic and satisfies $f \geq F$ in U.

Hints: Evidently F is continuous on U. Also if $a \in V$ [respectively, $a \in U\backslash\bar{V}$], then for some $r_a > 0$ we have $\bar{D}(a, r_a) \subset V$ [respectively, $\bar{D}(a, r_a) \subset U\backslash\bar{V}$]. Since g is harmonic [respectively, f is superharmonic], it follows that

$$F(a) = g(a) = \frac{1}{2\pi}\int_0^{2\pi} g(a + re^{i\theta})d\theta = \frac{1}{2\pi}\int_0^{2\pi} F(a + re^{i\theta})d\theta$$

[respectively,

$$F(a) = f(a) \geq \frac{1}{2\pi}\int_0^{2\pi} f(a + re^{i\theta})d\theta = \frac{1}{2\pi}\int_0^{2\pi} F(a + re^{i\theta})d\theta]$$

for every $r < r_a$. On the other hand, $f \geq g$ in $\bar{V}$ by the harmonic majorant property (5.22(i)). Therefore, $f \geq F$ in U and if $a \in \partial V$, $\bar{D}(a, r) \subset U$, then

$$F(a) = g(a) = f(a) \geq \frac{1}{2\pi}\int_0^{2\pi} f(a + re^{i\theta})d\theta \geq \frac{1}{2\pi}\int_0^{2\pi} F(a + re^{i\theta})d\theta.$$

We conclude from 5.23(i) that F is superharmonic in U.

(ii) *By imitating the above procedure establish the following: V, U open $\subset \mathbb{C}$, $f\colon \bar{V} \cap U \to (-\infty, 1]$ continuous, superharmonic in $V \cap U$ and $f(U \cap \partial V) = 1$. Then the extension of f to U given by $f(U\backslash\bar{V}) = 1$ is superharmonic in U.*

Exercise 5.26 (i) *If f is subharmonic in $D = D(0, 1)$, then $I(r) = \int_0^{2\pi} f(re^{i\theta})d\theta$ is a non-decreasing function of $r \in [0, 1)$.* (Cf. 11.60 (ii).)

Hints: Given $0 \le r \le R < 1$, let h be the solution of the Dirichlet problem in $D(0, R)$ with boundary data f. By the harmonic majorant property 5.22(i)

(1) $f \le h$ throughout $\bar{D}(0, R)$.

Moreover,

$$\int_0^{2\pi} f(Re^{i\theta})d\theta = \int_0^{2\pi} h(Re^{i\theta})d\theta, \quad \text{since } f = h \text{ on } C(0, R),$$

$$= 2\pi h(0)$$

(2) $$= \int_0^{2\pi} h(re^{i\theta})d\theta,$$

the last two equalities coming from 5.24. From (1) and (2) it follows, as desired, that

$$\int_0^{2\pi} f(Re^{i\theta})d\theta \ge \int_0^{2\pi} f(re^{i\theta})d\theta.$$

(ii) *If f is holomorphic in the open subset U of $\mathbb{C}$, then $|f|^p$ for each $p > 0$ and* $\log^+|f| = \log(\max\{|f|, 1\})$ *are each subharmonic in U.*

Hints: For each $a \in U$ we seek (after 5.23(i)) an $r_a > 0$ such that $D(a, r_a) \subset U$ and

$$h(a) \le \frac{1}{2\pi}\int_0^{2\pi} h(a + re^{i\theta})d\theta \quad \forall 0 \le r < r_a,$$

$h = |f|^p$ and $h = \log^+|f|$. There is no problem if $f(a) = 0$. If $f(a) \ne 0$, then there is an $r_a > 0$ such that $D(a, r_a) \subset U$ and f is zero-free in $D(a, r_a)$. Therefore $f = e^g$ for some g holomorphic in $D(a, r_a)$ [4.59 and 4.60] and $|f|^p = |e^{pg}|$, $\log^+|f| = \max\{\operatorname{Re} g, 0\} = \frac{1}{2}[\operatorname{Re} g + |\operatorname{Re} g|]$ are then subharmonic in $D(a, r_a)$ by 5.7 and its remark.

(iii) *If f is holomorphic in $D(0, 1)$ and $p > 0$, then $\int_0^{2\pi} |f(re^{i\theta})|^p d\theta$ is a non-decreasing function of $r \in [0, 1)$.*

Exercise 5.27 (Cf. 6.30) *Let $D = D(0, 1)$, $f \in H(D)$ be continuous in $\bar{D}$ and satisfy $f(0) = 0$, $|\operatorname{Re} f| \le A$. Show that*

$$|\operatorname{Im} f| \le \frac{2A}{\pi}\log\left(\frac{1 + r}{1 - r}\right) \quad \text{on } \bar{D}(0, r) \quad \forall 0 \le r < 1.$$

Hint: From the Cauchy–Schwarz formula 5.18 deduce that

$$\operatorname{Im} f(re^{i\psi}) = \frac{1}{2\pi}\int_{-\pi}^{\pi} \frac{2r \sin t}{1 - 2r\cos t + r^2} \operatorname{Re} f(e^{i(\psi - t)})dt, \quad 0 \le r < 1, \psi \in \mathbb{R}.$$

Write $h(t) = 1 - 2r\cos t + r^2$ and have

$$|\operatorname{Im} f(re^{i\psi})| \le \frac{1}{\pi}\int_0^\pi \frac{h'}{h} A = \frac{A}{\pi}[\log(h(\pi)) - \log(h(0))].$$

Exercise 5.28 (HARNACK [1886]) *If f is holomorphic and* $\operatorname{Re} f \ge 0$ *in* $D(0, 1)$, *then for any* $0 \le R \le 1$

$$(*)\qquad \frac{R - |z|}{R + |z|}\operatorname{Re} f(0) \le \operatorname{Re} f(z) \le \frac{R + |z|}{R - |z|}\operatorname{Re} f(0) \quad \forall z \in D(0, R).$$

Hint: For $R < 1$ use an appropriately scaled Poisson integral formula and obvious appraisals of the kernel $\operatorname{Re}[(Re^{it} + z)/(Re^{it} - z)]$ which occurs therein. Once (*) is established for $R < 1$, its validity for $R = 1$ follows by letting $R \uparrow 1$. If, in addition, $\operatorname{Im} f(0) = 0$, then this argument yields for the second inequality the stronger relation in which $\operatorname{Re} f$ is replaced by $|f|$. See also 6.35.

Exercise 5.29 (TSUJI [1955]) *Let* $H = \{z \in \mathbb{C}: \operatorname{Re} z > 0\}$, $\gamma: (0, 1] \to H$ *a continuous function such that* $\operatorname{Im}\gamma/\operatorname{Re}\gamma$ *is bounded and* $\lim_{t\downarrow 0}\gamma(t) = 0$. *Let h be a positive harmonic function in H.*

(i) *Show that if h is bounded on γ, then h is bounded in any triangular set* $\{z \in H: c^{-1}|\operatorname{Im} z| < \operatorname{Re} z < c\}$ *for each* $c > 0$.

Hints: Let $X = \sup[\operatorname{Re}\gamma]$, $M = \sup[h(\gamma)]$. We may confine attention to those c which satisfy

$$(1)\qquad \frac{|\operatorname{Im}\gamma|}{\operatorname{Re}\gamma} < c.$$

Furthermore, it obviously suffices to show that h is bounded in each set $S_c = \{z \in H: c^{-1}|\operatorname{Im} z| < \operatorname{Re} z < X\}$. To this end, let $n = n_c$ denote the least integer not less than $4c$ and consider any $x \in (0, X)$. There is a $z \in \gamma$ with $\operatorname{Re} z = x$. Write $y = \operatorname{Im} z$. By (1) we have $y \in [-cx, cx]$ and consequently by the choice of n

$$(2)\qquad \left[y - n\frac{x}{2}, y + n\frac{x}{2}\right] \supset [-cx, cx].$$

Successive applications of Harnack's inequality 5.28 tell us that

$$h(x + it) \le 2h(x + iy) \le 2M \quad \forall t \in \left[y - \frac{x}{2}, y + \frac{x}{2}\right],$$

whence

$$h(x + it) \le 2\cdot 2M = 4M \quad \forall t \in [y - x, y + x]$$

and, inductively, for each $k \ge 1$

$$(3.k)\qquad h(x + it) \le 2^k M \quad \forall t \in \left[y - \frac{kx}{2}, y + \frac{kx}{2}\right].$$

It follows from (2) and (3.n) that

$$h(x + it) \leq 2^n M \quad \forall t \in [-cx, cx].$$

This holds for each $x \in (0, X)$ and $S_c = \{x + it : 0 < x < X, |t| < cx\}$, so we have shown that $h \leq 2^n M$ in S_c.

(ii) *Show that if* $\lim_{t \downarrow 0} h(\gamma(t)) = 0$, *then* $\lim_{z \to 0, |\mathrm{Im}\, z| < c \mathrm{Re}\, z} h(z) = 0$ *for each* $c > 0$.

Hints: Given $\varepsilon > 0$ and $c > 0$, select $T \in (0, 1]$ such that

$$h(\gamma(t)) < \varepsilon \cdot 2^{-4c-1} \quad \forall t \in (0, T]$$

and then apply (i) to γ on $(0, T]$. With $X = \sup \mathrm{Re}\, \gamma(0, T]$, the proof in (i) shows that

$$h(z) < 2^{n_c} \cdot \varepsilon \cdot 2^{-4c-1} \quad \forall z \in S_c.$$

But by definition $n_c \leq 4c + 1$, so we achieve $h < \varepsilon$ on S_c.

§ 4 Existence of Power Series Expansions

Returning to the fundamental 5.18 we derive at once (cf. 5.3):

Corollary 5.30 *If U is an open subset of $\mathbb{C}$ and f is differentiable in U, then f is infinitely differentiable in U and for each $z_0 \in U$ the series*

$$\sum_{n=0}^{\infty} \frac{f^{(n)}(z_0)}{n!}(z - z_0)^n$$

converges to $f(z)$ in every open disk centered at z_0 which lies in U.

Proof: Given such a z_0 and such a disk $D(z_0, r)$ we just take an $0 < R < r$ and consider the function $h(z) = f(z_0 + Rz)$. This function is holomorphic in $D(0, 1)$ and continuous in $\bar{D}(0, 1)$. By the scholium to 5.18 we have then

$$h(z) = \sum_{n=0}^{\infty} a_n z^n \quad \forall z \in D(0, 1)$$

for some $a_n \in \mathbb{C}$. Therefore

$$(1) \qquad f(z) = \sum_{n=0}^{\infty} c_n (z - z_0)^n \quad \forall z \in D(z_0, R)$$

with $c_n = a_n R^{-n}$. By 3.6 we get from (1) that f is infinitely differentiable in $D(z_0, R)$ and that $f^{(n)}(z_0) = n!\, c_n$. In particular, the c_n do not depend on R and so, as R in (1) was arbitrary in $(0, r)$, we get from (1)

$$(2) \qquad f(z) = \sum_{n=0}^{\infty} \frac{f^{(n)}(z_0)}{n!}(z - z_0)^n \quad \forall z \in D(z_0, r).$$

Exercise 5.31 (Cauchy Product of Series) *Show that if $f(z) = \sum_{n=0}^{\infty} a_n z^n$ and $g(z) = \sum_{n=0}^{\infty} b_n z^n$ have radii of convergence $R_1 > 0$ and $R_2 > 0$, then $h(z) =$*

$\sum_{n=0}^{\infty} c_n z^n$ converges for all $|z| < R = \min\{R_1, R_2\}$ and $h(z) = f(z)g(z)$ for all such z, where $c_n = \sum_{k=0}^{n} a_k b_{n-k}$ for $n = 0, 1, 2, \ldots$.

Hint: The functions f and g are both holomorphic in $D(0, R)$ and $f^{(n)}(0)/n! = a_n$, $g^{(n)}(0)/n! = b_n$ by 3.6. Therefore $f \cdot g$ is differentiable in $D(0, R)$ and

$$f(z)g(z) = (f \cdot g)(z) = \sum_{n=0}^{\infty} \frac{(f \cdot g)^{(n)}(0)}{n!} z^n \text{ in } D(0, R)$$

by the last result. Finally, verify by induction on n the so-called *Leibniz formula:*

$$(f \cdot g)^{(n)} = \sum_{k=0}^{n} \binom{n}{k} f^{(k)} \cdot g^{(n-k)}.$$

Exercise 5.32 *If $\mathscr{F}$ is a family of functions holomorphic in the open set U, K is a compact subset of U and M is a constant such that $|f| \leq M$ on U for every $f \in \mathscr{F}$, show that there is a constant $C = C(M, K)$ such that $|f'| \leq C$ on K for every $f \in \mathscr{F}$.*

Hint: Use the Cauchy estimates (3.34) locally and take a finite cover of K.

Theorem 5.33 (Cauchy's Formulae for Starlike Regions) *Let Ω be an open, starlike subset of $\mathbb{C}$, f holomorphic in Ω, γ a piecewise smooth loop in Ω. Then for every non-negative integer k and every $z \in \Omega \backslash \gamma$ we have*

$$(**) \quad \frac{\operatorname{Ind}_\gamma(z)}{k!} f^{(k)}(z) = \frac{1}{2\pi i} \int_\gamma \frac{f(\xi)}{(\xi - z)^{k+1}} \, d\xi.$$

Proof: Fix $z_0 \in \Omega \backslash \gamma$. Then pick $r > 0$ so that $D(z_0, r) \subset \Omega$. Then in this disk f is given by a power series (5.30): for some $c_n \in \mathbb{C}$

$$f(z) = \sum_{n=0}^{\infty} c_n (z - z_0)^n \qquad \forall z \in D(z_0, r);$$

$$\frac{f(z) - f(z_0)}{z - z_0} = \sum_{n=1}^{\infty} c_n (z - z_0)^{n-1} \quad \forall z \in D(z_0, r) \backslash \{z_0\}.$$

It is therefore consistent to define a function F in Ω by

$$F(z) = \begin{cases} \displaystyle\sum_{n=0}^{\infty} c_{n+1} (z - z_0)^n, & z \in D(z_0, r) \\ \displaystyle\frac{f(z) - f(z_0)}{z - z_0} & z \in \Omega \backslash \{z_0\}. \end{cases}$$

F is evidently differentiable (3.6) throughout Ω, so we apply to it the Cauchy Theorem for Starlike Regions to get

$$0 = \int_\gamma F(z) dz = \int_\gamma \frac{f(z) - f(z_0)}{z - z_0} dz = \int_\gamma \frac{f(z)}{z - z_0} dz - f(z_0) \int_\gamma \frac{1}{z - z_0} dz.$$

This is the case $k = 0$ of (**) (recall 4.4). Using this, 2.14 and induction [recall that the function Ind_γ is locally constant], we get (**) for all k.

As a further consequence of 5.30 we have

Corollary 5.34 *Let Ω be a region in which every holomorphic function has a primitive. Then every zero-free holomorphic function in Ω has a holomorphic logarithm.*

Proof: Let $f: \Omega \to \mathbb{C}\backslash\{0\}$ be holomorphic. By 5.30 f' is also holomorphic and so therefore is f'/f. Let F be a primitive for this quotient: $F' = f'/f$. From this and the Chain Rule we see that the function fe^{-F} has derivative 0 in Ω, hence by 2.10 is constant in Ω. This constant is of course non-zero and hence (3.14) of the form e^c for some $c \in \mathbb{C}: fe^{-F} \equiv e^c$. Then $f = e^{F+c}$ and $F + c$ is the desired logarithm.

Corollary 5.35 *Every zero-free holomorphic function in an open starlike subset of $\mathbb{C}$ has a holomorphic logarithm there.*

Proof: 5.2 plus 5.34. Alternatively, 4.20 plus 4.60.

Exercise 5.36 (i) *Let F be holomorphic and zero-free in $D(0, 1)$ and satisfy $|F| \leq 1$. Show that*

$$\sup_{|z| \leq 1/5} |F(z)|^2 \leq \inf_{|z| \leq 1/7} |F(z)|.$$

Hints: There is a holomorphic function f in $D(0, 1)$ such that $F = e^{-f}$. From $e^{-\mathrm{Re} f} = |F| \leq 1$ we see $\mathrm{Re} f \geq 0$. From 5.28 (with $R = 1$)

$$\mathrm{Re}\, f(z) \geq \tfrac{2}{3} \mathrm{Re}\, f(0) \quad \forall |z| \leq \tfrac{1}{5},$$

$$\mathrm{Re}\, f(z) \leq \tfrac{4}{3} \mathrm{Re}\, f(0) \quad \forall |z| \leq \tfrac{1}{7}.$$

Therefore

$$\sup_{|z| \leq 1/5} [-2 \,\mathrm{Re}\, f(z)] \leq -\tfrac{4}{3} \mathrm{Re}\, f(0) \leq \inf_{|z| \leq 1/7} [-\mathrm{Re}\, f(z)].$$

(ii) *For each region Ω and each compact subset K of Ω there is a positive integer $n = n(K, \Omega)$, dependent only on K and Ω, such that*

(*) $\quad |f(z)|^{1/n} \leq |f(w)| \leq |f(z)|^n \quad \forall z, w \in K$

holds for all $f \in H(\Omega)$ which satisfy $|f| \geq 1$ in Ω.

Hints: From (i) we see that

(1) $\quad |F(z)|^2 \leq |F(w)| \leq |F(z)|^{1/2} \quad \forall z, w \in \bar{D}(0, \tfrac{1}{7})$

holds for all F which are holomorphic in $D(0, 1)$ and satisfy $0 < |F| \leq 1$. Consequently

(2) $\quad |f(z)|^{1/2} \leq |f(w)| \leq |f(z)|^2 \quad \forall z, w \in \bar{D}(a, \tfrac{1}{7}r)$

holds for all f which are holomorphic in $D(a, r)$ and satisfy $|f| \geq 1$. Use (2) and a covering compactness argument to get (*) [for an n of the form 2^m]. To this end one should first replace K by a *connected* compact subset of Ω. For the (easy) details see pp. 25 ff. of OSTROWSKI [1931]. See also the hints to 5.38(iii) below. (The harmonic function version of this inequality occurs as lemma 1, p. 174 in MANDELBROJT [1929].)

Exercise 5.37 *Let f be entire and suppose that $f(z)e^{-c|z|}$ is bounded by M for some $c > 0$. Show that $f'(z)e^{-c|z|}$ is bounded by ceM.*

Hints: By Cauchy's formula for disks, for all $z \in \mathbb{C}$ and all $r > 0$

$$f'(z) = \frac{1}{2\pi i}\int_{C(z,r)} \frac{f(\xi)}{(\xi - z)^2}\,d\xi,$$

$$|f'(z)| \le r \sup_{\xi\in C(z,r)} \frac{|f(\xi)|}{|\xi - z|^2} = \frac{1}{r}\sup_{|\xi - z| = r}|f(\xi)|$$

$$\le \frac{1}{r}\sup_{|\xi - z| = r} Me^{c|\xi|} = \frac{1}{r}Me^{c(|z|+r)} = \frac{e^{cr}}{r}Me^{c|z|}.$$

The function $\phi(r) = e^{cr}/r$ has its minimum over $(0, \infty)$ at the point $r = 1/c$.

Exercise 5.38 (i) *Let F be holomorphic in $S = (0, 1) \times \mathbb{R}$ and continuous in $\bar{S}$. Suppose that F is bounded in $\bar{S}$ and set*

$$M(x) = \sup\{|F(x + iy)| : y \in \mathbb{R}\}$$

for each $x \in [0, 1]$. Show that

$$M(x) \le M(0)^{1-x}M(1)^x \quad \forall x \in [0, 1].$$

Hints: Consider any $M_0 > M(0)$, $M_1 > M(1)$ and form $G(z) = M_0^{1-z}M_1^z$, $f = F/G$. Evidently f is holomorphic in S and bounded and continuous in $\bar{S}$, since $|G(z)| = M_0^{1-\operatorname{Re} z}M_1^{\operatorname{Re} z} \ge \min\{M_0, M_1\} > 0$ when $0 \le \operatorname{Re} z \le 1$. Also $|f| < 1$ on ∂S. It follows from 5.13 (or more elementarily by the introduction of the appropriate auxiliary function as in the hints to 5.16) that $|f| < 1$ throughout S, i.e.,

$$|F(z)| < M_0^{1-\operatorname{Re} z}M_1^{\operatorname{Re} z} \quad \forall z \in S.$$

Let $M_0 \downarrow M(0)$, $M_1 \downarrow M(1)$.

(ii) (Hadamard Three Circles Theorem) *Let $0 \le R_1 < R_2$, f holomorphic in $A(R_1, R_2)$. For each $R_1 < r < R_2$ set $M(r) = \sup\{|f(z)| : |z| = r\}$. Then*

(*) $$M(r)^{\log(r_2/r_1)} \le M(r_1)^{\log(r_2/r)}M(r_2)^{\log(r/r_1)}$$

for all $R_1 < r_1 < r < r_2 < R_2$.

Hints: Let such r_1, r, r_2 be given and consider the map

$$\phi(z) = \exp(\log r_1 + z\log(r_2/r_1)) = \exp((1 - z)\log r_1 + z\log r_2)$$

of $[0, 1] \times \mathbb{R}$ onto $\bar{A}(r_1, r_2)$ and the composite $F = f \circ \phi$. Evidently F fulfills the hypothesis of (i) and $F(x + iy) = f(r_1^{1-x}r_2^x e^{iy \log(r_2/r_1)})$, so

$$\sup\{|F(x + iy)| : y \in \mathbb{R}\} = \sup\{|f(r_1^{1-x}r_2^x e^{i\theta})| : \theta \in \mathbb{R}\} = M(r_1^{1-x}r_1^x).$$

For any $x \in (0, 1)$ we get then from (i)

$$M(r_1^{1-x}r_2^x) \le M(r_1)^{1-x}M(r_2)^x.$$

If we take $x = [\log(r/r_1)]/[\log(r_2/r_1)]$, we have $r_1^{1-x}r_2^x = r$ and (*) ensues when both sides of the last inequality are raised to power $\log(r_2/r_1)$.

(iii) *Let Ω be a region, U a non-empty open subset of Ω, K a non-empty compact subset of Ω. Show that there exists a positive constant $c = c(\Omega, U, K) < 1$ such that*

(*) $\sup|f(K)| \le [\sup|f(\Omega)|]^{1-c}[\sup|f(U)|]^c \quad \forall f \in H(\Omega).$

Hints: Since (*) holds for any $c \in (0, 1)$ if $\sup|f(\Omega)|$ is 0 or ∞, we can exclude such f and then, upon dividing f by this number, (*) is seen to be equivalent to

(**) $\sup|f(K)| \le [\sup|f(U)|]^c \quad \forall f \in H(\Omega)$ with $\sup|f(\Omega)| = 1.$

If the open disks $D_1, \ldots, D_n$ with $\bar{D}_j \subset \Omega$ cover K and the constants $c(\Omega, U, \bar{D}_j)$ have been produced satisfying (**) with $K = \bar{D}_j$, then the choice $c(\Omega, U, K) = \min\{c(\Omega, U, \bar{D}_j) : 1 \le j \le n\}$ will satisfy (**). Therefore it suffices to prove that for each point $z \in \Omega$ there is an $r > 0$ such that $\bar{D}(z, r) \subset \Omega$ and an appropriate constant $c(\Omega, U, \bar{D}(z, r))$ exists. To this end construct a sequence of points $z_1, \ldots, z_k = z$ and positive numbers $r_1, \ldots, r_k$ such that

$$D(z_1, r_1) \subset U$$

$$D(z_j, r_j) \subset D(z_{j-1}, 2r_{j-1}), \quad j = 2, \ldots, k$$

$$D(z_j, 4r_j) \subset \Omega, \quad j = 1, \ldots, k.$$

Given f as in (**), set $m = \sup|f(U)|$. Since f is bounded by 1 on $\Omega \supset D(z_1, 4r_1)$ and by m on $U \supset D(z_1, r_1)$, part (i) (plus the Maximum Modulus Principle) implies that

$$\sup|f(D(z_1, 2r_1))| \le m^\lambda, \quad \text{where } \lambda = \frac{\log 2}{\log 4} \in (0, 1).$$

Continuing through successive disks, we get, by successive applications of (ii)

$$\sup|f(D(z_k, 2r_k))| \le m^{\lambda^k}.$$

(iv) (STIELTJES [1894]) *Let V be a region, U a non-empty open subset of V, $\{f_n\} \subset H(V)$. If $\{f_n\}$ is uniformly convergent on U and uniformly bounded on each compact subset of V, then $\{f_n\}$ is in fact uniformly convergent on each compact subset of V.*

Hints: Let D be a non-empty open disk such that $\bar{D} \subset U$. Given compact $K \subset V$, it is not hard to construct a bounded region Ω such that $\bar{D} \cup K \subset$

$\Omega \subset \bar{\Omega} \subset V$. According to (iii) there is then a constant $c = c(\Omega, D, K) \in (0, 1)$ such that

$$(*) \quad \sup|f_n - f_m|(K) \le [\sup|f_n - f_m|(\Omega)]^{1-c}[\sup|f_n - f_m|(D)]^c.$$

By hypothesis there is a constant $M = M(\bar{\Omega})$ such that $\sup|f_n(\Omega)| \le M$ for all n. Therefore (*) yields

$$\sup|f_n - f_m|(K) \le (2M)^{1-c}[\sup|f_n - f_m|(D)]^c.$$

Exercise 5.39 (MORERA [1886]) *Prove the converse of the Cauchy–Goursat theorem: If f is continuous in the open subset U of $\mathbb{C}$ and satisfies $\int_{[a,b,c,a]} f = 0$ whenever* co$\{a, b, c\} \subset U$, *then f is holomorphic in U.*

Hint: The problem is a local one: if Ω is any open disk in U, then f has a primitive in Ω by the *proof* of 5.2 and so is holomorphic in Ω by 5.30.

Exercise 5.40 *Let $D = D(0, 1)$, $f\colon D \to \mathbb{C}$ a continuous function such that $\int_{\partial R} f = 0$ for every $R = [0, x] \times [0, y]$ (x, y real) which lies in D. Show that f is holomorphic in D.*

Hint: Form F in D by $F(z) = \int_{[0,\operatorname{Re} z, z]} f$ and use the hypothesis on f to show that the partial derivatives of F exist and satisfy $D_1F = f = -iD_2F$. Conclude from 2.16 that F is holomorphic. Then $F' = D_1F = f$ is too, by 5.30.

Theorem 5.41 (Removable Singularity) *Let U be an open subset of $\mathbb{C}$, f a holomorphic function in $U\backslash\{a\}$, for some $a \in U$. If f satisfies $\lim_{z\to a}(z - a)f(z) = 0$ [in particular, if f is bounded near a], then f is the restriction to $U\backslash\{a\}$ of some F which is holomorphic in U.*

Proof: Let $r > 0$ be such that $D_r = D(a, r) \subset U$. Define h on U by

$$(1) \quad h(z) = \begin{cases} 0 & z = a \\ (z - a)^2 f(z), & z \in U\backslash\{a\}. \end{cases}$$

Then h is differentiable in $U\backslash\{a\}$. Moreover $(h(z) - h(a))/(z - a) = (z - a)f(z)$ ($z \in U\backslash\{a\}$) and so the hypothesis shows that h is differentiable at a also, with $h'(a) = 0$. Therefore by 5.30 there is a series which represents h throughout D_r:

$$h(z) = \sum_{n=0}^{\infty} c_n(z - a)^n, \quad z \in D_r.$$

And $c_0 = h(a) = 0$, $c_1 = h'(a) = 0$ so we have

$$(2) \quad h(z) = \sum_{n=2}^{\infty} c_n(z - a)^n = (z - a)^2 \sum_{k=0}^{\infty} c_{k+2}(z - a)^k, \quad z \in D_r.$$

It follows from (1) and (2) that

$$f(z) = \sum_{k=0}^{\infty} c_{k+2}(z - a)^k \quad \forall z \in D_r\backslash\{a\}$$

and therefore we may (consistently) define the desired F by

$$F(z) = \begin{cases} f(z) & z \in U\backslash\{a\} \\ \sum_{k=0}^{\infty} c_{k+2}(z-a)^k, & z \in D_r. \end{cases}$$

(A different proof is suggested in 11.7.)

Exercise 5.42 *Let U be an open subset of $\mathbb{C}$, f holomorphic in U. Define $g\colon U \times U \to \mathbb{C}$ by*

$$g(z, w) = \begin{cases} \dfrac{f(z) - f(w)}{z - w}, & z, w \in U, z \neq w \\ f'(w) & z = w \in U. \end{cases}$$

Show that g is continuous and that for each $w \in U$, $g(z, w)$ is a holomorphic function of $z \in U$.

Hints: Let $(z_0, w_0) \in U \times U$ be given. If $z_0 \neq w_0$, then in a whole neighborhood of (z_0, w_0) the first formula in the definition of g prevails, so the continuity of g at (z_0, w_0) and the differentiability of $g(z, w_0)$ at z_0 are immediate. If $z_0 = w_0 \in U$, pick $r > 0$ so that $\bar{D}(w_0, r) \subset U$ and use the Cauchy Formulas to write

$$g(z, w) = \frac{1}{2\pi i}\int_{C(w_0,r)} \frac{f(\xi)d\xi}{(\xi - z)(\xi - w)} \quad \forall z, w \in D(z_0, r).$$

Differentiability of $g(z, w_0)$ at z_0 then follows from 2.14 and continuity of g at (z_0, w_0) is an easy exercise in the uniform continuity of the integrand.

Exercise 5.43 *Let V be an open subset of $\mathbb{C}$, $f\colon [0, 1] \times V \to \mathbb{C}$ a bounded function. Suppose $f(t, z)$ is a Riemann integrable (resp., continuous) function of $t \in [0, 1]$ for each $z \in V$ and a holomorphic function of $z \in V$ for each $t \in [0, 1]$. Show that then $D_2 f(t, z)$ is a Riemann integrable (resp., continuous) function of $t \in [0, 1]$ for each z. (See 7.18 for more on this theme.)*

Hints: Given $z \in V$, let $r > 0$ be such that $\bar{D}(z, r) \subset V$ and for each integer $n > 2/r$ define a Riemann integrable (resp., continuous) function f_n on $[0, 1]$ by

$$f_n(t) = n\left[f\left(t, z + \frac{1}{n}\right) - f(t, z)\right].$$

Then the Cauchy Integral Formulas give for all $t \in [0, 1]$

$$f_n(t) - D_2 f(t, z) = \frac{1}{2\pi i}\int_{C(z,r)} \left[\frac{nf(t, \xi)}{\xi - z - \frac{1}{n}} - \frac{nf(t, \xi)}{\xi - z} - \frac{f(t, \xi)}{(\xi - z)^2}\right] d\xi$$

$$= \frac{1}{2\pi i n}\int_{C(z,r)} \frac{f(t, \xi)d\xi}{\left(\xi - z - \frac{1}{n}\right)(\xi - z)^2}.$$

It follows from this, 2.9(i), and the fact $n \geq 2/r$ that

$$|f_n(t) - D_2 f(t, z)| \leq \frac{2M}{nr^2},$$

holding for all $t \in [0, 1]$. Here M is a bound for f on $[0, 1] \times V$. Thus $D_2 f(t, z)$ is the limit, uniformly in t, of the Riemann integrable (resp., continuous) functions $f_n(t)$.

Here is an elementary preview of a theme to be developed at length in Chapter VII. (Cf. also 5.38(iv), 5.45(iii) and 5.74.)

Exercise 5.44 (i) (WEIERSTRASS *Werke I*, 67–74; *II*, 201–233.) *A uniform limit of holomorphic functions is holomorphic.*

Hint: Morera (5.39).

(ii) *If f_n are holomorphic in the open set U and $f_n \to f$ uniformly in each compact subset of U, then not only is f holomorphic in U, but for each positive integer j, $f_n^{(j)} \to f^{(j)}$ uniformly in each compact subset of U.*

Hint: The problem is a local one. Express f_n via Cauchy's integral formula for a disk and take the limit under the integral. Then cite 2.14.

(iii) *If harmonic functions h_n converge uniformly on compacta to a (finite) limit function h, then h is harmonic. Moreover, all the partial derivatives of h_n converge uniformly on compacta to the corresponding partial derivatives of h.*

Hints: In analogy with (i), use the integral characterization of harmonicity in 5.24 to prove the first assertion. To prove both assertions at once, in analogy with (ii), exploit the Poisson formula. (Everything also follows from (ii) via 5.21(i) and 7.12(i).)

(iv) *Let $f_n: \mathbb{R} \to \mathbb{C}$ be differentiable, f_n' continuous. Suppose $f_n \to f$, $f_n' \to F$ uniformly on each compact subset of $\mathbb{R}$. Show that f is differentiable and $f' = F$.*

Hint: $f_n(x) = f_n(a) + \int_a^x f_n' \to f(a) + \int_a^x F$ for each x. Therefore $f(x) = f(a) + \int_a^x F$ for all $x \in \mathbb{R}$ (and any $a \in \mathbb{R}$).

§ 5 Harmonic Majorization

A simple but primitive form of the theme of this section is illustrated in the following exercise.

Exercise 5.45 (OSTROWSKI [1922/23]) (i) *Let f be holomorphic in $D = D(0, 1)$ and continuous on $\bar{D}$. Let A be a subarc of $C(0, 1)$ of length $\alpha > 0$. Define*

$$\lambda(z) = \frac{\alpha}{2\pi} \frac{1 - |z|}{1 + |z|}, \quad z \in D.$$

Show that

(*) $|f(z)| \le [\sup|f(A)|]^{\lambda(z)}[\sup|f(D)|]^{1-\lambda(z)} \quad \forall z \in D.$

Hints: Consider any $m > \sup|f(A)|$, $M > \sup|f(D)|$ and $M \ge m$. Define

(1) $$\phi(z) = \log(\max\{|f(z)|, m\}) = \log m + \log^+\left|\frac{f(z)}{m}\right|, \quad z \in \bar{D}.$$

By 5.26(ii) ϕ is subharmonic in D. Therefore if we define

(2) $$h(z) = \frac{1}{2\pi}\int_0^{2\pi} \operatorname{Re}\left[\frac{e^{i\theta}+z}{e^{i\theta}-z}\right]\phi(e^{i\theta})d\theta, \quad z \in D,$$

and $h = \phi$ on ∂D, then we have from 5.8 (applied to the difference $\phi - h$)

(3) $\phi \le h$ in $\bar{D}$.

After a rotation we can assume that

$$A = \{e^{i\theta} : 0 \le \theta \le \alpha\}.$$

On A we have $|f| \le m$, so $\phi = \log m$. Therefore

$$h(z) \le \frac{1}{2\pi}\int_0^{\alpha} \operatorname{Re}\left[\frac{e^{i\theta}+z}{e^{i\theta}-z}\right]\log m\, d\theta + \frac{1}{2\pi}\int_{\alpha}^{2\pi} \operatorname{Re}\left[\frac{e^{i\theta}+z}{e^{i\theta}-z}\right]\log M\, d\theta$$

(since, by equation (2) in the proof of 5.20, the kernel

$$\operatorname{Re}\left[\frac{e^{i\theta}+z}{e^{i\theta}-z}\right] = \frac{1-|z|^2}{|e^{i\theta}-z|^2} \text{ is non-negative)}$$

$$= \psi(z)\log m + (1-\psi(z))\log M$$

(4) $$= \log M - \psi(z)[\log M - \log m],$$

where

(5) $$\psi(z) = \frac{1}{2\pi}\int_0^{\alpha} \operatorname{Re}\left[\frac{e^{i\theta}+z}{e^{i\theta}-z}\right]d\theta \ge \frac{1}{2\pi}\int_0^{\alpha} \frac{1-|z|}{1+|z|}\, d\theta = \lambda(z)$$

and (taking $h = 1$ in 5.20)

$$\frac{1}{2\pi}\int_0^{2\pi} \operatorname{Re}\left[\frac{e^{i\theta}+z}{e^{i\theta}-z}\right]d\theta = 1.$$

From (5) and (4) we get

(6) $$h(z) \le \log M - \lambda(z)[\log M - \log m] = \lambda(z)\log m + (1-\lambda(z))\log M.$$

Since $|f| \le \max\{|f|, m\} = e^{\phi}$, (6) and (3) yield

$$|f(z)| \le m^{\lambda(z)}M^{1-\lambda(z)}$$

and (*) follows from this by letting $m \downarrow \sup|f(A)|$, $M \downarrow \sup|f(D)|$.

(ii) *Let A, B be complementary, non-degenerate arcs of the unit circle. Show that for each $r \in (0, 1)$ there exists $\lambda(r) = \lambda_A(r) \in (0, 1)$ such that*

(*) $$\sup_{|z|\le r}|f(z)| \le M_A^{\lambda(r)}M_B^{1-\lambda(r)} \quad \forall 0 < r < 1$$

holds for all numbers $0 \le M_A \le M_B < \infty$ *and all* $f \in H(D)$ *which satisfy*

$$(**)\qquad \overline{\lim_{z\to u}}|f(z)| \le \begin{cases} M_A & \forall u \in \bar{A} \\ M_B & \forall u \in \bar{B}. \end{cases}$$

Hints: Define $\lambda_A(r) = \alpha(1 - r)/2\pi(1 + r)$, α the length of A. Consider any f of the described kind. Since $M_A \le M_B$, it follows from (**) and 5.12 that

$$(7)\qquad |f| \le M_B \quad \text{throughout } D.$$

If we define f_n on $\bar{D}$ by $f_n(z) = f((n - 1)z/n)$, then (i) applied to f_n yields

$$\left|f\left(\frac{n-1}{n}z\right)\right| = |f_n(z)| \le [\sup|f_n(\bar{A})|]^{\lambda(r)}[\sup|f_n(D)|]^{1-\lambda(r)}$$
$$\le [\sup|f_n(\bar{A})|]^{\lambda(r)}M_B^{1-\lambda(r)} \quad \forall |z| = r.$$

So by the Maximum Modulus Principle

$$(8)\qquad \left|f\left(\frac{n-1}{n}z\right)\right| \le [\sup|f_n(\bar{A})|]^{\lambda(r)}M_B^{1-\lambda(r)} \quad \forall z \in \bar{D}(0, r).$$

Choose $u_n \in \bar{A}$ such that $|f((n - 1)u_n/n)| = \sup|f_n(\bar{A})|$. Then choose $1 \le n_1 < n_2 < \cdots$ such that $\sup|f_{n_j}(\bar{A})| \to \overline{\lim}_{n\to\infty} \sup|f_n(\bar{A})|$ and also such that $u_{n_j} \to u \in \bar{A}$. By (**) we have then

$$(9)\qquad \overline{\lim_{n\to\infty}} \sup|f_n(\bar{A})| = \lim_{j\to\infty}\left|f\left(\frac{n_j-1}{n_j}u_{n_j}\right)\right| \le M_A.$$

From (8) and (9) we get

$$|f(z)| \le M_A^{\lambda(r)}M_B^{1-\lambda(r)} \quad \forall z \in \bar{D}(0, r).$$

(iii) (Cf. § 90, Montel [1927]) *Let* D, A *be as in* (ii), F_n *holomorphic functions in* D *which are continuous on* $D \cup A$. *Suppose that* $\{F_n\}$ *is uniformly bounded on* D *and uniformly convergent on* A. *Show that* $\{F_n\}$ *is then uniformly convergent on each compact subset of* D.

Hints: It suffices to show that for each $r < 1$ and each subsequence $1 \le n_1 < n_2 < \cdots$ the functions $f_j = F_{n_{j+1}} - F_{n_j}$ converge to 0 uniformly in $D(0, r)$. If we set $M_j = \sup|f_j(A)|$, then $M_j \to 0$ by hypothesis and by (ii)

$$|f_j(z)| \le M_j^{\lambda(r)}(2M)^{1-\lambda(r)} \quad \forall z \in D(0, r),$$

where M is some bound for all the F_n in D. Since $\lambda(r) > 0$, the result follows.

To exploit the full potential of the idea in 5.45(i) we need to have available for various standard sets (like slit disks, half-disks, etc.) harmonic functions with very tightly specified boundary behavior and for which some appraisals of the interior values can be gotten. These are used as comparison functions to provide upper bounds for holomorphic functions via the various maximum theorems. In many important applications we can manufacture the relevant harmonic function by composing elementary transformations but usually a few "corner" points on the boundary will be points of discontinuity and therefore the first order of business is a technical result: we extend 5.12 to allow some singular points on the boundary.

Exercise 5.46 *Show that a finite set of $z_0 \in \partial U$ can be exempted from the hypothesis in* 5.12, *as long as $\overline{\lim}_{z \to z_0} h(z) < \infty$ for such z_0, and the conclusion still holds.*

Hints: Let d denote the diameter of U, $z_1, \ldots, z_N$ the exceptional points of ∂U. Each function $z - z_n$ is holomorphic and zero-free in U and so has local holomorphic logarithms (3.19). Consequently, $\log|z - z_n|$ is harmonic in U. The sum $u(z) = \sum_{n=1}^{N} 2^{-n} \log(|z - z_n|/d)$ is then also a harmonic function in U. It is non-positive in U and $\lim_{z \to z_n} u(z) = -\infty$ for each n. Consequently, for each $\varepsilon > 0$ the function $h_\varepsilon = h + \varepsilon u$ fulfills all the hypotheses of 5.12 and therefore $h_\varepsilon \le M$ throughout U, i.e., $h(z) + \varepsilon u(z) \le M$ for each $z \in U$. This holds for every $\varepsilon > 0$, so let $\varepsilon \downarrow 0$.

Remark: It would have cost no more effort to allow a countable number of exceptional points: the relevant series for u in the above proof converges to a harmonic function by 5.44(iii). However, even this version of the result is to be subsumed later under 7.15.

Lemma 5.47 *There is a non-negative harmonic function h in $D(0, 1)\backslash(-1, 0]$ which is bounded by* 1 *and satisfies*

(i) $$\overline{\lim_{z \to \zeta}}\, h(z) = \begin{cases} 0, & \zeta \in C(0, 1)\backslash\{-1\} \\ 1, & \zeta \in (-1, 0) \end{cases}$$

and

(ii) $$h(z) \ge 1 - \frac{4}{\pi} \tan^{-1} \sqrt{|z|} \ge \frac{1}{\pi}(1 - |z|) \quad \forall z \in D(0, 1)\backslash(-1, 0].$$

Proof: Let L be the holomorphic logarithm in $\mathbb{C}\backslash(-\infty, 0]$ which satisfies

(1) $L(re^{i\theta}) = \log r + i\theta, \quad r > 0, |\theta| < \pi.$

Let $D = D(0, 1)$, $C = C(0, 1)$, $H = (0, \infty) \times \mathbb{R}$, $U = \mathbb{R} \times (0, \infty)$ and consider first

$$\phi(z) = \frac{1 + z}{1 - z}, \quad z \in \bar{D}\backslash\{1\}.$$

We know (cf. 2.5) that ϕ is conformal in D and that

(2) $\phi(D \cap U) = H \cap U,$

(3) $\phi(C \cap U) = i(0, \infty),$

(4) $\phi(-1, 1) = (0, \infty).$

Therefore the composite

$$(5) \qquad v = \frac{2}{\pi} \operatorname{Im} L \circ \phi$$

is defined in $\bar{D} \cap U$, is harmonic in $D \cap U$ and satisfies

$(2)' \quad v(D \cap U) = (0, 1) \quad$ by (1), (5) and (2)

$(3)' \quad v(C \cap U) = \{1\} \quad$ by (1), (5) and (3)

$(4)' \quad v(-1, 1) = \{0\} \quad$ by (1), (5) and (4).

Now look at the rotated square-root map $z \to ie^{L(z)/2}$. Because of (1) it evidently maps $D \setminus (-1, 0]$ onto $D \cap U$, $C \setminus \{-1\}$ onto $C \cap U$ and $(0, 1)$ onto $i(0, 1)$. Therefore for the composite

$$(6) \qquad h(z) = 1 - v(ie^{L(z)/2}), \quad z \in \bar{D} \setminus [-1, 0]$$

we have from (2)′ and (3)′ that

$$(7) \qquad h(D \setminus (-1, 0]) \subset (0, 1)$$

$$(8) \qquad h(C \setminus \{-1\}) = \{0\}.$$

But from (1) it is clear that for any $x \in (-\infty, 0)$

$$\operatorname{Im} L(z) \to \begin{cases} \pi & \text{if } \operatorname{Im} z > 0 \text{ and } z \to x \\ -\pi & \text{if } \operatorname{Im} z < 0 \text{ and } z \to x. \end{cases}$$

Consequently, for any $x \in (-\infty, 0)$

$$ie^{L(z)/2} \to \begin{cases} -\sqrt{|x|} & \text{if } \operatorname{Im} z > 0 \text{ and } z \to x \\ \sqrt{|x|} & \text{if } \operatorname{Im} z < 0 \text{ and } z \to x. \end{cases}$$

It follows then from (6) and (4)′ that

$$(9) \qquad \lim_{z \to x} h(z) = 1 \quad \forall x \in (-1, 0).$$

All that remains is to check (ii). To this end note that by direct computation, for any $w \in \mathbb{C} \setminus \{-1\}$

$$(10) \qquad \frac{\operatorname{Im}\left(\dfrac{1 + w}{1 - w}\right)}{\operatorname{Re}\left(\dfrac{1 + w}{1 - w}\right)} = \frac{2 \operatorname{Im} w}{1 - |w|^2}.$$

We look at $z = re^{i\theta} \in D \setminus (-1, 0]$ $(0 < r < 1, |\theta| < \pi)$. Then

$$w = ie^{L(z)/2} = i\sqrt{r}e^{i\theta/2} \in D \cap U$$

and so by (2)

$$(11) \qquad \phi(ie^{L(z)/2}) = Re^{i\tilde{\theta}}$$

with $R > 0$ and $0 < \tilde{\theta} < \pi/2$. According to (10) we have

$$(12)\quad \tan\tilde{\theta} = \frac{2\sqrt{r}\cos\theta/2}{1-r}.$$

From (1) and (11)

$$\begin{aligned}\operatorname{Im} L(\phi(ie^{L(z)/2})) &= \tilde{\theta} \\ &\overset{(12)}{=} \tan^{-1}\left(\frac{2\sqrt{r}\cos\theta/2}{1-r}\right) \\ &\le \tan^{-1}\left(\frac{2\sqrt{r}}{1-r}\right).\end{aligned}$$

It follows from (5), (6) and the last inequality that

$$h(z) \ge 1 - \frac{2}{\pi}\tan^{-1}\left(\frac{2\sqrt{r}}{1-r}\right) = 1 - \frac{2}{\pi}\tan^{-1}\left(\frac{2\sqrt{|z|}}{1-|z|}\right).$$

If we use the "double angle" formula for the function tan, the last inequality becomes

$$\begin{aligned}h(z) &\ge 1 - \frac{4}{\pi}\tan^{-1}\sqrt{|z|} \\ &= \frac{4}{\pi}\left[\frac{\pi}{4} - \tan^{-1}\sqrt{|z|}\right], \quad z \in D\backslash(-1,0].\end{aligned}$$

Recalling the discussion in 3.22, we have then

$$\begin{aligned}h(z) &\ge \frac{4}{\pi}\int_{\sqrt{|z|}}^{1}\frac{dt}{1+t^2} \ge \frac{4}{\pi}\int_{\sqrt{|z|}}^{1}\frac{dt}{2} = \frac{2}{\pi}(1-\sqrt{|z|}) \\ &= \frac{2}{\pi}\frac{1-|z|}{1+\sqrt{|z|}} > \frac{1}{\pi}(1-|z|).\end{aligned}$$

Corollary 5.48 *Let f be holomorphic and bounded by 1 in $D(0,1)\backslash(-1,0]$ and for some $0 < \delta < 1$ satisfy*

$$\overline{\lim_{z\to\zeta}}|f(z)| \le \delta \quad \forall\zeta \in (-1,0).$$

Then

$$|f(z)| \le \delta^{(1-|z|)/\pi} \quad \forall z \in D(0,1)\backslash(-1,0].$$

Proof: Let $D = D(0,1)$, $C = C(0,1)$. The function

$$\phi = \log(\max\{|f|,\delta\}) = \log\delta + \log^+\left|\frac{f}{\delta}\right|$$

is subharmonic in $D\backslash(-1,0]$, by 5.26(ii). It is bounded and satisfies

$$(1)\quad \overline{\lim_{z\to\zeta}}\,\phi(z) \le \begin{cases}0 & \text{if } \zeta \in C \\ \log\delta & \text{if } \zeta \in (-1,0).\end{cases}$$

Let h be the harmonic function of 5.47. According to 5.47(i) and (1) we have (remember that $-\log \delta > 0$)

$$\overline{\lim_{z\to\zeta}}[\phi - (\log \delta)h](z) \le 0$$

for all points $\zeta \in (C\backslash\{-1\}) \cup (-1, 0) = \partial(D\backslash(-1, 0])\backslash\{-1, 0\}$. Therefore by 5.46 the subharmonic function $\phi - (\log \delta)h$ is non-positive throughout $D\backslash(-1, 0]$:

$$\phi \le (\log \delta)h$$

(2) $\quad |f| \le e^{\phi} \le e^{(\log \delta)h} = \delta^h \quad$ in $D\backslash(-1, 0]$.

Since $0 < \delta < 1$ and $h(z) \ge (1 - |z|)/\pi$ for all $z \in D\backslash(-1, 0]$ [5.47(ii)], the desired upper bound on $|f(z)|$ follows from (2).

Exercise 5.49 *Let γ: $[0, 1] \to \mathbb{C}$ be a curve such that $\gamma(0), \gamma(1) \in \mathbb{R}$ and $|\mathrm{Im}\, \gamma| > 0$ on $(0, 1)$. Form*

$$\Gamma(t) = \begin{cases} \gamma(t) & 0 \le t \le 1 \\ \bar{\gamma}(2 - t) & 1 \le t \le 2. \end{cases}$$

Notice that Γ is a loop. Show that $|\mathrm{Ind}_\Gamma(x)| = 1$ for every real x between $\gamma(0)$ and $\gamma(1)$.

Hints: Replacing γ by $\bar{\gamma}$ if necessary, we can suppose that $\mathrm{Im}\, \gamma > 0$ in $(0, 1)$. By reversing the parameter direction if necessary, we can suppose that $\gamma(0) \le \gamma(1)$. Therefore we are to consider points x such that

(1) $\quad \gamma(0) < x < \gamma(1)$.

Let L be the holomorphic logarithm in $\mathbb{C}\backslash i(-\infty, 0]$ which satisfies

(2) $\quad L(re^{i\theta}) = \log r + i\theta \quad$ when $r > 0, \dfrac{-\pi}{2} < \theta < \dfrac{3\pi}{2}$.

(Cf. 3.43.) Form

$$\phi(t) = \begin{cases} L(\gamma(t) - x) & 0 \le t \le 1 \\ \bar{L}(\gamma(2 - t) - x) & 1 < t \le 2. \end{cases}$$

Notice that ϕ is continuous at $t = 1$ (and trivially elsewhere) because by (1) and (2), $L(\gamma(1) - x)$ is real. Since $\overline{e^w} = e^{\bar{w}}$ for any $w \in \mathbb{C}$, we see that

(3) $\quad e^{\phi} = \Gamma - x$.

It follows that

$$\mathrm{Ind}_{\Gamma - x}(0) = \frac{1}{2\pi i}[\phi(2) - \phi(0)] = \frac{1}{2\pi i}[\bar{L}(\gamma(0) - x) - L(\gamma(0) - x)]$$

$$= \frac{-1}{\pi}\mathrm{Im}\, L(\gamma(0) - x).$$

Since $\gamma(0) - x$ is a negative real number by (1), (2) shows that $\mathrm{Im}\, L(\gamma(0) - x) = \pi$.

Theorem 5.50 (MILLOUX [1925]) *Let $D = D(0, 1)$ and let γ be a curve with initial point 0, terminal point on $C(0, 1)$ and all other points in D. Let $f \in H(D)$ be bounded by 1 and for some $0 < \delta < 1$ satisfy*

$$|f(\zeta)| \le \delta \quad \forall \zeta \in \gamma.$$

Then

$$|f(z)| \le \delta^{(1-|z|)/2\pi} \quad \forall z \in D.$$

Proof: (PÓLYA [1929], p. 633) After a rotation we may suppose that the terminal point of γ is -1. Let bar denote complex conjugation and consider the function

$$(1) \qquad F(z) = f(z)\bar{f}(\bar{z}), \quad z \in D.$$

By 2.4 it is holomorphic. Of course it is bounded by 1 in D and it satisfies

$$(2) \qquad |F(\zeta)| \le \delta \quad \forall \zeta \in \gamma \cup \bar{\gamma}.$$

Consider any $x \in (-1, 0)$. If $x \in \gamma$, then $|F(x)| \le \delta$ by (2). If $x \notin \gamma$, there is an appropriate portion γ_1 of γ which, except for having a different parameter interval, answers to the description of γ in 5.49. Therefore $x \in C$, a bounded component of $\mathbb{C}\backslash\gamma_1 \cup \bar{\gamma}_1$. Necessarily $C \subset D$, since $\mathbb{C}\backslash(D \cup \{-1\})$ is connected and disjoint from $\gamma \cup \bar{\gamma}$. Since C is a component of $\mathbb{C}\backslash\gamma_1 \cup \bar{\gamma}_1$, we have $\partial C \subset \gamma_1 \cup \bar{\gamma}_1 \subset \gamma \cup \bar{\gamma}$. From (2) then

$$|F| \le \delta \quad \text{on } \partial C.$$

Therefore by the Maximum Modulus Principle $|F| \le \delta$ holds throughout C, in particular at the point x. We have therefore shown that

$$|F(x)| \le \delta \quad \forall x \in (-1, 0),$$

i.e.,

$$|f(x)|^2 \le \delta \quad \forall x \in (-1, 0).$$

We are therefore in a position to apply 5.52 with $\delta^{1/2}$ in the role of the δ there.

Remarks 5.51 For the history of this important theorem (which derives from anterior work of Carleman) see the notes in LANDAU [1930] and R. NEVANLINNA [1933]. Landau proves that the curve γ can be abandoned in favor of the weaker hypothesis

$$(*) \qquad \sup_{0 \le r < 1} \inf_{\theta \in \mathbb{R}} |f(re^{i\theta})| \le \delta,$$

that is, on each circle centered at 0 there be at least one point at which $|f|$ does not exceed δ. Landau's clever proof of this does not use the harmonic majorant technique. V. I. LEVIN [1934] replaced $\sup_{0 \le r < 1}$ in (*) with $\sup_{\rho < r < R}$ and found a similar inequality as Milloux but with the constant $1/2\pi$ replaced by one dependent on ρ and R. Though he seems to rely on Landau's result, his proof is independent and different. The question of the largest admissible constant in the role of $1/2\pi$ is also discussed in Landau's paper. Also, in the curve version of the theorem, we might relax $f \in H(D)$ to $f \in H(D\backslash\gamma)$. [The inequalities about $|f|$ on γ become $\overline{\lim}\,|f|$ as γ is approached through $D\backslash\gamma$.] In addition to Landau's and Nevanlinna's references see MILLOUX [1930], [1932], AHLFORS [1932b], BEURLING [1933], MAITLAND [1939], HEINS [1945], LEWIS [1958], the treatment in CARATHÉODORY's book [1960] (pp. 107–109) and the definitive treatment in NEVANLINNA's book [1970] (pp. 102–113). MILLOUX [1948a] is an elegant qualitative version of Landau's result above with a very elementary *reductio ad absurdum* proof based on 7.5 or 7.6.

Here is a related example of the harmonic majorization method. It is convenient in 5.52 through 5.54 to use the abbreviations: $D = D(0, 1)$, $C = C(0, 1)$, $U = \mathbb{R} \times (0, \infty)$, $D^+ = D \cap U$, $C^+ = C \cap U$.

Lemma 5.52 *There exists a harmonic function h_+ in D^+ which is bounded by* 1 *and satisfies*

(i) $$\lim_{z \to \zeta} h_+(z) = \begin{cases} 0 & \text{if } \zeta \in C^+ \\ 0 & \text{if } \zeta \in (-1, 0) \\ 1 & \text{if } \zeta \in (0, 1), \end{cases}$$

(ii) $$\lim_{r \downarrow 0} \left[1 - \frac{\theta}{\pi} - h_+(re^{i\theta})\right] = 0 \quad \textit{uniformly for } \theta \in (0, \pi).$$

Proof: Consider first the function

$$\phi_1(z) = \frac{1 + z}{1 - z}, \quad z \in \bar{D}\backslash\{1\}$$

which was examined in 5.47. From the results of 5.47 and by direct inspection it is easy to see that

$$\phi_1(-1, 0) = (0, 1)$$
$$\phi_1(0, 1) = (1, \infty)$$
$$\phi_1(C^+) = i(0, \infty)$$
$$\phi_1(D^+) = (0, \infty) \times (0, \infty).$$

Therefore

$$\phi_2 = \phi_1^2$$

satisfies

$$\phi_2(-1, 0) = (0, 1)$$
$$\phi_2(0, 1) = (1, \infty)$$
$$\phi_2(C^+) = (-\infty, 0)$$
$$\phi_2(D^+) = U.$$

Finally the map

$$\phi_3(z) = \phi_2(z) - 1 = \frac{4z}{1 - z^2}$$

satisfies

$$\phi_3(-1, 0) = (-1, 0)$$
$$\phi_3(0, 1) = (0, \infty)$$
$$\phi_3(C^+) = (-\infty, -1)$$
$$\phi_3(D^+) = U.$$

Now let L be the holomorphic logarithm in $\mathbb{C}\backslash i(-\infty, 0]$ which satisfies

$$L(re^{i\theta}) = \log r + i\theta \quad \forall r > 0,\ \theta \in (-\pi/2, 3\pi/2).$$

(Cf. 3.43.) Look at

$$h(z) = 1 - \frac{1}{\pi} \operatorname{Im} L(z), \quad z \in \mathbb{C}\backslash i(-\infty, 0].$$

It is a harmonic function, is bounded by 1 in U, and satisfies

$$h(-\infty, 0) = 0$$
$$h(0, \infty) = 1.$$

If we set $h_+ = h \circ \phi_3$, property (i) is clear from all the foregoing.

Now for $r > 0$ and $\theta \in (0, \pi)$ we have

$$-1 + \frac{\theta}{\pi} + h_+(re^{i\theta}) = \frac{\theta}{\pi} - \frac{1}{\pi} \operatorname{Im} L\left[\frac{4re^{i\theta}}{1 - (re^{i\theta})^2}\right]$$
$$= \frac{1}{\pi} \operatorname{Im} L(e^{i\theta}) - \frac{1}{\pi} \operatorname{Im} L\left[\frac{e^{i\theta}}{1 - (re^{i\theta})^2}\right].$$

Since L is uniformly continuous on a neighborhood of $\overline{C^+}$, the assertion (ii) is now obvious.

Exercise 5.53 *Modify the last argument appropriately to obtain a harmonic function h_- on D^+ which is bounded by 1 and satisfies*

(i) $$\lim_{z \to \zeta} h_-(z) = \begin{cases} 0 & \text{if } \zeta \in C^+ \\ 1 & \text{if } \zeta \in (-1, 0) \\ 0 & \text{if } \zeta \in (0, 1), \end{cases}$$

(ii) $$\lim_{r \downarrow 0} \left[\frac{\theta}{\pi} - h_-(re^{i\theta})\right] = 0 \quad \textit{uniformly for } \theta \in (0, \pi).$$

Lemma 5.54 *Let* $\gamma\colon [0,1] \to \overline{D^+}$ *be an arc with* $\gamma(0) = 0$, $\gamma(1) \in C^+$ *and* $\gamma(0,1) \subset D^+$. *Let* w *be a positive superharmonic function on* D^+ *such that* $w \geq 1$ *on* $\gamma(0,1)$. *Then for each* $\varepsilon > 0$ *there exists an* $0 < r_\varepsilon < 1$ *such that*

$$w(re^{i\theta}) \geq \min\left\{\frac{\theta}{\pi}, 1 - \frac{\theta}{\pi}\right\} - \varepsilon \quad \textit{for all } 0 < r < r_\varepsilon,\ \theta \in (0, \pi).$$

Proof: $\gamma(1)$ determines two arcs on C^+ and each of them together with γ itself and one of the intervals $(-1, 0)$, $(0, 1)$ determines a Jordan-curve. Let them be J_-, J_+ with $(-1,0) \subset J_-$, $(0,1) \subset J_+$. The regions $\Omega_- = \mathscr{I}(J_-)$, $\Omega_+ = \mathscr{I}(J_+)$ have the property that

$$D^+ = \Omega_- \cup \gamma(0,1) \cup \Omega_+ \quad \text{(disjoint union)}.$$

For all these assertions see 4.47. Since $w \geq 1 \geq h_+$ on $\gamma(0,1)$ and $w \geq 0$ throughout D^+, it follows from 5.52(i) that

$$\varlimsup_{z \to \zeta} (h_+ - w)(z) \leq 0 \quad \forall \zeta \in C^+ \cup (-1, 0) \cup \gamma(0,1).$$

This union contains $J_- \backslash \{-1, 0\} = \partial\Omega_- \backslash \{-1, 0\}$ and so from the Maximum Principle in the form 5.46 we conclude that

$$w \geq h_+ \quad \text{on } \Omega_-.$$

Similarly

$$w \geq h_- \quad \text{on } \Omega_+.$$

Therefore

$$w \geq \min\{h_+, h_-\} \quad \text{on } \Omega_- \cup \Omega_+ = D^+ \backslash \gamma.$$

Since $h_+ \leq 1$, $h_- \leq 1$ and $w \geq 1$ on γ, we then actually have

$$w \geq \min\{h_+, h_-\} \quad \text{on } D^+.$$

The assertion of the lemma follows easily from this inequality and part (iii) of 5.52 and 5.53.

Remark 5.55 Use of the Jordan curve theorem (in the form of 4.47) is only a convenience. This can be avoided (and "arc" relaxed to "curve" in the hypothesis) by using an argument like that in 4.9 to show that $(-1, 0)$ and $(0, 1)$ each lie on the boundary of a different component of $D^+ \backslash \gamma$. It is easy to see that the boundaries of these components lie in $[-1, 0] \cup C^+ \cup \gamma$ and $[0, 1] \cup C^+ \cup \gamma$, respectively.

Corollary 5.56 (LINDELÖF [1915]) *Let* f *be holomorphic and bounded in* $D^+ = \{z \in \mathbb{C}: |z| < 1 < \operatorname{Im} z + 1\}$ *and* $\Gamma\colon [0,1] \to \mathbb{C}$ *be an arc such that* $\Gamma(0) = 0$, $\Gamma(0,1) \subset D^+$ *and* $\lim_{t \downarrow 0} f(\Gamma(t)) = a$, *for some* $a \in \mathbb{C}$. *Then for each* $\varepsilon > 0$

$$(*) \qquad \lim_{r \downarrow 0} f(re^{i\theta}) = a \quad \textit{uniformly for } \theta \in (\varepsilon, \pi - \varepsilon).$$

If $\tilde{\Gamma}$ and $\tilde{a}$ are another such arc and limit, then from (*) *and its tilde analog it follows that $\tilde{a} = a$.*

Proof: We can assume $a = 0$ and $|f| \leq 1$. Let $0 < \varepsilon < 1$ be given and set $\delta(\varepsilon) = e^{(2/\varepsilon)\log \varepsilon}$. Then choose $0 < T < 1$ sufficiently small that

(1) $\quad |f(\Gamma(t))| \leq \delta(\varepsilon) \quad \forall 0 < t \leq T.$

The function

$$h = \log(\max\{|f|, \delta(\varepsilon)\}) = \log \delta(\varepsilon) + \log^+\left|\frac{f}{\delta(\varepsilon)}\right|$$

is subharmonic in D^+ by 5.26(ii) and is non-positive. Since $\Gamma(0) = 0$ and Γ is continuous, there is a least $t \in (0, T]$, call it t^*, such that $|\Gamma(t^*)| = |\Gamma(T)|$ ($= R$, say). Then (1) yields

$$h(\Gamma(t^*t)) = \log \delta(\varepsilon) = \frac{2}{\varepsilon}\log \varepsilon \quad \forall 0 < t \leq 1.$$

Because of the minimality of t^*, the arc $\gamma(t) = R^{-1}\Gamma(t^*t)$, $t \in [0, 1]$, and the superharmonic (remember, $\log \varepsilon < 0$) function

(2) $\quad w(z) = \dfrac{\varepsilon h(Rz)}{2 \log \varepsilon}, \quad z \in D^+,$

fulfill the hypotheses of 5.54. Therefore there exists an $0 < r_\varepsilon < 1$ such that

$$w(re^{i\theta}) \geq \min\left\{\frac{\theta}{\pi}, 1 - \frac{\theta}{\pi}\right\} - \frac{\varepsilon}{2} \quad \forall 0 < r < r_\varepsilon,\ \theta \in (0, \pi)$$

$$> \varepsilon - \frac{\varepsilon}{2} \quad \forall 0 < r < r_\varepsilon,\ \theta \in (\varepsilon\pi, (1 - \varepsilon)\pi)$$

(3) $\quad = \dfrac{\varepsilon}{2}.$

From (2), (3) and the fact that $\log \varepsilon < 0$

$$h(re^{i\theta}) \leq \log \varepsilon \quad \forall 0 < r < Rr_\varepsilon,\ \theta \in (\varepsilon\pi, (1 - \varepsilon)\pi),$$

whence

$$|f(re^{i\theta})| \leq e^{h(re^{i\theta})} \leq \varepsilon \quad \forall 0 < r < Rr_\varepsilon,\ \theta \in (\varepsilon\pi, (1 - \varepsilon)\pi).$$

Remarks 5.57 Lindelöf proved 5.56 under the weaker hypothesis that there be at least two complex numbers absent from the range of f. (See also Tanaka [1964], [1976].) We will secure a version of this in Chapter XII. (See 12.30 and 12.35.) For a beautiful geometric proof of 5.56 which does not use the Jordan curve theorem (or any surrogate) see §§ 307–308 of Carathéodory [1960].

There are many generalizations and extensions of 5.56. E.g., if one uses the fact, mentioned in the Chapter I notes, that every curve is arcwise connected, the arc in 5.56 can be allowed to be any curve and the conclusion still follows (cf. Remark 5.55). However, the hypotheses can be weakened much further: HALL [1937] shows that if f is a bounded holomorphic function in $\operatorname{Im} z > 0$ and $\inf\{|f(re^{i\theta})| : 0 < \theta < \pi\} \to 0$ as $r \to \infty$ through an appropriately dense set of r values, then $\lim_{r\to\infty} f(re^{i\theta}) = 0$ uniformly in $\varepsilon \le \theta \le \pi - \varepsilon$, for each $\varepsilon > 0$. (Cf. 12.35 and CARTWRIGHT [1935].) An account of Hall's work may also be found in FUCHS [1967], pp. 81–87 and an extension in KAWAKAMI [1956]. For a generalization of 5.56 of a different kind see GEHRING and LOHWATER [1958] (or p. 20 of COLLINGWOOD and LOHWATER [1966]).

We offer one final interesting application of harmonic majorization.

Exercise 5.58 (i) *Let $a > 0$, $R = (-\pi/2, \pi/2) \times (0, a)$, $B = \{\pm\pi/2\} \times [0, a] \cup [-\pi/2, \pi/2]$. Prove that there exists a continuous function $g: \bar{R} \to [0, 1]$ which is harmonic and positive in R and 0 on B.*

Hints: Let $D = D(0, 1)$, $U = \{z \in \mathbb{C}: \operatorname{Im} z > 0\}$, $\psi(w) = (w - i)/(w + i)$ $(w \in \bar{U})$. Then ψ maps $\bar{U}$ onto $\bar{D}\backslash\{1\}$ and $\mathbb{R}$ onto $C(0, 1)\backslash\{1\}$. The function sin maps $\bar{R}$ into $\bar{U}$ and B into a compact subset of $\mathbb{R}$. Therefore the composite $\psi \circ \sin$ maps $\bar{R}$ into $\bar{D}\backslash\{1\}$ and B onto a compact subset K of $C(0, 1)\backslash\{1\}$. Let $f: C(0, 1) \to [0, 1]$ be any continuous function satisfying $f(K) = 0$, $f(1) = 1$ and let F solve the Dirichlet problem in D for f. Since $F = f \ge 0$ on $C(0, 1)$, it follows from 5.10 that $F > 0$ in D. Let $g = F \circ \psi \circ \sin$.

(ii) *Let $r > 1$, $S_r = \{z \in \mathbb{C}: \operatorname{Re} z > 0, 1/r < |z| < r\}$. Prove that there is a continuous function $f_r: \bar{S}_r \to [0, 1]$ which is harmonic and positive in S_r and 0 on $\partial S_r\backslash[i/r, ir]$.*

Hint: $z \to (1/r)e^{-iz}$ is a conformal map of $(-\pi/2, \pi/2) \times (0, 2\log r)$ onto S_r and a homeomorphism between their closures.

Exercise 5.59 *Let $r > 1$, $S_r = \{z \in \mathbb{C}: \operatorname{Re} z > 0, 1/r < |z| < r\}$. Prove that there is a positive constant $c(r)$, dependent only on r, such that $u(1) \ge c(r)$ holds for all continuous functions u on $\bar{S}_r$ which are harmonic and positive in S_r and for which there exists a curve $\gamma: [0, 1] \to \{z \in \mathbb{C}: \operatorname{Re} z > 0\}$ with $|\gamma(0)| = 1/r$, $|\gamma(1)| = r$ and $u \ge 1$ on γ.*

Hints: Let such a function u and such a curve γ be given. It suffices to deal with the case where $1 \notin \gamma$. Let Ω be the component of $S_r\backslash\gamma$ which contains 1. According to 4.9 $\bar{\Omega}$ cannot meet both the positive and negative y-axis. Therefore either $\partial\Omega \subset \gamma \cup \partial S_r\backslash[i/r, ir]$ or $\partial\Omega \subset \gamma \cup \partial S_r\backslash[-i/r, -ir]$. Say, the former holds. If f_r is as provided by 5.58(ii), then $u - f_r \ge 1 - f_r \ge 0$ on γ and $u - f_r = u \ge 0$ on $\partial S_r\backslash[i/r, ir]$. Therefore $u - f_r \ge 0$ on $\partial\Omega$. It follows from the Maximum Principle that $u - f_r \ge 0$ throughout Ω. In particular, $u(1) \ge f_r(1)$.

Theorem 5.60 (JØRGENSEN [1939]) *Let* $H = \{z \in \mathbb{C}: \operatorname{Re} z > 0\}$, $f: H \to \overline{H}$ *holomorphic. Suppose that*

(i) $$\lim_{\substack{x\to\infty\\ x\in\mathbb{R}}} \frac{\operatorname{Re} f(x)}{x} = 0$$

and that for some finite constant M

(ii) $$\overline{\lim_{z\to iy}} \operatorname{Re} f(z) \le M \quad \forall y \in \mathbb{R}.$$

Then $\operatorname{Re} f \le M$ *throughout* H.

Proof: From (i) and (ii) it follows easily that for each $\delta > 0$ and each $\varepsilon > 0$ there is a finite constant $M(\varepsilon) \ge M$ such that

(1) $$\overline{\lim_{z\to\zeta}}[\operatorname{Re} f(z) - \varepsilon \operatorname{Re} z - \delta|\operatorname{Im} z|] \le M(\varepsilon) \quad \forall \zeta \in i\mathbb{R} \cup (0, \infty).$$

We want first to establish the inequality

(2) $$\operatorname{Re} f(z) - \varepsilon \operatorname{Re} z - \delta|\operatorname{Im} z| \le M(\varepsilon) \quad \forall z \in H.$$

The arguments for $H_+ = \{z \in H: \operatorname{Im} z > 0\}$ and $H_- = \{z \in H: \operatorname{Im} z < 0\}$ are the same, so we (only) look at H_+ and argue by contradiction, i.e., we assume that the (open) set

$$U = \{z \in H_+: \operatorname{Re} f(z) - \varepsilon \operatorname{Re} z - \delta \operatorname{Im} z > M(\varepsilon)\}$$

is not void. Let C be any component of U. We have $\partial C \subset \partial U \subset \overline{H}_+\backslash U$. At points z of $H_+\backslash U$ we have (2). The other points ζ of $\overline{H}_+\backslash U$ lie in $\overline{H}_+\backslash H_+ \subset i\mathbb{R} \cup (0, \infty)$ and for them we have (1). Therefore

$$\overline{\lim_{z\to\zeta}}[\operatorname{Re} f(z) - \varepsilon \operatorname{Re} z - \delta \operatorname{Im} z] \le M(\varepsilon) \quad \forall \zeta \in \partial C.$$

If C were bounded, then this inequality would imply (by 5.12) the validity of the inequality in (2) for all $z \in C \subset U$, in contradiction with the definition of U. Therefore C must be unbounded. Since C is open (1.30(i)′) and connected, it follows readily from 1.28 that there exists a continuous $\gamma: [0, 1) \to C$ such that

(3) $$\overline{\lim_{t\uparrow 1}}|\gamma(t)| = \infty.$$

We have then

$$\operatorname{Re} f(z) - \varepsilon \operatorname{Re} z - \delta \operatorname{Im} z > M(\varepsilon) \quad \forall z \in \gamma$$

and so *a fortiori*

(4) $$\operatorname{Re} f(z) > \eta[\operatorname{Re} z + \operatorname{Im} z] \ge \eta|z| \quad \forall z \in \gamma,$$

where $\eta = \min\{\varepsilon, \delta\}$. For each sufficiently large r, e.g., $r > r_0 = \inf|\gamma|[0, 1)$, the curve γ meets $C(0, r)$. Set $S_2 = \{z \in \mathbb{C}: \operatorname{Re} z > 0, \frac{1}{2} < |z| < 2\}$ and consider $x > 2r_0$. (4) implies that

$$\operatorname{Re} f(xz) > \eta|xz| > \tfrac{1}{2}\eta x \quad \text{whenever } xz \in \gamma \cap xS_2.$$

Therefore if we set $u(z) = 2 \operatorname{Re} f(xz)/\eta x$, then

(5) $\quad u(z) > 1 \quad \text{whenever } z \in \dfrac{1}{x}\gamma \cap S_2.$

We cite 5.59 to conclude therefore that $u(1) \geq c(2)$, that is,

$$\frac{2 \operatorname{Re} f(x)}{\eta x} \geq c(2)$$

$$\frac{\operatorname{Re} f(x)}{x} \geq \frac{\eta c(2)}{2}.$$

This holds for all sufficiently large positive x, in violation of (i). It follows that U is void and (2) holds throughout H_+. Similarly (2) holds throughout H_-, thus throughout H. Let $\delta \downarrow 0$ in (2) to conclude that

(6) $\quad \operatorname{Re} f(z) - \varepsilon \operatorname{Re} z \leq M(\varepsilon) \quad \forall z \in H.$

To the function $e^{f(z) - \varepsilon z}$ we apply 5.13 and conclude from (ii) and (6) that

(7) $\quad \operatorname{Re} f(z) - \varepsilon \operatorname{Re} z \leq M \quad \forall z \in H.$

Let $\varepsilon \downarrow 0$ in (7) and the proof is complete.

Remark 5.61 Of course 5.60 is a theorem about positive harmonic functions in the right half-plane H. To secure this (apparent) generalization, the reader can either repeat the above proof, citing 7.15 instead of 5.13, or he can note that any positive harmonic function in H is the real part of a single holomorphic function there (5.21(i) + 2.5) and apply 5.60 as proven. [Another proof of 5.60 occurs in 8.36(iv).]

§ 6 Uniqueness Theorems

The general theme of this section, well illustrated by the first theorem, is the complete determination of a holomorphic function from its behavior on a small subset of its domain. The Cauchy integral formula already illustrates this, since it re-captures all the function values inside a disk in terms of those on the boundary.

Theorem 5.62 (Fundamental Uniqueness Theorem of Riemann) *If f is holomorphic in the region Ω and $f^{-1}(0)$ has a limit point in Ω, then $f \equiv 0$.*

Proof: Let Ω_0 be the set of limit points *in* Ω of $f^{-1}(0)$. Thus $\Omega_0 \neq \varnothing$ by hypothesis. We will show that the hypothesis of the Basic Connectedness Lemma is fulfilled with $\theta = 1$ and conclude therefrom that $\Omega_0 = \Omega$. Since f is continuous, $\Omega_0 \subset f^{-1}(0)$ and we shall be finished. So suppose $z_0 \in \Omega_0$ and $D(z_0, r) \subset \Omega$. Use 5.30 to write

$$f(z) = \sum_{n=0}^{\infty} c_n(z - z_0)^n \quad \forall z \in D(z_0, r).$$

We claim that $D(z_0, r) \subset f^{-1}(0)$, whence $D(z_0, r) \subset \Omega_0$. For if not, there is a first non-zero coefficient in this series, say c_k:

$$f(z) = (z - z_0)^k \sum_{n=0}^{\infty} c_{n+k}(z - z_0)^n = (z - z_0)^k h(z),$$

where h is defined in $D(z_0, r)$ by the obvious power series and $h(z_0) = c_k \neq 0$. Since h is continuous, it is non-zero sufficiently near z_0 and so $f(z) = (z - z_0)^k h(z)$ shows that f has no zeros near z_0 except z_0 itself, contrary to z_0 being a limit point of $f^{-1}(0)$.

Remark 5.63 This result is best possible in the sense that if A is any subset of Ω with no limit points in Ω, then there is a holomorphic function on Ω whose zero set is exactly A. This will be proved in 7.32.

Example 5.64 *If f is holomorphic in the region Ω and for each $z \in \Omega$ there is a non-negative integer $n = n(z)$ such that $f^{(n)}(z) = 0$, then f is a polynomial.*

Proof: Setting $\Omega_n = \{z \in \Omega : f^{(n)}(z) = 0\}$ ($n = 0, 1, 2, \ldots$), the zero set of the holomorphic function $f^{(n)}$, the hypothesis is that $\Omega = \bigcup_{n=0}^{\infty} \Omega_n$. Since Ω is uncountable and a countable union of compacta (by 1.31), some Ω_n meets some compact subset of Ω in an infinite set, hence has a limit point in Ω. Then it follows that $f^{(n)}$ is identically zero in Ω. Finally, n-successive applications of 2.10(iv) show that f is a polynomial of degree at most $n - 1$.

Remark: DONOGHUE [1969], p. 53 gives a neat but intricate Baire Category argument to establish the generalization of this result to infinitely differentiable functions on an open interval of the real line, due to COROMINAS and SUNYER BALAGUER [1954].

Exercise 5.65 (Extension of 4.60) *Let U be an open subset of $\mathbb{C}$, $f \in H(U)$, $g: U \to \mathbb{C}$ a continuous function such that $g^n = f$ for some positive integer n. Show that g is holomorphic in U.*

Hints: We may assume U is connected. All is trivial if f is the 0 function. Otherwise $A = f^{-1}(0)$ consists of isolated points, by 5.62, so $U \backslash A$ is open. g is holomorphic there by 4.60 and, being continuous, it is holomorphic near each point of A as well by 5.41.

Exercise 5.66 *Let f be holomorphic in $D(0, 1)$ and suppose that $|f(re^{i\theta})|$ is constant for $r \in [0, 1)$ and also that $|f(re^{i\psi})|$ is constant for such r. If $(\theta - \psi)/\pi$ is irrational, then f is constant.*

Hints: If the constant value of $|f|$ on (either of) the rays is 0, then f is 0 by 5.62. Otherwise $f(0) \neq 0$ and by considering $f(\rho e^{i\psi}z)/f(0)$ for sufficiently small positive ρ, we may suppose that f has no zeros in $D = D(0, 1)$, that $\psi = 0$ and that $f(0) = 1$. 5.35 provides a $g \in H(D)$ such that $f = e^{ig}$. Write $g(z) = \sum_{n=0}^{\infty} c_n z^n$. Since $|f(t)| = |f(0)| = 1$ for all $0 \leq t < 1$, we see that g is real-valued in $[0, 1)$ and so

(1) $\quad c_n \in \mathbb{R}, \quad n = 0, 1, 2, \ldots.$

Similarly $g(e^{i\theta}t)$ is real for all $0 \leq t < 1$, so we conclude that

(2) $\quad c_n e^{in\theta} \in \mathbb{R}, \quad n = 0, 1, 2, \ldots.$

If for some $n > 0$, $c_n \neq 0$ we conclude from dividing (2) by (1) that $e^{in\theta} \in \mathbb{R}$. This means $e^{in\theta} = \pm 1$ and so there is an integer m such that either $n\theta - \pi = 2\pi m$ or $n\theta = 2\pi m$. In either case θ is a rational multiple of π (since $n \neq 0$).

Exercise 5.67 *$a_n \in \mathbb{R}$ are distinct, $a_n \to 0$, f is holomorphic in $D(0, r)$ for some $r > 0$ and $f(a_n)$ is real for all large n.*

(i) *Show that $f(\bar{z}) = \bar{f}(z)$ for all $z \in D(0, r)$.*

Hint: Form $F(z) = \bar{f}(\bar{z})$, which is holomorphic in $D(0, r)$ [by 2.4], and consider $g = F - f$. It is holomorphic and vanishes at a_n for all large n.

(ii) *If all a_n are positive and $f(a_{2n}) = f(a_{2n+1})$ for all large n, then f is constant.*

Hint: Since by (i), f is real (and differentiable) on $(-r, r)$, the Mean Value Theorem provides (for each $n \geq N$, say) b_n between a_{2n} and a_{2n+1} such that

$$0 = \frac{f(a_{2n+1}) - f(a_{2n})}{a_{2n+1} - a_{2n}} = f'(b_n).$$

Since $a_n \to 0$, the set $\{b_N, b_{N+1}, \ldots\}$ is not finite and accumulates at 0. Conclude that $f' \equiv 0$.

Definition 5.68 If f is holomorphic and not constant in a connected neighborhood of a, it follows from 5.30 that $f^{(n)}(a) \neq 0$ for some non-negative integer n. The non-negative integer $n(a,f) = \min\{n : f^{(n)}(a) \neq 0\}$ is called the *multiplicity* (*or order*) *of f at a* or the *multiplicity* (*or order*) *of a as a zero of f.*

Exercise 5.69 *Let f be holomorphic and not constant in the region Ω. Let $a_1, \ldots, a_k \in \Omega$. Using 5.30 and induction on k show that there is a holomorphic function F in Ω such that*

$$f(z) = F(z) \prod_{j=1}^{k} (z - a_j)^{n(a_j,f)} \quad \forall z \in \Omega$$

and that $F(a_j) \neq 0$ for each $j = 1, \ldots, k$.

Exercise 5.70 (CAUCHY [1844]) *An entire function F such that $|F(z)| \to \infty$ as $|z| \to \infty$ is a polynomial.*

Hints: There is an $r_1 > 0$ such that $|F(z)| \geq 1$ for all $|z| \geq r_1$. Therefore the zeros of F are all in the compact set $\bar{D}(0, r_1)$, hence there are only finitely many, say $a_1, \ldots, a_l$ with multiplicities $m_1, \ldots, m_l$. Let

$$P(z) = \prod_{j=1}^{l} (z - a_j)^{m_j},$$

a monic polynomial of degree $n = m_1 + \cdots + m_l$. Evidently P/F is an entire function f and satisfies $|f(z)| \leq 2|z|^n$ for all $|z| \geq R$ and some $R \geq r_1$. (See (1) in the proof of 4.49.) For each $k > n$ and each $r \geq R$, 3.34 shows that

$$\frac{|f^{(k)}(0)|}{k!} \leq \frac{M(r)}{r^k} \leq \frac{2r^n}{r^k} = \frac{2}{r^{k-n}} \to 0 \quad \text{as } r \to \infty.$$

Therefore $f^{(k)}(0) = 0$ for $k > n$ and f is a polynomial. But by definition of P, the function P/F has no zeros. Thus the polynomial f has no zeros and so is constant, say $f \equiv c \neq 0$. Then $F = (1/c)P$.

Exercise 5.71 (i) *Let f be holomorphic in the region Ω and for some $a \in \Omega$ satisfy $f'' + f \equiv 0, f'(a) = f(a) = 0$. Show that $f \equiv 0$.*

Hint: $[(f')^2 + f^2]' \equiv 0$ so $(f' + if)(f' - if) = (f')^2 + f^2 \equiv (f'(a))^2 + (f(a))^2 = 0$. From 5.62 then follows that either $f' + if \equiv 0$ or $f' - if \equiv 0$. Now use 3.24 and the "boundary" conditions. Alternatively, form $g = f - if'$ and $h = f + if'$, check that $g' = ig$, $h' = -ih$ and then cite 3.24.

(ii) *Let F be a holomorphic function in the open starlike set Ω and set $g = F'' + F$. Show that $F \equiv A \sin + B \cos$, where A and B are primitives in Ω of the (holomorphic) functions $g \cos$ and $-g \sin$, respectively, which for some $a \in \Omega$ satisfy*

$$A(a) = F(a) \sin a + F'(a) \cos a, \qquad B(a) = F(a) \cos a - F'(a) \sin a.$$

(Note that such primitives exist by 5.2.)

Hint: Apply (i) to the function $f = F - A \sin - B \cos$.

(iii) *Modify the above techniques to solve the differential equation $F'' + cF = g$ (F as before, c a non-zero complex constant).*

Exercise 5.72 (i) *Let $f_1, \ldots, f_n$ be holomorphic in the region Ω. Show that if $\phi = |f_1| + \cdots + |f_n|$ assumes a maximum value in Ω, then each f_j is constant.*

Hints: If $\sup \phi(\Omega) = \phi(z_0)$, $z_0 \in \Omega$, then by replacing each f_j by the appropriate scalar multiple of itself, we can suppose that $f_j(z_0) = |f_j(z_0)|$. Then the maximum of $|f_1 + \cdots + f_n|$ over Ω is achieved at z_0 and is $\phi(z_0)$. It follows from 5.10 that $f_1 + \cdots + f_n \equiv f_1(z_0) + \cdots + f_n(z_0) = \phi(z_0) \geq \phi$. Taking real parts, we have

$\sum \operatorname{Re} f_j \geq \phi \geq \sum |f_j|$. Since $\operatorname{Re} f_j \leq |f_j|$ for each j, we infer that $\operatorname{Re} f_j = |f_j|$ for each j. Thus $\operatorname{Im} f_j = 0$ and f_j is constant by 2.17(ii).

(ii) *Let $F_1, \ldots, F_N$ be holomorphic in the region Ω, $p_1, \ldots, p_N$ positive real numbers and suppose that $\phi = |F_1|^{p_1} + \cdots + |F_N|^{p_N}$ assumes a maximum value in Ω. Show that each F_j is constant.*

Hints: Let $z_0 \in \Omega$ and $\phi \leq \phi(z_0)$. If $\phi(z_0) = 0$, all is trivial. Otherwise, after re-naming, we can suppose that $F_j(z_0) \neq 0$ for $1 \leq j \leq n \leq N$ and $F_j(z_0) = 0$ for $n < j \leq N$. Then for some $r > 0$, $D = D(z_0, r) \subset \Omega$ and $|F_j| > 0$ in D for $j = 1, 2, \ldots, n$. By 5.35 there exist $g_j \in H(D)$ such that $F_j = e^{g_j}$. It follows that $|F_j|^{p_j} = [e^{\operatorname{Re} g_j}]^{p_j} = e^{\operatorname{Re} p_j g_j} = |e^{p_j g_j}|$. Setting $f_j = e^{p_j g_j}$, we therefore have $\sum_{j=1}^{n} |f_j| \leq \phi \leq \phi(z_0) = \sum_{j=1}^{n} |f_j(z_0)|$ in D. According to (i), each f_j is constant in D. So then is each $|F_j| = |f_j|^{1/p_j}$ and then by 5.10 each F_j is constant in D, thence (5.62) constant in Ω. It follows that

$$\phi(z_0) = \sum_{j=1}^{n} |F_j(z_0)|^{p_j} \equiv \sum_{j=1}^{n} |F_j|^{p_j} \leq \sum_{j=1}^{N} |F_j|^{p_j} = \phi \leq \phi(z_0),$$

so $\sum_{j=n+1}^{N} |F_j|^{p_j} \equiv 0$.

According to 5.62 a holomorphic function in a region is uniquely determined by its values on any sequence with a limit point in the region. But how specifically can the values of the function at an arbitrary point be constructed from its values on the sequence? The next exercise answers this nicely.

Exercise 5.73 *Let f be holomorphic in a neighborhood of $\bar{D}(z_0, r)$, $z_n \in D(z_0, r)$ and $\lim_{n \to \infty} z_n = z_0$. Suppose the z_n are all distinct. Define $\psi_0 \equiv 1$,*

$$\psi_k(z) = \prod_{j=1}^{k} (z - z_j), \quad z \in \bar{D}(z_0, r), k = 1, 2, \ldots,$$

$$M_k(z) = \sup_{\xi \in C(z_0, r)} \left| \frac{z - z_k}{\xi - z_k} \right|, \quad z \in D(z_0, r), k = 0, 1, \ldots,$$

$$M = \sup_{\xi \in C(z_0, r)} |f(\xi)|.$$

(i) *Show that*

$$\psi'_{k+1}(z_l) = \prod_{\substack{j=1 \\ j \neq l}}^{k+1} (z_l - z_j) \neq 0, \quad l = 1, \ldots, k+1; k = 0, 1, \ldots$$

(ii) *Show that*

$$\frac{1}{\psi_{k+1}(z)} = \sum_{j=1}^{k+1} \frac{1}{(z - z_j)\psi'_{k+1}(z_j)}, \quad z \in C(z_0, r), k = 0, 1, \cdots$$

Hint: After a few algebraic maneuvers (using (i)), the desired equation is equivalent to

$$1 - \sum_{j=1}^{k+1} \prod_{\substack{l=1 \\ l \neq j}}^{k+1} \frac{z - z_l}{z_j - z_l} = 0.$$

Since this is a polynomial of degree at most k with the $k + 1$ distinct zeros $z_1, \ldots, z_{k+1}$, it is identically zero (4.50).

(iii) *Deduce from* (ii) *that*

$$\sum_{j=1}^{k+1} \frac{f(z_j)}{\psi'_{k+1}(z_j)} = \frac{1}{2\pi i} \int_{C(z_0,r)} \frac{f(\xi)}{\psi_{k+1}(\xi)}\, d\xi, \quad k = 0, 1, \ldots.$$

Call this number A_k.

(iv) (Cauchy, 1826. See also HERMITE [1878]) *Verify that*

$$f(z) - \sum_{k=0}^{n} A_k \psi_k(z) = \frac{1}{2\pi i} \int_{C(z_0,r)} \frac{f(\xi)\psi_{n+1}(z)\,d\xi}{\psi_{n+1}(\xi)(\xi - z)}$$

for all $z \in D(z_0, r)$ *and all positive integers* n.

Hint: (iii) and induction on n.

(v) *Deduce from* (iv) *that*

$$\left| f(z) - \sum_{k=0}^{n} A_k \psi_k(z) \right| \leq \sup_{\xi \in C(z_0,r)} \left| \frac{\psi_{n+1}(z)}{\psi_{n+1}(\xi)} \right| \frac{rM}{r - |z - z_0|}$$

$$\leq \frac{rM}{r - |z - z_0|} \prod_{k=1}^{n+1} M_k(z)$$

for all $z \in D(z_0, r)$ *and all positive integers* n.

(vi) *For each* $z \in D(z_0, r)$, $\xi \in C(z_0, r)$, $k = 0, 1, \ldots$

$$\left| \frac{z - z_k}{\xi - z_k} \right| \leq \frac{|z - z_0| + |z_k - z_0|}{\big||\xi - z_0| - |z_k - z_0|\big|} = \frac{|z - z_0| + |z_k - z_0|}{r - |z_k - z_0|}.$$

It follows that $\overline{\lim}_{k\to\infty} M_k(z) \leq |z - z_0|/r < 1$.

(vii) *Deduce from* (v) *and* (vi) *that*

$$\lim_{n\to\infty} \left| f(z) - \sum_{k=0}^{n} A_k \psi_k(z) \right| = 0, \quad z \in D(z_0, r).$$

Therefore, recalling (iii), *we have*

$$f(z) = \sum_{k=0}^{\infty} \left[\sum_{j=1}^{k+1} \frac{f(z_j)}{\psi'_{k+1}(z_j)} \right] \psi_k(z), \quad z \in D(z_0, r).$$

(viii) *Show that the convergence in* (vii) *is uniform for* z *in any compact subset of* $D(z_0, r)$.

Exercise 5.74 *Let* $\{f_m\}_{m=1}^{\infty}$ *be holomorphic and uniformly bounded in a neighborhood of* $\overline{D}(z_0, r)$ *and let* $z_1, z_2, \ldots$ *be distinct elements of* $D(z_0, r)$ *convergent to* z_0.

Suppose that $\lim_{m\to\infty} f_m(z_j) = 0$ *for each* $j = 1, 2, \ldots$. *Show that then* $\lim_{m\to\infty} f_m = 0$ *uniformly in each closed subdisk of* $D(z_0, r)$. (Cf. 7.5 and its notes.)

Hints: Let M be a bound for the f_m in $\bar{D}(z_0, r)$. Given $R < r$ and $\varepsilon > 0$, use (vi) above to choose n so large that

$$\frac{rM}{r - |z - z_0|} \prod_{k=1}^{n+1} M_k(z) \leq \frac{\varepsilon}{2} \quad \forall z \in D(z_0, R).$$

Deduce from (v) that then

$$\left| f_m(z) - \sum_{k=0}^{n} \left[\sum_{j=1}^{k+1} \frac{f_m(z_j)}{\psi'_{k+1}(z_j)} \right] \psi_k(z) \right| \leq \frac{\varepsilon}{2} \quad \forall z \in D(z_0, R).$$

Finally, choose m_ε so large that this double sum has modulus less than $\varepsilon/2$ for every $m \geq m_\varepsilon$ and conclude that

$$|f_m(z)| \leq \varepsilon \quad \forall z \in D(z_0, R), \forall m \geq m_\varepsilon.$$

The simple idea in the next exercise will experience several profound extensions in the sequel. 5.76 is an outright generalization. In Chapter XIII the same conclusion is reached by entirely different means without the simple-connectivity requirement. A weak version of 5.76 may be found in E. H. TAYLOR [1913].

Exercise 5.75 *Let* f *be holomorphic in* $D(0, 1)$ *and continuous on* $\bar{D}(0, 1)$. *Suppose that* $0 \leq \theta_0 < \theta_1 \leq 2\pi$ *and* $f(e^{i\theta}) = 0$ *for all* $\theta_0 \leq \theta \leq \theta_1$. *Show that* $f \equiv 0$.

Hint: Without loss of generality, $\theta_0 = 0$. Consider then the function

$$F(z) = f(z)f(ze^{i\theta_1})\cdots f(ze^{in\theta_1})$$

where $n\theta_1 \geq 2\pi$. Show that $F = 0$ in $C(0, 1)$. Alternatively, cite 5.45(i).

Theorem 5.76 (RADÓ [1924]) *Let* Ω *be a simply-connected region,* $f \in H(\Omega)$. *If* f *is bounded and for some neighborhood* V *of some boundary point of* Ω, f *satisfies*

$$\lim_{z\to z'} f(z) = 0 \quad \forall z' \in V \cap \partial\Omega,$$

then $f \equiv 0$.

Proof: We may assume that the boundary point in question is 0. We first deal with the case in which Ω satisfies the supplementary hypothesis that 0 is not an isolated boundary point. That is, we first assume

(*) every neighborhood of 0 meets $\mathbb{C}\backslash\bar{\Omega}$.

Pick $r_0 > 0$ so that $D(0, 2r_0) \subset V$ and consider any $z_0 \in \Omega \cap D(0, r_0)$. Then $D(z_0, r_0)$ is a neighborhood of 0 and so by (*) contains a point $z_1 \in \mathbb{C}\backslash\bar{\Omega}$. Therefore, for a sufficiently large k, the points $z_0 + (z_1 - z_0)e^{i\theta}$ lie in $D(z_0, r_0)$ and in $\mathbb{C}\backslash\bar{\Omega}$ whenever $|\theta| \leq 2\pi/k$. We can therefore apply Lindelöf's theorem 5.14, with $r = |z_1 - z_0| < r_0$ and $\varepsilon = 0$, since $D(z_0, r) \subset D(0, 2r_0) \subset V$. The conclusion is that $f(z_0) = 0$. This is true for every z_0 is the non-empty open subset $\Omega \cap D(0, r_0)$ of Ω and so by the Fundamental Uniqueness Theorem and connectedness of Ω the desired conclusion $f = 0$ is obtained.

Now to get out from under (*). According to 4.65 there is a holomorphic function ϕ on Ω such that

(1) $\phi^2(z) = z \quad \forall z \in \Omega.$

Then ϕ is univalent:

$$\phi(z) = \phi(z') \Rightarrow z = \phi^2(z) = \phi^2(z') = z'.$$

Moreover 0 is a boundary point of the simply-connected (by 4.68) region $\phi(\Omega)$, since $z_n \in \Omega$ and $z_n \to 0$ imply $\phi(z_n) \to 0$ [by (1)]. Of course 0 is then also a boundary point of $-\phi(\Omega)$. Now it follows from (1) that the open set $-\phi(\Omega)$ is disjoint from $\phi(\Omega)$. [$z, w \in \Omega$ and $-\phi(z) = \phi(w)$ imply

$$z \overset{(1)}{=} [-\phi(z)]^2 = \phi^2(w) \overset{(1)}{=} w$$

and then $-\phi(z) = \phi(w)$ entails $z = w = 0$, yet $0 \notin \Omega$.] Therefore every neighborhood of 0 contains a non-empty open set (viz., its intersection with $-\phi(\Omega)$) which is disjoint from $\phi(\Omega)$. This shows that the condition (*) is satisfied by $\Omega_0 = \phi(\Omega)$ in the role of Ω. We will apply the conclusion of the first paragraph to $f_0 = f \circ \phi^{-1}$ and this will finish the proof. The relevant boundary neighborhood for f_0 will of course be $V_0 = \{z \in \mathbb{C}: z^2 \in V\}$. To see this, what we have to show is that

(2) if $z_n \in \Omega_0$ and $z_n \to z \in V_0 \cap \partial\Omega_0$,

(3) then $\{\phi^{-1}(z_n)\}$ converges to some boundary point of Ω lying in V.

Let $w_n = \phi^{-1}(z_n)$. Then $z_n^2 = w_n$. Therefore $\{w_n\}$ does converge to some $w[=z^2] \in \bar{\Omega}$. If $w \in \Omega$, then $z_n = \phi(w_n) \to \phi(w) \in \phi(\Omega) = \Omega_0$, contradicting $z_n \to z \in \partial\Omega_0$. Hence

(4) $w_n \to w \in \partial\Omega.$

Finally, $z \in V_0$ means that

(5) $z^2 = w \in V.$

With (4) and (5), (3) is established and the proof complete.

§ 7 Local Theory

Theorem 5.77 (Open Map Theorem) *Let f be holomorphic and not constant in the region Ω. Then $f(\Omega)$ is an open set in $\mathbb{C}$.*

Proof: Let $w_0 \in f(\Omega)$ be given, say $w_0 = f(z_0)$. Since f is not constant, the zeros of the function $f - w_0$ do not accumulate at z_0, by 5.62. That is, $f - w_0$ has no zero in $D(z_0, r)\backslash\{z_0\}$ for some $r > 0$ with $D(z_0, r) \subset \Omega$. Consequently, the number $m = \min\{|f(z) - w_0|: |z - z_0| = r/2\}$ is positive. Consider any $w \in \mathbb{C}$ satisfying

$$|w - w_0| < m/2.$$

Then the function $f - w$ satisfies on $|z - z_0| = r/2$

(1) $\quad |f(z) - w| \geq |f(z) - w_0| - |w_0 - w| \geq m - m/2 = m/2,$

while

(2) $\quad |f(z_0) - w| = |w_0 - w| < m/2.$

It follows that $f - w$ must have a zero in $\bar{D}(z_0, r/2)$, for if not then the Maximum Modulus Principle may be applied to the function $1/(f - w)$, which would be holomorphic in $D(z_0, r/2)$ and continuous in $\bar{D}(z_0, r/2)$, and would yield

$$\frac{2}{m} \overset{(2)}{<} \frac{1}{|f(z_0) - w|} \leq \sup\left\{\frac{1}{|f(z) - w|}: |z - z_0| = r/2\right\} \overset{(1)}{\leq} \frac{2}{m},$$

a contradiction. Thus $f(z) - w = 0$ for some $z \in \bar{D}(z_0, r/2) \subset \Omega$, so $w \in f(\Omega)$. That is, $D(w_0, m/2) \subset f(\Omega)$.

Corollary 5.78 (Differentiability of the Inverse) *Let U be an open subset of $\mathbb{C}$, f holomorphic and one-to-one in U. Then f' is never zero, $f(U)$ is open, f^{-1} is holomorphic therein and $(f^{-1})' = 1/f' \circ f^{-1}$.*

Proof: U is a union of open disks and to f on each such disk we may apply the Open Map Theorem to learn that f is an open map: $f(U)$ is open and f^{-1} is continuous. Let g denote f^{-1}. At every point $w_0 \in f(U)$ such that $f'(g(w_0)) \neq 0$ we use the fact that $g(w) \neq g(w_0)$ for $w \neq w_0$ (g being one-to-one) to see that

$$\frac{g(w) - g(w_0)}{w - w_0} = \frac{1}{\dfrac{w - w_0}{g(w) - g(w_0)}} = \frac{1}{\dfrac{f(g(w)) - f(g(w_0))}{g(w) - g(w_0)}}.$$

Since g is continuous at w_0 and f is differentiable at $g(w_0)$, we see from this that g is differentiable at w_0 and that $g'(w_0) = 1/f'(g(w_0))$.

It remains only to show that f' never vanishes. Since f is not constant on any open disk in U, f' is not identically zero on any such disk so (5.62) the zeros of the holomorphic (5.30) function f' cannot accumulate in U. Since f is a homeomorphism, the f images of these zeros do not accumulate in $f(U)$. Therefore by the above, the set of points in $f(U)$ where g does not have a derivative does not accumulate in $f(U)$, i.e., is composed of isolated points. Since g is continuous on $f(U)$, it is bounded in a neighborhood of each of these points and so we can conclude from 5.41 that g is differentiable at these points as well. Therefore by 5.30 g' is continuous. In summary, $(g' \circ f)\cdot f'$ is a continuous function on U which equals 1 except perhaps at the points of an isolated set. By continuity then $(g' \circ f)\cdot f' = 1$ throughout U, so f' is never 0.

We have the following local converse to this last result. 5.30 and 3.7 constitute one proof and here is another, which is more constructive and proves a little more. A third proof occurs in 5.89.

Theorem 5.79 *Let U be an open subset of $\mathbb{C}$, f a holomorphic function in U, $z_0 \in U$ and $f'(z_0) \neq 0$. Then f is one-to-one near z_0.*

Proof: We lose no generality by supposing that $z_0 = f(z_0) = 1 - f'(z_0) = 0$. Let $\rho \in (0, 1)$ and set $\phi(z) = f(z) - z$. Use continuity of ϕ' at 0 (5.30) to select $r > 0$ such that

(1) $\quad |\phi'| < \rho \quad \text{in } \bar{D}(0, r) \subset U.$

Notice that if $w \in D(0, r(1 - \rho))$ and $z \in D(0, r)$, then

$$|w - \phi(z)| = \left| w - \int_{[0,z]} \phi' \right| \leq |w| + |z| \sup|\phi'[0, z]| < r(1 - \rho) + r\rho,$$

using (1), and so $w - \phi(z) \in D(0, r)$. Therefore the recursive definition $g_0 \equiv 0$, $g_{n+1}(w) = w - \phi(g_n(w))$ $(n \geq 0)$ always makes sense for $w \in D(0, r(1 - \rho))$. These functions map into $D(0, r)$ and satisfy

$$\begin{aligned} |g_{n+1}(w) - g_n(w)| &= |\phi(g_n(w)) - \phi(g_{n-1}(w))| = \left| \int_{[g_n(w), g_{n-1}(w)]} \phi' \right| \\ &\leq \rho |g_n(w) - g_{n-1}(w)| \leq \cdots \\ &\leq \rho^n |g_1(w) - g_0(w)| = \rho^n |w|, \end{aligned}$$

$$\begin{aligned} |g_{N+k}(w) - g_N(w)| &\leq \sum_{j=1}^{k} |g_{N+j}(w) - g_{N+j-1}(w)| \\ &\leq \sum_{j=1}^{k} \rho^{N+j-1}|w| < |w|\rho^N \sum_{j=1}^{\infty} \rho^{j-1} \\ &= |w| \frac{\rho^N}{1 - \rho} < r\rho^N. \end{aligned}$$

Therefore $\{g_N\}$ converges uniformly in $D(0, r(1 - \rho))$ and the limit function g maps into $\bar{D}(0, r) \subset U$ and satisfies

$$g(w) = \lim_{n \to \infty} g_{n+1}(w) = \lim_{n \to \infty} [w - \phi(g_n(w))] = w - \phi\left(\lim_{n \to \infty} g_n(w) \right) = w - \phi(g(w)),$$

that is,

(2) $\quad w = g(w) + \phi(g(w)) = f(g(w)) \quad \forall w \in D(0, r(1 - \rho)).$

By 5.44 g is holomorphic in $D(0, r(1 - \rho))$ and evidently (2) implies that f is one-to-one in the set $g(D(0, r(1 - \rho)))$. We could cite 5.77 to affirm that the latter is a neighborhood of $g(0) = 0$ and the proof would be complete. Alternatively, we can proceed without 5.77 thus: (2) implies that $f(U)$ is a neighborhood of $f(0) = 0$ and, together with the Chain Rule, provides the equality

$g'(0) = 1$. Therefore the function g fulfills all the hypotheses which f does and we can attribute to it the conclusions so far gained for f, to learn that $g(D(0, r(1 - \rho)))$ is a neighborhood of $g(0) = 0$.

Exercise 5.80 (Cf. 9.17(ii)) *Prove the following result from* POMPEIU [1911]: *If f is holomorphic in the open subset U of $\mathbb{C}$ and $z_0 \in U$, then in every neighborhood of z_0 there is a pair of distinct points z_1, z_2 such that*

$$\frac{f(z_2) - f(z_1)}{z_2 - z_1} = f'(z_0).$$

Hint: $g(z) = f(z) - zf'(z_0)$ has $g'(z_0) = 0$ and so by 5.79 is not one-to-one in any neighborhood of z_0.

Exercise 5.81 *Show that there is a function A defined by an absolutely convergent power series in $\bar{D}(0, 1)$ such that* $\sin(A(z)) = z$ *for all $z \in \bar{D}(0, 1)$.*

Hints: Since $\sin'(0) = 1$, there is (by 5.78 and 5.79) a neighborhood $U = D(0, r)$ of 0 (with $0 < r < 1$, say) and a holomorphic function A on U such that $A(0) = 0$ and

(1) $\quad \sin(A(z)) = z$

for all $z \in U$. Differentiate this relation and get

(2) $\quad 1 = \sin'(A(z)) \cdot A'(z) = \cos(A(z)) \cdot A'(z)$.

But $\sin'(x) = \cos(x) > 0$ for $x \in (-\pi/2, \pi/2)$, so sin maps $(-\pi/2, \pi/2)$ in a strictly monotone fashion onto $(-1, 1)$. There is therefore an inverse function $f: (-1, 1) \to (-\pi/2, \pi/2)$. Since sin is one-to-one near 0 and A is continuous with $A(0) = 0$, it follows that for sufficiently small $0 < \delta < r$ we have

$$A(x) = f(x) \quad \forall x \in (-\delta, \delta).$$

For such x, $A(x) = f(x) \in (-\pi/2, \pi/2)$ so $\cos(A(x)) > 0$ and consequently $\cos(A(x)) = \sqrt{1 - \sin^2(A(x))} = \sqrt{1 - x^2}$. From this and (2) we have

(3) $\quad A'(x) = \dfrac{1}{\sqrt{1 - x^2}} \quad \forall x \in (-\delta, \delta).$

The holomorphic function A' thus satisfies $A'(-x) = A'(x)$ for all $x \in (-\delta, \delta)$ and so by the Fundamental Uniqueness Theorem A' is an even function. It follows easily that the coefficients of all odd powers in its series expansion are 0: $A'(z) = \sum_{n=0}^{\infty} a_{2n} z^{2n}$. Therefore

(4) $\quad S(z) = \sum_{n=0}^{\infty} a_{2n} z^n$

defines a holomorphic function for $z \in D(0, r^2)$ such that

(5) $\quad A'(z) = S(z^2) \quad \forall z \in D(0, r).$

From this equation and (3) we get $S(x^2) = (1 - x^2)^{-1/2}$ for all $x \in (-\delta, \delta)$, whence

$$S(t) = (1 - t)^{-1/2} \quad \forall 0 \le t \le \delta^2.$$

From this it follows by induction, taking right-hand derivatives at 0 along the positive real axis, that

$$S^{(k)}(0) = \tfrac{1}{2}(\tfrac{1}{2} + 1)\cdots(\tfrac{1}{2} + k - 1) = \frac{1\cdot 3\cdots(2k-1)}{2^k}$$

$$(6) \qquad \frac{S^{(k)}(0)}{k!} = \frac{1\cdot 3\cdots(2k-1)}{2\cdot 4\cdots(2k)}, \quad k = 1, 2, \ldots.$$

Now from (4) and (5)

$$A'(z) = \sum_{k=0}^{\infty} \frac{S^{(k)}(0)}{k!} z^{2k} = 1 + \sum_{k=1}^{\infty} \frac{1\cdot 3\cdots(2k-1)}{2\cdot 4\cdots(2k)} z^{2k},$$

whence, since $A(0) = 0$,

$$(7) \qquad A(z) = z + \sum_{k=1}^{\infty} \frac{1\cdot 3\cdots(2k-1)}{2\cdot 4\cdots(2k)} \frac{z^{2k+1}}{2k+1} \quad \forall z \in D(0, r).$$

This series evidently converges throughout $D(0, 1)$ (its coefficients are each less than 1). If we let it define A in $D(0, 1)\backslash D(0, r)$, then the Fundamental Uniqueness Theorem shows that (1) holds for all $z \in D(0, 1)$. From (1) holding for all real z in $(0, 1)$, it follows that $\sin(A(x))$ is never 0 or ± 1 for $x \in (0, 1)$. Consequently $A(x)$, which is real and positive for such x by (7), is not an integer multiple of $\pi/2$. Therefore A maps the connected interval $(0, 1)$ continuously into $(0, \infty)\backslash\{n\pi/2 : n = 0, 1, 2, \ldots\}$, and so for some n maps it wholly into $(n\pi/2, (n+1)\pi/2)$. $A(0) = 0$ shows that $n = 0$. It follows that $A(x) < \pi/2$ for all $x \in (0, 1)$, whence

$$x + \sum_{k=1}^{N} \frac{1\cdot 3\cdots(2k-1)}{2\cdot 4\cdots(2k)} \frac{x^{2k+1}}{2k+1} < A(x) < \frac{\pi}{2}, \quad N = 1, 2, \ldots.$$

Let $x \uparrow 1$ here and conclude that

$$1 + \sum_{k=1}^{N} \frac{1\cdot 3\cdots(2k-1)}{2\cdot 4\cdots(2k)} \frac{1}{2k+1} \le \frac{\pi}{2}, \quad N = 1, 2, \ldots.$$

This shows that in fact the series (7) is absolutely convergent throughout $\bar{D}(0, 1)$. By continuity then the equality (1), known to hold for $z \in D(0, 1)$, holds throughout $\bar{D}(0, 1)$. In particular $\sin(A(1)) = 1$. Since $A(1) \in [0, \pi/2]$, it follows that $A(1) = \pi/2$, giving

Scholium 5.82 (i) $\displaystyle \frac{\pi}{2} = 1 + \sum_{k=1}^{\infty} \frac{1\cdot 3\cdots(2k-1)}{2\cdot 4\cdots(2k)} \cdot \frac{1}{2k+1}.$

(ii) Take $z = \frac{1}{2}$ in formula (1) of 5.81, note that $\sin(\pi/6) = \frac{1}{2}$ (proof?) and deduce that

$$\frac{\pi}{6} = \frac{1}{2} + \frac{1}{2}\cdot\frac{1}{3\cdot 2^3} + \frac{1\cdot 3}{2\cdot 4}\cdot\frac{1}{5\cdot 2^5} + \cdots.$$

(This series converges more rapidly than that in (i).)

Theorem 5.83 *Let f be holomorphic and not constant in a connected neighborhood Ω of z_0. Then there is a positive R, a positive integer n, an open neighborhood V of z_0 in Ω and a one-to-one holomorphic function F in V such that $f \equiv f(z_0) + F^n$ in V and $f^{-1}(w) \cap V$ consists of n distinct points for each $w \in D(f(z_0), R)\backslash\{f(z_0)\}$.*

Proof: We may suppose $f(z_0) = 0$, for otherwise we can work with $f - f(z_0)$. According to 5.69 there is a positive integer n and a holomorphic function h in Ω such that

(1) $\quad f(z) = (z - z_0)^n h(z) \quad \text{and} \quad h(z_0) \neq 0, \quad \forall z \in \Omega.$

There is therefore an $r_0 > 0$ such that $D(z_0, r_0) \subset \Omega$ and $0 \notin h(D(z_0, r_0))$ and then (5.35) a holomorphic function g in $D(z_0, r_0)$ such that

(2) $\quad h = e^g \quad \text{in } D(z_0, r_0).$

Form

(3) $\quad F(z) = (z - z_0)e^{g(z)/n}, \quad z \in D(z_0, r_0).$

Since $F'(z_0) = e^{g(z_0)/n} \neq 0$, it follows from 5.79 that there is an $0 < r < r_0$ such that F is one-to-one in $D(z_0, r)$. By the Open Map Theorem, $F(D(z_0, r))$ is an open neighborhood of $F(z_0) = 0$ and so there is an $R > 0$ such that $D(0, R^{1/n}) \subset F(D(z_0, r))$. Let $V = D(z_0, r) \cap F^{-1}(D(0, R^{1/n}))$. Then V is an open neighborhood of z_0, F is one-to-one in V and $f \equiv F^n$ in V [from (1), (2) and (3)]. Finally, given non-zero $w \in D(0, R)$, let $w_1, \ldots, w_n$ be its n distinct nth roots. These points lie in $D(0, R^{1/n})$ so their F-pre-images in V are distinct, say $z_1, \ldots, z_n$. Then $w = w_j^n = [F(z_j)]^n = f(z_j)$. Conversely, if $z \in V$ and $f(z) = w$, then $w = [F(z)]^n$ so $F(z) \in \{w_1, \ldots, w_n\}$, whence $z \in F^{-1}\{w_1, \ldots, w_n\} \cap V = \{z_1, \ldots, z_n\}$.

There is another approach to the local theory which we will now explore. It utilizes the index and leads to other interesting and useful facts, as well as different proofs of some results already secured.

Theorem 5.84 *If f and h are holomorphic in $D(a, R)$, then for each $0 < r < R$ and each $w \in \mathbb{C}\backslash f(C(a, r))$ the set $f^{-1}(w) \cap D(a, r)$ is finite and*

$$\frac{1}{2\pi i}\int_{C(a,r)} \frac{hf'}{f - w} = \sum_{z \in f^{-1}(w) \cap D(a,r)} n(z, f - w)h(z).$$

Proof: First note that since $w \in \mathbb{C}\backslash f(C(a, r))$,

(1) $f^{-1}(w) \cap D(a, r) = f^{-1}(w) \cap \bar{D}(a, r)$.

If this set were infinite, then it would have a limit point in $\bar{D}(a, r)$ and by 5.62 $f \equiv w$ in $D(a, r)$. In particular, $f(C(a, r)) = \{w\}$, a contradiction. Let then $a_1, \ldots, a_m$ be the (distinct) points of $f^{-1}(w) \cap D(a, r)$, if any. 5.69 provides a holomorphic function F in $D(a, R)$ such that (void products being 1)

(2) $$f(z) - w = F(z)\prod_{j=1}^{m}(z - a_j)^{n_j}, \quad z \in D(a, R)$$

and

(3) $F(a_j) \neq 0, \ j = 1, \ldots, m.$

Here

(4) $n_j = n(a_j, f - w), \ j = 1, \ldots, m.$

Since $a_1, \ldots, a_m$ are all the zeros of $f - w$ in $\bar{D}(a, r)$, it follows from (2) and (3) that F is never zero in $\bar{D}(a, r)$, hence by continuity never zero in some (convex) open neighborhood V of $\bar{D}(a, r)$ in $D(a, R)$. From (2) then

$$\frac{f'(z)}{f(z) - w} = \frac{F'(z)}{F(z)} + \sum_{j=1}^{m} \frac{n_j}{z - a_j} \quad \forall z \in V\backslash\{a_1, \ldots, a_m\},$$

whence

$$\int_{C(a,r)} \frac{hf'}{f - w} = \int_{C(a,r)} \frac{hF'}{F} + \sum_{j=1}^{m} n_j \int_{C(a,r)} \frac{h(z)}{z - a_j}\, dz.$$

By Cauchy's Theorem and Integral Formula for Convex Regions this gives

(5) $$\int_{C(a,r)} \frac{hf'}{f - w} = \sum_{j=1}^{m} 2\pi i n_j h(a_j) \operatorname{Ind}_{C(a,r)}(a_j).$$

But $a_j \in D(a, r)$, which is a connected subset of $\mathbb{C}\backslash C(a, r)$, so

$$\operatorname{Ind}_{C(a,r)}(a_j) = \operatorname{Ind}_{C(a,r)}(a).$$

The latter is trivially 1, so (5) and (4) give

$$\frac{1}{2\pi i}\int_{C(a,r)} \frac{hf'}{f - w} = \sum_{j=1}^{m} n(a_j, f - w)h(a_j).$$

When we recall that $\{a_1, \ldots, a_m\} = f^{-1}(w) \cap D(a, r)$, the desired formula is at hand.

Corollary 5.85 *If f is holomorphic in a neighborhood V of a, then for each $r > 0$ such that $\bar{D}(a, r) \subset V$ and such that f has no zero in $C(a, r)$, the number of zeros* *of f in $D(a, r)$ and the sum of these zeros, counted according to multiplicity, are respectively*

$$\frac{1}{2\pi i}\int_{C(a,r)}\frac{f'(\xi)}{f(\xi)}\,d\xi \quad \textit{and} \quad \frac{1}{2\pi i}\int_{C(a,r)}\frac{\xi f'(\xi)}{f(\xi)}\,d\xi.$$

Proof: Take $w = 0$ and h in the last theorem successively equal to the constant function 1 and the identity function.

Corollary 5.86 (Cf. 9.15) *If f is holomorphic in $D(a, R)$ and if for each $0 < r < R$ we let $\gamma_r = C(a, r)$, then for every $w \in \mathbb{C}\backslash f(C(a, r))$ the number of zeros of $f - w$ in $D(a, r)$, counted according to multiplicity, is* $\mathrm{Ind}_{f\circ\gamma_r}(w)$.

Proof: The number of these zeros is by the last corollary (with $f - w$ in the role of the f there)

$$\begin{aligned}\frac{1}{2\pi i}\int_{\gamma_r}\frac{f'}{f-w} &= \frac{1}{2\pi i}\int_0^1\frac{f'\circ\gamma_r}{f\circ\gamma_r - w}\cdot\gamma_r'\\ &= \frac{1}{2\pi i}\int_0^1\frac{(f\circ\gamma_r)'}{f\circ\gamma_r - w} = \frac{1}{2\pi i}\int_{f\circ\gamma_r}\frac{1}{\xi - w}\,d\xi,\end{aligned}$$

which is $\mathrm{Ind}_{f\circ\gamma_r}(w)$.

Exercise 5.87 (Cf. 8.18 and its proof.) *Let f and g be holomorphic in a neighborhood of $\bar{D}(0, 1)$ and satisfy $|f + g| < |f| + |g|$ on $C(0, 1)$. Show that f and g have the same number of zeros (multiplicities counted) in $D(0, 1)$.*

Hints: The strict inequality entails that g and $F_t = f - tg$ are each zero-free on $\gamma = C(0, 1)$ for every real $t \geq 0$. If we form $G_t = f/g - t$, then $F_t = gG_t$, so

(1) $\quad \mathrm{Ind}_{F_t\circ\gamma}(0) = \mathrm{Ind}_{g\circ\gamma}(0) + \mathrm{Ind}_{G_t\circ\gamma}(0)$.

Now by 4.6(i), $\mathrm{Ind}_{F_t\circ\gamma}(0)$ is a continuous function of $t \in [0, \infty)$, hence is constant. Since $F_0 = f$, it follows therefore from (1) that

(2) $\quad \mathrm{Ind}_{f\circ\gamma}(0) = \mathrm{Ind}_{g\circ\gamma}(0) + \mathrm{Ind}_{G_t\circ\gamma}(0) \quad \forall t \geq 0$.

Finally note that for $t > \sup_\gamma|f/g|$, we have $[0, \infty) \subset \mathbb{C}\backslash G_t \circ \gamma$, causing the last index in (2) to be 0. Now cite 5.86.

Alternatively we can argue thus: In the hints to 4.27(ii) it is shown that the quotient f/g has a continuous logarithm on $\gamma = C(0, 1)$, and so $\mathrm{Ind}_{(f/g)\circ\gamma}(0) = 0$. Since $f = g\cdot(f/g)$, it follows that

$$\mathrm{Ind}_{f\circ\gamma}(0) = \mathrm{Ind}_{g\circ\gamma}(0) + \mathrm{Ind}_{(f/g)\circ\gamma}(0) = \mathrm{Ind}_{g\circ\gamma}(0)$$

and we again conclude by citing 5.86.

Corollary 5.88 *Let f be holomorphic and not constant in the region Ω. If $z_0 \in \Omega$, $w_0 = f(z_0)$ and m is the multiplicity of z_0 as a zero of $f - w_0$, then there exist* *open sets V and W such that* $z_0 \in V \subset \Omega$, $W = f(V)$ *and for each* $w \in W\backslash\{w_0\}$ *the set* $f^{-1}(w) \cap V$ *consists of exactly m distinct points.*

Proof: Since $f - w_0$ is not identically 0, f' is not either (2.10) and so (5.62) neither the zeros of $f - w_0$ nor those of f' accumulate at z_0, i.e., there exists $r > 0$ such that

(1) $\quad |f - w_0| > 0 \quad \text{and} \quad |f'| > 0 \quad \text{in } \bar{D}(z_0, r)\backslash\{z_0\} \subset \Omega.$

Therefore $w_0 \in \mathbb{C}\backslash f(C(z_0, r))$ and it follows from the last corollary that $\text{Ind}_{f\circ\gamma_r}(w_0)$ is the number of zeros of $f - w_0$ in $D(z_0, r)$, counted according to multiplicities. By (1) this is just the multiplicity of z_0 as a zero of $f - w_0$, viz., m:

$$\text{Ind}_{f\circ\gamma_r}(w_0) = m.$$

Since $\text{Ind}_{f\circ\gamma_r}$ is a continuous integer-valued function on the open set $\mathbb{C}\backslash f(C(z_0, r))$, there is an $R > 0$ such that $W = D(w_0, R) \subset \mathbb{C}\backslash f(C(z_0, r))$ and

(2) $\quad \text{Ind}_{f\circ\gamma_r}(w) = m \quad \forall w \in W.$

Set $V = f^{-1}(W) \cap D(z_0, r)$. It follows from (2) and the last corollary that for each $w \in W$, $f - w$ has exactly m zeros in $D(z_0, r)$, hence in V. In particular, since $m > 0$, it follows that $f(V) = W$. Moreover if $w \in W\backslash\{w_0\}$ and $z \in f^{-1}(w) \cap D(z_0, r)$, then $z \neq z_0$ and so $(f - w)'(z) = f'(z) \neq 0$ by (1). Therefore the multiplicity of each zero of $f - w$ in V is 1 and the m zeros are therefore distinct.

Corollary 5.89 *If f is holomorphic in a neighborhood of* z_0, *then f is one-to-one near* z_0 *if and only if* $f'(z_0) \neq 0$.

Proof: f is one-to-one near z_0 if and only if the integer m of the last corollary is 1. Consulting the definition of multiplicity, we see that this is equivalent to $f'(z_0) \neq 0$.

Remarks 5.90 The Open Map Theorem is an immediate consequence of 5.88 and the differentiability of the inverse of a one-to-one holomorphic function follows at once from this and from the last corollary. For the calculation at the beginning of the proof of 5.78 needed only the facts that f^{-1} is continuous (f is open) and f' never vanishes. Alternatively, yet another proof (5.92) follows from the next corollary.

Corollary 5.91 *If f is holomorphic and one-to-one in the open set U, then for every* $a \in U$ *and every* $r > 0$ *such that* $\bar{D}(a, r) \subset U$ *we have*

$$f^{-1}(w) = \frac{1}{2\pi i}\int_{C(a,r)} \frac{zf'(z)}{f(z) - w}\, dz \quad \forall w \in f(D(a, r)).$$

Proof: There exists $R > r$ such that $D(a, R) \subset U$. Apply 5.84 with $h(z) \equiv z$ and get

$$\frac{1}{2\pi i}\int_{C(a,r)} \frac{zf'(z)}{f(z) - w}\,dz = \sum_{z \in f^{-1}(w) \cap D(a,r)} n(z, f - w)z \quad \forall w \in \mathbb{C}\backslash f(C(a, r)).$$

Now if $w \in f(D(a, r))$, then $w \in \mathbb{C}\backslash f(C(a, r))$ since f is one-to-one. Also because f is one-to-one, $f^{-1}(w) \cap D(a, r)$ consists just of the unique $z \in U$ such that $f(z) = w$ and for this z we have $n(z, f - w) = 1$ because $f'(z) \neq 0$ (5.89). The result follows.

Exercise 5.92 (i) *Deduce from the last corollary that* f^{-1} *is holomorphic and exhibit integral formulas for* $(f^{-1})^{(k)}$ *analogous to the one there for* f^{-1}.

(ii) *Using the formula*

$$(f^{-1})'(w) = \frac{1}{2\pi i}\int_{C(a,r)} \frac{zf'(z)}{(f(z) - w)^2}\,dz, \quad w \in f(D(a, r))$$

from (i), *integrate by parts to get*

$$(f^{-1})'(w) = \frac{1}{2\pi i}\int_{C(a,r)} \frac{1}{f(z) - w}\,dz, \quad w \in f(D(a, r)).$$

(iii) *For simplicity let* $a = 0 = f(a)$. *Then* $0 \notin f(C(0, r))$ *and consequently* $m = \inf|f(C(0, r))| > 0$. *If* $r > R > 0$ *is sufficiently small,* f *maps* $D(0, R)$ *into* $D(0, m)$. *Consider* w *in the neighborhood* $W = f(D(0, R))$ *of* 0 *and use* (ii) *to show that*

$$(f^{-1})'(w) = \frac{1}{2\pi i}\int_{C(0,r)} \frac{1}{f(z)}\left[\frac{1}{1 - w/f(z)}\right]dz$$

$$= \sum_{n=1}^{\infty} w^{n-1}\frac{1}{2\pi i}\int_{C(0,r)} \frac{1}{f^n(z)}\,dz.$$

$$f^{-1}(w) = \sum_{n=1}^{\infty} w^n \frac{1}{2n\pi i}\int_{C(0,r)} \frac{1}{f^n}.$$

We can now present an elegant proof of a useful (e.g., in Chapter XIII) Implicit Function Theorem for functions of two complex variables. The reader will observe that the last corollary is (essentially) the special case $F(w, z) = f(z) - w$ of this theorem.

Theorem 5.93 *Let* V *and* W *be open, connected neighborhoods in* $\mathbb{C}$ *of* z_0, w_0 *and let* $F\colon W \times V \to \mathbb{C}$ *be continuous and have a continuous second partial derivative on* $W \times V$ *and satisfy* $F(w_0, z_0) = 0$, $D_2F(w_0, z_0) \neq 0$. *Then there* *exist $r > 0$ and $R > 0$ such that F is never zero in $D(w_0, R) \times C(z_0, r) \subset W \times V$ and the function*

$$G(w) = \frac{1}{2\pi i}\int_{C(z_0,r)} \frac{\xi D_2F(w, \xi)}{F(w, \xi)}\, d\xi, \quad w \in D(w_0, R)$$

maps $D(w_0, R)$ into $D(z_0, r) \subset V$, is continuous and satisfies for

$$(w, z) \in D(w_0, R) \times D(z_0, r)\colon F(w, z) = 0 \Leftrightarrow z = G(w).$$

Proof: For each $w \in W$ let $F_w\colon V \to \mathbb{C}$ be the holomorphic function $F_w(z) = F(w, z)$. We have

(1) $\quad F'_{w_0}(z_0) = D_2F(w_0, z_0) \neq 0,$

so there is an $r > 0$ such that

(2) $\quad F_{w_0}$ has no zero in $\bar{D}(z_0, r)\setminus\{z_0\} \subset V$.

Moreover since D_2F is continuous, it follows from (1) that $|D_2F| > 0$ on some neighborhood of (w_0, z_0), so we may also take r small enough that

(3) $\quad |D_2F| > 0 \quad$ on $U \times \bar{D}(z_0, r)$

for some neighborhood U of w_0 in W. From (2) it follows that $\{w_0\} \times C(z_0, r)$ is a compact subset of $W \times V$ disjoint from the (relatively) closed subset $F^{-1}(0)$, and so there is a neighborhood of this set which is disjoint from $F^{-1}(0)$. In particular, we may take the U of (3) sufficiently small that

(4) $\quad |F| > 0 \quad$ on $U \times C(z_0, r)$.

Then according to 5.85

(5) $$N(w) = \frac{1}{2\pi i}\int_{C(z_0,r)} \frac{F'_w(\xi)}{F_w(\xi)}\, d\xi = \frac{1}{2\pi i}\int_{C(z_0,r)} \frac{D_2F(w, \xi)}{F(w, \xi)}\, d\xi, \quad w \in U$$

is the number of zeros of F_w in $D(z_0, r)$, counted according to multiplicity. But $F'_w(z) = D_2F(w, z) \neq 0$ by (3) for each $z \in D(z_0, r)$, so each zero of F_w has multiplicity 1. Then from 5.85 also follows that

(6) $$G(w) = \frac{1}{2\pi i}\int_{C(z_0,r)} \frac{\xi F'_w(\xi)}{F_w(\xi)}\, d\xi = \frac{1}{2\pi i}\int_{C(z_0,r)} \frac{\xi D_2F(w, \xi)}{F(w, \xi)}\, d\xi, \quad w \in U$$

is the sum of the zeros of F_w in $D(z_0, r)$. Now by (2) we have

(7) $\quad N(w_0) = 1.$

Also it is clear that

$$\frac{D_2F(w, \gamma(t))\gamma'(t)}{F(w, \gamma(t))} \quad \text{and} \quad \frac{\gamma(t) D_2F(w, \gamma(t))\gamma'(t)}{F(w, \gamma(t))}$$

are uniformly continuous on $U \times [0, 1]$, where $\gamma = C(z_0, r)$, and so from (5) and (6) it follows that N and G are continuous. Being integer-valued, N is therefore constantly 1 near w_0, say in $D(w_0, R) \subset U$. Thus for each $w \in D(w_0, R)$ the equation

$$F(w, z) = F_w(z) = 0, \quad z \in D(z_0, r)$$

has a unique solution z and that z is $G(w)$.

Remark 5.94 In fact the function G is holomorphic in $D(w_0, R)$, by 7.18.

Notes to Chapter V

Versions of 5.1 were found by Gauss in 1811 and by Cauchy in 1814 and in 1825. (See JOURDAIN [1905b], CAUCHY [1846], [1900] and ETTLINGER [1921–22].) Continuity of f' seemed necessary for the validity of these proofs. (E.g., one of Cauchy's involved iterated integration of the partial derivatives of $\operatorname{Re} f$ and $\operatorname{Im} f$, which, *prima facie*, would seem to require some hypothesis on f' beyond its mere existence!) GOURSAT [1900] (see also E. H. MOORE [1900]) succeeded in modifying earlier proofs he had given so as to eliminate the need for continuity of f'. (Actually Cauchy had vacillated on the matter of continuity of f': see pp. 1011–1012 of BURKHARDT [1914] and § IV of PRINGSHEIM [1900].) For the technique of triangulation see PRINGSHEIM [1901] and [1903]. Far-reaching extensions of 5.1 will be developed in Chapters IX and X; this is the most central and important single result of the whole theory. For history and references to original memoirs and the many subsequent proofs and extensions, see the monograph of HEFFTER [1960] (esp. pp. 55 ff.) and the bibliography and notes of FALLIN and GOULD [1965]. See also corollary 9.27 and the notes thereon at the end of Chapter IX, STÄCKEL [1900], section II of BRILL and NOETHER [1894], pp. 27–36 of MITTAG-LEFFLER [1923] and § 35 of BURKHARDT [1914]. The latter is a good history of Cauchy's many fundamental contributions; see also VALSON [1970]. ETTLINGER [1921–22] is an invaluable "explication de texte."

It is sometimes maintained that the use of integration theory to secure 5.3 is a methodological adulteration. The renunciation of integration theory here is debilitating but not fatal; circuitous alternative paths to 5.3 have been hewn out. See CONNELL [1965] and G. T. WHYBURN [1961–62], [1964], [1968] and their bibliographies for one approach; (the program there is however much broader than merely eliminating integration) and PRINGSHEIM's book [1925], [1932] for another. In the latter, limits of Riemann sums based on (2^n)th roots of 1 replace integrals over circles. These roots can easily be constructed inductively and so this procedure has the added advantage of freeing the local theory from dependence on the exponential function. A completely different approach is offered in ADEL'SON-VEL'SKIĬ and KRONROD [1945] and another in LELAND [1964–65], [1971].

Each of the (improper) integrals in 5.4 can be computed by various contour integrals or by the Residue Theorem of Chapter XI, but the development presented has a certain unity and is perhaps minimally arduous. For more on these kinds of integrals see HAYASHI [1922], [1923], A. N. SINGH [1925], DEAUX and DELCOURTE [1957], LAGRANGE [1957] and BUTLER [1960]. In connection with 5.4(i) see 10.44 of RUDIN [1974].

The concept of subharmonicity was introduced by Hartogs and greatly elaborated by F. Riesz. See the little monograph of RADÓ [1937] for history and relevant papers. These functions have many analogies with convex functions on open subintervals of $\mathbb{R}$. They certainly enter function theory naturally and provide just the appropriate level of generality in which to prove many results for the most efficient application to function theory, as the results of §§ 2 and 3 illustrate. See MONTEL [1928], [1940], BRELOT [1939] and HAYMAN and KENNEDY [1976] for more applications of them to function theory. By contrast to the integral (mean value) properties of harmonic and subharmonic functions which occupy center stage in §§ 2 and 3, we will look at differential properties (e.g., Laplace's Equation) in § 7 of Chapter IX and we will show that the Dirichlet problem is solvable for simply-connected regions.

It is customary to require only upper semi-continuity in the definition of subharmonicity and to allow the function to take the value $-\infty$. While a bit more delicacy is then called for in the early stages of the theory, there are many advantages. For example, now $\log|f|$ is subharmonic whenever f is holomorphic and artifices like that of considering $\log(\max\{|f|, \delta\})$ and letting $\delta \downarrow 0$, used often in this chapter, can be eliminated.

The maximum principle for harmonic functions was proved in RIEMANN's dissertation [pp. 22–23 of his *Werke*] by a brief reference to the mean value property 5.21(ii) (the heart of the matter, to be sure). Subsequently more extensive treatments were given by NEUMANN [1871] (pp. 340–344) and H. A. SCHWARZ (pp. 203–204, vol. II of his *Gesammelte Mathematische Abhandlungen*).

The device of taking powers in the proof of 5.13 is due to E. Landau. That the line between 5.12 and 5.13 is not illusory is illustrated by the example $f(z) = \exp(\exp(z))$, $z \in \Omega = (0, \infty) \times (-\pi/2, \pi/2)$: f is bounded (by 1) on $\partial\Omega$ but unbounded on Ω. Though the proof of 5.12 in the text is taken from LINDELÖF [1915], the result itself pre-dates this; e.g., it is mentioned as a known result in PHRAGMÉN and LINDELÖF [1908]. If h, U (connected) are each unbounded but otherwise fulfill the hypotheses of 5.12, then there exists continuous γ: $[0, 1) \to U$ such that $\lim_{t\to 1}|\gamma(t)| = \lim_{t\to 1} h(\gamma(t)) = \infty$. The version of this in which $U = \mathbb{C}$ (due to IVERSEN [1914]) will be presented in Chapter XV.

The rotation trick used in 5.14 (and also in 9.10) can be seen in its most pristine form in 5.75 and is due to Painlevé (in an appendix to a 1908 memoir on differential equations by Boutroux). For an extension of 5.14 see exercise IV.134 of PÓLYA and SZEGÖ [1976] and for another proof SAKS [1928/29].

The proof of 5.17 in the text is adapted from H. A. SCHWARZ (pp. 181–185, vol. II, *op. cit.*), to whom is also due the first rigorous proof (that in the text—see pp. 186–188, *op. cit.*) of the important continuity of $\operatorname{Re} f$ on $\bar{D}$ in 5.20. See § 34 of BURKHARDT [1914] for early history. Geometric proofs of 5.20 appear in BÔCHER [1906], DARBOUX [1910], PERKINS [1928] and DUFFIN [1957]. A clever but "natural" analytic proof is offered by VILLAT [1911]. FEJÉR [1904] solved the Dirichlet problem in the disk as follows: His *Hauptsatz* of this paper affirms that the arithmetic means of the partial sums of the Fourier series of any continuous, 2π-periodic function h converge uniformly to it. These sums have natural harmonic extensions in $D(0, 1)$ and they converge there too, by the maximum principle, to the desired harmonic extension f. This method is quite versatile and can be adapted to an annulus, as was done by GRONWALL [1912/13] and PICONE [1926a]. See also S. N. BERNSTEIN [1909], 9.38 and its proof, 11.54, 11.55 and the notes to Chapter XI. Finally, one can get the Poisson formula (and indeed the Cauchy integral formula too) from the so-called Green's theorem. For this see OSGOOD [1928], pp. 658–666. For a seven-line deduction of the Poisson formula from Cauchy's integral formula see ÅKERBERG [1961]. On 5.18 see also BOGGIO [1911–12].

It is an important fact that in 5.20 h may be permitted to be discontinuous at the points of certain subsets S of ∂D and the conclusion that $\operatorname{Re} f$ is continuous on $\bar{D} \backslash S$ still obtains. [h should, of course, be at least (Lebesgue) integrable.] E.g., this is true for any finite set S (with h bounded and continuous in $\partial D \backslash S$) and the proof in the text needs only trifling modifications to cover this case. (Cf. 8.32(i).) This fact is very useful in constructing harmonic majorants—one takes for h the characteristic function of an appropriate set. The use of such a function occurs implicitly in the hints to 5.45(i), but its explicit use was circumvented with the ϕ constructed there.

The result 5.21(ii) is called Gauss' mean value theorem for harmonic functions. 5.22(i) is due to F. RIESZ [1922/23b]. For a further strengthening of 5.23 see READE [1943] and RUDIN [1951]. For 5.24 see KOEBE [1906b], LEVI [1909], TONELLI [1909], VOLTERRA [1909] and KELLOGG [1934]. In the latter it is shown that if U is a bounded open subset of $\mathbb{C}$, f is continuous on $\bar{U}$ and at each $a \in U$ satisfies the mean value equality in (CM) for some $r = r(a) > 0$, then f is harmonic in U. For an easy proof of this last result in the special case $U = D(0, 1)$, see p. 250 of ZALCMAN [1972] and § 10 of that work for related matters and references. (See also FENTON [1976].) There is an obvious areal mean value property (see Chapter XVIII) which also characterizes harmonicity. See NICOLESCO [1945] and *Mathematical Reviews* vol. 52 #14330, 53 #8454 and the references there for recent work on this. In a sense the circumferential and the areal mean value properties also characterize circles and disks respectively. See GREEN [1956] and KURAN [1972]. (In Chapter XVIII Kuran's beautiful result will be presented.)

5.26(i) is from F. RIESZ [1922/23b]. The result 5.26(iii) itself is due to HARDY [1915]. See also LANDAU [1916], HILLE [1916–17], JULIA [1927d], MONTEL

[1928], p. 41, and A. E. TAYLOR [1950b]. There is a generalization due to LITTLEWOOD [1925] (see also F. RIESZ [1925]): If $w \in H(D(0, 1))$ satisfies $|w(z)| \leq |z|$ for all z, then

$$\int_0^{2\pi} |f(w(re^{i\theta}))|^p d\theta \leq \int_0^{2\pi} |f(re^{i\theta})|^p d\theta$$

for all $0 \leq r < 1$ and all $f \in H(D(0, 1))$. [To recover 5.26(iii) take $w(z) = r_1 z/r_2$, given $0 \leq r_1 < r_2 < 1$.] More references on this matter may be found on p. 135 of the bibliography BERNARDI [1966]. For a nice proof of 5.26(iii) based on 5.72 see exercise III.310 of PÓLYA and SZEGÖ [1972]. For Littlewood's extension with an arbitrary subharmonic function in the role of $|f|^p$ see PRIVALOV [1935].

If u is a continuous real-valued function on [0, 1] and ϕ is a convex, increasing (whence continuous) function on $\mathbb{R}$, then a well-known inequality due to JENSEN [1906] asserts that $\phi(\int_0^1 u) \leq \int_0^1 (\phi \circ u)$. It follows at once that $\phi \circ u$ is subharmonic if u is. This subsumes all the examples in 5.26(ii). Moreover, the two assertions in 5.26(ii) are essentially equivalent *a priori*: MONTEL [1928] shows that if $f > 0$, then f^p is subharmonic for all positive p if and only if $\log \circ f$ is subharmonic.

5.27 is due to Carathéodory and Plemelj. (See p. 21 of CARATHÉODORY [1914] and also 6.30.) For an extension of 5.28 see p. 16 of LINDELÖF [1909]. For other proofs of 5.29 see OU [1957], lemma 1 of GEHRING and LOHWATER [1958] and also theorem 7 of GOLDBERG [1962]. For the holomorphic function version of it see 5.56 and 5.57 (also 12.30–12.36).

The particular series in 5.30 is called the Taylor series of f at z_0. (See PRINGSHEIM [1900].) The all important fact 5.30 (due to Cauchy) constitutes the essential difference between differentiable (complex-valued) functions on an open subset of $\mathbb{R}$ and those on an open subset of $\mathbb{C}$. The demand that the difference quotients have a limit is so much more restrictive when the approach can be made through a disk rather than merely along one line, that the existence of such limits entails the existence of all higher derivatives as well as the convergence to the function of its Taylor series. No such conclusions hold on the line. We all know differentiable functions on $\mathbb{R}$ whose derivatives are not differentiable. (A primitive of Weierstrass' everywhere continuous and nowhere differentiable function on $\mathbb{R}$—see BANACH [1931] and MAZURKIEWICZ [1931]—even provides a function which is everywhere differentiable but whose derivative is nowhere differentiable.) Furthermore there exist infinitely differentiable functions on (0, 1) (a plenitude of them—see MORGENSTERN [1954], SALZMANN and ZELLER [1955], and DARST [1973]) which do not coincide with their Taylor series in any subinterval of (0, 1). A concrete example will be constructed in Chapter XVII.

The derivation of 5.30 (thence 5.33) from the Cauchy–Schwarz formula 5.18, "the road less traveled by" of Robert Frost, is the path followed by H. A. SCHWARZ on p. 189, vol. II of his *Gesammelte Mathematische Abhandlungen*. See also footnotes 889, 888 of BURKHARDT [1914]. There are more economical routes. For example, one can prove 5.1 under the (*prima facie*) weaker hypothesis that f is continuous in U, holomorphic in $U\backslash\{z_0\}$ and satisfies $(z - z_0)f(z) \to 0$ as $z \to z_0$, for some $z_0 \in U$. This makes the result immediately applicable to $(f(z) - f(z_0))/(z - z_0)$ in the role of f and gives the Cauchy integral formula (as in the proof of 5.33) as well as the removable singularity theorem (5.41) in one stroke. For such a development see SCHOTTKY [1916] or RUDIN [1974].

5.30 can be secured without the intervention of an integral representation; see chapter VI of G. T. WHYBURN [1964]. One can also deduce 5.30, under the additional assumption that f' is locally bounded, directly from the Cauchy-Riemann equation by Fourier analysis. See YOUNG [1912].

Notice that the existence of all the higher complex derivatives insures, without appeal to any other theorems of analysis, that all partial derivatives of a holomorphic or harmonic function exist and that these are independent of the order in which they are carried out. For example, if $h = \operatorname{Re} f$ and f is holomorphic, then $D_1(D_2h) = \operatorname{Re} D_1(D_2f) = \operatorname{Re} D_1(if') = \operatorname{Re}(if'')$, while $D_2(D_1h) = \operatorname{Re} D_2(D_1f) = \operatorname{Re} D_2(f') = \operatorname{Re}(if'')$, thanks to the Cauchy–Riemann equation. For the same reason all these partial derivatives are again harmonic functions.

I took 5.36(i) from HOFFMAN [1962].

The Three Circles Theorem (5.38(ii)) [so christened by E. Landau] occurs without proof in HADAMARD [1896]. See footnote 13 of OSTROWSKI [1922] for history. For an interesting alternative proof and more history see COLLINGWOOD [1932]. See also p. 14 of F. and R. NEVANLINNA [1922], CARLSON [1938], ROBINSON [1943] and § 3.4 of AHLFORS [1973] for sharper versions. For the "three lines" theorem in 5.38(i) see DOETSCH [1920], WALTHER [1921], HARDY and ROGOSINSKI [1946] and BONSALL [1949]. The boundedness requirement here can be weakened with Phragmén–Lindelöf techniques from Chapter XIV. (See, e.g., theorem 12.9 of RUDIN [1974].) It is interesting to note that 5.38(i) is a quantitative extension (for strip regions) of the result 5.13. For a proof of 5.38(i) and (ii) involving harmonic functions (cf. the following remarks) see MONTEL [1928] (pp. 45, 55), HERVÉ [1970] (pp. 11, 12) and BEAR [1974]. Modulo problems with 0, the Three Circles Theorem asserts that $\log M(r) = \max_{|z|=r} \log|f(z)|$ is a convex function of $\log r$, $r \in (0, 1)$. In this form it generalizes: if $h\colon A(R_1, R_2) \to \mathbb{R}$ is subharmonic, then $M(r) = \max_{|z|=r} h(z)$ is a convex function of $\log r$ for $r \in (R_1, R_2)$. The proof is effected by comparing $h(z)$ to $\alpha \log|z| + \beta$ via 5.22(i), for appropriate α and β. The case $h = \log|f|$ is 5.38(ii). (Cf. the earlier remarks on $-\infty$ as a value of subharmonic functions.) From this result a subharmonic Liouville theorem follows: if h is a subharmonic function defined in $\mathbb{C}$ which satisfies

$$\lim_{r\to\infty} \frac{1}{\log r}\left[\sup_{|z|=r} h(z)\right] = 0,$$

then h is constant. See § 2.7.1 of HAYMAN and KENNEDY [1976].

The result 5.38(iii) is from OSTROWSKI [1922]. When applied to uniformly convergent sequences of holomorphic functions [as was done in (iv)], it implies that the "rates of convergence on any two compacta with interiors are comparable." (See also MILLOUX [1924] and MANDELBROJT [1929].) Even more: convergence in one region induces convergence in another, and so the phenomenon in 5.38(iv) [and that in 5.45(iii) and 5.74 too] is called "induced convergence." Many more striking examples of it occur in Chapters VII and XII.

The result 5.39 is (essentially) contained in MORERA [1886] (see also his paper [1902]). However, several subsequent writers prove 2.11 without acknowledging any priority. See, e.g., DE LA VALLÉE POUSSIN [1893], OSGOOD [1895/96] and POMPEIU [1905a]. All these writers point out how easily Weierstrass' theorem 5.44(i) follows from this result and in addition de la Vallée Poussin shows how, using the Cauchy estimates, 5.44(ii) can be deduced from 2.11 too. The stronger form of 2.11 in 5.40, involving only rectangles, seems to have first occurred in the second edition (1912) of the book by OSGOOD [1928]. Much smaller classes of loops suffice in the role of the triangles and rectangles of 5.39 and 5.40 to imply the differentiability of f. See RADEMACHER [1919b], WOLFF [1924], RIDDER [1930a], G. SPRINGER [1957], DELSARTE [1958], [1961], ROYDEN [1962], ZALCMAN [1972] (especially its notes and bibliography), [1974]; BROWN, SCHREIBER and TAYLOR [1973] and BERENSTEIN and TAYLOR [1977]. See also exercise III.171 of PÓLYA and SZEGŐ [1972] and p. 333 of RICHARDS [1970] for two proofs that circles suffice. For an elementary generalization of Morera's theorem of a different kind see PERRY and YOUNGS [1947] and C. MÜLLER [1954].

5.41 appears in Riemann's dissertation [pp. 23–24 of his *Werke*], but was known to Weierstrass earlier (see OSGOOD [1895/96] for history); the proof given is from D. R. CURTISS [1905–06], inspired, it would seem, by HÖLDER [1882]. For different treatments see CHESSIN [1895–96], PRINGSHEIM [1896a] and SCHOTTKY [1916]; LANDAU [1905/06] is a treatment similar to Curtiss'.

5.44(iv) is from DARBOUX [1875], p. 83. Notice that only convergence of $\{f_n(a)\}$, and not that of the function sequence $\{f_n\}$ need be hypothesized. Moreover, use of the Mean Value Theorem instead of integrals allows the continuity of the f_n' to be dispensed with. (See, e.g., RUDIN [1976], p. 152.) For an elementary extension to holomorphic functions see FERRAR [1930].

There is a version of 5.46 in which the hypothesis $\overline{\lim}_{z\to z_0} h(z) < \infty$ is replaced by $h = |f|, f \in H(U)$ and f univalent near z_0. See MINAMI [1937]. For a different kind of extension see ZAREMBA [1909]. In general, some boundedness hypothesis at the exceptional points is needed in 5.46, as the following striking example of Picard shows: $h(z) = \mathrm{Re}(1/z) - 1$ in $\Omega = D(\frac{1}{2}, \frac{1}{2})$ has $h = 0$ at every boundary point of Ω save 0, yet h is not negative in Ω. On this example see BOULIGAND [1932].

The method of harmonic majorization is very powerful and of broad applicability. It goes back to Lindelöf, Carleman, Ostrowski and the Nevanlinnas. The general setting, of which the examples in the text are special cases, is this: a region Ω and a portion B of $\partial\Omega$ are given. For any $f \in H(\Omega)$ bounded by M and satisfying $\overline{\lim}|f| \leq m(\leq M)$ on B we have $|f| \leq m^{\omega}M^{1-\omega}$, where ω is the harmonic function with values in [0, 1] which approaches 1 (resp., 0) as the variable tends to B (resp., $\partial\Omega\backslash B$). This is known as the two-constants theorem of Ostrowski and F. and R. NEVANLINNA [1922]. (See also SAKS [1928/29].) The problem is to ensure that such an ω exists and to find useful bounds for it in Ω. The function ω is traditionally called (after R. Nevanlinna) the harmonic measure of B with respect to Ω, and one writes $\omega(z, B, \Omega)$ for its value at $z \in \Omega$. (For the geometric significance see § 158 of CARATHÉODORY [1958].) For more applications of the method see R. NEVANLINNA [1935], BRELOT [1939], § 18.3 of HILLE [1962], FUCHS [1967], R. NEVANLINNA [1970] and AHLFORS [1973].

5.66 is from ROSENBLATT [1945]; the proof given was suggested by S. Saeki. On 5.62 see BURDICK and LESLEY [1975]. 5.67(i) is from p. 66, vol. II of H. A. SCHWARZ' *Gesammelte Mathematische Abhandlungen.*

In contrast to 5.70 the entire function $f(z) = z + e^z$ clearly satisfies $\lim_{r\to\infty}|f(re^{i\theta})| = \infty$ for each fixed θ, yet f is not a polynomial. (See KOCH [1903–04].) 5.72 is exercise III.299 and III.302 of PÓLYA and SZEGÖ [1972].

5.73 is from BENDIXSON [1886/87], H. LAURENT [1902] and JOURDAIN [1905a]; a version for strips occurs in DEMIN [1946]. A different treatment of the problem is given by FRANKLIN [1927] (cf. ECHOLS [1893]). For related matter see ABIAN [1974] and SMITH [1975]. A more general problem (see, e.g., IZUMI [1929]), that of "interpolation," is to find necessary and sufficient conditions on complex numbers w_n which will insure the existence of a holomorphic function f satisfying $f(z_n) = w_n$ for all $n \geq 1$ and, as in 5.73, to find formulas for the values of $f(z)$ in terms of z and the w_n. The literature on this and allied problems in which supplemental requirements are imposed on the interpolating function f (e.g., that it be bounded, or univalent) is too enormous to attempt to survey, but the monograph of WHITTAKER [1935] can be mentioned as a modest starting point for the interested reader. See also WALSH [1969]. Isolated special cases will be treated from time to time in the sequel. (See e.g., 6.9(iii), 7.33 and Chapter XIV, where VALIRON [1925c] will be discussed.)

5.75 has many generalizations. Here is one (FATOU [1906], p. 395): if f is bounded and holomorphic in $D(0, 1)$ and $\lim_{r\uparrow 1} f(re^{i\theta}) = 0$ for each θ in $[\theta_0, \theta_1]$, then $f = 0$. See also Chapter XIII, WOLF [1940] and RIDDER [1967]. The ultimate theorem of this type is due to Lusin and Privalov: if f is holomorphic in $D(0, 1)$ and for some Lebesgue measurable subset S of positive measure in $[0, 2\pi)$ and for each $\theta \in S$ we have $f(z)\to 0$ as z converges to $e^{i\theta}$ within any angle (less than π) in $D(0, 1)$ whose vertex is $e^{i\theta}$ (so-called non-tangential approach), then $f = 0$. Moreover there is such a set S and an f holomorphic in $D(0, 1)$ such that S has measure 2π, f is not the zero function and yet $\lim_{r\uparrow 1} f(re^{i\theta}) = 0$ for each $\theta \in S$. For the (complicated) details see chapter 4 of PRIVALOV [1956] or pp. 145 ff. of

COLLINGWOOD and LOHWATER [1966]. For the last quoted result see WOLFF [1928] and BURGERS [1929]. (Cf. also § 9 of CARATHÉODORY [1913b].) In 9.10 we will see another kind of generalization of 5.75. There are several explicit formulas which recover the function values inside $D(0, 1)$ from those on a part of the boundary. For example, (Carleman), if A is an arc of the unit circle of length $\alpha\pi$ and f is holomorphic in $D(0, 1)$ and continuous on $A \cup D(0, 1)$, then

$$f(z) = \lim_{k\to\infty} \frac{e^{-k}}{2\pi i} \int_A \frac{f(\xi) \exp\left[k\left(\frac{\xi}{z}\right)^{1/\alpha}\right]}{\xi - z}\, d\xi \quad \forall z \in D(0, 1)\backslash\{0\}.$$

See also GOLUZIN and KRYLOW [1933], ZIN [1953], PICONE [1954] and PATIL [1972].

For an extension of 5.76 to subharmonic functions see SAKS [1928/29] and BRELOT [1939], p. 469.

The proof of 5.78 comes from POMPEIU [1911] (see also [1933]). The converse of 5.78, i.e., f' zero-free implies f univalent, is obviously false. (Take $f(z) = e^z$.) When U is a region, U or $f(U)$ is simply-connected, and f has the additional property that $f^{-1}(K)$ is compact for each compact subset K of $f(U)$, then the non-vanishing of f' does imply the univalence of f. See *Satz* II of CARATHÉODORY and RADEMACHER [1917] or *Satz* 5 and 5a, § 1, chapter IV of BEHNKE and SOMMER [1965]. (See also § 13 of ZALCMAN [1974]!) In fact in this situation, if the condition f' zero-free is replaced by the condition that f be a local homeomorphism (see 5.79), then the differentiability of f can be dropped from the hypothesis! For other extensions see BROWDER [1954], GORDON [1972], PLASTOCK [1974], and HO [1975]. On 5.79 see the historical note in HILLE [1959], p. 87.

5.80 is a sort of converse to the Mean Value Theorem for differentiable functions on the line. For complex versions of the latter see POMPEIU [1905b], D. R. CURTISS [1906–07]; pp. 1021, 1022, vol. 59 of the *Jahrbuch*, HUBER [1948], RUBINSTEIN [1969], SAMUELSSON [1973].

For further elaboration on 5.83 and 5.88 see HILLE [1959], p. 267 (the Weierstrass *Vorbereitungssatz*). For an integral-free treatment of 5.85 see VIVANTI [1904].

5.87 is a version of Rouché's theorem, to be discussed in detail in Chapter VIII. (See 8.18 and 11.51.) The first proof suggested for 5.87 is from GLICKSBERG [1976]; for this method see the historical note on p. 158 of SAKS and ZYGMUND [1971]. Either method yields a geometrically very satisfying version of the theorem for Jordan regions once the Argument Principle 9.15, generalizing 5.86, is available. But such a result is subsumed by 8.18, in which no special assumptions about the boundary occur. The reader with more background may be titillated by the exotic proof in VAN DULST [1971] or that in ABIAN [1977]. 5.92(iii) is from HAMEL [1940].

For detailed history of many of the results of this chapter see OSGOOD's encyclopedia article.

Chapter VI
Schwarz' Lemma and its Many Applications

§ 1 Schwarz' Lemma and the Conformal Automorphisms of Disks

This chapter is fairly tightly unified around the following simple result:

Theorem 6.1 (Schwarz' Lemma) *Let f be holomorphic and bounded by* 1 *in* $D = D(0, 1)$ *and* $f(0) = 0$. *Then*
(i) $|f'(0)| \le 1$
(ii) $|f(z)| \le |z| \quad \forall z \in D$
and equality in (i) *or in* (ii) *for some non-zero z occurs if and only if*
(iii) $f(z) \equiv cz$ *for some unimodular complex constant c.*

Proof: Write the power series expansion about 0 of f in D: with $c_n = f^{(n)}(0)/n!$

(1) $$f(z) = \sum_{n=0}^{\infty} c_n z^n \quad \forall z \in D.$$

Set

(2) $$g(z) = \sum_{n=1}^{\infty} c_n z^{n-1}, \quad z \in D.$$

Thus, since $c_0 = f(0) = 0$,

(3) $$zg(z) = f(z).$$

If $z \in D$ and we pick $1 > r > |z|$, then by the Maximum Modulus Principle

$$|g(z)| \le \sup_{|w|=r} |g(w)| \overset{(3)}{=} \sup_{|w|=r} \frac{|f(w)|}{r} \le \frac{1}{r}.$$

Let $r \uparrow 1$ to get

(4) $$|g(z)| \le 1 \quad \forall z \in D.$$

From (4) and (3) follows (ii). But also, taking $z = 0$ in (4) and noting that $g(0) = c_1 = f'(0)$, gives (i).

If equality holds in (i), then (4) shows that $|g|$ attains a maximum in D at 0 and so by the Maximum Modulus Principle g is constant in D. If c is that constant, then $|c| = |g(0)| = |f'(0)| = 1$ and (iii) holds: $f(z) \equiv zg(z) \equiv cz$.

If equality occurs in (ii) for some non-zero z, then $|g(z)| = 1$ and so (4) says that $|g|$ attains a maximum in D at z, making g constant and (iii) follows as before.

Remarks: A less general version of 6.1 (for univalent f only) will be found on pp. 109–111 of H. A. SCHWARZ' *Gesammelte Mathematische Abhandlungen*, vol. II. The full scope of the result was recognized by CARATHÉODORY [1912a], who gave the above proof (based on a verbal suggestion of E. Schmidt) and instituted the designation. However, a less transparent variant of this proof occurs already in an 1884 paper of H. Poincaré. See footnote 427 of LICHTENSTEIN [1919] for history and KOEBE [1920] for refinements. The reader will soon appreciate the keystone character of this deceptively simple result.

Theorem 6.2 *Let $D = D(0, 1)$ and for each $a \in D$ define f_a by*

$$f_a(z) = \frac{z - a}{1 - \bar{a}z}, \quad z \in \mathbb{C}\backslash\{1/\bar{a}\}.$$

Then

(i) *f_a is one-to-one, continuous on $\bar{D}$, holomorphic in D and $f_a(D) = D$, $f_a(\bar{D}) = \bar{D}$. Furthermore the inverse map to f_a on $\bar{D}$ is f_{-a}. Conversely:*

(ii) *If f is a one-to-one holomorphic map of D onto D, then $f = cf_a$ for some $a \in D$ and some unimodular complex constant c. In particular, f extends to be holomorphic in a neighborhood of $\bar{D}$ and to a homeomorphism of $\bar{D}$ onto $\bar{D}$.*

Proof: Part (i) is quite elementary and was completely attended to in 2.6. In proving (ii), consider first the case $f(0) = 0$. Schwarz' lemma may then be applied to each of the holomorphic functions f and f^{-1} to get

$$|f(z)| \leq |z| = |f^{-1}(f(z))| \leq |f(z)| \quad \forall z \in D.$$

It follows that $|f(z)| = |z|$, so again from Schwarz, $f(z) \equiv cz$ for some unimodular constant c.

In the general case, set $a = f^{-1}(0)$ and form $F = f \circ f_{-a}$. By part (i) this function satisfies all the hypotheses which f does and moreover $F(0) = f(f_{-a}(0)) = f(a) = 0$. By what was just shown, $F(z) = cz$, whence $f(z) = F(f_{-a}^{-1}(z)) = F(f_a(z)) = cf_a(z)$.

Exercise 6.3 (i) *If f is a holomorphic function in $D(0, 1)$ which satisfies $|f| < 1$, then*

$$\left|\frac{f(z) - f(w)}{1 - \overline{f(w)}f(z)}\right| \leq \left|\frac{z - w}{1 - \bar{w}z}\right| \quad \forall z, w \in D(0, 1).$$

Hints: Let

$$\phi(z) = \frac{z - w}{1 - \bar{w}z}, \qquad \psi(\xi) = \frac{\xi - f(w)}{1 - \overline{f(w)}\xi}, \quad z, \xi \in D(0, 1)$$

and apply 6.1 to the composite $\psi \circ f \circ \phi^{-1}$.

(ii) *If F is a holomorphic function in $D(0, R)$ bounded by M, then*

$$\left|\frac{F(z) - F(w)}{z - w}\right| \leq \frac{2MR}{|R^2 - \bar{w}z|} \quad \forall z, w \in D(0, R).$$

Hint: Take $f(z) = F(Rz)/M$ in part (i).

(iii) *Show that*

$$\left|\frac{a-b}{1-\bar{a}b}\right| \geq \frac{|a|-|b|}{1-|a||b|} \quad \forall a, b \in D(0, 1).$$

Hint: As in the proof of 2.6,

$$\left|\frac{a-b}{1-\bar{a}b}\right|^2 = 1 - \frac{(1-|a|^2)(1-|b|^2)}{|1-\bar{a}b|^2} \geq 1 - \frac{(1-|a|^2)(1-|b|^2)}{(1-|a||b|)^2}$$

$$= \frac{(|a|-|b|)^2}{(1-|a||b|)^2}.$$

(iv) (LINDELÖF [1909], p. 11.) *Let f be holomorphic and bounded by* 1 *in* $D(0, 1)$. *Then*

$$|f(z)| \leq \frac{|f(0)|+|z|}{1+|f(0)||z|} \quad \forall z \in D(0, 1).$$

Hint: By (iii) and (i)

$$\frac{|f(z)|-|f(0)|}{1-|f(0)||f(z)|} \leq \left|\frac{f(z)-f(0)}{1-\overline{f(0)}f(z)}\right| \leq |z|.$$

Solve the extreme inequality algebraically for $|f(z)|$.

Exercise 6.4 *If a_1, a_2, b_1, b_2, are complex numbers of modulus* 1 *with $a_1 \neq a_2$ and $b_1 \neq b_2$, then there is a conformal map f of $D(0, 1)$ onto itself such that $f(a_1) = b_1$, $f(a_2) = b_2$.*

Hints: If f maps the pair $(1, -1)$ to the pair (a_1, a_2) and g maps the pair $(1, -1)$ to the pair (b_1, b_2) then $g \circ f^{-1}$ maps the pair (a_1, a_2) to the pair (b_1, b_2). So it suffices to deal with $a_1 = 1$, $a_2 = -1$. Also it suffices to map the pair $(1, -1)$ to the pair $(1, b_1^{-1}b_2)$, for b_1 times any such map sends the pair $(1, -1)$ to the pair (b_1, b_2). Thus we may assume $b_1 = 1$, $b_2 = b \neq 1$. Let c be the square root of $-b$ which has *positive* real part. Let $a = (c-1)/(c+1)$ and $\alpha = (c+1)/(\bar{c}+1)$. Then $|a| < 1 = |\alpha|$ and the function $f(z) = \alpha(z-a)/(1-\bar{a}z)$ satisfies $f(1) = 1$, $f(-1) = b$.

Exercise 6.5 *If f is a holomorphic function on $D = D(0, 1)$ with $f(D) \subset D$ and if f fixes more than one point in D, then f is the identity.*

Hints: For an appropriate conformal map ϕ of D the function $F = \phi^{-1} \circ f \circ \phi$ maps D into D and fixes at least two points, one of which is 0. Apply Schwarz' lemma to F.

Exercise 6.6 *Show that every conformal map of $D(0, 1)$ onto itself maps each open subdisk of $D(0, 1)$ onto another. (For the converse of this see* CARATHÉODORY [1937b] *and* HARUKI [1972].)

Hint: Given $b \in D(0, 1)\backslash\{0\}$, the map $f(z) = (z + b)/(1 + \bar{b}z)$ is the composite $f_1 \circ f_3 \circ f_2 \circ f_1$, where $f_1(z) = z + 1/\bar{b}$, $f_2(z) = 1/z$ and $f_3(z) = z(|b|^2 - 1)/\bar{b}^2$.

Exercise 6.7 (JENSEN [1898–99] and PETERSEN [1899]) *Let f be holomorphic in a neighborhood of $\bar{D}(0, r)$ and bounded by M. If $z_1, \ldots, z_k$ are zeros of f in $D(0, r)\backslash\{0\}$, show that*

$$\frac{r^k|f(0)|}{|z_1 \cdots z_k|} \le M.$$

Hints: (CARATHÉODORY and FEJÉR [1907]) The function $f_j(z) = r(z - z_j)/(r^2 - \bar{z}_j z)$ is holomorphic in a neighborhood of $\bar{D}(0, r)$ and has a first order zero at z_j only. Consequently $F = f/(f_1 \cdots f_k)$ is holomorphic in a neighborhood of $\bar{D}(0, r)$. On $C(0, r)$, $|f_j| = 1$ (by 6.2) so $|F| = |f| \le M$ there. Cite the Maximum Modulus Principle to conclude that $|F(0)| \le M$.

Remark: When $r = 1$ the product $f_1 \cdots f_k$ above is called a (finite) *Blaschke product*. Because these products have modulus 1 on the unit circle, zeros can be removed from a function, without increasing its upper bound, by dividing off such a product. This is often more advantageous than removing the zeros by merely dividing off the factors $z - z_j$. Cf. 6.9 and 6.12 below.

Exercise 6.8 *Let f be an entire function with $|f(0)| = 1$. For each $\rho > 0$ let $M(\rho)$ denote $\sup|f(\bar{D}(0, \rho))|$ and $n(\rho)$ the number of zeros, counted according to multiplicity, of f in $\bar{D}(0, \rho)$. Show that*

$$2^{n(\rho)} \le M(2\rho) \qquad \forall \rho > 0.$$

Hints: Let $z_1, z_2 \cdots$ be an enumeration of the zeros of f repeated according to multiplicity and ordered by increasing modulus. Thus for each $R > 0$

(*) $\quad |z_j| \le R \quad$ for $1 \le j \le n(R)$.

With $\rho > 0$ fixed, consider $r = 2\rho$, $k = n(2\rho)$, $M = M(2\rho)$ in 6.7. Then

$$2^{n(\rho)} \overset{(*)}{\le} \prod_{j=1}^{n(\rho)} \left|\frac{2\rho}{z_j}\right| \overset{(*)}{\le} \prod_{j=1}^{n(2\rho)} \left|\frac{2\rho}{z_j}\right| = \frac{r^k|f(0)|}{|z_1 \cdots z_k|} \overset{(6.7)}{\le} M = M(2\rho).$$

Exercise 6.9 (*Following* LÖWNER *and* RADÓ [1923]; *see also* M. M. LAVRENTIEFF [1956].)

(i) *For all $a, z \in D(0, 1)$*

$$\left|\frac{z - a}{1 - \bar{a}z}\right| \le e^{(1 - |a|)(|z| - 1)/(|z| + 1)}.$$

Hints: According to 6.3(iv)

$$\left|\frac{z - a}{1 - \bar{a}z}\right| \le \frac{|z| + |a|}{1 + |a||z|} = 1 - (1 - |z|)\frac{1 - |a|}{1 + |a||z|}$$

$$< 1 - (1 - |z|)\frac{1 - |a|}{1 + |z|}.$$

The desired inequality follows from this when we recall that $1 - t < e^{-t}$ for positive t.

(ii) *For each set of k distinct numbers* $z_1, \ldots, z_k \in D(0, 1)$ *there is a positive finite constant* $C_k = C(z_1, \ldots, z_k)$ *such that*

$$|f(z)| \leq 2mC_k + Me^{[(|z|-1)/(|z|+1)]\sum_{j=1}^{k}(1-|z_j|)}, \quad z \in D(0, 1)$$

for all f which are holomorphic in $D(0, 1)$, *bounded by M there and by m on* $\{z_1, \ldots, z_k\}$.

Hints: Set $C_k = \sum_{j=1}^{k} \prod_{l=1, l\neq j}^{k} 2|z_l - z_j|^{-1}$. Let such an f be given and form the polynomial

$$L(z) = \sum_{j=1}^{k} f(z_j) \prod_{\substack{l=1 \\ l\neq j}}^{k} (z - z_l)(z_j - z_l)^{-1}.$$

Then we evidently have

(1) $\quad |L(z)| \leq C_k \max_{1\leq j\leq k} |f(z_j)| \leq C_k m, \quad z \in D(0, 1)$

and

(2) $\quad L(z_j) = f(z_j), \quad j = 1, 2, \ldots, k.$

Form the Blaschke product $B(z) = \prod_{j=1}^{k} (z - z_j)/(1 - \bar{z}_j z)$. Then because of (2), $(L - f)/B$ is holomorphic in $D(0, 1)$. Compute its maximum on $C(0, r)$, cite the Maximum Modulus Principle, let $r \uparrow 1$ and notice that $|B(z)| \to 1$ as $|z| \to 1$. We learn that

$$\left|\frac{L - f}{B}\right| \leq \sup_{D(0,1)} |L - f|$$

$$\leq mC_k + M, \quad \text{by (1) and the definition of } M.$$

It follows that

$$|f| \leq (mC_k + M)|B| + |L| \leq (mC_k + M)|B| + mC_k, \quad \text{by (1) again,}$$

$$\leq 2mC_k + M|B|, \text{ since } |B| \leq 1.$$

Finally, use (i) to majorize the Blaschke product in this last inequality.

(iii) *Let f be holomorphic and bounded by M in* $D(0, 1)$ *and let* $z_1, z_2, \ldots$ *be zeros of f repeated according to multiplicity. Show that*

$$|f(z)| \leq Me^{[(|z|-1)/(|z|+1)]\sum_{j=1}^{k}(1-|z_j|)} \quad \forall z \in D(0, 1), \forall k \geq 1.$$

Deduce that if $\sum_{j=1}^{\infty} (1 - |z_j|) = \infty$, *then* $f = 0$. (Cf. 12.6.)

Hint: Let B be the Blaschke product on $z_1, \ldots, z_k$ and argue as in the hints to (ii): form f/B, apply the Maximum Modulus Principle and use (i).

(iv) *F is holomorphic and bounded in* $H = (0, \infty) \times \mathbb{R}$ *and has zeros at points* $z_1, z_2, \ldots$ *with* $\inf_n |z_n| > 0$. *(Here a* z_n *may recur as often as its multiplicity.) Show that either* $F \equiv 0$ *or* $\sum_{n=1}^{\infty} \text{Re}(1/z_n) < \infty$.

Hints: Let $\delta = \inf_n|z_n|$. Map H onto $D = D(0, 1)$ by $\phi(z) = (z - 1)/(z + 1)$. (See 2.5.) $f = F \circ \phi^{-1}$ is a bounded holomorphic function in D with zeros at the points $w_n = \phi(z_n)$. We have

$$1 - |w_n|^2 = 1 - \left|\frac{z_n - 1}{z_n + 1}\right|^2 = \frac{4}{\left|1 + \dfrac{1}{z_n}\right|^2} \frac{z_n + \bar{z}_n}{2z_n\bar{z}_n} = \frac{4}{\left|1 + \dfrac{1}{z_n}\right|^2} \operatorname{Re}\left(\frac{1}{z_n}\right)$$

$$\geq \frac{4\delta^2}{(\delta + 1)^2} \operatorname{Re}\left(\frac{1}{z_n}\right), \quad \text{since } |z_n| \geq \delta \text{ entails } \left|1 + \frac{1}{z_n}\right| \leq 1 + \frac{1}{\delta}.$$

Since $|w_n| < 1$, it follows that

$$1 - |w_n| = \frac{1 - |w_n|^2}{1 + |w_n|} \geq \frac{1 - |w_n|^2}{2} \geq \frac{2\delta^2}{(\delta + 1)^2} \operatorname{Re}\left(\frac{1}{z_n}\right).$$

Now cite (iii).

Exercise 6.10 *Let f be holomorphic and satisfy $|f| < 1$ in $D(0, 1)$. Define g in $D(0, 1)$ by $g(z) = (f(0) - f(z))/(z - z\bar{f}(0)f(z))$. Suppose that g^* is holomorphic and bounded by* 1 *in $D(0, 1)$ and that $g^* - g$ has a zero of order at least n at* 0 *($n \geq 1$). Show that for the function $f^*(z) = (f(0) - zg^*(z))/(1 - \bar{f}(0)zg^*(z))$ the difference $f^* - f$ has a zero of order at least $n + 1$ at* 0.

Hint: Direct computation reveals that for all $z \in D(0, 1)$

$$f(z) - f^*(z) = z[g^*(z) - g(z)]\left[\frac{1 - \bar{f}(0)f(z)}{1 - \bar{f}(0)zg^*(z)}\right].$$

The third factor on the right is holomorphic in $D(0, 1)$, since $|\bar{f}(0)| < 1$ and $|g^*| \leq 1$.

Exercise 6.11 *Let f, g be entire functions which for some $c < 1$ satisfy*
(i) *$|f(z)| \leq e^{c|z|}$ & $|g(z)| \leq e^{c|z|}$ for all (sufficiently large) z,*
(ii) *$f(n) = g(n)$ for all positive integers n.*
Show that then $f = g$.

Hints: (PÓLYA [1915]) Suppose not. Then (5.62) there exists at least one $a \in (0, 1)$ such that $f(a) \neq g(a)$. Thus

$$h(z) = f(z + a) - g(z + a)$$

is entire, not identically 0 and satisfies

$$h(a_n) = 0, \quad n = 1, 2, 3, \ldots,$$

where $a_n = n - a$. For any $c < C < 1$ we have

(1) $|h(z)| \leq e^{C|z|}$ for all sufficiently large z.

Notice that $0 < a_j < j$ for every $j = 1, 2, \ldots$. Therefore for large n we may apply 6.7 to h with $r = n$ and, because of (1), $M = e^{Cn}$. There results

$$\frac{n^n|h(0)|}{n!} < \frac{n^n|h(0)|}{|a_1 \cdots a_n|} \le e^{Cn},$$

(2) $$\frac{n}{(n!)^{1/n}} |h(0)|^{1/n} \le e^C.$$

Notice that if we set $c_n = n^n/n!$, then $c_{n+1}/c_n = (1 + 1/n)^n$ and so $\lim_{n\to\infty} c_{n+1}/c_n = e$ by 3.21. It follows then from 3.9 that $\lim_{n\to\infty}(c_n)^{1/n}$ exists and equals e. Insert this fact into (2), recall that $h(0) \neq 0$ and cite 3.1. We conclude that $e \le e^C$, whence $C \ge 1$, contrary to the definition of C.

§ 2 Many-to-one Maps of Disks onto Disks

Exercise 6.12 (i) (FATOU [1923]) *If f is holomorphic in $D(0, 1)$ and*

$$\lim_{r\uparrow 1}|f(re^{i\theta})| = 1$$

uniformly for $\theta \in \mathbb{R}$, then there exist $a_1, \ldots, a_n \in D(0, 1)$ and $u \in C(0, 1)$ such that

$$f(z) = u\prod_{j=1}^{n} \frac{z - a_j}{1 - \bar{a}_j z} \quad \forall z \in D(0, 1).$$

Hints: f can have only finitely many zeros in $D(0, 1)$. If these are $a_1, \ldots, a_n$, repeated according to multiplicity, then let $g(z) = \prod_{j=1}^{n} (z - a_j)/(1 - \bar{a}_j z)$ and note that g/f, f/g are each holomorphic in $D(0, 1)$, and their moduli are continuous on $\bar{D}(0, 1)$. Apply the Maximum Modulus Principle to each to conclude that $|f/g| \equiv 1$.

(ii) *Deduce from* (i) *that an entire function of constant modulus on the unit circle is a monomial.*

Exercise 6.13 *Let f: $D \to D$ be holomorphic and exactly m-to-1. By this is meant that for each $w \in D$ the number of zeros of $f - w$, counted according to multiplicity, is m. Here $D = D(0, 1)$.*

(i) *Show that* $\lim_{|z|\to 1}|f(z)|$ *exists and equals* 1.

Hints: (RADÓ [1922/23c]) In the contrary case we can find $z_n \in D$ such that

(1) $|z_n| \to 1$

(2) $\overline{\{f(z_n)\}} \subset D$

(3) $f(z_n) \to w$

(4) infinitely many $f(z_n)$ are different from w.

By (2) and (3) we have $w \in D$. Let the distinct zeros of $f - w$ be $a_1, \ldots, a_k$ with respective multiplicities $m_1, \ldots, m_k$. Thus

(5) $m_1 + \cdots + m_k = m.$

Use 5.83 to select $\varepsilon > 0$ sufficiently small and for each j an open neighborhood N_j of a_j whose closure lies in D and such that

(6) each point of $D(w, \varepsilon)\backslash\{w\}$ has m_j distinct pre-images in N_j.

We can demand that the N_j are disjoint. Select $R \in (0, 1)$ sufficiently near 1 that

(7) $\bar{N}_1 \cup \cdots \cup \bar{N}_k \subset D(0, R)$.

Now use (1), (4) and (3) to select n so large that

$$|z_n| > R, \quad f(z_n) \neq w \quad \text{and} \quad f(z_n) \in f(N_1) \cap \cdots \cap f(N_k) \cap D(w, \varepsilon).$$

(Recall, $f(N_j)$ is an open neighborhood of $f(a_j) = w$.) Then by (6) the point $f(z_n)$ has m_j distinct pre-images in N_j, hence $m_1 + \cdots + m_k = m$ distinct pre-images in $N_1 \cup \cdots \cup N_k \subset D(0, R)$, while it also has the pre-image $z_n \in D\backslash D(0, R)$, since $|z_n| > R$. The point $f(z_n)$ thus has $m + 1$ distinct pre-images.

(ii) (FATOU [1920]) *Deduce that for some (not necessarily distinct) points* $c_1, \ldots, c_m \in D$ *and* $c \in C(0, 1)$

$$f(z) = c\prod_{j=1}^{m} \frac{z - c_j}{1 - \bar{c}_j z}.$$

Hint: Because of (i) the conclusion of 6.12 is applicable.

(iii) *Establish the converse of* (ii): *every such product f is an exactly m-to-1 map of D onto D.*

Hint: Given $b \in D$, apply 5.87 with $g = b - f$.

Exercise 6.14 (i) *State and prove the analog of* 6.13(i) *for arbitrary regions.*

(ii) (RADÓ, op. cit. *and* CARATHÉODORY *and* RADEMACHER [1917]; *also lemma* 3.1 *of* LANDAU *and* OSSERMAN [1959–60].) *Prove the following converse of* (i): *If $f: \Omega_1 \to \Omega_2$ is holomorphic, Ω_1, Ω_2 bounded regions, and there is no sequence $\{z_n\} \subset \Omega_1$ which converges to $\partial\Omega_1$ and on which f converges to a point of Ω_2 (equivalently, $f^{-1}(K)$ is compact for each compact $K \subset \Omega_2$), then f is exactly m-to-1 for some integer m.*

Hint: Show first that the number $N(w)$ of zeros of $f - w$ is finite for each $w \in \Omega_2$, that $N \geq 1$ and finally that N is continuous, i.e., $N^{-1}(\{n\})$ is open for every integer $n \geq 1$.

§ 3 Applications to Half-planes, Strips and Annuli

Exercise 6.15 *Let $H = \{z \in \mathbb{C}: \operatorname{Im} z > 0\}$ and $f: H \to H$ be holomorphic. Show that*

$$\left|\frac{f(z) - f(z_0)}{f(z) - \overline{f(z_0)}}\right| \leq \left|\frac{z - z_0}{z - \bar{z}_0}\right| \quad \forall z, z_0 \in H.$$

Hints: For each $w_0 = x_0 + iy_0 \in H$ define $\phi = \phi_{w_0}: H \to \mathbb{C}$ by

$$\phi(w) = \frac{w - w_0}{w - \overline{w}_0}, \quad w \in H$$

$$= \frac{w' - i}{w' + i}, \quad w' = \frac{w - x_0}{y_0} \in H.$$

Show (as in 2.5) that ϕ is a conformal map of H onto $D(0, 1)$. Then form $\phi_{f(z_0)} \circ f \circ \phi_{z_0}^{-1}$ and apply Schwarz' lemma to it.

Exercise 6.16 *Show that the conformal maps of $H = \mathbb{R} \times (0, \infty)$ onto itself are exactly all maps of the form*

$$(*) \qquad \phi(z) = \frac{az + b}{cz + d}, \quad \textit{where } a, b, c, d \in \mathbb{R} \textit{ and } ad - bc > 0.$$

Hints: $f(z) = (z - i)/(z + i)$ maps H conformally onto $D = D(0, 1)$. (See the hints to 6.15.) Observe that if the most general conformal map of D onto D is written (after 6.2) as

$$T(z) = e^{i\theta} \cdot \frac{z - a}{1 - \bar{a}z} \quad (\theta \in \mathbb{R}, a, z \in D),$$

and if we set $A = e^{i\theta/2}$, $B = -Aa$, then

$$T(z) = \frac{Az + B}{\overline{B}z + \overline{A}}.$$

Using this representation of T, form $f^{-1} \circ T \circ f$ to arrive at the most general conformal map of H onto itself.

Exercise 6.17 *Let ϕ be a conformal map of $H = \mathbb{R} \times (0, \infty)$ onto itself which has no fixed point in H.*

(i) *Show that then ϕ fixes one or two points in $\mathbb{R}$.*

(ii) *Show that there is another conformal map ψ of H onto H such that*

$$\psi(\phi(\psi^{-1}(z))) = \alpha z + \beta \quad \forall z \in H,$$

for some $\alpha > 0$, $\beta \in \mathbb{R}$.

Hints: Let ϕ have the form (*) in 6.16 and let x be one of the fixed points of ϕ on $\mathbb{R}$, i.e., one of the (real) roots of

$$cz^2 + (d - a)z - b = 0,$$

and consider

$$\psi(z) = \frac{xz + B}{z - x}, \quad B \in \mathbb{R} \text{ and } -x^2 - B > 0.$$

(iii) *Show that if $\alpha, \beta \in \mathbb{R}$, $\alpha > 0$ and $\Phi(z) = \alpha z + \beta$, then there is a conformal map Ψ of H onto itself such that*

$$\begin{cases} \text{either} & \Psi(\Phi(\Psi^{-1}(z))) = z + \beta \quad \forall z \in H \\ \text{or} & \Psi(\Phi(\Psi^{-1}(z))) = \alpha z \quad \forall z \in H. \end{cases}$$

Hint: If $\alpha = 1$ or $\beta = 0$ there is nothing to prove. Otherwise, choose $\beta' = \pm 1$ so that $(\alpha - 1)\beta'/\beta > 0$, set $\alpha' = (\alpha - 1)\beta'/\beta$ and take $\Psi(z) = \alpha' z + \beta'$.

(iv) *There is a conformal map θ of H onto itself such that*

$$\begin{cases} \text{either} & \theta(\phi(\theta^{-1}(z))) = z + \beta \quad \forall z \in H \\ \text{or} & \theta(\phi(\theta^{-1}(z))) = \alpha z \quad \forall z \in H, \end{cases}$$

for some $\alpha > 0$, $\beta \in \mathbb{R}$.

Hint: Combine (ii) and (iii).

Theorem 6.18 *Let $\phi: \mathbb{R} \times (-\pi/2, \pi/2) \to \mathbb{R} \times (-\pi/2, \pi/2)$ be holomorphic. Then*

(i) $|\phi'(x)| \le 1 \quad \forall x \in \mathbb{R}$,

and

(ii) $|\phi'(x_0)| = 1$ *for some $x_0 \in \mathbb{R}$ implies that $\phi(x_0)$ is real.*

Proof: Fix $x_0 \in \mathbb{R}$ and let $x_1 = \operatorname{Re} \phi(x_0)$, $y_1 = \operatorname{Im} \phi(x_0) \in (-\pi/2, \pi/2)$. Define

$$\phi_1(z) = \frac{1+z}{1-z} \qquad z \in D(0, 1)$$

$$\phi_3(z) = e^{z - x_1} \qquad z \in \mathbb{C}$$

$$\phi_4(z) = \frac{z - e^{iy_1}}{z + e^{-iy_1}}, \quad z \in \mathbb{R} \times (-\pi/2, \pi/2)$$

and let $\phi_2 - x_0$ be a holomorphic logarithm in the right half-plane which is real-valued on $(0, \infty)$ (cf. 3.43). Then $\phi_5 = \phi_4 \circ \phi_3 \circ \phi \circ \phi_2 \circ \phi_1$ maps $D(0, 1)$ into itself [by 2.5 and 2.6] and $\phi_5(0) = 0$. From Schwarz' lemma, $|\phi_5'(0)| \le 1$. Computing via the Chain Rule,

$$\phi_5'(0) = \frac{e^{iy_1}}{\cos y_1} \phi'(x_0).$$

It follows that

$$|\phi'(x_0)| \le \cos y_1 \le 1.$$

Moreover, we see that $|\phi'(x_0)| = 1$ entails $\cos y_1 = 1$, whence $y_1 = 0$.

Theorem 6.19 *Let $\phi: \mathbb{R} \times (-\pi/2, \pi/2) \to \mathbb{R} \times (-\pi/2, \pi/2)$ be holomorphic. Then*

(i) $|\phi(x_2) - \phi(x_1)| \le |x_2 - x_1| \quad \forall x_1, x_2 \in \mathbb{R}$.

(ii) *Equality in* (i) *for some pair $x_1 \ne x_2$ implies that*

$$\phi(z) \equiv z + c \quad \text{or} \quad \phi(z) \equiv -z + c \quad \text{for some real constant } c.$$

Proof: To get (i), integrate the inequality of the last theorem: if $x_2 \geq x_1$,

$$|\phi(x_2) - \phi(x_1)| = \left|\int_{x_1}^{x_2} \phi'\right| \leq \int_{x_1}^{x_2} |\phi'| \leq \int_{x_1}^{x_2} 1 = |x_2 - x_1|.$$

Equality here for some $x_1 < x_2$ implies that $|\phi'| \equiv 1$ in $[x_1, x_2]$. Then by (ii) of the last theorem ϕ, whence also ϕ', is real in $[x_1, x_2]$ and so $|\phi'| \equiv 1$ and connectedness mean (via 1.7) that $\phi' \equiv 1$ or $\phi' \equiv -1$ in $[x_1, x_2]$. By the Fundamental Uniqueness Theorem (5.62), $\phi' \equiv 1$ or $\phi' \equiv -1$ holds throughout $\mathbb{R} \times (-\pi/2, \pi/2)$. It follows that for some constant c, $\phi(z) \equiv z + c$ or $\phi(z) \equiv -z + c$. In either case, the fact that the range of ϕ lies in $\mathbb{R} \times (-\pi/2, \pi/2)$ clearly forces c to be real.

Theorem 6.20 *Let* $0 < r_j < R_j < \infty$, $A_j = \{z \in \mathbb{C}: r_j < |z| < R_j\}$ $(j = 1, 2)$, $r = \sqrt{R_1 r_1}$ *and* $\gamma = C(0, r)$. [*Note:* $r_1 < r < R_1$. *In fact, the following conclusions are valid for any such* r.] *Suppose* $f\colon A_1 \to A_2$ *is holomorphic and set* $n = \operatorname{Ind}_{f\circ\gamma}(0)$. *Then*

(i) $|n| \leq \dfrac{\log(R_2/r_2)}{\log(R_1/r_1)}$.

(ii) *If equality holds in* (i), *then* $f(z) \equiv cz^n$ *for some complex constant* c.

Proof: Let $M_j = \log(R_j/r_j)$, $F(z) = f(re^{iM_1 z/\pi})$. Then F is a holomorphic map of the strip $S = \mathbb{R} \times (-\pi/2, \pi/2)$ into A_2. By 5.35 it has a holomorphic logarithm L:

(1) $L \in H(S)$ & $e^L = F$.

Since $F(S) = f(A_1) \subset A_2$, $\operatorname{Re} L(S) \subset (\log r_2, \log R_2)$. Therefore, if we set

(2) $\phi = \dfrac{i\pi}{M_2}[-L + \log\sqrt{R_2 r_2}]$,

then ϕ is a holomorphic map of S into S. Now for $t \in [0, 1]$

$$(f\circ\gamma)(t) = f(re^{2\pi it}) = F\left(\frac{2\pi^2}{M_1}t\right) \overset{(1)}{=} e^{L((2\pi^2/M_1)t)}.$$

It follows from this and the definition (4.2) of the index that

$$\begin{aligned} n &= \operatorname{Ind}_{f\circ\gamma}(0) \\ &= \frac{1}{2\pi i}\left[L\left(\frac{2\pi^2}{M_1}\right) - L(0)\right] \\ &\overset{(2)}{=} \frac{M_2}{2\pi^2}\left[\phi\left(\frac{2\pi^2}{M_1}\right) - \phi(0)\right]. \end{aligned} \tag{3}$$

Since ϕ meets all the hypotheses of the last theorem, we conclude that

$$|n| = \frac{M_2}{2\pi^2}\left|\phi\left(\frac{2\pi^2}{M_1}\right) - \phi(0)\right| \leq \frac{M_2}{2\pi^2}\left|\frac{2\pi^2}{M_1} - 0\right| = \frac{M_2}{M_1}. \tag{4}$$

This proves inequality (i). Moreover, if equality holds in (i), then from (4) $|\phi(2\pi^2/M_1) - \phi(0)| = 2\pi^2/M_1$, so from part (ii) of the last theorem we infer that for some real constant a

(5) $\phi(z) \equiv z + a$ or $\phi(z) \equiv -z + a$.

In fact, from (3) we see that the sign in (5) is that of n, i.e.,

(5)′ $\phi(z) \equiv \frac{n}{|n|} z + a.$

Insert (5)′ into (2) and (1) to see that

$$F(z) \equiv e^{M_2 i\phi(z)/\pi + \log\sqrt{R_2 r_2}} = \sqrt{R_2 r_2}\, e^{M_2 i(nz/|n| + a)/\pi}.$$

For any $\rho \in (r_1, R_1)$ and any real θ then

$$\begin{aligned} f(\rho e^{i\theta}) &= F\left(\frac{\pi}{M_1}\theta + \frac{i\pi}{M_1}\log\frac{r}{\rho}\right) \\ &= \sqrt{R_2 r_2}\, e^{M_2 i a/\pi}\, e^{(n/|n|)(M_2/M_1)(\log(\rho/r) + i\theta)}. \end{aligned}$$

Remembering that $|n| = M_2/M_1$, this gives finally

$$\begin{aligned} f(\rho e^{i\theta}) &= \sqrt{R_2 r_2}\, e^{M_2 i a/\pi} e^{n(\log(\rho/r) + i\theta)} \\ &= \frac{\sqrt{R_2 r_2}}{r^n} e^{M_2 i a/\pi} (\rho e^{i\theta})^n. \end{aligned}$$

If we christen this ugly constant c, assertion (ii) is proven.

Remarks 6.21 This result will be quite important to us later on. Here is an alternative proof of part of it. (Cf. REICH [1966] and the review thereof in *Zentralblatt*; also p. 207 of MARDEN, RICHARDS and RODIN [1967].) Use 4.61 to write

$$f(z) = z^m e^{\phi(z)}, \quad z \in A_1$$

for some integer m and holomorphic function ϕ in A_1. Evidently then $m = n$. Take logarithms of the moduli of both sides of this equation and have, since $f(A_1) \subset A_2$,

$$\log r_2 \leq \log|f(z)| = n \log|z| + \operatorname{Re}\phi(z) \leq \log R_2.$$

For each $r_1 < r < R_1$, set $z = re^{i\theta}$ and integrate in θ:

(1.r) $\log r_2 \leq n \log r + \frac{1}{2\pi}\int_0^{2\pi} \operatorname{Re}\phi(re^{i\theta})d\theta \leq \log R_2.$

The integral here is independent of r, by 5.5(i). Therefore if we subtract (1.r) from (1.R) and let $r \downarrow r_1$, $R \uparrow R_1$, we get

$$\log r_2 - \log R_2 \leq n \log R_1 - n \log r_1 \leq \log R_2 - \log r_2.$$

Exercise 6.22 *Use an argument similar to that in the remarks above to prove the following* (A. J. MACINTYRE [1939]): *if* $f\colon \bar{A}(r, R) \to \mathbb{C}\backslash\{0\}$ *is continuous and holomorphic in* $A(r, R)$, *then*

$$\sup_{\theta\in\mathbb{R}}\left|\frac{f(Re^{i\theta})}{f(re^{i\theta})}\right| \geq \left(\frac{R}{r}\right)^n,$$

where $n = \operatorname{Ind}_{f\circ C(0,\rho)}(0)$ *for any* $r \leq \rho \leq R$.

§ 4 The Theorem of Carathéodory, Julia, Wolff *et al.*

Theorem 6.23 *Let* $H = \{z \in \mathbb{C}\colon \operatorname{Re} z > 0\}$ *and* $f\colon H \to \bar{H}$ *be holomorphic. Then there exists a (non-negative) real constant c such that*

(i) $\operatorname{Re}[f(z) - cz] \geq 0 \quad \forall z \in H,$

and

(ii) $$\lim_{\substack{|z|\to\infty\\ z\in S_\varepsilon}} \frac{f(z)}{z} = \lim_{\substack{|z|\to\infty\\ z\in S_\varepsilon}} \frac{\operatorname{Re} f(z)}{\operatorname{Re} z} = c \quad \forall \varepsilon > 0,$$

where $S_\varepsilon = \{z \in H\colon \operatorname{Re} z > \varepsilon|z|\}$. *Moreover,*

(iii) $$\lim_{\substack{|z|\to\infty\\ z\in S_\varepsilon}} f'(z) = c \quad \forall \varepsilon > 0.$$

Proof: (LANDAU and VALIRON [1929]) Define

(1) $$c = \inf\left\{\frac{\operatorname{Re} f(z)}{\operatorname{Re} z}\colon z \in H\right\}, \qquad g(z) = f(z) - cz \quad (z \in H).$$

Then $\operatorname{Re} g \geq 0$ in H and by the Maximum Principle we can assume strict inequality

(2) $\operatorname{Re} g > 0$ in H.

(Otherwise $g(z) \equiv ia$, for some $a \in \mathbb{R}$, and all is clear.) Using 6.15 *mutatis mutandis*, we have

(3) $$\left|\frac{g(z) - g(z_0)}{g(z) + \bar{g}(z_0)}\right| \leq \left|\frac{z - z_0}{z + \bar{z}_0}\right| \quad \forall z, z_0 \in H.$$

For the moment fix z and z_0 and denote the right side of (3) by r. Evidently $r < 1$. Set

$$w = \frac{g(z) - i\operatorname{Im} g(z_0)}{\operatorname{Re} g(z_0)} \quad \text{[recalling (2)]}.$$

Then (3) says

$$\frac{|w| - 1}{|w| + 1} \leq \left|\frac{w - 1}{w + 1}\right| \leq r.$$

Solve the extreme inequality for $|w|$, remembering that $r < 1$:

$$|w| \le \frac{1+r}{1-r} = \frac{(1+r)^2}{1-r^2} = \frac{(|z+\bar{z}_0| + |z-z_0|)^2}{|z+\bar{z}_0|^2 - |z-z_0|^2}$$

$$(4) \qquad = \frac{(|z+\bar{z}_0| + |z-z_0|)^2}{4(\operatorname{Re} z)(\operatorname{Re} z_0)} \le \frac{(|z| + |z_0|)^2}{(\operatorname{Re} z)(\operatorname{Re} z_0)}.$$

Then

$$\left|\frac{g(z)}{z}\right| \le \left|\frac{i \operatorname{Im} g(z_0)}{z}\right| + \left|\frac{g(z) - i \operatorname{Im} g(z_0)}{z}\right| = \left|\frac{\operatorname{Im} g(z_0)}{z}\right| + |w| \left|\frac{\operatorname{Re} g(z_0)}{z}\right|.$$

$$\overset{(4)}{\le} \left|\frac{\operatorname{Im} g(z_0)}{z}\right| + \frac{(|z| + |z_0|)^2}{(\operatorname{Re} z)(\operatorname{Re} z_0)} \cdot \frac{\operatorname{Re} g(z_0)}{|z|}.$$

If now $z \in S_\varepsilon$, this inequality yields

$$\left|\frac{g(z)}{z}\right| \le \left|\frac{\operatorname{Im} g(z_0)}{z}\right| + \frac{(|z| + |z_0|)^2}{\varepsilon |z|^2} \cdot \frac{\operatorname{Re} g(z_0)}{\operatorname{Re} z_0}$$

$$(5) \qquad = \left|\frac{\operatorname{Im} g(z_0)}{z}\right| + \frac{1}{\varepsilon}\left(1 + \left|\frac{z_0}{z}\right|\right)^2 \frac{\operatorname{Re} g(z_0)}{\operatorname{Re} z_0}.$$

Given $\delta > 0$, the definition of c insures that there exists a $z_0 \in H$ such that $\operatorname{Re} g(z_0)/\operatorname{Re} z_0 < \varepsilon\delta$. It is then clear that if ($z \in S_\varepsilon$ and) $|z|$ is sufficiently large, the right side of (5) is less than δ. This proves half of the claim (ii). For the other half just note that

$$\left|\frac{\operatorname{Re} g(z)}{\operatorname{Re} z}\right| \le \left|\frac{g(z)}{z}\right| \left|\frac{z}{\operatorname{Re} z}\right| < \frac{1}{\varepsilon}\left|\frac{g(z)}{z}\right| \quad \forall z \in S_\varepsilon.$$

It is routine to check that for $0 < \varepsilon < 1$, $z \in S_{3\varepsilon} \Rightarrow \bar{D}(z, \varepsilon|z|) \subset S_\varepsilon$. Then note that for such z

$$f'(z) - c = \frac{1}{2\pi i} \int_{C(z,\varepsilon|z|)} \frac{f(\xi) - c\xi}{(\xi - z)^2}\, d\xi,$$

whence

$$|f'(z) - c| \le \varepsilon|z| \cdot \sup_{|\xi - z| = \varepsilon|z|} \left|\frac{f(\xi) - c\xi}{(\xi - z)^2}\right|$$

$$= \sup_{|\xi - z| = \varepsilon|z|} \left|\frac{f(\xi)}{\xi} - c\right| \frac{|\xi|}{\varepsilon|z|}$$

$$\le \sup_{|\xi - z| = \varepsilon|z|} \left|\frac{f(\xi)}{\xi} - c\right| \cdot \frac{|z| + \varepsilon|z|}{\varepsilon|z|}$$

$$\le \frac{1+\varepsilon}{\varepsilon} \sup_{\substack{|\xi| \ge (1-\varepsilon)|z| \\ \xi \in S_\varepsilon}} \left|\frac{f(\xi)}{\xi} - c\right|.$$

Exercise 6.24 (Angular Derivative) *Let F be holomorphic in $D = D(0, 1)$ and $F(D) \subset D$.*

(i) *Let $\phi(z) = (1 + z)/(1 - z)$ $(z \in D)$ and note that ϕ maps conformally onto H. For each $0 < \varepsilon < 1$, let*

$$S_\varepsilon = \{z \in H: \operatorname{Re} z > \varepsilon|z|\} \quad \text{and} \quad C_\varepsilon = \{z \in D: 1 - |z| > \varepsilon|1 - z|\}.$$

Show that $\phi(C_\varepsilon) \subset S_\varepsilon$ and interpret geometrically.

(ii) *Define $f = \phi \circ F \circ \phi^{-1}$ and show that*

$$(1) \qquad \frac{f(\phi(z))}{\phi(z)} = \frac{1 - z}{1 - F(z)} \cdot \frac{1 + F(z)}{1 + z} \quad \forall z \in D.$$

(iii) *Let c be as in the last theorem. Show that for each $0 < \varepsilon < 1$*

$$(2) \qquad \lim_{\substack{z \to 1 \\ z \in C_\varepsilon}} \frac{1 - F(z)}{1 - z} = \frac{1}{c} \quad \text{if } c > 0,$$

whereas

$$(3) \qquad \lim_{\substack{z \to 1 \\ z \in C_\varepsilon}} \left| \frac{1 - F(z)}{1 - z} \right| = \infty \quad \text{if } c = 0.$$

Hints: Fix $0 < \varepsilon < 1$. Evidently $|\phi(z)| \to \infty$ as $z \to 1$. Therefore (i), (1) and 6.23 yield

$$(4) \qquad \lim_{\substack{z \to 1 \\ z \in C_\varepsilon}} \frac{1 - z}{1 - F(z)} \cdot \frac{1 + F(z)}{1 + z} = c.$$

Consider two cases.

Case I: $\lim_{z \to 1, z \in C_\varepsilon} F(z) = 1$. In this case (2) and (3) follow immediately from (4).

Case II: There is a sequence in C_ε convergent to 1 on which $|F - 1|$ remains bounded away from 0. In this case it follows from (4) (and the boundedness of F) that $c = 0$. Now consider any $z_n \in C_\varepsilon$ such that $z_n \to 1$. We have for each n

$$(5) \qquad \text{either } |1 - F(z_n)| \geq 1 \quad \text{or} \quad |1 + F(z_n)| \geq 1.$$

And by (1)

$$(6) \qquad \left| \frac{1 - F(z_n)}{1 - z_n} \right| = \left| \frac{\phi(z_n)}{f(\phi(z_n))} \right| \left| \frac{1 + F(z_n)}{1 + z_n} \right| \geq \left| \frac{\phi(z_n)}{f(\phi(z_n))} \right| \frac{|1 + F(z_n)|}{2}.$$

From (5) and (6) we have for each n

$$(7) \qquad \begin{cases} \text{either} & \left| \dfrac{1 - F(z_n)}{1 - z_n} \right| \geq \dfrac{1}{|1 - z_n|} \\[2ex] \text{or} & \left| \dfrac{1 - F(z_n)}{1 - z_n} \right| \geq \dfrac{1}{2} \cdot \left| \dfrac{\phi(z_n)}{f(\phi(z_n))} \right|. \end{cases}$$

Since $z_n \to 1$, we have $|\phi(z_n)| \to \infty$ and therefore (recalling (i))

$$(8) \qquad \left|\frac{\phi(z_n)}{f(\phi(z_n))}\right| \to \left|\frac{1}{c}\right| = \infty.$$

From (7) and (8) we learn that

$$\left|\frac{1 - F(z_n)}{1 - z_n}\right| \to \infty.$$

(iv) *Show that if $c > 0$, then for each $0 < \varepsilon < 1$*

$$\lim_{\substack{z \to 1 \\ z \in C_\varepsilon}} F'(z) = \frac{1}{c}.$$

How do these derivatives behave when $c = 0$?

Hints: By the Chain Rule

$$F'(z) = \left[\frac{1 - F(z)}{1 - z}\right]^2 \cdot f'(\phi(z)).$$

Now cite 6.23(iii) and (2) above. As to the query, let α be *any* complex number. Choose integer $n > |\alpha|$ and form $F(z) = \alpha z^n/n$. For this F the associated c is 0, since $|F - 1|$ is bounded away from 0. However, $F'(z) \to \alpha$ as $z \to 1$. Therefore if $c = 0$, any number is a possible value for the limit in (iv).

(v) *Show that if $F(0) = 0$, then $c \leq 1$. (See* UNKELBACH [1938], [1940] *and* HERZIG [1940] *for refinements of this inequality.)*

Hint: By 6.1 we have

$$(9) \qquad |F(z)| \leq |z| \quad \forall z \in D.$$

For $z \in (0, 1)$ then

$$\left|\frac{1 - F(z)}{1 - z}\right| \geq \frac{1 - |F(z)|}{|1 - z|} \geq \frac{1 - |z|}{|1 - z|} = \frac{1 - z}{1 - z} = 1.$$

It follows from (2) that $1/c \geq 1$.

Exercise 6.25 *Let f, H and S_ε be as in 6.23. Show that if f is not identically 0, then either*

(i) *there exists a positive finite constant a such that*

$$\lim_{\substack{|z| \to \infty \\ z \in S_\varepsilon}} zf(z) = a \quad \forall \varepsilon > 0$$

or

(ii) $$\lim_{\substack{|z| \to \infty \\ z \in S_\varepsilon}} |zf(z)| = \infty \quad \forall \varepsilon > 0.$$

(iii) *Deduce from (i) and (ii) that if $\varliminf_{r \to \infty} |rf(re^{i\theta})| = 0$ for some $\theta \in (-\pi/2, \pi/2)$, then f is identically 0.*

Hint: If f is not identically 0, then it has no zeros (Maximum Principle applied to $-\operatorname{Re} f$), so $1/f$ is a candidate for 6.23.

We close this section with an application of 6.24 to a technical result which will be needed in the analysis of convex functions in Chapter XVIII.

Exercise 6.26 *Let h be non-constant and holomorphic in a neighborhood of $\bar{D}(0, r)$, $h(0) = 0$ and let $z_0 \in C(0, r)$ satisfy*

$$|h(z_0)| = \sup\{|h(z)| : z \in D(0, r)\}.$$

Show that

$$\frac{z_0 h'(z_0)}{h(z_0)} \quad (\textit{is real and}) \geq 1.$$

Hints: Since h is non-constant, $h(z_0) \neq 0$ and we can form

$$\text{(1)} \qquad F(z) = \frac{h(z_0 z)}{h(z_0)}.$$

Then F is holomorphic and non-constant in a neighborhood of $\bar{D}(0, 1)$, $F(0) = 0$ and

$$\text{(2)} \qquad F(1) = 1 = \sup\{|F(z)| : z \in \bar{D}(0, 1)\}.$$

Since F is not constant in $D(0, 1)$, from (2) and the Maximum Modulus Principle we infer that

$$F(D(0, 1)) \subset D(0, 1).$$

Since F is differentiable at 1 and $F(1) = 1$,

$$\frac{1 - F(z)}{1 - z} = \frac{F(z) - F(1)}{z - 1} \to F'(1), \text{ which is finite.}$$

But then by 6.24(iii) and (v) we have

$$\frac{1 - F(z)}{1 - z} \to \frac{1}{c} \in [1, \infty)$$

as $z \in (0, 1)$ converges to 1. Thus

$$\text{(3)} \qquad F'(1) \text{ is real and } \geq 1.$$

Since a trivial calculation gives $z_0 h'(z_0)/h(z_0) = F'(1)$, the desired conclusion is contained in (3).

§ 5 Subordination

Let $f, g \colon D = D(0, 1) \to \mathbb{C}$ be holomorphic, $f(0) = g(0)$, $g(D) \subset f(D)$ and f univalent. Then there exists a holomorphic $\phi \colon D \to D$ such that $g = f \circ \phi$. We have only to take $\phi = f^{-1} \circ g$. Now (if g is not constant) Schwarz (6.1) gives $\phi(D(0, r)) \subset D(0, r)$ and so (in any case) $g(D(0, r)) \subset f(D(0, r))$, for all $0 < r < 1$. We codify this in the useful

Principle of Subordination 6.27: If $f, g\colon D(0, 1) \to \mathbb{C}$ are holomorphic, $f(0) = g(0)$ and f is univalent, then

$$g(D(0, 1)) \subset f(D(0, 1)) \Rightarrow g(D(0, r)) \subset f(D(0, r)) \quad \forall 0 < r \leq 1.$$

When f and g are so related, we say g *is subordinate to* f. Variations on this simple idea lead to many useful results:

Theorem 6.28 *If f is a one-to-one holomorphic function in $D(0, 1)$ and $f(D(0, 1))$ is convex, then $f(D(0, r))$ is also convex for each $0 < r < 1$.*

Proof: Consider any pair of points $w_1, w_2 \in f(D(0, r))$, say $w_j = f(z_j)$ and $|z_1| \leq |z_2| < r$. We are to show that for each fixed $0 < t_0 < 1$ the point

$$w_0 = t_0 w_1 + (1 - t_0) w_2$$

belongs to $f(D(0, r))$. This is trivial if $z_2 = 0$ (for then $z_1 = 0$ too and all $w_j = f(0)$). If $z_2 \neq 0$, define

$$g(z) = t_0 f(z z_1/z_2) + (1 - t_0) f(z), \quad z \in D(0, 1).$$

Since $z z_1/z_2 \in D(0, 1)$ for every $z \in D(0, 1)$, g is well-defined and holomorphic in $D(0, 1)$. Moreover $g(D(0, 1)) \subset f(D(0, 1))$ by convexity of the latter, and $f(0) = 0$ implies $g(0) = 0$. Thus by subordination

$$w_0 = g(z_2) \in g(D(0, r)) \subset f(D(0, r)).$$

Exercise 6.29 *Deduce the starlike analog of the last result: that is, if f is a one-to-one holomorphic function in $D(0, 1)$ and $f(D(0, 1))$ is starlike with respect to $f(0)$, then $f(D(0, r))$ is also starlike with respect to $f(0)$ for each $0 < r \leq 1$.*

Hint: Without loss of generality, $f(0) = 0$. Given $0 < r < 1$ and $|z_0| < r$ and $0 < t < 1$ we are to see that $tf(z_0) \in f(D(0, r))$. Form

$$g(z) = tf(z), \quad z \in D(0, 1)$$

and apply the Subordination Principle, as was done in 6.28.

Exercise 6.30 *Use 3.20 and Schwarz' lemma to improve 5.27 thus: If $f \in H(D)$, $f(0) = 0$ and $|\mathrm{Re} f| < A$, then*

(i) $$|f| \leq \frac{2A}{\pi} \log\left(\frac{1 + r}{1 - r}\right) \quad \textit{on } D(0, r), \quad \forall 0 \leq r < 1.$$

Hints: Consider the map

(1) $$L(z) = \frac{4A}{\pi i} \sum_{k=1}^{\infty} \frac{z^{2k-1}}{2k - 1}, \quad z \in D.$$

According to 3.20 it maps D conformally onto $(-A, A) \times \mathbb{R} \supset f(D)$. Since $L(0) = 0$, all the conditions are met for an application of the Principle of Subordination. This supplies us with the inclusions

$$(2) \qquad f(\bar{D}(0, r)) \subset L(\bar{D}(0, r)) \quad \forall 0 \leq r < 1.$$

Now it was further shown in 3.20 that

$$(3) \qquad e^{(\pi i/2A)L(z)} = \frac{1 + z}{1 - z} \quad \forall z \in D,$$

whence

$$(4) \qquad -\frac{\pi}{2A} \operatorname{Im} L(z) = \log\left|\frac{1 + z}{1 - z}\right| \quad \forall z \in D.$$

For $z \in \bar{D}(0, r)$ two applications of (1) yield

$$|L(z)| \leq \frac{4A}{\pi} \sum_{k=1}^{\infty} \frac{|z|^{2k-1}}{2k - 1} = -\operatorname{Im} L(|z|)$$

$$\overset{(4)}{=} \frac{2A}{\pi} \log\left|\frac{1 + |z|}{1 - |z|}\right| \leq \frac{2A}{\pi} \log\left(\frac{1 + r}{1 - r}\right).$$

From the last inequality and (2) follows the claim (i).

Using the same method, deduce that

$$\text{(ii)} \qquad |\operatorname{Re} f(z)| \leq \frac{4A}{\pi} \tan^{-1}(|z|) \quad \forall z \in D.$$

Hints: Since $\operatorname{Re} L$ maps into $(-A, A)$, $\operatorname{Im}[(\pi i/2A)L]$ maps into $(-\pi/2, \pi/2)$. It follows therefore from (3) that

$$\operatorname{Im}\left[\frac{\pi i}{2A} L(z)\right] = \tan^{-1}\left[\frac{\operatorname{Im}\left(\frac{1 + z}{1 - z}\right)}{\operatorname{Re}\left(\frac{1 + z}{1 - z}\right)}\right] = \tan^{-1}\left[\frac{2 \operatorname{Im} z}{1 - |z|^2}\right]$$

$$|\operatorname{Re} L(z)| = \frac{2A}{\pi} \tan^{-1}\left[\frac{2|\operatorname{Im} z|}{1 - |z|^2}\right] \leq \frac{2A}{\pi} \tan^{-1}\left[\frac{2|z|}{1 - |z|^2}\right]$$

$$= \frac{4A}{\pi} \tan^{-1}(|z|).$$

[The "double angle" formula for the tangent function intervened in the last step.] Consequently

$$|\operatorname{Re} L| \leq \frac{4A}{\pi} \tan^{-1}(r) \quad \text{in } \bar{D}(0, r),$$

and so from (2) the same inequality can be asserted for $|\operatorname{Re} f|$.

If only $\operatorname{Re} f$ and not $|\operatorname{Re} f|$ is bounded above, useful bounds on $|f|$ can still be found:

Theorem 6.31 (Hadamard–Borel–Carathéodory Inequalities) *Let f be holomorphic in $D(0, R)$ and continuous on $\bar{D}(0, R)$. (Actually it is enough that* $\operatorname{Re} f$ *be bounded above on $D(0, R)$.) For each $0 \leq r \leq R$ let $M(r) = \sup|f(C(0, r))|$ and $A(r) = \sup \operatorname{Re} f(C(0, r))$. Then*

(i) $$A(r) \leq \frac{R - r}{R + r} \operatorname{Re} f(0) + \frac{2r}{R + r} A(R),$$

(ii) $$M(r) \leq \frac{R + r}{R - r} |f(0)| + \frac{2r}{R - r} A(R).$$

Proof: We have $A(R) \geq \operatorname{Re} f(0)$ by the Maximum Principle and equality holds only if f is constant, in which case (i) and (ii) are obvious. So assume in what follows that

(1) $$A(R) > \operatorname{Re} f(0).$$

We are going to use the Principle of Subordination. Recall that by 2.5 $z \to (1 + z)/(1 - z)$ maps $D(0, 1)$ conformally onto $(0, \infty) \times \mathbb{R}$, and so $z \to (1 + z)/(1 - z) - 1 = 2z/(1 - z)$ maps $D(0, 1)$ conformally onto $(-1, \infty) \times \mathbb{R}$. Therefore, because of (1),

(2) $$\psi(z) = f(0) - [A(R) - \operatorname{Re} f(0)] \frac{2z}{1 - z}$$

maps $D(0, 1)$ conformally onto $(-\infty, A(R)) \times \mathbb{R}$. By definition of $A(R)$ [plus the Maximum Principle] the latter set contains $f(D(0, R))$. After re-scaling the variable under f, the stage is set to invoke the Principle of Subordination to conclude that $f(Rz) \in \psi(D(0, \rho))$ for all $z \in D(0, \rho)$ and all $0 \leq \rho < 1$. That is,

(3) $$f(z) \in \psi(D(0, r/R)) \qquad \forall z \in D(0, r), 0 \leq r < R.$$

From (3) we learn that

$$\begin{aligned} M(r) &\leq \sup|\psi(D(0, r/R))| \\ &\overset{(2)}{=} \sup\left\{\left| f(0) - [A(R) - \operatorname{Re} f(0)] \frac{2z}{1 - z}\right| : |z| < r/R\right\} \\ &\leq |f(0)| + [A(R) - \operatorname{Re} f(0)] \frac{2r/R}{1 - r/R} \\ &\leq |f(0)| + [A(R) + |f(0)|] \frac{2r/R}{1 - r/R} \\ &= \frac{R + r}{R - r} |f(0)| + A(R) \frac{2r}{R - r}. \end{aligned}$$

A similar deduction of (i) from (3) is possible. Only one step will be non-trivial, viz., checking that

$$\operatorname{Re}\left(\frac{-z}{1-z}\right) \leq \frac{\rho}{1+\rho} \quad \forall z \in D(0, \rho),\ 0 \leq \rho < 1.$$

This can be accomplished by computing that

$$\frac{|z|}{1+|z|} + \operatorname{Re}\left(\frac{z}{1-z}\right) = \frac{(1-|z|)(|z|+\operatorname{Re} z)}{(1+|z|)|1-z|^2} \geq 0$$

and noting that $|z|/(1+|z|) = 1/(1+1/|z|)$ is an increasing function of $|z|$. Alternatively, (i) is an almost immediate consequence of the lower Harnack inequality (5.28) applied to the function $A(R) - f$.

Corollary 6.32 *If f is holomorphic and zero-free in a neighborhood of $\bar{D}(0, R)$, then for every $0 \leq r \leq R$*

$$M(r) \leq |f(0)|^{(R-r)/(R+r)} M(R)^{2r/(R+r)}.$$

Proof: Choose $\rho > R$ such that f is holomorphic in $D(0, \rho)$ and then use 5.35 to select an F holomorphic in $D(0, \rho)$ such that $f = e^F$. Apply the previous theorem to F:

$$\operatorname{Re} F(z) \leq \frac{2r}{R+r} \sup_t \operatorname{Re} F(Re^{it}) + \frac{R-r}{R+r} \operatorname{Re} F(0) \quad \text{for } |z| = r.$$

Since the real exponential is monotone, we may exponentiate this inequality and permute all sups and exps to get

$$\begin{aligned} |f(z)| = e^{\operatorname{Re} F(z)} &\leq \left[\exp \sup_t \operatorname{Re} F(Re^{it})\right]^{2r/(R+r)} [\exp \operatorname{Re} F(0)]^{(R-r)/(R+r)} \\ &= \left[\sup_t \exp \operatorname{Re} F(Re^{it})\right]^{2r/(R+r)} [\exp \operatorname{Re} F(0)]^{(R-r)/(R+r)} \\ &= \left[\sup_t |f(Re^{it})|\right]^{2r/(R+r)} |f(0)|^{(R-r)/(R+r)} \\ &= M(R)^{2r/(R+r)} |f(0)|^{(R-r)/(R+r)} \quad \text{for all } |z| = r. \end{aligned}$$

Corollary 6.33 (HADAMARD [1892a]) *If f is an entire function which for some real R, B and K satisfies*

(*) $\quad \operatorname{Re} f(z) \leq B|z|^K \quad \forall |z| \geq R,$

then f is a polynomial of degree not greater than K.

Proof: In the notation of 6.31 we have from (*)

$$A(r) \leq Br^K \quad \forall r \geq R,$$

whence

$$M(r) \le \frac{2r}{2r-r}A(2r) + \frac{2r+r}{2r-r}|f(0)| = 2A(2r) + 3|f(0)|$$

$$\text{(1)} \qquad \le 2B(2r)^K + 3|f(0)| \quad \forall r \ge R/2.$$

From the Cauchy estimates (3.34) and (1) we get

$$\text{(2)} \qquad \frac{|f^{(n)}(0)|}{n!} \le 2^{K+1}Br^{K-n} + 3|f(0)|r^{-n},$$

holding for all positive integers n and all $r \ge R/2$. For each fixed $n > K$ let $r \uparrow \infty$ to see from (2) that $f^{(n)}(0) = 0$. Since (5.30) we have

$$f(z) = \sum_{n=0}^{\infty} \frac{f^{(n)}(0)}{n!} z^n \quad \text{for all } z \in \mathbb{C},$$

the vanishing of $f^{(n)}(0)$ for all $n > K$ proves the assertion.

Remark: The proof shows that inequality (*) need only by hypothesized to hold on a set of concentric circles whose radii are unbounded, and the conclusion still follows.

Theorem 6.34 *Let $\phi(z) = \sum_{n=1}^{\infty} a_n z^n$ and $\psi(z) = \sum_{n=1}^{\infty} b_n z^n$ be holomorphic in $D = D(0, 1)$. Suppose that ϕ is one-to-one, $\phi(D)$ is convex and $\psi(D) \subset \phi(D)$. Then*

$$|b_k| \le |a_1| \quad \forall k \ge 1.$$

The conclusion applies in particular to $\psi = \phi$.

Proof: We may form $h = \phi^{-1} \circ \psi$ and apply Schwarz' lemma to it:

$$\text{(1)} \qquad |h'(0)| \le 1.$$

Since $\phi \circ h = \psi$, the Chain Rule gives

$$\psi'(0) = \phi'(h(0))h'(0),$$

$$b_1 = a_1 h'(0).$$

Hence from (1)

$$\text{(2)} \qquad |b_1| \le |a_1|.$$

Next we create a new function to play the role of ψ in (2) and thereby parlay (2) into the final conclusion.

Given positive integer k we have, by convexity of $\phi(D) \supset \psi(D)$, that

$$\text{(3)} \qquad \frac{1}{k}\sum_{j=1}^{k} \psi(e^{2\pi ij/k} r^{1/k} e^{i\theta/k}) \in \phi(D) \quad \forall (r, \theta) \in [0, 1) \times [0, 2\pi).$$

Using the power series for ψ, this convex sum is

$$(4)\qquad \sum_{n=1}^{\infty} b_n\left[\frac{1}{k}\sum_{j=1}^{k} e^{2\pi i j n/k}\right] r^{n/k} e^{in\theta/k}.$$

Now if $e^{2\pi i n/k} \neq 1$, that is, if k does not divide n, then

$$\sum_{j=1}^{k} (e^{2\pi i n/k})^j = \frac{e^{2\pi i n/k} - (e^{2\pi i n/k})^{k+1}}{1 - e^{2\pi i n/k}} = 0,$$

while if k divides n this sum is clearly k. Therefore the (outer) sum in (4) need only be extended over the n which are multiples of k, and so that sum is

$$\sum_{m=1}^{\infty} b_{mk} r^m e^{im\theta}$$

and the assertion (3) is therefore that

$$(5)\qquad \Psi(z) = \sum_{m=1}^{\infty} b_{mk} z^m \in \phi(D) \quad \forall z \in D.$$

The function Ψ so defined satisfies all the hypotheses which ψ does, so from (2) we are entitled to conclude that

$$|b_k| \leq |a_1|.$$

Corollary 6.35 (Borel [1896–97]) *If $f(z) = 1 + \sum_{n=1}^{\infty} c_n z^n$ maps the unit disk into the right half-plane, then $|c_n| \leq 2$ for all n. It follows* (cf. 5.28) *that $|f(z)| \leq (1 + |z|)/(1 - |z|)$ for all $|z| < 1$.*

Proof: The function $z \to (1 + z)/(1 - z)$ maps $D = D(0, 1)$ conformally onto $(0, \infty) \times \mathbb{R}$, according to 2.5. Therefore the function

$$(*)\qquad \phi(z) = \frac{1+z}{1-z} - 1 = \frac{2z}{1-z} = 2z\sum_{n=0}^{\infty} z^n = \sum_{n=1}^{\infty} 2z^n, \quad z \in D$$

maps D conformally onto $H = (-1, \infty) \times \mathbb{R}$. By hypothesis $\psi(z) = f(z) - 1 = \sum_{n=1}^{\infty} c_n z^n$ also maps D into H. Apply 6.34 to conclude from (*) that $|c_n| \leq 2$ for all n.

Exercise 6.36 (Rogosinski [1945]) *If f is holomorphic in $D(0, 1)$ and satisfies $0 < |f| \leq 1$ there, then $|f'(0)| \leq 2/e$.*

Hints: (M. S. Robertson) We may assume that $|f| < 1$ throughout $D = D(0, 1)$; for if equality is attained, f is constant by the Maximum Modulus Principle, and then $f'(0) = 0$. We may also assume that $f(0) = |f(0)|$, multiplying f by a unimodular constant if necessary. Let (5.35) L be a holomorphic logarithm for $1/f$. We may assume $L(0) = -\log|f(0)| = -\log f(0)$. Since

$|f| < 1$, $\operatorname{Re} L > 0$. Thus $L/L(0)$ maps into the right half-plane and the case $n = 1$ of 6.35 implies that

$$\left|\frac{L'(0)}{L(0)}\right| \leq 2,$$

(1) $\quad |L'(0)| \leq 2|L(0)|.$

However, $e^{-L} = f$, so

$$-L'(0)e^{-L(0)} = f'(0)$$

$$-L'(0)f(0) = f'(0)$$

$$\left|\frac{f'(0)}{f(0)}\right| = |L'(0)| \overset{(1)}{\leq} 2|L(0)| = 2\log\frac{1}{|f(0)|},$$

(2) $\quad |f'(0)| \leq -2|f(0)| \log|f(0)|.$

Now compute the maximum of the function $-x \log x$ for $0 < x < 1$, remembering that $x \log x \to 0$ as $x \downarrow 0$ (by 3.18(4)).

Exercise 6.37 *Use the Poisson formula* (5.17) *to establish the following variant of* 6.31(ii).

$$M(r) \leq \frac{R - r}{R + r}|f(0)| + \frac{2r}{R + r}M(R) \qquad \forall 0 \leq r \leq R,$$

whenever f is holomorphic in $D(0, R)$, continuous on $\bar{D}(0, R)$ and $M(r) = \sup|f(C(0, r))|$ for each $0 \leq r \leq R$.

Hints: Without loss of generality, $R = 1$. Set $P(\theta, a) = (1 - |a|^2)/|e^{i\theta} - a|^2$, $P(a) = (1 - |a|)/(1 + |a|)$ for $\theta \in \mathbb{R}$, $a \in D(0, 1)$. Then

$$P(\theta, a) \geq \frac{1 - |a|^2}{(1 + |a|)^2} = \frac{1 - |a|}{1 + |a|} = P(a).$$

Consequently, from 5.17

$$\begin{aligned}
|f(a)| &= \left|\frac{1}{2\pi}\int_0^{2\pi} P(\theta, a)f(e^{i\theta})d\theta\right| \\
&\leq \left|\frac{1}{2\pi}\int_0^{2\pi} P(a)f(e^{i\theta})d\theta\right| + \left|\frac{1}{2\pi}\int_0^{2\pi} [P(\theta, a) - P(a)]f(e^{i\theta})d\theta\right| \\
&\leq P(a)\left|\frac{1}{2\pi}\int_0^{2\pi} f(e^{i\theta})d\theta\right| + \frac{1}{2\pi}\int_0^{2\pi} [P(\theta, a) - P(a)]|f(e^{i\theta})|d\theta \\
&\leq P(a)\left|\frac{1}{2\pi}\int_0^{2\pi} f(e^{i\theta})d\theta\right| + \frac{M(1)}{2\pi}\int_0^{2\pi} [P(\theta, a) - P(a)]d\theta \\
&= P(a)|f(0)| + M(1)[1 - P(a)] \\
&= \frac{1 - |a|}{1 + |a|}|f(0)| + M(1)\frac{2|a|}{1 + |a|}, \quad \text{by definition of } P(a).
\end{aligned}$$

Take the supremum over $a \in C(0, r)$.

Exercise 6.38 (Cf. 5.38(iii)) *Let Ω be a region, $a \in \Omega$, K compact $\subset \Omega$ and $0 \le r < 1$. Show that there exists a constant $C = C(\Omega, K, a, r) < 1$ such that $\sup|f(K)| \le C$ for all $f \in H(\Omega)$ which satisfy $|f(a)| \le r$, $|f| \le 1$.*

Hints: Pick $R > 0$ so that $\bar{D}(a, R) \subset \Omega$ and let $U = D(a, R/2)$. According to 6.37 and the Maximum Modulus Principle we have for any $f \in H(\Omega)$ with $|f| \le 1$

$$|f(z)| \le \frac{R - R/2}{R + R/2}|f(a)| + \frac{2R/2}{R + R/2} \quad \forall z \in U,$$

so

$$\sup|f(U)| \le \tfrac{1}{3}|f(a)| + \tfrac{2}{3} \le \tfrac{1}{3}r + \tfrac{2}{3}$$

whenever $|f(a)| \le r$. Now use 5.38(iii).

Notes to Chapter VI

For more references to literature on Schwarz' lemma see p. 148 of the bibliography Bernardi [1966]. Some generalizations are given by Julia [1920] and R. Nevanlinna [1929a]. The results in 6.3 are also from Carathéodory [1912a].

There is an elegant generalization of 6.4 in Cantor and Phelps [1965]. See also the Nevanlinna–Pick theorem (e.g., in R. Nevanlinna [1920], [1922b], [1929c], Pick [1915b], [1920], chapter X of Walsh [1969], or Marshall [1974]).

A remarkable relative of 6.5 is proved in Zinterhof [1972]: for every $\varepsilon > 0$ there is a conformal $f \in H(D)$, not the identity function, such that $f(D)$ is a convex subset of $D(0, 1 + \varepsilon)$ and yet f fixes infinitely many points in D.

On 6.7 see § 61 of Bieberbach's encyclopedia article. An equality refining this inequality may be obtained by taking $z = 0$ in equation (1) of the proof of 12.4.

6.8 is implicit in Jensen [1898–99] and in Petersen [1899]; a slightly weaker version is explicit in Schou [1897]. The uniqueness part of 6.9(iii) is due to Blaschke [1915]. For a sectorial version of it see Tsuji [1926] and Ou [1957].

The idea for 6.10 comes from Gronwall [1914–15b], Schur [1917] and R. Nevanlinna [1920]. 6.10 itself is only ancillary to 8.23.

The requirement $c < 1$ in 6.11 can be considerably relaxed. See Wigert [1916] and the results in Chapter XIV.

For results related to 6.12 see 11.14 and 11.48. For extensions of 6.12(ii) see Shimizu [1931b] and Hejhal [1969] and its references. A considerable extension of 6.12(i) was achieved by Lohwater [1952b].

For a systematic study of the ideas underlying 6.13 and 6.14 see pp. 31–36 of Heins [1962]. For a different proof and extension of 6.14 see Gordon [1972].

Cf. also STOÏLOW [1934], NOSHIRO [1937], ROBERTS [1940] and MIODUSZEWSKI [1961]. For regions of finite connectivity greater than 1 and for $m > 1$ no holomorphic self-maps exist which are exactly m-to-1 in the sense of 6.13. See HEINS [1941c].

6.18–6.20 are from HUBER [1951]. (Cf. also KHAJALIA [1937], SCHIFFER [1946] and TSUJI [1960].) 6.20 will be a key ingredient in the description of all the conformal maps of one annulus onto another. (See 10.13 and 10.14.) For another proof of 6.20 see LANDAU and OSSERMAN [1959].

6.23 and 6.24 are the work of many hands: JULIA [1920], WOLFF [1926a], CARATHÉODORY [1929a], LANDAU and VALIRON [1929], WOLFF and DE KOK [1932]. See also R. NEVANLINNA [1929c], AHLFORS [1930] (chapter III), M. RIESZ [1931], DINGHAS [1938], TSUJI [1939], HERZIG [1940] and VERBLUNSKY [1949]. For unified treatments, see SERRIN [1956], KOMATU [1961] and GOLDBERG [1962]. The development in the text is straight out of LANDAU and VALIRON [1929]; a similar account appears in VALIRON [1931]. Wolff and deKok, Verblunsky, Komatu, and Serrin proceed from an integral representation formula. We shall explore this in § 6 of Chapter VIII. In 6.23 it is also the case that for each fixed $y \in \mathbb{R}$, $\operatorname{Re} f(x+iy)/x$ is a non-increasing function of $x \in (0, \infty)$ and it is important that y/x remain bounded as $|x+iy| \to \infty$ (see VISSER [1932] and WOLFF [1930b]). Moreover, for F as in 6.24(v) CARATHÉODORY [1929a] showed that

$$\frac{|1 - F(z)|^2}{1 - |F(z)|^2} \leq \frac{1}{c} \frac{|1 - z|^2}{1 - |z|^2} \quad \text{for all } z \in D.$$

His treatment of these matters may also be found in volume II of his book [1960]. It is quite geometric. 6.25(iii) was originally proved in R. NEVANLINNA [1922a] by a somewhat different (though basically the same!) method.

The idea of subordination originates with LINDELÖF [1909]. For a revealing discussion of it and several more applications the reader is strongly encouraged to look at the expository paper of MACGREGOR [1972]. See also 18.5 of HILLE [1962] and the important basic papers of LITTLEWOOD [1925] (who coined the term) and ROGOSINSKI [1939], [1943]; also ROBINSON [1947]. On p. 134 of the bibliography BERNARDI [1966] the reader will find more references to the literature on subordination.

Theorem 6.28 is from STUDY [1913], p. 109 (though only after the analytic characterization of range-convexity in Chapter XVIII will the reader recognize it there); the proofs given in the text are due to RADÓ [1930], for 6.28 and TAKAHASHI [1930], for 6.29. See also SEIDEL [1931] and NABETANI [1935–36]. For a related result see GRUNSKY [1938]. 6.26 is from PÓLYA and SZEGÖ [1972], exercise III.291; they attribute it to K. Löwner. (It was recently re-discovered—see *Mathematical Reviews* vol. 43 #7611.)

The result 6.30(ii) is due to H. A. SCHWARZ (p. 190 and pp. 361–362, vol. II of his *Gesammelte Mathematische Abhandlungen*). For more on it see KOEBE [1920].

For other proofs of the Hadamard–Borel–Carathéodory inequality 6.31 see pp. 275–277 of LANDAU [1906], pp. 191–194 of LANDAU [1908b], p. 15 of LINDELÖF [1908] (which actually contains a generalization), exercises III.280–289 of PÓLYA and SZEGÖ [1972], ELKINS [1971] and ZALCMAN [1978]. A short proof can be based on the Poisson integral formula. See, e.g., HOLLAND [1973] or LANDAU (*loc. cit.*). For history and extensive generalizations of 6.1, 6.3, 6.7, 6.15 and 6.31 see JENSEN [1919–20] and for material of a related kind, MACINTYRE and ROGOSINSKI [1945].

Hadamard's proof of 6.33 is a simple deduction from the Poisson formula. Likewise, COWLING [1956] gives a straightforward deduction of 6.35 from Poisson. (See also CARATHÉODORY [1911].) Compare the hints to 6.37. Using similar techniques (a Poisson–Stieltjes, instead of a simple Poisson formula) HERZIG [1940] gives a complete description of the f for which equality holds in 6.35 for at least one n or for a relatively prime pair of n. For more on functions with positive real part see the literature references on p. 135 of BERNARDI [1966]. For more on the theme of 6.36 see RAJAGOPAL [1953]. The weak form of 6.33 in which $M_f(r) = \max\{|f(z)|: |z| = r\} \leq Br^K$ has the following interesting generalization (I. N. BAKER [1958]): If there are positive constants a, b and positive integer n such that $M_f(ar^n) \leq b[M_f(r)]^n$ for all large r, then f is a polynomial of degree no higher than n.

Chapter VII
Convergent Sequences of Holomorphic Functions

§ 1 Convergence in $H(U)$

We have already seen three amazing results on the convergence of sequences of holomorphic functions [5.38(iv), 5.45(iii) and 5.74] which have no analog for differentiable functions on the line. In this chapter we want to explore this theme more thoroughly. We naturally start with a convenient definition.

Definition 7.1 A sequence (set) of complex-valued functions on an open subset U of $\mathbb{C}$ is said to converge or be Cauchy (be bounded) *locally uniformly on* U, if U is a union of open sets in each of which the sequence (set) is uniformly convergent or uniformly Cauchy (uniformly bounded).

Remark 7.2 A routine covering-compactness argument shows that "locally uniformly on U" is equivalent to "uniformly on each compact subset of U" for each of the three phenomena above. We use this equivalence frequently in the sequel without further comment.

Exercise 7.3 *Let U be an open subset of* $\mathbb{C}$. *According to* 1.31 *there exists a sequence $\mathscr{K}$ of compact sets $K_1 \subset K_2 \subset \cdots \subset U$ such that every compact subset of U lies in some K_n. For $f, g \in H(U)$ define*

$$\|f\|_{K_n} = \sup|f(K_n)|$$

$$d_{\mathscr{K}}(f, g) = \sum_{n=1}^{\infty} 2^{-n} \min\{1, \|f - g\|_{K_n}\}.$$

(i) *Show that $d_{\mathscr{K}}$ is a metric in $H(U)$.*

(ii) *Show that $d_{\mathscr{K}}$ is translation invariant, i.e.,*

$$d_{\mathscr{K}}(f - h, g - h) = d_{\mathscr{K}}(f, g) \quad \forall f, g, h \in H(U).$$

(iii) *Show that for $f, f_1, f_2, \ldots \in H(U)$ the following are equivalent:*

(a) $\lim_{k\to\infty} f_k = f$ *locally uniformly in U,*
(b) $\lim_{k\to\infty} d_{\mathscr{K}}(f, f_k) = 0$.

(iv) *Show that any metric in $H(U)$ which enjoys properties* (ii) *and* (iii) *is complete.*

Hint: 5.44(i).

(v) *Show that if d is any metric in $H(U)$ which enjoys property* (iii), *then $D(f) = f'$ is a continuous map of the metric space $(H(U), d)$ into itself.*

Hint: 5.44(ii).

Theorem 7.4 (BLASCHKE [1915]) *Let $f_1, f_2, \ldots$ be holomorphic and uniformly bounded in $D(0, 1)$. Let $z_1, z_2, \ldots$ be* distinct *elements of $D(0, 1)$ satisfying*

$$\text{(i)} \qquad \sum_{j=1}^{\infty} (1 - |z_j|) = \infty$$

and suppose that

$$\text{(ii)} \qquad \lim_{n \to \infty} f_n(z_j) \quad \textit{exists for each } j.$$

Then $\{f_n\}$ is locally uniformly convergent on $D(0, 1)$. (Cf. 12.7(ii).)

Proof: (LÖWNER and RADÓ [1923]) Let $M/2$ be a uniform bound for the f_n in $D(0, 1)$ and consider any $0 < r < 1$ and any $\varepsilon > 0$. Use hypothesis (i) to select k so large that

$$\text{(1)} \qquad M \exp\left[\frac{r-1}{r+1} \sum_{j=1}^{k} (1 - |z_j|)\right] < \varepsilon/2.$$

With this k in hand let C_k be the constant furnished for $\{z_1, \ldots, z_k\}$ by 6.9(ii). Then use (ii) to select N so large that

$$\text{(2)} \qquad \max_{1 \le j \le k} |f_n(z_j) - f_l(z_j)| < \frac{\varepsilon}{4C_k} \quad \forall n, l \ge N.$$

For each pair of n, $l \ge N$ apply the inequality from 6.9(ii) to the function $f = f_n - f_l$, with $m = \varepsilon/4C_k$. It follows, using (1), that

$$|f_n(z) \quad f_l(z)| \le \varepsilon \quad \forall z \subset \bar{D}(0, r).$$

This establishes the local uniform Cauchyness of $\{f_n\}$ and finishes the proof.

Remark: Notice that hypothesis (i) is (dramatically) fulfilled if $\{z_j\}$ has a cluster point in $D(0, 1)$. That case, however, was handled already in 5.74. It suffices for most of what follows.

Easy connectedness arguments free us from the disk:

Corollary 7.5 (Vitali–Porter Theorem on Induced Convergence) *Let Ω be a region, $\{F_n\} \subset H(\Omega)$ be locally uniformly bounded in Ω. If $\{F_n(z)\}$ converges for a set of z having a limit point in Ω, then $\{F_n\}$ converges locally uniformly on Ω to some $F \in H(\Omega)$.*

Proof: Let C denote the set of $z \in \Omega$ such that $\{F_n(z)\}$ converges and let Ω_0 denote the set of limit points of C *in* Ω. Note that $\Omega_0 \neq \varnothing$ by hypothesis. 7.4 (or 5.74) shows that

$$(*) \qquad z_0 \in \Omega_0 \;\&\; \bar{D}(z_0, r) \subset \Omega \Rightarrow D(z_0, r) \subset \Omega_0 \cap C.$$

It follows from (*) and the Basic Connectedness Lemma that $\Omega_0 = \Omega$ and then a second citation of (*) shows that $C = \Omega$. A return to 7.4 (or 5.74) shows that in fact convergence is locally uniform in Ω. The limit function is holomorphic by 5.44.

Alternatively, a complete one-line proof is provided by citing 5.74 and 5.38(iv).

Corollary 7.6 (MONTEL [1907]) *If Ω is a region, $\{F_n\}$ a locally uniformly bounded sequence in $H(\Omega)$, then there exists a subsequence of $\{F_n\}$ which is locally uniformly convergent on Ω.*

Proof: Let K be a countable subset of Ω with a limit point in Ω. [For example, select $z_0 \in \Omega, r > 0$ with $D(z_0, r) \subset \Omega$ and let $K = \{z_0 + r/m: m = 2, 3, \ldots\}$.] Now $\{F_n(z)\}$ is a bounded set for each $z \in \Omega$. Therefore an application of the diagonal process to the countable set K produces a subsequence $\{F_{n_k}\}$ such that $\{F_{n_k}(z)\}$ converges for each $z \in K$. The asserted local uniform convergence follows upon applying 7.5 to the sequence $\{F_{n_k}\}$.

Exercise 7.7 *Show that the connectedness hypothesis in the last corollary is unnecessary.*

Hint: The components of Ω are open and disjoint, hence only countable in number. Apply 7.6 successively to each and take the diagonal subsequence.

Corollary 7.8 (MONTEL [1910], p. 21) *Let U be an open subset of $\mathbb{C}$, $\{f_n\}$ a locally uniformly bounded sequence in $H(U)$ with the property that every one of its convergent subsequences converges to g. Then the sequence $\{f_n\}$ itself converges to g, locally uniformly on U.*

Proof: If for some $z \in U$ the sequence of complex numbers $\{f_n(z)\}$ does not converge to the complex number $g(z)$, then there exists a $\delta > 0$ and $1 \le n_1 < n_2 < \cdots$ such that

$$|f_{n_j}(z) - g(z)| > \delta \quad \forall j = 1, 2, \ldots$$

By 7.7, $\{f_{n_j}\}$ contains a convergent subsequence. Its limit cannot be g. This contradiction shows that $\{f_n\}$ converges to g *at* each point of U. Local uniformity of convergence follows from Vitali–Porter.

Corollary 7.9 *Let $f_n(z) = \sum_{k=0}^{\infty} c_{n,k} z^k$ $(n = 0, 1, 2, \ldots)$ be holomorphic in $D(0, 1)$ and locally uniformly bounded there . Then $f_n \to f_0$ locally uniformly if and only if $c_{n,k} \to c_{0,k}$ for each $k = 0, 1, 2, \ldots$*

Proof: $(\Rightarrow)$ Since $c_{n,k} = f_n^{(k)}(0)/k!$, this assertion follows from 5.44(ii).

$(\Leftarrow)$ If g is a subsequential limit of $\{f_n\}$, then by 5.44(ii) and the condition $c_{n,k} \to c_{0,k}$, we see that $g^{(k)}(0)/k! = c_{0,k}$, i.e., by 5.30 $g = f_0$. Now cite 7.8.

Theorem 7.10 *Let Ω be a region, f_n zero-free and holomorphic in Ω and locally uniformly bounded on Ω. If* $\lim_{n\to\infty} f_n(z_0) = 0$ *for* some $z_0 \in \Omega$, *then* $\lim_{n\to\infty} f_n(z) = 0$ *for all $z \in \Omega$ and the convergence is locally uniform.*

Proof: Let C denote the set of $z \in \Omega$ such that $\{f_n(z)\}$ converges to 0 and let U denote the set of $z \in \Omega$ such that $\{f_n\}$ converges uniformly to 0 in some neighborhood of z. Note that $C \neq \varnothing$ by hypothesis and that $U \subset C$. We will show that

$$(*) \qquad z_1 \in C \;\&\; \bar{D}(z_1, 2r) \subset \Omega \Rightarrow D(z_1, r) \subset U \subset C.$$

Indeed let $B = B(z_1, r)$ be a bound for $\{f_n\}$ in $D(z_1, 2r)$ and take $R = 2r$ in 6.32, getting for each n

$$\sup_{|z-z_1|=r} |f_n(z)| \leq |f_n(z_1)|^{(R-r)/(R+r)}\left[\sup_t |f_n(z_1 + Re^{it})|\right]^{2r/(R+r)}$$
$$\leq |f_n(z_1)|^{1/3}B^{2/3}.$$

This clearly shows that if $\{f_n(z_1)\}$ converges to 0, then $\{f_n\}$ converges to 0 uniformly in $D(z_1, r)$, whence (*). In the light of (*) the Basic Connectedness Lemma is applicable and asserts that $C = \Omega$. But then (*) says that $U = \Omega$.

Corollary 7.11 (HURWITZ [1889]) *Let Ω be a region, F_n holomorphic and one-to-one in Ω and suppose that $\{F_n\}$ converges to F locally uniformly on Ω. Then F is either one-to-one or constant.* (Cf. 12.29(vi).)

Proof: Suppose F is not one-to-one, that is, $F(z) = F(w)$ for two distinct points $z, w \in \Omega$. Pick $r > 0$ so that $D(z, 2r) \subset \Omega\backslash\{w\}$. Form $f = F - F(w)$ and for each n form $f_n = F_n - F_n(w)$. Then the functions f_n are holomorphic and never zero in $D(z, 2r)$ (since F_n is one-to-one) and $\{f_n\}$ converges to f uniformly in $D(z, r)$. Since $f(z) = 0$, it follows from 7.10 that $f = 0$ throughout $D(z, r)$. But f is holomorphic in Ω [by 5.44], so by the Fundamental Uniqueness Theorem $f(D(z, r)) = 0$ implies $f(\Omega) = 0$. Thus $F \equiv F(w)$, a constant.

Theorem 7.12 *Let Ω be a region, f_n holomorphic in Ω such that $\{f_n(z_0)\}$ converges for some $z_0 \in \Omega$. Suppose that either*

(i) $\{\text{Re } f_n\}$ *converges locally uniformly on* Ω

or

(ii) $\text{Re } f_n(z) \leq \text{Re } f_{n+1}(z)$ *for all n and all $z \in \Omega$*

or

(iii) $|f_n(z)| \leq |f_{n+1}(z)|$ *for all n and all $z \in \Omega$* and $\lim_{n\to\infty} f_n(z_0) \neq 0$.

Then $\{f_n\}$ converges locally uniformly in Ω.

Proof: Let C denote the set of $z \in \Omega$ such that $\{f_n(z)\}$ converges and let U denote the set of $z \in \Omega$ such that $\{f_n\}$ converges uniformly in some neighborhood of z. Note that $z_0 \in C$ so $C \neq \varnothing$. Also $U \subset C$. For each $z_1 \in \Omega$ and each $r > 0$ such that

$$(1) \qquad D(z_1, 2r) \subset \Omega$$

we may apply the second Hadamard–Borel–Carathéodory Inequality to $f_m - f_n$ for every pair (m, n) with $n \geq m$ to get

$$\sup_{|z-z_1|=r} |f_n - f_m|(z) \leq \frac{2r}{2r-r} \sup_{|z-z_1|=2r} \operatorname{Re}(f_m - f_n)(z) + \frac{2r+r}{2r-r} |f_m - f_n|(z_1)$$

and, applying the Maximum Principles,

(2) $$\sup_{z \in \bar{D}(z_1, r)} |f_n(z) - f_m(z)| \leq 2 \sup_{|z-z_1|=2r} \operatorname{Re}(f_m - f_n)(z) + 3|f_n(z_1) - f_m(z_1)|.$$

If hypothesis (i) prevails, then it follows at once from (2) that

(3) $z_1 \in C$ & $\bar{D}(z_1, 2r) \subset \Omega \Rightarrow D(z_1, r) \subset U \subset C.$

On the other hand, if hypothesis (ii) prevails, then (2) yields

$$\sup_{z \in \bar{D}(z_1, r)} |f_n(z) - f_m(z)| \leq 3|f_n(z_1) - f_m(z_1)| \quad \forall n \geq m,$$

from which (3) again follows. From (3) and the Basic Connectedness Lemma it follows that $C = \Omega$. But then (3) says that $U = \Omega$.

Now consider hypothesis (iii). Because $\lim_{n\to\infty} f_n(z_0) \neq 0$, we can discard finitely many f_n and simply suppose that $f_1(z_0) \neq 0$. Then the set $A = f_1^{-1}(0)$ consists of isolated points (5.62) and $\Omega_0 = \Omega \backslash A$ is a region (1.24). We have $z_0 \in \Omega_0$, so $\Omega_0 \neq \varnothing$. As before, form the pointwise convergence set C and the locally uniform convergence set U, in Ω_0. We have $z_0 \in C$, so $C \neq \varnothing$. If $z_1 \in C$ and $D(z_1, r) \subset \Omega_0$, then each f_n has a holomorphic logarithm g_n in $D(z_1, r)$

(4) $f_n = e^{g_n}$ in $D(z_1, r)$.

Because $\{f_n(z_1)\}$ converges, it is easy to see that appropriate integer multiples of $2\pi i$ may be added to each g_n so that (4) still holds and in addition

(5) $\{g_n(z_1)\}$ converges.

Because of (4), (5) and (iii) the hypotheses of (ii) are met by the g_n in $D(z_1, r)$, so $\{g_n\}$ converges locally uniformly in $D(z_1, r)$. It follows that $D(z_1, r) \subset U$ and by the Basic Connectedness Lemma we again get $C = U = \Omega_0$.

Now consider any $a \in A$. Since the points of A are isolated, there exists an $r > 0$ such that $\bar{D}(a, r) \subset \Omega$ and $\bar{D}(a, r) \cap A = \{a\}$. In particular, $C(a, r) \subset \Omega \backslash A = \Omega_0$ and so by the result of the last paragraph $\{f_n\}$ is uniformly convergent on the compact set $C(a, r)$. By the Maximum Modulus Principle then, $\{f_n\}$ is uniformly convergent throughout $\bar{D}(a, r)$.

We have now shown that each point of Ω_0 and each point of $\Omega \backslash \Omega_0 = A$ has a neighborhood in which $\{f_n\}$ is uniformly convergent; that is, $\{f_n\}$ is locally uniformly convergent in Ω.

Scholium 7.13 If h_n are harmonic in a region Ω, $h_n \leq h_{n+1}$ for all n and $\{h_n(z_0)\}$ converges for some $z_0 \in \Omega$, then $\{h_n\}$ is locally uniformly convergent on Ω.

Remark: The scope of all these theorems will be greatly extended when we prove in Chapter XII the remarkable fact that a family $\mathscr{F}$ of holomorphic functions is locally uniformly bounded if there are two distinct complex numbers not in the range of any function in $\mathscr{F}$ and the set of function values at some one point is bounded.

Exercise 7.14 (i) *Let U be an open subset of $\mathbb{C}$, $h_n: U \to \mathbb{R}$ harmonic functions such that $\{h_n\}$ is locally uniformly bounded in U. Show that $\{h_n\}$ contains a locally uniformly convergent subsequence.*

Hints: It suffices to consider the case in which U is a disk and $\{h_n\}$ is uniformly bounded in U. This is because U is a countable union of such disks and we can apply the weaker conclusion *successively* in each disk and take a diagonal subsequence. Thus suppose h_n are harmonic in the disk D and $|h_n| \leq M < \infty$ in D for all n. According to 5.21(i) there exist $f_n \in H(D)$ such that $h_n = \operatorname{Re} f_n$. Then the functions $F_n = e^{f_n} \in H(D)$ satisfy $e^{-M} \leq e^{h_n} = |F_n| \leq e^M$. According to 7.6 there then exist $1 \leq n_1 < n_2 < \cdots$ and $F \in H(D)$ such that $\lim_{j\to\infty} F_{n_j} = F$ locally uniformly in D. Then $|F_{n_j}| \to |F|$ and $e^{-M} \leq |F| \leq e^M$. It follows from this and the uniform continuity of log on $[e^{-M}, e^M]$ that $h_{n_j} = \log|F_{n_j}| \to \log|F|$ locally uniformly in D.

(ii) *Let $h: D = D(0, 1) \to \mathbb{R}$ be subharmonic. Show that the following two properties of h are equivalent:*

$$\text{(1.h)} \quad \sup_{r<1} \int_0^{2\pi} h(re^{i\theta})d\theta < \infty,$$

(2.h) $h \leq H$ *for some harmonic function H in D.*

Hints: If (2.h) holds, then the integral in (1.h) is majorized by $\int_0^{2\pi} H(re^{i\theta})d\theta$, which equals $2\pi H(0)$.

Suppose (1.h) holds. Take a sequence $0 < r_1 < r_2 < \cdots \to 1$ and let h_n be the function which is harmonic in $D(0, r_n)$ and satisfies

$$\text{(3)} \qquad h_n = h \quad \text{on } C(0, r_n), \quad n = 1, 2, \ldots$$

Then by 5.22 we have

$$\text{(4)} \qquad h \leq h_{n+1} \quad \text{in } D(0, r_{n+1}), \quad n = 1, 2, \ldots$$

From (3), (4) and $r_n < r_{n+1}$ we have

$$h_n = h \leq h_{n+1} \quad \text{on } C(0, r_n)$$

and so, application of the Maximum Principle to $h_{n+1} - h_n$ in $D(0, r_n)$ gives

$$\text{(5)} \qquad h_n \leq h_{n+1} \quad \text{in } D(0, r_n), \quad n = 1, 2, \ldots$$

By the mean value property of harmonic functions

$$2\pi h_n(0) = \int_0^{2\pi} h_n(r_n e^{i\theta})d\theta \overset{(3)}{\le} \int_0^{2\pi} h(r_n e^{i\theta})d\theta, \quad n = 1, 2, \ldots.$$

From this equation and (1.h) we see that the sequence $\{h_n(0)\}$ is bounded above, hence, by (5), convergent. Apply 7.13 in $D(0, r_k)$ to the sequence $\{h_n\}_{n>k}$ for each k to come up with a harmonic function H in D such that $h_n \to H$ throughout D. Go to the limit in (4) to see that $h \le H$ throughout D.

(iii) *Show that if h is harmonic in $D = D(0, 1)$, then*

(3.h) $h = h_1 - h_2$, *for some non-negative harmonic functions h_j in D*

is a third condition equivalent to (1.|h|) *and* (2.|h|).

Hint: Given h_j, set $H = h_1 + h_2$. Given H, set $h_j = H - (-1)^j h/2$.

Here is a nice application of 7.13: an extension, which we will need later, of 5.13 and 5.46.

Theorem 7.15 (Phragmén and Lindelöf [1908]) *Let Ω be a region, $u: \Omega \to \mathbb{R}$ a subharmonic function which is bounded above. Let A be a countable, proper subset of $\partial\Omega$ and M a finite constant such that*

(0) $\quad \overline{\lim_{z\to\zeta}}\, u(z) \le M \quad \forall \zeta \in \partial\Omega \backslash A.$

Then $u \le M$ throughout Ω.

Proof: Let $A = \{a_1, a_2, \ldots\}$. Consider a fixed point $w \in \Omega$ and $r > 0$ such that $\bar{D}(w, 2r) \subset \Omega$. Select a finite number K such that

(1) $\quad \max\limits_{z\in\bar{D}(w,2r)} u(z) < K.$

Now the map $z \to r/(z - w)$ sends $\mathbb{C}\backslash\bar{D}(w, r)$ conformally onto $D(0, 1)\backslash\{0\}$ and for each n, $a_n \in \partial\Omega \subset \mathbb{C}\backslash\bar{D}(w, 2r)$, so $r/(a_n - w) \in D(0, 1)$. There is a conformal automorphism of $D(0, 1)$ which sends this point to 0. Thus (see 2.6) the composite

(2) $$f_n(z) = \frac{\dfrac{r}{z - w} - \dfrac{r}{a_n - w}}{1 - \dfrac{r}{\bar{a}_n - \bar{w}}\dfrac{r}{z - w}}, \quad z \in \mathbb{C}\backslash D(w, r),$$

maps $\mathbb{C}\backslash\bar{D}(w, r)$ conformally into $D(0, 1)$ and sends a_n into 0. Since $a_n \notin \Omega$, this function is zero-free on $\Omega\backslash\bar{D}(w, r)$ and so has local holomorphic logarithms. Therefore $\log|f_n|$ is harmonic in $\Omega\backslash\bar{D}(w, r)$; it is clearly also non-positive. Selecting any base point $b \in \Omega\backslash\bar{D}(w, r)$, we can choose coefficients $c_n > 0$ sufficiently small that

(3) $$\sum_{n=1}^{\infty} -c_n \log|f_n(b)| < \infty.$$

According to 7.13 then, the sequence of functions $s_N = \sum_{n=1}^{N} - c_n \log|f_n|$ converges (as $N \to \infty$) throughout the region (see 1.25(ii)) $\Omega \backslash \bar{D}(w, r)$ to a harmonic function h:

(4) $$h = \sum_{n=1}^{\infty} - c_n \log|f_n| \quad \text{in } \Omega \backslash \bar{D}(w, r).$$

If we define

(5) $$v(z) = \frac{\log\left|\frac{z-w}{r}\right|}{\log 2}, \quad z \in \bar{D}(w, 2r) \backslash D(w, r),$$

then v is harmonic in $D(w, 2r) \backslash \bar{D}(w, r)$ and satisfies

(6) $$v = 0 \quad \text{on} \quad C(w, r), \qquad v = 1 \quad \text{on} \quad C(w, 2r).$$

Now (2) shows that each f_n maps $C(w, r)$ into $C(0, 1)$, so each $\log|f_n|$, hence also each s_N, vanishes on $C(w, r)$. Therefore (6) shows that

$$s_N \le \left[\max_{z \in C(w,2r)} s_N(z)\right] v \le \left[\max_{z \in C(w,2r)} h(z)\right] v$$

obtains on the boundary of $D(w, 2r) \backslash \bar{D}(w, r)$, hence throughout this region by the Maximum Principle. Going to the limit on N then yields

(7) $$h \le \left[\max_{z \in C(w,2r)} h(z)\right] v \quad \text{on } D(w, 2r) \backslash \bar{D}(w, r).$$

Returning to (6), (7) shows that

$$\overline{\lim_{z \to \zeta}}\, h(z) \le 0 \quad \forall \zeta \in C(w, r).$$

Since $h \ge 0$, this means that

(8) $$\lim_{z \to \zeta} h(z) = 0 \quad \forall \zeta \in C(w, r).$$

Since $f_n(a_n) = 0$, $c_n > 0$ and all the summands in h are non-negative, it is clear that

(9) $$\lim_{z \to a_n} h(z) = +\infty \quad \forall n = 1, 2, \ldots$$

For each $\varepsilon > 0$ we form

(10) $$u_\varepsilon = u - \varepsilon h - \max\{K, M\} \quad \text{on } \Omega \backslash \bar{D}(w, r).$$

Then we have

(11) $$\overline{\lim_{z \to \zeta}}\, u_\varepsilon(z) \le 0 \quad \forall \zeta \in C(w, r) \cup \partial\Omega = \partial(\Omega \backslash \bar{D}(w, r)).$$

For $\zeta \in C(w, r)$ this follows from (8), (10) and (1). For $\zeta \in \partial\Omega \backslash A$ it follows from (0), (10) and $h \ge 0$. For $\zeta \in A$ it follows from (9), (10) and the fact that u is bounded above in Ω.

Next form the function

$$f_0(z) = \frac{r}{z - w}, \quad z \in \Omega\backslash\bar{D}(w, r).$$

Then f_0 is holomorphic and satisfies $0 < |f_0| \le 1$, so $h_0 = \log|f_0|$ is harmonic and non-positive. It clearly satisfies

$$(12) \qquad \lim_{\substack{|z|\to\infty \\ z\in\Omega\backslash\bar{D}(w,r)}} h_0(z) = -\infty.$$

Because $h_0 \le 0$ we get from (11)

$$(13) \qquad \overline{\lim_{z\to\zeta}}(u_\varepsilon + \varepsilon h_0)(z) \le 0 \quad \forall\zeta \in \partial(\Omega\backslash\bar{D}(w, r))$$

and because $h \ge 0$ and u is bounded above, u_ε is bounded above, and therefore (12) shows that

$$(14) \qquad \overline{\lim_{\substack{|z|\to\infty \\ z\in\Omega\backslash\bar{D}(w,r)}}} (u_\varepsilon + \varepsilon h_0)(z) \le 0.$$

With (13) and (14) in hand we can cite the remark to 5.12 to assert that

$$u_\varepsilon + \varepsilon h_0 \le 0 \quad \text{in } \Omega\backslash\bar{D}(w, r),$$

that is,

$$u(z) \le \varepsilon h(z) + \varepsilon h_0(z) + \max\{K, M\} \quad \forall z \in \Omega\backslash\bar{D}(w, r).$$

Here $\varepsilon > 0$ is arbitrary so

$$u(z) \le \max\{K, M\} \quad \forall z \in \Omega\backslash\bar{D}(w, r).$$

Because of the definition (1), this clearly implies that

$$(15) \qquad u(z) \le \max\{K, M\} \quad \forall z \in \Omega.$$

If we let $r \downarrow 0$, we can select K arbitrarily close to $u(w)$, so (15) yields

$$(16) \qquad u(z) \le \max\{u(w), M\} \quad \forall z, w \in \Omega.$$

If some $u(w) > M$, then this shows that

$$u(z) \le u(w) \quad \forall z \in \Omega,$$

whence by the Maximum Principle (5.8), $u \equiv u(w)$. That, however, would contradict (0), since $M < u(w)$ and by hypothesis there is at least one ζ for which (0) holds.

Remarks: A different arrangement of this proof is possible which is perhaps a little more transparent. The frontal assault above is replaced by three separate campaigns: I, Ω is bounded. II, Ω is unbounded but still not dense in $\mathbb{C}$. III is the general case. I is covered by the remark following 5.46. (For this 5.44(iii) suffices and 7.13 is not needed at all in the proof.) II is reduced to I by an inversion argument (whose details involve elementary topological considerations—cf. with the hints to 9.43(i)). Then III is reduced to I ∪ II by excising from Ω a subdisk $\bar{D}(w, r)$ and replacing M by $\max\{M, K\}$, then letting $r \downarrow 0$ at the end, as in the final phase of the above proof. Moreover, the *a priori* boundedness assumption on u can be relaxed a little to the requirements that $\overline{\lim}_{z\to\zeta} u(z) < \infty$ for every $\zeta \in \partial\Omega$ and, in case Ω is unbounded, $\overline{\lim}_{|z|\to\infty} u(z) < \infty$. It is a worthwhile exercise for the student to write out the details of these arguments.

Exercise 7.16 (OSGOOD [1901–02]) *Let U be an open subset of $\mathbb{C}$, f_n holomorphic in U. Suppose that $\lim_{n\to\infty} f_n(z)$ exists (as a complex number) for each $z \in U$. Call it $f(z)$. Show that there is a dense open subset U_0 of U such that the convergence is locally uniform on U_0 and (so) f is holomorphic in U_0.*

Hints: For every positive integer n and N the sets $|f_n|^{-1}[0, N]$ are relatively closed in U and each z belongs to $\bigcap_{n=1}^{\infty} |f_n|^{-1}[0, N]$ for sufficiently large N, since $\lim_{n\to\infty} f_n(z)$ exists. Therefore, if V is an open set with $\bar{V} \subset U$, then $\bar{V} = \bigcup_{N=1}^{\infty} \bigcap_{n=1}^{\infty} |f_n|^{-1}[0, N] \cap \bar{V}$. By Baire's Category Theorem there must be an $N = N(V)$ for which the set $\bigcap_{n=1}^{\infty} |f_n|^{-1}[0, N] \cap \bar{V}$ has non-void interior in $\bar{V}$ and so contains an open disk $D(V)$. Thus $|f_n| \le N$ in $D(V)$ for all n. Let $U_0 = \bigcup \{D(V): V \text{ open}, \bar{V} \subset U\}$. Evidently U_0 is open and dense. Each compact subset of U_0 lies in a finite union of the $D(V)$ and so $\{f_n\}$ is uniformly bounded on such a compactum.

Exercise 7.17 (CARATHÉODORY [1932], *Satz* 2) *Let f_n $(1 \le n \le \infty)$ be holomorphic in an open neighborhood V of z_0, $f_n \to f_\infty$ uniformly in V, and $f'_\infty(z_0) \neq 0$. Then there is a neighborhood U of z_0 in V, a neighborhood W of $f_\infty(z_0)$, and an integer N such that f_n is one-to-one in U and $W \subset f_n(U)$ for every $N \le n \le \infty$ and $f_n^{-1} \to f_\infty^{-1}$ uniformly on W.*

Hints: The condition $f'_\infty(z_0) \neq 0$ implies that f_∞ is one-to-one in some smaller neighborhood U_0 of z_0 (5.79). Consider $r > 0$ such that $\bar{D}(z_0, r) \subset U_0$. Since f_∞ is one-to-one in $\bar{D}(z_0, r)$, the compact sets $f_\infty(C(z_0, r))$ and $f_\infty(\bar{D}(z_0, r/2))$ are disjoint. Since the convergence is uniform, there is an $n_1(r)$ such that the sets $f_n(C(z_0, r))$ and $f_\infty(\bar{D}(z_0, r/2))$ are disjoint for every n satisfying $n_1(r) \le n \le \infty$. Thus for each $w \in f_\infty(\bar{D}(z_0, r/2))$, $f_n - w$ has no zeros on $C(z_0, r)$, so by 5.85 the number of zeros of $f_n - w$ in $D(z_0, r)$ is

$$(1) \qquad N_n(w) = \frac{1}{2\pi i} \int_{C(z_0,r)} \frac{f'_n}{f_n - w}, \qquad n_1(r) \le n \le \infty.$$

Now $f_n \to f_\infty$ uniformly on $C(z_0, r)$ and (5.44) $f'_n \to f'_\infty$ uniformly on $C(z_0, r)$. It follows easily from (1) that $N_n(w)$ converges to $N_\infty(w)$ uniformly on the compact set $f_\infty(\bar{D}(z_0, r/2))$. Since $N_\infty(w)$ is the number of zeros of $f_\infty - w$ in $D(z_0, r)$ and f_∞ is one-to-one in $D(z_0, r)$, we have $N_\infty(w) = 1$ for all $w \in f_\infty(D(z_0, r/2))$. Thus the sequence of integers $\{N_n(w)\}$ converges to 1 uniformly on the set $f_\infty(D(z_0, r/2))$. Therefore there is an $n_2(r) \ge n_1(r)$ such that

$$N_n(w) = 1 \quad \forall n_2(r) \le n \le \infty, \forall w \in f_\infty(D(z_0, r/2)),$$

that is,

$$\text{(2)} \qquad \begin{cases} f_n - w \text{ has exactly one zero in } D(z_0, r) \text{ for each } w \in f_\infty(D(z_0, r/2)) \\ \text{and each } n \text{ satisfying } n_2(r) \le n \le \infty. \end{cases}$$

Now because the convergence is uniform there is an $n_3(r) \ge n_2(r)$ such that

$$\text{(3)} \qquad f_n(\bar{D}(z_0, r/3)) \subset f_\infty(D(z_0, r/2)) \text{ for every } n \text{ satisfying } n_3(r) \le n \le \infty.$$

Indeed, we let $\varepsilon > 0$ be the distance from the compact subset $f_\infty(\bar{D}(z_0, r/3))$ of $f_\infty(D(z_0, r/2))$ to the complement of the latter open set and choose $n_3(r)$ so that $|f_n - f_\infty| < \varepsilon$ on $\bar{D}(z_0, r/3)$ for all $n \ge n_3(r)$.

It follows from (3) and (2) that

$$\text{(4)} \qquad f_n \text{ is one-to-one on } \bar{D}(z_0, r/3) \text{ for every } n \text{ satisfying } n_3(r) \le n \le \infty.$$

From (2) and (4) we see that

$$\text{(5)} \qquad \begin{cases} f_n \text{ is one-to-one on } \bar{D}(z_0, r/3) \text{ and takes each value in } f_\infty(D(z_0, r/6)) \\ \text{on } D(z_0, r/3) \text{ for each } n \text{ satisfying } \max\{n_3(r), n_2(r/3)\} \le n \le \infty. \end{cases}$$

We set $U = D(z_0, r/3)$, $W = f_\infty(D(z_0, r/6))$.

Finally, because of (5) and 5.91 we have

$$\text{(6)} \qquad f_n^{-1}(w) = \frac{1}{2\pi i}\int_{C(z_0, r/3)} \frac{z f_n'(z)}{f_n(z) - w}\, dz \quad \forall w \in W,$$

holding for all n such that $\max\{n_3(r), n_2(r/3)\} \le n \le \infty$. Since $f_n \to f_\infty$ and $f_n' \to f_\infty'$ uniformly on $C(z_0, r/3)$, it follows from (6) that $f_n^{-1} \to f_\infty^{-1}$ uniformly on W.

§ 2 Applications of the Convergence Theorems; Boundedness Criteria

Exercise 7.18 *Let U be an open subset of $\mathbb{C}$, let $f\colon U \times [0, 1] \to \mathbb{C}$ be bounded on each compact subset of $U \times [0, 1]$ and let $f(z, t)$ be a holomorphic function of z for each t and a Riemann integrable (for example, continuous) function of t for each z. Show that*

$$F(z) = \int_0^1 f(z, t)dt$$

is holomorphic in U, that for each positive integer j, $D_1^{(j)}f(z, t)$ is a Riemann integrable function for every $z \in U$ and that

$$F^{(j)}(z) = \int_0^1 D_1^{(j)}f(z, t)dt.$$

Hints: For each positive integer n the function

$$F_n(z) = \frac{1}{n}\sum_{k=1}^{n} f(z, k/n)$$

is holomorphic in U, the family $\{F_n\}$ is locally uniformly bounded on U and, by definition of the Riemann integral, $\lim_{n\to\infty} F_n(z) = F(z)$ for each $z \in U$. Apply the Vitali–Porter theorem in each component of U. 5.32 and 5.43 are also relevant.

The next result will recur in a more natural habitat in Chapter VIII where a proof based on rational approximation will be given. But it is hard to resist presenting here the recent and elegant proof of DIXON [1971], especially because it utilizes facts developed in this chapter.

Theorem 7.19 (A General Cauchy Theorem and Formula. Cf. 8.14.) *Let U be an open subset of $\mathbb{C}$, $\gamma_1, \ldots, \gamma_k$ piecewise smooth loops in U, $n_1, \ldots, n_k \in \mathbb{Z}$ such that*

$$(*) \qquad \sum_{j=1}^{k} n_j \operatorname{Ind}_{\gamma_j} = 0 \quad \text{in } \mathbb{C}\backslash U.$$

Then for each $f \in H(U)$

$$\text{(i)} \qquad \sum_{j=1}^{k} n_j \int_{\gamma_j} f = 0,$$

and

$$\text{(ii)} \qquad f(z) \sum_{j=1}^{k} n_j \operatorname{Ind}_{\gamma_j}(z) = \sum_{j=1}^{k} \frac{n_j}{2\pi i} \int_{\gamma_j} \frac{f(\xi)}{\xi - z}\, d\xi \quad \forall z \in U\backslash\gamma_1 \cup \cdots \cup \gamma_k.$$

Proof: Obviously (i) follows from (ii) by selecting any fixed $z \in U\backslash\gamma_1 \cup \cdots \cup \gamma_k$ and applying (ii) to the function $\xi \to (\xi - z)f(\xi)$ in the role of f.

Define $g: U \times U \to \mathbb{C}$ by

$$g(z, \xi) = \begin{cases} \dfrac{f(z) - f(\xi)}{z - \xi} & z, \xi \in U, z \neq \xi \\ f'(z) & z = \xi \in U. \end{cases}$$

Then $g(z, \xi)$ is a holomorphic function of z for each fixed ξ and (so) a holomorphic function of ξ for each fixed z. If $z \in U\backslash\gamma_1 \cup \cdots \cup \gamma_k$, then the first formula for g prevails for all $\xi \in \gamma_1 \cup \cdots \cup \gamma_k$ and we have

$$\sum_{j=1}^{k} n_j \int_{\gamma_j} g(z, \xi)d\xi = -f(z) \sum_{j=1}^{k} n_j \int_{\gamma_j} \frac{d\xi}{\xi - z} - \sum_{j=1}^{k} n_j \int_{\gamma_j} \frac{f(\xi)}{z - \xi}\, d\xi$$

$$(1) \qquad = -f(z)2\pi i \sum_{j=1}^{k} n_j \operatorname{Ind}_{\gamma_j}(z) + \sum_{j=1}^{k} n_j \int_{\gamma_j} \frac{f(\xi)}{\xi - z}\, d\xi \quad \text{(by 4.4).}$$

Now define $V = \{z \in \mathbb{C}\backslash\gamma_1 \cup \cdots \cup \gamma_k: \sum_{j=1}^{k} n_j \operatorname{Ind}_{\gamma_j}(z) = 0\}$. Since the index is integer-valued, the function $\sum_{j=1}^{k} n_j \operatorname{Ind}_{\gamma_j}$ is integer-valued and continuous and so the set V is open. Set

$$(2) \qquad G(z) = \begin{cases} \displaystyle\sum_{j=1}^{k} n_j \int_{\gamma_j} g(z, \xi)d\xi, & z \in U \\ \displaystyle\sum_{j=1}^{k} n_j \int_{\gamma_j} \frac{f(\xi)}{\xi - z} d\xi, & z \in V. \end{cases}$$

Then G is well-defined. For if $z \in U \cap V$, then on the one hand $z \in U\backslash\gamma_1 \cup \cdots \cup \gamma_k$, so (1) holds and on the other hand by the definition of V,

$$\sum_{j=1}^{k} n_j \operatorname{Ind}_{\gamma_j}(z) = 0,$$

so from (1) it follows that the two integrals in (2) are equal. The formulas for G show that it is holomorphic in U (5.42 and 7.18) and in V (2.14). Now hypothesis (*) means that $\mathbb{C}\backslash U \subset V$, so $U \cup V = \mathbb{C}$ and G is entire. Since $V \supset \mathbb{C}\backslash D(0,r)$ for all large r (4.3), use of the second formula for G and obvious estimates show that $\lim_{|z|\to\infty} G(z) = 0$. By Liouville G is therefore identically 0 in $\mathbb{C}$. For all $z \in U\backslash\gamma_1 \cup \cdots \cup \gamma_k$ equation (1) and the first formula in (2) therefore show that (ii) holds.

Exercise 7.20 (Homotopy Versions of Cauchy's Theorem) *Let U be an open subset of $\mathbb{C}$, $f \in H(U)$.*

(i) *Show that if γ_1, γ_2 are piecewise smooth loops which are U-loophomotopic, then $\int_{\gamma_1} f = \int_{\gamma_2} f$.*

Hints: By 4.12, γ_1 and γ_2 are U-homologous, so the assertion follows from part (i) of the last theorem, with $n_1 = 1$, $n_2 = -1$.

(ii) *Deduce from* (i) *the following fixed-endpoint homotopy version of Cauchy's theorem: If $h: [0, 1] \times [0, 1] \to U$ is continuous, $h(0, \tau) = h(0, 0)$, $h(1, \tau) = h(1, 0)$ for all $\tau \in [0, 1]$ and the curves $\gamma_0(t) = h(t, 0)$, $\gamma_1(t) = h(t, 1)$ $(0 \le t \le 1)$ are piecewise smooth, then $\int_{\gamma_0} f = \int_{\gamma_1} f$.*

Hints: Define $H: [0, 2] \times [0, 1] \to \mathbb{C}$ by

$$H(t, \tau) = \begin{cases} \gamma_0(t) & 0 \le t \le 1, 0 \le \tau \le 1 \\ h(2 - t, \tau), & 1 \le t \le 2, 0 \le \tau \le 1 \end{cases}$$

and let $\Gamma_0(t) = H(t, 0)$, $\Gamma_1(t) = H(t, 1)$ for $t \in [0, 2]$. Check that Γ_0, Γ_1 are piecewise smooth loops and that H is a loophomotopy between them in U. Apply (i) and observe that $\int_{\Gamma_0} f = 0$ and $\int_{\Gamma_1} f = \int_{\gamma_0} f - \int_{\gamma_1} f$.

The next definition is convenient in what follows; the designation is traditional.

Definition 7.21 The class $\mathcal{S}$ (for *schlicht*) consists of all holomorphic and univalent functions f on $D = D(0, 1)$ with the normalizations $f(0) = 0$ and $f'(0) = 1$.

Exercise 7.22 *Show that $\mathscr{S}$ is closed in $H(D)$ in the metric of* 7.3. *Consequently, by* 7.6 *and* 7.24 *below, $\mathscr{S}$ is compact.*

Exercise 7.23 *For each $f \in \mathscr{S}$ the number $\alpha = \inf|\mathbb{C}\backslash f(D)|$ satisfies $0 < \alpha \le 1$.*

Hints: Since $f(D)$ is open and contains 0, the number α is clearly positive and $D(0, \alpha) \subset f(D)$. Therefore f^{-1} is defined in $D(0, \alpha)$ and the function $F(w) = f^{-1}(\alpha w)$ is a holomorphic map of D into D which fixes 0. Apply the Schwarz lemma to F.

Theorem 7.24 *The family $\mathscr{S}$ is locally uniformly bounded on $D = D(0, 1)$.*

Proof: Let an $f \in \mathscr{S}$ be given and let $\alpha_f = \alpha$ be as defined in the previous exercise. Evidently there exists an $a_f \in \mathbb{C}$ with $|a_f| = \alpha_f$ and $a_f \in \partial f(D)$. Form

(1) $\quad g_f = \dfrac{1}{a_f} f - 1$

and have

(2) $\quad D(-1, 1) \subset g_f(D), \quad 0 \notin g_f(D), \quad g_f(0) = -1.$

As g_f is a conformal map, it follows from 5.35 that there exists on $g_f(D)$ a holomorphic square root h_f of the identity. Since $h_f^2(-1) = h_f^2(g_f(0)) = g_f(0) = -1$, we have $h_f(-1) = \pm i$. Replacing h_f with $-h_f$ if necessary, we can suppose $h_f(-1) = i$. Let I denote the identity function on D. Then $I \in \mathscr{S}$ and $h_f^2(z) = z = h_I^2(z)$ for all $z \in D(-1, 1)$ and $h_f(-1) = i = h_I(-1)$. It follows from 1.7 that $h_f = h_I$ throughout $D(-1, 1)$. Now $i = h_I(-1) \in h_I(D(-1, 1))$, so there is an $0 < r < 1$ such that

$$D(i, r) \subset h_I(D(-1, 1)) = h_f(D(-1, 1)).$$

Therefore from (2)

(3) $\quad D(i, r) \subset h_f(g_f(D)).$

But then $D(-i, r) \cap h_f(g_f(D)) = \varnothing$. Indeed, if $z \in D$ and $h_f(g_f(z)) = w \in D(-i, r)$, then use (3) to choose $z' \in D$ such that $h_f(g_f(z')) = -w$. Squaring, we get $g_f(z') = [h_f(g_f(z'))]^2 = (-w)^2 = w^2 = [h_f(g_f(z))]^2 = g_f(z)$. Since g_f is one-to-one, it follows that $z' = z$ and then $w = -w$, $w = 0$, a contradiction: $r < 1$ so $0 \notin D(-i, r)$. Thus $h_f(g_f(D)) \subset \mathbb{C}\backslash D(-i, r)$, that is,

(4) $\quad |h_f \circ g_f + i| \ge r \quad$ on D.

Now for each $f \in \mathscr{S}$ we have $0 < \alpha_f \le 1$ by the last exercise, so by (1) it suffices to show that the family $\{g_f : f \in \mathscr{S}\}$ is uniformly bounded on each compact subset of D. If this fails for some compact $K \subset D$, then there exist $f_n \in \mathscr{S}$, $g_n = g_{f_n}$ and $h_n = h_{f_n}$ such that

(5) $\quad \sup|g_n(K)| > n.$

The sequence of functions $F_n = 1/(h_n \circ g_n + i)$ is uniformly bounded in D by (4) and so by 7.6 some subsequence of it, say without loss of generality $\{F_n\}$ itself, converges locally uniformly on D to a holomorphic limit F. We have

$$F_n(0) = \frac{1}{h_n(g_n(0)) + i} = \frac{1}{h_n(-1) + i} = \frac{1}{2i},$$

so $F(0) = 1/2i$ and F is not identically zero. Since none of the functions F_n is ever zero on D, it follows from 7.10 that F is zero-free. Therefore $1/F$ is holomorphic in D and $1/F_n$ converges to $1/F$ uniformly on K. Since

$$\left[\frac{1}{F_n} - i\right]^2 = (h_n \circ g_n)^2 = g_n,$$

we have a contradiction to (5).

Corollary 7.25 (KOEBE [1907]) *There exists a positive constant* α *such that* $D(0, \alpha) \subset f(D)$ *for each* $f \in \mathscr{S}$. (Cf. 12.21.)

Proof: For $f \in \mathscr{S}$ let $\alpha_f = \inf|\mathbb{C}\backslash f(D)|$. The claim is that $\inf\{\alpha_f : f \in \mathscr{S}\}$ is positive. If this is not the case, pick $f_j \in \mathscr{S}$ such that $\alpha_j = \alpha_{f_j} < 1/j$ and set $g_j = f_j/\alpha_j$. In the proof of 7.24 it was shown that $\{g_j\}$ is locally uniformly bounded on D. It follows that $f_j = \alpha_j g_j$ converges to 0 locally uniformly on D. But then $f_j'(0)$ converges to 0 by 7.3(v), contrary to $f_j \in \mathscr{S}$.

Remarks: The largest α for which the conclusion above is true is $\frac{1}{4}$. This was conjectured by Koebe and proved by BIEBERBACH [1916] and by FABER [1916]. The proof is based on an area method and will be developed in Chapter XVIII. It is possible to parlay 7.25 into a much broader covering theorem (BLOCH [1925]): if f is holomorphic in a neighborhood of $\bar{D}$ and $|f'(0)| = 1$, then $f(D)$ contains "schlichtly" a disk of radius $\alpha/6$, for any α which satisfies the conclusion of 7.25. (See, e.g., VEECH [1967], pp. 149–150.) In particular, after $\alpha = \frac{1}{4}$ is shown to be admissible, we get that some open subset U_f of D is mapped conformally by f onto some open disk D_f of radius $\frac{1}{24}$. However, one can proceed *ab initio* with similar elementary means and get a better constant than $\frac{1}{24}$:

Exercise 7.26 (i) *Let* $f \in H(D)$ *satisfy* $f(0) = 0, f'(0) = 1$ *and* $|f'| < M < \infty$ *(so that* $M > |f'(0)| = 1$*). Show that* f *maps* $D(0, 1/M)$ *conformally onto a set which contains* $D(0, 1/2M)$.

Hints: Let

$$(1) \qquad g = \frac{M(f'-1)}{M^2 - f'} = \frac{\dfrac{f'}{M} - \dfrac{1}{M}}{1 - \dfrac{f'}{M}\cdot\dfrac{1}{M}}$$

and apply Schwarz' lemma (6.1) to learn that

(2) $\quad |g(z)| \le |z| \quad \forall z \in D.$

By the usual algebraic maneuvers involving "completing the square" show that for $r \le 1/M$ we have

$$(3) \qquad \left|\frac{M(w-1)}{M^2 - w}\right| \le r \Leftrightarrow \left|w - \frac{M^2(1-r^2)}{M^2 - r^2}\right| \le \frac{rM(M^2-1)}{M^2 - r^2}$$

and consequently, since $M^2(1 - r^2)/(M^2 - r^2) \le 1$, the point w in the disk (3) which is farthest from 1 is

$$\frac{M^2(1-r^2)}{M^2-r^2} - \frac{rM(M^2-1)}{M^2-r^2} = \frac{M(1-rM)}{M-r}.$$

Deduce therefore from (1), (2) and (3) that

(4) $$|f'(z) - 1| \leq 1 - \frac{M(1 - rM)}{M - r} = \frac{r(M^2 - 1)}{M - r} \quad \text{whenever } |z| = r \leq 1/M.$$

Since $r \leq 1/M$, the right side of inequality (4) is majorized by rM and so we have

(5) $$|f'(z) - 1| \leq |z|M \quad \forall z \in D(0, 1/M).$$

It follows by integration that for $z = |z|e^{i\theta} \in D(0, 1/M)$

$$|f(z) - z| = \left|\int_{[0,z]} (f' - 1)\right| = \left|\int_0^{|z|} (f' - 1)(\rho e^{i\theta})d\rho\right|$$
$$\leq \int_0^{|z|} |f' - 1|(\rho e^{i\theta})d\rho \leq \int_0^{|z|} \rho M d\rho = \frac{|z|^2}{2} M < \frac{1}{2M},$$

(6) $$|f(z) - z| < \frac{1}{2M} \quad \forall z \in D(0, 1/M).$$

Let j denote the identity function. If $w \in D(0, 1/2M)$, then (6) implies that

$$|j - w| > \frac{1}{2M} \geq |f - j| \quad \text{on } C(0, 1/M).$$

Therefore by Rouché's theorem (5.87), $j - w$ and $(f - j) + (j - w) = f - w$ have the same number of zeros in $D(0, 1/M)$. Thus $f - w$ has exactly one zero in $D(0, 1/M)$ for every $w \in D(0, 1/2M)$. Furthermore, (5) shows that f' maps $D(0, 1/M)$ into $D(1, 1)$, so that $\operatorname{Re} f' > 0$. For $z_1, z_2 \in D(0, 1/M)$ with $z_1 \neq z_2$ we therefore have

$$\operatorname{Re}[(f(z_2) - f(z_1))/(z_2 - z_1)] = \operatorname{Re}\left[\int_0^1 f'(z_1 + t(z_2 - z_1))dt\right] > 0.$$

In particular, $f(z_2) \neq f(z_1)$.

(ii) *Let F be holomorphic in $D(0, r)$ and satisfy $F'(0) \neq 0$, $|F'| < K < \infty$. Show that for some $\rho \leq r$, f maps $D(0, \rho)$ conformally onto a set which contains $D(F(0), r|F'(0)|^2/2K)$.*

Hint: Apply (i) to $f(z) = (F(rz) - F(0))/rF'(0)$ with $M = K/|F'(0)|$.

(iii) *Let f be holomorphic in a neighborhood of $\bar{D}$ with $|f'(0)| = 1$. Show that f maps some open subdisk D_f of D conformally onto a set which contains a disk of radius $1/3\sqrt{3}$.*

Hints: The continuous function $\phi(z) = (1 - |z|^2)|f'(z)|$ attains its maximum k over $\bar{D}$ at some point z_0 of D, since $\phi(0) = 1 > \phi(C(0, 1)) = 0$. Let $\psi(z) = (z + z_0)/(1 + \bar{z}_0 z)$ and $F = f \circ \psi$, a function on D. Check that

$$(1 - |z|^2)|F'(z)| = (1 - |\psi(z)|^2)|f'(\psi(z))| \quad \forall z \in D.$$

Consequently,

$$(1 - |z|^2)|F'(z)| \leq k,$$

$$|F'(z)| \leq \frac{k}{1 - |z|^2} \quad \forall z \in D.$$

For each $0 \leq r < 1$ we can therefore apply (ii) with $K = k/(1 - r^2)$ to learn that for some $\rho \leq r$, F maps $D(0, \rho)$ conformally onto a set which contains $D(f(z_0), kr(1 - r^2)/2)$. But $k \geq \phi(0) = 1$, so this implies that

(7) F maps $D(0, \rho)$ conformally onto a set which contains $D(f(z_0), r(1 - r^2)/2)$.

This holds for any $0 < r < 1$, so choose r to maximize the radius $r(1 - r^2)/2$, viz., $r = 1/\sqrt{3}$, $r(1 - r^2)/2 = 1/3\sqrt{3}$. Since ψ maps D conformally onto D and $D(0, \rho)$ onto some subdisk of D (see 6.6), and since $f = F \circ \psi^{-1}$, the assertion (iii) follows from (7) with this choice of r.

Theorem 7.27 *The sets $\mathcal{S}_1 = \{f' : f \in \mathcal{S}\}$ and $\mathcal{S}_2 = \{1/f' : f \in \mathcal{S}\}$ are locally uniformly bounded on $D = D(0, 1)$.*

Proof: For $\mathcal{S}_1$ use 7.24 and 5.32. If $\mathcal{S}_2$ fails to be uniformly bounded on some compact $K \subset D$, then there exist $f_n \in \mathcal{S}$ and $z_n \in K$ such that

(*) $|f_n'(z_n)| < 1/n.$

A subsequence of $\{z_n\}$ converges and by 7.6 and 7.24 a subsequence of $\{f_n\}$ converges locally uniformly on D. Without loss of generality then, $z_n \to z_0 \in K$ and $f_n \to f_0$ locally uniformly on D. Then $f_n' \to f_0'$ locally uniformly on D, whence $f_0'(0) = 1$ and f_0 is not constant. From Hurwitz' theorem f_0 is therefore one-to-one. However, $f_0'(z_0) = \lim_{n\to\infty} f_n'(z_n)$ follows easily from the uniform convergence of f_n' to f_0' on K. Therefore from (*) we have $f_0'(z_0) = 0$. This contradicts the fact that f_0 is one-to-one (5.78).

In more modern dress our proof goes thus: $(z, f) \to f'(z)$ is a continuous function Ψ on the metric space $D \times H(D)$ (easy to check using the Cauchy integral formula) and for each compact $K \subset D$ the set $K \times \mathcal{S}$ is a compact subset of $D \times H(D)$ (7.22). Ψ maps this compact set to a compact subset of $\mathbb{C}\backslash\{0\}$. The interested reader may formulate a proof of 7.25 in this language too.

Exercise 7.28 (i) *Let Ω be a region, $a \in \Omega$ and $\mathcal{F}$ the set of all univalent $f \in H(\Omega)$ which satisfy $|f(a)| \leq 1$ and $|f'(a)| \leq 1$. Show that both $\mathcal{F}$ and $\mathcal{F}' = \{f' : f \in \mathcal{F}\}$ are locally uniformly bounded in Ω.*

Hints: Let Ω_0 denote the subset of points $z \in \Omega$ such that $\mathcal{F}$ and $\mathcal{F}'$ are uniformly bounded in some neighborhood of z. Show that

(*) $\sup\{|f(z_0)| + |f'(z_0)| : f \in \mathcal{F}\} < \infty \;\&\; D(z_0, r) \subset \Omega \Rightarrow D(z_0, r/4) \subset \Omega_0$

and conclude with 1.27. To this end note the above supremum by $A < \infty$. The functions

$$\tilde{f}(z) = \frac{f(z_0 + rz) - f(z_0)}{rf'(z_0)}, \quad z \in D(0, 1), f \in \mathcal{F}$$

all belong to $\mathscr{S}$. Therefore, by 7.24 there is a constant B such that

$$|\tilde{f}| \leq B \quad \text{in } D(0, \tfrac{1}{2}) \text{ for each } f \in \mathscr{F}.$$

This says that

$$|f(z)| \leq r|f'(z_0)|B + |f(z_0)| \leq rAB + A \quad \forall z \in D(z_0, r/2), f \in \mathscr{F}.$$

From this inequality and Cauchy's integral (around $C(z_0, r/2)$) for f' we infer that

$$|f'(z)| \leq (rAB + A)8/r \quad \forall z \in D(z_0, r/4), f \in \mathscr{F}$$

and (*) is established.

(ii) *Let Ω be a region, $a \in \Omega$ and $\mathscr{F}$ the set of all univalent $f \in H(\Omega)$ which satisfy $|f(a)| \leq 1$ and $|f'(a)| \geq 1$. Show that $\{1/f' : f \in \mathscr{F}\}$ is locally uniformly bounded in Ω.*

Hint: Use 7.27 and an argument like that in (i) above.

(iii) (MONTEL [1925], p. 253) *Let Ω be a region, $a \in \Omega$ and $\{F_n\}$ univalent holomorphic functions in Ω such that $\{F_n(a)\}$ is convergent and $\{F_n'(a)\}$ converges to 0. Show that $\{F_n\}$ converges to a constant, locally uniformly in Ω.*

Hints: Let $L = \lim_{n \to \infty} F_n(a)$. Given $\varepsilon > 0$, pick $n(\varepsilon)$ so that $|F_n(a) - L| \leq \varepsilon$ and $|F_n'(a)| \leq \varepsilon$ for all $n \geq n(\varepsilon)$. Then the functions $f_n = (F_n - L)/\varepsilon$, $n \geq n(\varepsilon)$, belong to the family $\mathscr{F}$ of (i) and so for each compact $K \subset \Omega$ there is a constant $c(K) < \infty$ such that $|f_n| \leq c(K)$ on K, for all $n \geq n(\varepsilon)$.

(iv) *Let Ω be a region. Show that for each compact $K \subset \Omega$ there is a finite positive constant $C = C(\Omega, K)$, dependent only on Ω and K, such that*

$$\frac{1}{C} \leq \left|\frac{f'(z)}{f'(w)}\right| \leq C \quad \forall z, w \in K$$

holds for every univalent $f \in H(\Omega)$. [This is Koebe's Distortion Theorem.]

Hints: Let $\mathscr{F}$ denote the set of all univalent $f \in H(\Omega)$ and fix an $a \in \Omega$. For $f \in \mathscr{F}$ define $\tilde{f} = (f - f(a))/f'(a)$. According to (ii) and (i) there is a positive finite constant $c = c(\Omega, K)$ such that

$$\frac{1}{c} \leq |\tilde{f}'| \leq c \quad \text{on } K \text{ for each } f \in \mathscr{F},$$

that is,

$$\frac{1}{c} \leq \left|\frac{f'(z)}{f'(a)}\right| \leq c \quad \forall z \in K, f \in \mathscr{F}.$$

Multiplying a pair of such inequalities gives

$$\frac{1}{c^2} \leq \left|\frac{f'(z)}{f'(w)}\right| \leq c^2 \quad \forall z, w \in K, f \in \mathscr{F}.$$

Therefore we may take $C(\Omega, K) = c^2(\Omega, K)$.

Exercise 7.29 (Cf. 12.29(v).) *Let Ω be a region, A a subset of Ω with a limit point in Ω. Let $\mathscr{F}$ be a set of univalent holomorphic functions in Ω such that $\{f(a): f \in \mathscr{F}\}$ is bounded for each $a \in A$.*

(i) *Show that $\mathscr{F}$ is locally uniformly bounded on Ω.*

Hints: For each $a \in A$ let $M(a) = \sup\{|f(a)|: f \in \mathscr{F}\} < \infty$. Fix an $a_0 \in A$ and consider the functions

$$\tilde{f} = \frac{f - f(a_0)}{f'(a_0)}, \quad f \in \mathscr{F}.$$

The family $\tilde{\mathscr{F}} = \{\tilde{f}: f \in \mathscr{F}\}$ is locally uniformly bounded on Ω by 7.28(i). Also $\{f(a_0): f \in \mathscr{F}\}$ is bounded by $M(a_0)$. Since

$$f = f(a_0) + f'(a_0)\tilde{f},$$

the local uniform boundedness of $\mathscr{F}$ will follow if we prove that the set $\{f'(a_0): f \in \mathscr{F}\}$ is bounded. If this set is not bounded, pick $f_n \in \mathscr{F}$ such $|f_n'(a_0)| > n$. Then we have

$$|\tilde{f}_n(a)| \leq \frac{M(a) + M(a_0)}{n} \quad \forall a \in A, n = 1, 2, \ldots.$$

and so $\tilde{f}_n(a) \to 0$ for each $a \in A$. From the Vitali–Porter theorem we then have $\tilde{f}_n \to 0$ locally uniformly on Ω and so from 7.3(v), $\tilde{f}_n'(a_0) \to 0$, a manifest absurdity since $\tilde{f}'(a_0) = 1$ for every $f \in \mathscr{F}$.

(ii) *If $\{f_1, f_2, \ldots\} \subset \mathscr{F}$ and $\{f_n(a)\}$ converges for each $a \in A$, then $\{f_n\}$ converges locally uniformly on Ω.*

(iii) *If $\Omega = D(0, 1)$ and $A = \{a_1, a_2, \ldots\}$ are* distinct *points of Ω with $\sum_{j=1}^{\infty} (1 - |a_j|) = \infty$, draw the same conclusions as in* (i) *and* (ii).

Hint: Argue as in (i) but cite 7.4 instead of Vitali–Porter.

Exercise 7.30 *Let Ω be a region, $a_n \in \Omega$ and $a_n \to a \in \Omega$. Let $f_n: \Omega \to D(0, 1)$ be holomorphic and satisfy $|f_n(a_n)| \to 1$. Show that $|f_n| \to 1$ locally uniformly on Ω.*

Hint: Assume not. Then, by passing to a subsequence, we may assume that $\sup_n |f_n(z_n)| \leq 1 - \delta < 1$ for some $z_1, z_2, \ldots$ in a compact subset K of Ω. We may suppose a and all a_n lie in K. Use 7.6 to select $1 = n_1 < n_2 < \cdots$ such that $\{f_{n_j}\}$ converges to some holomorphic function f locally uniformly on Ω. Then $f_{n_j}(a_j) \to f(a)$, so $|f(a)| = 1$ and by the Maximum Modulus Principle f is constant. In particular, $|f| \equiv |f(a)| = 1$. Then

$$\delta \leq 1 - |f_{n_j}(z_{n_j})| \leq |f(z_{n_j}) - f_{n_j}(z_{n_j})| \leq \sup|f - f_{n_j}|(K)$$

contradicts the uniform-on-K convergence.

Exercise 7.31 *Let $a, b, c \in \mathbb{R}$ with $a < b$ and $f: S = (a, b) \times (c, \infty) \to D(0, 1)$ a holomorphic function. Suppose that the numbers $x_n + iy_n \in S$ satisfy*

(1) $x_n \to d \in (a, b)$ & $y_n \to \infty$

(2) $\overline{\lim}_{n\to\infty}(y_{n+1} - y_n) < \infty$

(3) $\lim_{n\to\infty}|f(x_n + iy_n)| = 1.$

Show that $\lim_{y\to\infty}|f(x + iy)| = 1$ *uniformly for x in any compact subset of* (a, b).

Hints: There exists $\delta > 0$ such that $\{x_n\} \subset (a + \delta, b - \delta)$ and $\{y_n\} \subset (c + \delta, \infty)$. Choose $0 < M < \infty$ such that $y_{n+1} - y_n < M$ for all n. Then form the functions

$$f_n(z) = f(z + i(y_n - c)), \quad z \in \Omega = (a, b) \times (c - \delta, c + \delta + M).$$

They satisfy

$$|f_n(x_n + ic)| = |f(x_n + iy_n)| \to 1.$$

Since $x_n + ic \to d + ic \in \Omega$, the last exercise gives

$$\lim_{n\to\infty} |f_n| = 1 \quad \text{uniformly on compact subsets of } \Omega.$$

In particular, given $\varepsilon > 0$ and compact $K \subset (a, b)$, there exists N such that

$$|f_n(x + iy)| > 1 - \varepsilon \quad \forall n \geq N, (x, y) \in K \times [c, c + M].$$

If $x \in K$ and $y > y_N$, let n be maximal such that $y_n < y$. Then $y_{n+1} \geq y > y_n$ and $n \geq N$. Consequently $y = y_n + t$ for some $t \in (0, y_{n+1} - y_n] \subset (0, M)$ and

$$f(x + iy) = f_n(x + i(t + c)) \in f_n(K \times [c, c + M]).$$

§ 3 Prescribing Zeros

I shall now deliver on a promise made in 5.63 and show the reader how to construct a holomorphic function with arbitrarily prescribed limit-point-free zero set. The theorem is very useful in investigations of the ideal structure of the algebra $H(U)$ (see Chapter XI) and in constructing functions with natural boundaries (see Chapter XVI). The exposition below follows Heins [1968]. Another proof, for simply-connected regions, is offered in Chapter XI, based on the (independently proved) Mittag–Leffler theorem.

Theorem 7.32 (Weierstrass [1876]). *Let U be an open subset of* $\mathbb{C}$, *A a subset of U with no limit point in U. With each* $a \in A$ *associate a positive integer* $n(a)$. *Then there exists a holomorphic function on U which has a zero of order* $n(a)$ *at each* $a \in A$ *and no others.*

Proof: Since A has no limit point in U, it meets each compact subset of U in only a finite set. Since (1.31) U is a countable union of compact subsets, it follows that A is at most countable. If A is finite, the polynomial $\prod_{a\in A} (z - a)^{n(a)}$ is evidently a function of the desired kind. So we shall assume that A is not finite. Let $\{a_1, a_2, \ldots\}$ be an (one-to-one) enumeration of A. For each positive integer k, write d_k for $n(a_k)$.

Consider first the case in which

(1) $\quad U \neq \mathbb{C}$ and A is bounded.

For each $a_n \in A$ there is, by a routine compactness argument, a point $b_n \in \mathbb{C}\backslash U$ such that

(2) $\quad |a_n - b_n| \leq |a_n - b| \quad \forall b \in \mathbb{C}\backslash U.$

Next notice that

(3) $\quad \lim_{n\to\infty} |a_n - b_n| = 0.$

For if not, then $n_1 < n_2 < \cdots$ and $\varepsilon > 0$ exist such that $|a_{n_k} - b_{n_k}| \geq \varepsilon$. Using the boundedness of A, we can assume in addition, by passing to a subsequence of $\{n_k\}$ if necessary, that $\lim_{k\to\infty} a_{n_k} = a$ exists. We have for any $b \in \mathbb{C}\backslash U$ and all k

$$|a - b| \geq |a_{n_k} - b| - |a - a_{n_k}| \overset{(2)}{\geq} |a_{n_k} - b_{n_k}| - |a - a_{n_k}|$$
$$\geq \varepsilon - |a - a_{n_k}|.$$

Therefore $|a - b| \geq \varepsilon$, $a \neq b$ and $a \notin \mathbb{C}\backslash U$. That is, the limit point a of $\{a_n\}$ lies in U, a contradiction. This establishes (3).

For each positive integer n define

(4) $\quad \phi_n(z) = \dfrac{a_n - b_n}{z - b_n}, \quad z \in \mathbb{C}\backslash\{b_n\} \supset U$

(5) $\quad \lambda_n(z) = -\displaystyle\sum_{k=n+1}^{\infty} \frac{z^k}{k}, \quad z \in D(0, 1)$

(6) $\quad E_n(z) = (1 - z)\exp\left[\displaystyle\sum_{k=1}^{n} \frac{z^k}{k}\right], \quad z \in \mathbb{C}.$

Now equation (2) in the proof of 3.19 says that

$$1 - z = \exp\left[-\sum_{k=1}^{\infty} \frac{z^k}{k}\right] \qquad \forall z \in D(0, 1)$$
$$= \exp\left[-\sum_{k=1}^{n} \frac{z^k}{k}\right] \exp \lambda_n(z) \quad \forall z \in D(0, 1).$$

Therefore

(7) $\quad E_n(z) = \exp \lambda_n(z) \quad \forall z \in D(0, 1).$

For each n choose $m(n)$ so that

(8) $\quad 2^{-m(n)} < d_n^{-1} 2^{-n}.$

(Notice that then $m(n) > n$.) Now let K be any compact subset of U. The compact set K and the closed set $\mathbb{C}\backslash U$ are disjoint, so by a routine compactness argument they are a positive distance apart. That is, there exists an $\varepsilon = \varepsilon_K > 0$ such that

(9) $\quad |z - b| \geq \varepsilon \quad \forall z \in K, b \in \mathbb{C} \backslash U.$

There exists by (3) an n_K such that

$$|a_n - b_n| \leq \varepsilon/2 \quad \forall n \geq n_K$$

and together with (9) this gives

(10) $\quad |\phi_n(z)| = \left|\dfrac{a_n - b_n}{z - b_n}\right| \leq \tfrac{1}{2} \quad \forall n \geq n_K, z \subset K.$

We therefore have for all m, all $n \geq n_K$ and all $z \in K$

$$|\lambda_m(\phi_n(z))| \leq \sum_{k=m+1}^{\infty} \frac{1}{k} |\phi_n(z)|^k \leq \sum_{k=m+1}^{\infty} |\phi_n(z)|^k \leq \sum_{k=m+1}^{\infty} 2^{-k} = 2^{-m}.$$

Therefore from (8)

(11) $\quad d_n|\lambda_{m(n)}(\phi_n(z))| < 2^{-n} \quad \forall z \in K, n \geq n_K.$

It follows that for every positive integer $m \geq n_K$ the series $\sum_{n=m+1}^{\infty} d_n \lambda_{m(n)} \circ \phi_n$ converges uniformly on K. The limit function, call it $g_{m,K}$, is holomorphic in $\mathring{K}$ by 5.44. Define

(12) $\quad F_N(z) = \prod_{n=1}^{N} E_{m(n)}^{d_n}(\phi_n(z)), \quad z \in \mathbb{C} \backslash \{b_1, \ldots, b_N\}, N = 1, 2, \ldots.$

Then for $N > M \geq n_K$ we have for $z \in K$

$$F_N(z) = F_M(z) \prod_{n=M+1}^{N} E_{m(n)}^{d_n}(\phi_n(z))$$

$$= F_M(z) \prod_{n=M+1}^{N} [\exp(\lambda_{m(n)}(\phi_n(z)))]^{d_n} \quad \text{by (10) and (7)}$$

(13) $$= F_M(z) \exp\left(\sum_{n=M+1}^{N} d_n \lambda_{m(n)} \circ \phi_n\right)(z).$$

Therefore we see that the sequence $\{F_N\}$ converges at each point of U and uniformly on each compact subset K of U, with

$$\lim_{N \to \infty} F_N(z) = F_M(z) e^{g_{M,K}(z)} \quad \forall z \in K, M \geq n_K.$$

Call this limit F. Thus, F is holomorphic in U and

(14) $\quad F(z) = F_M(z) e^{g_{M,K}(z)} \quad \forall z \in K, M \geq n_K.$

This shows on the one hand that the zeros of F are among those of the F_M ($M = 1, 2, \ldots$). In fact, each $a \in U$ lies in the interior of some compact $K \subset U$

and then (14) shows that the order of a as a zero of F is its order as a zero of F_M for any $M \geq n_K$. But an examination of the relevant definitions (4), (6), (12) shows that each F_M has a zero of order d_n at a_n for each $1 \leq n \leq M$ and no others. It follows that F has a zero of order d_n at each a_n, that is, a zero of order $n(a)$ at each $a \in A$, and no others.

Because of (3) and the boundedness of A there exists an $R > 0$ such that

$$|a_n|, |b_n| < R \quad \forall n.$$

Then for $|z| \geq 5R$ and all n

$$|z - b_n| \geq |z| - |b_n| > 4R > 2|a_n| + 2|b_n| \geq 2|a_n - b_n|$$

$$(15) \quad |\phi_n(z)| = \left|\frac{a_n - b_n}{z - b_n}\right| < \tfrac{1}{2} \quad \forall n, \forall |z| \geq 5R.$$

As in the derivation of (11) it follows from (15) that

$$(16) \quad d_n|\lambda_{m(n)}(\phi_n(z))| \leq 2^{-n} \quad \forall n, \forall |z| \geq 5R.$$

From (15), (7) and (12)

$$F_N(z) = \exp\left[\sum_{n=1}^{N} d_n\lambda_{m(n)} \circ \phi_n\right](z) \quad \forall N, \forall |z| \geq 5R.$$

Therefore (16) shows that

$$|F_N(z)| \leq \exp\left|\sum_{n=1}^{N} d_n\lambda_{m(n)}(\phi_n(z))\right| \quad \forall N, \forall |z| \geq 5R.$$

$$\leq \exp \sum_{n=1}^{N} 2^{-n} < e \qquad \forall N, \forall |z| \geq 5R.$$

Let $N \to \infty$ and conclude that

$$(17) \quad |F(z)| \leq e \quad \forall z \in U, |z| \geq 5R.$$

Now in the general case we may suppose, after a translation, that

$$(18) \quad 0 \in U \backslash A.$$

Let $\phi(z) = 1/z$ and consider

$$U_0 = \phi(U\backslash\{0\}), A_0 = \phi(A).$$

Since $0 \in U$, it is not a limit point of A and since $0 \notin A$ we infer that A is bounded away from 0. Therefore A_0 is bounded and U_0, A_0 satisfy the conditions in (1). Therefore there is a function F which is holomorphic in U_0 with zeros of order $n(a)$ at $\phi(a)$, for each $a \in A$, and no others. Form $f = F \circ \phi$ on $U\backslash\{0\}$. Then f has zeros only at the $a \in A$. In fact if we write

$$F(w) - F(\phi(a)) = [w - \phi(a)]^{n(a)}G_a(w), \quad w \in U_0,$$

where G_a is holomorphic in U_0 and $G_a(\phi(a)) \neq 0$, then

$$\begin{aligned} f(z) - f(a) &= [\phi(z) - \phi(a)]^{n(a)} G_a(\phi(z)) \\ &= (z - a)^{n(a)} \frac{G_a(1/z)}{(-az)^{n(a)}}, \quad z \in U \backslash \{0\}, \end{aligned}$$

so a is a zero of f of order $n(a)$. Since $0 \in U$ and U is open, there is an $r > 0$ such that $D(0, r) \subset U$. We may suppose that $r < 1/5R$. Then for $0 < |z| < r$ we have by (17)

$$|f(z)| = |F(1/z)| \leq e.$$

Therefore according to 5.41, f extends to be differentiable at 0, i.e., holomorphic throughout U. If 0 is a zero of (the extension of) f, say of order n_0, then $f(z)/z^{n_0}$ is the desired function.

In specific cases it is often possible to exhibit the above function rather explicitly. Due to the press of other matters we will only look at one such case now.

Exercise 7.33 *If* $a_n \in D(0, 1) \backslash \{0\}$ *and* $\sum_{n=1}^{\infty} (1 - |a_n|) < \infty$, *then*

$$\lim_{N \to \infty} \prod_{n=1}^{N} \frac{|a_n|}{a_n} \frac{z - a_n}{\bar{a}_n z - 1}$$

exists throughout $D(0, 1)$ *and defines a holomorphic function* (*so-called* Blaschke product) *which is bounded by* 1 *and whose zero set is* $\{a_1, a_2, \ldots\}$. (*See* 12.6 *and* 6.9(iii) *for the converse of this.*)

Hints: Suppose first that $|a_n| > r$ for all n and some $r > 0$. Then the functions

$$B_N(z) = \prod_{n=1}^{N} \frac{|a_n|}{a_n} \frac{z - a_n}{\bar{a}_n z - 1}, \quad N = 1, 2, \ldots$$

are holomorphic in $D(0, 1)$ and zero-free in $D(0, r)$. There exist then holomorphic L_N in $D(0, r)$ such that $e^{L_N} = B_N$ in $D(0, r)$. Since

$$B_N(0) = \prod_{n=1}^{N} |a_n| \in (0, \infty),$$

we can suppose that $L_N(0) = \log B_N(0)$. Notice that $\operatorname{Re} L_N = \log|B_N|$ and $|B_{N+1}| \leq |B_N| < 1$, so $0 > \operatorname{Re} L_1 \geq \operatorname{Re} L_2 \geq \cdots$ in $D(0, r)$. Since $\log x = \int_1^x (1/t)dt < x - 1$ for $x > 1$, we have

$$\log \frac{1}{|a_n|} < \frac{1}{|a_n|} - 1 = \frac{1 - |a_n|}{|a_n|} < \frac{1 - |a_n|}{r},$$

$$\begin{aligned} L_N(0) = \log B_N(0) &> -\sum_{n=1}^{N} \frac{1 - |a_n|}{r} = -\frac{1}{r} \sum_{n=1}^{N} (1 - |a_n|) \\ &\geq -\frac{1}{r} \sum_{n=1}^{\infty} (1 - |a_n|) > -\infty. \end{aligned}$$

Therefore the decreasing sequence $\{L_N(0)\}$ converges in $\mathbb{R}$. It follows then from 7.12(ii) that $\{L_N\}$ converges throughout $D(0, r)$, then from 7.5 that $\{B_N\}$ converges throughout $D(0, 1)$. Call the latter limit B. Since, as just noted, $|B(0)| > 0$, we have from 7.10 that B is zero-free in $D(0, r)$.

In general, since $|a_n| \to 1$, given $r < 1$ we can select n_r such that $|a_n| > r$ for all $n > n_r$ and then conclude from the above that

$$\lim_{N\to\infty} \prod_{n=n_r+1}^{N} \frac{|a_n|}{a_n} \frac{z - a_n}{\bar{a}_n z - 1}$$

is zero-free in $D(0, r)$, i.e., B/B_{n_r} is zero-free in $D(0, r)$. It follows that the zeros of B are precisely the points $\{a_1, a_2, \ldots\}$.

§ 4 Elementary Iteration Theory

Definition 7.34 If $S \subset \mathbb{C}$ and $f: S \to S$, then we define $f^{[0]}$ to be the identity map of S onto S and inductively $f^{[n+1]} = f \circ f^{[n]}$ for each non-negative integer n. We call $f^{[n]}$ the nth *iterate of f*.

We investigate in this section elementary results, derivable from the convergence criteria of previous sections, about the sequence $\{f^{[n]}\}$ when f is holomorphic. Points fixed by f and the convergence of $f^{[n]}$ to such a point are on center stage, as the first example illustrates:

Exercise 7.35 (i) (Cf. 7.38, 7.40) *Let U be an open subset of $\mathbb{C}$, $f \in H(U)$, $f(U) \subset U$. Suppose $z_0 \in U$, $f(z_0) = z_0$ and $|f'(z_0)| < 1$. Show that there is an open disk about z_0 in which the iterates of f converge uniformly to z_0. (Such a point z_0 is consequently called an* attractive fixed point.) *Hence, if in addition U is bounded and connected, the iterates converge everywhere in U to z_0.*

Hints: Let $R = \frac{1}{2}[1 + |f'(z_0)|]$. By definition of $f'(z_0)$ then there exists $r > 0$ such that $D(z_0, r) \subset U$ and

$$\left|\frac{f(z) - f(z_0)}{z - z_0}\right| \leq R \quad \forall z \in D(z_0, r)\backslash\{z_0\}.$$

Since $f(z_0) = z_0$, this says that

(*) $\quad |f(z) - z_0| \leq R|z - z_0| \quad \forall z \in D(z_0, r).$

Since $R < 1$, (*) shows that $f(D(z_0, r)) \subset D(z_0, r)$. Therefore we may iterate to get for any positive integer n

$$|f^{[n]}(z) - z_0| \leq R^n|z - z_0| \quad \forall z \in D(z_0, r).$$

If U is bounded and connected, cite 7.5 to conclude that $\{f^{[n]}\}$ converges locally uniformly on U to the constant z_0.

(ii) *Prove the following converse of* (i): *if U is a region, $f: U \to U$ holomorphic, $z_0 \in U$ and $f^{[n]}$ converges locally uniformly in U to z_0, then $f(z_0) = z_0$ and $|f'(z_0)| < 1$.*

Hints: From $f^{[n]}(z_0) \to z_0$ and continuity of f at z_0 follow $f(f^{[n]}(z_0)) \to f(z_0)$. On the other hand, $f(f^{[n]}(z_0)) = f^{[n+1]}(z_0) \to z_0$. We have $f^{[n]'} \to 0$ by 7.3(v) and $f^{[n]'}(z_0) = (f'(z_0))^n$ by the Chain Rule and the fact that $f(z_0) = z_0$. It must therefore be that $|f'(z_0)| < 1$.

(iii) (Ritt [1920–21]) *Let Ω be a region, f a holomorphic function in Ω such that $\overline{f(\Omega)}$ is a compact subset of Ω. Show that the iterates of f converge uniformly in Ω to a constant.*

Hints: For $n \in \mathbb{N}$ let $\Omega_n = f^{[n]}(\Omega)$, $K = \bigcap_{n=1}^{\infty} \Omega_n$. By induction we establish (with $\Omega_0 = \Omega$): $\overline{\Omega}_{n+1} = \overline{f(\Omega_n)} \subset f(\overline{\Omega}_n)$ compact $\subset f(\Omega_{n-1}) = \Omega_n$, for all $n \in \mathbb{N}$. Hence $K = \bigcap_{n=2}^{\infty} \overline{\Omega}_n$ is compact and not empty. Consider any locally uniformly convergent subsequence $\{f^{[n_j]}\}$ of iterates, and its limit function g. Evidently $g(\Omega) \subset K$. On the other hand, if $w \in K$, then $w \in \Omega_{n+1} = f^{[n+1]}(\Omega) = f^{[n]}(f(\Omega)) = f^{[n]}(\Omega_1)$ for all n, so $w = f^{[n]}(w_n)$ for some $w_n \in \Omega_1$. A subsequence $\{w_{n_{j_k}}\}$ of $\{w_{n_j}\}$ converges to some w_0 in the compact set $\overline{\Omega}_1$. Since $\{f^{[n_j]}\}$ converges to g uniformly on this compact set, it follows easily that $w = f^{[n_{j_k}]}(w_{n_{j_k}})$ converges to $g(w_0)$, i.e., $w \in g(\overline{\Omega}_1) \subset g(\Omega)$. Thus $g(\Omega) = K$. By the open map theorem (5.77) g is therefore constant and the compact set K is a single point. By 7.8 then the whole sequence $\{f^{[n]}\}$ converges to this point.

Exercise 7.36 *Let f be holomorphic in the open set U and satisfy $f(U) \subset U$. Show that if some subsequence of $\{f^{[n]}\}$ converges locally uniformly to the identity function, then f is one-to-one and onto.*

Hints: Suppose $\{f^{[n_j]}\}$ converges locally uniformly to the identity function ϕ on U. If $z, z' \in U$ and $f(z) = f(z')$, then $f^{[n]}(z) = f^{[n]}(z')$ for all n so $\phi(z) = \phi(z')$, that is, $z = z'$. Thus f is one-to-one. If $w \in U \backslash f(U)$ then $w \in U \backslash f^{[n]}(U)$ for all n, since evidently $f^{[n]}(U) \subset f(U)$, and so $f^{[n_j]} - w$ has no zero in U. However $f^{[n_j]}(w) - w \to \phi(w) - w = 0$. It follows from 7.10 that $f^{[n_j]} - w$ converges to 0 throughout any connected neighborhood of w in U, a contradiction.

Exercise 7.37 *Let Ω be a bounded region, f a holomorphic function on Ω, $f(\Omega) \subset \Omega$. If f is not an onto homeomorphism, then every convergent subsequence of its iterates has a constant limit.*

Hints: Let $n_1 < n_2 < \cdots$ be such that $\{f^{[n_j]}\}$ converges throughout Ω to a non-constant function ϕ. Set $m_j = n_{j+1} - n_j$. Because Ω is bounded, $\{f^{[n]}\}$ is uniformly bounded, so by 7.5 ϕ is holomorphic in Ω. There exist $j_1 < j_2 < \cdots$ such that $\{f^{[m_{j_k}]}\}$ converges locally uniformly on Ω to a holomorphic limit Φ.

This follows from the boundedness of $\{f^{[m_j]}\}$ and 7.6. Evidently $\phi(\Omega) \subset \bar{\Omega}$ and since ϕ is non-constant, $\phi(\Omega)$ is open (5.77). Hence $\phi(\Omega)$ meets Ω non-voidly. Consider $w \in \phi(\Omega) \cap \Omega$. Say $w = \phi(z)$. Then $f^{[n_j]}(z) \to \phi(z) = w$, so $f^{[n_{j_k}+1]}(z) \to w$ and $f^{[n_{j_k}]}(z) \to w$. Since $f^{[m_{j_k}]} \to \Phi$ uniformly on each compact subset of Ω, it follows easily that $f^{[m_{j_k}]}(f^{[n_{j_k}]}(z)) \to \Phi(w)$. Therefore from $f^{[n_j+1]} = f^{[m_j]} \circ f^{[n_j]}$ we deduce that $w = \Phi(w)$. Thus Φ coincides with the identity function in the non-void open subset $\phi(\Omega) \cap \Omega$ of Ω. It follows from the Fundamental Uniqueness Theorem that Φ is the identity function on Ω and then from 7.36 f is one-to-one and onto.

Exercise 7.38 *Let Ω be a bounded region, f a holomorphic function in Ω, $f(\Omega) \subset \Omega$. Suppose that f has a fixed point α in Ω but that f is not an onto homeomorphism. Show that $\{f^{[n]}\}$ converges to α, locally uniformly on Ω.*

Hints: Any convergent subsequence of the iterates of f converges to the constant α. For the limit is a constant function by 7.37 and the limit at the point α is α, since $f^{[n]}(\alpha) = \alpha$ for all n. By corollary 7.8 then the whole sequence $\{f^{[n]}\}$ converges to α.

Exercise 7.39 (Cf. 6.5) *If Ω is a bounded region, f holomorphic in Ω, $f(\Omega) \subset \Omega$ and f has at least two fixed points in Ω, then f is a conformal map of Ω onto Ω.*

Hint: In the contrary case the iterates would have to converge to each of the distinct fixed points, by the last exercise, a manifest impossibility.

Theorem 7.40 *Let Ω be a bounded region, f holomorphic in Ω, $f(\Omega) \subset \Omega$ and $f(a) = a$ for some $a \in \Omega$. Then*
(i) $|f'(a)| \leq 1$
(ii) $f'(a) = 1 \Leftrightarrow f$ *is the identity map*
(iii) $|f'(a)| = 1 \Leftrightarrow f$ *is one-to-one and onto.*
In particular, if f is a conformal map of Ω onto itself, $f(a) = a$ and $f'(a) > 0$, then f is the identity map.

Proof: (i) We may translate Ω by $-a$ and so suppose without loss of generality that $a = 0$. It is notationally convenient to set $f_n = f^{[n]}$. Then

$$f_n'(0) = f'(f_{n-1}(0)) \cdot f_{n-1}'(0).$$

But $f(0) = 0$ implies $f_n(0) = 0$ for all n, so

$$f_n'(0) = f'(0) f_{n-1}'(0).$$

This is the inductive step in a proof that

(1) $\quad f_n'(0) = (f'(0))^n \quad$ for all n.

If M is a bound for Ω and $r > 0$ is such that $\bar{D}(0, r) \subset \Omega$, then Cauchy's Formula gives for each positive integer k

$$\frac{f_n^{(k)}(0)}{k!} = \frac{1}{2\pi i}\int_0^{2\pi} \frac{f_n(re^{i\theta})}{(re^{i\theta})^{k+1}}\, ire^{i\theta}d\theta$$

(2.k) $$\frac{|f_n^{(k)}(0)|}{k!} \leq \frac{1}{2\pi}\int_0^{2\pi} \frac{|f_n(re^{i\theta})|}{r^k}\, d\theta \leq \frac{M}{r^k}, \quad n = 1, 2, \cdots$$

(since $f_n(\Omega) \subset \Omega$). From (1) and (2.1) follow that $|f'(0)| \leq 1$.

(ii) Now assume $f'(0) = 1$. If there are any $k \geq 2$ such that $f^{(k)}(0) \neq 0$, let m be the smallest and set $c_k = f^{(k)}(0)/k!$ for $k \geq m$.

$$f(z) = z + \sum_{k=m}^{\infty} c_k z^k$$

(3) $$= z + c_m z^m g(z), \quad \text{where } g(0) = 1.$$

Let us show that for each n there is an $r_n > 0$ and a holomorphic function g_n in $D(0, r_n)$ such that for all $z \in D(0, r_n)$

(3.n) $$f_n(z) = z + c_m z^m g_n(z) \quad \text{and} \quad g_n(0) = n.$$

The equation (3) is just (3.1). If (3.n) holds, then since $f_n(0) = 0$ we may choose $r_n > r_{n+1} > 0$ so that $f_n(D(0, r_{n+1})) \subset D(0, r_n)$. Then from (3) and (3.n) will follow for $z \in D(0, r_{n+1})$

$$\begin{aligned} f_{n+1}(z) &= f(f_n(z)) = f_n(z) + c_m f_n^m(z) g(f_n(z)) \\ &= z + c_m z^m g_n(z) + c_m[z + c_m z^m g_n(z)]^m g(f_n(z)) \\ &= z + c_m z^m[g_n(z) + (1 + c_m z^{m-1} g_n(z))^m g(f_n(z))]. \end{aligned}$$

Let $g_{n+1}(z)$ denote the square-bracketed term and note that since $m - 1 \geq 1$, the term z^{m-1} is 0 when $z = 0$, so

$$\begin{aligned} g_{n+1}(0) &= g_n(0) + g(f_n(0)) \\ &= g_n(0) + 1 \quad \text{by (3)} \\ &= n + 1 \quad \text{by (3.n)}, \end{aligned}$$

which establishes (3.n + 1).

It follows from (3.n) that

$$f_n^{(m)}(0) = m!\, c_m g_n(0) = nm!\, c_m, \quad n = 1, 2, \ldots$$

and since $c_m \neq 0$ this contradicts inequality (2.m) for sufficiently large n.

The conclusion is that $f^{(k)}(0) = 0$ for all $k \geq 2$, so $f(z) = z$ in $D(0, r)$, hence throughout the connected superset Ω.

(iii) Suppose $|f'(0)| = 1$. The sequence $\{f_n\}$ of iterates of f is bounded on Ω and so has a convergent subsequence $\{f_{n_j}\}$ (by 7.6). If the limit is ϕ, then $(f_{n_j})'(0) \to \phi'(0)$. But, as noted above, $f_n'(0) = (f'(0))^n$ for all n, so

$|f'_{n_j}(0)| = 1$, whence $|\phi'(0)| = 1$. Therefore ϕ is not constant and it follows from 7.37 that f is one-to-one and onto.

For the converse direction in (iii), apply (i) to both f and f^{-1}, noting that, by the Chain Rule, $(f^{-1})'(0) = 1/f'(f^{-1}(0)) = 1/f'(0)$.

Exercise 7.41 *Let Ω be a bounded region, $a \in \Omega$, $\mathscr{G} = \mathscr{G}_a$ the set of all conformal maps of Ω onto itself which fix a.*

(i) *$\mathscr{G}$ is a group under composition and is isomorphic to a subgroup of the unit circle. Hence $\mathscr{G}$ is commutative.*

Hint: The isomorphism is $f \to f'(a)$. Compare part (iii) of 7.40.

(ii) *If $\{f_n\} \subset \mathscr{G}$ and $\{f_n\}$ converges on Ω to f, then $f \in \mathscr{G}$.*

Hints: By 7.5 the convergence is locally uniform on Ω and f is holomorphic, since $\{f_n\}$ is bounded. Then $f'_n(a) \to f'(a)$. As $|f'_n(a)| = 1$, it follows that $|f'(a)| = 1$, so f is not constant. For $w \in \mathbb{C}\backslash\Omega$ the functions $f_n - w$ have no zero in Ω, so by 7.10 the non-constant limit function $f - w$ does not either. That is, $f(\Omega) \subset \Omega$. Then since $|f'(a)| = 1$, it follows from 7.40 that f is one-to-one and onto.

(iii) *If $\{f_n\} \subset \mathscr{G}$ and $\{f'_n(a)\}$ converges, then $\{f_n\}$ converges to an $f \in \mathscr{G}$.*

Hints: According to 7.8 it suffices to demonstrate that all convergent subsequences of $\{f_n\}$ have the same limit. (This limit will be in $\mathscr{G}$ by (ii).) If $\{f_{n_j}\}$, $\{f_{m_j}\}$ are two such convergent subsequences with limits ϕ and ψ, say, then $f'_{n_j}(a) \to \phi'(a)$ and $f'_{m_j}(a) \to \psi'(a)$. Therefore $\phi'(a) = \psi'(a)$. Since $\phi, \psi \in \mathscr{G}$ by (ii), it follows from (i) that $\phi = \psi$.

(iv) *The map $f \to f'(a)$ of $\mathscr{G}$ into the circle is bicontinuous and the image of $\mathscr{G}$ is a compact subgroup of the circle. Hence $\mathscr{G}$ is either finite and cyclic or isomorphic to the circle.*

Remarks 7.42 The group $\mathscr{G}$ is the whole circle if Ω is simply-connected. This follows easily (see 10.6) from the Riemann Mapping Theorem. AUMANN and CARATHÉODORY [1934] demonstrated the converse: If Ω is not simply-connected, then every element of $\mathscr{G}$ has finite order, hence, by the above, $\mathscr{G}$ is a finite (cyclic) group. The proof has points of similarity with 10.6 and 10.20; one needs a generalization of the Riemann map: a locally univalent map f of $D(0, 1)$ onto Ω with a special property—analytic continuability of the local inverses of f along curves in Ω. The construction of this function is very similar to the Koebe-Carathéodory construction of the mapping function for Riemann's theorem which will be presented in Chapter IX. The reader will find it carried out in §§ 315–321 of CARATHÉODORY's book [1960] and in § 7, chapter 3 of VEECH [1967]. In exercises 12–19 of the latter he will also find the elegant details of Aumann and Carathéodory's proof. A less constructive existence proof for f along the lines of 9.8 appears in § 1, chapter VI of GOLUZIN [1969].

For more about groups of conformal automorphisms of a region see RADÓ [1924–26], BRUNÉ [1932], HEINS [1946b], and MARDEN, RICHARDS and RODIN [1967] and their bibliographies.

Exercise 7.43 *Let* $\Omega = \{z \in \mathbb{C}: \operatorname{Re} z > 0,\ \operatorname{Im} z > 0,\ |z| < 1\}$, $f(z) = e^{\frac{1}{2}\pi i z}$. *Show that* $f(\Omega) \subsetneq \Omega$ *and that* f *has a fixed point in* Ω.

Hint: If $0 < \theta < \pi/2$ and $r > 0$, then $f(re^{i\theta}) = e^{-\frac{1}{2}\pi r \sin\theta} \exp(\frac{1}{2}\pi i r \cos\theta)$. Since $\frac{1}{2}\pi r \cos\theta \in (0, \pi/2)$ and $\sin\theta > 0$, it is therefore clear that $f(re^{i\theta}) \in \Omega$. Also we see that this point is fixed by f if and only if $r = e^{-\frac{1}{2}\pi r \sin\theta}$ and $\theta = \frac{1}{2}\pi r \cos\theta$, that is, if and only if the function $F(\theta) = \exp(-\theta\tan\theta) - 2\theta/(\pi\cos\theta)$ has a zero in $(0, \pi/2)$. The existence of such a zero follows from the Intermediate Value Theorem for continuous functions, since $F(0) = 1$ and $F(\pi/4) = e^{-\pi/4} - 1/\sqrt{2} < 0$, since $e^{\pi/4} = \sqrt{e^{\pi/2}} > \sqrt{e} > \sqrt{2}$. Finally note that $\frac{1}{2}i \in \bar{\Omega}$ but $\frac{1}{2}i \notin f(\bar{\Omega})$. Indeed if $re^{i\theta} \in \bar{\Omega}$ and $\frac{1}{2}\exp(\frac{1}{2}\pi i) = \frac{1}{2}i = f(re^{i\theta}) = e^{-\frac{1}{2}\pi r\sin\theta}\exp(\frac{1}{2}\pi i r\cos\theta)$, then $\frac{1}{2}\pi = \frac{1}{2}\pi r \cos\theta$ and it follows that $r = 1$, $\cos\theta = 1$, whence $re^{i\theta} = 1$ and $f(re^{i\theta}) = i$, a contradiction. Since $\frac{1}{2}i \in \bar{\Omega}\backslash f(\bar{\Omega}) \subset \bar{\Omega}\backslash \overline{f(\Omega)}$, certainly $f(\Omega) \neq \Omega$.

Exercise 7.44 (i) *Show that* $i^{i^{\cdot^{\cdot^{\cdot}}}}$ *converges, that is, the sequence* i_n *defined inductively by* $i_1 = i$, $i_{n+1} = i^{i_n} = e^{\frac{1}{2}\pi i_n}$, *converges.*

Hints: Let Ω and f be as in the last exercise. The conclusion of 7.38 is valid for Ω, since by the last exercise f is a holomorphic map of Ω into itself which is not onto but which fixes some point $\alpha \in \Omega$. It follows that the sequence $\{f^{[n]}\}$ of iterates of f converges throughout Ω to α. Now check that $\beta = f^{[4]}(1) \in \Omega$ and that $i_n = f^{[n]}(1) = f^{[n-4]}(\beta)$ for all $n > 4$ and we're done.

(ii) *Let* $a > 0$, $b > 0$, $a + b = 1$ *and* $f(z) = az + b/z$ $(z \in \mathbb{C}\backslash\{0\})$. *Then the iterates of* f *converge to* 1 *in the open right half-plane and to* -1 *in the open left half-plane. What happens on the imaginary axis?*

Hints: Note that each half-plane is conformal to a disk [see 2.5] and is invariant under f. Since $f(1) = 1$, $f(-1) = -1$, f has a fixed point in each half-plane but since $f'(\sqrt{b/a}) = f'(-\sqrt{b/a}) = 0$, f is not one-to-one in either half-plane (cf. 5.78). The assertion follows then from 7.38.

In the next four exercises we specialize Ω to the unit disk $D = D(0, 1)$, where much more detailed conclusions are possible.

Exercise 7.45 *If* $0 < b < 1$ *and for any* $a \in D$ *we set*

$$f_a(z) = \frac{z - a}{1 - \bar{a}z} \qquad (|z| \leq 1),$$

show that the iterates of f_b *converge to* -1 *at all points of* D.

Hints: Define $F_1 = f_b$, $F_n = f_b \circ F_{n-1}$. For any $0 < B < 1$ we have that

$$\text{(1)} \quad f_b \circ f_B = f_c \quad \text{where } c = -f_b(-B) = \frac{b + B}{1 + bB} > B.$$

It follows by induction that

$$F_n = f_{b_n}$$

where

(2) $$b_n = \frac{b + b_{n-1}}{1 + bb_{n-1}} \quad \text{and} \quad b = b_1 < b_2 < \cdots b_n < 1.$$

Let $B = \lim_{n\to\infty} b_n \in (0, 1]$. Letting $n \to \infty$ in (2) and solving for B, we see that $B = 1$. Then $F_n(z) = (z - b_n)/(1 - b_n z) \to (z - 1)/(1 - z) = -1$ for all $z \in D$.

Exercise 7.46 *Show that if f is a conformal map of $D = D(0, 1)$ onto itself and is not the identity map, then either f has a unique fixed point in $\bar{D}$, or else f has exactly two fixed points in $\bar{D}$ and these are of modulus 1. In the latter case the iterates of f converge on D to one of the fixed points.*

Hints: By 6.2(ii) there exist $a \in D$ and α of modulus 1 such that

$$f(z) = \alpha\cdot\frac{z - a}{1 - \bar{a}z}, \quad z \in D.$$

If $a = 0$, then $\alpha \neq 1$ (since f is not the identity), so 0 is clearly the unique fixed point. If $a \neq 0$, then $f(z) = z$ is equivalent to

(*) $$\bar{a}z^2 - (1 - \alpha)z - \alpha a = 0.$$

This is a quadratic and so has at most two roots. If z is a root, then $z \neq 0$ and $1/\bar{z}$ is also a root (recalling that $\bar{\alpha} = 1/\alpha$). Therefore (*) either has exactly one root in $\bar{D}$ or two distinct roots in $\bar{D}$, each of modulus 1. In the latter case, calling the fixed points b_1 and b_2, there is by 6.4 a conformal self-map h of D such that $h(1) = b_1$, $h(-1) = b_2$. Thus the composite $h^{-1} \circ f \circ h$ fixes 1 and -1. Since (by 6.2(ii)) $h^{-1} \circ f \circ h = \beta f_b$ for some $b \in D$ and some $|\beta| = 1$, we have

$$\beta\cdot\frac{1 - b}{1 - \bar{b}} = 1 \quad \text{and} \quad \beta\cdot\frac{-1 - b}{1 - \bar{b}(-1)} = -1.$$

Thus

$$\frac{1 - \bar{b}}{1 - b} = \frac{1 + \bar{b}}{1 + b}$$

and we see that $b = \bar{b}$, b is real. Then $\beta = 1$. Note that $b \neq 0$, since f is not the identity. Let $g = h^{-1} \circ f \circ h = f_b$ if $b > 0$, and if $b < 0$ let $j(z) \equiv -z$ and $g = j^{-1} \circ h^{-1} \circ f \circ h \circ j = j^{-1} \circ f_b \circ j = f_{-b}$. Then by the previous exercise $g^{[n]}(z) \to -1$ for all $z \in D$. Since $g^{[n]} = h^{-1} \circ f^{[n]} \circ h$ or $j^{-1} \circ h^{-1} \circ f^{[n]} \circ h \circ j$, we have

$$h^{-1}(f^{[n]}(h(z))) \to -1 \quad \text{or} \quad j^{-1} \circ h^{-1}(f^{[n]}(h(-z))) \to -1 \quad \text{for all } z \in D,$$

that is,

$$h^{-1}(f^{[n]}(w)) \to -1 \quad \text{or} \quad -h^{-1}(f^{[n]}(w)) \to -1 \quad \text{for all } w \in D.$$

Therefore,

$$f^{[n]}(w) \to h(-1) = b_2 \quad \text{or} \quad f^{[n]}(w) \to h(1) = b_1 \quad \text{for all } w \in D.$$

Exercise 7.47 (SHIELDS [1964]) *Let $\mathscr{F}$ be a family of continuous self-maps of the closed unit disk which are holomorphic in the open unit disk and commute with one another under composition. Show that the functions in $\mathscr{F}$ have a common fixed point.*

Hints: Let $D = D(0, 1)$. If $\mathscr{F}$ contains a constant function, then this is a common fixed point. Otherwise by the Maximum Modulus Principle $f(D) \subset D$ for each $f \in \mathscr{F}$. If some $f \in \mathscr{F}$ is not a homeomorphism of D onto D, then, on the one hand some subsequence of the iterates has a limit (7.6) and by 7.37 the limit is constant. Say, $f^{[n_k]} \to z_0$ (on D). Then for any $g \in \mathscr{F}$, $g(0) \in D$ so $f^{[n_k]}(g(0)) \to z_0$. Since $f^{[n_k]}(g(0)) = g(f^{[n_k]}(0)) \to g(z_0)$ (continuity of g on $\bar{D}$), $g(z_0) = z_0$, for all $g \in \mathscr{F}$.

Finally suppose that every $f \in \mathscr{F}$ is a homeomorphism of D onto D. We may exclude from $\mathscr{F}$ the identity function. Then each $f \in \mathscr{F}$ has one or two fixed points (7.46). If some $f \in \mathscr{F}$ has exactly one fixed point z_0, then

$$g(z_0) = g(f(z_0)) = f(g(z_0)) \quad \forall g \in \mathscr{F}$$

shows that $g(z_0)$ is fixed by f, so $g(z_0) = z_0$. Thus z_0 is a common fixed point. If some $f \in \mathscr{F}$ has two distinct fixed points, then by 7.46 $f^{[n]} \to z_0$ on D, where z_0 is one of the fixed points. As before, it follows that z_0 is a common fixed point of the functions in $\mathscr{F}$.

Remarks: When $\Omega = D(0, 1)$ much more than the conclusion of 7.37 is true: the whole sequence of iterates of f converges to a constant. We develop the special case of this when f is continuous on $\bar{D}(0, 1)$ in the next exercise. The general case is a bit messy: See WOLFF [1926b], DENJOY [1926] or VALIRON [1931], [1954]. A modified version of these proofs is sketched in 7.49.

In what follows f is a holomorphic function in $D = D(0, 1)$, $f(D) \subset D$ and f is *not* a conformal map of D onto D. We have $|f^{[n]}| \le 1$ for all n and so by 7.6 every subsequence of $\{f^{[n]}\}$ contains a convergent subsequence. Here and in the sequel convergence means local uniform convergence on D.

Exercise 7.48 (i) *Either f has a fixed point $\alpha \in D$ and then $\{f^{[n]}\}$ converges to α locally uniformly on D and the fixed point is unique, or there is no fixed point and then $\{|f^{[n]}|\}$ converges to 1, locally uniformly on D.*

Hints: After 7.39 and 7.38 only the last assertion needs proof. If it were false, then (7.2) there would be a $\delta < 1$ and a compact set $K \subset D$ such that $|f^{[n]}(\alpha_n)| \le \delta$ for infinitely many n and points $\alpha_n \in K$. Using 7.6 and the compactness of K we can select $n_1 < n_2 < \cdots$ so that $\{f^{[n_j]}\}$ converges uniformly on compact subsets of D to some $g \in H(D)$ and such that $\{\alpha_{n_j}\}$ converges to some $\alpha \in K$. Uniform convergence of $\{f^{[n_j]}\}$ to g on K evidently implies that $f^{[n_j]}(\alpha_{n_j}) \to g(\alpha)$. Thus $|g(\alpha)| \le \delta < 1$. By 7.37, g is constant. Thus $g(D) = g(\alpha) \in D$. Therefore, on the one hand, since $f(\alpha) \in D$

$$f^{[n_j]}(f(\alpha)) \to g(f(\alpha)) = g(\alpha)$$

and on the other hand

$$f^{[n_j]}(f(\alpha)) = f(f^{[n_j]}(\alpha)) \to f(g(\alpha)).$$

This makes $g(\alpha)$ a fixed point for f, contrary to the hypothesis on f.

In the sequel we suppose in addition to the hypotheses preceding (i) that f has no fixed point in D.

(ii) *If $E(z)$ denotes the set of limit points of $\{f^{[n]}(z)\}$ for each $z \in D$, then $E(z) = E(z')$ for all $z, z' \in D$. This common set (which lies on the unit circle by* (i)) *will be denoted simply E.*

Hints: If $\alpha \in E(z)$, there is a subsequence of $\{f^{[n]}\}$ which converges at z to α and a further subsequence $\{f^{[n_j]}\}$ which converges uniformly on each compact subset of D to a holomorphic function ϕ. By 7.37 ϕ is constant. Therefore $\phi \equiv \alpha$. In particular, $\{f^{[n_j]}(z')\}$ converges to $\phi(z') = \alpha$, so $\alpha \in E(z')$ for every $z' \in D$.

(iii) *The set E is closed and connected.*

Hints: Closure is clear. Suppose E is not connected: let E_1, E_2 be two non-void, disjoint, closed subsets of E with $E_1 \cup E_2 = E$. Let V_1, V_2 be disjoint open neighborhoods of E_1, E_2. Since $E \subset V_1 \cup V_2$ and E is all the limit points of $\{f^{[n]}(z)\}$ (fix $z \in D$), there are two mutually disjoint sequences $n_1 < n_2 < \cdots$ and $m_1 < m_2 < \cdots$, comprising all integers beyond some point, such that $\{f^{[n_j]}(z)\} \subset V_1, \{f^{[m_k]}(z)\} \subset V_2$. For infinitely many j we must have $n_j + 1 \in \{m_k\}$, otherwise $\{m_k\}$ would be finite. Let this be the case for $j_1 < j_2 < \cdots$. The sequence $\{f^{[n_{j_i}]}\}$ contains a subsequence which converges locally uniformly on D. Call this subsequence $\{F_i\}$ and the limit ϕ. By 7.37 ϕ is constant. Since $\{F_i(z)\} \subset \{f^{[n_j]}(z)\} \subset V_1$, this constant is an element α_1 of E_1. Therefore $F_i(w) \to \alpha_1$ for every $w \in D$. In particular for $w = f(z)$. Thus α_1 is a limit point of $\{F_i(f(z))\} \subset \{f^{[n_{j_i}]}(f(z))\} = \{f^{[n_{j_i}+1]}(z)\} \subset \{f^{[m_k]}(z)\} \subset V_2$, a contradiction.

For the next two parts we suppose in addition to the hypotheses preceding (i) that f has no fixed point in D and that f is continuous on $\bar{D}$.

(iv) *$f(\alpha) = \alpha$ for all $\alpha \in E$.*

Hints: If $\alpha \in E$, then as in (ii) there is a subsequence $\{f^{[n_j]}\}$ which converges throughout D to α. Fixing $z \in D$, we have

$$f^{[n_j]}(z) \to \alpha \quad \text{and} \quad f^{[n_j]}(f(z)) \to \alpha.$$

Since f is continuous on $\bar{D}$, it follows from the first of these however that $f^{[n_j]}(f(z)) = f(f^{[n_j]}(z)) \to f(\alpha)$.

(v) *$\{f^{[n]}\}$ converges on D to a constant.*

Hints: By 7.8 it suffices to prove that E is a single point. But if this is not the case, then E is an arc and $f(z) = z$ on E by (iv). Therefore by 5.75 $f(z) = z$ for all $z \in D$, contrary to f being fixed-point-free in D.

Final Remarks 7.49 When continuity of f on $\bar{D}$ is not available, one can set $L = [0, f(0)]$ and by a routine connectedness argument prove that for every $u \in E$ with at most two exceptions the interval $[ru, u]$ and the connected set $\bigcup_{n=N}^{\infty} f^{[n]}(L)$ have non-void intersection, for every $r < 1$ and every $N \in \mathbb{N}$. This fact, plus a little argument like that in (iv) above, enables one to select $r_k = r_k(u) \uparrow 1$ such that $f(r_k u) \to u$. Hence for all those $u \in E$, with at most two exceptions, for which $\lim_{r \uparrow 1} f(ru)$ exists, the value of the limit is u. This existence set is Lebesgue-almost-all of E, according to the famous radial limit theorem of FATOU [1906], pp. 364–367 (mentioned also in the Chapter IX notes). Since f is not the identity function, it follows from Fatou's equally famous uniqueness theorem (pp. 394–395, *op. cit.*) that the arc E has measure 0, i.e., is a single point. The conclusion (v) follows from this fact and 7.8.

Notice that the removal of the continuity assumption makes the conclusion valid for any simply-connected region $\Omega \neq \mathbb{C}$ in the role of D, thanks to the Riemann mapping theorem (10.2).

Notes to Chapter VII

The convergence in 7.1 is also often called "almost uniform convergence" and the topology defined by any of the equivalent metrics of 7.3 is called the "c.c." (compact convergence) or the "compact-open" topology.

LANDAU [1918] points out that 7.4 is a simple consequence of well-known earlier results of Mittag–Leffler and Jensen and generally downplays BLASCHKE [1915] by giving elementary proofs of the other results there. (History, however, has dealt more kindly with Blaschke on this point.) The use of Blaschke products to get bounds as in 6.9(ii) and to secure convergence as in 7.4 has been discovered independently by other writers. See especially the formula on p. 57 of WHITTAKER [1935] and the way this is used by BOWEN and MACINTYRE [1954] and WHITTAKER [1970b]. This device is also used very effectively by V. I. LEVIN [1934]. A related technique, expansion of a function into a series of Blaschke products with remainder, occurs in LAMMEL [1958].

The important convergence theorem 7.5 is due independently to VITALI [1903], [1904] and PORTER [1904–1905b]. There is a beautiful proof based on 7.9 (itself proved *en route*) in LINDELÖF [1913] and, essentially the same proof, in JENTZSCH [1918]. These proofs were probably inspired by MONTEL [1907], pp. 300–301 and [1910], pp. 21 ff. For a similar and comparably elegant proof see CARATHÉODORY and LANDAU [1911]. For other proofs see JACOBSTHAL [1927], BOWEN and MACINTYRE [1954], LAMMEL [1958] and for interesting related results LINDWART and PÓLYA [1914] and SHISHA [1965]. For history and references to the precursors of Vitali and Porter (like STIELTJES [1894]) see § 58 of BIEBERBACH's encyclopedia article. A good historical discussion also may be found in OSTROWSKI [1923b].

There are extensions of the Vitali–Porter theorem in which the set on which convergence is hypothesized, instead of having a limit point in the region, is allowed to cluster on the boundary (e.g., 7.4) or even lie wholly on the boundary (e.g., 5.45(iii)): See MONTEL [1907], p. 309, [1917], KHINTCHINE [1922–24], [1923], OSTROWSKI [1922/23], F. RIESZ [1922/23a], KUNUGUI [1942]. The conclusion also follows if convergence (in a certain sense) is hypothesized on a subset of the boundary having positive measure. See KOVANKO [1924], HARTOGS [1928], pp. 157 ff. of BIEBERBACH's text [1931] or p. 456 of DIENES [1931]; also pp. 397 ff. of GOLUZIN [1969]. For related matter see WHITTAKER [1970b].

In Chapter XIV we will see that the positive integers are a uniqueness set for entire functions with certain growth restrictions. (Cf. 6.11.) When these conditions are fulfilled uniformly for a sequence, then convergence of the sequence on the positive integers implies its local uniform convergence in the whole plane. (See, e.g., 15.3.5 of HILLE [1962].)

From 12.25 and 12.24 follows a dramatic strengthening of the Vitali–Porter convergence theorem: It suffices for the conclusion in 7.5 that there be two distinct complex numbers which are not in the range of any of the functions F_n. See *Satz* VI of CARATHÉODORY and LANDAU [1911].

7.6 is due to MONTEL [1907] (pp. 298–302), with an independent proof by KOEBE [1908], p. 341. Though this fact appears here as a trivial glissade from the Vitali–Porter theorem, in fact the two are easily seen to be equivalent via the Fundamental Uniqueness Theorem (5.62). See, for example, GOLUZIN [1969], p. 18. An *ab initio* proof of 7.6 is then desirable and such exist (*op. cit.*, pp. 15 ff.) Montel himself followed this path in his [1910] monograph and prior to LINDELÖF [1913] (see p. 178) it was customary to deduce 7.5 from 7.6. Generally, 7.8 can be coupled with any uniqueness theorem to yield a convergence theorem. E.g., DENJOY [1929] proved that if $a_n, b_n \in D = D(0, 1)$, then a necessary and sufficient condition that there be *at most one* $f \in H(D)$ which is bounded by 1 and takes the value b_n at a_n for each $n = 1, 2, \ldots$, is that

$$(*) \qquad \sum_{n=1}^{\infty} \frac{(1 - |a_n|)}{(1 - |b_n|)} = \infty;$$

and R. NEVANLINNA [1929d] (last paragraph) pointed out that this entails the following extension of the Vitali–Porter theorem: the condition (*) on numbers $a_n, b_n \in D$ is necessary and sufficient that every sequence $\{f_k\} \subset H(D)$ with $|f_k| \leq 1$ which converges at a_n to b_n for each n, actually converge locally uniformly throughout D.

The aforementioned direct proof of 7.6 is also probably the most perspicuous. It is based on ideas of C. Arzelà and G. Ascoli and runs thus: Use boundedness to get equicontinuity (via Cauchy's integral formula), the diagonal process to get a subsequence which converges at each point of a countable dense set D and the equicontinuity to insure that this subsequence converges throughout $\bar{D}$. (This kind of argument appears in the second paragraph of the proof of 9.24 and in the latter half of the hints to 12.35 below.)

7.10 is from MONTEL [1912] (see also [1925], p. 247). Alternative proofs of it may be based on 5.36, 5.38 or 5.87. 7.13 is due to HARNACK [1886]. Notice that the positive integer multiples of a fixed holomorphic function illustrate the need for the hypothesis that the limit at z_0 be *non-zero* in 7.12(iii). 7.14(i) is asserted in KOEBE [1909], IV *Mitteilung*, p. 329 as being a consequence of 7.6. A proof occurs in OSGOOD [1928], p. 697. See footnote 524 of LICHTENSTEIN [1919] for the geneology of this result. For the genesis of 7.14(ii) and (iii) the reader should examine G. C. EVANS [1923] and F. and R. NEVANLINNA [1922]. PRIVALOV [1935] contains a subharmonic function version of 7.14(iii). A beautiful half-plane version of 7.14(ii) and one for strips may be found in BRAWN [1973] and the references cited there.

The exposition of 7.15 in the text is from HEINS [1962]. For generalizations see pp. 15 ff. of F. and R. NEVANLINNA [1922], ASCOLI [1928b], PRIVALOV [1935] and BRUCKNER, LOHWATER and RYAN [1969]. The Phragmén–Lindelöf device in 5.46 and 7.15 will be explored further in Chapter XIV. HARTOGS and ROSENTHAL [1928], [1931] make an exhaustive topological study of the set U_0 of 7.20. It is, e.g., simply-connected if U is. (To see this use the Maximum Modulus Principle and the criterion 4.65(viii) for simple-connectivity.

For another proof of 7.16 see MONTEL [1932a] and pp. 315 ff. of MONTEL [1907]. On 7.18 and 5.43 see DE LA VALLÉE POUSSIN [1893], SEVERINI [1901], YOUNG [1910], BÔCHER [1910–11] and HARDY [1914]. The statement and proof of 7.18 are from MONTEL [1910], pp. 28, 29. For a nice related result see CUNNINGHAM [1967]. ŚWIĄTKOWSKI [1967] also deals with related matters.

For an interesting alternative proof of 7.20, based on approximation by Bernstein polynomials in two variables, see FLATTO and SHISHA [1973] and for yet another proof EBERLEIN [1975]. A nice self-contained treatment occurs in REDHEFFER [1969] and one based on the paving technique of 4.1 is given on pp. 83–84 of CONWAY [1973].

7.17, 7.23, 7.28 and 7.36, 7.37. 7.40 and 7.41 were culled from the pages of BEHNKE and SOMMER [1965]. For 7.24 and 7.25 see also pp. 29–31 of MONTEL [1933b]. A sharp quantitative version of 7.24 and sharp, explicit bounds in 7.27 and 7.28(iv) will be produced in Chapter XVIII. (See p. 111 of LANDAU [1929a].) Though 7.25 occurs in KOEBE [1907] (pp. 204–205), it is contained in an earlier stronger result of HURWITZ [1904] (namely 12.21 below) in which moreover a lower bound for the constant α occurs. With $\alpha = \frac{1}{16}$ the result occurs in PLEMELJ [1912]. KOEBE [1909b], p. 215 conjectured that the largest value for α is $\frac{1}{4}$ and that this is attained (only) when $f(D)$ is the plane slit along $(-\infty, -\frac{1}{4}]$ or a rotation thereof. The result 7.26 (with an unspecified universal constant replacing $1/3\sqrt{3}$) is from BLOCH [1924b] and [1925]. Many proofs and refinements followed:

FEKETE [1925], [1927b], VALIRON [1926], [1930], LANDAU [1926], [1929], A. J. MACINTYRE [1938] (whose bibliography mentions other references). The proof in the text (essentially that on pp. 616–618 of LANDAU [1929b]) was taken from MILLOUX [1953]. Cf. HEINS [1962] for a slightly better constant obtained by comparably elementary means. Other covering theorems will be presented and discussed in Chapter XVIII and its notes. See also MONTEL [1929] and chapter VII of [1933b] for covering theorems and the paper of LANDAU [1929b] and its detailed historical notes.

7.28(iv), proved via 7.6, occurs first in KOEBE [1909a], p. 73, then later with a different proof in KOEBE [1910], pp. 46–52. See also PICK [1916] and LÖWNER [1917]. For related harmonic function results see OSGOOD [1923–24], MANDELBROJT [1929] and PLEMELJ [1933], for locally univalent functions see MONTEL [1937].

7.29(i) is from MONTEL [1912], p. 508 (see also p. 70 of his book [1927]) and 7.29(ii) is from MONTEL [1912], p. 533. Both results were rediscovered by PRIVALOV [1924d], whose proofs are those suggested in the text. 7.29(iii) was proposed as a problem in vol. 32 (1923), p. *16* of *Jahresbericht der Deutschen Mathematiker Vereinigung* by LÖWNER and RADÓ.

7.30 is from BOWEN [1964–65a], though his proof is different; 7.31 is from the same paper and represents an extension of a result of CARTWRIGHT [1962], who took the hypothesis $\lim_{y\to\infty}|f(x_0 + iy)| = 1$ for some $x_0 \in (a, b)$. See 12.39 and the Chapter XII notes for related material. Cartwright's proof is a slight variation on that of 12.38(i). For nice extensions of 7.30 and 7.31 to subharmonic functions see BEARDON [1971].

In connection with 7.32, see 11.40 and the paper by BARTH and SCHNEIDER [1972]. Actually Weierstrass only did the case where $A = \{a_1, a_2, \ldots\}$ and $|a_n| \to \infty$. The general case occurs in MITTAG–LEFFLER [1884a] and it is essentially the latter's proof which is given in the text. 7.33 is from BLASCHKE [1915]; for another proof see p. 133 of SCHUR [1918] and FATOU [1923]. See the penultimate paragraph of the Chapter XII notes.

7.35(i) is due to SCHRÖDER [1870] and the proofs of (i) and (ii) are from pp. 344–345 of KOENIGS [1883] and from RITT [1920–21]. 7.35(iii) is stated in FARKAS [1884] but the proof there is inadequate. A multi-dimensional version occurs in WAVRE [1926]. The special case in the last sentence of 7.40 is due to BIEBERBACH [1913a]. The proof given for (i) and (ii) is from RADÓ [1924–26]. See also RADÓ [1922/23a] and SAKS and ZYGMUND [1971], p. 232 (also 12.45 below) for a generalization to unbounded regions. For extensions of 7.40(iii) see CARATHÉODORY [1932], pp. 779–786, AUMANN and CARATHÉODORY [1934], HEINS [1941b], and HERVÉ [1951]. Proofs of 7.39 and 7.40 also occur in KAMOWITZ [1976]. Notice that when $\Omega = D(0, 1)$, 7.38 and 7.40 follow very easily from Schwarz' lemma, without using compactness of the sequence of iterates (JULIA [1918]). The hypothesis in 7.37–7.41 that Ω be bounded can be dramatically weakened; see 12.45.

7.44(i) is due to SHELL [1962], the proof in the text is from A. J. MACINTYRE [1966]. For an alternative treatment of 7.44(ii) see exercise III.262 in PÓLYA and SZEGÖ [1972] and for related material Shell's paper and that of BARROW [1936]. The latter paper contains, for example, a proof that if a is positive, the sequence $a^{a^{\cdot^{\cdot}}}$ converges (to a finite real number) if and only if $e^{-e} \leq a \leq e^{1/e}$. See also BROMWICH [1926], p. 23, MITCHELMORE [1974], STROMBERG [1980], *Jahrbuch über die Fortschritte der Mathematik* vol. 29, p. 343, vol. 44, p. 126 and *Mathematical Reviews* vol. 15, p. 950, vol. 48 #2615. For iterated square roots see SCHUSKE and THRON [1962], GERBER [1973], and ROHDE [1974].

The proof of 7.47 in SHIELDS [1964] uses fewer results from function theory but instead relies in an interesting way on elementary facts about compact topological semigroups. An extension of 7.47 to two complex dimensions is given by EUSTICE [1972].

Returning momentarily to 7.49, notice that with E reduced to a single point, $u = u_f$ say, the set $\Gamma = \bigcup_{n=0}^{\infty} f^{[n]}(L)$ is a path from 0 to u and as $z \to u$ in Γ we have $f(z) \to u$. (Such a path is produced in a different way by CRAIG and MACINTYRE [1967].) Using this fact and a version of 5.56, one can easily show that $f(z)$ converges to u when z converges to u from inside any (open) triangle in D which has a vertex at u. This says that the extension of f to $D \cup \{u\}$ which fixes u is "almost continuous" and makes clear that the following result of BEHAN [1973] is a generalization of 7.47: If g is another function answering to the description of f in 7.49 and if $f \circ g = g \circ f$, then $u_f = u_g$. For generalizations of 7.49 itself (to multiply-connected regions) the reader should consult the prize-winning paper of HEINS [1941b]. In both these latter papers the results of 6.24 come into play in an interesting way. On the question of locating u_f and the fixed point in 7.48(i), some information may be found in CRAIG and MACINTYRE [1967].

For other interesting techniques and problems in iteration theory see RÅDSTRÖM [1953], the papers of I. N. BAKER (only a few of which are listed in the bibliography), VALIRON [1931], JULIA [1918], [1919b], FATOU [1919], [1920], [1926], CREMER [1925] (survey), WOLFF [1927], [1929], LIEBECK [1961], ELJOSEPH [1968] (the latter two being relevant to 7.45 and 7.46), the book of MONTEL [1957] and chapter 8 of GROSS [1972]. Incidentally, it is his 1927 paper which is alluded to without coordinates in the opening words "J'ai démontré ailleurs la proposition suivante ..." of WOLFF [1926b], p. 200.

Chapter VIII
Polynomial and Rational Approximation—Runge Theory

§ 1 The Basic Integral Representation Theorem

The power of "integral representations" was demonstrated in 3.8: they lead to local power series expansions. If the (Riemann) integrals involved are approximated by Riemann sums, we get global rational approximations to the function. (See 8.8 below.) If further the domains are properly disposed in $\mathbb{C}$, the "poles" in these rational functions can be "shoved to infinity" and the rational functions thereby approximated by polynomials. Global polynomial approximants are thus produced. They are a very powerful tool for investigating holomorphic functions: often a theorem is easy for polynomials and remains valid under local uniform convergence, hence its validity passes over to holomorphic functions generally.

So our first task is to get a global integral representation for a holomorphic function, to replace the local ones of Chapter V. The development is routine but is almost completely combinatorial and so rather unpleasant to write out: certain integrals occur in oppositely oriented pairs and cancel each other; allegedly all that remains are integrals of such-and-such a kind. Our main task is to keep track of the cancellations and confirm this. Perhaps the reader can devise a more felicitous exposition once he has seen the ideas.

Theorem 8.1 (The Basic Integral Representation Theorem) *Let U be an open subset of $\mathbb{C}$, K a compact subset of U. Then there exist intervals $\gamma_1, \gamma_2, \ldots, \gamma_k$ lying in $U \backslash K$ such that*

$$(1) \qquad f(z) = \sum_{j=1}^{k} \frac{1}{2\pi i} \int_{\gamma_j} \frac{f(\xi)}{\xi - z}\, d\xi$$

holds for every $f \in H(U)$ and every $z \in K$.

Proof: Let

$$0 < \delta = \tfrac{1}{2} \operatorname{dist}(K, \mathbb{C} \backslash U)$$

and for every pair of integers n and m set

$$Q(n, m) = [n\delta, (n + 1)\delta] \times [m\delta, (m + 1)\delta]$$
$$a(n, m) = n\delta + im\delta.$$

Also define (coefficients)

$$c(n, m) = \begin{cases} 1 & \text{if } Q(n, m) \cap K \neq \varnothing \\ 0 & \text{if } Q(n, m) \cap K = \varnothing \end{cases}$$

$$c'(n, m) = c(n, m) - c(n, m - 1)$$
$$c''(n, m) = c(n - 1, m) - c(n, m).$$

Since

$$[a(n, m), a(n + 1, m)] \subset Q(n, m) \cap Q(n, m - 1),$$

it follows from the definitions of c' and c that

$$(1)' \quad [a(n, m), a(n + 1, m)] \cap K \neq \varnothing \Rightarrow c'(n, m) = 0.$$

Also if $c'(n, m) \neq 0$, then $Q(n, m) \cup Q(n, m - 1)$ meets K and since the interval $[a(n, m), a(n + 1, m)]$ lies in each of these rectangles and each has diameter $\sqrt{2}\delta$, all points of $[a(n, m), a(n + 1, m)]$ are within $\sqrt{2}\delta < \operatorname{dist}(K, \mathbb{C}\backslash U)$ of K, hence lie in U:

$$(2)' \quad c'(n, m) \neq 0 \Rightarrow [a(n, m), a(n + 1, m)] \subset U.$$

Now if, say, $\max\{|n|, |m|\} > 1 + \delta^{-1} \sup|K|$, then $\inf|Q(n, m)| > \sup|K|$, so $c(n, m) = 0$. Similarly, $c(n - 1, m) = c(n, m - 1) = 0$. Thus,

$$(3) \quad \max\{|n|, |m|\} > 1 + \delta^{-1} \sup|K| \Rightarrow c'(n, m) = c''(n, m) = 0.$$

Therefore there are only finitely many pairs (n, m) for which $c'(n, m) \neq 0$, say

$$(3)' \quad c'(n, m) \neq 0 \Leftrightarrow (n, m) \in \{(n_1, m_1), \ldots, (n_l, m_l)\}.$$

Then (1)′ and (2)′ assert that

$$(4) \quad [a(n_j, m_j), a(n_j + 1, m_j)] \subset U\backslash K, \quad j = 1, 2, \ldots, l.$$

Similarly there exist $(n_{l+1}, m_{l+1}), \ldots, (n_k, m_k)$ such that

$$(3)'' \quad c''(n, m) \neq 0 \Leftrightarrow (n, m) \in \{(n_{l+1}, m_{l+1}), \ldots, (n_k, m_k)\}$$

and

$$(4) \quad [a(n_j, m_j), a(n_j, m_j + 1)] \subset U\backslash K, \quad j = l + 1, \ldots, k.$$

Consider now any g which is continuous on $\bigcup_{c(n,m)=1} \partial Q(n, m)$. By 2.9

$$\begin{aligned}
&\sum_{(n,m)} c(n, m) \int_{\partial Q(n,m)} g \\
&\quad = \sum_{(n,m)} c(n, m) \left\{ \int_{[a(n,m), a(n+1,m)]} g + \int_{[a(n+1,m), a(n+1,m+1)]} g \right. \\
&\qquad\qquad \left. + \int_{[a(n+1,m+1), a(n,m+1)]} g + \int_{[a(n,m+1), a(n,m)]} g \right\} \\
&\quad = \sum_{(n,m)} c(n, m) \int_{[a(n,m), a(n+1,m)]} g + \sum_{(n,m)} c(n, m) \int_{[a(n+1,m), a(n+1,m+1)]} g \\
&\qquad + \sum_{(n,m)} c(n, m) \int_{[a(n+1,m+1), a(n,m+1)]} g + \sum_{(n,m)} c(n, m) \int_{[a(n,m+1), a(n,m)]} g
\end{aligned}$$

$$= \sum_{(n,m)} c(n,m) \int_{[a(n,m),a(n+1,m)]} g + \sum_{(n,m)} c(n-1,m) \int_{[a(n,m),a(n,m+1)]} g$$

$$+ \sum_{(n,m)} c(n,m-1) \int_{[a(n+1,m),a(n,m)]} g + \sum_{(n,m)} c(n,m) \int_{[a(n,m+1),a(n,m)]} g$$

$$= \sum_{(n,m)} [c(n-1,m) - c(n,m)] \int_{[a(n,m),a(n,m+1)]} g$$

$$+ \sum_{(n,m)} [c(n,m) - c(n,m-1)] \int_{[a(n,m),a(n+1,m)]} g$$

$$(5) \quad = \sum_{j=1}^{l} c'(n_j,m_j) \int_{[a(n_j,m_j),a(n_j+1,m_j)]} g + \sum_{j=l+1}^{k} c''(n_j,m_j) \int_{[a(n_j,m_j),a(n_j,m_j+1)]} g.$$

Define therefore

$$\gamma_j = \begin{cases} [a(n_j,m_j), a(n_j+1,m_j)] & \text{if } 1 \le j \le l \text{ and } c'(n_j,m_j) = 1 \\ [a(n_j+1,m_j), a(n_j,m_j)] & \text{if } 1 \le j \le l \text{ and } c'(n_j,m_j) = -1 \\ [a(n_j,m_j), a(n_j,m_j+1)] & \text{if } l < j \le k \text{ and } c''(n_j,m_j) = 1 \\ [a(n_j,m_j+1), a(n_j,m_j)] & \text{if } l < j \le k \text{ and } c''(n_j,m_j) = -1 \end{cases}$$

and have from (4)

$$(6) \quad \gamma_j \subset U \backslash K, \quad 1 \le j \le k$$

and from (5)

$$(7) \quad \sum_{(n,m)} c(n,m) \int_{\partial Q(n,m)} g = \sum_{j=1}^{k} \int_{\gamma_j} g.$$

Now let $f \in H(U)$ be given. For all $z \in \mathbb{C} \backslash \gamma_1 \cup \cdots \cup \gamma_k$ (thus by (6) for all $z \in K$ at least) define

$$(8) \quad F(z) = \sum_{j=1}^{k} \frac{1}{2\pi i} \int_{\gamma_j} \frac{f(\xi)}{\xi - z} \, d\xi.$$

It is clear that F is continuous on its domain, hence on K. Let now $z \in \mathring{Q}(n', m')$ for some (n', m') with $c(n', m') = 1$. If $c(n, m) = 1$ then, as noted before, $Q(n, m) \subset U$. Therefore U contains a convex neighborhood of the closed rectangle $Q(n, m)$. Since the function $\xi \to (f(\xi) - f(z))/(\xi - z)$ is holomorphic in U, we may cite Cauchy's Theorem for Convex Regions (5.2) to conclude that

$$\int_{\partial Q(n,m)} \frac{f(\xi) - f(z)}{\xi - z} \, d\xi = 0$$

$$(9) \quad f(z) \cdot \int_{\partial Q(n,m)} \frac{1}{\xi - z} \, d\xi = \int_{\partial Q(n,m)} \frac{f(\xi)}{\xi - z} \, d\xi.$$

Again Cauchy's Theorem for Convex Regions shows that

$$\int_{\partial Q(n,m)} \frac{1}{\xi - z}\, d\xi = 0$$

if $z \notin Q(n, m)$, i.e., if $(n, m) \neq (n', m')$; and by 4.5 this integral is $2\pi i$ if $z \in \mathring{Q}(n, m)$, i.e., if $(n, m) = (n', m')$. From these remarks and (9) follows

$$f(z) \cdot 2\pi i = f(z) \cdot \sum_{(n,m)} c(n, m) \int_{\partial Q(n,m)} \frac{d\xi}{\xi - z}$$

$$(10) \qquad = \sum_{(n,m)} c(n, m) \int_{\partial Q(n,m)} \frac{f(\xi)}{\xi - z}\, d\xi.$$

The function $g(\xi) = f(\xi)/(\xi - z)$ is eligible for an application of (7), thus transforming (10) into

$$(11) \quad f(z) = \sum_{j=1}^{k} \frac{1}{2\pi i} \int_{\gamma_j} \frac{f(\xi)}{\xi - z}\, d\xi = F(z).$$

Now $K \subset \bigcup_{c(n,m)=1} Q(n, m)$, so if $z \in K$, then $z \in Q(n', m')$ for some (n', m') with $c(n', m') = 1$. Pick $z_1, z_2, z_3, \ldots \in \mathring{Q}(n', m')$ so that $\lim_{r \to \infty} z_r = z$. From (11) we have $f(z_r) = F(z_r)$ and from this and the continuity of f in $U \supset Q(n', m')$ and of F in $\mathbb{C} \backslash \gamma_1 \cup \cdots \cup \gamma_k \supset K \cup \mathring{Q}(n', m')$ we conclude that $f(z) = F(z)$.

Exercise 8.2 *Let $a_1, \ldots, a_k, b_1, \ldots, b_k \in \mathbb{C}$ satisfy*

$$(\text{k.N}) \quad \sum_{j=1}^{k} a_j^N = \sum_{j=1}^{k} b_j^N$$

for all positive integers N. Show that then there is a permutation ϕ of $\{1, 2, \ldots, k\}$ such that $b_j = a_{\phi(j)}$ for each $j = 1, 2, \ldots, k$.

Hints: To construct a proof by induction on k it clearly suffices to show that some a_j lies in the set $\{b_1, b_2, \ldots, b_k\}$. To this end, choose the notation so that

$$a_l = a_{l+1} = \cdots = a_k \quad \text{but } a_j \neq a_k \text{ if } j < l,$$

and suppose, in order to reach a contradiction, that $a_k \notin \{b_1, b_2, \ldots, b_k\}$. There is then a polynomial p which has zeros at $\{a_1, \ldots, a_{l-1}\} \cup \{b_1, b_2, \ldots, b_k\}$ but $p(a_k) \neq 0$. For example,

$$p(z) = \prod_{j=1}^{l-1} (z - a_j) \prod_{n=1}^{k} (z - b_n).$$

Because of equation (k.N), $N = 0, 1, \ldots, k + l - 1$, we have

$$\sum_{j=1}^{k} p(a_j) = \sum_{j=1}^{k} p(b_j),$$

that is,

$$(k - l + 1)p(a_k) = 0.$$

As $k \geq l$ and $p(a_k) \neq 0$, this is a contradiction.

Exercise 8.3 *Here is a variant of* 8.1. *Let K be compact $\subset U$ open $\subset \mathbb{C}$, as there. Then there exist closed polygons $\Gamma_1, \ldots, \Gamma_n$ lying in $U \backslash K$ such that*

$$f(z) = \sum_{l=1}^{n} \frac{1}{2\pi i} \int_{\Gamma_l} \frac{f(\xi)}{\xi - z}\, d\xi$$

for every $f \in H(U)$ and every $z \in K$.

Hints: Let $\gamma_1, \ldots, \gamma_k$ be the intervals provided by 8.1, say $\gamma_j = [a_j, b_j]$. For every non-negative integer n we then have

$$(1) \qquad \sum_{j=1}^{k} \frac{1}{2\pi i} \int_{\gamma_j} \xi^n d\xi = 0.$$

For we may pick a fixed $z \in K$ and apply 8.1 to the function $f(\xi) = (\xi - z)\xi^n$. From (1) and 2.10 we get

$$\sum_{j=1}^{k} \frac{1}{n+1} [b_j^{n+1} - a_j^{n+1}] = 0,$$

that is,

$$(2) \qquad \sum_{j=1}^{k} b_j^N = \sum_{j=1}^{k} a_j^N$$

for every integer $N \geq 1$. Then according to the last exercise there exists a permutation ϕ of $\{1, 2, \ldots, k\}$ such that $b_j = a_{\phi(j)}$ for each $j = 1, 2, \ldots, k$. Write ϕ as a product of disjoint cycles $\phi_1, \phi_2, \ldots, \phi_n$ where, say, ϕ_l is the cycle $(j_l, \phi(j_l), \ldots, \phi^{N_l}(j_l), j_l)$. Then

$$\begin{aligned}\{\gamma_1, \gamma_2, \ldots, \gamma_k\} &= \{[a_1, a_{\phi(1)}], \ldots, [a_k, a_{\phi(k)}]\} \\ &= \bigcup_{l=1}^{n} \{[a_{j_l}, a_{\phi(j_l)}], \ldots, [a_{\phi^{N_l}(j_l)}, a_{j_l}]\},\end{aligned}$$

a disjoint union. Now set

$$\Gamma_l = [a_{j_l}, a_{\phi(j_l)}, \ldots, a_{\phi^{N_l}(j_l)}, a_{j_l}].$$

§ 2 Applications to Approximation

Definition 8.4 (i) A *rational* function is a quotient of two polynomials. When the common factors are removed from these polynomials (via 4.50(ii)), the zeros of the denominator are called the *poles* of the rational function. The domain of the rational function is the plane minus these poles.

(ii) For $w \in \mathbb{C}$ we will, in this section at least, use the special designation F_w for the rational function $F_w(z) = 1/(z - w)$ $(z \in \mathbb{C} \backslash \{w\})$.

Remarks: The reader can easily check that poles are well-defined by this procedure (cf. 11.17). However this definition will be subsumed in Chapter XI by the more general notion of a pole of a meromorphic function.

Theorem 8.5 *Let K be a compact subset of $\mathbb{C}$, Ω a component of $\mathbb{C}\backslash K$ and $a \in \Omega$. Then for each $w \in \Omega$, the function F_w can be uniformly approximated on K by polynomials in F_a.*

Proof: Let $C(K)$ denote the set of all continuous complex functions on K and $A(K)$ the uniform closure in $C(K)$ of the polynomials in F_a. Let S be the set of $w \in \Omega$ for which $F_w \in A(K)$. Note that Ω is open (by 1.30(i)′). If $w_0 \in S$ and $D(w_0, r) \subset \Omega$, then for any $w \in D(w_0, r)$ and all $z \in K$

$$\begin{aligned} F_w(z) &= (z - w)^{-1} = [(z - w_0) - (w - w_0)]^{-1} \\ &= (z - w_0)^{-1}[1 - (w - w_0)(z - w_0)^{-1}]^{-1} \\ &= (z - w_0)^{-1} \sum_{k=0}^{\infty} [(w - w_0)(z - w_0)^{-1}]^k \\ &= \sum_{k=0}^{\infty} (w - w_0)^k (z - w_0)^{-k-1} \\ &= \sum_{k=0}^{\infty} (w - w_0)^k F_{w_0}^{k+1}(z) \end{aligned}$$

and the convergence is uniform in K because for $z \in K$, $|(w - w_0)(z - w_0)^{-1}| \le |w - w_0|r^{-1} < 1$, by definition of r. Therefore the functions

$$R_n = \sum_{k=0}^{n} (w - w_0)^k F_{w_0}^{k+1}$$

approximate F_w uniformly on K. Since $w_0 \in S$, we have $F_{w_0} \in A(K)$, which is clearly a uniformly closed algebra, so $R_n \in A(K)$ and then $F_w \in A(K)$. That is, $w \in S$. We have shown that the non-void (a belongs) subset S of Ω satisfies

$$w_0 \in S \ \& \ D(w_0, r) \subset \Omega \Rightarrow D(w_0, r) \subset S,$$

so by the Basic Connectedness Lemma $S = \Omega$.

Theorem 8.6 *Notation as above, the set S of w in $\mathbb{C}\backslash K$ for which F_w is uniformly approximable on K by polynomials is the unbounded component Ω of $\mathbb{C}\backslash K$.*

Proof: We first show that $S \subset \Omega$, that is, S does not meet any bounded component C of $\mathbb{C}\backslash K$. Suppose otherwise, say $w \in S \cap C$. Let

$$M = \max\{|z - w| : z \in K\}$$

and use the fact $w \in S$ to choose a polynomial p such that

$$|p(z) - F_w(z)| < 1/M \quad \text{for all } z \in K,$$

whence, recalling the definition of F_w,

$$\text{(1)} \qquad |(z - w)p(z) - 1| < |z - w|/M \le 1 \quad \text{for all } z \in K.$$

Next note that the boundary of C lies entirely in K. For C is open by 1.30(i)′, so $\partial C = \bar{C}\backslash C$. If $z \in \partial C$ and also $z \in \mathbb{C}\backslash K$, then some disk D about z lies wholly in $\mathbb{C}\backslash K$. Then D and C are connected sets with points in common (recalling that D is a neighborhood of z and $z \in \bar{C}$) and so their union is connected. As this set lies in $\mathbb{C}\backslash K$, the maximality of C entails that $D \cup C = C$. Consequently $z \in D \subset C$, contrary to $z \in \partial C = \bar{C}\backslash C$. The first inequality in (1) now implies that

(2) $\quad |(z - w)p(z) - 1| < 1 \quad$ for all $z \in \partial C$.

Now the polynomial $1 - (z - w)p(z)$ is bounded on $\bar{C}$ because C is a bounded set. Its maximum modulus on $\bar{C}$ must occur (at least once) on ∂C and so is less than 1 by (2). In particular, this polynomial has modulus less than 1 at $z = w$, an evident contradiction. Thus $S \subset \Omega$.

Let $P(K)$ be the uniform closure of the polynomials in $C(K)$ and set $M = \sup|K|$. If $|a| \geq M + 1$, then

$$F_a(z) = -a^{-1}(1 - za^{-1})^{-1} = -a^{-1} \sum_{k=0}^{\infty} (za^{-1})^k$$

uniformly in $z \in K$, since for such z we have $|za^{-1}| \leq M/(M + 1)$. The partial sums of this series are polynomials, so $F_a \in P(K)$ and $a \in S$ (whence $a \in \Omega$). As $P(K)$ is a uniformly closed algebra, it follows that the uniform closure of the polynomials in F_a lies in $P(K)$. But by the last theorem F_w belongs to the closure of the algebra of polynomials in F_a, for every $w \in \Omega$, so $F_w \in P(K)$ for all $w \in \Omega$. That is, $\Omega \subset S$; so from the result of the first paragraph, $S = \Omega$, as claimed.

Exercise 8.7 *Modify the proof of* 8.6 *to show that if a belongs to a bounded component of $\mathbb{C}\backslash K$, then F_w is not uniformly approximable on K by polynomials in F_a if w belongs to a different bounded component of $\mathbb{C}\backslash K$. What is the situation if w belongs to the unbounded component of $\mathbb{C}\backslash K$?*

Corollary 8.8 *If K is a compact subset of $\mathbb{C}$ and f is holomorphic in a neighborhood U of K, then f is uniformly approximable on K by linear combinations of the functions $F_w(z) = 1/(z - w)$ with $w \in \mathbb{C}\backslash K$.*

Proof: Because of 8.1 it suffices to consider f of the form

$$f(z) = \int_{[a,b]} \frac{\phi(\xi)}{\xi - z}\, d\xi = \int_0^1 \frac{\phi(tb + (1 - t)a)(b - a)}{tb + (1 - t)a - z}\, dt$$

for some $[a, b] \subset U\backslash K$ and some continuous function ϕ on $[a, b]$. The definition of the integral as a limit of Riemann sums and a standard compactness argument show that, given $\varepsilon > 0$, there exist $0 = t_0 < t_1 < \cdots < t_n = 1$ such that

$$\left| f(z) - \sum_{j=1}^{n} \frac{\phi(t_j b + (1 - t_j)a)(b - a)(t_j - t_{j-1})}{t_j b + (1 - t_j)a - z} \right| < \varepsilon \quad \text{for all } z \in K$$

or, setting $w_j = t_j b + (1 - t_j)a$, $c_j = \phi(w_j)(a - b)(t_j - t_{j-1})$,

$$\left| f(z) - \sum_{j=1}^{n} c_j(z - w_j)^{-1} \right| < \varepsilon \quad \text{for all } z \in K$$

$$\left| f(z) - \sum_{j=1}^{n} c_j F_{w_j}(z) \right| < \varepsilon \qquad \text{for all } z \in K.$$

Noting that $w_j \in [a, b] \subset \mathbb{C}\backslash K$, completes the proof.

Corollary 8.9 (The Polynomial Runge Theorem I) *Let K be a compact subset of $\mathbb{C}$ such that $\mathbb{C}\backslash K$ is connected. Let f be holomorphic in an open neighborhood U of K. Then on K, f is a uniform limit of polynomials.*

Proof: Because of the last result, it suffices to approximate each F_w ($w \in \mathbb{C}\backslash K$) uniformly on K with polynomials. This is possible by 8.6, since $\mathbb{C}\backslash K$ is connected and unbounded.

Corollary 8.10 (The Polynomial Runge Theorem II) *Let U be an open, simply-connected subset of $\mathbb{C}$. If f is holomorphic in U, then f is uniformly approximable by polynomials on each compact subset of U.*

Proof: Given compact $C \subset U$, 1.31 furnishes a compact K such that $C \subset K \subset U$ and $\mathbb{C}\backslash K$ is connected. Now apply the last corollary.

Corollary 8.11 (The Rational Runge Theorem I) *Let K be a compact subset of $\mathbb{C}$ and $A \subset \mathbb{C}$. Suppose that A meets each bounded component of $\mathbb{C}\backslash K$. If f is holomorphic in an open neighborhood U of K, then f is uniformly approximable on K by rational functions with poles in A.*

Proof: First approximate f uniformly on K via 8.8 by linear combinations of the functions F_w, $w \in \mathbb{C}\backslash K$. For a given w let C be the component of $\mathbb{C}\backslash K$ which contains w. If C is unbounded, then F_w is uniformly approximable on K by polynomials (8.6), while if C is bounded, there is by hypothesis an $a \in C \cap A$ and then (8.5) F_w is uniformly approximable on K by polynomials in F_a.

Corollary 8.12 (The Rational Runge Theorem II) *Let U be an open subset of $\mathbb{C}$ and let $A \subset \mathbb{C}$. Suppose that A meets each bounded component of $\mathbb{C}\backslash U$. If f is holomorphic in U, then f is uniformly approximable on each compact subset of U by rational functions with poles in A.*

Proof: This follows from 8.11 via 1.31 just as 8.10 was deduced from 8.9 via 1.31.

Exercise 8.13 *Deduce directly from the Polynomial Runge Theorem Cauchy's Theorem for Simply-Connected Regions in the following form: let U be an open, simply-connected subset of $\mathbb{C}$. Then $\int_\gamma f = 0$ for every f holomorphic in U and every piecewise smooth loop γ in U.*

Hint: The result is clear for any f which is a derivative, in particular for all polynomials, whence for any f which is a uniform limit of polynomials on the range of γ. That range being a compact subset of U, all f holomorphic in U qualify by 8.10.

Exercise 8.14 *Deduce directly from the Rational Runge Theorem the General Cauchy Theorem in the following form: Let U be an open subset of $\mathbb{C}$, $c_1, \ldots, c_k \in \mathbb{C}$ and $\gamma_1, \ldots, \gamma_k$ piecewise smooth loops in U. Then $\sum_{j=1}^k c_j \int_{\gamma_j} F = 0$ for every $F \in H(U)$ if and only if $\sum_{j=1}^k c_j \int_{\gamma_j} F_w = 0$ for all $w \in \mathbb{C}\backslash U$, that is, if and only if $\sum_{j=1}^k c_j \operatorname{Ind}_{\gamma_j}$ vanishes identically on $\mathbb{C}\backslash U$.*

Hints: Since $F_w \in H(U)$ for every $w \in \mathbb{C}\backslash U$, one implication is trivial. Suppose $\sum_{j=1}^k c_j \int_{\gamma_j} F_w = 0$ for all $w \in \mathbb{C}\backslash U$, and let $F \in H(U)$ be given. On the compact subset $K = \gamma_1 \cup \cdots \cup \gamma_k$ of U, F is uniformly approximable by polynomials, and linear combinations of powers of the functions F_w with $w \in \mathbb{C}\backslash U$. This follows from the *proof* of 8.12. We have $\int_{\gamma_j} g = 0$ for every g which is a polynomial or a power F_w^n with $n > 1$ because such g are derivatives and each γ_j is a closed curve. On the other hand, $\sum_{j=1}^k c_j \int_{\gamma_j} g = 0$ by hypothesis if $g = F_w$ for any $w \in \mathbb{C}\backslash U$.

Remark: One can get 8.14 directly from the representation 8.1 by an elementary Fubini (8.26(vi)) argument. See, e.g., ČERNÝ [1976].

Example 8.15 Let

$$S_n = [-n, n] \times \left[\frac{1}{n}, n\right]$$

$$U_n = (-n - 1, n + 1) \times \left(\frac{1}{2n}, n + 1\right)$$

$$V_n = (-n - 1, n + 1) \times \left(-\frac{1}{3n}, \frac{1}{3n}\right).$$

The characteristic function χ_n of V_n is holomorphic in the open set $U_n \cup V_n \cup -U_n$ and the compact subset $S_n \cup [-n, n] \cup -S_n$ has a connected complement in $\mathbb{C}$. Therefore Polynomial Runge I provides polynomials P_n such that

$$|P_n - \chi_n| < 1/n \quad \text{on } S_n \cup [-n, n] \cup -S_n, \quad \text{that is,}$$

$$\begin{cases} |P_n| < 1/n & \text{on } S_n \cup -S_n \\ |P_n - 1| < 1/n & \text{on } [-n, n]. \end{cases}$$

It follows that

$$\lim_{n\to\infty} P_n(z) = \begin{cases} 0 & \text{for all } z \in \mathbb{C}\backslash\mathbb{R} \\ 1 & \text{for all } z \in \mathbb{R}. \end{cases}$$

This example shows that the local uniform boundedness hypotheses of Chapter VII cannot (in general) be weakened to pointwise boundedness.

Exercise 8.16 *Produce a sequence of polynomials p_n such that* $\lim_{n\to\infty} p_n(z) = 0$ *for every* $z \in \mathbb{C}$ *but such that this convergence is not locally uniform on* $\mathbb{C}$.

Hint: Set $p_n = P_n^2 - P_n$, where P_n is the polynomial of 8.15. Then by 7.5 and the result of 8.15 it follows that there is a compact subset K of $\mathbb{C}$ such that $\{P_n\}$ is not uniformly bounded on K. Since $|p_n| \geq |P_n|^2 - |P_n| = |P_n|(|P_n| - 1)$, the sequence $\{p_n\}$ is not uniformly bounded on K either.

Exercise 8.17 (i) *Show that in the two Rational Runge Theorems it is enough that the closure of A meet each bounded component of $\mathbb{C}\backslash K$ (resp., $\mathbb{C}\backslash U$).*

(ii) *Show that in the various Runge theorems interpolation at any finite set of points can be achieved by the approximating functions.* (Cf. SINCLAIR [1956].)

Hints: Consider 8.9 and a finite subset A of K together with a positive integer $n(a)$ for each $a \in A$. Let p be any polynomial such that $f - p$ has a zero of order at least $n(a)$ at each $a \in A$. Then the function $F(z) = \dfrac{f(z) - p(z)}{\prod_{a\in A} (z - a)^{n(a)}}$ is holomorphic in U and given $\varepsilon > 0$, 8.9 provides a polynomial P_ε such that $|F - P_\varepsilon| \leq \varepsilon'$ in K, where $\varepsilon' = \varepsilon/\max_{z\in K} \prod_{a\in A} |z - a|^{n(a)}$. The polynomial $p_\varepsilon(z) = p(z) + P_\varepsilon(z) \prod_{a\in A} (z - a)^{n(a)}$ satisfies $|f - p_\varepsilon| \leq \varepsilon$ on K and $f - p_\varepsilon$ has a zero of order at least $n(a)$ at each $a \in A$.

(iii) *Prove the following relative Runge theorem: If U is an open subset of an open subset V of $\mathbb{C}$ and $V\backslash U$ has no bounded component, then every function in $H(U)$ can be uniformly approximated on compact subsets of U by (rational) functions in $H(V)$.*

Hint: $\mathbb{C}\backslash V$ is eligible for the role of A in 8.12.

§ 3 Other Applications of the Integral Representation

Theorem 8.18 (ROUCHÉ [1862]) *Let U be a bounded open subset of $\mathbb{C}$, Φ, Ψ continuous on $\overline{U}$ and holomorphic in U. Suppose that*

(*) $\quad |\Phi + \Psi| < |\Phi| + |\Psi| \quad$ *on ∂U.*

Then, counting multiplicities, the functions Φ and Ψ have the same number [*which is finite*] *of zeros in U.*

Proof: Consider first compact $K \subset U$ and use 8.3 to select polygonal loops $\Gamma_1, \ldots, \Gamma_n \subset U\backslash K$ with the property that

$$(1) \quad f(z) = \sum_{i=1}^{n} \frac{1}{2\pi i} \int_{\Gamma_i} \frac{f(\xi)d\xi}{\xi - z} \quad \forall z \in K$$

holds for every $f \in H(U)$. If $F \in H(U)$ and z is any point in K, we may apply (1) to the holomorphic function $f(\xi) = (\xi - z)F(\xi)$, which is 0 at z, to get

$$(2) \quad \sum_{i=1}^{n} \frac{1}{2\pi i} \int_{\Gamma_i} F(\xi)d\xi = 0.$$

Next suppose $f \in H(U)$ has no zeros in $U \backslash K$ and only finitely many in K, say at $a_1, \ldots, a_N$ with respective multiplicities $m_1, \ldots, m_N$. Then the function

$$(3) \qquad F(z) = \frac{f'(z)}{f(z)} - \sum_{j=1}^{N} \frac{m_j}{z - a_j}$$

is (extendable to be) holomorphic in U. Indeed, according to 5.69 there is a zero-free holomorphic function h in U such that

$$f(z) = h(z) \prod_{j=1}^{N} (z - a_j)^{m_j} \quad \forall z \in U$$

and so the right side of (3) is $h'(z)/h(z)$. Apply (2) to this F to conclude that

$$(4) \qquad \sum_{l=1}^{n} \frac{1}{2\pi i} \int_{\Gamma_l} \frac{f'(\xi)}{f(\xi)}\, d\xi = \sum_{j=1}^{N} m_j \sum_{l=1}^{n} \frac{1}{2\pi i} \int_{\Gamma_l} \frac{1}{\xi - a_j}\, d\xi.$$

But an application of (1) with the constant function 1 in the role of f there (and $z = a_1, a_2, \ldots, a_N$) shows that

$$1 = \sum_{l=1}^{n} \frac{1}{2\pi i} \int_{\Gamma_l} \frac{1}{\xi - a_j}\, d\xi, \quad j = 1, 2, \ldots, N$$

and so (4) becomes

$$(5) \qquad \sum_{l=1}^{n} \frac{1}{2\pi i} \int_{\Gamma_l} \frac{f'(\xi)}{f(\xi)}\, d\xi = \sum_{j=1}^{N} m_j.$$

Now to prove the theorem, let $K = \{z \in \bar{U} : |\Phi(z) + \Psi(z)| = |\Phi(z)| + |\Psi(z)|\}$. This is a closed, hence compact, subset of $\bar{U}$ and by (*) it does not meet ∂U, hence lies in U. By definition we have

$$(**) \qquad |\Phi + \Psi| < |\Phi| + |\Psi| \quad \text{in } \bar{U} \backslash K.$$

To apply the considerations of the first paragraph with either Φ or Ψ in the role of f, we have to check that neither Φ nor Ψ has infinitely many zeros in K. If, say, Φ has infinitely many zeros in K, then these zeros accumulate at some point of K and by the Fundamental Uniqueness Theorem Φ is 0 throughout the component C of U which contains that point and by continuity on $\bar{U}$, $\Phi(\bar{C}) = 0$. C is open (by 1.30(i)′) and relatively closed in U. We cannot therefore have $\bar{C} \subset U$, else $C = \bar{C}$ would be a non-void open-and-closed subset of the connected set $\mathbb{C}$. So $\bar{C} \cap (\bar{U} \backslash U) \neq \varnothing$. But this means that Φ has a zero in $\bar{U} \backslash U = \partial U$, contrary to (*). The same argument shows that Ψ has only finitely many zeros in K. Therefore the argument of the first paragraph of the proof applies with either Φ or Ψ in the role of f. Consequently, to prove the theorem it suffices by (5) to confirm that

$$(6) \qquad \sum_{l=1}^{n} \frac{1}{2\pi i} \int_{\Gamma_l} \frac{\Phi'}{\Phi} = \sum_{l=1}^{n} \frac{1}{2\pi i} \int_{\Gamma_l} \frac{\Psi'}{\Psi}.$$

But (**) shows that $\Phi(z)$ is not a non-negative multiple of $\Psi(z)$ for any $z \in U \backslash K$, so Φ/Ψ maps $U\backslash K$ into $\mathbb{C}\backslash[0, \infty)$. Letting L be the Principal Branch of the Logarithm in $\mathbb{C}\backslash[0, \infty)$ (see 3.43), we can therefore form

$$g = L \circ \left(\frac{\Phi}{\Psi}\right) \quad \text{in } U\backslash K.$$

A little calculation with the Chain Rule gives

$$(7) \quad g' = \frac{\Phi'}{\Phi} - \frac{\Psi'}{\Psi} \quad \text{in } U\backslash K.$$

Since Γ_l is a piecewise smooth loop in $U\backslash K$, we have from 2.10(ii)

$$(8) \quad \int_{\Gamma_l} g' = 0, \quad l = 1, 2, \ldots, N.$$

If we add in (8) and use (7), the equation (6) is confirmed.

Exercise 8.19 *Use Rouché's theorem to prove the following relative of* 7.11. *If* f_n *are holomorphic in the open subset* U *of* $\mathbb{C}$, $f_n \to f_0$ *locally uniformly on* U *and* f_0 *has a zero of order* $n_0 < \infty$ *at* $z_0 \in U$, *then for every* $r > 0$ *there exists* $n(r)$ *such that for each* $n \geq n(r)$ *the function* f_n *has at least* n_0 *zeros (counted according to multiplicity) in* $U \cap D(z_0, r)$.

Hints: Since the zero of f_0 at z_0 is of finite order, f_0 is not identically 0 in any neighborhood of z_0, so the zeros of f_0 do not accumulate at z_0. Given $r > 0$, we may therefore choose $0 < R < r$ so that $\bar{D}(z_0, R) \subset U$ and such that f_0 has no zeros on $C(z_0, R)$. Thus $\delta = \inf|f_0(C(z_0, R))| > 0$. Choose $n(r)$ so large that $|f_n - f_0| < \delta$ on $\bar{D}(z_0, R)$ for all $n \geq n(r)$. Then for each such n the hypotheses of Rouché's theorem are fulfilled by f_n and $-f_0$ in the role of Φ and Ψ respectively.

Remarks: The method of this exercise, that of 7.17 and that of 7.11 are three different ways of proving that the number $N(f)$ of zeros of a (non-zero) holomorphic function f is lower semicontinuous, that is, if $f_n \to f$, then

$$\varliminf_{n\to\infty} N(f_n) \geq N(f).$$

Exercise 8.20 *Let* U *be an open subset of* $\mathbb{C}$, $C_0, C_1, \ldots, C_m$ *distinct bounded components of* $\mathbb{C}\backslash U$; *if* $m > 0$, *suppose* U *is connected. Then there exists a loop* Γ *in* U *such that* $\mathrm{Ind}_\Gamma(C_0) \neq 0$ *and* $\mathrm{Ind}_\Gamma(C_j) = 0$ *for* $j = 1, 2, \ldots, m$. (Cf. 10.10.)

Hints: Use 1.39 to produce unbounded connected closed subsets $K_1, \ldots, K_m$ of $\mathbb{C}\backslash C_0$ which contain $C_1, \ldots, C_m$ respectively. Then $\mathbb{C}\backslash K_1 \cup \cdots \cup K_m$ is an open neighborhood of the compact set C_0, so 1.34 provides a compact set V which is relatively open in $\mathbb{C}\backslash U$, contains C_0 and lies in $\mathbb{C}\backslash K_1 \cup \cdots \cup K_m$. It follows easily that $U \cup V$ is open. Apply 8.3 with the U there equal to the set $(U \cup V)\backslash K_1 \cup \cdots \cup K_m$, $K = V$ (and $f = 1$) to come up with piecewise smooth loops $\Gamma_1, \ldots, \Gamma_n$ in $\mathbb{C}\backslash K$ such that

$$1 = \sum_{l=1}^{n} \frac{1}{2\pi i} \int_{\Gamma_l} \frac{1}{\xi - z}\, d\xi = \sum_{l=1}^{n} \mathrm{Ind}_{\Gamma_l}(z) \quad \forall z \in V.$$

For any fixed $z_0 \in C_0 \subset V$ there is then some l such that

$$\mathrm{Ind}_{\Gamma_l}(z_0) \neq 0,$$

that is,

$$\mathrm{Ind}_{\Gamma_l}(C_0) \neq 0.$$

However, $\Gamma_l \subset \mathbb{C}\backslash K_1 \cup \cdots \cup K_m$ so that by 4.3

$$\mathrm{Ind}_{\Gamma_l}(K_j) = 0, \quad j = 1, 2, \ldots, m,$$

as the K_j are unbounded and connected. Since $C_j \subset K_j$, we are finished.

Exercise 8.21 *Show that an open subset U of $\mathbb{C}$ is homologically-connected if and only if it is simply-connected.* (Cf. 4.65.)

Hints: ($\Rightarrow$) If $\mathbb{C}\backslash U$ has a bounded component C, then the last exercise provides a loop Γ in U such that $\mathrm{Ind}_\Gamma(C) \neq 0$. In particular, Γ is not U-nullhomologous.

($\Leftarrow$) If Γ is a loop in U, then Ind_Γ is constant in each component C of $\mathbb{C}\backslash U$ and vanishes at infinity. Since each C is unbounded, it follows that $\mathrm{Ind}_\Gamma(C) = 0$.

Exercise 8.22 *Let U be an open subset of $\mathbb{C}$ with the property:*

$$(*) \quad \begin{cases} \textit{For any two loops } \gamma_0, \gamma_1 \textit{ in } U \textit{ there exist integers } n_0, n_1 \textit{ not both } 0 \\ \textit{such that } n_0\bullet\gamma_0 \textit{ is loophomotopic to } n_1\bullet\gamma_1 \textit{ in } U. \end{cases}$$

Show that U is connected and that $\mathbb{C}\backslash U$ has at most one bounded component.

Hints: If c_0, c_1 are two points of U, we may apply (*) with γ_0: $[0, 1] \to c_0$ and γ_1: $[0, 1] \to c_1$ constant curves. If H: $[0, 1] \times [0, 1] \to U$ is the loophomotopy, then $H(\tau, 0)$ $(0 \leq \tau \leq 1)$ is a curve in U from c_0 to c_1. Thus U is connected. If C_0, C_1 are two different bounded components of $\mathbb{C}\backslash U$, choose loops γ_0, γ_1 in U after 8.20 so that

$$\mathrm{Ind}_{\gamma_0}(C_0)\,\mathrm{Ind}_{\gamma_1}(C_1) \neq 0 \;\&\; \mathrm{Ind}_{\gamma_0}(C_1) = \mathrm{Ind}_{\gamma_1}(C_0) = 0.$$

It follows easily from 4.14 and 4.12 that (*) fails for these loops.

§ 4 Some Special Kinds of Approximation

Exercise 8.23 *Let F be continuous on $\bar{D}$, holomorphic in $D = D(0, 1)$ and $|F| \leq 1$. Then there exists a sequence of functions E_n such that*

(i) *E_n is continuous on $\bar{D}$ and holomorphic in D,*

(ii) *$|E_n| = 1$ on $C = C(0, 1)$* (cf. 6.12),

(iii) *$E_n \to F$ locally uniformly on D.*

Hints: Let us call any function which has properties (i) and (ii) a *unit function.* (We know from 6.12 that these are precisely the finite Blaschke products.) Take as inductive hypothesis

$$(H_n) \quad \begin{cases} \text{If } f \text{ is continuous on } \bar{D}, \text{ holomorphic in } D \text{ and} \\ |f| \leq 1, \text{ then there is a unit function } E \text{ such that} \\ f - E \text{ has a zero at 0 of order at least } n. \end{cases}$$

If $|f(0)| = 1$, then we may take $E = f(0)$ and if $|f(0)| < 1$, we may take $E(z) = (f(0) - z)/(1 - \bar{f}(0)z)$, in order to establish that (H_1) holds. If (H_n) holds and f is given which is continuous on $\bar{D}$, holomorphic in D and $|f| < 1$ in D, then the function

$$g(z) = \frac{1}{z}\frac{f(0) - f(z)}{1 - \bar{f}(0)f(z)}$$

is also continuous on $\bar{D}$, holomorphic in D and $|g| \leq 1$. Apply (H_n) to g and come up with a unit function g^* such that $g^* - g$ has a zero at 0 of order at least n. Evidently the unit functions are a semigroup under composition so

$$f^*(z) = \frac{f(0) - zg^*(z)}{1 - \bar{f}(0)zg^*(z)}$$

is a unit function, and according to 6.10 for it we know that $f^* - f$ has a zero at 0 of order at least $n + 1$. With (H_n) in hand for all n, we can cite 7.9 to get (iii).

Exercise 8.24 (Fisher [1968]) *Let f be continuous on $\bar{D}$, holomorphic in $D = D(0, 1)$ and $|f| \leq 1$. Then f is a uniform limit on $\bar{D}$ of convex combinations of unit functions.*

Hints: f is uniformly continuous on $\bar{D}$ so, setting $f_r(z) = f(rz)$ $(z \in \bar{D}, r \in [0, 1])$, we have that $\lim_{r\uparrow 1} f_r = f$ uniformly on $\bar{D}$. Given $\varepsilon > 0$, choose $r < 1$ so that $|f - f_r| < \varepsilon/2$ in $\bar{D}$ and then use the last exercise to choose a unit function E such that $|f - E| < \varepsilon/2$ in $\bar{D}(0, r)$. It follows that $|f_r - E_r| < \varepsilon/2$, whence $|f - E_r| < \varepsilon$, in $\bar{D}$. Therefore, it suffices that we show E_r is a convex combination of unit functions. As $(gh)_r = g_r h_r$ for any functions g, h on $\bar{D}$ and as the convex combinations of unit functions are clearly a multiplicative semigroup, it suffices in view of 6.12 to restrict attention to E of the form $E(z) = (z - a)/(1 - \bar{a}z)$, $|a| < 1$. Then we have for all $z \in \bar{D}$

$$E_r(z) = \frac{r(1 - |a|^2)}{1 - |a|^2 r^2}\left[\frac{z - ar}{1 - \bar{a}rz}\right] + \frac{|a|(1 - r^2)}{1 - |a|^2 r^2}\left[\frac{-a}{|a|}\right]$$
$$+ \frac{(1 - r)(1 - |a|)}{2(1 + |a|r)}[1] + \frac{(1 - r)(1 - |a|)}{2(1 + |a|r)}[-1].$$

On the right side the square bracketed terms are unit functions and their coefficients are non-negative and add up to 1.

Exercise 8.25 *By iterating the approximation of the last exercise, show that for each function f which is continuous on $\bar{D}$, holomorphic in $D = D(0, 1)$ and satisfies $\sup|f(D)| < 1$, there exist unit functions E_n and positive numbers c_n such that $\sum_{n=1}^{\infty} c_n = 1$ and $\sum_{n=1}^{\infty} c_n E_n = f$.*

The next exercise will be useful later on. It represents Weierstrass' own *Ansatz* on his famous polynomial approximation theorem. Later we will examine other derivations. The proof here is another "convolution approximate identity" argument similar to that in the solution of the Dirichlet problem for a disk. The actual numerical value of the integral in part (i) is clearly of no relevance. The literature abounds with elementary computations of it (see, e.g., KRAFFT [1955], [1956] or pp. 328–329 of SPIVAK [1967]) and we will deduce its value as a by-product of some residue calculations in Chapter XI. It is of special significance in probability theory.

Exercise 8.26 (i) *Prove that* $\int_{-\infty}^{\infty} e^{-x^2}dx = \lim_{R\to\infty}\int_{-R}^{R} e^{-x^2}dx$ *exists. Calling it* I, *show that* $0 < I < 1 + 2e^{-1/4}$.

Hint: For $x \geq \frac{1}{2}$, $e^{-x^2} \leq 2xe^{-x^2} = \dfrac{d}{dx}[-e^{-x^2}]$. Therefore for $R \geq \frac{1}{2}$

$$\phi(R) = \int_{-R}^{R} e^{-x^2}dx$$

$$= \int_{-1/2}^{1/2} e^{-x^2}dx + 2\int_{1/2}^{R} e^{-x^2}dx < 1 + 2[-e^{-x^2}]_{1/2}^{R} < 1 + 2e^{-1/4}.$$

The function ϕ is therefore bounded. As it is evidently increasing, the desired limit exists.

(ii) *Let* $a, b \in \mathbb{R}$, $a < b$ *and* $f\colon [a, b] \to \mathbb{C}$ *a continuous function. For each positive integer* k *form*

$$f_k(x) = \frac{k}{I}\int_a^b e^{-k^2(x-t)^2}f(t)dt, \quad x \in \mathbb{R}.$$

Show that for each $\varepsilon > 0$

$$\lim_{k\to\infty} f_k(x) = \begin{cases} f(x) & \textit{uniformly for } x \in [a + \varepsilon, b - \varepsilon] \\ 0 & \textit{uniformly for } x \in \mathbb{R}\backslash[a - \varepsilon, b + \varepsilon]. \end{cases}$$

Hints: Fix $\varepsilon > 0$. For $x \in \mathbb{R}\backslash[a - \varepsilon, b + \varepsilon]$ and $t \in [a, b]$ we have $(x - t)^2 \geq \varepsilon^2$ so, upon setting $M = \sup|f[a, b]|$,

$$|f_k(x)| \leq \frac{k}{I}\int_a^b Me^{-k^2\varepsilon^2}dt = \frac{(b - a)M}{I} ke^{-k^2\varepsilon^2}$$

$$< \frac{(b - a)M}{I} k\frac{1}{k^2\varepsilon^2} = \frac{(b - a)M}{\varepsilon^2 Ik}$$

(since $e^{k^2\varepsilon^2} > k^2\varepsilon^2$).

Now extend f to $\mathbb{R}$ by making $f(-\infty, a - 1] = f[b + 1, \infty) = 0$ and f linear on $[a - 1, a]$ and on $[b, b + 1]$. Note that $\sup|f(\mathbb{R})|$ is still M. Changing the variable of integration, we see that

$$f_k(x) = \frac{1}{I}\int_{k(a-x)}^{k(b-x)} e^{-\tau^2}f\left(x + \frac{\tau}{k}\right)d\tau.$$

Remember that

$$(1)\qquad \int_{-\infty}^{\infty} e^{-\tau^2} d\tau = I.$$

Next write

$$(2)\qquad f_k(x) - f(x) = \frac{1}{I}\int_{\sqrt{k}(a-x)}^{\sqrt{k}(b-x)} e^{-\tau^2}\left[f\left(x+\frac{\tau}{k}\right) - f(x)\right] d\tau$$

$$+ \frac{1}{I}\int_{k(a-x)}^{\sqrt{k}(a-x)} e^{-\tau^2}\left[f\left(x+\frac{\tau}{k}\right) - f(x)\right] d\tau$$

$$+ \frac{1}{I}\int_{\sqrt{k}(b-x)}^{k(b-x)} e^{-\tau^2}\left[f\left(x+\frac{\tau}{k}\right) - f(x)\right] d\tau$$

$$+ f(x)\left[\frac{1}{I}\int_{k(a-x)}^{k(b-x)} e^{-\tau^2} d\tau - 1\right].$$

Since f is continuous and compactly supported, it is uniformly continuous and so, given $\delta > 0$, there exists k_δ such that

$$(3)\qquad \left|f\left(x+\frac{\tau}{k}\right) - f(x)\right| \le \tfrac{1}{2}\delta \quad \forall k \ge k_\delta, x \in [a, b], \tau \in [\sqrt{k}(a-x), \sqrt{k}(b-x)].$$

In view of (1), we may also require of k_δ that

$$(4)\qquad 1 - \frac{1}{I}\int_{-\varepsilon\sqrt{k}}^{\varepsilon\sqrt{k}} e^{-\tau^2} d\tau < \frac{1}{8M}\delta \quad \forall k \ge k_\delta.$$

For $x \in [a, b]$ and $k \ge k_\delta$ we have from (2) and (3)

$$|f_k(x) - f(x)| \le \frac{1}{I}\int_{\sqrt{k}(a-x)}^{\sqrt{k}(b-x)} \frac{\delta}{2} e^{-\tau^2} d\tau$$

$$+ \left[\frac{1}{I}\int_{k(a-x)}^{\sqrt{k}(a-x)} 2Me^{-\tau^2} d\tau + \frac{1}{I}\int_{\sqrt{k}(b-x)}^{k(b-x)} 2Me^{-\tau^2} d\tau\right]$$

$$+ M\left[\frac{1}{I}\int_{k(a-x)}^{k(b-x)} e^{-\tau^2} d\tau - 1\right]$$

$$\overset{(1)}{\le} \frac{\delta}{2} + \left[2M\frac{1}{I}\int_{-\infty}^{\sqrt{k}(a-x)} e^{-\tau^2} d\tau + \frac{1}{I}\int_{\sqrt{k}(b-x)}^{\infty} e^{-\tau^2} d\tau\right]$$

$$+ M\left[\frac{1}{I}\int_{k(a-x)}^{k(b-x)} e^{-\tau^2} d\tau - 1\right]$$

$$\overset{(4)}{\le} \frac{\delta}{2} + 2M\left[\frac{1}{I}\int_{-\infty}^{-\varepsilon\sqrt{k}} e^{-\tau^2} d\tau + \frac{1}{I}\int_{\varepsilon\sqrt{k}}^{\infty} e^{-\tau^2} d\tau\right] + \frac{\delta}{8},$$

$$\text{for } x \in [a + \varepsilon, b - \varepsilon],$$

$$\overset{(1)}{=} \frac{\delta}{2} + 2M\left[1 - \frac{1}{I}\int_{-\varepsilon\sqrt{k}}^{\varepsilon\sqrt{k}} e^{-\tau^2} d\tau\right] + \frac{\delta}{8}$$

$$\overset{(4)}{\leq} \frac{\delta}{2} + \frac{\delta}{4} + \frac{\delta}{8} < \delta, \quad x \in [a + \varepsilon, b - \varepsilon].$$

This holds for all $k \geq k_\delta$.

(iii) (Weierstrass [1885]) *If $A, B \in \mathbb{R}$, $A < B$ and $f: [A, B] \to \mathbb{C}$ is continuous, then for each $\varepsilon > 0$ there is a polynomial p such that $|f - p| < \varepsilon$ on $[A, B]$.*

Hints: Let $a = A - 1$, $b = B + 1$ and extend f continuously to $[a, b]$ (as was done in (ii)). Then pick k so that $|f - f_k| < \varepsilon/2$ on $[A, B]$. Finally choose N so large that

$$(*) \quad \left| f_k(x) - \int_a^b \frac{k}{I} \sum_{n=0}^{N} \frac{[-k^2(x - t)^2]^n}{n!} f(t)dt \right|$$

$$\leq \frac{k}{I} \sup|f[a, b]| \cdot \sup_{(x,t)\in[a,b]\times[a,b]} \left| e^{-k^2(x-t)^2} - \sum_{n=0}^{N} \frac{[-k^2(x - t)^2]^n}{n!} \right| < \frac{\varepsilon}{2}.$$

This is possible since the series converges to the exponential uniformly on compact subsets of $\mathbb{R}$. The integral on the left of (*) is a linear combination of terms of the form $x^i \int_a^b t^j f(t)dt$, which one sees by multiplying out the various terms $(x - t)^{2n}$.

(iv) *Formulate and prove the two-dimensional version of* (ii).

(v) *Formulate and prove the two-dimensional analog of* (iii). (*Approximation will be by linear combinations of $x^k y^j$, $k, j \geq 0$.*)

Hints: From (iv) just as (iii) was got from (ii). Alternatively, one can proceed independently of (iv). Let $F: [A, B] \times [A, B] \to \mathbb{C}$ be continuous and extend to $[a, b] \times [a, b]$ as in (iii). Form

$$F_k(x, y) = \frac{k}{I} \int_a^b e^{-k^2(x-t)^2} F(t, y)dt.$$

Examine the proof of (ii) to see that, because F is uniformly continuous on $[a, b] \times [a, b]$, all convergence there is uniform in the parameter y. Apply the procedure of (iii) to F_k to come up with a polynomial P such that

$$(*) \quad \left| F_k(x, y) - \frac{k}{I} \int_a^b P(x, t)F(t, y)dt \right| < \varepsilon \quad \forall (x, y) \in [A, B] \times [A, B].$$

Each integral

$$f(y) = \int_a^b t^n F(t, y)dt$$

involved in (*) is uniformly approximable by polynomials in y, by (iii).

(vi) *Using* (v) *prove Fubini's theorem in the form*

$$\int_a^b \int_c^d f(x, y)dydx = \int_c^d \int_a^b f(x, y)dxdy$$

whenever $a, b, c, d \in \mathbb{R}$ *and* $f: [a, b] \times [c, d] \to \mathbb{C}$ *is continuous.*

Hints: The claim is evidently true when f is a linear combination of multinomials $x^k y^m$ (k, m non-negative integers) and if f_n converges to f uniformly on $[a, b] \times [c, d]$, each iterated integral of f_n converges to the corresponding integral of f.

§ 5 Carleman's Approximation Theorem

Carleman [1927] discovered the remarkable fact that a continuous function on $\mathbb{R}$ can be approximated by an entire function *uniformly on* $\mathbb{R}$. This generalizes the Weierstrass theorem 8.26(iii), since on any *compact* portion of $\mathbb{R}$ the entire function can in turn be approximated uniformly by polynomials (partial sums of its power series). The proof given below is based on Hoischen [1967] and Scheinberg [1976]. Carleman's original proof is also quite beautiful and the reader is urged to look at it. It uses an integral representation like that used to prove the Runge theorems. Unfortunately the Runge theorem itself is not strong enough because the compacta on which approximation is to take place touch the boundary. This necessitates some delicate maneuvering in Carleman's proof. The techniques below circumvent this problem.

Exercise 8.27 (Poincaré [1892]) *Let* $f: \mathbb{R} \to \mathbb{C}$ *be continuous. Construct a zero-free entire function g which is real-valued on* $\mathbb{R}$ *and satisfies*

$$g(x) > |f(x)| \quad \forall x \in \mathbb{R}.$$

Hints: For each positive integer n let

(1) $M_n = \max\{|f(x)| : |x| \leq n + 1\}$

and choose integer k_n to satisfy both

(2) $k_n \geq n$

and

(3) $\left[\dfrac{n^2}{n + 1}\right]^{k_n} > M_n.$

If $|z| \leq N$, then $|z^2/(n + 1)| < \frac{1}{2}$ for all $n \geq 2N^2$, so from (2) we see that the series

(4) $M_0 + \displaystyle\sum_{n=1}^{\infty} \left[\frac{z^2}{n + 1}\right]^{k_n}$

converges uniformly for z in $\bar{D}(0, N)$. This is the case for every N, so (4) defines an entire function h. Evidently h is positive on $\mathbb{R}$ and for $|x| < 1$, $h(x) \geq M_0 \geq |f(x)|$; while for $1 \leq n \leq |x| < n + 1$, we have

$$h(x) \geq \left[\frac{x^2}{n+1}\right]^{k_n} \geq \left[\frac{n^2}{n+1}\right]^{k_n} > M_n \geq |f(x)|.$$

Therefore $h(x) \geq |f(x)|$ for all real x. It remains only to set $g = e^h$ to get a zero-free such majorant.

Exercise 8.28 (i) *Let $f\colon \mathbb{R} \to \mathbb{C}$ be continuous. Then for each integer n there exists a continuous function $f_n\colon \mathbb{R} \to \mathbb{C}$ which is supported in $[-1, 1]$ and*

$$f(x) = \sum_{n=-\infty}^{\infty} f_n(x - n) \quad \forall x \in \mathbb{R}.$$

Hints: Let ϕ be the function which is 0 outside $(-1, 1)$, is 1 on $[-\frac{1}{2}, \frac{1}{2}]$ and is piecewise linear. Then set

$$\Phi(x) = \sum_{n=-\infty}^{\infty} \phi(x - n) \quad \forall x \in \mathbb{R}.$$

We have

$$\Phi(x) = \sum_{n=-N-1}^{N+1} \phi(x - n) \quad \forall x \in [-N, N],$$

so Φ is continuous at every point. Also, given t, we can select an integer n so that $n - \frac{1}{2} \leq t \leq n + \frac{1}{2}$. Then $\phi(t - n) = 1$, so $\Phi(t) \geq 1$. Finally set

$$f_n(x) = \frac{\phi(x)f(x + n)}{\Phi(x + n)}, \quad x \in \mathbb{R}.$$

(ii) *Let $f\colon \mathbb{R} \to \mathbb{C}$ be continuous and have support in $[-1, 1]$. Set $S = \{z \in \mathbb{C}\colon |\text{Re } z| > 3 \;\&\; |\text{Re } z| > 2|\text{Im } z|\}$. Given $\varepsilon > 0$, there exists an entire function F such that*

(1) $|f(x) - F(x)| < \varepsilon \quad \forall x \in \mathbb{R}$

(2) $|F(z)| < \varepsilon \quad \forall z \in S.$

Hints: For each positive integer k form

$$f_k(z) = \frac{k}{I}\int_{-1}^{1} e^{-k^2(z-t)^2} f(t)\,dt, \quad z \in \mathbb{C}, I = \int_{-\infty}^{\infty} e^{-x^2}dx.$$

Upon expanding the exponential here into a series and interchanging that sum with the integral, we have a power series for f_k. Thus f_k is entire. By 8.26(ii) with $a = -2$, $b = 2$ we have $f_k \to f$ uniformly in $[-\frac{3}{2}, \frac{3}{2}]$ and with $a = -1$, $b = 1$ we have $f_k \to 0$ uniformly in $\mathbb{R}\backslash[-\frac{3}{2}, \frac{3}{2}]$. Thus

(3) $f_k \to f$ uniformly on $\mathbb{R}$.

For $z = x + iy \in S$ and $t \in [-1, 1]$ we have

$$\begin{aligned}\operatorname{Re} k^2(z - t)^2 &= k^2((x - t)^2 - y^2)\\ &= k^2x^2\left(1 - \frac{2t}{x} + \frac{t^2}{x^2} - \frac{y^2}{x^2}\right)\\ &\geq k^2x^2\left(1 - \frac{2}{|x|} - \left|\frac{y}{x}\right|^2\right)\\ &\geq k^2x^2(1 - \tfrac{2}{3} - (\tfrac{1}{2})^2) = \frac{k^2x^2}{12} > \frac{3k^2}{4}\end{aligned}$$

and so

$$\begin{aligned}|f_k(z)| &\leq \frac{k}{I}\int_{-1}^{1} |e^{-k^2(z-t)^2}f(t)|\,dt = \frac{k}{I}\int_{-1}^{1} e^{-\operatorname{Re} k^2(z-t)^2}|f(t)|\,dt\\ &\leq \frac{k}{I}e^{-3k^2/4}\int_{-1}^{1} |f(t)|\,dt.\end{aligned}$$

Since $e^{3k^2/4} > 3k^2/4$, this gives

$$(4) \qquad |f_k(z)| \leq \frac{k}{I}\cdot\frac{4}{3k^2}\cdot\int_{-1}^{1} |f| = \frac{1}{k}\cdot\frac{4}{3I}\cdot\int_{-1}^{1} |f| \quad \forall z \in S.$$

We get (1) and (2) from (3) and (4) by taking $F = f_k$ for sufficiently large k.

(iii) *Let $f\colon \mathbb{R} \to \mathbb{C}$ be continuous. Then there exists an entire function F such that*

$$|f(x) - F(x)| < 1 \quad \forall x \in \mathbb{R}.$$

Hints: Let f_n be as furnished by (i). For each n, part (ii) provides an entire function F_n such that

$$(1.n) \quad |f_n(t) - F_n(t)| < 2^{-|n|-2} \quad \forall t \in \mathbb{R}$$

$$(2.n) \quad |F_n(z)| < 2^{-|n|} \quad \forall z \in S.$$

If $|z| \leq N$ and $|n| > 3N + 3$, then

$$|\operatorname{Re}(z - n)| \geq |n| - |\operatorname{Re} z| \geq |n| - N > 3$$

and

$$|\operatorname{Im}(z - n)| = |\operatorname{Im} z| \leq N \leq \tfrac{1}{2}(|n| - N) < \tfrac{1}{2}|\operatorname{Re}(z - n)|.$$

It follows that $z - n \in S$, and so from (2.n)

$$|F_n(z - n)| < 2^{-|n|}, \quad \text{if } |z| \leq N \;\&\; |n| > 3N + 3.$$

Therefore the series $\sum_{n=-\infty}^{\infty} F_n(z - n)$ converges uniformly for $z \in \bar{D}(0, N)$, for each positive integer N, and consequently defines an entire function F. For this F and all real x we have

$$\begin{aligned}|F(x) - f(x)| &= \left|\sum_{n=-\infty}^{\infty} F_n(x-n) - \sum_{n=-\infty}^{\infty} f_n(x-n)\right| \\ &\leq \sum_{n=-\infty}^{\infty} |F_n(x-n) - f_n(x-n)| \\ &< \sum_{n=-\infty}^{\infty} 2^{-|n|-2} \quad \text{by (1.n)} \\ &= \tfrac{3}{4}.\end{aligned}$$

Theorem 8.29 (Carleman [1927]) *Let* $f\colon \mathbb{R} \to \mathbb{C}$, $\varepsilon\colon \mathbb{R} \to (0, \infty)$ *be continuous. Then there exists an entire function g such that*

$$|f(x) - g(x)| < \varepsilon(x) \quad \forall x \in \mathbb{R}.$$

Proof: 8.27 provides a zero-free entire function h such that

$$h(x) > \frac{1}{\varepsilon(x)} \quad \forall x \in \mathbb{R}$$

and the last exercise provides an entire function F such that

$$|(hf)(x) - F(x)| < 1 \quad \forall x \in \mathbb{R}.$$

Consequently

$$\left|f(x) - \frac{F(x)}{h(x)}\right| < \frac{1}{h(x)} < \varepsilon(x) \quad \forall x \in \mathbb{R}$$

and we take for g the entire function F/h.

§ 6 Harmonic Functions in a Half-plane

Corollary 8.30 (Dirichlet Problem in a Half-plane) *If* $g\colon \mathbb{R} \to \mathbb{R}$ *is continuous, then there exists f holomorphic in* $U = \{z \in \mathbb{C}\colon \operatorname{Im} z > 0\}$ *such that the function G defined by*

$$G(z) = \begin{cases} \operatorname{Re} f(z), & z \in U \\ g(z) & z \in \mathbb{R} \end{cases}$$

is continuous on $\bar{U} = U \cup \mathbb{R}$.

Proof: First set $\varepsilon(x) = 1/(1 + |x|)$ and apply Carleman's theorem to come up with an entire function G_1 such that

$$|G_1(x) - g(x)| < \varepsilon(x) \quad \forall x \in \mathbb{R}.$$

Therefore, by replacing g with the function $g_1(x) = [\operatorname{Re} G_1 - g](x)$, it suffices to suppose in addition that the given function g satisfies

(*) $$\lim_{|x| \to \infty} g(x) = 0.$$

The function

$$\phi(z) = i\frac{1+z}{1-z}$$

maps $\bar{D}(0, 1)\backslash\{1\}$ onto $\bar{U}$, $C(0, 1)\backslash\{1\}$ onto $\mathbb{R}$ and has inverse

$$\psi(w) = \frac{w-i}{w+i}.$$

Since $\lim_{z\to 1}|\phi(z)| = \infty$, the continuous function $g \circ \phi$ on $C(0, 1)\backslash\{1\}$ has by (*) an extension to a continuous function h on $C(0, 1)$. Solve the Dirichlet Problem in the Disk for h [5.20] and then compose this solution with ψ.

Remark: Using ϕ, ψ above we can carry the Poisson formula for $D(0, 1)$ into one for U (see pp. 121–123 of HOFFMAN [1962] for details), motivating and providing an alternative development of

Exercise 8.31 (Poisson Formula for a Half-plane) *Let f: $\mathbb{R} \to \mathbb{R}$ be continuous and bounded, $U = \mathbb{R} \times (0, \infty)$. For each positive integer n define*

$$F_n(z) = \frac{1}{\pi}\int_{-n}^{n} \frac{\operatorname{Im} z}{|z-t|^2} f(t)dt, \quad z \in U.$$

(i) *Show that each F_n is harmonic in U.*

Hint: $\dfrac{\operatorname{Im} z}{|z-t|^2} = \operatorname{Re}\left(\dfrac{i}{z-t}\right)$.

(ii) *Show that for all $x \in \mathbb{R}$, $y \in (0, \infty)$*

$$\frac{1}{\pi}\int_{-\infty}^{\infty} \frac{y}{(x-t)^2 + y^2}\, dt = 1.$$

Hint: Make appropriate changes of variable in 3.22.

(iii) *Show that* $\lim_{n\to\infty} F_n$ *exists uniformly on each compact subset of U.*

Hint: Note that $|z - t|^2 = (\operatorname{Re} z - t)^2 + (\operatorname{Im} z)^2$ and use (ii).

(iv) *Letting F denote the limit in* (iii), *show that F is harmonic in U.*

Hint: 5.44(iii).

(v) *Show that* $\lim_{z\to x, z\in U} F(z) = f(x)$ *for each $x \in \mathbb{R}$, and so F is (extendable to be) continuous on $\bar{U}$.*

Hint: Modify the techniques used in the proof of 5.20, using now (ii). (See the hints to 8.32(vi) below.)

(vi) *Show that F is bounded: in fact* $\sup_{z\in U}|F(z)| = \sup_{x\in\mathbb{R}}|f(x)|$.

Hint: Use (ii) again.

(vii) *Let g: $\bar{U} \to \mathbb{R}$ be bounded and continuous and be harmonic in U. Form*

(1) $$G(z) = \frac{1}{\pi}\int_{-\infty}^{\infty} \frac{\operatorname{Im} z}{|z-t|^2} g(t)dt = \lim_{n\to\infty} \frac{1}{\pi}\int_{-n}^{n} \frac{\operatorname{Im} z}{|z-t|^2} g(t)dt, \quad z \in U$$

and for convenience set $G = g$ on $\mathbb{R}$. Let S denote the holomorphic square root in $\mathbb{C}\backslash(-\infty, 0]$ which is positive on the positive real-axis. (*See* 3.43.) *Extend S continuously to $\mathbb{C}\backslash(-\infty, 0)$ by setting $S(0) = 0$. Show that for each $\varepsilon > 0$ the function*

(2) $$F_\varepsilon(z) = G(z) - g(z) - \varepsilon \operatorname{Re} S(-iz), \quad z \in \bar{U}$$

satisfies

(3) $$\lim_{\substack{|z| \to \infty \\ z \in \bar{U}}} F_\varepsilon(z) = -\infty.$$

Hints: If $z \in \bar{U}$ is written $z = re^{i\theta}$, where $0 \le \theta \le \pi$ and $r \ge 0$, then $S(-iz) = S(re^{i(\theta - \pi/2)}) = \sqrt{r}e^{i(\theta/2 - \pi/4)}$. Consequently $\operatorname{Re} S(-iz) = \sqrt{r}\cos(\theta/2 - \pi/4) \ge \sqrt{r}\cos(\pi/2 - \pi/4) = \sqrt{r/2} = \sqrt{|z|/2}$. By (vi) $G - g$ is bounded and so (3) follows.

(viii) *Show that $F_\varepsilon \le 0$ throughout U.*

Hints: If $F_\varepsilon(z) > 0$ for some $z \in U$, use (3) to select $R > 0$ so large that

$$|F_\varepsilon| < F_\varepsilon(z) \quad \text{on } [-R, R] \times \{R\} \cup \{\pm R\} \times [0, R].$$

Since $G = g$ on $\mathbb{R}$ and $\operatorname{Re} S(-iw) \ge 0$ for all $w \in \bar{U}$, it follows that

$$F_\varepsilon < F_\varepsilon(z) \quad \text{on the boundary of } [-R, R] \times [0, R],$$

in violation of the Maximum Principle.

(ix) *Show that $G - g \le 0$ in U.*

Hint: For each fixed $z \in U$, let $\varepsilon \downarrow 0$ in the inequality $F_\varepsilon(z) \le 0$.

(x) *Show that $G = g$ in U.*

Hint: Apply (ix) with $-g$ in the role of g to secure the opposite inequality.

(xi) *Explore integrability hypotheses on the continuous function f, weaker than the boundedness assumption above, under which all the conclusions* (i)–(v) *remain valid. One such is $\int_{-\infty}^{\infty} |f(t)|/(1 + |t|)dt < \infty$.* (For other conditions see, e.g., § 2 chapter V of B. Levin [1964].) We will see in the next exercise that a sufficient condition for (i)–(v) to hold is that f be the restriction of a non-negative harmonic function in the upper half-plane.

Exercise 8.32 (i) *Let $f: \mathbb{R} \to \mathbb{R}$ be bounded and continuous save at the points of a finite set S. Show that* (i)–(vi) *of* 8.31 *remain valid if in* (v) *only $x \in \mathbb{R}\backslash S$ are considered and F is extended only to $\bar{U}\backslash S$, while in* (vi) *only less-than-or-equal is asserted.*

(ii) *Let $g: \bar{U} \to [0, \infty)$ be continuous and harmonic in U. Show that for positive a and b the function*

(*) $$G_{a,b}(z) = \frac{1}{\pi}\int_{-a}^{a} \frac{\operatorname{Im} z}{|z - t|^2} g(t + ib)dt, \quad z \in U,$$

satisfies

(1) $G_{a,b}$ *is non-negative and harmonic in* U,

(2) $$\lim_{\substack{z \to x \\ z \in U}} G_{a,b}(z) = \begin{cases} g(x + ib), & x \in (-a, a) \\ 0 & x \in \mathbb{R} \setminus [-a, a], \end{cases}$$

(3) $$\lim_{\substack{|z| \to \infty \\ z \in U}} G_{a,b}(z) = 0.$$

Hints: (1) and (2) are from (i) with $f(t) = g(t + ib)$ for $t \in [-a, a]$ and $f(t) = 0$ for $t \in \mathbb{R} \setminus [-a, a]$. For the proof of (3) consider $z \in U$ with $|z| > 2\sqrt{2}a$. Then either $\operatorname{Im} z \geq |z|/\sqrt{2}$ or $|\operatorname{Re} z| \geq |z|/\sqrt{2}$. In the first case (*) shows that

$$G_{a,b}(z) \leq \frac{1}{\pi} \int_{-a}^{a} \frac{\operatorname{Im} z}{(\operatorname{Im} z)^2} g(t + ib)dt = \frac{1}{\pi \operatorname{Im} z} \int_{-a}^{a} g(t + ib)dt$$
$$\leq \frac{1}{|z|} \frac{\sqrt{2}}{\pi} \int_{-a}^{a} g(t + ib)dt$$

and in the second case

$$G_{a,b}(z) \leq \frac{1}{\pi} \int_{-a}^{a} \frac{|z|}{(\operatorname{Re} z - t)^2} g(t + ib)dt$$
$$\leq \frac{1}{\pi} \frac{|z|}{(|\operatorname{Re} z| - a)^2} \int_{-a}^{a} g(t + ib)dt$$
$$\leq \frac{1}{\pi} \frac{|z|}{\left(\frac{|z|}{\sqrt{2}} - a\right)^2} \int_{-a}^{a} g(t + ib)dt$$
$$\leq \frac{1}{\pi} \frac{|z|}{\left(\frac{|z|}{\sqrt{2}} - \frac{|z|}{2\sqrt{2}}\right)^2} \int_{-a}^{a} g(t + ib)dt$$
$$= \frac{1}{|z|} \frac{8}{\pi} \int_{-a}^{a} g(t + ib)dt.$$

(iii) *With* g *and* $G_{a,b}$ *as in* (ii) *show that*

$$G_{a,b}(z) \leq g(z + ib) \quad \forall z \in U.$$

Hints: Let $z_0 \in U$ and $\varepsilon > 0$ be given. Use (3) to select $r > |z_0|$ such that $G_{a,b} \leq \varepsilon$ on $C(0, r) \cap U$. From this, the non-negativity of g and (2) above we have

$$\varlimsup_{\substack{z \to \zeta \\ z \in U}} [G_{a,b}(z) - g(z + ib)] \leq \varepsilon \quad \forall \zeta \in \partial[D(0, r) \cap U] \setminus \{\pm a\}.$$

Infer via 5.46 that

$$G_{a,b}(z) - g(z + ib) \leq \varepsilon \quad \forall z \in D(0, r) \cap U.$$

Since $z_0 \in D(0, r) \cap U$, this yields

$$G_{a,b}(z_0) \leq g(z_0 + ib) + \varepsilon$$

and as $\varepsilon > 0$ is arbitrary here, the desired conclusion follows.

(iv) *Show that*

$$G_a(z) \stackrel{\text{def.}}{=} \frac{1}{\pi}\int_{-a}^{a} \frac{\operatorname{Im} z}{|z-t|^2} g(t)dt \le g(z) \quad \forall z \in U, a > 0.$$

Hint: Let $b \downarrow 0$ in (iii).

(v) *Show that*

$$G(z) = \lim_{a\to\infty} \frac{1}{\pi}\int_{-a}^{a} \frac{\operatorname{Im} z}{|z-t|^2} g(t)dt = \frac{1}{\pi}\int_{-\infty}^{\infty} \frac{\operatorname{Im} z}{|z-t|^2} g(t)dt$$

exists for each $z \in U$ and defines a harmonic function in U.

Hint: The integrand being non-negative, the integral increases with a. Cite 7.13 and (iv) above.

(vi) *Show that*

$$\lim_{\substack{z\to x\\ z\in U}} G(z) = g(x) \quad \forall x \in \mathbb{R}.$$

Hints: Let $x \in \mathbb{R}$ and $\varepsilon > 0$ be given. Set $g^* = g + g(x)$. This is a non-negative harmonic function in U for which the conclusions (ii)–(v) above hold. Choose $\delta > 0$ so that $|g(t) - g(x)| < \varepsilon$ whenever $t \in \mathbb{R}$ and $|t - x| \le 8\delta$. Then, using 8.31(ii), we have for $z \in U$

$$\begin{aligned}
|G(z) - g(x)| &= \frac{1}{\pi}\left|\int_{-\infty}^{\infty} \frac{\operatorname{Im} z}{|z-t|^2}[g(t) - g(x)]dt\right| \\
&\le \frac{1}{\pi}\int_{x-8\delta}^{x+8\delta} \frac{\operatorname{Im} z}{|z-t|^2} |g(t) - g(x)|\, dt \\
&\quad + \frac{1}{\pi}\int_{|t-x|\ge 8\delta} \frac{\operatorname{Im} z}{|z-t|^2} g^*(t)dt \\
&\le \frac{\varepsilon}{\pi}\int_{x-8\delta}^{x+8\delta} \frac{\operatorname{Im} z}{|z-t|^2}dt + \frac{\operatorname{Im} z}{\pi}\int_{|t-x|\ge 8\delta} \frac{g^*(t)}{|z-t|^2}dt
\end{aligned}$$

(5) $$\le \varepsilon + \frac{\operatorname{Im} z}{\pi}\int_{|t-x|\ge 8\delta} \frac{g^*(t)}{|z-t|^2}dt, \quad \text{by 8.31(ii)}, \quad \forall z \in U.$$

Now consider $z \in D(x, \delta) \cap U$ and $t \in \mathbb{R}\backslash[x - 8\delta, x + 8\delta]$. We have

(6) $$|x + 2i\delta - t| \ge |x - t| - 2\delta \ge 6\delta$$

so

$$\begin{aligned}
|z - t| &\ge |x + 2i\delta - t| - |z - x - 2i\delta| \\
&\ge |x + 2i\delta - t| - |z - x| - 2\delta \\
&\ge |x + 2i\delta - t| - 3\delta
\end{aligned}$$

(7) $$\ge \tfrac{1}{2}|x + 2i\delta - t|, \quad \text{by (6).}$$

From (7) and (5) we get

$$|G(z) - g(x)| \leq \varepsilon + \frac{\operatorname{Im} z}{\pi} \int_{|t-x| \geq 8\delta} \frac{4g^*(t)}{|x + 2i\delta - t|^2} dt$$

$$\leq \varepsilon + \frac{\operatorname{Im} z}{\delta} \cdot \frac{2}{\pi} \int_{-\infty}^{\infty} \frac{2\delta g^*(t)}{|x + 2i\delta - t|^2} dt$$

$$= \varepsilon + \frac{\operatorname{Im} z}{\delta} 2G^*(x + 2i\delta)$$

$$\leq \varepsilon + \frac{\operatorname{Im} z}{\delta} 2g^*(x + 2i\delta), \quad \text{by (iv) and (v) applied to } g^*.$$

Since x is real, it follows that

$$\overline{\lim_{\substack{z \to x \\ z \in U}}} |G(z) - g(x)| \leq \varepsilon.$$

The next exercise broaches a theme which will be examined further in Chapter XIII, the "reflection principle."

Exercise 8.33 *Let* $a \in \mathbb{R}$, $r > 0$, $D = D(a, r)$, $D^+ = \{z \in D: \operatorname{Im} z > 0\}$. *Let* $h: D^+ \to \mathbb{R}$ *be harmonic and satisfy* $\lim_{z \to x} h(z) = 0$ *for each* $x \in (a - r, a + r)$. *Define*

$$H(z) = \begin{cases} h(z) & z \in D^+ \\ 0 & z \in (a - r, a + r) \\ -h(\bar{z}), & \bar{z} \in D^+ \end{cases}$$

and show that H *is harmonic in* D.

Hints: Evidently H is continuous in D and harmonic in each of the open subsets D^+ and $\{z: \bar{z} \in D^+\}$. Consequently, for any z in either of these sets we have $2\pi H(z) = \int_0^{2\pi} H(z + \rho e^{it})dt$ for all sufficiently small $\rho > 0$. (See 5.21(ii).) On the other hand, if $x \in (a - r, a + r)$ and $\rho > 0$ is small enough that $D(x, \rho) \subset D$, then

$$\int_0^{2\pi} H(x + \rho e^{it})dt = \int_0^{\pi} H(x + \rho e^{it})dt + \int_{\pi}^{2\pi} H(x + \rho e^{i\tau})d\tau$$

$$= \int_0^{\pi} H(x + \rho e^{it})dt - \int_{\pi}^{0} H(x + \rho e^{i(2\pi - t)})dt$$

$$(\tau = 2\pi - t)$$

$$= \int_0^{\pi} [H(x + \rho e^{it}) + H(x + \rho e^{-it})]dt$$

$$= 0, \quad \text{since } H(\bar{z}) = -H(z) \text{ for all } z \in D,$$

$$= H(x).$$

It now follows from 5.23 (i) that H is harmonic in D.

Theorem 8.34 *If* $h\colon U = \{z \in \mathbb{C}\colon \operatorname{Im} z > 0\} \to [0, \infty)$ *is a harmonic function such that* $\lim_{z\to x} h(z) = 0$ *for every real* x, *then* $h(z) \equiv c \operatorname{Im} z$ *for some non-negative constant* c.

Proof: (LOOMIS and WIDDER [1942]) According to the last exercise the definition

$$H(z) = \begin{cases} h(z) & \text{if } \operatorname{Im} z > 0 \\ 0 & \text{if } \operatorname{Im} z = 0 \\ -h(\bar{z}) & \text{if } \operatorname{Im} z < 0 \end{cases}$$

produces a harmonic function in $\mathbb{C}$. As an easy consequence of 5.21(i) there is an entire F such that $H = \operatorname{Im} F$. We have $H(\bar{z}) = -H(z)$ for all z, so $F(z) - \bar{F}(\bar{z})$ is an entire function (2.4) with 0 imaginary part and so it is constant (2.17). This real constant is seen to be 0 by considering $z = 0$. Thus

(1) $\quad F(z) = \bar{F}(\bar{z}) \quad \forall z \in \mathbb{C}.$

If we write the power series

$$F(z) = \sum_{n=0}^{\infty} c_n z^n,$$

then (1) and the uniqueness of the c_n show that $c_n = \bar{c}_n$ for all n, so all c_n are real and we have

$$H(re^{i\theta}) = \operatorname{Im} F(re^{i\theta}) = \sum_{n=0}^{\infty} c_n \operatorname{Im}(re^{i\theta})^n$$

(2)
$$= \sum_{n=0}^{\infty} c_n r^n \sin(n\theta) \quad \forall r > 0,\ \theta \in \mathbb{R}.$$

Using the orthogonality relations

$$\int_0^{\pi} \sin(n\theta)\sin(k\theta)d\theta = \begin{cases} \pi/2 & \text{if } n = k \\ 0 & \text{if } n \neq k \end{cases}$$

and integrating term-by-term the uniformly (for each fixed r) convergent series arising from (2), we get

(3)
$$\int_0^{\pi} h(re^{i\theta})\sin(k\theta)d\theta = \sum_{n=0}^{\infty} c_n r^n \int_0^{\pi} \sin(n\theta)\sin(k\theta)d\theta = \frac{\pi}{2} c_k r^k,$$

$$|c_k| \le \frac{2}{\pi} r^{-k} \int_0^{\pi} |h(re^{i\theta})\sin(k\theta)|d\theta.$$

Since $h \ge 0$, this reads

(4)
$$|c_k| \le \frac{2}{\pi} r^{-k} \int_0^{\pi} h(re^{i\theta})|\sin(k\theta)|d\theta \quad \forall r > 0,\ k \in \mathbb{N}.$$

But an easy induction confirms that

(5) $\quad |\sin(k\theta)| \leq k|\sin\theta| \quad \forall \theta \in \mathbb{R},\ k \in \mathbb{N}.$

Indeed, equality holds in (5) when $k = 1$ and if (5) holds for $k = m$, then

$$\begin{aligned}|\sin(m+1)\theta| &= |\sin(m\theta)\cos\theta + \cos(m\theta)\sin\theta| \\ &\leq |\sin(m\theta)| + |\sin\theta| \leq m|\sin\theta| + |\sin\theta|,\end{aligned}$$

giving (5) for $k = m + 1$. Use of (5) in (4) gives

$$\begin{aligned}|c_k| &\leq \frac{2}{\pi} r^{-k} \int_0^\pi kh(re^{i\theta}) \sin\theta d\theta \\ &= \frac{2}{\pi} r^{-k} k\left(\frac{\pi}{2} c_1 r\right) \quad \text{by (3)} \\ &= \frac{kc_1}{r^{k-1}}.\end{aligned}$$

This inequality holds for all $k \in \mathbb{N}$ and all $r > 0$. For $k \geq 2$ we let $r \to \infty$ and conclude that $c_k = 0$. It follows that

$$F(z) = c_0 + c_1 z$$

with real coefficients c_0 and c_1. Consequently

$$h(z) = \operatorname{Im} F(z) = c_1 \operatorname{Im} z \quad \forall z \in U.$$

Corollary 8.35 *Let $U = \mathbb{R} \times (0, \infty)$ and g a continuous function on $\overline{U}$ which is harmonic and non-negative in U. Then for some $c \geq 0$*

$$g(z) = c \operatorname{Im} z + \frac{1}{\pi} \int_{-\infty}^{\infty} \frac{\operatorname{Im} z}{|z - t|^2} g(t) dt \quad \forall z \in U.$$

Proof: According to 8.32(iv)–(vi) the integral

$$G(z) = \frac{1}{\pi} \int_{-\infty}^{\infty} \frac{\operatorname{Im} z}{|z - t|^2} g(t) dt$$

is finite for every $z \in U$ and defines a harmonic function which satisfies

$$G \leq g \quad \text{in } U,$$

$$\lim_{\substack{z \to x \\ z \in U}} G(z) = g(x) \quad \forall x \in \mathbb{R}.$$

Therefore $h = g - G$ fulfills all the hypotheses of 8.34.

Exercise 8.36 *Let g, G be as* 8.32(ii)–(v). *For each $\varepsilon > 0$ set*

$$S_\varepsilon = \{z \in \mathbb{C} : \operatorname{Im} z > \varepsilon |z|\}.$$

(i) *Show that*

$$\lim_{\substack{|z|\to\infty \\ z\in S_\varepsilon}} \frac{G(z)}{\operatorname{Im} z} = 0 \quad \forall 0 < \varepsilon < 1.$$

(ii) *Use* (i) *and the representation* 8.35 *to prove*, mutatis mutandis, *the special case of the Carathéodory–Julia–Wolff theorem* 6.23 *in which the function is continuous on the closed half-plane. I.e., with the notation and hypotheses of* 8.35 *use* (i) *above to show that for each* $0 < \varepsilon < 1$

$$(*) \qquad \lim_{\substack{|z|\to\infty \\ z\in S_\varepsilon}} \frac{g(z)}{\operatorname{Im} z} = \inf\left\{\frac{g(z)}{\operatorname{Im} z} : \operatorname{Im} z > 0\right\}.$$

(iii) *Conserving all the notation and hypotheses of* (ii) *except the continuity of* g *on the closed half-plane, prove that* (*) *still holds.*

Hint: For each fixed (small) positive b, apply (ii) to the function $g_b(z) = g(z + ib)$.

(iv) *Use the representation* 8.35 *to secure the version of* 5.60 *in which* $\operatorname{Re} f$ *is continuous on the closed half-plane but the hypothesis* 5.60(i) *is weakened to*

$$(**) \qquad \varliminf_{\substack{x\to\infty \\ x\in\mathbb{R}}} \frac{\operatorname{Re} f(x)}{x} = 0.$$

Hint: (**) and the non-negativity of everything in sight insure that the c of 8.35 is 0. Now recall 8.31(vi).

Exercise 8.37 *Let L denote the Principal Branch of the Logarithm in* $\mathbb{C}\backslash(-\infty, 0]$, $U = \{z \in \mathbb{C}: \operatorname{Im} z > 0\}$, $D = D(0, 1)$, $C = C(0, 1)$ *and*

$$\phi(z) = \frac{1 + z}{1 - z}, \quad z \in \mathbb{C}\backslash\{1\}.$$

(i) *Show that* ϕ *maps* $\bar{D}\backslash\{1\}$ *into* $\mathbb{C}\backslash(-\infty, 0]$ *and that the function* $g = \operatorname{Im} L \circ \phi$ *satisfies*

(1) g *is bounded and continuous in* $\bar{D}\backslash\{\pm 1\}$ *and harmonic in* D,

(2) $g(-1, 1) = 0$,

(3) $g(U \cap C) = \pi/2$.

Hint: Examine the proof of 5.47, up to the introduction of the square root transformation used therein.

(ii) *Show that*

$$g(z) = \tan^{-1}\left(\frac{2 \operatorname{Im} z}{1 - |z|^2}\right) \quad \forall z \in D.$$

Hints: If $-\pi/2 < \theta < \pi/2$, $r > 0$ and $w = re^{i\theta}$, then

$$\tan\theta = \frac{\sin\theta}{\cos\theta} = \frac{\operatorname{Im} w}{\operatorname{Re} w}.$$

Since by definition of L, $\operatorname{Im} L(w) = \theta$, it follows that

$$\operatorname{Im} L(w) = \tan^{-1}\left(\frac{\operatorname{Im} w}{\operatorname{Re} w}\right).$$

Finally, note that by direct computation

$$\frac{\operatorname{Im}\phi(z)}{\operatorname{Re}\phi(z)} = \frac{2\operatorname{Im} z}{1 - |z|^2} \quad \forall z \in D.$$

(iii) *Show that*

$$\lim_{r\to\infty} rg(z/r) = 2\operatorname{Im} z \quad \forall z \in \mathbb{C}.$$

Hints: Fix $z \in \mathbb{C}$. Of course, only $r > |z|$ will then be considered. By (ii)

$$rg(z/r) = r\tan^{-1}\left(\frac{2\operatorname{Im} z/r}{1 - |z/r|^2}\right)$$

$$= \frac{\tan^{-1}\left(\dfrac{2\operatorname{Im} z/r}{1 - |z/r|^2}\right) - \tan^{-1}(0)}{\dfrac{2\operatorname{Im} z/r}{1 - |z/r|^2}} \cdot \frac{2\operatorname{Im} z}{1 - |z/r|^2}.$$

As r grows, the expression $2(\operatorname{Im} z/r)/(1 - |z/r|^2)$ approaches 0, so the first term on the right approaches $(\tan^{-1})'(0) = 1/\tan'(0) = \cos^2(0) = 1$, while the second factor on the right obviously approaches $2\operatorname{Im} z$.

Exercise 8.38 (F. and R. NEVANLINNA [1922]) *Let* $U = \{z \in \mathbb{C}: \operatorname{Im} z > 0\}$ *and* h *a harmonic function in* U *which satisfies*

(*) $\displaystyle\varliminf_{z\to x} h(z) \geq 0 \quad \forall x \in \mathbb{R}.$

For each $r > 0$ *set* $m(r) = \inf\{h(z): z \in U \cap C(0, r)\}$ *and suppose that*

(**) $\displaystyle b = \varlimsup_{r\to\infty} \frac{m(r)}{r} > -\infty.$

(i) *Use the last exercise and harmonic majorization to show that*

$$h(z) \geq \frac{4b}{\pi}\operatorname{Im} z \quad \forall z \in U.$$

In particular, b *is finite.*

Hints: Notice first that on account of (*), $m(r)$ is finite for every r. Fix $r > 0$ and in $U \cap D(0, r)$ consider the function

$$u(z) = \frac{2}{\pi} m(r) g(z/r) - h(z),$$

where g is the function constructed in 8.37. Thanks to 8.37(i), the definition of $m(r)$ and (*) above, we have

$$\overline{\lim_{z \to \zeta}}\, u(z) \leq 0 \quad \forall \zeta \in [U \cap C(0, r)] \cup (-r, r) = \partial[U \cap D(0, r)] \setminus \{\pm r\}.$$

Because of (*) and the boundedness of g we are therefore entitled to cite 5.46 in order to conclude that

$$u(z) \leq 0 \quad \forall z \in U \cap D(0, r),$$

that is,

$$h(z) \geq \frac{2}{\pi} m(r) g(z/r) \quad \forall z \in U \cap D(0, r).$$

Given $z \in U$, select a sequence $r_j > |z|$ such that

$$\frac{m(r_j)}{r_j} \to b.$$

Then the last inequality gives

$$h(z) \geq \frac{2}{\pi} \cdot \frac{m(r_j)}{r_j} \cdot r_j g(z/r_j)$$

and from 8.37(iii) it follows that

$$h(z) \geq \frac{2}{\pi} \cdot b \cdot 2 \operatorname{Im} z.$$

(ii) *Assuming that in addition to the hypotheses above, h is continuous on $\overline{U}$, show that for some real constant a*

$$h(z) = a \operatorname{Im} z + \frac{1}{\pi} \int_{-\infty}^{\infty} \frac{\operatorname{Im} z}{|z - t|^2} h(t)\, dt \quad \forall z \in U.$$

Hint: Because of (i) the function $g(z) = h(z) - (4b/\pi) \operatorname{Im} z$ fulfills the hypotheses of 8.35.

(iii) *Under the hypotheses of* (ii) *show that*

$$\lim_{r \to \infty} \frac{m(r)}{r} \quad \textit{exists and equals } \min\{0, a\}.$$

Hints: From (ii) above and 8.36(i) we have that

$$a = \lim_{\substack{|z| \to \infty \\ z \in S_\varepsilon}} \frac{h(z)}{\operatorname{Im} z} \quad \forall \varepsilon > 0,$$

where $S_\varepsilon = \{z \in \mathbb{C}: \operatorname{Im} z > \varepsilon|z|\}$. In particular,

$$a = \lim_{r\to\infty} \frac{h(re^{i\theta})}{r \sin\theta} \quad \forall\theta \in (0, \pi).$$

Since by definition of m we have

$$\frac{m(r)}{r} \le \frac{h(re^{i\theta})}{r} = \sin\theta \cdot \frac{h(re^{i\theta})}{r\sin\theta},$$

it follows that

$$\overline{\lim_{r\to\infty}} \frac{m(r)}{r} \le \sin\theta \cdot \overline{\lim_{r\to\infty}} \frac{h(re^{i\theta})}{r\sin\theta} = \sin\theta\cdot a \quad \forall\theta \in (0, \pi),$$

whence

$$\text{(1)} \qquad \overline{\lim_{r\to\infty}} \frac{m(r)}{r} \le \inf\{a \sin\theta: 0 < \theta < \pi\}.$$

On the other hand, since $h(t) \ge 0$ for all $t \in \mathbb{R}$ (by hypothesis (*) and continuity of h on $\bar{U}$), it follows from (ii) that

$$h(z) \ge a \operatorname{Im} z \quad \forall z \in U,$$

that is,

$$ar \sin\theta \le h(re^{i\theta}) \quad \forall r > 0,\ \theta \in (0, \pi)$$

and so

$$r\cdot\inf\{a \sin\theta: 0 < \theta < \pi\} \le \inf\{h(re^{i\theta}): 0 < \theta < \pi\} = m(r).$$

It follows that

$$\text{(2)} \qquad \inf\{a \sin\theta: 0 < \theta < \pi\} \le \underline{\lim_{r\to\infty}} \frac{m(r)}{r}.$$

Finally, note that the infimum featuring in (1) and (2) equals 0 if $a \ge 0$ and equals a if $a < 0$.

Exercise 8.39 *Let h be harmonic and non-negative in $U = \{z \in \mathbb{C}: \operatorname{Im} z > 0\}$.*
(i) *Show that if h is continuous in $\bar{U}$, then*

$$\int_{-\infty}^{\infty} h(x + iy)dx \ge \int_{-\infty}^{\infty} h(t)dt \quad \forall y > 0.$$

[*The finiteness of these integrals is not being claimed.*]

Hints: Fix $y > 0$. Then for any $n, m \in \mathbb{N}$, $t \in \mathbb{R}$

$$\frac{1}{\pi}\int_{-n}^{n} \frac{y\,dx}{|x + iy - t|^2} = \frac{1}{\pi}\int_{t-n}^{t+n} \frac{y\,d\tau}{|iy - \tau|^2}$$

and so it follows from 8.31(ii) that

$$\text{(1)}\qquad \lim_{n\to\infty}\frac{1}{\pi}\int_{-n}^{n}\frac{y\,dx}{|x+iy-t|^2}=1 \quad \text{uniformly for } t\in[-m,m].$$

The non-negative constant c supplied us by 8.35 can be assumed to be 0, because $h(x+iy)\geq cy$ and if $c>0$ the left-hand integral in the desired inequality is ∞. With $c=0$ it follows from 8.35 that for all $n, m\in\mathbb{N}$

$$\begin{aligned}\int_{-n}^{n}h(x+iy)dx &= \int_{-n}^{n}\left(\frac{1}{\pi}\int_{-\infty}^{\infty}\frac{yh(t)dt}{|x+iy-t|^2}\right)dx\\ &\geq \int_{-n}^{n}\left(\frac{1}{\pi}\int_{-m}^{m}\frac{yh(t)dt}{|x+iy-t|^2}\right)dx\\ &= \int_{-m}^{m}h(t)\left(\frac{1}{\pi}\int_{-n}^{n}\frac{y\,dx}{|x+iy-t|^2}\right)dt, \quad \text{by 8.26(vi).}\end{aligned}$$

From this inequality and (1) infer that

$$\int_{-\infty}^{\infty}h(x+iy)dx=\lim_{n\to\infty}\int_{-n}^{n}h(x+iy)dx\geq\int_{-m}^{m}h(t)dt,$$

holding for all $m\in\mathbb{N}$.

(ii) *Show that if h is continuous on $\bar{U}$ and $\int_{-\infty}^{\infty}h(x+iy_0)dx$ is finite for some $y_0>0$, then*

$$\int_{-\infty}^{\infty}h(x+iy)dx=\int_{-\infty}^{\infty}h(t)dt \quad \forall y>0.$$

Hints: Taking $y=y_0$ in (i) shows that $\int_{-\infty}^{\infty}h(t)dt<\infty$. Consequently, if we fix $y>0$ and note that for any $x\in\mathbb{R}$, $0<m<M$

$$\int_{m<|t|<M}\frac{h(t)dt}{|x+iy-t|^2}\leq\frac{1}{y^2}\int_{m<|t|<M}h(t)dt,$$

it follows that the latter approaches 0 uniformly in x as $m\to\infty$. Therefore

$$\lim_{m\to\infty}\frac{1}{\pi}\int_{-m}^{m}\frac{yh(t)dt}{|x+iy-t|^2} \quad \text{exists uniformly for } x\in\mathbb{R}$$

and consequently for any $n\in\mathbb{N}$

$$\begin{aligned}\int_{-n}^{n}\left(\frac{1}{\pi}\int_{-\infty}^{\infty}\frac{yh(t)dt}{|x+iy-t|^2}\right)dx &= \int_{-n}^{n}\lim_{m\to\infty}\left(\frac{1}{\pi}\int_{-m}^{m}\frac{yh(t)dt}{|x+iy-t|^2}\right)dx\\ &= \lim_{m\to\infty}\int_{-n}^{n}\left(\frac{1}{\pi}\int_{-m}^{m}\frac{yh(t)dt}{|x+iy-t|^2}\right)dx\\ &= \lim_{m\to\infty}\int_{-m}^{m}h(t)\left(\frac{1}{\pi}\int_{-n}^{n}\frac{y\,dx}{|x+iy-t|^2}\right)dt,\\ &\qquad \text{by 8.26(vi),}\\ \text{(2)}\qquad &\leq \lim_{m\to\infty}\int_{-m}^{m}h(t)dt, \quad \text{by 8.31(ii).}\end{aligned}$$

As in (i), the constant c supplied by 8.35 is 0 and so according to the representation 8.35 the left side of (2) is just $\int_{-n}^{n} h(x + iy)dx$.

(iii) *Show that*

$$\int_{-\infty}^{\infty} h(x + iy_1)dx = \int_{-\infty}^{\infty} h(x + iy_0)dx \quad \forall y_1, y_0 > 0.$$

Hints: We may assume that the integral is finite for some $y_0 > 0$. That being the case, (ii) may be applied to the function $h(z + ib)$ in the role of $h(z)$, for any $0 < b < y_0$. We get

$$\int_{-\infty}^{\infty} h(x + iy)dx = \int_{-\infty}^{\infty} h(t + ib)dt \quad \forall y > b.$$

Notes to Chapter VIII

Our development of Runge theory follows SAKS and ZYGMUND [1971] and RUDIN [1966], with some modifications, especially in the matter of "shoving the poles to infinity" (see pp. 177 ff. of SAKS and ZYGMUND); though composed earlier than GRABINER [1976], it is quite similar to his treatment. Actually RUNGE [1885] only proved rational approximation, but the pole-shoving technique occurs in this paper and, as 8.9 shows, the step from rational functions to polynomials via this technique is pretty small. (See pp. 218–220 of MITTAG–LEFFLER [1900] on this point.) Polynomial approximation occurs explicitly (proved by other means) in HILBERT [1897].

Another proof of Polynomial Runge II follows from BIEBERBACH [1914] (pp. 107–112). He shows that for (certain) simply-connected Ω the conformal map ϕ of Ω onto $D(0, 1)$ (see next chapter) can be locally uniformly approximated in Ω by polynomials P_n. Given $f \in H(\Omega)$, let S_k be the partial sums of the power series of $f \circ \phi^{-1}$ in $D(0, 1)$. Then the polynomials $S_k \circ P_n$ provide the desired locally uniform in Ω approximation to f. Cf. 9.26.

For more on rational approximation see the treatises of WALSH [1935] and [1969]. In connection with 8.8, WOLFF [1921] shows that one can choose $a_k \in \mathbb{C}\backslash K$ and $A_k \in \mathbb{C}$ such that $f = \sum_{k=1}^{\infty} A_k F_{a_k}$ and $\sum_{k=1}^{\infty} |A_k| < \infty$. See also DENJOY [1922]. For a discussion of the pre-1920 progeny of Runge's approximation theorem (e.g., PAINLEVÉ [1886], [1888], [1898], HILBERT [1897], FABER [1903], FEJÉR [1918]), see § 59 of BIEBERBACH's encyclopedia article and for details of some of the methods, MONTEL's monograph [1910]. The problems of polynomial and rational approximation and those of analytic continuation are closely connected. E.g., both ends are served by summability methods which alter a power series to make it converge in regions other than disks. Also, from the Basic Integral Representation 8.1 [or from the various general Cauchy Integral Formulas like 7.19] follow, by the method of 8.8, other polynomial approximations to the function, if polynomial approximations to $1/(1 - z)$

other than $\sum_{n=0}^{\infty} z^n$ are used. These may converge in larger regions and provide holomorphic extensions of the given function. See §§ 37–44 of BIEBERBACH's encyclopedia article, chapters 10 and 11 of HILLE [1962] and BUHL [1925]. Yet another source of polynomial approximants is subsequences of the full sequence of partial sums of the power series of the function. Such subsequences may converge outside the disk of convergence of the full series and thus lead to polynomial approximations and holomorphic extensions to larger regions. This is the phenomenon of overconvergence to be discussed in Chapter XVI and its notes. There is also the phenomenon of expansion in series of Faber polynomials, a direct analog of power series expansions in disks. See the notes to Chapter IX. Finally, polynomial and rational approximants arise in interpolation theory. Here one is attempting to find explicit expressions for a function in terms of its values at certain points. 5.73 is a good example; as are the results in VALIRON [1925c], which will be developed in Chapter XIV. I refer the reader to the work of Fejér, Fekete and Leja. See also chapters 1 and 2 of SMIRNOV and LEBEDEV [1968].

Saks and Zygmund were the first to discern the elegant path to the general Cauchy theorem (8.14) via Runge's theorem. (But note WALSH's use of a similar technique in [1933]; see 9.27.) Many subsequent expositors, however, inexplicably chose not to follow this *camino real*, until HEINS [1962] and RUDIN [1966] took up and extended the theme. (See also HEINS [1968] and CONWAY [1973].)

There is a remarkable strengthening of Polynomial Runge I (8.9) due to MERGELYAN [1952]: if K is a compact, simply-connected set, then every f which is continuous on K and holomorphic in the interior of K is uniformly approximable on K by polynomials. Note that this includes every continuous function on K if K has a void interior. (M. A. LAVRENTIEFF [1936], pp. 25–35 and HARTOGS and ROSENTHAL [1931]. Thus this result subsumes Weierstrass' classical polynomial approximation theorem, 8.26(iii).) For a readable account of this profound theorem, see the last chapter of RUDIN [1974]. A good survey of the precursors of this theorem (one of which is proved in 9.26 below) is offered in the first few pages of MERGELYAN [1952]. (The reader who looks there for Mergelyan's own proof should also take advantage of the simplification wrought in it by BROWDER [1960].) With such a theorem in hand, general Cauchy theorems can be proved in which the curve of integration lies on the boundary of the domain of the holomorphic function. Cf. 9.27. In [1952] MERGELYAN also proved a corresponding strengthening (i.e., f continuous on K but holomorphic only in the interior of K) of 8.11 for the case where $\mathbb{C}\backslash K$ has only finitely many components.

Mergelyan's theorem, together with CARLESON's functional analytic proof of it [1965], has spawned a big industry, wedding functional analysis techniques with hard classical analysis. See chapters II and VIII of GAMELIN [1969], chapters five and six of STOUT [1971] and the notes of ZALCMAN [1968a] and GARNETT [1972]; for the classical approach see SMIRNOV and LEBEDEV [1968].

The proof of 8.18 is adapted from those in BIEBERBACH [1934], SAKS and ZYGMUND [1971] and ESTERMANN [1962]. (See also the remark of KARAMATA and FENCHEL, p. 365, vol. 5 of *Zentralblatt*.) For related material see BANK and ORLAND [1968].

For the last word on 8.24 and 8.25 see MARSHALL [1976].

In Chapter XIII we will approach the important approximation theorem 8.26(iii) from a different angle and discuss at length its history and generalizations. 8.26(v) was asserted on p. 797 of WEIERSTRASS [1885], with a proof supplied on pp. 27 ff., vol. 3 of his *Werke*. Proofs were also supplied by LEBESGUE [1898], [1908], MITTAG–LEFFLER [1900], BOREL [1928] (see chapter 4), LANDAU [1908a], TONELLI [1910], TAKENAKA [1919] and FRANKLIN [1925b]. For references to other proofs see pp. 1186–1187 of BOREL and ROSENTHAL [1924]. 8.26(vi) is also valid for bounded functions f such that $f(x, y)$ is Riemann integrable in x for each fixed y and Riemann integrable in y for each fixed x. In this case $\int_a^b f(x, y)dx$ is a Riemann integrable function of y and $\int_c^d f(x, y)dy$ is a Riemann integrable function of x and the equality in 8.26(vi) holds. See YOUNG [1910], LICHTENSTEIN [1910], [1911], FICHTENHOLZ [1913], GILLESPIE [1918–19], BESICOVITCH [1918–1919], ETTLINGER [1926–27], RIDDER [1930b,c] and RENNIE [1974]. For more on 8.27 see the *American Mathematical Monthly* 60 (1953), 427.

As noted in the text, Runge's theorems are inadequate to Carleman's proof of his theorem 8.29. However, with the aid of the more potent theorem of Mergelyan mentioned above, his proof simplifies greatly; compare, for example, the proof of KAPLAN [1955–56]. It is from the latter paper that 8.30 is taken. The satisfyingly constructive approach in 8.31 is not available in the more general case 8.30 but FINKELSTEIN and SCHEINBERG [1975] have found a constructive proof of a somewhat different nature for 8.30, and a similar solution of 8.30 was given by R. NEVANLINNA [1925]. SINCLAIR [1958] showed that the g in 8.29 could be chosen to satisfy in addition $g(n) = f(n)$ for all non-zero integers n. Finally HOISCHEN [1975] has shown that if $0 \leq c_n \leq c_{n+1} \to \infty$, $\{x_k\} \subset \mathbb{R}$ has no limit points and f is infinitely differentiable on $\mathbb{R}$, then an entire function g exists such that $|f^{(n)} - g^{(n)}| < \varepsilon$ in $\mathbb{R}\backslash[-c_n, c_n]$ for all $n = 0, 1, 2, \ldots$ and $g(x_k) = f(x_k)$ for all $k = 1, 2, \ldots$ See also RUBEL and VENKATESWARAN [1976] and GAUTHIER and HENGARTNER [1977]. In other directions Carleman's theorem has been extended by ROTH [1938], [1976], KELDYCH and LAVRENTIEFF [1939], MERGELYAN [1952], ARAKÉLIAN [1964], NERSESJAN [1971] and BROWN, GAUTHIER and SEIDEL [1975]. See FUCHS [1968] for a detailed exposition of Arakélian's work and ARAKÉLIAN [1971] for a survey.

The suggested proof of 8.32 is adapted from LOOMIS and WIDDER [1942], as is the proof of 8.34. The result 8.34 itself goes back at least to TSCHEBOTAREFF [1928] (p. 675), who credits Krein and Grandjot, and may also be found in BOULIGAND [1932]. A proof based on the argument principle appears in B. LEVIN's book [1964], p. 230. Another beautiful proof based on the reflection principle is given by TIDEMAN [1954]. See also BALK [1957]. The reflection principle originates with H. A. SCHWARZ; see pp. 12, 149–151 of vol. I and p. 66 of vol. II of his *Gesammelte Mathematische Abhandlungen*. There will be more on it in Chapter XIII. KURAN [1970] contains the following result, to which 8.34 is an easy corollary: If $h: \bar{U} \to \mathbb{R}$ is continuous, harmonic in U and $h(\mathbb{R}) = 0$, then $\sup_{y>0} \int_{-\infty}^{\infty} \frac{h^+(x + iy)}{x^2 + y^2}\, dx < \infty$ is equivalent to $h(x + iy) \equiv cy$ for some real constant c. Bearing in mind the extension to $\mathbb{C}$ provided by 8.33, the main result of KURAN [1966b] is also a generalization of 8.34.

The importance of the (Poisson) integral representation 8.35 is clear from the consequences we derived from it. Such a representation can be obtained under less restrictive hypotheses than those of 8.35 or 8.38(ii). In particular, continuity on the closed half-plane can be dispensed with if the Riemann integral is replaced by a Riemann–Stieltjes integral. A compactness argument under the boundedness provided by 8.32(iii) and (iv), in the form of Helly's Selection Theorem, is used to get the integrator function. See R. NEVANLINNA [1925], WOLFF and DE KOK [1932], CAUER [1932], VERBLUNSKY [1935], TSUJI [1939], [1959] (pp. 149–152) and, especially, LOOMIS and WIDDER [1942] and F. NEVANLINNA and NIEMINEN [1955]. The basic fact that positive harmonic functions admit a representation as a (Poisson) kernel integrated against a positive measure, which is so central to the modern theory, goes back to HERGLOTZ [1911], and R. NEVANLINNA [1922a]. The reader unable to fashion a proof of his own for 8.36(i) will find one on p. 96 of BOAS [1954] and on p. 525 of SERRIN [1956]. The latter paper is the source of the ideas in 8.38(ii) and (iii). The proof of 8.38(i) is from p. 19 of AHLFORS [1930] and the result 8.38(iii) is due to HEINS [1946a]. See also AHLFORS [1937], [1966b], DINGHAS [1938], HEINS [1948], AHLFORS and HEINS [1949], TSUJI [1956] and pp. 412–420 of HILLE's book [1962]. In the last two papers it is shown that not only does $\lim m(r)/r$ exist, but even $\lim h(re^{i\theta})/r$ exists for "most" θ. HEINS (*op. cit.*) shows that if h is only superharmonic and satisfies 8.38(*), then $\lim m(r)/r$ exists in $[-\infty, 0)$. ESSÉN [1975] represents the present state of the art. Finally, KURAN [1966a] was my source for 8.39; for related results see HARDY, INGHAM and PÓLYA [1927], [1928] and YÛJÔBÔ [1952].

As remarked before 8.31, some of the results of § 6 can be proved a little more easily by transforming them into statements about the unit disk. This is a good exercise program for the reader. The situation in the disk shows at a glance why all the improper integrals (which were tamed by various *ad hoc* devices) in 8.32 are finite: they are mean values of the appropriate positive harmonic function there, hence all are equal to the value of the function at 0. Notice that the disk analog of 8.34 is contained in LOHWATER [1952]. (The actual result proved there suggests that the boundary continuity hypothesis in 8.34 may be weakened to $\lim_{y\downarrow 0} h(x + iy) = 0$ for all real x. The interested reader might try to confirm this. On this point compare with 5.29(ii) and WOLF [1947].) See also BRUCKNER, LOHWATER and RYAN [1969].

Chapter IX
The Riemann Mapping Theorem

§ 1 Introduction

In his Göttingen dissertation of 1851 Bernard Riemann enunciated (p. 40 of his *Werke*) and attempted to prove the famous theorem that now bears his name:

Theorem 9.1 *Every simply-connected region other than* $\mathbb{C}$ *itself is conformal to an open disk.*

The gap in Riemann's proof consisted in the assumption that a certain extremal problem in the calculus of variations necessarily had a solution. (See MONNA [1975], pp. 32–34.) Hilbert and others subsequently filled this gap (MONNA [1975], chapter IV). In essence the matter comes down to the compactness criterion of Montel (7.6). A remnant of this method of proof persists in what has become the canonical (for concision and simplicity) proof, that of L. Fejér and F. Riesz, which is sketched in exercise 9.8 below. Here we propose to explore another proof. This is constructive in that it produces more or less explicit sequences of maps that converge to the desired conformal equivalence. Moreover, the maps are elementary: repeated compositions of square roots and conformal automorphisms of the unit disk. Explicit estimates are obtained for the rate of convergence and even the induced convergence theorems of Chapter VII can be avoided, so this proof is truly elementary (in its ingredients, if not in its genesis and execution!). This proof evolved at the hands of Paul Koebe and Constantin Carathéodory around 1910 and will be presented in all its glorious detail in section 2 below. The reader is reminded of 4.65, according to which property (HS_1) is equivalent to simple-connectivity. However, no use is made of this equivalence in the proof; we simply take (HS_1) as hypothesis, since only that property of the region is used. The grand synthesis occurs in 10.2.

Definition 9.2 (i) A *Koebe region* is an open, connected, proper subset of the open unit disk D which contains 0 and has property (HS_1).

(ii) The *Koebe radius* of a Koebe region Ω is the number

$$R = \inf\{|z| : z \in \mathbb{C}\backslash\Omega\}.$$

[Notice that $D(0, R) \subset \Omega$ and $0 < R < 1$.]

(iii) If Ω is a Koebe region and R is its Koebe radius, then a *Koebe map* on Ω is any conformal map k of Ω with the following properties

(a) $k(\Omega) \subsetneq D$

(b) $k(0) = 0$

(c) $k'(0) > 1 + \frac{1}{32}(1 - R)^2$

(d) $|k(z)| \geq |z| \quad \forall z \in \Omega.$

Lemma 9.3 *If Ω is a Koebe region with Koebe radius R and if k is a Koebe map on Ω, then $k(\Omega)$ is a Koebe region and its Koebe radius is not less than R.*

Proof: Only the last assertion needs proof. Let R^* denote the Koebe radius of $k(\Omega)$. From the definition of R^* it is obvious that there is a boundary point b of $k(\Omega)$ with $|b| = R^*$. Select $z_n \in \Omega$ such that $k(z_n) \to b$. Since k is a homeomorphism, the points z_n cannot accumulate in Ω. In particular, since $D(0, R) \subset \Omega$ by definition of R, we must have

$$\overline{\lim_{n\to\infty}}|z_n| \geq R.$$

But then since $|k(z)| \geq |z|$ for all $z \in \Omega$,

$$R^* = |b| = \lim_{n\to\infty} |k(z_n)| \geq \overline{\lim_{n\to\infty}}|z_n| \geq R.$$

Lemma 9.4 *Every Koebe region possesses a Koebe map.*

Proof: Let Ω be a Koebe region, with Koebe radius R. Note that $0 < R < 1$. From the definition of R it is obvious that there is a boundary point a of Ω with

(1) $|a| = R.$

Thus $a \in D$ and we may form f_a:

(2) $f_a(z) = \dfrac{z - a}{1 - \bar{a}z}, \quad z \in D,$

a conformal map of D onto D (see 2.6) which is zero-free in Ω. By property (HS_1) of Ω, f_a has a holomorphic square root g *on* Ω. Evidently, $\Omega \subsetneq D$ and f_a a one-to-one map of D onto D mean that $f_a(\Omega) \subsetneq D$, that is, $g^2(\Omega) \subsetneq D$. It follows that $g(\Omega) \subsetneq D$. So we may set

(3) $b = g(0),$

form

(4) $G = f_b \circ g$

and have

(5) $G(\Omega) \subsetneq D$

and

(6) $G(0) = 0.$

Notice that $g^2 = f_a$ and (3) yield

(7) $b^2 = -a.$

By the Chain Rule and a direct calculation on (2)

$$2g(0)g'(0) = [g^2]'(0) = f_a'(0) = 1 - |a|^2$$

$$g'(0) = \frac{1 - |a|^2}{2g(0)}$$

$$(8) \qquad = \frac{1 - |a|^2}{2b} \quad \text{by (3).}$$

Therefore another application of the Chain Rule gives

$$G'(0) \overset{(4)}{=} f_b'(g(0))g'(0) \overset{(3)}{=} f_b'(b)g'(0)$$

$$= \frac{1 - |b|^2}{(1 - \bar{b}b)^2} \cdot g'(0)$$

$$\overset{(8)}{=} \frac{1}{1 - |b|^2} \cdot \frac{1 - |a|^2}{2b}$$

$$\overset{(7)}{=} \frac{1 - |a|^2}{1 - |a|} \cdot \frac{1}{2b}$$

$$(9) \qquad \overset{(1)}{=} \frac{1 + R}{2b}.$$

We finally set

$$k = \frac{|G'(0)|}{G'(0)} \cdot G$$

and we see from (5) and (6) that k has properties (a) and (b) of 9.2(iii), while (9) shows that

$$k'(0) = |G'(0)| = \frac{1 + R}{2|b|} = \frac{1 + R}{2\sqrt{R}} = 1 + \frac{(1 - \sqrt{R})^2}{2\sqrt{R}}$$

$$> 1 + \frac{(1 - \sqrt{R})^2}{2}$$

$$(10) \qquad > 1 + \tfrac{1}{2}\left(1 - \sqrt{\frac{1 + R}{2}}\right)^2.$$

From $2(3 + R)^2 - 16(1 + R) = 2 - 4R + 2R^2 = 2(1 - R)^2 > 0$, we infer that $2(3 + R)^2 > 16(1 + R)$,

$$\left[\frac{3 + R}{4}\right]^2 > \frac{1 + R}{2}.$$

Using this in (10) gives

$$k'(0) > 1 + \tfrac{1}{2}\left(1 - \frac{3 + R}{4}\right)^2 = 1 + \tfrac{1}{32}(1 - R)^2,$$

thus establishing property (c).

Let s denote the square map on $\mathbb{C}$: $s(z) = z^2$. Then the above definitions show that

$$f_b^{-1} \circ G = g \qquad \text{on } \Omega$$

$$(11) \qquad s \circ f_b^{-1} \circ G = g^2 = f_a \quad \text{on } \Omega.$$

Now $B = f_a^{-1} \circ s \circ f_b^{-1}$ is a holomorphic map of D into D and since $f_a^{-1} = f_{-a}$, $f_b^{-1} = f_{-b}$ (see 2.6), we see that

$$B(0) = f_{-a}((f_{-b}(0))^2) = f_{-a}(b^2) \overset{(7)}{=} f_{-a}(-a) = 0.$$

Therefore by Schwarz

$$|B(w)| \le |w| \quad \forall w \in D$$

and combining this with (11) gives

$$|z| = |B(G(z))| \le |G(z)| = |k(z)| \quad \forall z \in \Omega,$$

which is property (d) of a Koebe map.

Lemma 9.5 *Let $r > 0$, $1 > \varepsilon > 0$ and F be holomorphic in a neighborhood of $\bar{D}(0, r)$ with $|F| \le 1$ and $F(0) \ge 1 - \varepsilon$. Then*

$$|F(z) - 1| \le \frac{\sqrt{2\varepsilon}}{1 - |z|/r} \quad \forall z \in D(0, r).$$

Proof: We may assume $r = 1$. For $|w| \le 1$ we have

$$|w - 1|^2 = 1 - w - \bar{w} + w\bar{w} \le 2 - w - \bar{w}.$$

Therefore

$$\begin{aligned}\int_0^{2\pi} |F(e^{i\theta}) - 1|^2 d\theta &\le \int_0^{2\pi} [2 - F(e^{i\theta}) - \bar{F}(e^{i\theta})] d\theta \\ &= 2\pi[2 - F(0) - \bar{F}(0)] \quad \text{by 5.5(ii)} \\ &\le 4\pi\varepsilon, \quad \text{since } F(0) \ge 1 - \varepsilon.\end{aligned}$$

The Cauchy–Schwarz inequality for integrals then yields

$$\begin{aligned}\int_0^{2\pi} |F(e^{i\theta}) - 1| d\theta &\le \left[\int_0^{2\pi} |F(e^{i\theta}) - 1|^2 d\theta \int_0^{2\pi} 1 \, d\theta\right]^{1/2} \\ &\le 2\pi\sqrt{2\varepsilon}.\end{aligned}$$

It follows from the last inequality and Cauchy's formula that

$$\begin{aligned}|F(z) - 1| &= \frac{1}{2\pi}\left|\int_0^{2\pi} \frac{F(e^{i\theta}) - 1}{e^{i\theta} - z} ie^{i\theta} d\theta\right| \quad \forall z \in D(0, 1) \\ &\le \frac{1}{2\pi} \cdot \frac{1}{1 - |z|} \int_0^{2\pi} |F(e^{i\theta}) - 1| d\theta \\ &\le \frac{\sqrt{2\varepsilon}}{1 - |z|}.\end{aligned}$$

Corollary 9.6 *Let* $0 < R < 1$, Ω *a region containing* 0 *and* f *a conformal map of* Ω *such that* $f(0) = 0, f'(0) > 0$. *Suppose that*

$$D(0, R) \subset \Omega \cap f(\Omega) \subset \Omega \cup f(\Omega) \subset D(0, 1).$$

Then

$$|f(z) - z| \leq \frac{3\sqrt{1 - R}}{R^2 - |z|} \quad \forall z \in D(0, R^2).$$

Proof: Let $g = f^{-1}$ and apply Schwarz' lemma to both $f(Rz)$ and $g(Rz)$, $z \in D(0, 1)$, to get ($w = Rz$)

(1) $\quad |f(w)| \leq |w|/R, \quad |g(w)| \leq |w|/R \quad \forall w \in D(0, R).$

Consider $z \in D(0, R^2)$. Then $|f(z)| < R$ by (1), so

(2) $\quad |z| = |g(f(z))| \leq |f(z)|/R$ by (1).

Combining (1) and (2) we have

$$R^2 \leq |Rf(z)/z| \leq 1 \quad \forall z \in D(0, R^2).$$

If we define $r = R^2$, $\varepsilon = 1 - R^2$ and $F(z) = Rf(z)/z$, for $z \in \Omega$, then the last inequality becomes

(3) $\quad 1 - \varepsilon = R^2 \leq |F(z)| \leq 1 \quad \forall z \in D(0, r).$

Furthermore, if we go to the limit $z \to 0$ in the definition of F, we get $F(0) = Rf'(0) > 0$, so (3) also yields

$$F(0) \geq 1 - \varepsilon.$$

Therefore F fulfills the hypotheses of the last lemma. Consequently, for $z \in D(0, r)$

$$\begin{aligned}
|f(z) - z| &\leq |f(z) - z/R| + |z/R - z| \\
&= \frac{|z|}{R}\left|\frac{Rf(z)}{z} - 1\right| + \frac{|z|}{R}(1 - R) \\
&\leq \left|\frac{Rf(z)}{z} - 1\right| + (1 - R), \quad \text{since } |z| < r = R^2 < R, \\
&= |F(z) - 1| + 1 - R \\
&\leq \frac{\sqrt{2\varepsilon}}{1 - |z|/r} + 1 - R, \quad \text{by the last lemma,} \\
&= \frac{R^2\sqrt{2(1 - R)(1 + R)}}{R^2 - |z|} + 1 - R, \quad \text{by definition of } \varepsilon \text{ and } r, \\
&< \frac{2R^2\sqrt{1 - R}}{R^2 - |z|} + 1 - R, \quad \text{since } 1 + R < 2, \\
&= \frac{2R^2 + \sqrt{1 - R}(R^2 - |z|)}{R^2 - |z|}\sqrt{1 - R} \\
&< \frac{3}{R^2 - |z|}\sqrt{1 - R}.
\end{aligned}$$

This holds for all $z \in D(0, r) = D(0, R^2)$ and completes the proof.

§ 2 The Proof of Carathéodory and Koebe

The proof below is synthesized from ideas of Koebe, Carathéodory and Ostrowski and follows somewhat the exposition in BISHOP [1967].

Theorem 9.7 (Riemann Mapping Theorem) *Every region having property* (HS_1) *other than* $\mathbb{C}$ *itself is conformal to a disk.*

Proof: We first show that it is enough to treat Koebe regions, that is, every region $\Omega \neq \mathbb{C}$ having property (HS_1) is conformally equivalent to a Koebe region. Pick $w_0 \in \mathbb{C}\backslash\Omega$ and use property (HS_1) of Ω to produce a holomorphic function ϕ on Ω such that

$$\phi^2(z) = z - w_0 \quad \forall z \in \Omega.$$

Then ϕ is one-to-one and

(1) $\quad \phi(\Omega) \cap -\phi(\Omega) = \varnothing.$

These facts are simple to check. (See the proof of corresponding facts in 5.76.) Since ϕ is an open map, $\phi(\Omega)$ contains an open disk, hence an open disk of the form $D(a, r)$ where $a \neq 0$ and $0 < r < |a|$. From $D(a, r) \subset \phi(\Omega)$ and (1) follows that $-D(a, r) = \{-z: z \in D(a, r)\} = D(-a, r)$ does not meet $\phi(\Omega)$:

$$D(-a, r) \subset \mathbb{C}\backslash\phi(\Omega)$$

[cf. the proof of 7.24] or, since $\mathbb{C}\backslash\phi(\Omega)$ is closed,

(2) $\quad \bar{D}(-a, r) \subset \mathbb{C}\backslash\phi(\Omega).$

So we form $\psi = r/(\phi + a)$ and we have a conformal map of Ω into the open unit disk D. Indeed by (2) $|\phi(z) - (-a)| > r$ for all $z \in \Omega$. Note that $\psi(\Omega) \neq D$, since by its very form the function ψ is never 0. We look at $f_c(\psi(\Omega))$ for some $c \in \psi(\Omega)$, where f_c is the conformal map of D onto D given by

$$f_c(z) = \frac{z - c}{1 - \bar{c}z}, \quad z \in D.$$

Since $f_c(c) = 0$, we have $0 \in f_c(\psi(\Omega))$ and since $\psi(\Omega) \subsetneq D$, we have $f_c(\psi(\Omega)) \subsetneq D$. Therefore $f_c(\psi(\Omega))$ is a Koebe region. (For clearly property (HS_1) passes over to this set from Ω under the conformal map $f_c \circ \psi$.)

So we are reduced to the case of a Koebe region Ω. We make an inductive construction thus: let $\Omega_0 = \Omega$ and let f_1 be a Koebe map of Ω_0. Set $\Omega_1 = f_1(\Omega_0)$. Suppose defined Koebe regions $\Omega_0, \ldots, \Omega_n$ and functions $f_1, \ldots, f_n$ such that f_j is a Koebe map of Ω_{j-1} and $\Omega_j = f_j(\Omega_{j-1})$ for each $j = 1, 2, \ldots, n$. Then let f_{n+1} be a Koebe map on Ω_n (such exist by 9.4) and let Ω_{n+1} be the Koebe region $f_{n+1}(\Omega_n)$. This completes the inductive definition.

Let R_n be the Koebe radius of Ω_n ($n = 0, 1, 2, \ldots$). Notice that by 9.3 we have then

(3) $\quad 0 < R_0 \le R_1 \le R_2 \le \cdots < 1$

and by definition of the Koebe radius

(4) $\quad D(0, R_n) \subset \Omega_n \quad \forall n \ge 0.$

For each $n \ge 0$ and each $p \ge 1$ set

(5) $\quad \psi_{n,p} = f_{n+p} \circ f_{n+p-1} \circ \cdots \circ f_{n+1}$ from Ω_n onto Ω_{n+p}.

By the Chain Rule, the fact that $f_k(0) = 0$ for all k and induction we have

$$\psi'_{n,p}(0) = f'_{n+p}(0) \cdots f'_{n+1}(0).$$

So by property (c) of Koebe maps

$$\psi'_{n,p}(0) \ge \prod_{j=n+1}^{n+p} [1 + \tfrac{1}{32}(1 - R_j)^2]$$

(6) $$\qquad \ge [1 + \tfrac{1}{32}(1 - R_{n+p})^2]^p > 0 \quad \text{by (3)}.$$

Now all Ω_{n+p} lie in D, so

(7) $\quad |\psi_{n,p}(z)| < 1 \quad \forall z \in \Omega_n, \quad \forall n \ge 0, p \ge 1.$

If we apply Schwarz' lemma to each $\psi_{n,p}$ in the disk $D(0, R_0)$, which lies in each Ω_n by (3) and (4), we get from (7)

$$|\psi'_{n,p}(0)| \le 1/R_0 \quad \forall n \ge 0, p \ge 1,$$

which together with (6) yields

$$1/R_0 \ge [1 + \tfrac{1}{32}(1 - R_p)^2]^p,$$

whence

$$1 - R_p \le 4\sqrt{2}\sqrt{R_0^{-1/p} - 1}.$$

From this and 3.1 we infer that

(8) $\quad R_p \to 1.$

Now let K compact $\subset \Omega_0$ be given. Since $\psi_{0,n}(0) = 0$ and $|\psi_{0,n}| < 1$ for each n, it follows from 6.38 that for some $c = c(K) < 1$ we have

(9) $\quad |\psi_{0,n}(z)| \le c \quad \forall z \in K, \forall n = 1, 2, \ldots$

Since $c < 1$, (8) insures that there exists N such that

(10) $\quad c < R_n^2 \quad \forall n \ge N.$

From (9) and (10) we have

(11) $\quad \psi_{0,n}(K) \subset D(0, R_n^2) \quad \forall n \ge N.$

Now by (3) and (4) we have

$$\begin{aligned} D(0, R_n) \subset D(0, R_{n+p}) \subset \Omega_{n+p} &= f_{n+p}(\Omega_{n+p-1}) = \cdots \\ &= f_{n+p} \circ f_{n+p-1} \circ \cdots \circ f_{n+1}(\Omega_n) \\ &= \psi_{n,p}(\Omega_n). \end{aligned} \tag{12}$$

From (4), (7) and (12) we thus have

$$D(0, R_n) \subset \Omega_n \cap \psi_{n,p}(\Omega_n) \subset \Omega_n \cup \psi_{n,p}(\Omega_n) \subset D(0, 1).$$

These inclusions together with (6) show that the hypotheses of 9.6 are fulfilled by $f = \psi_{n,p}$ and we can therefore assert that

$$|\psi_{n,p}(w) - w| \leq \frac{3\sqrt{1 - R_n}}{R_n^2 - |w|} \quad \forall w \in D(0, R_n^2).$$

In conjunction with (11) this yields

$$|\psi_{n,p}(\psi_{0,n}(z)) - \psi_{0,n}(z)| \leq \frac{3\sqrt{1 - R_n}}{R_n^2 - |\psi_{0,n}(z)|} \quad \forall z \in K, n \geq N, p \geq 1.$$

From (5) however we see that $\psi_{n,p} \circ \psi_{0,n} = \psi_{0,n+p}$. Using this and (9) in the last inequality yields

$$|\psi_{0,n+p}(z) - \psi_{0,n}(z)| \leq \frac{3\sqrt{1 - R_n}}{R_n^2 - c} \quad \forall z \in K, n \geq N, p \geq 1.$$

In view of (8) this last inequality asserts that the sequence $\{\psi_{0,k}\}$ is uniformly Cauchy on K. Therefore we have shown that

(13) $\psi = \lim_{k \to \infty} \psi_{0,k}$ exists locally uniformly in Ω_0.

This limit is a holomorphic function (5.44) and it further follows from (9) that $|\psi(z)| \leq c(K) < 1$ for each $z \in K$ and each compact $K \subset \Omega_0$. That is,

(14) $\psi(\Omega_0) \subset D(0, 1)$.

Next we show that ψ is surjective. Let $w \in D(0, 1)$ be given. Use (8) to select N so that

(15) $R_n > |w|$ for all $n \geq N$.

Then by (4) we have for $p \geq 1$

$$\begin{aligned} w \in D(0, R_{N+p}) \subset \Omega_{N+p} &= f_{N+p}(\Omega_{N+p-1}) = \cdots \\ &= f_{N+p} \circ f_{N+p-1} \circ \cdots \circ f_{N+1}(\Omega_N) \\ &= f_{N+p} \circ f_{N+p-1} \circ \cdots \circ f_{N+1} \circ \psi_{0,N}(\Omega_0). \end{aligned}$$

Therefore there exists $z_p \in \Omega_0$ such that

$$w = f_{N+p} \circ f_{N+p-1} \circ \cdots \circ f_{N+1} \circ \psi_{0,N}(z_p) \tag{16}$$

$$= \psi_{0,N+p}(z_p). \tag{17}$$

By (16) and successive applications of property (d) of Koebe maps we have

(18) $|w| \geq |\psi_{0,N}(z_p)| \quad \forall p \geq 1.$

From (4) and (15) we have that $\bar{D}(0, |w|)$ is a compact subset of Ω_N, so $\psi_{0,N}^{-1}(\bar{D}(0, |w|))$ is a compact subset C of Ω_0. It follows from (18) that z_p belongs to this compact subset of Ω_0 for all $p \geq 1$. Therefore some subsequence $\{z_{p_j}\}$ converges to a $z \in C$. Since $\{\psi_{0,k}\}$ converges to ψ *uniformly* on C, it follows easily that $\psi_{0,N+p_j}(z_{p_j}) \to \psi(z)$. That is, recalling (17), $w = \psi(z)$. As w is any point of $D(0, 1)$, this establishes the surjective character of ψ.

The fact that ψ is one-to-one, which completes the proof, is now a consequence of Hurwitz' theorem 7.11. For ψ, being surjective, is not constant but it is the limit of one-to-one functions.

We can however prove the univalence of ψ directly without appeal to 7.11 thus. Define

(5)′ $\phi_{n,p} = \psi_{n,p}^{-1} = f_{n+1}^{-1} \circ \cdots \circ f_{n+p-1}^{-1} \circ f_{n+p}^{-1}$ from Ω_{n+p} onto Ω_n.

Since all maps fix 0 and are one-to-one, 5.78 gives

(6)′ $$\phi_{n,p}'(0) = \frac{1}{\psi_{n,p}'(0)} \overset{(6)}{>} 0.$$

Let compact $L \subset D(0, 1)$ be given. Choose $R' < R < 1$ so that $L \subset D(0, R')$ and then choose N so large that

(19) $R < R_n^2 \quad \forall n \geq N.$

By (4) then

(20) $L \subset D(0, R') \subset D(0, R) \subset D(0, R_n^2) \subset \Omega_n \quad \forall n \geq N.$

We have

$$D(0, R_n) \subset D(0, R_{n+p}) \subset \Omega_{n+p}$$

and

$$D(0, R_n) \subset \Omega_n = \phi_{n,p}(\Omega_{n+p}),$$

whence

$$D(0, R_n) \subset \Omega_{n+p} \cap \phi_{n,p}(\Omega_{n+p}) \subset \Omega_{n+p} \cup \phi_{n,p}(\Omega_{n+p}) \subset D(0, 1)$$

for all $n \geq 0$, $p \geq 1$. From these inclusions and (6)′ we infer via 9.6 that

(21) $$|\phi_{n,p}(w) - w| \leq \frac{3\sqrt{1 - R_n}}{R_n^2 - |w|} \quad \forall w \in D(0, R_n^2),\ n \geq 0, p \geq 1.$$

Then (20) teams up with (21) to give

$$|\phi_{n,p}(w) - w| \le \frac{3\sqrt{1-R_n}}{R_n^2 - R'} \quad \forall w \in L,\ n \ge N,\ p \ge 1,$$

$$(22) \qquad \overset{(19)}{\le} \frac{3\sqrt{1-R_n}}{R - R'} \quad \forall w \in L,\ n \ge N,\ p \ge 1.$$

Property (d) of a Koebe map k says that $|k^{-1}(w)| \le |w|$ for all w in the range of k. It follows by induction from (5)′ that

$$(23) \quad |\phi_{n,p}(w)| \le |w| \quad \forall w \in \Omega_{n+p},\ n \ge 0,\ p \ge 1,$$

hence from (20) follows also

$$\phi_{n,p}(L) \subset D(0, R) \quad \forall n \ge N,\ p \ge 1.$$

For $w \in L$, $n \ge N$ and $p \ge 1$ the point $z = \phi_{n,p}(w)$ thus lies in $D(0, R)$ which also contains w and which lies in the domain Ω_n of $\phi_{0,n}$, by (20). We are therefore in a position to apply 6.3(ii) with the F there equal to $\phi_{0,n}$ and the M there equal to 1. We get

$$|\phi_{0,n+p}(w) - \phi_{0,n}(w)| = |\phi_{0,n}(\phi_{n,p}(w)) - \phi_{0,n}(w)|$$

$$\le \frac{2R}{|R^2 - \bar{w}\phi_{n,p}(w)|} |\phi_{n,p}(w) - w|$$

$$\le \frac{2R}{R^2 - |w|\,|\phi_{n,p}(w)|} \cdot \frac{3\sqrt{1-R_n}}{R - R'} \quad \text{by (22)},$$

$$(24) \qquad \le \frac{2R}{R^2 - (R')^2} \cdot \frac{3\sqrt{1-R_n}}{R - R'} \quad \text{by (23) and (20)},$$

an inequality which holds for all $w \in L$, all $n \ge N$ and all $p \ge 1$. Since the right side of (24) converges to 0 as $n \to \infty$, we see that the sequence $\{\phi_{0,n}\}_{n=N}^{\infty}$ is uniformly Cauchy on L. Therefore

$$(13)' \quad \phi = \lim_{k\to\infty} \phi_{0,k} \quad \text{exists locally uniformly in } D(0, 1).$$

Now consider any $z_0 \in \Omega_0$. If in (9) we take for K the set $\{z_0\}$, we will have

$$(25) \quad \{\psi_{0,n}(z_0): n = 1, 2, 3, \dots\} \subset \bar{D}(0, c).$$

As the latter is a compact subset of $D(0, 1)$, the sequence $\{\phi_{0,k}\}$ converges to ϕ *uniformly* on it. It follows easily from (13) and the inclusion (25) that therefore

$$\phi_{0,k}(\psi_{0,k}(z_0)) \to \phi(\psi(z_0))$$

and so, since $\phi_{0,k}$ is the inverse of $\psi_{0,k}$,

(26) $z_0 = \phi(\psi(z_0)).$

This holds for all $z_0 \in \Omega_0$. It establishes at once the univalence of ψ.

Remark: CARATHÉODORY [1914] proves the univalence and the surjectivity of ϕ with the zero-counting integral of 5.85 (see the hints to 9.22); the last vestige of non-constructivity (I use the term informally, not in the strict sense of Bishop, *et al.*) can thereby be eliminated from the proof of the mapping theorem. On the other hand, a pure existence proof can be fashioned very expeditiously thus:

§ 3 Fejér and Riesz' Proof

Exercise 9.8 *Let Ω be a Koebe region and $\mathscr{F}$ the family of all conformal maps f of Ω into $D(0, 1)$ with $f(0) = 0$.*

(i) *Show that $\mathscr{F}$ contains a function f_0 which maximizes $|f'(0)|$ among all f in $\mathscr{F}$.*

Hint: Use 7.6 and 7.11.

(ii) *Show that any such f_0 provides the sought-for conformal map of Ω* onto $D(0, 1)$.

Hint: If f_0 is not surjective, it can be followed by a Koebe map to produce an $f_1 \in \mathscr{F}$ with $|f_1'(0)| > |f_0'(0)|$, by the Chain Rule and property (c) of Koebe maps.

Exercise 9.9 (Poincaré's Uniqueness Theorem) *Let Ω be a simply-connected region other than $\mathbb{C}$ itself, $w_0 \in \Omega$, $z_0 \in D = D(0, 1)$ and $\theta \in \mathbb{R}$. Show that there is exactly one conformal map f of D onto Ω which satisfies $f(z_0) = w_0$ and $f'(z_0) = e^{i\theta}|f'(z_0)|$.*

Hints: (p. 111 of CARATHÉODORY [1912a]) For existence, just move the disk D around by an appropriate conformal automorphism under any Riemann map of D onto Ω. For uniqueness, consider any other function g answering to this same description and form $\phi^{-1} \circ f^{-1} \circ g \circ \phi$ and $\phi^{-1} \circ g^{-1} \circ f \circ \phi$, where $\phi(z) = (z + z_0)/(1 + \bar{z}_0 z)$. These are conformal automorphisms of D which fix 0. Their derivatives at 0 are positive and reciprocals of one another, so one of them is not less than 1. To that one apply the uniqueness clause in Schwarz' lemma. (Uniqueness also follows from 7.40.)

§ 4 Boundary Behavior for Jordan Regions

A Jordan region is evidently simply-connected (see 4.46) hence has property (HS_1), by 4.65. OSGOOD conjectured in the encyclopedia article [1901] (p. 56) that any conformal map of a Jordan region onto $D(0, 1)$ extends to a homeomorphism of its closure. This was confirmed more or less simultaneously by

Osgood and Taylor [1913] and Carathéodory [1913b]. We present here one solution of the problem and later discuss some other approaches. All are quite geometrical, presenting formidable problems of exposition. The reason for singling out Jordan regions is simple: if $g: D(0, 1) \to \Omega$ is conformal and extends to a homeomorphism of $\bar{D}(0, 1)$, then Ω is *necessarily* a Jordan region, the inside of the Jordan-curve $g(C(0, 1))$.

Lemma 9.10 (Koebe [1913a]) *Let a, b be two distinct numbers of modulus 1, γ_n arcs in $D = D(0, 1)$ with endpoints a_n, b_n satisfying*

(i) $a_n \to a$, $b_n \to b$

(ii) $|\gamma_n| \geq \delta > 0$ *for all n.*

Let f be a bounded holomorphic function in D such that

(iii) $\sup|f(\gamma_n)| \to 0$ *as* $n \to \infty$.

Then $f = 0$.

Proof: It is enough to show that these conditions imply $f(0) = 0$. For then, if f is not the 0 function, and $0 \leq N < \infty$ is the order of 0 as a zero of f, we can observe (using (ii)) that $F(z) = z^{-N}f(z)$ fulfills all the hypotheses f does and that consequently the contradiction $F(0) = 0$ ensues.

Next note that we can augment (ii) with the assumption

(ii)′ $\inf|\gamma_n| \to 1$ as $n \to \infty$.

For if (ii)′ fails, there is an $r_1 < 1$ and $1 < n_1 < n_2 < \cdots$ such that γ_{n_j} meets $D(0, r_1)$ for each j. By (i) we can suppose n_1 so large that $|a_{n_j}|, |b_{n_j}| \geq (1 + r_1)/2 = r_2$ for all j. Connectedness considerations then show that γ_{n_j} meets $C(0, r)$ for each j and each $r \in (r_1, r_2)$. From compactness considerations and (iii) it follows that f has a zero on $C(0, r)$ for each $r \in (r_1, r_2)$, hence from 5.62, $f = 0$.

Next note that none of the hypotheses (i), (ii)′, (iii) is altered by a conformal movement of D under f, since such a map is a homeomorphism of D onto D and $\bar{D}$ onto $\bar{D}$. There is such a map which moves a to i and b to $-i$. (See 6.4.) Therefore with no loss of generality we take

(i)′ $a_n \to i$, $b_n \to -i$.

This says that γ_n visits both the upper half-plane and the lower half-plane for all large n, so such γ_n must cut the real axis. Replacing $f(z)$ by $f(-z)$ and γ_n by $-\gamma_n$ if necessary, and going to a subsequence again, we can simply suppose

(1) γ_n meets $(0, 1)$ for every n.

We may also suppose, from (i)′, that

(2) $\operatorname{Im} a_n > |\operatorname{Re} a_n|$ for all n.

We form

$$S = \{z \in D: \operatorname{Re} z > |\operatorname{Im} z|\}.$$

Let the parameter interval for each γ_n be $[0, 1]$. We have $\gamma_n(0) = a_n$ (or this can be arranged by replacing $\gamma_n(t)$ with $\gamma_n(1 - t)$) and $\gamma_n(\tau_n) \in (0, 1)$ for some $\tau_n > 0$. Thus $\operatorname{Re} \gamma - |\operatorname{Im} \gamma|$ is (by (2)) negative at 0 and positive at τ_n. Hence there is a $t \in (0, \tau_n)$ such that $\operatorname{Re} \gamma(t) = |\operatorname{Im} \gamma(t)|$. Let t_n denote the largest such t. Then

$$(3) \qquad \operatorname{Re} \gamma_n(t_n) = |\operatorname{Im} \gamma_n(t_n)|, \quad \gamma_n(t_n, \tau_n] \subset S.$$

Finally, let s_n be the least $t \in (t_n, \tau_n]$ such that $\gamma_n(t) \in (0, 1)$. Then γ_n either maps $[t_n, s_n)$ into the upper half-plane or into the lower half-plane. One of these alternatives prevails for infinitely many n and by discarding the others and renumbering, we can have

$$(4) \qquad \operatorname{Re} \gamma_n(t_n) = |\operatorname{Im} \gamma_n(t_n)|, \gamma_n(s_n) \in (0, 1), \gamma_n(t_n, s_n) \subset S_0 \quad \forall n,$$

where S_0 is the intersection of S with one of the upper or lower half-planes. Since γ_n is a one-to-one curve, it is clear from (4) that the set (bar denoting conjugation) $\gamma_n[t_n, s_n] \cup \bar{\gamma}_n[t_n, s_n]$ is the range of a one-to-one curve. Let $\Gamma_n\colon [0, 1] \to \mathbb{C}$ be an arc with

$$(5) \qquad \Gamma_n(0) = \gamma_n(t_n), \qquad \Gamma_n(1) = \bar{\gamma}_n(t_n)$$

$$(6) \qquad \Gamma_n(0, 1) = \gamma_n(t_n, s_n] \cup \bar{\gamma}_n(t_n, s_n] \subset S.$$

By (ii) and 4.1 we can find continuous $\phi_n\colon [0, 1] \to \mathbb{C}$ such that

$$(7) \qquad \Gamma_n = e^{2\pi i \phi_n}.$$

According to (4) we have

$$\Gamma_n(0) = \gamma_n(t_n) = |\gamma_n(t_n)| e^{\pm i\pi/4}, \qquad \Gamma_n(1) = \bar{\gamma}_n(t_n) = |\gamma_n(t_n)| e^{\mp i\pi/4}.$$

Therefore (7) shows that

$$(8) \qquad \phi_n(1) - \phi_n(0) = \pm\tfrac{1}{4} + k_n, \quad \text{for some integer } k_n.$$

If we define

$$(9) \qquad \psi_n(t) = \begin{cases} \phi_n(t) & 0 \le t \le 1 \\ \phi_n(t - 1) + \phi_n(1) - \phi_n(0) & 1 \le t \le 2 \\ \phi_n(t - 2) + 2[\phi_n(1) - \phi_n(0)] & 2 \le t \le 3 \\ \phi_n(t - 3) + 3[\phi_n(1) - \phi_n(0)] & 3 \le t \le 4, \end{cases}$$

then ψ_n is (well-defined and) continuous and by (7) and (8) it satisfies

$$(10) \qquad e^{2\pi i \psi_n(t)} = \begin{cases} \Gamma_n(t) & 0 \le t \le 1 \\ \pm i\Gamma_n(t - 1) & 1 \le t \le 2 \\ -\Gamma_n(t - 2) & 2 \le t \le 3 \\ \mp i\Gamma_n(t - 3) & 3 \le t \le 4. \end{cases}$$

We have

$$\psi_n(4) - \psi_n(0) \overset{(9)}{=} 4[\phi_n(1) - \phi_n(0)] \overset{(8)}{=} \pm 1 + 4k_n.$$

Therefore $e^{2\pi i\psi_n}$ is a *closed* curve and

(11) $\operatorname{Ind}_{e^{2\pi i\psi_n}}(0) = \pm 1 + 4k_n \neq 0.$

From (10) and the inclusion (6) it follows that $e^{2\pi i\psi_n}$ is one-to-one on $(0, 4)$ and so (recall 3.26) its range is a Jordan-curve J_n. From (11) and 4.42 it follows that $0 \in \mathscr{I}(J_n)$. Since $J_n \subset D$, $\mathbb{C}\backslash D$ is a connected unbounded subset of $\mathbb{C}\backslash J_n$, consequently it lies in $\mathscr{O}(J_n)$. Thus

(12) $0 \in \mathscr{I}(J_n) \subset D.$

Form the function

$$F(z) = f(z)\bar{f}(\bar{z})f(iz)\bar{f}(-i\bar{z})f(-z)\bar{f}(-\bar{z})f(-iz)\bar{f}(i\bar{z}), \quad z \in D.$$

It is holomorphic by 2.4. For any $z \in J_n$, (10) shows that one of the numbers $\pm z$, $\pm iz$ belongs to the range of Γ_n. Therefore, glancing back to (6), we see that one of the factors in $F(z)$ is majorized by

$$\varepsilon_n \overset{\text{def.}}{=} \sup|f(\gamma_n)|.$$

If M is a majorant for $|f|$ on D, this means that

(13) $|F| \leq \varepsilon_n M^7$ on J_n.

From (12), (13) and the Maximum Modulus Principle we infer that

$$|F(0)| \leq \varepsilon_n M^7.$$

This holds for every n and by (iii), $\varepsilon_n \to 0$. We conclude that $0 = F(0) = |f(0)|^8$.

Lemma 9.11 *Let Ω be a Jordan region, g a conformal map of Ω onto $D = D(0, 1)$. If $w_n \in \Omega$ and $w_n \to w \in \partial\Omega$, then $z_n = g(w_n)$ converges to some point of ∂D.*

Proof: Let $f = g^{-1}$. We have

$$|z_n| \to 1.$$

For otherwise $\{z_n\}$ has a cluster point z in D and then $f(z)$ is a cluster point of $f(\{z_n\}) = \{w_n\}$ in Ω, contrary to $w_n \to w \in \partial\Omega$. By compactness of $\bar{D}$, it suffices to show that $\{z_n\}$ has only one cluster point. (Cf. 7.8.) So we suppose a, b two distinct cluster points of $\{z_n\}$ and try to deduce a contradiction.

Thus we have two sequences $\{a_n\}, \{b_n\} \subset D$ such that

(1) $a_n \to a, \quad b_n \to b, \quad |a| = |b| = 1,$

(2) $f(a_n) \to w, \quad f(b_n) \to w.$

We may suppose that no a_n equals b_n. Then $f(a_n) \neq f(b_n)$. Because of (2) and 4.48 there are polygons P_n joining $f(a_n)$ to $f(b_n)$ in Ω such that $P_n \to w$. By 1.14 each P_n may be replaced with an arc Γ_n with the same properties. Thus

(3) for any $\varepsilon > 0$, $\Gamma_n \subset D(w, \varepsilon)$ for all sufficiently large n.

Now form $\gamma_n = g \circ \Gamma_n$. These are arcs in D and their endpoints are a_n, b_n. Also, for any $r < 1$, γ_n is disjoint from $D(0, r)$ for all sufficiently large n, because of (3): if γ_n visits $D(0, r)$ for large n, then $f \circ \gamma_n = \Gamma_n$ visits the compact subset $f(\bar{D}(0, r))$ of Ω for large n, against (3) and the fact $w \in \partial\Omega$. Since $f(\gamma_n) = \Gamma_n \to w$, all the hypotheses of Koebe's arc lemma above are met (with $f - w$ in the role of the function f there). Conclude that $f - w \equiv 0$, a contradiction.

Lemma 9.12 *Let Ω be a Jordan region, g a conformal map of Ω onto $D = D(0, 1)$. Then g extends to a continuous map of $\bar{\Omega}$ onto $\bar{D}$.*

Proof: For each $w \in \partial\Omega = \bar{\Omega}\backslash\Omega$ the last lemma provides a unique $z \in \partial D$ such that

(1) $g(w_n) \to z$ whenever $w_n \in \Omega$ and $w_n \to w$.

Define $g(w)$ to be this z. This extends g to $\bar{\Omega}$. To check continuity of the extended g on $\bar{\Omega}$, we only have to see that

$$w_n' \in \bar{\Omega} \;\&\; w_n' \to w \in \partial\Omega \quad \text{imply } g(w_n') \to g(w).$$

To this end select $w_n \in \Omega$ such that

(2) $|w_n - w_n'| < 1/n$

and

(3) $|g(w_n) - g(w_n')| < 1/n$.

If $w_n' \in \Omega$ we can just take $w_n = w_n'$. Otherwise, $w_n' \in \partial\Omega$ and the choice of such w_n is possible because of the way $g(w_n')$ is defined. From (2) and $w_n' \to w$, we have $w_n \to w$. Then from (1) we get $g(w_n) \to g(w)$. Finally then, from (3) follows $g(w_n') \to g(w)$.

Since $\bar{\Omega}$ is compact, the set $g(\bar{\Omega})$ is compact and contains $g(\Omega) = D$. Hence $g(\bar{\Omega}) \supset \bar{D}$, and we have $g(\bar{\Omega}) = \bar{D}$.

Lemma 9.13 *Let Ω be a Jordan region, g a conformal map of Ω onto $D = D(0, 1)$. Then the extension of the last lemma is one-to-one on $\bar{\Omega}$.*

Proof: The proof of the last lemma shows that g maps $\partial\Omega$ into ∂D, so only univalence on $\partial\Omega$ has to be checked. (Here we are using the name g for the extension too.) We can argue by contradiction and suppose that

$$a, b \in \partial\Omega, \quad a \neq b \quad \text{and} \quad g(a) = g(b).$$

We may rotate D and assume that $g(a) = 1$. Let ϕ be a homeomorphism of ∂D onto $\partial\Omega$ (which is a Jordan-curve). By preceding ϕ with an appropriate homeomorphism of $\bar{D}$ onto $\bar{D}$ (see 6.4) we can suppose

$$\phi(+i) = a, \qquad \phi(-i) = b.$$

Let

$$\Gamma(t) = \phi(e^{it}), \quad 0 \le t \le 2\pi.$$

Then

$$\partial\Omega = \Gamma[0, 2\pi], \qquad \Gamma\left(\frac{\pi}{2}\right) = a, \qquad \Gamma\left(\frac{3\pi}{2}\right) = b.$$

Since the accessible points of $\partial\Omega$ are dense in $\partial\Omega$ (1.16) and Γ is a local homeomorphism, we can select a_n, b_n accessible points of $\partial\Omega$ sufficiently near a and b respectively that

$$(1) \qquad \begin{cases} a_n \in \Gamma\left(\frac{\pi}{4}, \frac{3\pi}{4}\right) \ \& \ |g(a_n) - 1| < 1/n \\ b_n \in \Gamma\left(\frac{5\pi}{4}, \frac{7\pi}{4}\right) \ \& \ |g(b_n) - 1| < 1/n. \end{cases}$$

Let, say

$$(2) \qquad \begin{cases} a_n = \Gamma(\alpha_n), \quad \frac{\pi}{4} < \alpha_n < \frac{3\pi}{4} \\ b_n = \Gamma(\beta_n), \quad \frac{5\pi}{4} < \beta_n < \frac{7\pi}{4}. \end{cases}$$

Since a_n, b_n are distinct accessible points of $\partial\Omega$, there is by 1.29 a cross-cut C_n in Ω joining a_n to b_n. Form

$$(3) \qquad \begin{cases} \Gamma_n' = \Gamma[\alpha_n, \beta_n] \supset \Gamma\left[\frac{3\pi}{4}, \pi\right] = A' \\ \Gamma_n'' = \Gamma[\beta_n, \alpha_n + 2\pi] \supset \Gamma\left[\frac{7\pi}{4}, 2\pi\right] = A''. \end{cases}$$

According to 4.47, $C_n \cup \Gamma_n'$ and $C_n \cup \Gamma_n''$ are Jordan-curves which satisfy

$$(4) \qquad \Omega \backslash C_n = \mathscr{I}(C_n \cup \Gamma_n') \cup \mathscr{I}(C_n \cup \Gamma_n''), \text{ disjoint union.}$$

We write $\gamma(t) = e^{it}$, $-\pi \le t \le \pi$, and for each n select $s_n, t_n \in [-\pi, \pi]$ such that $\gamma(s_n) = g(a_n)$, $\gamma(t_n) = g(b_n)$. According to (1) we then have

$$(5) \qquad s_n, t_n \in \left(-\frac{\pi}{2}, \frac{\pi}{2}\right) \quad \text{and} \quad s_n \to 0, t_n \to 0.$$

Let

$$(6) \qquad \gamma_n' = \gamma[s_n, t_n], \ \gamma_n'' \ \text{ the complementary arc in } \partial D.$$

Notice that, since g maps Ω univalently onto D, $g(C_n)$ is a cross-cut c_n in $g(\Omega) = D$ with endpoints $g(a_n) = \gamma(s_n)$ and $g(b_n) = \gamma(t_n)$. Therefore, citing 4.47 once again, we have from (6) that $c_n \cup \gamma_n'$ and $c_n \cup \gamma_n''$ are Jordan-curves which satisfy

(7) $\quad D\backslash c_n = \mathscr{I}(c_n \cup \gamma_n') \cup \mathscr{I}(c_n \cup \gamma_n''),$ disjoint union.

Since g is a homeomorphism of $\Omega\backslash C_n$ onto $g(\Omega\backslash C_n) = g(\Omega)\backslash g(C_n) = D\backslash c_n$, connectedness considerations show that each of the two sets on the right of (4) is mapped onto one of those on the right of (7). In particular, for every $n = 1, 2, \ldots$

(8) $\quad$ either $g(\mathscr{I}(C_n \cup \Gamma_n')) = \mathscr{I}(c_n \cup \gamma_n')$ or $g(\mathscr{I}(C_n \cup \Gamma_n'')) = \mathscr{I}(c_n \cup \gamma_n')$.

Consider an n for which the first half of the alternative prevails:

(8)′ $\quad g(\mathscr{I}(C_n \cup \Gamma_n')) = \mathscr{I}(c_n \cup \gamma_n')$.

For any $w \in \Gamma_n'$ there is (by 4.41) a sequence $\{w_k\} \subset \mathscr{I}(C_n \cup \Gamma_n')$ such that $w_k \to w$. Then from (8)′ we have

$$g(w_k) \in \mathscr{I}(c_n \cup \gamma_n') \;\&\; g(w_k) \to g(w) \in \partial D.$$

Therefore

$$\begin{aligned} g(w) \in \overline{\mathscr{I}(c_n \cup \gamma_n')} \cap \partial D &= [\mathscr{I}(c_n \cup \gamma_n') \cup (c_n \cup \gamma_n')] \cap \partial D \\ &\subset (D \cup \gamma_n') \cap \partial D = \gamma_n'. \end{aligned}$$

This all shows that

$$g(\Gamma_n') \subset \gamma_n'.$$

In particular, from (3)

(9)′ $\quad g(A') \subset \gamma_n'$.

We show in the same way that if for a given n the second half of the alternative (8) prevails, then

(9)″ $\quad g(A'') \subset \gamma_n'$.

Now one of (9)′, (9)″ must prevail for infinitely many n. Let $n \to \infty$ through these values and remember (5) and (6). It follows that

$$\text{either } g(A') \subset \{1\} \quad \text{or } g(A'') \subset \{1\}.$$

Consequently, by 5.76, g is constant. Contradiction!

Theorem 9.14 (Osgood and Taylor [1913], Carathéodory [1913b]) *Let Ω be a Jordan region and f a conformal map of $D = D(0, 1)$ onto Ω. Then f extends to a homeomorphism of $\bar{D}$ onto $\bar{\Omega}$.*

Proof: Combine the last two lemmas.

As a complement to this result we have 9.16 below.

§ 5 A Few Applications of the Osgood–Taylor–Carathéodory Theorem

Corollary 9.15 (Argument Principle) *Let Γ be a simple loop, f a continuous function on $\overline{\mathscr{I}(\Gamma)}$ which is holomorphic in $\mathscr{I}(\Gamma)$ and zero-free on Γ. Then the number of zeros of f in $\mathscr{I}(\Gamma)$, counted according to multiplicity, is finite and equals* $|\mathrm{Ind}_{f\circ\Gamma}(0)|$. [This generalizes 5.86.]

Proof: Let ϕ be a conformal map of $D = D(0, 1)$ onto $\mathscr{I}(\Gamma)$ and extend it to a homeomorphism of $\bar{D}$ onto $\overline{\mathscr{I}(\Gamma)}$ with $C(0, 1)$ mapping to Γ. Define $\Gamma_r(t) = re^{2\pi it}$ for $0 \le r, t \le 1$. Then $\phi \circ \Gamma_1$ and Γ have the same range, so

(1) $|\mathrm{Ind}_{f\circ\Gamma}(0)| = |\mathrm{Ind}_{f\circ\phi\circ\Gamma_1}(0)|$

by 4.7(ii). Consider $F = f \circ \phi$. It is easy to see that

(2) $\begin{cases} F \text{ has a zero of order } n \text{ at } z \in D \text{ if and only if} \\ f \text{ has a zero of order } n \text{ at } \phi(z). \end{cases}$

We essentially checked this near the end of the proof of 7.32. Therefore, the number of zeros of f in $\mathscr{I}(\Gamma)$, counted according to multiplicity, equals the number of zeros of F in D, counted according to multiplicity. Call this number N. Since the zero set of F in $\bar{D}$ is closed and lies wholly in D [the zero set of f lies wholly in $\mathscr{I}(\Gamma) = \phi(D)$ by hypothesis], this set lies in $D(0, r)$ for all r sufficiently near 1. Consequently we have from 5.86

(3) $N = \mathrm{Ind}_{F\circ\Gamma_r}(0)$

for all $r \in (0, 1)$ sufficiently close to 1. Let $r \uparrow 1$ and cite 4.6 to assert that

(4) $\mathrm{Ind}_{F\circ\Gamma_r}(0) \to \mathrm{Ind}_{F\circ\Gamma_1}(0)$.

It follows from (3), (4) and (1) that

$$N = \mathrm{Ind}_{F\circ\Gamma_1}(0) = |\mathrm{Ind}_{f\circ\Gamma}(0)|.$$

Corollary 9.16 (Darboux–Picard) *Let D be an open disk, $f\colon \bar{D} \to \mathbb{C}$ be continuous and satisfy*

(i) *f is holomorphic in D.*
(ii) *f is one-to-one on ∂D.*

Then f is one-to-one throughout $\bar{D}$ and $f(D)$ is the inside of the Jordan-curve $f(\partial D)$.

Proof: Apply the last corollary to $\Gamma = \partial D$: if $w \in \mathbb{C}\backslash f(\partial D)$, then $f - w$ is zero-free on ∂D and so

(1) the number of zeros of $f - w$ in D equals $|\mathrm{Ind}_{f\circ\partial D}(w)|$.

But by hypothesis $f \circ \partial D$ is a simple loop, so by 4.42 this index is always ± 1 or 0, according as w belongs to the inside of the Jordan-curve $J = f(\partial D)$ or to its outside. We learn therefore from (1) that

(2) $f - w$ has a unique zero in D for all $w \in \mathscr{I}(J)$

(3) $f - w$ has no zero in D for all $w \in \mathscr{O}(J)$.

In particular

(4) $\mathscr{I}(J) \subset f(D) \subset \mathbb{C}\backslash\mathscr{O}(J) = \mathscr{I}(J) \cup J$.

Since (5.77) $f(D)$ is an open set, it cannot meet J without meeting $\mathcal{O}(J)$ [see 4.41] and therefore (4) entails

(5) $\quad \mathscr{I}(J) = f(D).$

Then

$$\varnothing \overset{(5)}{=} J \cap f(D) = f(\partial D) \cap f(D)$$

shows that f takes no value in D which it also takes on ∂D. By (2) and (5) f is one-to-one in D and by hypothesis f is one-to-one on ∂D. It follows that f is one-to-one throughout $\bar{D}$ and the proof is complete.

Exercise 9.17 (i) *With the aid of the Osgood–Taylor–Carathéodory theorem formulate and prove a version of the last corollary in which D is any Jordan region.* (Cf. § 11 *of* CARATHÉODORY *and* RADEMACHER [1917] *and Satz* 3 *of* RADÓ [1922/23c].)

(ii) (*Global version of* 5.80 *from* MONTEL [1932d].) *Let* Γ *be a Jordan-curve,* $F: \overline{\mathscr{I}(\Gamma)} \to \mathbb{C}$ *continuous and holomorphic in* $\mathscr{I}(\Gamma)$. *Show that for each* $z_0 \in \mathscr{I}(\Gamma)$ *there exist two distinct points* $z_1, z_2 \in \Gamma$ *such that*

$$\frac{F(z_2) - F(z_1)}{z_2 - z_1} = F'(z_0).$$

Hint: The function $f(z) = F(z) - F'(z_0)z$ has $f'(z_0) = 0$ and so (5.78) is not univalent in $\mathscr{I}(\Gamma)$. By (i) it cannot be univalent on Γ either.

(iii) *Show that every point inside of a Jordan-curve is the midpoint of some chord of the curve.*

Hint: Consider $F(z) = z^2$ in (ii).

Typical applications of the Argument Principle are offered in the next exercise. For more complicated examples see KRZYŻ [1971], § 3.9.

Exercise 9.18 (i) *Show that for any positive numbers a, b, c the polynomial* $P(z) = z^8 + az^3 + bz + c$ *has exactly two zeros in the (open) first quadrant.*

Hints: Evidently there exists an $r > 0$ such that

$$|z^{-8}P(z) - 1| < 1 \quad \forall |z| \geq r,$$

so that if we set $g(z) = z^{-8}P(z)$, then

(1) $\quad g(z) \in D(1, 1) \quad \forall |z| \geq r,$

(2) $\quad P(z) = z^8 g(z) \quad \forall |z| \geq r.$

Now $P(it) = (t^8 + c) + i(-at^3 + bt)$ for real t and so

(3) $\quad P[0, r] \subset [c, \infty) \subset (0, \infty)$

(4) $\quad P[0, ir] \subset [c, \infty) \times \mathbb{R} \subset (0, \infty) \times \mathbb{R}.$

Let γ be the piecewise smooth simple loop defined by

$$\text{(5)}\qquad \gamma(t) = \begin{cases} -it & -r \le t \le 0 \\ t & 0 \le t \le r \\ re^{\pi i(t-r)/2} & r \le t \le r+1 \end{cases}$$

and set $\Gamma = P \circ \gamma$. Thus by (2)

$$\text{(6)}\qquad \Gamma(t) = \begin{cases} P(\gamma(t)) & -r \le t \le r \\ \gamma^8(t)g(\gamma(t)) & r \le t \le r+1. \end{cases}$$

Let L be the holomorphic logarithm in $\mathbb{C}\backslash(-\infty, 0]$ which satisfies

$$\text{(7)}\qquad -\pi < \operatorname{Im} L < \pi.$$

According to (3), (4) and (1) we may form $L(P[0, r])$, $L(P[0, ir])$ and $L(g(z))$ for all $|z| \ge r$. Therefore we may define a continuous function $\phi: [-r, r+1] \to \mathbb{C}$ by

$$\phi(t) = \begin{cases} L(P(\gamma(t))) & -r \le t \le r \\ L(r^8) + 4\pi i(t-r) + L(g(\gamma(t))) & r \le t \le r+1. \end{cases}$$

We must check that this is well-defined; then continuity is obvious. So we must check that

$$L(P(\gamma(r))) = L(r^8) + L(g(\gamma(r))),$$

that is,

$$L(P(r)) = L(r^8) + L(g(r)).$$

Since the exponentials of both sides of this putative equation are equal (by (2)), we know that the real parts of both sides are equal and the imaginary parts differ by an integer multiple of 2π, say

$$\operatorname{Im}[L(g(r)) + L(r^8) - L(P(r))] = 2\pi n.$$

Since L is real on $(0, \infty)$, this means

$$\operatorname{Im}[L(g(r)) - L(P(r))] = 2\pi n$$

and from (7) it follows that $n = 0$.

Because of (6) and (5)

$$\text{(8)}\qquad \Gamma = e^{\phi}.$$

Now we have

$$e^{L(r^8) + L(g(ir))} = r^8 g(ir) = (ir)^8 g(ir) \overset{(2)}{=} P(ir) = e^{L(P(ir))}$$

and therefore the reasoning above shows that

$$\text{(9)}\qquad L(r^8) + L(g(ir)) = L(P(ir)).$$

From (8) and the definition of index we have

$$\begin{aligned}\mathrm{Ind}_{P\circ\gamma}(0) = \mathrm{Ind}_{\Gamma}(0) &= \frac{1}{2\pi i}[\phi(r+1) - \phi(-r)] \\ &= \frac{1}{2\pi i}[L(r^8) + 4\pi i + L(g(\gamma(r+1))) - L(P(\gamma(-r)))] \\ &= \frac{1}{2\pi i}[L(r^8) + 4\pi i + L(g(ir)) - L(P(ir))] \\ &= 2, \quad \text{by (9)}.\end{aligned}$$

Consequently, by 9.15 P has exactly two zeros in $\mathscr{I}(\gamma)$. Of course, by (1) and (2) P has no zeros of modulus greater-than-or-equal-to r. Thus P has exactly two zeros in the open first quadrant.

(ii) *Show that the polynomial $P(z) = z^4 + z^3 + 5z^2 + 2z + 4$ has no zeros in the first quadrant.*

Hints: Define $g(z) = z^{-4}P(z)$ and select $r > 2$ sufficiently large that

(1) $\quad g(z) \in D(1, 1) \quad \forall |z| \geq r.$

Notice that

(2) $\quad P(it) = (t^2 - 4)(t^2 - 1) - it(t^2 - 2) \quad \forall t \in \mathbb{R}.$

Therefore

(3) $\quad P(it) \in \mathbb{C}\backslash[0, \infty) \quad \text{for } t > 0$

(indeed $t > 0 \;\&\; P(it) \in \mathbb{R} \Rightarrow t = \sqrt{2} \Rightarrow P(it) = -2$) and

(4) $\quad P(t) \in [4, \infty) \quad \text{for } t \geq 0.$

Let L_1 be the holomorphic logarithm in $\mathbb{C}\backslash[0, \infty)$ with $0 < \operatorname{Im} L_1 < 2\pi$, L the holomorphic logarithm in $\mathbb{C}\backslash(-\infty, 0]$ with $-\pi < \operatorname{Im} L < \pi$. Define γ as in (i). Then by (3), (4) and (1) we may define

$$\phi(t) = \begin{cases} L_1(P(\gamma(t))) & -r \leq t < 0 \\ L(P(\gamma(t))) & 0 \leq t \leq r \\ L(r^4) + 2\pi i(t - r) + L(g(\gamma(t))) & r \leq t \leq r + 1. \end{cases}$$

As in (i), ϕ is well-defined at r. Moreover

$$\lim_{t\uparrow 0} L_1(P(\gamma(t))) = \lim_{t\downarrow 0} L_1(P(it)) = \log 4$$

because by (2), $P(it)$ approaches 4 from the first quadrant as $t \downarrow 0$. Since L is real on $(0, \infty)$,

$$L(P(\gamma(0))) = L(P(0)) = L(4) = \log 4.$$

Therefore ϕ is continuous at 0, hence throughout $[-r, r+1]$. As in (i),

$$e^{L(r^4)+L(g(ir))} = r^4 g(ir) = (ir)^4 g(ir) = P(ir) = e^{L_1(P(ir))}$$

implies that for some integer n

$$L(r^4) + L(g(ir)) - L_1(P(ir)) = 2\pi i n,$$
$$\operatorname{Im}[L(g(ir)) - L_1(P(ir))] = 2\pi n.$$

Now $-\pi < \operatorname{Im} L < \pi$ and from (2) and the fact $r > 2$, $P(ir)$ is in the fourth quadrant, so $\pi < \operatorname{Im} L_1(P(ir)) < 2\pi$. Consequently

$$-3\pi = -\pi - 2\pi < \operatorname{Im} L(g(ir)) - \operatorname{Im} L_1(P(ir)) < \pi - \pi = 0.$$

Therefore n above is -1, and we compute

$$\begin{aligned}\operatorname{Ind}_{P\circ\gamma}(0) &= \frac{1}{2\pi i}[\phi(r+1) - \phi(-r)] \\ &= \frac{1}{2\pi i}[L(r^4) + 2\pi i + L(g(ir)) - L_1(P(ir))] \\ &= 0.\end{aligned}$$

Conclude as before.

Exercise 9.19 (SCHÖNFLIES [1906]. *See* 4.44.) *Let J be a Jordan-curve, h any homeomorphism of the unit circle C onto J.*

(i) *If Ω is the inside of J, then h extends to a homeomorphism of $\bar{D}$ onto $\bar{\Omega}$, D the open unit disk. If $0 \in \Omega$, then it can be arranged that the extension map 0 to 0.*

Hints: There is by the Osgood–Taylor–Carathéodory theorem a homeomorphism F of $\bar{D}$ onto $\bar{\Omega}$ which maps D onto Ω and (hence) maps C onto J. Furthermore we can have $F(0) = 0$ if $0 \in \Omega$. 4.43 provides an extension H of $F^{-1} \circ h$ to a homeomorphism of $\bar{D}$ onto $\bar{D}$ with, moreover, $H(0) = 0$. Then $F \circ H$ is an extension of h of the desired kind.

(ii) *If U is the outside of J, then h extends to a homeomorphism of $\mathbb{C}\backslash D$ onto $\bar{U}$.*

Hints: We may assume that $0 \in \Omega$. Let ϕ be the homeomorphism $\phi(z) = 1/z$ of $\mathbb{C}\backslash\{0\}$ onto $\mathbb{C}\backslash\{0\}$. It is easy to see that $\phi(J)$ is a Jordan-curve $\tilde{J}$ and that $\tilde{\Omega} = \phi(U) \cup \{0\}$ is its inside. Consequently by part (i), the homeomorphism $\tilde{h} = \phi \circ h \circ \phi^{-1}$ of C onto $\tilde{J}$ extends to a homeomorphism $\tilde{H}$ of $\bar{D}$ onto $\bar{\tilde{\Omega}} = \phi(\bar{U}) \cup \{0\}$ which satisfies $\tilde{H}(0) = 0$. Then $\phi^{-1} \circ \tilde{H} \circ \phi$ maps $\phi^{-1}(\bar{D}\backslash\{0\}) = \mathbb{C}\backslash D$ homeomorphically onto $\phi^{-1}(\phi(\bar{U})) = \bar{U}$ and extends h.

(iii) *h extends to a homeomorphism of $\mathbb{C}$ onto $\mathbb{C}$.*

Hint: Put (i) and (ii) together.

Remarks: If $f(z) = \sum_{n=0}^{\infty} c_n z^n$ is a conformal map of D onto the inside of some Jordan-curve J, then not only does f extend continuously to $\bar{D}$, but the series converges uniformly in $\bar{D}$. This beautiful result of FEJÉR [1914] will be proved (by an area method) in Chapter XVIII. One consequence of it is that every Jordan-curve J is the range of a homeomorphism h of $C(0, 1)$ which has the form $h(z) = \sum_{n=0}^{\infty} c_n z^n$, convergence uniform in $z \in C(0, 1)$.

Exercise 9.20 (Extension of 5.16) *Let Ω be a Jordan region, $z_0 \in \partial\Omega$, F: $\bar{\Omega}\backslash\{z_0\} \to \mathbb{C}$ bounded and continuous, holomorphic in Ω. Pick any* other *point $z_1 \in \partial\Omega$ and let A_1, A_2 be the two arcs of $\partial\Omega$ determined by z_0 and z_1.*

(i) *Suppose that*

$$\overline{\lim_{\substack{z \to z_0 \\ z \in A_j}}} |F(z)| \le m \quad (j = 1, 2).$$

Show that then

$$\overline{\lim_{\substack{z \to z_0 \\ z \in \Omega}}} |F(z)| \le m.$$

(ii) *Suppose that $\lim_{z \to z_0, z \in A_j} F(z)$ each exist $(j = 1, 2)$. Show that F is (extendable to be) continuous on $\bar{\Omega}$.*

Hints: There is a conformal map ϕ of Ω onto $D = D(0, 1)$. This map extends to a homeomorphism of $\bar{\Omega}$ onto $\bar{D}$. We may suppose (after a rotation) that $\phi(z_0) = 1$. The map $\psi(z) = (z + 1)/(z - 1)$ sends D conformally onto $S = (-\infty, 0) \times \mathbb{R}$ (see 2.5) and the upper unit semi-circle onto the negative imaginary axis, the lower unit semi-circle onto the positive imaginary axis. Apply 5.16 to the composite $f = F \circ \phi^{-1} \circ \psi^{-1}$.

§ 6 More on Jordan Regions and Boundary Behavior

The object here is to use the Osgood–Taylor–Carathéodory theorem to get uniform polynomial approximation on the whole closure of a Jordan region, thus improving the Polynomial Runge theorems. Along the way we look at some elementary instances of the following phenomenon which CARATHÉODORY [1912a] explored in depth: if the simply-connected regions Ω_n converge in some geometric sense to Ω, do the associated (appropriately normalized) Riemann maps of Ω_n onto $D(0, 1)$ converge to that of Ω onto $D(0, 1)$? None of the material of this admittedly rather heavy section is needed anywhere else in the book and it may be skipped by the less ardent reader.

Exercise 9.21 *Let Γ_n, Γ: $C = C(0, 1) \to \mathbb{C}$ be homeomorphisms such that*

(i) $|\Gamma_n - \Gamma| \to 0$ *uniformly on C*

(ii) $\overline{\mathscr{I}(\Gamma)} \subset \mathscr{I}(\Gamma_{n+1}) \subset \mathscr{I}(\Gamma_n)$ *for all n.*

Show that

$$\overline{\mathscr{I}(\Gamma)} = \bigcap_{n=1}^{\infty} \overline{\mathscr{I}(\Gamma_n)} = \bigcap_{n=1}^{\infty} \mathscr{I}(\Gamma_n).$$

Hints: It suffices to show that $\bigcap_{n=1}^{\infty} \overline{\mathscr{I}(\Gamma_n)} \subset \overline{\mathscr{I}(\Gamma)}$. Let $w \in \bigcap_{n=1}^{\infty} \overline{\mathscr{I}(\Gamma_n)}$. If $w \in \Gamma$, there is nothing further to prove, since $\Gamma \subset \overline{\mathscr{I}(\Gamma)}$. If $w \notin \Gamma$, then for all large n, $w \notin \Gamma_n$, by (i). Consequently $w \in \overline{\mathscr{I}(\Gamma_n)} \backslash \Gamma_n = \mathscr{I}(\Gamma_n)$ and [by 4.42] $|\text{Ind}_{\Gamma_n}(w)| = 1$. From (i) and 4.6 we know that $\text{Ind}_{\Gamma_n}(w) \to \text{Ind}_{\Gamma}(w)$. Consequently, $|\text{Ind}_{\Gamma}(w)| = 1$. This means that w is not in the (unique) unbounded component of $\mathbb{C}\backslash\Gamma$, so $w \in \mathscr{I}(\Gamma)$, the unique bounded component of $\mathbb{C}\backslash\Gamma$.

Exercise 9.22 *Let Ω_n be bounded, simply-connected regions such that $\Omega_{n+1} \subset \Omega_n$ for all $n = 1, 2, \ldots$. Let $\Omega = \bigcap_{n=1}^{\infty} \Omega_n$ and suppose that 0 is an interior point of Ω. Let Ω_0 be the component of the interior of Ω which contains 0. Let f_n be the unique* [9.9] *conformal map of $D = D(0, 1)$ onto Ω_n which satisfies $f_n(0) = 0$, $f_n'(0) > 0$. Show that*

(i) *$f = \lim_{n\to\infty} f_n$ exists locally uniformly on D*

(ii) *f is the unique conformal map of D onto Ω_0 which satisfies $f(0) = 0$, $f'(0) > 0$.*

Hints: Since Ω_1 is bounded, the family $\{f_n\}$ is bounded. Therefore by 7.6 there are plenty of convergent subsequences. We will show that the limit function of any one of them is conformal, takes the value 0 at the point 0, has positive derivative there and has range equal to Ω_0. Since the function answering to this description is unique, we can invoke 7.8 to learn that the whole sequence $\{f_n\}$ converges to this function.

Thus suppose f is the local uniform limit of a subsequence of $\{f_n\}$. We may, for notational simplicity, suppose that subsequence is the whole sequence $\{f_n\}$. Choose $r > 0$ so that $\bar{D}(0, r) \subset \Omega$. Such r exist by hypothesis. Then apply Schwarz' lemma, appropriately scaled, to f_n^{-1} in $D(0, r)$ to get $|(f_n^{-1})'(0)| \leq 1/r$. From $f_n(0) = 0$ and the Chain Rule we have $(f_n^{-1})'(0) = 1/f_n'(0)$. Thus $f_n'(0) \geq r$ for all n. Since $f_n' \to f'$ (by 5.44(ii)), we have $f'(0) \geq r$. Hence f is not constant and so by Hurwitz' theorem f is conformal.

Given $z_0 \in D$, since f is one-to-one, $f - f(z_0)$ has a zero of order 1 in D at z_0. Since $f_n - f(z_0)$ converges to $f - f(z_0)$ locally uniformly in D, it follows from 8.19 (or from 7.10) that for all sufficiently large n, $f_n - f(z_0)$ has a zero in D, i.e., $f(z_0) \in f_n(D) = \Omega_n$. This shows that $f(D) \subset \bigcap \Omega_n$. Since $0 = f(0) \in f(D)$ and $f(D)$ is connected, we get $f(D) \subset \Omega_0$.

Use the Basic Connectedness Lemma to show that this inclusion is not proper. To this end show that

$$(1) \qquad w_0 \in f(D) \;\&\; D(w_0, 4\delta) \subset \Omega_0 \Rightarrow D(w_0, \delta) \subset f(D).$$

Write $w_0 = f(z_0)$. For each n the function $f_n^{-1}(w_0 + \xi)$, for $\xi \in D(0, 4\delta)$, is holomorphic and maps into D, hence is bounded by 1; consequently, for $\xi = 0$ it satisfies $|f_n^{-1}(w_0)| = |f_n^{-1} \circ f(z_0)| \leq |z_0|$, by an application of 6.1 to $f_n^{-1} \circ f$. Therefore an application of 6.3(iv) shows that

$$|f_n^{-1}(w_0 + \xi)| \le \frac{|z_0| + 1/2}{1 + |z_0|/2} = r, \quad \text{say, } \forall \xi \in D(0, 2\delta).$$

(2) $\quad |f_n^{-1}(w)| \le r < 1 \quad \forall w \in D(w_0, 2\delta).$

Set $R = \sqrt{r}$. For $z \in C(0, R)$, $f_n(z)$ cannot lie in $D(w_0, 2\delta)$, since otherwise by (2), $|z| = |f_n^{-1}(f_n(z))| < R$. Consequently,

$$|f_n(z) - w| \ge \delta \quad \forall w \in D(w_0, \delta), z \in C(0, R)$$

and so

(3) $$\frac{1}{2\pi i}\int_{C(0,R)} \frac{f_n'}{f_n - w} \to \frac{1}{2\pi i}\int_{C(0,R)} \frac{f'}{f - w} \quad \forall w \in D(w_0, \delta).$$

The left-hand expression in (3) is by 5.85 the number of zeros of $f_n - w$ in $D(0, R)$ and this is at least 1, since $f_n^{-1}(w) \in D(0, R)$ [by (2)] and $f_n(f_n^{-1}(w)) = w$. Consequently, the value of the expression on the right in (3) is at least 1. As this expression represents the number of zeros of $f - w$ in $D(0, R)$, we thus learn that $w \in f(D(0, R)) \subset f(D)$. This is the case for each $w \in D(w_0, \delta)$, thus confirming (1) and therewith the surjective character of f.

Theorem 9.23 (Carathéodory [1912a]) *Let $\Gamma, \Gamma_n: C = C(0, 1) \to \mathbb{C}$ be homeomorphisms such that*

(i) $\quad |\Gamma_n - \Gamma| \to 0$ *uniformly on C*

(ii) $\quad 0 \in \mathscr{I}(\Gamma) \subset \overline{\mathscr{I}(\Gamma)} \subset \mathscr{I}(\Gamma_{n+1}) \subset \mathscr{I}(\Gamma_n)$ *for all n.*

Let f_n be the unique [9.9] *conformal map of $D = D(0, 1)$ onto $\mathscr{I}(\Gamma_n)$ which satisfies $f_n(0) = 0$, $f_n'(0) > 0$. Its extension to a homeomorphism of $\bar{D}$ onto $\mathscr{I}(\Gamma_n)$ is also called f_n. Then the family $\{f_n\}$ is (uniformly) equicontinuous on C.*

Proof: (Radó [1922/23d]) We argue by contradiction. If $\{f_n\}$ is not equicontinuous on C then, invoking compactness several times, passing to subsequences and re-labelling to assume that the final subsequence is the original one, we find

(1) $\quad$ distinct points $z_n, w_n \in C$ such that

(2) $\quad z_n, w_n \to z_0,$

(3) $\quad |f_n(z_n) - f_n(w_n)| \ge 2\delta > 0 \quad \forall n,$

(4) $\quad f_n(z_n) \to a, \; f_n(w_n) \to b.$

It follows from (i) and (4) that $a, b \in \Gamma$. The points z_n, w_n divide C into two arcs, one of which, call it A_n, converges to z_0 in the obvious sense: every neighborhood of z_0 contains all but finitely many A_n. This one sees easily from (2) by using the exponential representation of C. Now $f_n(A_n)$ is an arc which meets both the disjoint [by (3)] open sets $D(f_n(z_n), \delta)$ and $D(f_n(w_n), \delta)$ [in their centers]. Being connected, it cannot lie wholly in their union. Therefore there exists a point

(5) $\quad c_n \in f_n(A_n)$ such that $|c_n - f_n(z_n)| \ge \delta$ and $|c_n - f_n(w_n)| \ge \delta.$

Passage to another subsequence lets us assume, in order to reach the ultimately sought contradiction, that

(6) $c_n \to c.$

From (5) and (4) then we have

(7) $|c - a| \geq \delta, \qquad |c - b| \geq \delta.$

From (i) and (6) we observe that $c \in \Gamma$. Let t_a, t_b, t_c be the points of C which satisfy

(8) $\Gamma(t_a) = a, \qquad \Gamma(t_b) = b, \qquad \Gamma(t_c) = c.$

Let B denote that (compact) arc of C with endpoints t_a, t_b which does not contain t_c. Let A denote a (compact) non-degenerate arc of C symmetric about t_c but disjoint from B. (Use the exponential representation of C to effect all this.) Now Γ is one-to-one and continuous. Since A and B are compact and disjoint, $\Gamma(A)$ and $\Gamma(B)$ are too. Since t_c and $\overline{C\backslash A}$ are compact and disjoint, so are $\Gamma(t_c) = c$ and $\Gamma(\overline{C\backslash A}) = \overline{\Gamma(C)\backslash\Gamma(A)}$. Hence there is an η such that

(9) $0 < 3\eta < \operatorname{dist}(c, \Gamma\backslash\Gamma(A))$

(10) $0 < 2\eta < \operatorname{dist}(\Gamma(A), \Gamma(B)).$

Using (i), (4) and (6) and discarding a few f_n, we can assume that

(11) $|\Gamma_n - \Gamma| < \eta$ on C for all n,

(12) $|f_n(z_n) - a| < \eta, |f_n(w_n) - b| < \eta, |c_n - c| < \eta$ for all n.

From these inequalities we want to deduce that

(13) $D(c, 2\eta) \cap \Gamma_n \subset f_n(A_n) \quad \forall n.$

Suppose for some n this is not the case. Then

(14) $|\Gamma_n(t) - c| < 2\eta \;\&\; \Gamma_n(t) \notin f_n(A_n)$ for some $t \in C$.

The point c_n belongs to $f_n(A_n) \subset f_n(C) = \Gamma_n$, so we can write

(15) $c_n = \Gamma_n(t_n)$, some $t_n \in C$.

We have

(16) $|\Gamma(t_n) - c| \leq |\Gamma(t_n) - \Gamma_n(t_n)| + |\Gamma_n(t_n) - c| \overset{(11)}{<} \eta + |c_n - c| \overset{(12)}{<} 2\eta,$

(17) $|\Gamma(t) - c| \leq |\Gamma(t) - \Gamma_n(t)| + |\Gamma_n(t) - c| \overset{(11)}{<} \eta + |\Gamma_n(t) - c| \overset{(14)}{<} 3\eta.$

Because of (9) the inequalities (16) and (17) imply that

$$\Gamma(t_n), \Gamma(t) \in \Gamma(A).$$

Since Γ is one-to-one, this means $t_n, t \in A$. Let A' denote the subarc of A having these endpoints. Now $\Gamma_n(t_n)$ and $\Gamma_n(t)$, the endpoints of $\Gamma_n(A')$, separate $f_n(z_n)$

and $f_n(w_n)$ in Γ_n, because $\Gamma_n(t_n) = c_n \in f_n(A_n)$ [by (5)] and $\Gamma_n(t) \in \Gamma_n \backslash f_n(A_n)$ [by (14)]. Therefore the arc $\Gamma_n(A')$ contains exactly one of $f_n(z_n)$ and $f_n(w_n)$, say $f_n(z_n) \in \Gamma_n(A')$:

(18) $\quad f_n(z_n) = \Gamma_n(t_n')$ for some $t_n' \in A'$.

But then we have

$$|\Gamma(t_n') - \Gamma(t_a)| \overset{(8)}{=} |\Gamma(t_n') - a| \overset{(18)}{\leq} |f_n(z_n) - a| + |\Gamma_n(t_n') - \Gamma(t_n')| < 2\eta$$

by (12), (11). Since $t_a \in B$ and $t_n' \in A' \subset A$, this inequality violates (10). We have thus finally confirmed (13).

Now define

(19) $\quad \varepsilon_n = \sup\{|f_n^{-1}(w) - z_0| : w \in f_n(A_n)\}$.

Since $A_n \to z_0$, we have

(20) $\quad \varepsilon_n \to 0$.

Our next goal is to use Lindelöf's estimate 5.14 to show that

(21) $\quad f_n^{-1}(w) - z_0 \to 0 \quad \forall w \in D(c, \eta) \cap \mathscr{I}(\Gamma)$.

Let $w_0 \in D(c, \eta) \cap \overline{\mathscr{I}(\Gamma)}$ be given. Consider the disk $D(w_0, \eta)$. It contains $c \in \Gamma$, the boundary of $\mathscr{I}(\Gamma)$. Therefore $D(w_0, \eta)$ contains points outside $\overline{\mathscr{I}(\Gamma)}$. If one such point has distance $r(<\eta)$ to w_0, then some non-degenerate arc of $C(w_0, r)$ lies outside $\overline{\mathscr{I}(\Gamma)}$. Let γ be such a (compact) arc with length $2\pi r/k$ for some positive integer k. γ is disjoint from $\overline{\mathscr{I}(\Gamma)}$ which by 9.21 equals $\bigcap_{n=1}^{\infty} \overline{\mathscr{I}(\Gamma_n)}$. Therefore by compactness γ is disjoint from $\overline{\mathscr{I}(\Gamma_n)}$ for all large n. Say,

(22) $\quad \gamma \cap \overline{\mathscr{I}(\Gamma_n)} = \varnothing \quad \forall n \geq n_0$.

Notice that $D(w_0, \eta) \subset D(c, 2\eta)$ and therefore by (13), $\Gamma_n \cap D(w_0, \eta) \subset f_n(A_n)$. It follows from this inclusion and from (19) that

(23) $\quad \overline{\lim_{w \to w'}} |f_n^{-1}(w) - z_0| \leq \varepsilon_n \quad \forall w' \in \Gamma_n \cap D(w_0, \eta)$.

The function $f_n^{-1} - z_0$ is bounded by 2 and Γ_n is the boundary of $\mathscr{I}(\Gamma_n)$. Therefore (22) and (23) affirm that the hypotheses of Lindelöf's theorem 5.14 are met. We conclude that

$$|f_n^{-1}(w_0) - z_0| \leq (\varepsilon_n 2^{k-1})^{1/k} \quad \forall n \geq n_0.$$

If we let $n \to \infty$ and recall (20), the claim (21) stands confirmed.

Now according to 9.22 the sequence $\{f_n\}$ converges locally uniformly in D to the conformal map f of D onto $\mathscr{I}(\Gamma)$ which satisfies $f(0) = 0, f'(0) > 0$. By 7.17 $\{f_n^{-1}\}$ converges locally uniformly in $\mathscr{I}(\Gamma)$ to f^{-1}. This contradicts (21).

Therefore (21), and therewith the assumption (of non-equicontinuity) from which (21) was deduced, is untenable. This proves the theorem.

Corollary 9.24 *Let the notation and hypotheses be as in the last theorem. Then* $\{f_n\}$ *converges uniformly in* $\bar{D}$ *to the unique function* f *which maps* D *conformally onto* $\mathscr{I}(\Gamma)$ *and satisfies* $f(0) = 0, f'(0) > 0$.

Proof: We first show that every subsequence of $\{f_n\}$ contains another which converges uniformly on $\bar{D}$ to the function f described above. Then a short little argument like that in 7.8 shows that the sequence $\{f_n\}$ itself must converge uniformly on $\bar{D}$. Here are the details.

For the first step, by a notational change it suffices to show that $\{f_n\}$ contains a subsequence which is uniformly convergent on $\bar{D}$ to f. By the diagonal process we select a subsequence $\{f_{n_k}\}$ which converges at each point of some countable dense subset of $C = C(0, 1)$. Since the family $\{f_n\}$ is equicontinuous on C, Cauchyness of $\{f_{n_k}(z)\}$ for a dense set of $z \in C$ and compactness lead via a finite-covering argument to the uniform Cauchyness of $\{f_{n_k}\}$ on C. The Maximum Modulus Principle propagates this into D. Thus $\{f_{n_k}\}$ converges uniformly on $\bar{D}$. The limit is a continuous function on $\bar{D}$. On D this limit is f, by 9.22. Consequently, by continuity, f is the limit everywhere in $\bar{D}$.

Now the argument of 7.8 shows that the whole sequence $\{f_n\}$ converges at *each* point of C. Then equicontinuity ensures that this convergence is uniform on C and Maximum Modulus makes it uniform throughout $\bar{D}$, as noted in the last paragraph.

Exercise 9.25 *Show that if* $\Gamma: C(0, 1) \to \mathbb{C}$ *is any homeomorphism with* $0 \in \mathscr{I}(\Gamma)$, *there exist* Γ_n *satisfying the hypotheses of* 9.23.

Hints: The result is trivial if Γ is the identity map. Extend Γ to a homeomorphism of $\mathbb{C}$ onto $\mathbb{C}$ [SCHÖNFLIES 9.19] and thereby reduce the general case to this trivial one.

We are now in a position to prove a significant extension of the Polynomial Runge Theorem 8.9, still considerably short of the Mergelyan extension, however.

Theorem 9.26 (WALSH [1927]) *Let* Ω *be a Jordan region,* $F: \bar{\Omega} \to \mathbb{C}$ *a continuous function which is holomorphic in* Ω. *Then* F *is uniformly approximable on* $\bar{\Omega}$ *by polynomials.*

Proof: Let $\Gamma: C = C(0, 1) \to \mathbb{C}$ be a homeomorphism such that $\Omega = \mathscr{I}(\Gamma)$ [definition of Jordan region]. We may assume $0 \in \Omega$. Then let Γ_n, f_n be the maps furnished by 9.25 and 9.23. According to 9.24 the functions f_n converge uniformly on $\bar{D}$ to the unique conformal map f of D onto Ω which satisfies $f(0) = 0$, $f'(0) > 0$. Given $\varepsilon > 0$, choose $\delta > 0$ so that

(1) $$z_1, z_2 \in \bar{D} \;\&\; |z_1 - z_2| < \delta \Rightarrow |F \circ f(z_1) - F \circ f(z_2)| < \varepsilon.$$

This is just the uniform continuity of $F \circ f$ on $\bar{D}$. Similarly, choose $\eta > 0$ such that

(2) $\quad w_1, w_2 \in \bar{\Omega} \;\&\; |w_1 - w_2| < \eta \Rightarrow |f^{-1}(w_1) - f^{-1}(w_2)| < \delta.$

Next choose an n such that

(3) $\quad |f_n(z) - f(z)| < \eta \quad \forall z \in \bar{D}.$

Apply f^{-1} in (3) to get via (2)

$$|f^{-1}(f_n(z)) - z| < \delta \quad \forall z \in f_n^{-1}(\bar{\Omega}) \subset \bar{D}.$$

That is, upon setting $w = f_n(z)$,

(4) $\quad |f^{-1}(w) - f_n^{-1}(w)| < \delta \quad \forall w \in \bar{\Omega}.$

From (4) and (1) we have

$$|F \circ f \circ f^{-1}(w) - F \circ f \circ f_n^{-1}(w)| < \varepsilon \quad \forall w \in \bar{\Omega},$$

i.e.,

(5) $\quad |F(w) - F \circ f \circ f_n^{-1}(w)| < \varepsilon \quad \forall w \in \bar{\Omega}.$

Since $\bar{\Omega}$ is a compact subset of the simply-connected region $\mathscr{I}(\Gamma_n)$, Runge's Polynomial Approximation Theorem 8.10 furnishes a polynomial P such that

(6) $\quad |P - F \circ f \circ f_n^{-1}| < \varepsilon \quad \text{on } \bar{\Omega}.$

Combining (5) and (6), we have finally

$$|F - P| < 2\varepsilon \quad \text{on } \bar{\Omega}.$$

Corollary 9.27 (Cauchy's Theorem on the Boundary) *Let Ω be a Jordan region, Γ a piecewise smooth loop in $\bar{\Omega}$, $f\colon \bar{\Omega} \to \mathbb{C}$ a continuous function which is holomorphic in Ω. Then $\int_\Gamma f = 0$.*

Proof: (WALSH [1933]) This follows from the last theorem just as 8.13 was deduced from the Polynomial Runge Theorem.

Exercise 9.28 (PAINLEVÉ [1888]) *Let Γ be a piecewise smooth, simple loop, C a piecewise smooth cross-cut* (1.15) *in $\mathscr{I}(\Gamma)$. Let f be continuous on $\overline{\mathscr{I}(\Gamma)}$ and holomorphic in $\mathscr{I}(\Gamma)\backslash C$. Show that in fact f is holomorphic in $\mathscr{I}(\Gamma)$.*

Hints: Let Γ_1, Γ_2 be piecewise smooth arcs such that $\Gamma_1 \cap C = \Gamma_2 \cap C$ and $\Gamma_1 \cup \Gamma_2 = \Gamma$. See 4.47. Then by the last corollary

$$\int_{\Gamma_1} f + \int_C f = \int_C f - \int_{\Gamma_2} f = 0.$$

Subtract to get

$$\int_\Gamma f = \int_{\Gamma_1} f + \int_{\Gamma_2} f = 0.$$

For each $z \in \mathscr{I}(\Gamma)\backslash C$ apply this conclusion to the function

$$F(\zeta) = \frac{f(\zeta) - f(z)}{\zeta - z}$$

in the role of f to get

$$(*) \qquad nf(z) = \frac{1}{2\pi i}\int_\Gamma \frac{f(\zeta)}{\zeta - z}\, d\zeta \quad \forall z \in \mathscr{I}(\Gamma)\backslash C$$

(where n is 1 or -1). Since C is a homeomorph of $[0, 1]$, the removal of any three points destroys its connectedness (since $[0, 1]$ evidently has this property). Therefore C cannot contain any disk. (See 1.24.) In particular, every neighborhood of a point of $C \cap \mathscr{I}(\Gamma)$ contains points of $\mathscr{I}(\Gamma)\backslash C$. In other words, $\mathscr{I}(\Gamma)\backslash C$ is dense in $\mathscr{I}(\Gamma)$. Now both sides of (*) are continuous functions of $z \in \mathscr{I}(\Gamma)$, hence the equality there holds throughout $\mathscr{I}(\Gamma)$. That formula proves that f is holomorphic in $\mathscr{I}(\Gamma)$. (See 2.14.)

§ 7 Harmonic Functions and the General Dirichlet Problem

In this section we will connect our definition of harmonic function with the traditional one and then solve the Dirichlet problem (5.19) for (bounded) simply-connected regions. This is appropriate here because, as we will see in 9.47, it is possible to use the solubility of the Dirichlet problem to provide yet another proof of the Riemann mapping theorem. In connection with which it should be noted that the development here does not utilize the Riemann mapping theorem. This approach to the Riemann mapping theorem is close to the historical one followed by Riemann, Schwarz, Neumann, Hilbert, Poincaré, Osgood and Lebesgue, i.e., via potential theory.

Exercise 9.29 *If U is an open subset of $\mathbb{C}$ and $h: U \to \mathbb{R}$ is harmonic, then $D_{11}h$, $D_{22}h$ (exist and are continuous and) satisfy $D_{11}h + D_{22}h = 0$.*

Hints: The problem is a local one so we can assume U is a disk and $h = \operatorname{Re} f$ for some $f \in H(U)$. Differentiate the Cauchy–Riemann equation.

Remark: We write $\nabla^2 h$ for $D_{11}h + D_{22}h$. This is called the *Laplacian of h* and $\nabla^2 h = 0$ is called *Laplace's equation*. (See pp. 17, 18 of Osgood [1901].)

Theorem 9.30 *Let U be an open subset of $\mathbb{C}$, f a continuous real-valued function on U for which $D_{11}f$, $D_{22}f$ exist in U and satisfy $\nabla^2 f \le 0$. Then f is superharmonic.*

Proof: Consider any disk $D = D(z_0, r)$ $(r > 0)$ whose closure lies in U. Let h be the solution of the Dirichlet problem in D with boundary data f. The function $h - f$ assumes a maximum in $\bar{D}$. If this maximum is positive, it must occur at some $z_1 \in D$, since $h - f = 0$ on ∂D. Then for sufficiently small $c > 0$ the function

$$(1) \qquad \phi(z) = c|z - z_0|^2 + h(z) - f(z)$$

is larger at z_1 than on ∂D. (We have only to take $0 < cr^2 < h(z_1) - f(z_1)$.) Hence its maximum over $\bar{D}$ occurs at some point $z_2 = (x_2, y_2) \in D$. Then the function $\phi(x, y_2)$ has a relative maximum at x_2, so $D_{11}\phi(x_2, y_2) \leq 0$. Similarly $\phi(x_2, y)$ has a relative maximum at y_2, so $D_{22}\phi(x_2, y_2) \leq 0$. Returning to (1), this means that

$$(2) \qquad 0 \geq 4c + \nabla^2(h - f)(x_2, y_2).$$

Since $\nabla^2 h = 0$ in D by the last exercise, and $\nabla^2 f \leq 0$ by hypothesis, it follows from (2) that $0 \geq 4c$. This is a contradiction. We conclude that the maximum of $h - f$ on $\bar{D}$ is non-positive, i.e., $h - f \leq 0$ throughout $\bar{D}$. Thus

$$f(z_0) \geq h(z_0) = \frac{1}{2\pi}\int_0^{2\pi} h(z_0 + re^{i\theta})d\theta \quad \text{(by 5.24)}$$

$$= \frac{1}{2\pi}\int_0^{2\pi} f(z_0 + re^{i\theta})d\theta.$$

Thus f satisfies the typical inequality in the definition of superharmonicity.

Corollary 9.31 (Converse of 9.29) *If f is a continuous real-valued function in the open subset U of $\mathbb{C}$ and $D_{11}f$, $D_{22}f$ exist and satisfy Laplace's equation in U, then f is harmonic.*

Proof: 5.24 plus the last theorem.

Exercise 9.32 (Converse of 9.30) *If f is a superharmonic function in an open subset U of $\mathbb{C}$ and $D_{11}f$, $D_{22}f$ exist and are continuous in U, then $\nabla^2 f \leq 0$.*

Hints: If $\nabla^2 f(a) > 0$ for some $a \in U$, then by continuity there is a whole open disk $D \subset U$ in which $\nabla^2 f > 0$. But then by the theorem f is subharmonic in D. Being also superharmonic, it is harmonic there (by 5.24). Hence by 9.29 $\nabla^2 f = 0$ in D, contradiction.

Exercise 9.33 (i) *Let f be holomorphic in the region Ω. Define $F(x, y) = |f(x + iy)|^2$ and show that*

$$D_{11}F(x, y) + D_{22}F(x, y) = 4|f'(x + iy)|^2$$

for all $x + iy \in \Omega$.

(ii) *Deduce from* (i) *that if g and h are holomorphic in Ω and* $\operatorname{Re} g = |h|$, *then g and h are each constant. (Thus a harmonic function is seldom the modulus of a holomorphic function.)*

Hint: If h is not constant, go to a disk $D \subset \Omega$ in which h has no zeros and in D write $h = f^2$ for some $f \in H(D)$ (5.35).

(iii) *Show that if u, v are (real-valued) harmonic functions in a region Ω and $f = u + iv$ is holomorphic in some non-empty open subset of Ω, then f is holomorphic throughout Ω.*

Hints: The result follows from the Basic Connectedness Lemma (1.27, with $\theta = 1$) together with the validity of the special case where Ω is a disk. To prove the latter, cite 5.21(i) twice to come up with $F, G \in H(\Omega)$ such that $u = \operatorname{Re} F$, $v = \operatorname{Im} G$ in Ω. Let D be a non-empty open disk in Ω in which f is holomorphic. Since $i(F - f)$ is real-valued in D, it is constant there (2.17(ii)) and by adding i times this constant to F we can retain $u = \operatorname{Re} F$ in Ω and have in addition $F = f$ in D. Do the same to G. Then $F = G = f$ in D, so by 5.62 $F = G$ in Ω. It follows that

$$f = u + iv = \operatorname{Re} F + i \operatorname{Im} G = \operatorname{Re} F + i \operatorname{Im} F = F \quad \text{in } \Omega.$$

(iv) *Show that if u is a (real-valued) harmonic function in a disk Ω and if the product $U(z) = u(z) \operatorname{Re} z$ is also harmonic in Ω, then $u(z) = a \operatorname{Im} z + b$ for some real constants a and b.*

Hints: Compute that

$$\nabla^2 U(x + iy) = 2D_1 u(x + iy) + x\nabla^2 u(x + iy) \quad \forall x + iy \in \Omega.$$

Since both U and u satisfy Laplace's equation, we learn that $2D_1 u(x + iy) = 0$, so $u(x + iy)$ is a function only of y and $0 = \nabla^2 u(x + iy) = D_{22} u(x + iy)$.

(v) *The product of two harmonic functions is seldom harmonic. Prove the following precise version of this assertion: if u and v are non-constant (real-valued) harmonic functions in a region Ω, then their product is harmonic in Ω if and only if there exists a non-zero real constant c such that $u + icv$ is holomorphic in Ω.*

Hints: One direction is trivial: if $f = u + icv$ is holomorphic in Ω and c is real and non-zero, then $uv = \operatorname{Re}(f^2/2ci)$ is harmonic in Ω. Now suppose that $U = uv$ is harmonic in Ω. If both $D_1 u$ and $D_2 u$ are identically 0, then by 2.16 u is holomorphic and $u' = D_1 u = 0$. The constancy of u in Ω would follow. Therefore not both $|D_1 u|$ and $|D_2 u|$ are identically 0 and there exists an open disk D in Ω where one of them is positive. In that disk there is a holomorphic function f such that $u = \operatorname{Re} f$. According to the Cauchy–Riemann equation $f' = D_1 u - iD_2 u$ and so f' is zero-free in D. In some smaller disk D_0, f is therefore one-to-one (5.79) and on the open set $V_0 = f(D_0)$ there is an inverse function g. Then

$$U \circ g = [v \circ g][u \circ g] = [v \circ g][(\operatorname{Re} f) \circ g] = [v \circ g][\operatorname{Re}(f \circ g)]$$

is harmonic in V_0. Since $f \circ g(w) = w$ for all $w \in V_0$, this says that

$$U \circ g(w) = v \circ g(w) \operatorname{Re} w, \quad w \in V_{,0}$$

is harmonic. By (iv) there are constants a, b such that

$$v \circ g(w) = a \operatorname{Im} w + b \quad \forall w \in V,$$

where V is an open disk in V_0; that is,

$$v(z) = a \operatorname{Im} f(z) + b \quad \forall z \in f^{-1}(V) \cap D_0.$$

Since $u = \operatorname{Re} f$ in D_0, the last equation gives

(*) $au + iv = af + ib$ in $f^{-1}(V) \cap D_0$.

The equation (*) affirms that $au + iv$ is holomorphic in $f^{-1}(V) \cap D_0$, so by (iii) it is holomorphic throughout Ω. But then $a \neq 0$, else by 2.17(ii) v would be constant in Ω. Set $c = 1/a$.

We now have all that is needed to solve the Dirichlet problem in a quite general setting. Start with

Definition 9.34 Let U be an open subset of $\mathbb{C}$, $b \in \partial U$. A *barrier* in U at b is a set $\{h_r : 0 < r < 1\}$ of non-negative superharmonic functions in U which each satisfy

(i) $\lim_{z \to b} h_r(z) = 0$,
(ii) $h_r(U \backslash D(b, r)) = 1$.

Remarks: The reader will encounter no difficulty confirming (via the method of 5.25) that if the Dirichlet problem is solvable for an open set U, then U has a barrier at each of its boundary points. The importance of the concept lies in the validity of the converse, whose proof begins with

The Basic Construction: Let U be an open subset of $\mathbb{C}$, f a bounded continuous function on $\bar{U}$ which is superharmonic in U. Let, say, $|f| \leq M$ in $\bar{U}$. Let $\{D_n\}_{n=1}^{\infty}$ be a sequence of open disks whose closures each lie in U. The sequence is to have one further property: each point z of U lies in an open disk D_z which lies wholly inside D_n for infinitely many n. As an example of such a sequence, take all the open disks with rational (complex) centers and rational radii, whose closures lie in U. Now the Dirichlet problem is solvable in each D_n, so we can define a sequence of functions F_n on $\bar{U}$ inductively thus: F_1 equals f in $\bar{U} \backslash D_1$ and coincides in D_1 with the solution of the Dirichlet problem with boundary function f. After $F_1, \ldots, F_{n-1}$ are defined, we let F_n equal F_{n-1} in $\bar{U} \backslash D_n$ and coincide in D_n with the solution of the Dirichlet problem with boundary function F_{n-1}. Evidently each F_n is continuous. Since, by the Maximum Principle, the solution of the Dirichlet problem in D_n with boundary function F_{n-1} attains its maximum modulus over $\bar{D}_n$ on $\partial D_n \subset \bar{U} \backslash D_n$, where it coincides with F_{n-1}, we have $\sup_{\bar{U}} |F_n| \leq \sup_{\bar{U}} |F_{n-1}|$, and so by induction

(1) $|F_n| \leq M$ on $\bar{U}$, $\forall n = 1, 2, \ldots$

Also from 5.25(i) we learn that each F_n is superharmonic in U and $F_n \leq F_{n-1}$ in U. Thus

(2) $f \geq F_1 \geq F_2 \geq \cdots \geq -M$ in U.

From (1) and (2) we see that $\lim_{n \to \infty} F_n(z)$ exists for every $z \in U$. Call this limit $F(z)$. Now let $z \in U$ be given and pick $1 \leq n_1 < n_2 < \cdots$ such that $D_z \subset \bigcap_{j=1}^{\infty} D_{n_j}$. Since each function F_n is harmonic in D_n, it follows from (2) and

7.13 that the convergence of F_{n_j} to F is locally uniform in D_z. We have proved that

(3) $\lim_{n\to\infty} F_n = F$ exists throughout U and defines a harmonic function there.

Lemma 9.35 *If U has a barrier at $z_0 \in \partial U$, then F satisfies*

(4) $\lim_{z\to z_0} F(z) = f(z_0)$.

Proof: Let $\{h_r\}$ be a barrier in U at z_0 and let $\varepsilon > 0$ be given. Choose $r > 0$ sufficiently small that

(5) $f(z_0) - \varepsilon < f(z) < f(z_0) + \varepsilon \quad \forall z \in \overline{U} \cap D(z_0, r)$.

We will first show that

(6) $f(z_0) - 2Mh_r(z) - \varepsilon \le f(z) \le f(z_0) + 2Mh_r(z) + \varepsilon \quad \forall z \in U$.

For $z \in U \cap D(z_0, r)$ this is a trivial consequence of (5), h_r being non-negative. For $z \in U\backslash D(z_0, r)$ we have

$$|f(z) - f(z_0)| \le 2M = 2Mh_r(z)$$

by property (ii) of a barrier, so again (6) follows. We claim that

(6.1) $f(z_0) - 2Mh_r(z) - \varepsilon \le F_1(z) \le f(z_0) + 2Mh_r(z) + \varepsilon \quad \forall z \in U$.

By (6) and the definition of F_1 this holds for all $z \in \partial D_1$. But the function on the left of (6.1) is subharmonic, that on the right is superharmonic. Therefore, F_1 being harmonic in D_1, both inequalities persist throughout D_1. For the validity of (6.1) in $U\backslash D_1$ we fall back on (6) and the fact that $F_1 = f$ in $U\backslash D_1$. It should be clear how to repeat this reasoning to establish inductively

(6.n) $f(z_0) - 2Mh_r(z) - \varepsilon \le F_n(z) \le f(z_0) + 2Mh_r(z) + \varepsilon \quad \forall z \in U$.

Go to the limit here and get

(7) $f(z_0) - 2Mh_r(z) - \varepsilon \le F(z) \le f(z_0) + 2Mh_r(z) + \varepsilon \quad \forall z \in U$.

Using the first property of a barrier, select $0 < r_\varepsilon < r$ sufficiently small that

(8) $h_r(z) \le \dfrac{\varepsilon}{2M} \quad \forall z \in U \cap D(z_0, r_\varepsilon)$.

From (7) and (8) conclude that

$$f(z_0) - 2\varepsilon \le F(z) \le f(z_0) + 2\varepsilon \quad \forall z \in U \cap D(z_0, r_\varepsilon),$$

which completes the proof.

Exercise 9.36 *Let U be an open subset of $\mathbb{C}$ which has a barrier at each point of ∂U. If f is a bounded, continuous function on $\overline{U}$ which is superharmonic in U, show* *that there exists a bounded continuous function F on $\overline{U}$ which is harmonic in U and coincides on ∂U with f.*

Hints: Extend the F of the above construction to $\overline{U}$ by the decree $F = f$ on ∂U. Then $\lim_{z \to z_0, z \in U} F(z) = F(z_0)$ for each $z_0 \in \partial U$ and what is at issue is that $\lim_{z \to z_0, z \in \overline{U}} F(z) = F(z_0)$. This is easily proven from the continuity of f $(= F)$ on ∂U. See the proof of 9.12 for details of this kind of argument.

Lemma 9.37 *Let U be a bounded open subset of $\mathbb{C}$ which has a barrier at each point of ∂U, $C_{\mathbb{R}}(\partial U)$ the continuous real-valued functions on ∂U and $\mathscr{H}$ the subset of all such functions which admit a continuous extension to $\overline{U}$ which is harmonic in U. Then $\mathscr{H}$ is a vector space which is closed in the uniform norm and contains all the functions $P_{n,m}(z) = (\operatorname{Re} z)^n(\operatorname{Im} z)^m$, n, m non-negative integers.*

Proof: It is obvious that $\mathscr{H}$ is a vector space. If $f_n \in \mathscr{H}$ and $f_n \to f$ uniformly on ∂U, let F_n be the corresponding extension of f_n. Then $F_n - F_m$ is a bounded harmonic function on U, so

$$\sup_{\overline{U}} |F_n - F_m| = \sup_{\partial U} |F_n - F_m| = \sup_{\partial U} |f_n - f_m| \to 0.$$

Therefore $\{F_n\}$ is uniformly convergent on $\overline{U}$ to a continuous function F which coincides on ∂U with f. Thanks to 5.44 the limit function F is also harmonic in U. Thus $f \in \mathscr{H}$. Finally, consider any non-negative integers n, m and let $P = P_{n,m}$. Set $P_1(z) = |z|^2 = |z^2|$. This function is subharmonic in $\mathbb{C}$, by 5.7. The continuous function $\nabla^2 P$ has a maximum value M, say, on the compact set $\overline{U}$. Since $\nabla^2 P_1 = 4$, it follows that $\nabla^2(P - \frac{1}{4}MP_1) \leq 0$ throughout $\overline{U}$. Therefore by 9.30 $P_2 = P - \frac{1}{4}MP_1$ is superharmonic in U. 9.36 ensures that both P_2 and $-P_1$ belong to $\mathscr{H}$. Since $\mathscr{H}$ is a vector space, it follows that $P = P_2 + \frac{1}{4}MP_1$ also belongs to $\mathscr{H}$. This proves the lemma.

Corollary 9.38 *The Dirichlet problem is solvable for any bounded open subset of $\mathbb{C}$ which has a barrier at each point of its boundary.*

Proof: In the language of 9.37 the problem is to see that $\mathscr{H} = C_{\mathbb{R}}(\partial U)$. But this follows from 9.37 and the two-dimensional Weierstrass approximation theorem 8.26(v).

We next examine conditions which ensure that the hypotheses of 9.38 are fulfilled.

Lemma 9.39 *Let I be a non-empty set of positive integers and (A_k, B_k) $(k \in I)$ be mutually disjoint, bounded open subintervals of $\mathbb{R}$. Then there exists a harmonic* *function h in the open right half-plane H such that*

(i) $0 \le h \le \pi$,

(ii) $h(z) \le \dfrac{1}{\operatorname{Re} z} \sum_{k \in I} (B_k - A_k) \quad \forall z \in H$,

(iii) $$\lim_{z \to ic} h(z) = \begin{cases} \pi & \forall c \in S = \bigcup_{k \in I} (A_k, B_k) \\ 0 & \forall c \in \mathbb{R} \backslash \bar{S}. \end{cases}$$

Proof: (Cf. 8.32(i).) Let l be the Principal Branch of the Logarithm in $\mathbb{C}\backslash(-\infty, 0]$ and define

$$h_k(z) = \operatorname{Im} l\left(\frac{z - iA_k}{z - iB_k}\right), \quad z \in H, k \in I.$$

These functions are harmonic in H and satisfy

(1) $$h_k(z) = \int_{A_k}^{B_k} \frac{\operatorname{Re} z}{|z - it|^2}\, dt, \quad z \in H, k \in I.$$

See 3.44. If F is any finite subset of I and $n \in \mathbb{N}$ satisfies

$$-n < A_k < B_k < n \quad \text{for all } k \in F,$$

then (1) and the disjointness of the (A_k, B_k) show that

$$0 \le \sum_{k \in F} h_k(z) \le \int_{-n}^{n} \frac{\operatorname{Re} z}{|z - it|^2}\, dt = \int_{(-n-y)/x}^{(n-y)/x} \frac{d\tau}{1 + \tau^2} \quad \forall z = x + iy \in H.$$

Let $n \to \infty$ and get

$$0 \le \sum_{k \in F} h_k(z) \le \int_{-\infty}^{\infty} \frac{d\tau}{1 + \tau^2} = 2\int_0^{\infty} \frac{d\tau}{1 + \tau^2}$$

(2) $$= 2\left(\int_0^1 \frac{d\tau}{1 + \tau^2} + \int_1^{\infty} \frac{d\tau}{1 + \tau^2}\right) = 4\int_0^1 \frac{d\tau}{1 + \tau^2} = \pi, \quad \text{by 3.22, } \forall z \in H.$$

Since F is any finite subset of I, we can cite 7.13 to conclude that

$$h(z) = \sum_{k \in I} h_k(z), \quad z \in H$$

defines a harmonic function in H which satisfies (i). If we return to (1) and write

$$h_k(z) = \frac{1}{x}\int_{A_k}^{B_k} \frac{dt}{1 + \left(\dfrac{t - y}{x}\right)^2} \le \frac{1}{x}\int_{A_k}^{B_k} 1 = \frac{B_k - A_k}{x} \quad \forall z = x + iy \in H, k \in I,$$

then

$$h(z) = \sum_{k \in I} h_k(z) \le \frac{1}{x} \sum_{k \in I} (B_k - A_k) \quad \forall z = x + iy \in H,$$

which gives (ii).

Now fix $k \in I$ and divide I into two subclasses

$$I_k = \{j \in I: A_j \geq B_k\}, \qquad {}_kI = \{j \in I: B_j \leq A_k\}.$$

The intervals (A_j, B_j) are disjoint and therefore ${}_kI$, I_k (obviously disjoint) comprise all of I except k. Hence

$$\begin{aligned} 0 \leq h(z) - h_k(z) &= \sum_{j \in I_k} h_j(z) + \sum_{j \in {}_kI} h_j(z) \\ &\leq \int_{B_k}^{+\infty} \frac{\operatorname{Re} z}{|z - it|^2}\, dt + \int_{-\infty}^{A_k} \frac{\operatorname{Re} z}{|z - it|^2}\, dt \end{aligned}$$

$$(3) \qquad = \int_{(B_k - y)/x}^{+\infty} \frac{1}{1 + \tau^2}\, d\tau + \int_{-\infty}^{(A_k - y)/x} \frac{1}{1 + \tau^2}\, d\tau,$$

for all $z = x + iy \in H$. If $y \in (A_k, B_k)$, then $(B_k - y)/x > 0$ and $(A_k - y)/x < 0$. Consequently, neither interval of integration in (3) contains 0 and we may make the change of variable $t = 1/\tau$ to get in (3)

$$(4) \qquad 0 \leq h(z) - h_k(z) \leq \int_{x/(A_k - y)}^{x/(B_k - y)} \frac{1}{1 + t^2}\, dt, \quad z = x + iy \in H, y \in (A_k, B_k).$$

From which we see at once that

$$(5) \qquad \lim_{z \to ic} [h(z) - h_k(z)] = 0 \quad \forall c \in (A_k, B_k), k \in I.$$

On the other hand

$$h_k(z) = \int_{A_k}^{B_k} \frac{\operatorname{Re} z}{|z - it|^2}\, dt = \int_{(A_k - y)/x}^{(B_k - y)/x} \frac{d\tau}{1 + \tau^2}, \quad z = x + iy \in H,$$

from which we see, as in the derivation of (2), that

$$(6) \qquad \lim_{z \to ic} h_k(z) = \int_{-\infty}^{\infty} \frac{d\tau}{1 + \tau^2} = \pi \quad \forall c \in (A_k, B_k), k \in I.$$

Next consider $c \in \mathbb{R} \backslash \bar{S}$. There exist $c_1 < c < c_2$ such that $(c_1, c_2) \cap S = \varnothing$. For $y \in (c_1, c_2)$ calculations isomorphic to those leading up to (4) give us

$$0 \leq h(z) \leq \int_{x/(c_1 - y)}^{x/(c_2 - y)} \frac{dt}{1 + t^2}$$

and consequently

$$\lim_{z \to ic} h(z) = 0.$$

This holds for each $c \in \mathbb{R} \backslash \bar{S}$ and together with (5) and (6) confirms property (iii) of h.

Theorem 9.40 *An open subset U of $\mathbb{C}$ has a barrier at each boundary point whose component in $\mathbb{C}\backslash U$ is not a single point. In particular, an open simply-connected subset of the plane has a barrier at each of its boundary points.*

Proof: It is convenient to treat first the case that U is simply-connected. After a translation, the generic boundary point of U can be supposed to be 0. In this case there is, by 4.65, a holomorphic logarithm L in U. Consider any $0 < r < 1$. If $U \cap C(0, r)$ is void, define

$$h_r = \begin{cases} 0 & \text{in } D(0, r) \\ 1 & \text{in } U\backslash D(0, r) \end{cases}$$

and have a harmonic function in U which satisfies (i) and (ii) of 9.34.

In what follows we suppose that $U \cap C(0, r)$ is not void. As this intersection is a relatively open subset of $C(0, r)$, its components are open arcs of the form $\{re^{i\theta}: \alpha < \theta < \beta\}$. They are disjoint and (so) there are at most countably many of them. Let them be

(1) $\quad \gamma_k = \{re^{i\theta}: \alpha_k < \theta < \beta_k\}, \quad k \in I \subset \mathbb{N},$

where

(2) $\quad 0 \leq \alpha_k < \beta_k \leq 2\pi.$

For $\theta \in (\alpha_k, \beta_k)$ we have $e^{L(re^{i\theta})} = re^{i\theta}$ and so

$$\begin{cases} \operatorname{Re} L(re^{i\theta}) = \log r \\ \dfrac{\operatorname{Im} L(re^{i\theta}) - \theta}{2\pi} \in \mathbb{Z}. \end{cases}$$

This last quotient is, however, a continuous function of θ in the connected set (α_k, β_k) and so it is constant, say N_k. It follows that

(3) $\quad L(re^{i\theta}) = \log r + i\theta + 2\pi i N_k \quad \forall \theta \in (\alpha_k, \beta_k).$

Define the function L_r by

(4) $\quad L_r(z) = \log r - L(z), \quad z \in U$

and set

$$A_k = -\beta_k - 2\pi N_k < -\alpha_k - 2\pi N_k = B_k.$$

It follows from (1), (3) and (4) that

(5.k) $\quad L_r(\gamma_k) = i(A_k, B_k).$

Of course, we also have from (4) that

(6) $\quad L_r(U \cap D(0, r))$ lies in the open right half-plane, H,

since $\operatorname{Re} L(z) = \log|z|$ for every $z \in U$. Moreover, since the γ_k are disjoint, it follows from the equalities (5.k) that the intervals (A_k, B_k) are mutually disjoint. Let h be the non-negative harmonic function in H which 9.39 furnishes for these intervals. By (1) and the disjointness of the γ_k the intervals (α_k, β_k) are disjoint and so (2) shows that

$$2\pi \geq \sum_{k \in I} (\beta_k - \alpha_k) = \sum_{k \in I} (B_k - A_k).$$

Consequently, by property (ii) of h we have

$$(7) \qquad 0 \leq h(z) \leq \frac{2\pi}{\operatorname{Re} z} \quad \forall z \in H.$$

Because of (6) we can define a harmonic function h_r in $U \cap D(0, r)$ by:

$$(8) \qquad h_r = \frac{1}{\pi} h \circ L_r \quad \text{in } U \cap D(0, r).$$

Thanks to (7) it satisfies

$$0 \leq h_r(z) \leq \frac{2}{\operatorname{Re} L_r(z)} = \frac{2}{\log r - \log|z|}$$

and so

$$(9) \qquad \lim_{z \to 0} h_r(z) = 0.$$

On the other hand, consider any $w \in U \cap C(0, r)$ and $z \in U \cap D(0, r)$ with $z \to w$. Then $L_r(z) \to L_r(w)$. There is some k such that $w \in \gamma_k$ and so by (5.k), $L_r(w) = ic$ for some $c \in (A_k, B_k)$. From property (iii) of h and $L_r(z) \to L_r(w) = ic$, it follows that $h(L_r(z)) \to \pi$ and so we've proved that

$$(10) \qquad \lim_{z \to w} h_r(z) = 1 \quad \forall w \in U \cap C(0, r).$$

Of course by (8) and property (i) of h we have

$$(11) \qquad 0 \leq h_r \leq 1 \quad \text{in } U \cap D(0, r).$$

We now extend h_r to U by the decree $h_r(U \backslash D(0, r)) = 1$. Because of (10), (11) and 5.25(ii) the extended h_r is superharmonic in U.

The family $\{h_r\}$ is evidently a barrier in U at the point 0.

Next consider the general case of an arbitrary open subset U of $\mathbb{C}$ and a boundary point b of U such that the component K of $\mathbb{C} \backslash U$ which contains b is not a single point. If K is unbounded, then $\mathbb{C} \backslash K$ is (open and) simply-connected and so by the result of the first paragraph possesses a barrier at b which (after restriction) is evidently also a barrier in U at b. Assume now that K is bounded, hence compact. By hypothesis $K \backslash \{b\}$ is not empty. Without loss of generality let $0 \in K \backslash \{b\}$ and consider the transformation $\psi(z) = 1/z$ in $\mathbb{C} \backslash \{0\} \supset U \cup K \backslash \{0\}$. Let

C be the component of $\psi(K\backslash\{0\})$ which contains $\psi(b)$. According to 1.41, C is unbounded. Moreover, $\psi(b)$ clearly lies on the boundary of $\mathbb{C}\backslash C$. Given $0 < r < 1$, choose $0 < R < 1$ sufficiently small that

(12) $\psi^{-1}(D(\psi(b), R)) \subset D(b, r)$.

By the result already proven there is a non-negative superharmonic function h in $\mathbb{C}\backslash C$ with the properties

(13) $\lim\limits_{\substack{w\to\psi(b)\\ w\in\mathbb{C}\backslash C}} h(w) = 0,$

(14) $h(\mathbb{C}\backslash C\backslash D(\psi(b), R)) = 1.$

From 5.22(ii) follows the superharmonicity of the composite $h \circ \psi$. This function is also non-negative and because of (12) and (14) satisfies

$$h \circ \psi(U\backslash D(b, r)) = 1.$$

Since ψ is continuous on the set $\mathbb{C}\backslash\{0\}$ which contains $U \cup \{b\}$, the convergence to b of points z in U implies the convergence to $\psi(b)$ of points $w = \psi(z)$ in $\psi(U) \subset \mathbb{C}\backslash C$ and therefore from (13) follows

$$\lim_{\substack{z\to b\\ z\in U}} h \circ \psi(z) = 0.$$

Corollary 9.41 *The Dirichlet problem is solvable for any bounded, open subset of $\mathbb{C}$ whose complement contains no one-point components; in particular, for any bounded, open and simply-connected subset of $\mathbb{C}$.*

Proof: 9.38 and 9.40.

Exercise 9.42 *Formulate and prove versions of* 9.36 *and* 9.37 *allowing a finite proper subset E of ∂U at each point of which no barrier exists.*

Hints: The extension F in 9.36 will be continuous only in $\overline{U}\backslash E$. In 9.37 the new $\mathscr{H}$ should comprise those $f \in C_{\mathbb{R}}(\partial U)$ which have a *bounded*, continuous extension F to $\overline{U}\backslash E$ which is harmonic in U. The crucial equality $\sup_{\overline{U}\backslash E}|F| = \sup_{\partial U}|f|$ still holds, thanks to 7.15 and the boundedness of F.

With 9.42 in hand, more mileage can be gotten from the inversion device used in the proof of 9.40 and a modest improvement of 9.41 can be eked out:

Exercise 9.43 *Let U be an unbounded, open subset of $\mathbb{C}$ such that $\mathbb{C}\backslash U$ has no one-point components. Let f: $\partial U \to \mathbb{R}$ be continuous and, in case ∂U is unbounded, satisfy the extra condition*

(*) $\lim\limits_{\substack{|z|\to\infty\\ z\in\partial U}} f(z)$ *exists in* $\mathbb{R}$.

(i) *Show that, if in addition U is not dense in $\mathbb{C}$, then f has a bounded, continuous extension to $\overline{U}$ which is harmonic in U.*

Hints: U not dense in $\mathbb{C}$ means that $\mathbb{C}\backslash U$ contains a non-empty open disk, say $D(0, 1)$. Let $\psi(z) = 1/z$. The set $U_0 = \psi(U)$ is an open subset of $\bar{D}(0, 1)$, $\mathbb{C}\backslash U_0 = \psi(\mathbb{C}\backslash\{0\}\backslash U) \cup \{0\}$ and $\partial U_0 = \psi(\partial U) \cup \{0\}$ (since U is unbounded). 0 belongs to just one component of $\mathbb{C}\backslash U$ and that component contains all of $D(0, 1)$. Hence there are no one-point components in $\mathbb{C}\backslash\{0\}\backslash U$, consequently, none in $\psi(\mathbb{C}\backslash\{0\}\backslash U) = (\mathbb{C}\backslash U_0)\backslash\{0\}$ and so the only possible one-point component in $\mathbb{C}\backslash U_0$ is $\{0\}$. By 9.40 U_0 therefore has a barrier at every point of its boundary except possibly 0. We define f_0 on ∂U_0 by $f_0 = f \circ \psi^{-1}$ on $\psi(\partial U)$, $f_0(0)$ equal 0 if ∂U is bounded, and $f_0(0)$ equal the value of the limit in (*) if ∂U is unbounded. Since 0 is an isolated point of ∂U_0 if ∂U is bounded, it follows that f_0 is continuous on ∂U_0. By the version of 9.37 stated in 9.42, coupled with 8.26(v), we are assured of the existence of a bounded, continuous function F_0 on $\bar{U}_0\backslash\{0\}$ which is harmonic in U_0 and agrees with f_0 on $\partial U_0\backslash\{0\}$. The sought-for extension of f is then evidently $F = F_0 \circ \psi$.

(ii) *Show that f has a bounded, continuous extension to $\bar{U}$ which is superharmonic in U.*

Hints: We may suppose that $\bar{D}(0, 1) \subset U$. From (*) it follows trivially that f is bounded. Let $M = \sup|f(\partial U)|$. Let $U_0 = U\backslash D(0, 1)$ and notice that $\partial U_0 = \partial U \cup C(0, 1)$ and no component of $\mathbb{C}\backslash U_0$ consists of a single point. Define f_0 on ∂U_0 by $f_0 = M$ on $C(0, 1)$, $f_0 = f$ on ∂U and apply (i) to come up with a bounded, continuous function F_0 on $\bar{U}_0$ which is harmonic in U_0 and coincides with f_0 on ∂U_0. Extend F_0 to F on $\bar{U}$ by the decree $F = M$ on $D(0, 1)$. Since $|F_0| = |f_0| \le M$ on ∂U_0 and F_0 is bounded in U_0, it follows from the remark after 5.12 that in fact $|F_0| \le M$ throughout U_0. A direct appeal to the definition of superharmonicity in terms of circumferential mean value inequalities (in the form 5.23(i)) then shows that F is superharmonic in U.

(iii) *Show that f has a bounded, continuous extension to $\bar{U}$ which is harmonic in U.*

Hints: Use (ii), 9.36 and 9.40.

Remarks 9.44 CONWAY [1973] modifies The Basic Construction somewhat and utilizes the barrier in a slightly different way, so as to avoid appeal to the two-dimensional Weierstrass approximation theorem. Moreover, a careful reading of his proof establishes that in 9.43 the hypothesis (*) can be replaced by the weaker requirement that f be bounded on ∂U and the conclusion (iii) still follows. However, the exceptional role of one-point components in this theory is unavoidable, as 11.59(iii) illustrates.

§ 8 The Dirichlet Problem and the Riemann Mapping Theorem

Definition 9.45 Let U be an open subset of $\mathbb{C}$, $u: U \to \mathbb{R}$ a harmonic function. A function $v: U \to \mathbb{R}$ such that $u + iv \in H(U)$ is called a *harmonic conjugate* of u. If for every such u there exists such a v, we will say that U has *property* (H).

Thus U has property (H) if and only if each (real-valued) harmonic function on U is the real part of some holomorphic function on U.

Exercise 9.46 *Property (H) implies property (HL).*

Hints: Evidently either property is possessed by an open set if and only if it is possessed by each component thereof. So consider a region Ω with property (H) and any zero-free $f \in H(\Omega)$. Then $\log|f|$ is harmonic in Ω since in any disk in Ω it coincides with the real part of a holomorphic logarithm of f. (See 5.35.) Consequently by property (H) there is a function $g \in H(\Omega)$ such that $\log|f| = \operatorname{Re} g$ throughout Ω. That is, $|fe^{-g}| = 1$ throughout Ω. From the Maximum Modulus Principle it follows that fe^{-g} is constant in Ω, say c. Since $c \neq 0$, it has the form e^a for some $a \in \mathbb{C}$. Then $g + a$ is a holomorphic logarithm for f in Ω.

Theorem 9.47 *Any proper subregion Ω of $\mathbb{C}$ which has property (H) is conformal to an open disk.*

Proof: According to 9.46, Ω has property (HL) and then, as we showed in the opening salvo of the proof of 9,7, Ω is conformal to a bounded region. Evidently that region inherits both property (H) and property (HL). Hence we may assume that Ω itself is bounded. An examination of the first part of the proof of 9.40 shows that the only property of the set U (there hypothesized to be simply-connected) which is used is property (HL). Consequently, Ω has a barrier at each of its boundary points and so by 9.38 the Dirichlet problem is solvable for Ω. Since for each $a \in \Omega$ the function $\log|z - a|$ is continuous on $\partial\Omega$, there is then a continuous function u_a on $\overline{\Omega}$ which is harmonic in Ω and satisfies

(1) $\quad u_a(z) = \log|z - a| \quad \forall z \in \partial\Omega.$

By property (H) of Ω we have

(2) $\quad u_a = \operatorname{Re} \phi_a \quad$ for some $\phi_a \in H(\Omega)$.

Consider the holomorphic function

(3.a) $\quad f_a(z) = (z - a)e^{-\phi_a(z)}, \quad z \in \Omega.$

We have

$$\begin{aligned} |f_a(z)| &= |z - a|e^{-\operatorname{Re}\phi_a(z)} \\ &= |z - a|e^{-u_a(z)} \quad \forall\, z \in \Omega. \end{aligned} \tag{4}$$

From (4) and (1) we see that

(5.a) $\quad \lim_{z \to \xi} |f_a(z)| = 1 \quad \forall \xi \in \partial\Omega.$

It follows from this and 5.12 that

(6) $\quad |f_a(z)| \leq 1 \quad \forall z \in \Omega.$

But equation (3.a) also shows that f_a is zero only at the point a and so, in particular, f_a is not constant. Therefore by the Maximum Modulus Principle 5.10, strict inequality prevails everywhere in (6):

(6.a) $|f_a(z)| < 1 \quad \forall z \in \Omega.$

We aim to show that f_a is one-to-one and $f_a(\Omega) = D(0, 1)$. Consider any $b \in \Omega\backslash\{a\}$. Since $|f_a(b)| < 1$, we may form

$$\Phi(w) = \frac{w - f_a(b)}{1 - \overline{f_a(b)}w}, \quad w \in D(0, 1).$$

This is a conformal map of $D(0, 1)$ onto itself which satisfies

(7) $\Phi(C(0, 1)) = C(0, 1).$

(See 2.6.) Since by (6.a), $f_a(\Omega) \subset D(0, 1)$, we may form

(8) $F = \Phi \circ f_a.$

Now $\Phi(f_a(b)) = 0$, so $F(b) = 0$. However, equation (3.b) shows that the holomorphic function f_b has a first order zero at b and no other zeros. Consequently, F/f_b is holomorphic in Ω. Now for any $\xi \in \partial\Omega$

$$\lim_{z\to\xi}\left|\frac{F(z)}{f_b(z)}\right| \overset{(5.b)}{=} \lim_{z\to\xi}|F(z)| \overset{(8)}{=} \lim_{z\to\xi}|\Phi(f_a(z))|$$
$$= 1, \quad \text{by (7) and equation (5.a).}$$

Therefore the Maximum Modulus Principle in the form 5.12 implies that

(9) $$\left|\frac{F(z)}{f_b(z)}\right| \le 1 \quad \forall z \in \Omega.$$

Since $F(a) = -f_a(b)$, it follows from (9) that

(10) $$\left|\frac{f_a(b)}{f_b(a)}\right| = \left|\frac{F(a)}{f_b(a)}\right| \le 1,$$

(11.a.b) $|f_a(b)| \le |f_b(a)|.$

But the roles of a and b may be interchanged and the last inequality together with the corresponding inequality (11.b.a) gives us

$$|f_a(b)| = |f_b(a)|.$$

This, however, says that equality holds in (10) and so by the Maximum Modulus Principle (9) becomes

(12) $$\left|\frac{F(z)}{f_b(z)}\right| = 1 \quad \forall z \in \Omega.$$

Since f_b vanishes only at b (see equation (3.b)), it follows from (12) that F is never zero in $\Omega\backslash\{b\}$, i.e. recalling (8),

(13.b) $f_a(z) \ne f_a(b) \quad \forall z \in \Omega\backslash\{b\}.$

Here b is any point of $\Omega\backslash\{a\}$. However, the inequality (13.a) is true too, as follows at once from (3.a). The inequalities (13.b) are thus valid for all $b \in \Omega$. They establish the univalence of f_a.

Now to prove the surjective character of f_a, suppose $w \in D(0, 1)\backslash f_a(\Omega)$. We may then form

$$G(z) = \frac{1 - \bar{w}f_a(z)}{f_a(z) - w}, \quad z \in \Omega.$$

Arguments like those used above show that

$$\lim_{z \to \xi} |G(z)| = 1 \quad \forall \xi \in \partial\Omega.$$

Therefore by the Maximum Modulus Principle in the form 5.12 we infer that

$$|G(z)| \leq 1 \quad \forall z \in \Omega.$$

Since, however, $G(a) = -1/w$ and $w \in D(0, 1)$, we have a contradiction.

Remarks 9.48 The converse direction, deducing the solubility of the Dirichlet Problem from the Riemann Mapping theorem, is evidently only possible when something is known about the boundary behavior of the latter. In case the Riemann map extends to a homeomorphism of the closure of the region Ω, i.e., in the case of Jordan regions, the solubility of the Dirichlet Problem becomes a banality: one transfers everything over to the disk, uses the Poisson formula there, then transfers back to $\bar{\Omega}$. When this is done for Ω bounded by a (sufficiently) smooth, simple, positively oriented loop Γ, the solution takes the form

$$(*) \qquad h(z) = \frac{1}{2\pi} \int_\Gamma u(\xi) D_N g_z(\xi) d|\xi|, \quad z \in \Omega,$$

where u is the boundary datum and for each $z \in \Omega$, g_z is the *Green's function* on Ω with singularity at z. This is a (the) function in $\Omega\backslash\{z\}$ such that $g_z(\xi) + \log|\xi - z|$ is harmonic in Ω, while $\lim_{\xi \to w} g_z(\xi) = 0$ for each $w \in \partial\Omega$. (Namely, $g_z(\xi) = \log|1 - \bar{f}(\xi)f(z)| - \log|f(\xi) - f(z)|$, where f is a conformal map of Ω onto $D(0, 1)$.) In (*), $D_N g_z(\xi)$ denotes the directional derivative of g_z at ξ in the direction inwardly normal to Γ at ξ, i.e.,

$$D_N g_z(\Gamma(t)) = \frac{1}{|\Gamma'(t)|} [-D_1 g_z(\Gamma(t)) \operatorname{Im} \Gamma'(t) + D_2 g_z(\Gamma(t)) \operatorname{Re} \Gamma'(t)].$$

The notation $d|\xi|$ is defined by

$$\int_\Gamma \phi(\xi) d|\xi| = \int_\alpha^\beta \phi(\Gamma(t))|\Gamma'(t)| dt$$

($[\alpha, \beta]$ the parameter interval of Γ, ϕ any continuous function on Γ). One has in (*) the exact analog of the Poisson formula in the disk. While the *solubility* of the problem in Ω is indeed trivial (after 9.14), this transferal to Ω of the Poisson integral formula via f is not trivial. To achieve it, smoothness of the extended f on the boundary is needed and this must somehow be wrung out of the smoothness properties of Γ. The integral one gets then actually involves $f^{-1}|C(0, 1)$ instead of Γ. This is changed to an integral involving Γ by means of an elementary argument like that in 4.7. (Cf. with the proof of 5.17.)

Often (*) is derived from the classical Green's–Stokes' formula which relates the integral of a certain differential form over a surface to that of another form over the boundary of the surface. However, we will not pursue such matters in this book; the interested reader may consult any book on potential theory. (It should be noted in passing, however, that if Lebesgue integration is used, less smoothness has to be demanded of Γ for (*) to hold. See, e.g., VERBLUNSKY [1951].)

Notes to Chapter IX

For some of the history of this famous theorem proved, modulo some gaps, by RIEMANN in his Göttingen dissertation in 1851 [pp. 1–48 of his *Werke*] see WALSH [1973], pp. 320–321 of HILLE [1962], § 74 of DINGHAS [1961], pp. 49–53 of SIMONART [1931] and the encyclopedia articles of LICHTENSTEIN [1919] (chapter IV) and BIEBERBACH [1921]. In CARATHÉODORY [1912a] appears the method (used in treating the $\phi_{n,p}$ in the proof of 9.7, as well as in the proof of 9.22) of "variable regions," showing that when the regions converge geometrically, the Riemann maps from them into $D(0, 1)$ converge locally uniformly. For more on this see BIEBERBACH [1913a] and pp. 283–285 of POMMERENKE [1975] and the references cited there. The result 9.6, a special case of the above, affirms roughly that if Ω is close to a disk, then its Riemann map is close to the identity. CARATHÉODORY [1914] gives a version of 9.6 and uses it in a beautiful, self-contained proof of 9.7. (See pp. 61–72 of SIMONART [1931] for a French exposition.) KOEBE's proof is in his paper [1915], having been announced and sketched in [1912]. He called this method the "Schmiegungsverfahren." For other constructive proofs see LEJA [1935], LEHTO [1949], POMMERENKE [1965], and JULIA [1927a]; compare also pp. 339 ff. of HILLE [1962] and the references there. OSTROWSKI [1929a] shows that not only does $R_n \to 1$ ((8) in the proof of 9.7), but in fact

$$1 - R_n < \frac{128|\log R_1|}{1 - R_1}\cdot\frac{1}{n}$$

for all $n \geq 512|\log R_1|$. See also VEECH [1967], p. 112. For another proof of 9.6 (based on area) see pp. 10–11 of BIEBERBACH [1931]; also VEECH [1967], pp. 99 ff., OSTROWSKI [1930], M. MÜLLER [1938], LAVRENTIEFF and KWASSELAVA [1940] and SAKASHITA [1956] for improvements on the majorant. For more on 9.3 see exercise IV.91 of PÓLYA and SZEGÖ [1976].

By property (d) of Koebe maps we have $|\psi_{0,k}| \leq |\psi_{0,k+1}|$ for all k. Hence the convergence of the sequence $\{\psi_{0,k}\}$ in the proof of 9.7 can also be secured quite expeditiously by applying 7.12(iii) to the functions $\psi_{0,k}(z)/z$.

The proof in § 3 appears in RADÓ [1922/23b]; see also § II of OSTROWSKI [1929a]. For other accounts of the fundamental theorem see BIEBERBACH [1914], LINDELÖF [1914], [1915], [1920], MONTEL [1917], FABER [1922], CARATHÉODORY [1928], [1948], chapter 10 of DE LA VALLÉE POUSSIN [1949], JULIA [1950] and GARABEDIAN [1976].

For some regions it is possible to write down the Riemann map explicitly. This is the case, for example, with the interior of a simple closed polygon. The formulas which accomplish this are known as the Schwarz–Christoffel formulas. For a development of them (using 9.16) the reader can consult SAKS and ZYGMUND [1971], pp. 233–237 or AHLFORS [1966a], pp. 227–233; and pp. 274–275 of LICHTENSTEIN [1919] for history. For some other useful explicit conformal maps see pp. 105–143 of SANSONE and GERRETSEN [1969].

The whole problem of conformal mapping, and Riemann's theorem in particular, can be treated from a Hilbert space [of areally square-integrable holomorphic functions] point of view. For a definitive account of this procedure see BERGMAN [1970] together with LEHTO [1949], and for short and very readable introductions STONE [1960], § 17.2 of HILLE [1962], or EPSTEIN [1965]. The latter is a completely *ab ovo* treatment requiring of the reader no knowledge of either Lebesgue integration or Hilbert space theory.

§ 48 of LICHTENSTEIN's encyclopedia article [1919] is a good source for the history of the fundamental mapping theorem 9.14. There are essentially four approaches centered around CARATHÉODORY [1913b,c], [1952]; COURANT [1914a] and KOEBE [1913a], [1915]; OSGOOD and TAYLOR [1913]; LINDELÖF [1914], [1915]. The first uses cross-cuts in Jordan regions and a famous theorem of Fatou that $\lim_{r\uparrow 1} f(re^{i\theta})$ exists in $\mathbb{C}$ for almost all (Lebesgue measure) $\theta \in [0, 2\pi]$ whenever f is a bounded holomorphic function in $D(0, 1)$. Actually only the case of univalent f is needed and for this CARATHÉODORY [1913b], § 2 gives an elegant proof using area ideas that avoids Lebesgue integration. Area arguments also dominate in the second method above in which it is shown by *reductio ad absurdum* that the Riemann map must be uniformly continuous. (See also DE LA VALLÉE POUSSIN [1930], WOLFF [1930a], TSUJI [1930a], [1959], MCSHANE [1937], LELONG-FERRAND [1955] and ARSOVE [1968a].) The third method is old-school potential theory arguments. Lindelöf's is closer in spirit to Carathéodory's. (See pp. 372–373 of LICHTENSTEIN *loc. cit.*) For other accounts see §§ 5–8 of STUDY [1913], MONTEL [1917] (also chapter IV of [1927]), FABER [1922], DE LA VALLÉE POUSSIN [1932], [1949], DOUGLAS [1931], FLOYD [1946], KLEINER [1955], [1962] and COLLINGWOOD and PIRANIAN [1964]. The treatment in the text is based on Carathéodory and on LINDELÖF's exposition [1920]. It uses, I believe, the fewest auxiliary results and no geometric facts which cannot be precisely formulated and proven. Many of the other proofs abound with vague arguments. The critical reader will undoubtedly find them unsatisfactory; it is hoped that in eliminating these deficiencies the present treatment has not compromised too grievously with esthetics. Carathéodory himself was critical of the cavalier proof which Koebe offered for his fundamental lemma 9.10 and gave (in [1913b]) another (KOEBE rejoins on pp. 212–213 of [1915]); that in the text is adapted from GOLUZIN [1969]. For another proof see MYLLER–LÉBÉDEFF [1938], TSUJI [1959], p. 302 and for generalizations see § 19 of MONTEL [1917], FLEET [1954], § 5 of BAGEMIHL and SEIDEL [1960], [1961], BAGEMIHL [1960], GAVRILOV [1965], CHOU [1967], RUNG [1968] and pp. 261–270 of POMMERENKE [1975]. A measure-theoretically more careful treatment of the main boundary argument in COURANT [1914] and FABER [1922] may be found on pp. 308–311 of RUDIN [1974]. (See also NOVINGER [1975a].)

There are several alternative proofs that g is univalent (lemma 9.13). See, e.g., chapter 6 of HEINS [1962], where the solubility of the Dirichlet problem is utilized. Montel and Lindelöf use a version of 5.56 to this end: the half-disk in 5.56 is replaced with $D = D(i, 1)$. However, in addition one needs to know that *every* point of $\partial\Omega$ is accessible from Ω (cf. 1.16). This is true, and trivial after the fact, i.e., after 9.14 is proved, but an *ab initio* proof is unpleasant. With this result in hand the proof of 9.13 is trivial: If $a, \tilde{a} \in \partial\Omega$ and $g(a) = w = g(\tilde{a})$, select arcs $\gamma, \tilde{\gamma}$ with terminal points $a, \tilde{a}$, and lying, except for these points, in Ω. Then $\Gamma = g \circ \gamma$, $\tilde{\Gamma} = g \circ \tilde{\gamma}$ are arcs with common terminal point w which lie, except for this point, in D. We have $f \to a$ on Γ and $f \to \tilde{a}$ on $\tilde{\Gamma}$, so by 5.56 $\tilde{a} = a$.

The reader may note that the maps f and $g = f^{-1}$ are treated by different techniques in the proof of the Osgood–Taylor–Carathéodory theorem. It should be clear, however, that if the arc lemma 9.10 is freed from the disk, then the technique of 9.11 could be used on g as well as f. This approach is offered in FLETT [1954]. It rests on an inequality of Carleman (see pp. 301–302 of TSUJI [1959]) and, even after the details are supplied, it is perhaps slightly shorter than the development in the text. The reader is encouraged to look at it and to supply those details.

Carathéodory (see also STUDY [1913]) in his theory of "prime ends" also made definitive investigations into the boundary behavior of non-Jordan regions. An account may be found in POMMERENKE [1975] or COLLINGWOOD and LOHWATER [1966]. DENJOY [1941], [1942] also investigated continuous, not necessarily univalent, extensions of the Riemann map to the boundary. See also SCHÖNHAGE [1969].

Another important problem is "conformality" of the Riemann map at the boundary (cf. 6.24) when various smoothness hypotheses are placed on the bounding curve. For this vast area of research I shall be content to cite chapter 10 of POMMERENKE [1975], LELONG–FERRAND [1955] and the two-volume work of GATTEGNO and OSTROWSKI [1949] and its bibliographies. Finally there is the problem of constructability of conformal maps, especially important in (extra-mathematical) applications. Again there is an enormous literature and I cite only two references: SEIDEL [1952] and GAIER [1964]. The bibliography of Gaier's book (consisting of 490 items, equipped with FM, Zbl and MR coordinates) covers the field very well. It also will inform the reader on other approaches to the mapping problem not discussed here.

For more literature on boundary behavior see the references on p. 155 of BERNARDI [1966], the book of COLLINGWOOD and LOHWATER [1966] and its bibliography.

The effect of 9.14 is to concentrate a lot of geometric difficulties in one place. Now to prove theorems involving boundary behavior of an arbitrary holomorphic function in an arbitrary Jordan region, instead of confronting all the unpleasantries of proof associated with an arbitrary Jordan region, one can prove the result for disks or some other geometrically simple region and transfer them to the general region by citing 9.14. E.g., the limiting behavior of a bounded function along two sides of an angular sector is easy to analyze but that along two abutting arcs of a Jordan-curve is not (see, e.g., pp. 226–228 of DINGHAS [1961]); so we reduce the latter to the former, as we did in 9.20. For an intriguing application of 9.14 to a problem in Fourier theory (theorem of Bohr and Pál), see p. 127 of ZALCMAN [1974].

9.14 also permits the introduction for a Jordan region Ω of a sequence of polynomials $\{P_n\}$, called Faber polynomials, which play for the region Ω a role analogous to that played by the monomials z^n in the unit disk D: for certain Ω, each $f \in H(\Omega)$ has a unique expansion $\sum c_n P_n$ ($c_n \in \mathbb{C}$) which converges locally uniformly in Ω to f. The literature on Faber polynomials is considerable; the papers of SZEGÖ [1921], HEUSER [1934], J. H. CURTISS [1971], SUETIN [1964] and their bibliographies should get the interested reader started.

9.14 together with the Riemann mapping theorem provides the following important characterization of Jordan regions: a subset Ω of $\mathbb{C}$ is a Jordan region if and only if it is the image of $D(0, 1)$ under a homeomorphism of $\bar{D}(0,1)$. R. L. MOORE [1918] has given a wholly internal characterization: Ω is a Jordan region if and only if it is a simply-connected region which is uniformly connected *im kleinen*, i.e., for every $\varepsilon > 0$ there is a $\delta(\varepsilon) > 0$ such that any two points of Ω closer together than $\delta(\varepsilon)$ lie in a connected subset of Ω of diameter less than ε. See also ARSOVE [1967].

For more applications of the Argument Principle see exercises III. 181–194 of PÓLYA and SZEGÖ [1972]. The last of these is Rouché's theorem for Jordan regions (weaker than our 8.18). See also p. 193 of SAKS and ZYGMUND [1971]. For a generalization of it see MONTEL [1932b] and [1933a]; LIPKA [1929] and TERASAKA [1929]; KERÉKJÁRTÓ [1934/35]. For a topological version of the Argument Principle see RADÓ [1936].

For an extension of 9.16 to complex-valued harmonic function see H. KNESER [1926]. For original credits see footnote 106 of OSGOOD [1901]. A significant topological extension (based on 4.57) was made by MEISTERS and OLECH [1963];

see also G. T. WHYBURN [1951], lemma 1. For a different extension see BAGEMIHL [1967]. A very elementary, purely topological version of 9.16 may be found in DERRICK [1973].

Besides the literature mentioned in 4.44, there are proofs of 9.19 in R. L. MOORE [1926], GEHMAN [1926] and CAIRNS [1951]. (The latter proves the Schönflies and Jordan Curve Theorems together.) See also BIEBERBACH [1913b], FLOYD [1946] and CHOQUET [1945]. (In the latter harmonic extensions are sought.) A series of papers by W. Huebsch and M. Morse (circa 1960) treats the higher dimensional case exhaustively.

For more material related to 9.26 see JULIA [1927], [1928], WALSH [1929], FARRELL [1932], LEJA [1936] and the monographs of WALSH. Of course the definitive result here is Mergelyan's theorem mentioned in the Chapter VIII notes. The reader may feel that we paid too high a price for theorem 9.26; especially if he is versed in abstract measure and integration theory, he could buy the full Mergelyan theorem with a comparable expenditure of energy. *Nolo contendere.*

The *raison d'être* for Cauchy's theorem in the form 9.27 is mainly esthetical; one seldom needs this general a version of the theorem in applications and for the special configurations in which one does, more elementary *ad hoc* arguments usually suffice. However, if one does want the theorem and is not willing to go all the way to Walsh's result 9.26, then the proof of BECKENBACH [1943] or DELANGE [1956] is the best. The theorem proved there is a little less general than 9.27, however: the loop Γ must be the boundary of Ω. The most general Cauchy theorem that I know of in the literature is NÖBELING [1949]: If Γ is a rectifiable loop and Ind_Γ is bounded in $\mathbb{C}\backslash\Gamma$, f is holomorphic in $\{z: \text{Ind}_\Gamma(z) \neq 0\}$ and continuous on the closure of this set, then $\int_\Gamma f = 0$. Many proofs of 9.27 have appeared in the literature; few are both artistically and logically satisfactory. Here is a partial roll call: PUCCIANO [1909], WATSON [1914], POLLARD [1923] (heavy), LEAU [1926], ČECH [1930], RIDDER [1930a], [1941], KAMKE [1932], HEILBRONN [1933], DENJOY [1933], [1955], ESTERMANN [1933], HU [1935], GONÇALVES [1940], [1941], [1942] (the *Zentralblatt* reviewer was never convinced), HESTENES [1941], REID [1941], MINAMI [1942], LOOMIS [1944], DEURING [1949], H. KNESER [1949], M. H. A. NEWMAN [1951], CRAVEN [1964], GRUNSKY [1966], ARSOVE [1968b], [1969]. For a more complete list (including some Chinese references!) and some comments see FALLIN and GOULD [1965].

9.28 is just the most prominent product of a big industry concerned with the study of "removable sets" (of singularities). See POMPEIU [1905a] and chapter V of ZORETTI [1911] for more on this theme. Cf. also VASILESCO [1930], the monograph of CARLESON [1967] and ZALCMAN [1968b].

The proof technique in 9.30 comes from ZAREMBA [1905]. In the 19th century, just as continuity of f' had to be hypothesized for holomorphic functions (see Chapter V notes), so too the continuity of the partial derivatives in 9.31 had to be hypothesized for harmonic functions. ZAREMBA [1905] showed that the latter hypothesis could be dropped. See WILKOSZ [1922], LOOMAN [1923–24], HOPF [1930] and ZAREMBA [1939]. However, WILKOSZ [1922] gives an example (due to Zaremba; see also ZAREMBA [1939]) of a discontinuous function h whose Laplacian exists and is identically zero: $h(x, y) = xy/(x^2 + y^2)^2$ if (x, y) is not $(0, 0)$, $h(0, 0) = 0$. If h is bounded, this cannot happen though (TOLSTOV [1951]).

On the other hand, relative to 9.31, the requirement that the Laplacian exist and equal zero throughout the region can be considerably weakened and the harmonicity of the function still inferred. See RIDDER [1940], [1946] and BECKENBACH [1945] as well as the references cited in the Chapter II notes.

In the proof of 9.39 I have followed CONWAY [1973], which the reader should consult for the geometric significance of the proof.

By now we have seen both integral and differential characterizations of harmonic functions. There are many more, including areal mean value properties and mixtures of integral and differential properties. See, e.g., BLASCHKE [1916], [1918] and SAKS [1932].

Laplace's equation is very prevalent in classical physics and the Dirichlet problem is one of several kinds of boundary value problem for partial differential equations which are important there. Laplace's equation arises as follows: $h(x, y)$ measures a quantity (energy, fluid mass, heat) at the point (x, y) and the net flow into any region of the plane equals the net flow out, i.e., a steady state or an equilibrium state exists. This flow across a curve is often proportional to the normal derivative of h with respect to the curve. If one records this information in the form of an integral over a square of the normal derivative of h (in the case of a square parallel to the axes this is D_1h or D_2h) being equal to 0, divides by the perimeter of the square and lets the side length decrease to 0, there emerges the equation $\nabla^2 h = 0$. Actually, harmonicity is completely characterized by the vanishing of all such integrals or by the vanishing of the integral of the normal derivative over all (small) circles. See KOEBE [1906b], BÔCHER [1906], SEVERINI [1911], G. C. EVANS [1921], [1928], SAKS [1932], and RIDDER [1946]. [The integrals here are of the kind discussed in 9.48.]

Laplace's equation and the whole subject of potential theory were intensely investigated in the second half of the 19th century, but not until early in this century were the methods of analytic function theory brought into the subject on a large scale. The pendulum has now swung even further and the subject is dominated by the methods of real analysis, probability and the theory of distributions. The Dirichlet problem for an arbitrary bounded, simply-connected region was finally disposed of by LEBESGUE [1907]. See also COURANT [1914b], LICHTENSTEIN [1916], pp. 734–788 of OSGOOD [1928], INOUE [1938], LEJA [1950]. The use of superharmonic functions as in the text (descendant from the "méthode de balayage" of Poincaré, 1890) was pioneered by PERRON [1923] and simplified by REMAK [1924], [1926], RADÓ and RIESZ [1925], and WHITNEY [1932]. Similar ideas were used by Wiener and Kellogg about this same time. The development in the text follows KELLOGG [1928], a similar account appearing in his book [1929]. The method of approximating the boundary data with harmonic polynomials (as in 9.37) also occurs in WALSH [1930] and JULIA [1930a]. For fine expository accounts of the problem see CARATHÉODORY [1937a] and BRELOT [1933], [1938]. For a very detailed history of the Dirichlet problem up to about 1915, including outlines of the proofs, the encyclopedia article of LICHTENSTEIN is excellent; the survey article of KELLOGG [1926b] is also valuable. For modern developments see ANGER [1961/62], with its bibliography of over 200 entrics, and BRELOT [1952], [1972], also equipped with extensive bibliographies.

For an alternative short proof of 9.47, at equation (5.a) one can cite 6.14(ii) to conclude that f_a is exactly m-to-1 for some m. Since f_a assumes the value 0 only at a and with multiplicity 1 (see (3.a)), it follows that $m = 1$.

Chapter X
Simple and Double Connectivity

§ 1 Simple Connectivity

We come now to the climax of much of our previous work. To state the result in its most awesome form we introduce some convenient definitions.

Definition 10.1 Let U be an open subset of $\mathbb{C}$. Say that U has

property (C) if the conclusion of Cauchy's theorem holds in U, that is, $\int_\gamma f = 0$ for every $f \in H(U)$ and every piecewise smooth loop γ in U;

property (P) if every holomorphic function in U has a primitive; that is, every function in $H(U)$ is the derivative of another;

property (AP) if each holomorphic function in U is locally uniformly approximable on U by polynomials; that is, for each $f \in H(U)$ there exist polynomials p_n such that $p_n \to f$ locally uniformly on U.

Theorem 10.2 (Hauptsatz of the Cauchy Theory) *For any proper subregion Ω of $\mathbb{C}$ the following properties are equivalent:*

(a) *Ω is conformal to an open disk*
(b) *Ω is homeomorphic to an open disk*
(c) *Ω is loophomotopically-connected*
(d) *Ω is homologically-connected*
(e) *Ω is simply-connected*
(f) *Ω contains the inside of every Jordan-curve in Ω*
(g) *Ω has property (C)*
(h) *Ω has property (P)*
(i) *Ω has property (HL)*
(j) *Ω has property (HS)*
(k) *Ω has property (HS_1)*
(l) *Ω has property (CL)*
(m) *Ω has property (CS)*
(n) *Ω has property (H)*
(o) *Ω has property (AP).*

Proof: Here is one possible cyclic tour, touching all bases:

$$\begin{array}{ccccccccccccccc}
(n) & \Longleftarrow & (a) & \overset{\text{trivial}}{\Longrightarrow} & (b) & \Longrightarrow & (c) & \overset{4.12}{\Longrightarrow} & (d) & \overset{4.65}{\Longrightarrow} & (e) & \overset{8.10}{\Longrightarrow} & (o) & \Longrightarrow & (g) \\
{\scriptstyle 9.46}\Big\Downarrow & & {\scriptstyle 9.7}\Big\Uparrow & & & & & & & & & & & & \Big\Downarrow{\scriptstyle 2.11} \\
(i) & \underset{4.65}{\Longrightarrow} & (k) & \underset{\text{trivial}}{\Longleftarrow} & (j) & \underset{4.65}{\Longleftarrow} & (m) & \underset{4.65}{\Longleftarrow} & (l) & \underset{4.65}{\Longleftarrow} & (f) & \underset{4.65}{\Longleftarrow} & (i) & \underset{5.34}{\Longleftarrow} & (h)
\end{array}$$

Substantiation of the three unlabelled implications follows.

(a) ⇒ (n) According to 5.21(i), every open disk has property (*H*) and that property is transmitted to Ω via the conformal map supplied by (a).

(b) ⇒ (c) Open disks are loophomotopically-connected by 4.16 and homeomorphisms clearly preserve this property.

(o) ⇒ (g) Examine the hints to 8.13.

Remarks: Part of the profundity of this theorem is that some of the properties which it establishes as equivalent (all but (d), (e) and (f)) are entirely internal to Ω, one is entirely external to Ω (namely (e), which concerns the connectedness of the complement of Ω) and one involves both the interior and the exterior of Ω (namely (d), which concerns how curves in Ω behave with respect to points outside Ω). Moreover, properties (a), (n), (o) and (h)–(k) are "analytic" while the rest are purely "topological." For bounded regions yet another (internal, analytic) characterization of simple-connectivity was described in 7.42 and still another is discussed in the last paragraph of the notes to this chapter.

In 10.7 below we will see (essentially) a direct proof that (h) implies (n), i.e., that property (*H*) follows from property (*P*). Moreover, 9.47 is a "direct" proof that (n) implies (a), while 7.19 provides a direct proof that (d) implies (g). A direct proof of the equivalence of (d) and (e) is 8.21.

Corollary 10.3 *For any open subset* Ω *of* $\mathbb{C}$ *statements* (c) *through* (o) *in* 10.2 *are all equivalent.*

Proof: It is easy to see that the whole plane $\mathbb{C}$ satisfies (c) through (o)—the reader can search for the relevant previous theorem or exercise to cite in each case. Therefore we put aside the case $\Omega = \mathbb{C}$.

The equivalence of (d), (e), (f) and (i)–(m) is averred in 4.65, which does not have connectedness in its hypothesis. As to the other properties, we shall show that each is possessed by Ω if and only if it is possessed by each component of Ω. [Our corollary then follows from 10.2.] For (h) and (n) this is obvious. Since a curve in Ω and the range of any homotopy in Ω involving it are connected sets, they lie wholly in one component of Ω, so (g) and (c) are clear. For (o) we can pass back and forth from 8.10 and 10.2 thus: Ω has (AP) ⇒ Ω has (C) by the proof of 8.13 ⇒ each component of Ω has (C) [as just noted] ⇒ each component of Ω has (AP) by 10.2 ⇒ each component of Ω simply-connected (10.2) ⇒ Ω simply-connected by 4.66 ⇒ Ω has (AP) by 8.10.

Exercise 10.4 (i) *Show that a region in* $\mathbb{C}$ *is simply-connected if and only if either* (a) *its boundary has no bounded component or* (b) *it is bounded and has a connected boundary.*

Hint: 10.2, 1.35 and 1.37(i).

(ii) *Show that an open subset* U *of* $\mathbb{C}$ *is simply-connected* (*if and*) *only if it contains every bounded* (*open*) *set* V *such that* $\partial V \subset U$.

Hints: 10.3(f) insures that this condition implies simple-connectivity. For the converse cite 1.37(ii).

Exercise 10.5 *Let n be a positive integer, U an open, simply-connected subset of $\mathbb{C}$, $f \in H(U)$. Show that f has a holomorphic nth root in U if and only if the order of each of its zeros is divisible by n.*

Hints: One direction is trivial. For the other we may assume U is connected and f is not identically 0. Then $A = f^{-1}(0)$ consists of isolated points and Weierstrass (7.32) provides an $h \in H(U)$ with a zero of order $n(a,f)/n$ at each $a \in A$ and no other zeros. Then f/h^n is zero-free and so by property (HL) it has the form e^F for some $F \in H(U)$. Then $g = he^{F/n}$ is a holomorphic nth root of f.

Exercise 10.6 *Use 10.2 to give a new proof of the simply-connected case of 7.41: if Ω is a simply-connected proper subregion of $\mathbb{C}$, $a \in \Omega$ and $\mathscr{G} = \mathscr{G}_a$ is the group (under composition) of all conformal self-maps of Ω which fix a, then $f \to f'(a)$ is an isomorphism of $\mathscr{G}$ onto the multiplicative group $C(0, 1)$.*

Hints: By 10.2 there is a conformal map ψ of Ω onto $D = D(0, 1)$. Following with the appropriate conformal self-map of D, we can suppose that $\psi(a) = 0$. Now 6.2(ii) shows that the conformal self-maps of D which fix 0 all have the form $z \to uz$ for some $u \in C(0, 1)$. Consequently the typical function in $\mathscr{G}$ has the form $f_u(z) = \psi^{-1}(u\psi(z))$. By the chain rule $f_u'(a) = u$, so the map $f_u \to u$ is well-defined. It is evidently one-to-one and homomorphic.

Exercise 10.7 *Let U be an open, simply-connected subset of $\mathbb{C}$, $h: U \to \mathbb{R}$ a harmonic function. According to 10.3 there is a holomorphic function f in U such that $h = \operatorname{Re} f$. Find an expression for f directly in terms of h. In fact, show that $D_1h - iD_2h = f'$ and that consequently*

$$f(z) = h(z_0) + \int_{\Gamma(z_0, z)} (D_1h - iD_2h)$$

for any $z_0, z \in U$ and any piecewise smooth curve $\Gamma(z_0, z)$ in U joining z_0 to z. (Note that such curves always exist if U is connected, thanks to 1.28.)

Hints: Given $z_0 \in U$, we may require of f that $\operatorname{Im} f(z_0) = 0$, i.e.,

$$f(z_0) = h(z_0).$$

From the Cauchy–Riemann equation we have $f' = D_1 f = -iD_2 f$. Equating imaginary parts shows that

$$D_1(\operatorname{Im} f) = -D_2(\operatorname{Re} f),$$

whence $f' = D_1 f = D_1(\operatorname{Re} f) - iD_2(\operatorname{Re} f) = D_1h - iD_2h$. Therefore from 2.10(i) we have

$$f(z) - h(z_0) = f(z) - f(z_0) = \int_{\Gamma(z_0, z)} f' = \int_{\Gamma(z_0, z)} (D_1h - iD_2h).$$

Exercise 10.8 (i) *Show that even without the simple-connectivity hypothesis, $D_1h - iD_2h$ is holomorphic in the open set U whenever h is harmonic there.*

Hint: Apply 10.7 in subdisks of U.

(ii) *Use* (i) *to prove a local maximum principle for harmonic functions; equivalently, show that a harmonic function which is constant in some non-empty open subset of a region is constant in the whole region.*

Hints: If the harmonic function h attains a local maximum in the region Ω, then it is constant in some disk. Therefore (5.62) the holomorphic function $D_1h - iD_2h$ is identically 0 in Ω, i.e., $D_1h = D_2h = 0$ throughout Ω. From 2.16 it follows that h is actually holomorphic in Ω and $h' \equiv 0$. Cite 2.10(iv). An alternative one-line proof can be based on 9.33(iii).

(iii) *Give a counter-example to the conclusion* (ii) *when the function is only subharmonic.*

Hint: $f(z) = \max\{0, \operatorname{Re} z\}$ is subharmonic in $\mathbb{C}$ (by 5.25(ii) or directly from definition 5.6).

Exercise 10.9 *Let U be an open subset of $\mathbb{C}$ such that $\mathbb{C}\backslash U$ has exactly one bounded component C. Then there exists a smooth simple loop Γ in U such that* $\operatorname{Ind}_\Gamma(C) = 1$.

Hints: If $U \cup C = \mathbb{C}$, then we take any $r > 2\sup|C|$, pick any point $w \in C$ and let $\Gamma = C(w, r)$. This curve lies in $\mathbb{C}\backslash C = U$. If $U \cup C \neq \mathbb{C}$, then by 1.38 $U \cup C$ is open and simply-connected. The connected component Ω of $U \cup C$ which contains C is open and is simply-connected by 4.66. Consequently by 10.2 there is a conformal map f of $D(0, 1)$ onto Ω. $f^{-1}(C)$ is a compact subset of $D(0, 1)$ and hence lies in $D(0, r)$ for some $r < 1$. Let $\gamma = C(0, r)$. Then $(f \circ \gamma) \cap C = \varnothing$ and if $w \in C \subset f(D(0, r))$, then f assumes the value w exactly once (counting multiplicity) in $D(0, r)$. By 5.86 then $\operatorname{Ind}_{f\circ\gamma}(w) = 1$. Let $\Gamma = f \circ \gamma$.

Exercise 10.10 *Let Ω be a region such that $\mathbb{C}\backslash\Omega$ has finitely many bounded components $C_1, \ldots, C_n$. Then for each $j = 1, 2, \ldots, n$ there exists a smooth simple loop γ_j in Ω such that* $\operatorname{Ind}_{\gamma_j}(C_j) = 1$ *and* $\operatorname{Ind}_{\gamma_j}(C_k) = 0$ *if $k \neq j$.* (Cf. 8.20.)

Hints: It suffices to construct γ_1. Use 1.40 to find an open $\Omega_1 \subset \Omega$ such that C_1 is the unique bounded component of $\mathbb{C}\backslash\Omega_1$. Then use the last exercise to construct a smooth simple loop γ_1 in Ω_1 such that $\operatorname{Ind}_{\gamma_1}(C_1) = 1$. By Corollary 4.3 $\operatorname{Ind}_{\gamma_1}$ vanishes in $\mathbb{C}\backslash(\Omega_1 \cup C_1) \supset C_2 \cup \cdots \cup C_n$, since $\mathbb{C}\backslash(\Omega_1 \cup C_1)$ is the union of the unbounded components of $\mathbb{C}\backslash\Omega_1$.

Exercise 10.11 *If Ω is a region such that $\mathbb{C}\backslash\Omega$ has finitely many bounded components $C_1, \ldots, C_n$, then there exist smooth simple loops $\gamma_1, \ldots, \gamma_n$ in Ω such that*

$$\int_\gamma f = \sum_{k=1}^{n} \operatorname{Ind}_\gamma(C_k) \int_{\gamma_k} f$$

for all $f \in H(\Omega)$ and all piecewise smooth loops γ in Ω.

Hints: If $\gamma_1, \ldots, \gamma_n$ are the loops provided by the last exercise, then

$$\text{Ind}_\gamma - \sum_{k=1}^{n} \text{Ind}_\gamma(C_k)\,\text{Ind}_{\gamma_k} = 0$$

in each unbounded component of $\mathbb{C}\backslash\Omega$ by 4.3 and in each bounded component $C_1, \ldots, C_n$ in virtue of the construction of γ_k. Therefore 8.14 (or 7.19) applies.

Exercise 10.12 *Consider the case of* 4.59 *in which the set U there is a region Ω and the function f there is holomorphic.*

(i) *Use* 10.11 *to evaluate the integers k_j featuring there.*

(ii) *Prove the existence of the representation* 4.59 *itself from* 10.11.

Hints: Let $C_1, \ldots, C_n$ and $\gamma_1, \ldots, \gamma_n$ be as in 10.11. Select a point p_j from each C_j. Now consider any zero-free $f \in H(\Omega)$. Define $k_j = \text{Ind}_{f\circ\gamma_j}(0)$ and set $F(z) = f(z)\prod_{j=1}^{n}(z - p_j)^{-k_j}$. For any piecewise smooth loop γ in Ω we have

$$\begin{aligned}\int_\gamma \frac{F'}{F} &= \sum_{j=1}^{n} \text{Ind}_\gamma(C_j)\int_{\gamma_j}\frac{F'}{F} \quad \text{by } 10.11\\ &= \sum_{j=1}^{n} \text{Ind}_\gamma(p_j)\left[\int_{\gamma_j}\frac{f'}{f} - \sum_{l=1}^{n} k_l \int_{\gamma_j}\frac{dz}{z - p_l}\right]\\ &= \sum_{j=1}^{n} \text{Ind}_\gamma(p_j)[2\pi i\,\text{Ind}_{f\circ\gamma_j}(0) - 2\pi i k_j] \quad \text{by } 10.10\\ &= 0 \quad \text{by definition of the } k_j.\end{aligned}$$

It follows from 2.11 that F'/F has a primitive ϕ in Ω. From $\phi' = F'/F$ deduce that the function $Fe^{-\phi}$ has derivative 0 in Ω, so the desired holomorphic logarithm for F is ϕ plus an appropriate constant.

§ 2 Double Connectivity

Theorem 10.13 *Let $0 < r < R < \infty$, $A = \{z \in \mathbb{C} : r < |z| < R\}$, γ, Γ piecewise smooth loops in A. Then*

(i) $\text{Ind}_\gamma(0)\int_\Gamma f = \text{Ind}_\Gamma(0)\int_\gamma f$ *for all $f \in H(A)$.*

(ii) *If* $\text{Ind}_\Gamma(0) \neq 0$, *$f \in H(A)$ is zero-free and* $\text{Ind}_{f\circ\Gamma}(0) = 0$, *then f has a holomorphic logarithm.*

(iii) *If $f \in H(A)$, $f(A) \subset A$ and* $\text{Ind}_{f\circ\Gamma}(0) \neq 0$, *then either $f(z) \equiv cz$ or $f(z) \equiv c/z$ for some constant c.*

Proof: (i) Ind_γ and Ind_Γ each vanish identically in $\mathbb{C}\backslash D(0, R)$ and in $\bar{D}(0, r)$ have the constant values $\text{Ind}_\gamma(0)$, $\text{Ind}_\Gamma(0)$ respectively. Therefore, $\text{Ind}_\gamma(0)\,\text{Ind}_\Gamma - \text{Ind}_\Gamma(0)\,\text{Ind}_\gamma$ vanishes identically in $\bar{D}(0, r) \cup [\mathbb{C}\backslash D(0, R)] = \mathbb{C}\backslash A$. Therefore (i) follows from 8.14 (or 7.19).

(ii) We have

$$(1) \quad 2\pi i\,\text{Ind}_{f\circ\Gamma}(0) = \int \frac{(f\circ\Gamma)'}{f\circ\Gamma} = \int\left(\frac{f'}{f}\circ\Gamma\right)\cdot\Gamma' = \int_\Gamma \frac{f'}{f}.$$

By hypothesis this in 0, so (i) with f'/f in the role of f gives

$$\mathrm{Ind}_\Gamma(0)\int_\gamma \frac{f'}{f} = 0.$$

Also, by hypothesis this latter index is non-zero, so we conclude that

$$\int_\gamma \frac{f'}{f} = 0.$$

As γ is any piecewise smooth loop in A, it follows from 2.11 that f'/f has a primitive F in A and then $f = e^{F-c}$ for some constant c, as shown in the proof of 5.34.

(iii) Let $\gamma = C(0, \sqrt{Rr})$. It follows from (i) that

(2) $$\mathrm{Ind}_\gamma(0)\int_\Gamma \frac{f'}{f} = \mathrm{Ind}_\Gamma(0)\int_\gamma \frac{f'}{f}.$$

Put (1) (and its analog for γ) into (2) to get

$$\mathrm{Ind}_\gamma(0)\,\mathrm{Ind}_{f\circ\Gamma}(0) = \mathrm{Ind}_\Gamma(0)\,\mathrm{Ind}_{f\circ\gamma}(0)$$

or, since $\mathrm{Ind}_\gamma(0) = 1$,

(3) $$\mathrm{Ind}_{f\circ\Gamma}(0) = \mathrm{Ind}_\Gamma(0)\,\mathrm{Ind}_{f\circ\gamma}(0).$$

Since the left side of (3) is non-zero by hypothesis, it follows that $n = \mathrm{Ind}_{f\circ\gamma}(0) \neq 0$ and therefore by 6.20(ii) (with $r_1 = r_2 = r$, $R_1 = R_2 = R$), f has the asserted form.

Corollary 10.14 *Let* $0 < r_j < R_j < \infty$, $A_j = \{z \in \mathbb{C}: r_j < |z| < R_j\}$ $(j = 1, 2)$ *and f a conformal map of A_1 onto A_2. Then f has the form* $f(z) \equiv cz$ *or* $f(z) \equiv c/z$ *for some constant c. In particular,* $R_2/r_2 = R_1/r_1$.

Proof: Replacing f by $(r_1/r_2)f$, we can suppose that $r_1 = r_2$. Then, considering f^{-1} instead of f if necessary, we can suppose that $R_2 \leq R_1$, which insures that f maps A_1 into $A_2 \subset A_1$. Note that f has no holomorphic logarithm. For if $g \in H(A_1)$ and $f = e^g$, then the fact that $z = f(f^{-1}(z)) = e^{g(f^{-1}(z))}$ for all $z \in A_2$ would contradict the result of 4.22(i). It follows then from 10.13(ii) that $\mathrm{Ind}_{f\circ\Gamma}(0) \neq 0$, where $\Gamma = C(0, \sqrt{r_1R_1})$, and consequently by 10.13(iii) f has the asserted form.

Theorem 10.15 *Let* $0 < r < R < \infty$, $A = A(r, R)$. *Let $\mathcal{U}$ denote the invertible elements in the ring $H^\infty(A)$ and $\mathcal{E}$ the subset of $H^\infty(A)$ consisting of functions which possess holomorphic nth roots for every positive integer n. [It follows trivially from 4.19 and 4.60 that $\mathcal{E}$ consists precisely of 0 and the exponentials of functions in $H(A)$ whose real parts are bounded above.] Let* $\|f\| = \sup|f(A)|$ *for* $f \in H^\infty(A)$. *Then we have*

(*) $$\inf_{f\in\mathcal{U}\setminus\mathcal{E}} \|f\|\,\|1/f\| = R/r.$$

Proof: According to 4.22(i) the function $f_1(z) = z$ has no holomorphic square root, so $f_1 \in \mathscr{U}\backslash\mathscr{E}$ and

(1) $$\inf_{f\in\mathscr{U}\backslash\mathscr{E}} \|f\|\,\|1/f\| \leq \|f_1\|\,\|1/f_1\| = Rr^{-1}.$$

Now consider any $f \in \mathscr{U}\backslash\mathscr{E}$. We propose to show by *reductio ad absurdum* that

(2) $$\|f\|\,\|1/f\| < R/r$$

is impossible. To this end, assume (2). Evidently we may scale f to have in fact

$$\|f\| < R \;\&\; \|1/f\| < 1/r.$$

Then f maps A into A. Setting $\gamma = C(0, \sqrt{Rr})$, we note that $f \notin \mathscr{E}$ and 10.13(ii) imply that

(3) $$\mathrm{Ind}_{f\circ\gamma}(0) \neq 0.$$

It follows from (3) and 10.13(iii) that either $f(z) = cz$ or $f(z) = c/z$. In either case, therefore, a contradiction to (2) is reached.

Remarks 10.16 Let $Q = \{r_1 + r_2 i : r_1, r_2 \text{ rational}\}$ and for each $f \in H^\infty(A)$ let $S(f)$ denote the closure of $f(A)$. If f is non-constant, set

(4) $$\sigma(f) = \{\lambda \in Q: \lambda 1 - f \notin \mathscr{U} \quad \text{or } \bar{\lambda} 1 - f \notin \mathscr{U}\}.$$

For any $\lambda \in \mathbb{C}$ and any $f \in H^\infty(A)$, $\lambda 1 - f \notin \mathscr{U}$ is clearly equivalent to $\lambda \in S(f)$. Moreover, if f is non-constant, $f(A)$ is open, so $Q \cap S(f)$ is dense in $S(f)$. Therefore, it follows from (4) that $\sigma(f)$ is not void and

$$\left.\begin{aligned} \sup|\sigma(f)| &= \sup|f(A)| \\ \inf|\sigma(f)| &= \inf|f(A)| \end{aligned}\right\} f \in H^\infty(A), \quad f \text{ non-constant.}$$

Since the functions in $\mathscr{U}\backslash\mathscr{E}$ are non-constant, (*) then says

(**) $$\frac{R}{r} = \inf_{f\in\mathscr{U}\backslash\mathscr{E}} \left[\frac{\sup|\sigma(f)|}{\inf|\sigma(f)|}\right].$$

Next let us observe that $\mathscr{U}$, $\mathscr{E}$ and σ involve only the ring structure of $H^\infty(A)$. That is, if $A_j = \{z \in \mathbb{C}: r_j < |z - c_j| < R_j\}$, where $c_j \in \mathbb{C}$, $0 < r_j < R_j < \infty$ ($j = 1, 2$) and Φ is a ring isomorphism of $H^\infty(A_1)$ onto $H^\infty(A_2)$ then, with the obvious notation, $\Phi(\mathscr{U}_1) = \mathscr{U}_2$, $\Phi(\mathscr{E}_1) = \mathscr{E}_2$ and $\sigma(\Phi(f)) = \sigma(f)$. The first two of these equalities are obvious. For the third, note that $\Phi(1) = 1$ and the additivity of Φ imply its rational homogeneity: $n\Phi(m/n\cdot 1) = \Phi(n\cdot m/n\cdot 1) = \Phi(m\cdot 1) = m\cdot 1$ for all integers m and n ($n \neq 0$). Also $(\Phi(i))^2 = \Phi(i^2) = \Phi(-1) = -1$, so by 1.7 $\Phi(i)$ is either the constant function i or the constant function $-i$. Therefore for rational r_1 and r_2, $\Phi\{r_1\cdot 1 \pm r_2\cdot i\} = \{r_1\cdot 1 \pm r_2\cdot i\}$. Together with $\Phi(\mathscr{U}_1) = \mathscr{U}_2$ and $\Phi(\mathscr{E}_1) = \mathscr{E}_2$, this implies via definition (4) that $\sigma(\Phi(f)) = \sigma(f)$ for all $f \in \mathscr{U}_1\backslash\mathscr{E}_1$. Thus from (**) $R_2/r_2 = R_1/r_1$. In particular, if ϕ is a conformal map of A_2 onto A_1, then $\Phi(f) = f\circ\phi$ clearly defines a ring isomorphism of $H^\infty(A_1)$ onto $H^\infty(A_2)$ and so we have a new proof (cf. 10.14) that the existence of such a ϕ entails the equality $R_2/r_2 = R_1/r_1$. However, we have done more than prove this equality; we have found the radius ratio R/r in the ring structure of $H^\infty(A)$!

Next we take up a study of doubly-connected regions, culminating in the beautiful classification theorem 10.20.

Theorem 10.17 *Let Ω be a region such that $\mathbb{C}\backslash\Omega$ is not a single point but is compact and connected. Then Ω is conformal to $D(0, 1)\backslash\{0\}$ and there is a conformal map F: $D(0, 1)\backslash\{0\} \to \Omega$ of the form $F(z) = c/z + \sum_{n=0}^{\infty} c_n z^n$.*

Proof: We may assume, after a translation, that $0 \in K = \mathbb{C}\backslash\Omega$. Consider then ϕ: $\mathbb{C}\backslash\{0\} \to \mathbb{C}\backslash\{0\}$ given by $\phi(z) = 1/z$. $\phi(\Omega)$ is an open, connected set. If $0 < r < \infty$ is such that $K \subset D(0, r)$, then $D(0, 1/r)\backslash\{0\} \subset \phi(\Omega)$. It follows that $\Omega_0 = \{0\} \cup \phi(\Omega)$ is open and connected. Moreover $\mathbb{C}\backslash\Omega_0 = \phi(K\backslash\{0\})$ and according to 1.41 each component of the latter is unbounded. This means that Ω_0 is simply-connected. By 10.2 there is therefore a conformal map f of $D = D(0, 1)$ onto Ω_0. Preceding f with the appropriate conformal self-map of D, we can suppose $f(0) = 0$. Then $\phi \circ f = 1/f$ maps $D\backslash\{0\}$ conformally onto Ω. Since f has a first order zero at 0 and no other zeros in D, we have $f(z) = zg(z)$, g holomorphic and zero-free in D; so in $D\backslash\{0\}$, $1/f(z) = h(z)/z$ where h is holomorphic and zero-free in D.

Lemma 10.18 *Let Ω be a region such that $\mathbb{C}\backslash\Omega$ has a unique bounded component C. Pick $z_0 \in C$ and let $E(w) = z_0 + e^w$ ($w \in \mathbb{C}$). Then $E^{-1}(\Omega)$ is connected.*

Proof: We obviously lose no generality by supposing $z_0 = 0$. Let Γ: $[0, 1] \to \Omega$ be a smooth loop as provided by 10.9 so that

(1) $\quad \operatorname{Ind}_\Gamma(0) = 1.$

According to 4.1 there exists a smooth ϕ: $[0, 1] \to \mathbb{C}$ such that $e^\phi = \Gamma$. We have then from (1) and definition 4.2

(2) $\quad 2\pi i = 2\pi i \operatorname{Ind}_\Gamma(0) = \phi(1) - \phi(0).$

For each $z \in \Omega$ let Γ_z: $[0, 1] \to \Omega$ be a curve from $\Gamma(0)$ to z (see 1.28). Again 4.1 provides continuous ϕ_z: $[0, 1] \to \mathbb{C}$ such that $e^{\phi_z} = \Gamma_z$. By adding an appropriate integer multiple of $2\pi i$ to ϕ_z, we may assume that it satisfies

(3) $\quad \phi_z(0) = \phi(0).$

Now consider any $w \in E^{-1}(\Omega)$ and let $z = E(w) = e^w$. We have

$$e^w = \Gamma_z(1) = e^{\phi_z(1)}$$

so

(4) $\quad w - \phi_z(1) = 2\pi i n,\quad$ some integer n.

Form the curve

$$\psi(t) = \begin{cases} \phi(t) & 0 \le t \le 1 \\ \phi(t-1) + 2\pi i & 1 \le t \le 2 \\ \vdots & \\ \phi(t-n+1) + 2\pi i(n-1) & n-1 \le t \le n \\ \phi_z(t-n) + 2\pi i n & n \le t \le n+1. \end{cases}$$

Because of (2) and (3) ψ is well-defined, hence continuous.

We have

$$e^{\psi(t)} = \left.\begin{cases} e^{\phi(t)} & = \Gamma(t) & 0 \le t \le 1 \\ \vdots & & \\ e^{\phi(t-n+1)} & = \Gamma(t-n+1) & n-1 \le t \le n \\ e^{\phi_z(t-n)} & = \Gamma_z(t-n) & n \le t \le n+1 \end{cases}\right\} \in \Omega$$

and therefore $\psi(t) \in E^{-1}(\Omega)$ for all $t \in [0, n+1]$. Since $\psi(0) = \phi(0)$ and $\psi(n+1) = \phi_z(1) + 2\pi i n \overset{(4)}{=} w$, this shows that any point w of $E^{-1}(\Omega)$ can be joined by a curve in $E^{-1}(\Omega)$ to the fixed point $\phi(0)$, proving (by 1.20 and 1.1) that $E^{-1}(\Omega)$ is connected.

Lemma 10.19 *Let C be a compact, connected subset of $D = D(0, 1)$ with $0 \in C$. Let $E(w) = e^w$ $(w \in \mathbb{C})$. Then $E^{-1}(D \backslash C)$ is simply-connected.*

Proof: $\mathbb{C} \backslash E^{-1}(D\backslash C) = E^{-1}(\mathbb{C}\backslash(D\backslash C)) = E^{-1}(\mathbb{C}\backslash D) \cup E^{-1}(C)$. We have

$$E^{-1}(\mathbb{C}\backslash D) = \{w : |e^w| \ge 1\} = \{w : \operatorname{Re} w \ge 0\},$$

an unbounded connected set. Therefore it suffices to show that $E^{-1}(C)$ has no bounded component. Argue by contradiction: if $E^{-1}(C)$ has a bounded component, then there is a compact, relatively open subset K of $E^{-1}(C)$ which contains that component (1.34). If U is an open subset of $\mathbb{C}$ such that $K = U \cap E^{-1}(C)$, then $E(K) = E(U) \cap C$. Since E is an open map (5.77), the set $E(U)$ is open, so $E(U) \cap C$ is relatively open in C. That is, $E(K)$ is relatively open in C. But K is compact, so $E(K)$ is too. Thus $E(K)$ is a non-void, proper $(0 \in C, 0 \notin E(K))$, clopen subset of the connected set C, a contradiction.

Theorem 10.20 *For a doubly-connected region Ω one of the following three situations obtains:*

(i) Ω *equals* $\mathbb{C}\backslash\{c\}$ *for some* $c \in \mathbb{C}$

(ii) Ω *is conformal to* $D(0, 1)\backslash\{0\}$

(iii) Ω *is conformal to* $A(r, 1)$, *some* $0 < r < 1$.

Proof: Let C be the unique bounded component of $\mathbb{C}\backslash\Omega$. The case where $C = \mathbb{C}\backslash\Omega$ is a single point, is (i) above. The case where $C = \mathbb{C}\backslash\Omega$ is not a single point, is dealt with in 10.17 and is (ii) above. So we suppose that

(1) $\Omega \cup C \neq \mathbb{C}$.

By 1.38 and 10.2 there is a conformal map of $\Omega \cup C$ onto $D = D(0, 1)$. Following this with an appropriate conformal self-map of D, we see that we may simply assume in addition to (1) that

(2) $0 \in C$ compact, connected $\subset D$ and $\Omega = D\backslash C$.

Let $E: \mathbb{C} \to \mathbb{C}\backslash\{0\}$ be the exponential map, $E(w) = e^w$. Cite the last two lemmas to assert that

(3) $E^{-1}(\Omega)$ is open, connected, and simply-connected.

Evidently this is a proper subset of $\mathbb{C}$, so by (two applications of) 10.2 we can find a conformal map f of $E^{-1}(\Omega)$ onto $H = \mathbb{R} \times (0, \infty)$. $T(w) = w + 2\pi i$ evidently maps $E^{-1}(\Omega)$ conformally onto itself and is fixed-point free. Then $\phi = f \circ T \circ f^{-1}$ is a fixed-point free conformal map of H onto H. According to 6.17(iv) there is a conformal map ψ of H onto H such that either for some $\beta \in \mathbb{R}$,

(4) $\psi(\phi(\psi^{-1}(z))) = z + \beta$ for all $z \in H$

or for some $\alpha > 0$,

(5) $\psi(\phi(\psi^{-1}(z))) = \alpha z$ for all $z \in H$.

Set then

(6) $F = \psi \circ f$,

a conformal map of $E^{-1}(\Omega)$ onto H such that either for some $\beta \in \mathbb{R}$,

(4)′ $F(T(F^{-1}(z))) = z + \beta$ for all $z \in H$

or for some $\alpha > 0$,

(5)′ $F(T(F^{-1}(z))) = \alpha z$ for all $z \in H$.

(I) We shall treat the case (4)′ in detail. Notice that $\beta \neq 0$, since ϕ is fixed-point free. Define

(7) $A = \{e^{2\pi i w/\beta}: w \in H\}$.

Thus A is either $D\backslash\{0\}$ (if $\beta > 0$) or the conformally equivalent (under $z \to 1/z$) set $\mathbb{C}\backslash\bar{D}$ (if $\beta < 0$). We will show that Ω is conformal to A. The map h is defined as follows: given $z \in \Omega$, pick $v \in E^{-1}(\Omega)$ so that $e^v = z$ and let $h(z) = E(2\pi i F(v)/\beta)$. First we need to show that h is well-defined. So suppose $v_1, v_2 \in E^{-1}(z)$. Then

$e^{v_1} = e^{v_2} \Rightarrow v_1 = v_2 + 2\pi i n$

for some integer n. Reversing the roles of v_1 and v_2 if necessary, we may suppose $n \geq 0$. Then

$$v_1 = v_2 + 2\pi in = T^{[n]}(v_2),$$

where $T^{[n]}$ is the nth iterate of T. It follows that

$$\begin{aligned} F(v_1) = F(T^{[n]}(v_2)) &= (F \circ T^{[n]} \circ F^{-1})(F(v_2)) \\ &= (F \circ T \circ F^{-1})^{[n]}(F(v_2)) \\ &= F(v_2) + n\beta \quad \text{by (4)}'. \end{aligned}$$

Therefore

$$E\left(\frac{2\pi i}{\beta} F(v_1)\right) = E\left(\frac{2\pi i}{\beta} F(v_2)\right).$$

Thus h is well-defined. It clearly maps Ω onto A, since F maps $E^{-1}(\Omega)$ onto H. The proof that h is one-to-one is similar to the above analysis. Finally, to prove that h is holomorphic is just a local matter: each point of Ω lies in a neighborhood N in which a holomorphic logarithm exists, i.e., a holomorphic function L on N such that $e^{L(z)} = z$ for all $z \in N$. Thus $L(N) \subset E^{-1}(\Omega)$ and for each $z \in N$ we may use $L(z)$ for the v above, giving

$$h(z) = E\left(\frac{2\pi i}{\beta} F(L(z))\right).$$

This exhibits h as a composition of holomorphic functions in the neighborhood N.

(II) The case (5)′ is not too different. Here is a sketch which leaves a few details to the reader. Let L be the Principal Branch of the Logarithm in $\mathbb{C}\backslash(-\infty, 0]$ (see 3.43). Then on $(0, \infty)$, L is real-valued and in fact is the natural logarithm of 3.17. Since T has no fixed points, it follows from (5)′ that $\alpha \neq 1$. Therefore, if we define $S = \mathbb{R} \times (0, \pi)$ and $A = \{e^{2\pi i w/L(\alpha)}: w \in S\}$, then A is an annulus and H is mapped conformally onto S by L. Define $h(z) = E(2\pi i L(F(v))/L(\alpha))$ for any $v \in E^{-1}(z)$ and verify as before that h is well-defined and maps Ω conformally onto A.

Exercise 10.21 *Show that the three conditions of the last theorem are mutually exclusive.*

Hint: If f: $D\backslash\{0\} \to A(r, 1)$ is conformal, use 5.41 and 5.35 to get $f = e^g$ for some $g \in H(D)$ and reach a contradiction as in the proof of 10.14.

Exercise 10.22 *Here is an internal characterization of double-connectivity (for regions). Show that a region Ω is doubly-connected if and only if*

(i) *Ω is not simply-connected, and*

(ii) *For any two loops γ_0, γ_1 in Ω there exist integers n_0, n_1 not both 0 such that $n_0 \bullet \gamma_0$ is loophomotopic in Ω to $n_1 \bullet \gamma_1$.*

Hints: From 4.15(ii) and 10.20 we see that (i) and (ii) are necessary conditions. On the other hand, condition (i) insures that $\mathbb{C} \backslash \Omega$ has at least one bounded component and condition (ii) and 8.22 insure that it has at most one.

Exercise 10.23 *Let U, V be homeomorphic regions in $\mathbb{C}$. Show that one is doubly-connected if and only if the other is.*

Hints: 4.68 gives one proof. Here is another. If ϕ is a homeomorphism of U onto V and U is doubly-connected, then U has properties (i) and (ii) above and these pass over to V under ϕ because of the obvious fact that $\phi^{-1} \circ (n \bullet \Gamma) = n \bullet (\phi^{-1} \circ \Gamma)$ for any integer n and any loop Γ in V.

Exercise 10.24 (Cf. 4.66.) *Show that if U is an open, doubly-connected subset of $\mathbb{C}$ and V is a component of U, then V is either simply- or doubly-connected and either alternative is possible. What can you say about an open set if its components are each doubly-connected?*

Hints: If $\mathbb{C} \backslash V$ has at least two different bounded components C_1, C_2, use 8.20 to come up with loops γ_1, γ_2 in V such that

$$\operatorname{Ind}_{\gamma_1}(C_1) \operatorname{Ind}_{\gamma_2}(C_2) \neq 0 \ \& \ \operatorname{Ind}_{\gamma_1}(C_2) = \operatorname{Ind}_{\gamma_2}(C_1) = 0.$$

If neither $C_1 \subset U$ nor $C_2 \subset U$, say $c_1 \in C_1 \backslash U$ and $c_2 \in C_2 \backslash U$, we will have

(1) $\operatorname{Ind}_{\gamma_1}(c_1) \operatorname{Ind}_{\gamma_2}(c_2) \neq 0,$

and

(2) $\operatorname{Ind}_{\gamma_1}(c_2) = \operatorname{Ind}_{\gamma_2}(c_1) = 0.$

Since $\operatorname{Ind}_{\gamma}$ is constant on components of $\mathbb{C} \backslash \gamma$, it follows from (1) and (2) that c_1, c_2 belong to different components of $\mathbb{C} \backslash U$ and from (1) and 4.3 that each of these components is bounded. This contradicts the double-connectivity of U. We conclude that one of C_1, C_2 lies in U, say $C_1 \subset U$. The connected set C_1 must be wholly in one component V_1 of U. Of course $V_1 \neq V$ so, distinct components being disjoint, $V_1 \cap V = \varnothing$. On the other hand, C_1 meets $\overline{V}$ and so V_1 meets $\overline{V}$. But then since V_1 is open, V_1 meets V, a contradiction.

Notes to Chapter X

As the reader can see, the *Hauptsatz* (my designation) 10.2 is just the final assembly of many subsections proved elsewhere; this final *Beweisordnung* has however become fairly standard (see the texts of Rudin and Conway). For a somewhat different treatment involving the Jordan Curve Theorem see HERVÉ [1970].

For any open subset U of $\mathbb{C}$ any $f \in H(U)$ is the *pointwise* limit in U of a sequence of polynomials. (See, for example, MONTEL [1910], pp. 101–103, M. A. LAVRENTIEFF [1936], p. 13, REIJNIERSE [1938] or MARKUSCHEWITSCH [1967], p. 97. In view of 8.9 the proof comes down to a geometric construction: find simply-connected, compact subsets K_n of U such that $U = \bigcup_{k=1}^{\infty} \bigcap_{n=k}^{\infty} K_n$. None of the cited references handles this construction very satisfactorily.) This is to be contrasted to the equivalence (e) $\Leftrightarrow$ (o) in 10.2. 10.4(i) is from HAUSDORFF [1914], p. 345. (Cf. also MAZURKIEWICZ [1922a].) Other proofs of 10.2(f), 10.4, 10.9 may be found in BASYE [1935].

10.10 and 10.11 are from SAKS and ZYGMUND [1971] and 10.13 is from HUBER [1951] (see also TSUJI [1960] and SCHIFFER [1946]). For far-reaching extensions of 10.13(iii) see LANDAU and OSSERMAN [1959–60] and MARDEN, RICHARDS and RODIN [1967]. The conclusion 10.14 that the radius ratio is a conformal invariant occurs already in SCHOTTKY [1877], p. 326. For a modern treatment of great generality see pp. 220–224 of AHLFORS and SARIO [1960]. For more on the interesting phenomenon of ring equivalence of $H^\infty(\Omega_1)$ with $H^\infty(\Omega_2)$ implying conformal equivalence of Ω_1 with Ω_2 see BERS [1948], KAKUTANI [1955], RUDIN [1955], NAGASAWA [1959], NAKAI [1963], [1975], ALLING [1968], SU [1972] and the nice exposition in HEINS [1968], pp. 200–204. The beautiful results 10.15, 10.16 are from BECK [1964] and RICHARDS [1968].

There are a variety of proofs for 10.14 but that in the text is the smoothest. Most others require some knowledge of the boundary behavior of f, which is arduous and unesthetic to confirm. (Cf. Chapter IX.) In his careful account RUDIN [1974] (cf. also SCHIFFER [1946]; a more cavalier version of this proof occurs already in KOEBE [1906a], p. 143) first establishes that as z approaches ∂A_1, $f(z)$ approaches ∂A_2 "in a consistent manner"; this is an adequate weak substitute for continuity of f on $\bar{A}_1$. Other authors are less candid, continuity on the boundary being used but not acknowledged. One alternative proof proceeds by normalizing the annuli appropriately and considering iterates of the given conformal map; another uses a reflection principle (see Chapter XIII) to extend the map to an entire function which is 0 at 0 and of constant modulus on some circle centered at 0. (Compare then 6.12(ii).) For these see exercises 24 and 25, p. 316 of RUDIN [1974] and RITT [1920–21], or GOLUZIN [1969], p. 208 or RADÓ [1924–26]. For a proof based on area (due to Carleman) see PÓLYA and SZEGÖ [1976], exercises IV.81–83. For a harmonic function analog of 10.14 see NITSCHE [1962].

The concept of a covering space (mentioned in the notes to Chapter IV) is again in the background in 10.18 and 10.19; these lemmas assert in effect that $E^{-1}(U)$ is a covering space of U. See, for example, JULIA [1955] or VEECH [1967] (esp.

pp. 103–113) or AHLFORS and SARIO [1960]. On the inversion technique in 10.17 (and in 9.40) see KNASTER and KURATOWSKI [1924]. The idea of the proof of 10.20 is from H. KNESER [1958], pp. 372–375. For an elementary treatment of 10.14 and 10.20 entirely in terms of covering surfaces see PESCHL [1967], pp. 226–227. For other treatments of the fundamental theorem on doubly-connected regions (10.20), the reader may want to look at LE VAVASSEUR [1902], pp. 195–205 of KOEBE [1915], CREMER [1930], GRAESER [1930], GIRONZA SOLANAS [1932], ZARANKIEWICZ [1934], KHAJALIA [1937], [1938], KOMATU [1945a], [1949], ROYDEN [1952], ALBRECHT [1954], LEJA [1954], HEINHOLD [1957] and GAIER [1957], [1964] (chapter V). See also the literature cited on p. 151 of BERNARDI [1966].

For regions with degree of connectivity n (i.e., $n - 1$ bounded components in the complement) there are also canonical conformal types. Generally these involve the extended plane $\mathbb{C}_\infty$ (see the Chapter I notes), e.g., $\mathbb{C}_\infty$ with some parallel line segments [i.e., compact intervals in $\mathbb{C}$] removed (parallel slit regions); some intervals on lines through a common point removed (radial slit regions); or some arcs of concentric circles removed (circular slit regions). Here is a representative result, a version of which, along with more references to the literature, will be presented in Chapter XVIII. Theorem: (i) For any open, connected subset Ω of $\mathbb{C}_\infty$ there exists a conformal map of Ω onto a parallel slit domain. (ii) If $c \in \Omega \cap \mathbb{C}$, then among all conformal maps f of Ω into $\mathbb{C}_\infty$ which near c have a representation of the form $1/(z - c) + \sum_{k=1}^{\infty} \alpha_k(z - c)^k$ there is one for which $\operatorname{Re} \alpha_1$ is maximal. (iii) The image of Ω under any such extremal function is a parallel slit region. (iv) If $\mathbb{C}_\infty \backslash \Omega$ has n components, then $\mathbb{C}_\infty \backslash f(\Omega)$ does too for any extremal f and for a given $c \in \Omega \cap \mathbb{C}$ there is only one function of the form (ii) which maps Ω onto a parallel slit region. This result and its radial slit and circular slit avatars is due variously to Hilbert, Koebe, Grötzsch, Rengel, dePossel. The proof details are elegant and elementary. There are excellent expositions in BIEBERBACH [1967], GOLUZIN [1969] and NEHARI [1952]. Distortion and area theorems from Chapter XVIII are used to establish the local uniform boundedness of the functions in (ii) and then 7.6 is exploited to get an extremal function. This proof, reminiscent of Fejér and Riesz's proof of the Riemann Mapping Theorem, is from DEPOSSEL [1931]. (See also RENGEL [1934], the various papers of KOEBE, and the long series of papers by GRÖTZSCH in the *Proceedings of the Saxon Academy* at Leipzig between 1928 and 1935. GOLUZIN [1969] and AHLFORS and SARIO [1960] contain the references on the latter.) In fact at one point of the proof Riemann's theorem is invoked. Later DEPOSSEL [1939] found a constructive (and short) way around this (see also GARABEDIAN [1976]) and the theorem thus proved subsumes Riemann's completely. There is, not surprisingly, an intimate connection between the solvability of the Dirichlet problem and the existence of conformal maps (a special case of which is examined in § 8 of Chapter IX). The mapping of a multiply-connected region onto one of the canonical ones above is treated via the Dirichlet problem in AHLFORS

[1966a], pp. 243–253. Several papers in the bibliography deal with mappings of multiply-connected regions but I have not tried to provide extensive coverage of it. I mention only the monographs of JULIA [1955], BERGMAN [1970] and GOLUZIN [1969] and the papers of LEHTO [1949], REICH and WARSCHAWSKI [1960], ROYDEN [1952], STONE [1962] and especially the survey GAIER [1978]. See also GAIER's book and its extensive bibliography as well as the annotated bibliography of SEIDEL [1952] and the recent notes of GRUNSKY [1978].

There is an important analytic characterization of simple-connectivity which was not treated in the text, the (*Holomorphic*) *Monodromy Theorem*. Its enunciation requires a short definition: For a curve $\Gamma: [0,1] \to \mathbb{C}$ and an open disk D_0 which contains the initial point of Γ, say that a function $f_0 \in H(D_0)$ is *analytically continuable along* Γ if there exist $0 = t_0 < t_1 < t_2 < \cdots < t_n = 1$, open disks $D_j \supset \Gamma[t_j, t_{j+1}]$ and functions $f_j \in H(D_j)$ such that $f_j = f_{j+1}$ in $D_j \cap D_{j+1}$ for $j = 0, 1, 2, \ldots, n-1$. The theorem asserts that a region Ω is simply-connected if and only if

$$(*) \quad \begin{cases} \text{for every open disk } D \subset \Omega, \text{ the set } H(\Omega)|D \text{ includes all } f \in H(D) \\ \text{which are analytically continuable along } every \text{ curve in } \Omega \text{ with} \\ \text{initial point in } D. \end{cases}$$

In spite of the difficulty of verifying that a function is analytically continuable along *every* curve in a region (and the widespread carelessness in doing so—see § 13 of ZALCMAN [1974] and § 3 of STYER and MINDA [1974]), the property (*) of simply-connected regions can be quite useful for getting globally defined functions from locally defined ones. E.g., it is essential in the proofs of the results described in 7.42. If $f \in H(\Omega)$ is zero-free, then it follows from 5.35 that f has local logarithms and that by the addition of appropriate constants these provide analytic continuations of one another along any curve in Ω. Therefore, a region Ω which satisfies (*) has property (*HL*). On the other hand, one verifies without difficulty that if f_0 is analytically continuable along every curve in a homotopy and if these curves all have the same initial point a and the same terminal point b, then the corresponding f_n functions above all agree in a neighborhood of b. Now if Ω is loophomotopically-connected, it is not hard to prove directly that any two curves in Ω with the same initial point and the same terminal point are part of a homotopy of such curves. [Alternatively, one can observe that disks trivially enjoy this property (the homotopy is a convex combination of the two curves) and use 10.2 to transfer it to Ω.] Consequently, if Ω is loophomotopically-connected and f_0 is analytically continuable along every curve in Ω, then any two continuations like f_n above agree in the overlap of their domains (recall 5.62) and these functions serve to define an extension of f_0 to an element of $H(\Omega)$. This is how property (*) is deduced from simple-connectivity. Though the details of these arguments are routine (and can reasonably be recommended to the reader as an exercise), textbook accounts of this theorem seldom achieve both perspicuity and precision. These attributes are, however, realized in RUDIN [1966], [1974], where interested readers will find the details of the above sketch interpolated.

Chapter XI
Isolated Singularities

§ 1 Laurent Series and Classification of Singularities

If f is holomorphic in $V\backslash\{z_0\}$ for some neighborhood V of z_0, we call z_0 an *isolated singularity* of f. These we propose to study in this chapter. The first step is the acquisition of a bilateral series expansion for f around such a point. Such an expansion flows naturally from the appropriate Cauchy integral formula (cf. the proof of 3.8), which we now derive.

Theorem 11.1 (A Cauchy Formula for the Annulus) *Let* $0 \leq r < R \leq \infty$, $c \in \mathbb{C}$, $A = \{z \in \mathbb{C}: r < |z - c| < R\}$, *$f$ holomorphic in A. Then for any* $r < a < b < R$

$$f(z) = \frac{1}{2\pi i}\int_{C(c,b)} \frac{f(\xi)}{\xi - z}\,d\xi - \frac{1}{2\pi i}\int_{C(c,a)} \frac{f(\xi)}{\xi - z}\,d\xi, \quad a < |z - c| < b.$$

Proof: Of the many avenues open to us at this point (e.g., 10.13(i) or 7.19), we choose the most elementary. Let $\gamma = C(c, a)$, $\Gamma = C(c, b)$ and let z satisfying $a < |z - c| < b$ be given. Form

$$F(\xi) = \begin{cases} \dfrac{f(\xi) - f(z)}{\xi - z}, & \xi \in A\backslash\{z\} \\ f'(z) & \xi = z. \end{cases}$$

As noted many times before, this function is holomorphic in A. Apply 5.5(i) to it and split the integrand apart, keeping in mind that

$$\int_\gamma \frac{d\xi}{\xi - z} = 0 \quad \text{while} \quad \int_\Gamma \frac{d\xi}{\xi - z} = 2\pi i.$$

Theorem 11.2 (Laurent Expansion) *Let* $0 \leq r < R \leq \infty$, $c \in \mathbb{C}$,

$$A = \{z \in \mathbb{C}: r < |z - c| < R\},$$

f holomorphic in A. Then there exist functions F, G holomorphic in $D(0, R)$, $D(0, 1/r)$ respectively such that

$$f(z) = F(z - c) + \frac{1}{z - c}G\left(\frac{1}{z - c}\right) \quad \forall z \in A.$$

It follows that there exist $c_n \in \mathbb{C}$ such that

$$f(z) = \sum_{n=-\infty}^{\infty} c_n(z - c)^n \overset{\text{def.}}{=} \lim_{N\to\infty} \sum_{n=-N}^{N} c_n(z - c)^n \quad \forall z \in A$$

with convergence absolute and uniform on compact subsets of A.

Proof: Without loss of generality, $c = 0$. In what follows consider $r < \rho < R$. For each such ρ define $\gamma_\rho = C(0, \rho)$ and

$$(1) \qquad f_\rho(z) = \frac{1}{2\pi i}\int_{\gamma_\rho} \frac{f(\xi)}{\xi - z}\, d\xi, \quad z \in \mathbb{C}\backslash\gamma_\rho.$$

If $|z| < \rho_1 < \rho_2 < R$, then apply 5.5 with $g(\xi) = f(\xi)/(\xi - z)$, $b = R$ and a any number between $\max\{|z|, r\}$ and ρ_1 to conclude that $f_{\rho_1}(z) = f_{\rho_2}(z)$. We can therefore consistently define a holomorphic function F in $D(0, R)$ by

$$(2) \qquad F(z) = f_\rho(z) \quad \text{if } |z| < \rho < R.$$

We can apply these considerations to the function $g(z) = (1/z)f(1/z)$ in the annulus $\{z \in \mathbb{C}: 1/R < |z| < 1/r\}$. The analog of (2) says that the equation

$$(3) \qquad G(z) = g_\rho(z) \quad \text{if } |z| < \rho < 1/r$$

defines a holomorphic function in $D(0, 1/r)$. Notice that if $r < \rho < |z|$, we have

$$\begin{aligned} g_{1/\rho}\left(\frac{1}{z}\right) &= \frac{1}{2\pi i}\int_{\gamma_{1/\rho}} \frac{g(\xi)}{\xi - 1/z}\, d\xi = \int_0^1 \frac{g(\rho^{-1}e^{2\pi i t})}{\rho^{-1}e^{2\pi i t} - 1/z}\cdot\frac{1}{\rho}\, e^{2\pi i t}dt \\ &= \int_0^1 \frac{f(\rho e^{-2\pi i t})}{\rho^{-1}e^{2\pi i t} - 1/z}\, dt = z\int_0^1 \frac{f(\rho e^{-2\pi i t})\rho e^{-2\pi i t}}{z - \rho e^{-2\pi i t}}\, dt \\ &= -z\int_0^1 \frac{f(\rho e^{2\pi i \tau})\rho e^{2\pi i \tau}}{\rho e^{2\pi i \tau} - z}\, d\tau \quad (\text{setting } \tau = 1 - t) \\ &= -\frac{z}{2\pi i}\int_{\gamma_\rho} \frac{f(\xi)}{\xi - z}\, d\xi = -z f_\rho(z), \end{aligned}$$

$$(4) \qquad f_\rho(z) = -\frac{1}{z}\, g_{1/\rho}\left(\frac{1}{z}\right), \quad r < \rho < |z|.$$

Finally, if $z \in A$ is given, pick $r < \rho_1 < |z| < \rho_2 < R$ and get

$$\begin{aligned} F(z) + \frac{1}{z}G\left(\frac{1}{z}\right) &\overset{(2)}{=} f_{\rho_2}(z) + \frac{1}{z}G\left(\frac{1}{z}\right) \overset{(3)}{=} f_{\rho_2}(z) + \frac{1}{z}\, g_{1/\rho_1}\left(\frac{1}{z}\right) \\ &\overset{(4)}{=} f_{\rho_2}(z) - f_{\rho_1}(z) \\ (5) \qquad &= f(z), \quad \text{by 11.1.} \end{aligned}$$

If the power series for F and G are respectively

$$F(z) = \sum_{n=0}^{\infty} c_n z^n \quad \forall |z| < R, \qquad G(z) = \sum_{n=0}^{\infty} c_{-n-1} z^n \quad \forall |z| < 1/r,$$

then (5) gives

$$f(z) = \sum_{n=-\infty}^{\infty} c_n z^n \quad \forall z \in A.$$

Exercise 11.3 *If f is holomorphic in $A = \{z \in \mathbb{C}: r < |z - c| < R\}$ for some $0 \le r < R \le \infty$, show that the functions F and G provided by* 11.2 *are unique. Find integral formulas for the coefficients c_n in* 11.2 *analogous to those in* 3.8 *and conclude that these coefficients are unique.*

We are thus enabled to make

Definition 11.4 Let f be holomorphic in $A = \{z \in \mathbb{C}: r < |z - c| < R\}$ for some $c \in \mathbb{C}$, $0 \le r < R \le \infty$. Then

(i) The bilateral series $\sum_{n=-\infty}^{\infty} c_n(z - c)^n$ provided by 11.2 is called the *Laurent series of f* in A and the c_n are called the *Laurent coefficients of f*.

(ii) When $r = 0$ we call c_{-1} the *residue of f at c*, noted $\operatorname{Res}(f, c)$, and the function $\sum_{n=1}^{\infty} c_{-n}(z - c)^{-n}$ (that is, $(z - c)^{-1}G((z - c)^{-1})$ is called the *principal part of f at c*. Note that by 11.2 this series converges for all $z \in \mathbb{C}\backslash\{c\}$.

(iii) When $r = 0$ we call c a *removable singularity* of f if $c_{-n} = 0$ for all $n > 0$, a *polar singularity* or a *pole of f of order k* (k positive) if $c_{-k} \ne 0$ but $c_{-n} = 0$ for all $n > k$, and an *essential singularity* of f if $c_{-n} \ne 0$ for infinitely many $n > 0$.

Remark: By 11.3,

$$c_{-1} = \frac{1}{2\pi i}\int_{C(c,\rho)} f(z)dz = \frac{\rho}{2\pi}\int_0^{2\pi} f(c + \rho e^{i\theta})e^{i\theta}d\theta$$

(for any $\rho \in (r, R)$). This is the sole surviving term when the Laurent series of f is integrated term-by-term. Hence the designation "residue". More generally, what remains after integrating a holomorphic function around a loop is the sum of the residues of f inside the loop, as will be proved in 11.29 below. The rationale for the terminology in (iii) is contained in 11.5(i) and 11.7(ii).

Exercise 11.5 *Let U be an open neighborhood of c, f holomorphic in $U\backslash\{c\}$. Then*

(i) *c is a removable singularity of f if and only if f has a holomorphic extension to U.*

(ii) *c is a pole of f if and only if for some non-constant polynomial P the function $f(z) - P(1/(z - c))$ has c as a removable singularity.*

(iii) *c is an essential singularity if and only if for some non-polynomial entire function g the function $f(z) - g(1/(z - c))$ has c as a removable singularity.*

Theorem 11.6 (Casorati–Weierstrass) *Let U be an open neighborhood of c, f holomorphic in $U\backslash\{c\}$. Then c is an essential singularity of f if and only if $f(D(c, R) \cap U\backslash\{c\})$ is dense in $\mathbb{C}$ for every $R > 0$.*

Proof: ($\Leftarrow$) If c is removable, then $\lim_{z\to c}|f(z)|$ exists and is finite, while if c is a pole, it is clear from 11.5(ii) that $\lim_{z\to c}|f(z)| = \infty$. In either case the density condition fails.

($\Rightarrow$) Suppose for some $R > 0$, $f(D(c, R) \cap U\backslash\{c\})$ is not dense in $\mathbb{C}$. We may suppose $D = D(c, R) \subset U$. There is accordingly a $w \in \mathbb{C}$ and $d > 0$ such that $|f(z) - w| \geq d$ for all $z \in D\backslash\{c\}$. The function $h = 1/(f - w)$ is therefore bounded (by $1/d$) and holomorphic in $D\backslash\{c\}$, so by 5.41 may be extended to a holomorphic function, call it h also, in D. Thus

$$(1) \qquad f = w + \frac{1}{h} \quad \text{in } D\backslash\{c\}.$$

Therefore if $h(c) \neq 0$, then (1) provides a holomorphic extension of f to D and by 11.5 c is a removable singularity. If $h(c) = 0$ and $m \geq 1$ is the order of c as a zero of h, then

$$(2) \qquad h(z) = (z - c)^m H(z), \quad z \in D,$$

for some function H which is holomorphic in D with $H(c) \neq 0$. Because $h = 1/(f - w)$ in $D\backslash\{c\}$, it follows from (2) that H has no zero in $D\backslash\{c\}$, so $g = 1/H$ is holomorphic in D. By 5.30 there are c_n such that

$$(3) \qquad g(z) = \sum_{n=0}^{\infty} c_n(z - c)^n \quad \forall z \in D.$$

Note that $c_0 = g(c) \neq 0$. From (1), (2) and (3)

$$\begin{aligned} f(z) &= \sum_{n=0}^{m-1} c_n(z - c)^{n-m} + c_m + w + \sum_{n=m+1}^{\infty} c_n(z - c)^{n-m} \\ &= P\left(\frac{1}{z - c}\right) + F(z), \quad z \in D\backslash\{c\}, \end{aligned}$$

where P is a polynomial of degree m and F is holomorphic in D. It follows from 11.5(ii) that in this case c is a pole of f.

Remark: In Chapter XII we prove a remarkable strengthening of this theorem due to Julia and Picard which implies that if c is an essential singularity, then in fact for every $R > 0$, $f(D(c, R) \cap U\backslash\{c\})$ contains every complex number with at most one exception.

Theorem 11.7 *Let U be an open neighborhood of c, f holomorphic in $U\backslash\{c\}$. Then*
(i) *c is a removable singularity of f if and only if f is bounded near c.*
(ii) *c is a pole of f if and only if* $\lim_{z\to c}|f(z)| = \infty$.

Proof: ($\Rightarrow$) clear, from 11.5, for example.

($\Leftarrow$) (i) 5.41 and 11.5. Alternatively, 11.3 and 3.34.

(ii) This condition clearly precludes $f(D(c, R)\backslash\{c\})$ being dense for all sufficiently small R, so 11.6 implies that c is not an essential singularity, i.e., is removable or a pole. But by (i) it is not removable.

Exercise 11.8 *For each integer n let $c_n \in \mathbb{C}$ and suppose that $a, b \in \mathbb{C}$ satisfy: $0 < |a| < |b|$. Show that if the series $\sum_{n=-\infty}^{\infty} c_n z^n$ converges (in the sense that $\lim_{N\to\infty} \sum_{n=-N}^{N} c_n z^n$ exists in $\mathbb{C}$) for $z = a$ and for $z = b$, then the series $\sum_{n=0}^{\infty} c_n z^n$ converges for all $|z| < |b|$ and the series $\sum_{n=0}^{\infty} c_{-n} z^{-n}$ converges for all $|z| > |a|$.*

Exercise 11.9 *Let U, V be open subsets of $\mathbb{C}$ with bounded complements. Let $g\colon U \to V$ and $h\colon V \to \mathbb{C}$ be holomorphic. Suppose that their Laurent expansions for large z are of the form*

$$g(z) = \sum_{j=-1}^{\infty} a_j z^{-j}, \quad a_{-1} \neq 0$$

$$h(z) = \sum_{j=-1}^{\infty} b_j z^{-j}, \quad b_{-1} \neq 0.$$

(i) *Show that for the composite $h \circ g$ the Laurent expansion for large z is of the form*

$$h \circ g(z) = \sum_{j=-1}^{\infty} c_j z^{-j},$$

where

$$c_{-1} = a_{-1}b_{-1}, c_0 = b_0 + a_0 b_{-1}, c_1 = \frac{b_1 + a_{-1}b_{-1}a_1}{a_{-1}}.$$

Hints: The hypotheses on g and h are that the functions

$$G(z) = g\left(\frac{1}{z}\right) - \frac{a_{-1}}{z}, \qquad H(z) = h\left(\frac{1}{z}\right) - \frac{b_{-1}}{z}$$

have removable singularities at 0, that $G(0) = a_0$ and $G'(0) = a_1$, while $H(0) = b_0$ and $H'(0) = b_1$. That is,

(1) $\lim_{|z|\to\infty} [g(z) - a_{-1}z]$ exists and equals a_0

(2) $\lim_{|z|\to\infty} z[g(z) - a_{-1}z - a_0] = a_1$

(3) $\lim_{|z|\to\infty} [h(z) - b_{-1}z]$ exists and equals b_0

(4) $\lim_{|z|\to\infty} z[h(z) - b_{-1}z - b_0] = b_1.$

Let $\phi(z) = z[g(z) - a_{-1}z - a_0]$ for large z. Then

$$g(z) = a_{-1}z + a_0 + \frac{\phi(z)}{z}$$

and

(2)′ $\lim_{|z|\to\infty} \phi(z) = a_1.$

Now for large z

$$h \circ g(z) - a_{-1}b_{-1}z = h\left(a_{-1}z + a_0 + \frac{\phi(z)}{z}\right) - b_{-1}\left(a_{-1}z + a_0 + \frac{\phi(z)}{z}\right) + \left(a_0 + \frac{\phi(z)}{z}\right)b_{-1}.$$

Since $|a_{-1}z + a_0 + \phi(z)/z| \to \infty$ as $|z| \to \infty$ [by (2)′ and the fact that $a_{-1} \neq 0$], it follows from (3) and (2)′ that

(5) $\lim_{|z|\to\infty} [h \circ g(z) - a_{-1}b_{-1}z]$ exists and equals $b_0 + a_0 b_{-1}$.

Then for large z

$$\begin{aligned} &z[h \circ g(z) - a_{-1}b_{-1}z - b_0 - a_0b_{-1}] \\ &= z\left[h\left(a_{-1}z + a_0 + \frac{\phi(z)}{z}\right) - b_{-1}\left(a_{-1}z + a_0 + \frac{\phi(z)}{z}\right) - b_0\right] + b_{-1}\phi(z) \\ &= \frac{1}{a_{-1}}\left(a_{-1}z + a_0 + \frac{\phi(z)}{z}\right) \\ &\quad \times \left[h\left(a_{-1}z + a_0 + \frac{\phi(z)}{z}\right) - b_{-1}\left(a_{-1}z + a_0 + \frac{\phi(z)}{z}\right) - b_0\right] \\ &\quad - \frac{1}{a_{-1}}\left(a_0 + \frac{\phi(z)}{z}\right) \\ &\quad \times \left[h\left(a_{-1}z + a_0 + \frac{\phi(z)}{z}\right) - b_{-1}\left(a_{-1}z + a_0 + \frac{\phi(z)}{z}\right) - b_0\right] \\ &\quad + b_{-1}\phi(z). \end{aligned}$$

Then let $|z| \to \infty$ and cite (4), (3) and (2)′ to see that

$$\begin{aligned} \lim_{|z|\to\infty} z[h \circ g(z) - a_{-1}b_{-1}z - b_0 - a_0b_{-1}] &= \frac{b_1}{a_{-1}} - \frac{a_0}{a_{-1}}\cdot 0 + b_{-1}a_1 \\ \text{(6)} \qquad &= \frac{b_1 + a_{-1}b_{-1}a_1}{a_{-1}}. \end{aligned}$$

From (5) and (6) we see that $h \circ g(1/z) - a_{-1}b_{-1}/z$ has a removable singularity at 0 and its Laurent series there looks like

$$h \circ g\left(\frac{1}{z}\right) - \frac{a_{-1}b_{-1}}{z} = \sum_{j=0}^{\infty} c_j z^j,$$

where

$$c_0 = b_0 + a_0b_{-1}, \qquad c_1 = \frac{b_1 + a_{-1}b_{-1}a_1}{a_{-1}}.$$

Therefore for all large z

$$h \circ g(z) = \sum_{j=-1}^{\infty} c_j z^{-j}$$

where c_{-1}, c_0 and c_1 have the asserted values.

(ii) *Show that if g is a conformal map of U onto V, then the Laurent expansion of g^{-1} for large z is of the form*

$$g^{-1}(z) = \sum_{j=-1}^{\infty} b_j z^{-j}, \quad \textit{where} \quad b_{-1} = \frac{1}{a_{-1}}, \qquad b_0 = \frac{-a_0}{a_{-1}}, \qquad b_1 = -a_1.$$

Hints: From (1) we see that $|g(z)| \to \infty$ if and only if $|z| \to \infty$. Consequently, setting $w = g(z)$,

$$\lim_{|w|\to\infty} \left[g^{-1}(w) - \frac{w}{a_{-1}} \right] = \lim_{|z|\to\infty} \frac{1}{a_{-1}} [a_{-1}z - g(z)] = \frac{-a_0}{a_{-1}}.$$

It follows as in the conclusion of (i) that $g^{-1}(1/z) - 1/(a_{-1}z)$ has a removable singularity at 0 and a Laurent expansion there of the form

$$g^{-1}\left(\frac{1}{z}\right) - \frac{1}{a_{-1}z} = \sum_{j=0}^{\infty} b_j z^j, \quad \text{where } b_0 = \frac{-a_0}{a_{-1}},$$

$$g^{-1}(z) = \sum_{j=-1}^{\infty} b_j z^{-j}, \quad \text{where } b_{-1} = \frac{1}{a_{-1}}, \ b_0 = \frac{-a_0}{a_{-1}}.$$

Finally, apply (i) with g^{-1} in the role of h to learn that b_1 must satisfy

$$\frac{b_1 + a_{-1}b_{-1}a_1}{a_{-1}} = 0,$$

the coefficient of $1/z$ in the Laurent expansion of the identity function.

Exercise 11.10 *Let f be holomorphic in $\mathbb{R} \times (a, b)$ for some real a, b with $a < b$ and suppose that f is periodic of period 2π. Show that there exist $c_k \in \mathbb{C}$ such that*

$$f(z) = \sum_{k=-\infty}^{\infty} c_k e^{ikz} \quad \forall z \in \mathbb{R} \times (a, b)$$

with convergence uniform in $\mathbb{R} \times K$ for any compact $K \subset (a, b)$.

Hints: Define F in $A = \{z \in \mathbb{C}: e^{-b} < |z| < e^{-a}\}$ by

$$F(re^{i\theta}) = f(\theta - i \log r), \quad e^{-b} < r < e^{-a}, \ \theta \in \mathbb{R}.$$

Then F is well-defined due to the 2π-periodicity of f. Alternatively viewed, we have

(1) $\quad f(z) = F(e^{iz}) \quad \forall z \in \mathbb{R} \times (a, b).$

Given $z_0 \in A$, there is a holomorphic logarithm iL_0 defined in a neighborhood V_0 of z_0. Thus each $z \in V_0$ satisfies $z = e^{iL_0(z)}$ and so by (1), $F(z) = f(L_0(z))$ for all $z \in V_0 \cap A$. Therefore $F = f \circ L_0$ is holomorphic in $V_0 \cap A$, i.e., F is holomorphic in A. Write the Laurent expansion of F in this annulus

$$(2) \qquad F(w) = \sum_{k=-\infty}^{\infty} c_k w^k \quad \forall w \in A.$$

From (1) and (2) the desired expansion follows.

Exercise 11.11 *Find integral formulas for the c_k of the last exercise.*

Hint: Fix $c \in (a, b)$. Then the series converges uniformly on $\mathbb{R} \times \{c\}$, so for every integer n

$$\int_0^{2\pi} f(x + ic)e^{-in(x+ic)}dx = \sum_{k=-\infty}^{\infty} c_k \int_0^{2\pi} e^{i(k-n)(x+ic)}dx = 2\pi c_n.$$

Exercise 11.12 *Let f be an entire function such that $f(z)e^{-c|z|}$ is bounded for some $c > 0$. Suppose that f is periodic of period 2π. Show that there exist $c_k \in \mathbb{C}$ with $|k| \le N \le c$ such that $f(z) = \sum_{k=-N}^{N} c_k e^{ikz}$ for all $z \in \mathbb{C}$.*

Hints: According to 11.10 and 11.11

$$(1) \qquad f(z) = \sum_{k=-\infty}^{\infty} c_k e^{ikz} \quad \forall z \in \mathbb{C},$$

where $c_k = \frac{1}{2\pi}\int_0^{2\pi} f(x)e^{-ikx}dx$. If M is a bound for $f(z)e^{-c|z|}$, then by 5.37

$$|f^{(n)}(z)| \le (ce)^n M e^{c|z|}$$

for all $z \in \mathbb{C}$ and all positive integers n. Integrate by parts n-times to get

$$c_k = \frac{1}{2\pi}\frac{1}{(ik)^n}\int_0^{2\pi} f^{(n)}(x)e^{-ikx}dx.$$

Then

$$|c_k| \le \frac{1}{|k|^n}(ce)^n M e^{2\pi c} = Me^{2\pi c}\left(\frac{ce}{|k|}\right)^n.$$

This holds for all positive integers n, so if $|k| > ce$, then $c_k = 0$. Thus if K is the greatest integer not greater than ce, we have from (1)

$$(2) \qquad f(z) = \sum_{k=-K}^{K} c_k e^{ikz} \quad \forall z \in \mathbb{C}.$$

Because $f(z)e^{-c|z|}$ is bounded on $i\mathbb{R}$, it follows from (2) that in fact no c_k with $|k| > c$ can be non-zero.

§ 2 Rational Functions

We have already encountered rational functions in Chapter VIII. Here we are going to examine some interesting examples and derive a criterion for a power series to represent a rational function. Notice that the definition of pole for a rational function given in 8.4 coincides with that in 11.4, as a consequence of 11.7(ii), for example. Our first result here is purely algebraic and could be proved accordingly. For an *ab initio* algebraic proof see, e.g., HAMILTON [1972]. For history see § 24 of NETTO and LEVAVASSEUR [1907].

Theorem 11.13 *If f is a rational function and its poles occur at $a_1, \ldots, a_n$ with orders $m_1, \ldots, m_n$, then there is a polynomial p and $c_{jk} \in \mathbb{C}$ such that*

$$f(z) = p(z) + \sum_{j=1}^{n} \sum_{k=1}^{m_j} \frac{c_{jk}}{(z - a_j)^k}, \quad z \in \mathbb{C} \backslash \{a_1, \ldots, a_n\}.$$

Proof: The limit $\lim_{|z| \to \infty} |f(z)|$ exists in $[0, \infty]$. In fact, if P and Q are polynomials such that $f = P/Q$, then this limit is infinite if the degree of P exceeds that of Q, is zero if the degree of Q exceeds that of P and is the ratio of the moduli of the leading coefficients of P and Q when they have the same degree. Since $\lim_{|z| \to \infty} 1/|z - a_j| = 0$, it follows that if $f_j(z) = \sum_{k=1}^{m_j} c_{jk}/(z - a_j)^k$ is the principal part of f at a_j, then the function $p = f - \sum_{j=1}^{n} f_j$ can be extended to an entire function and the limit $\lim_{|z| \to \infty} |p(z)|$ exists in $[0, \infty]$ (and equals $\lim_{|z| \to \infty} |f(z)|$). If this limit is infinite then p is a polynomial by 5.70, while if this limit is finite it is obvious that p is bounded and so by Liouville (3.33) is constant.

Example 11.14 *If f is a rational function and $|f(z)| = 1$ whenever $|z| = 1$, then*

$$f(z) = cz^m \prod_{k=1}^{n} \frac{z - a_k}{1 - \bar{a}_k z}$$

for some integer m and some c, $a_k \in \mathbb{C}$ with $|a_k| \neq 1$.

Proof: The proof is almost identical to that suggested for 6.12. By first dividing away the appropriate (positive or negative) power of z, we may assume without loss of generality that f has neither a pole nor a zero at 0.

Let $a_1, \ldots, a_N$ and $b_1, \ldots, b_M$ be the zeros and poles, respectively, of f in $D(0, 1) = D$, each counted according to its multiplicity. Form

$$R(z) = \prod_{k=1}^{N} \frac{z - a_k}{1 - \bar{a}_k z} \prod_{l=1}^{M} \frac{1 - \bar{b}_l z}{z - b_l}.$$

If $a \in D$ and $|z| = 1$, then

$$\left|\frac{z - a}{1 - \bar{a}z}\right| = \left|\frac{\bar{z}z - \bar{z}a}{1 - \bar{a}z}\right| = \left|\frac{\overline{1 - \bar{a}z}}{1 - \bar{a}z}\right| = 1,$$

so $|R(z)| = 1$. Moreover f/R and R/f are each (extendable to be) holomorphic in D. Applying the Maximum Modulus Principle to each, we learn that $|f/R| \equiv 1$ in D. Therefore for some $b \in \mathbb{C}$, $f/R \equiv b$, $f \equiv bR$ in D. Then (cf. 11.17 below)

$f \equiv bR$ throughout $\mathbb{C}$. Finally, set $n = N + M$, $a_{N+l} = 1/\bar{b}_l$ $(l = 1, 2, \ldots, M)$ and $c = (\bar{b}_1/b_1)\cdots(\bar{b}_M/b_M)\cdot b$ in order to have

$$f(z) = bR(z) = c\prod_{k=1}^{n}\frac{z - a_k}{1 - \bar{a}_k z}.$$

Example 11.15 *If f is a rational function and $f(z) > 0$ whenever $|z| = 1$, then*

$$f(z) \equiv c\prod_{k=1}^{n}\frac{(z - z_k)(1 - \bar{z}_k z)}{(z - p_k)(1 - \bar{p}_k z)}$$

for some p_k, $z_k \in D(0, 1)$ and some positive c.

Proof: By considering the rational function $1/f$ instead of f if necessary, we may assume that 0 is a pole of f of order, say, l with $l \geq 0$ (and possibly $l = 0$). Form the rational function

$$F(z) = \frac{1}{\bar{f}\left(\frac{1}{\bar{z}}\right)}.$$

We have $fF = 1$ on the boundary of $D = D(0, 1)$. For if $|z| = 1$, then $1/\bar{z} = z$ and $f(1/\bar{z}) = f(z)$ is real so $F(z) = 1/f(z)$. Conclude from the uniqueness theorem that $fF = 1$ throughout $\mathbb{C}$ (cf. 11.17 below):

$$f(z) \equiv \bar{f}\left(\frac{1}{\bar{z}}\right).$$

It follows from this equality that if $z_1, \ldots, z_n$ are the zeros of f in $D\backslash\{0\}$ and $p_1, \ldots, p_m$ are the poles of f in $D\backslash\{0\}$, then $1/\bar{z}_1, \ldots, 1/\bar{z}_n$ are the zeros of f in $\mathbb{C}\backslash D$ and $1/\bar{p}_1, \ldots, 1/\bar{p}_m$ are the poles of f in $\mathbb{C}\backslash D$. If therefore we form

(1) $$R(z) = \frac{(z - z_1)\cdots(z - z_n)(1 - \bar{z}_1 z)\cdots(1 - \bar{z}_n z)}{(z - p_1)\cdots(z - p_m)(1 - \bar{p}_1 z)\cdots(1 - \bar{p}_m z)}\cdot\frac{1}{z^l},$$

then f/R is a rational function with no zeros or poles, hence a constant:

(2) $f = cR$.

Now for $z \neq 0$

$$R(z) = \frac{(z - z_1)\cdots(z - z_n)\left(\frac{1}{z} - \bar{z}_1\right)\cdots\left(\frac{1}{z} - \bar{z}_n\right)}{(z - p_1)\cdots(z - p_m)\left(\frac{1}{z} - \bar{p}_1\right)\cdots\left(\frac{1}{z} - \bar{p}_m\right)}\cdot z^{n-m-l},$$

so that if $|z| = 1$, that is, $1/z = \bar{z}$, we get

(3) $$R(z) = \frac{|z - z_1|^2\cdots|z - z_n|^2}{|z - p_1|^2\cdots|z - p_m|^2}\cdot z^{n-m-l}.$$

Since $f(z)$ is positive for such z, it follows from (2) and (3) that

(4) cz^{n-m-l} is positive if $|z| = 1$.

It follows that $n - m - l = 0$ and $c > 0$. But set then

$$p_{m+1} = \cdots = p_{m+l} = p_n = 0$$

and the asserted result follows from (1) and (2).

Corollary 11.16 *If* $p(\theta) = \sum_{k=-n}^{n} a_k e^{ik\theta}$ $(a_k \in \mathbb{C})$ *and* $p(\theta) \geq 0$ *for all real* θ, *then there exists* $P(\theta) = \sum_{k=0}^{n} c_k e^{ik\theta}$ *such that* $p = |P|^2$.

Proof: Let $\varepsilon > 0$ and consider

$$(1) \qquad R_\varepsilon(z) = \varepsilon + \sum_{k=-n}^{n} a_k z^k.$$

This is a rational function which satisfies

$$R_\varepsilon(e^{i\theta}) = \varepsilon + p(\theta) > 0 \quad \text{for all real } \theta.$$

Therefore R_ε has the form prescribed by the last example: For some integer $n(\varepsilon)$, some $z_k(\varepsilon)$, $p_k(\varepsilon) \in D$ and some $c(\varepsilon) > 0$

$$(2) \qquad R_\varepsilon(z) = c(\varepsilon) \prod_{k=1}^{n(\varepsilon)} \frac{(z - z_k(\varepsilon))(1 - \bar{z}_k(\varepsilon)z)}{(z - p_k(\varepsilon))(1 - \bar{p}_k(\varepsilon)z)}.$$

Since 0 is the only possible pole of R_ε, after appropriate cancellations have been made in (2), we will have $p_k(\varepsilon) = 0$ for all k:

$$(3) \qquad R_\varepsilon(z) = c(\varepsilon) \prod_{k=1}^{n(\varepsilon)} (z - z_k(\varepsilon))\left(\frac{1}{z} - \bar{z}_k(\varepsilon)\right).$$

A comparison of (3) with (1) shows that by ignoring or creating factors with $z_k(\varepsilon) = 0$ we can simply suppose, as far as (3) is concerned, that the $n(\varepsilon)$ there equals n. Then

$$\varepsilon + p(\theta) = R_\varepsilon(e^{i\theta}) = c(\varepsilon) \prod_{k=1}^{n} |e^{i\theta} - z_k(\varepsilon)|^2.$$

Letting

$$P_\varepsilon(\theta) = \sqrt{c(\varepsilon)} \prod_{k=1}^{n} [e^{i\theta} - z_k(\varepsilon)] = \sum_{k=0}^{n} c_k(\varepsilon) e^{ik\theta},$$

we thus have

$$(4) \qquad \varepsilon + p = |P_\varepsilon|^2.$$

A simple computation shows that

$$\sum_{k=0}^{n} |c_k(\varepsilon)|^2 = \frac{1}{2\pi} \int_0^{2\pi} |P_\varepsilon|^2 = \varepsilon + \frac{1}{2\pi} \int_0^{2\pi} p$$

and so $\{(c_0(\varepsilon), \ldots, c_n(\varepsilon)) \colon 0 < \varepsilon < 1\}$ is a bounded set in $\mathbb{C}^{n+1}$ and there exist $\varepsilon_j \downarrow 0$ such that $\lim_{j \to \infty} c_k(\varepsilon_j)$ exists for each k. Calling this limit c_k and setting $P(\theta) = \sum_{k=0}^{n} c_k e^{ik\theta}$, it follows from (4) that $p = |P|^2$.

Exercise 11.17 *If two rational functions agree on an infinite set, they are identical.*

Hint: If P_1, P_2, Q_1, Q_2 are polynomials and $P_1/Q_1 = P_2/Q_2$ on a set S, then $P_1Q_2 - P_2Q_1$ is a polynomial which is zero on S. If S is infinite, this polynomial is identically zero.

Exercise 11.18 *If f is a rational function such that* $|f(z)| = 1$ *whenever* $\operatorname{Re} z = 0$, *then*

$$f(z) = c\prod_{k=1}^{n} \frac{z + \bar{z}_k}{z - z_k}$$

for some $z_k \in \mathbb{C}$ and some unimodular c.

Hint: Let $f = P/Q$. Since $\operatorname{Re} z = 0$ implies $z = -\bar{z}$,

$$1 = f(z)\bar{f}(z) = f(z)\bar{f}(-\bar{z}),$$

whence

$$P(z)\bar{P}(-\bar{z}) = Q(z)\bar{Q}(-\bar{z}) \quad \forall z \in \mathbb{R}.$$

Since both sides of this equation are polynomials, it follows that the equality in fact prevails throughout $\mathbb{C}$.

Exercise 11.19 *Show that a one-to-one entire function f is a first degree polynomial.*

Hints: The function $F(z) = f(1/z)$ $(z \in \mathbb{C}\backslash\{0\})$ cannot have an essential singularity at 0. Otherwise, by 11.6 $F(D(0, 1)\backslash\{0\})$ is dense in $\mathbb{C}$; in particular it meets the open set $F(\mathbb{C}\backslash\bar{D}(0, 1))$, i.e., $f(\mathbb{C}\backslash\bar{D}(0, 1))$ meets $f(D(0, 1))$, violating univalence. It follows from 11.5(i) and 11.7(ii) that $\lim_{|z|\to\infty}|f(z)| = \lim_{|z|\to 0}|F(z)|$ exists in $[0, \infty]$. From 5.70 and Liouville it follows that f is a polynomial, which must be of degree 1 since f is one-to-one.

Alternatively, notice that $f(\mathbb{C})$ is connected (1.3) and simply-connected (4.69). Consequently if $f(\mathbb{C}) \neq \mathbb{C}$, 10.2 provides a conformal $g: f(\mathbb{C}) \to D(0, 1)$ and the composite $g \circ f$ violates Liouville. Thus f is surjective and $f^{-1}: \mathbb{C} \to \mathbb{C}$ is continuous, so $f^{-1}(\bar{D}(0, M))$ is compact for every $M > 0$. Outside this compact set we have $|f| > M$. This means that $\lim_{|z|\to\infty}|f(z)| = \infty$ and we finish as before. (For yet another proof of the surjectivity of f, form the functions $f_r(z) = rf(z/(rf'(0))) - rf(0)$ with $z \in D(0, 1)$, note that $\{f_r : r > 0\}$ is a subset of $\mathscr{S}$ and apply 7.25.)

Exercise 11.20 *Show that if f is holomorphic in $\mathbb{C}$ with the exception of finitely many poles and $f(1/z)$ has an inessential singularity at 0, then f is a rational function.*

The next five exercises develop the rationality criterion mentioned at the beginning of this section. Besides its intrinsic interest, it will be essential to certain Chapter XVI developments.

Exercise 11.21 *Let $q_0, q_1, q_2, \ldots$ be rational numbers and $z_0, \ldots, z_n \in \mathbb{C}$ not all 0 satisfy*

(1) $$z_0 q_k + z_1 q_{k+1} + \cdots + z_n q_{k+n} = 0 \quad \forall k = 0, 1, 2, \ldots$$

Show that then there exist rational numbers $r_0, \ldots, r_n$ not all 0 satisfying

(2) $$r_0 q_k + r_1 q_{k+1} + \cdots + r_n q_{k+n} = 0 \quad \forall k = 0, 1, 2, \ldots$$

Hints: Let $Q_k = (q_k, q_{k+1}, \ldots, q_{k+n}) \in \mathbb{C}^{n+1}$ $(k = 0, 1, 2, \ldots)$. Then (1) says that the subspace spanned by these vectors has $\mathbb{C}$-dimension less than $n + 1$. Let $Q_{k_1}, \ldots, Q_{k_m}$ $(m \le n)$ be a maximal $\mathbb{C}$-linearly independent set. Then $Q_{k_1}, \ldots, Q_{k_m}$ are also linearly independent over the rationals. As column rank (over the rationals) does not exceed row rank (over the rationals) in the $m \times (n + 1)$ matrix

$$\begin{bmatrix} Q_{k_1} \\ \vdots \\ Q_{k_m} \end{bmatrix}$$

(for a painless and brief proof of which see LIEBECK [1966]), there exists a nontrivial rational linear relation among the $n + 1$ columns of this matrix. That is, rational $r_0, \ldots, r_n$ not all 0 exist such that

(3) $$\begin{bmatrix} Q_{k_1} \\ \vdots \\ Q_{k_m} \end{bmatrix} \begin{bmatrix} r_0 \\ \vdots \\ r_n \end{bmatrix} = 0 \in \mathbb{C}^m.$$

Now, given Q_k, there exist $a_1, \ldots, a_m \in \mathbb{C}$ such that

$$Q_k = a_1 Q_{k_1} + \cdots + a_m Q_{k_m}$$

(by maximality of $\{Q_{k_1}, \ldots, Q_{k_m}\}$). Consequently

$$Q_k \begin{bmatrix} r_0 \\ \vdots \\ r_n \end{bmatrix} = a_1 Q_{k_1} \begin{bmatrix} r_0 \\ \vdots \\ r_n \end{bmatrix} + \cdots + a_m Q_{k_m} \begin{bmatrix} r_0 \\ \vdots \\ r_n \end{bmatrix} \overset{(3)}{=} 0 \in \mathbb{C}.$$

Exercise 11.22 *Let $f(z) = \sum_{n=0}^{\infty} a_n z^n$ and $g(z) = \sum_{n=0}^{\infty} b_n z^n$ each have positive radius of convergence. Suppose the a's are all integers and have no common integer factors save ± 1. Assume also that the b's are all integers. Show that any common integer factor of all the coefficients of $f(z)g(z)$ is also a common integer factor of all the b's.*

Hints: The nth coefficient of the product series is (5.31)

$$c_n = \sum_{j=0}^{n} a_j b_{n-j}.$$

Suppose $p > 1$ is a prime factor of all the c's which nevertheless is not a common factor of all the b's. There exists then a least $l \geq 0$ such that p is not a factor of a_l and a least $n \geq l$ such that p is not a factor of b_{n-l}. Then p is not a factor of the product $a_l b_{n-l}$. Yet by minimality of l and n, p is a factor of each term in the sum $\sum_{j=0}^{l-1} a_j b_{n-j} + \sum_{j=l+1}^{n} a_j b_{n-j}$, as well as being a factor of c_n. Since the difference of c_n and these sums is $a_l b_{n-l}$, we have a contradiction.

Exercise 11.23 (FATOU [1904]) *Let the power series $f(z) = \sum_{n=0}^{\infty} c_n z^n$ have positive radius of convergence and integer coefficients. Suppose that f is a rational function. Show that then there exist polynomials P, Q with integer coefficients such that $f = P/Q$, $Q(0) = 1$ and P, Q have no common non-constant polynomial factor.*

Hints: By hypothesis there exist $a_0, \ldots, a_m; b_0, \ldots, b_k \in \mathbb{C}$ such that

(1) $$f(z) = \frac{a_0 + a_1 z + \cdots + a_m z^m}{b_0 + b_1 z + \cdots + b_k z^k} \quad (z \text{ near } 0).$$

We may assume that $b_k \neq 0$ and that

(*) $$\begin{cases} \text{No denominator in any representation of } f \text{ as a} \\ \text{quotient of polynomials has degree less than } k. \end{cases}$$

Multiplying away the denominator in (1) and equating coefficients, we have

(2) $$a_j = \sum_{l=0}^{\min(k,j)} b_l c_{j-l}, \quad j = 0, 1, \ldots, m$$

(3) $$0 = \sum_{l=0}^{\min(k,j)} b_l c_{j-l} \quad \forall j > m.$$

In particular then

(4) $$0 = \sum_{l=0}^{k} b_l c_{j-l} \quad \forall j > \max\{m, k\}.$$

It follows from 11.21 then that there exist rational numbers $q_0, \ldots, q_k$, not all 0, such that

(4)′ $$0 = \sum_{l=0}^{k} q_l c_{j-l} \quad \forall j > \max\{m, k\} = N.$$

We may clear the denominators in (4)′ and so suppose simply that the q's are all integers. We may also divide away any common integer factors and so suppose that

(5) $q_0, \ldots, q_k$ are integers with no common integer factors save ± 1.

If we define integers $p_0, \ldots, p_N$ by

(2)′ $$p_j = \sum_{l=0}^{\min(k,j)} q_l c_{j-l}, \quad 0 \leq j \leq N,$$

then reversing the computations above which led to (2)–(4), gives

$$(6) \quad f(z) = \frac{P(z)}{Q(z)},$$

where $P(z) = p_0 + p_1 z + \cdots + p_N z^N$ and $Q(z) = q_0 + q_1 z + \cdots + q_k z^k$.

Notice that, because the degree of Q is k or less, it follows from (6) and (*) that P and Q have no common non-constant polynomial factor.

Next, use 4.52 to select polynomials $A, B \in \mathscr{R}$ (i.e., with rational coefficients) such that $AP + BQ = 1$. If d is the product of the denominators of the coefficients in A and B, then $U = dA$ and $V = dB$ are polynomials with integer coefficients such that

$$UP + VQ = d,$$

whence from (6)

$$(7) \quad (Uf + V)Q = d.$$

Since f is a power series with integer coefficients and positive radius of convergence, so is $Uf + V$. Hence from (5), (7) and 11.22 it follows that d is a common integer factor of all the coefficients of $Uf + V$. If therefore we equate the constant terms in (7), we get

$$[U(0)f(0) + V(0)]q_0 = d,$$

where the square-bracketed term is an integer of which d is a factor. It follows that the square-bracketed term must in fact be $\pm d$ and so $q_0 = \pm 1$.

Exercise 11.24 *Let the power series $f(z) = \sum_{n=0}^{\infty} c_n z^n$ have positive radius of convergence. Show that f is a rational function if and only if there exist numbers $a_0, \ldots, a_p$ not all 0 and an integer N such that*

$$(*) \quad a_0 c_n + a_1 c_{n+1} + \cdots + a_p c_{n+p} = 0 \quad \forall n \geq N.$$

Hint: ($\Leftarrow$) Set $P(z) = a_0 z^p + a_1 z^{p-1} + \cdots + a_p$ and use (*) to show that $f(z)P(z)$ is a polynomial of degree not greater than $N + p$.

Exercise 11.25 (Kronecker [1881]) *Let $c_0, c_1, \ldots$ be complex numbers and for each non-negative integer m set*

$$C_m = \begin{bmatrix} c_0 & c_1 & \cdots & c_m \\ c_1 & c_2 & \cdots & c_{m+1} \\ \vdots & \vdots & & \vdots \\ c_m & c_{m+1} & \cdots & c_{2m} \end{bmatrix}.$$

Show that if the power series $\sum_{n=0}^{\infty} c_n z^n$ *has positive radius of convergence, then it represents a rational function if and only if* $\det C_m = 0$ *for all sufficiently large m.*

Hints: ($\Leftarrow$) Let p be the smallest non-negative integer with the property that

(1) $\quad \det C_m = 0 \quad \forall m \geq p.$

If $p = 0$, use induction on m to show that $c_m = 0$ for all $m \geq 0$ (hence $f = 0$, a rational function): we have $c_0 = \det C_0 = 0$. If $c_0 = c_1 = \cdots = c_{n-1} = 0$, then $0 \overset{(1)}{=} \det C_n = \pm c_n^{n+1}$, whence $c_n = 0$.

If $p > 0$, we have

(2) $\quad \det C_{p-1} \neq 0, \qquad \det C_p = 0.$

It follows that the columns of C_{p-1} are linearly independent while the columns of C_p are linearly dependent. Therefore the last column of C_p is a linear combination of the others, that is, there exist $a_0, \ldots, a_{p-1} \in \mathbb{C}$, $a_p = 1$ such that

(3) $$\sum_{j=0}^{p} a_j c_{m+j} = 0, \qquad 0 \leq m \leq p.$$

Make the definition

(4) $$s_m = \sum_{j=0}^{p} a_j c_{m+j}, \quad m = 0, 1, 2, \ldots$$

We want to show that $s_m = 0$ for all $m \geq 0$. Go by induction on m and suppose, in view of (3), that for some $q > 0$

(5) $\quad s_m = 0 \quad \forall 0 \leq m < p + q.$

Now consider C_{p+q}. Perform on it the following column operations: For each $m \geq p$ add to the $(m + 1)$st column (the column headed by c_m) the preceding p columns, each of these columns having been first multiplied by $a_0, \ldots, a_{p-1}$ respectively. The resulting matrix has the same determinant as C_{p+q} and, thanks to (5), has the form

$$\begin{bmatrix} c_0 & c_1 \cdots c_{p-1} & & & \\ c_1 & c_2 \cdots c_p & & 0 & \\ \vdots & \vdots \qquad \vdots & & & \\ c_{p-1} & c_p \cdots c_{2(p-1)} & & & \\ & & 0 & 0 \cdots 0 & s_{p+q} \\ & & 0 & \cdot\ 0\ s_{p+q} & \vdots \\ & & \vdots & 0 \ \cdot^{\cdot^{\cdot}} & \vdots \\ & & 0 & s_{p+q} & \vdots \\ & & s_{p+q} & \cdots\cdots\cdots & \vdots \end{bmatrix}$$

Therefore $0 \overset{(1)}{=} \det C_{p+q} = \pm s_{p+q}^{q+1} \det C_{p-1}$. Because of (2) this entails $s_{p+q} = 0$, thus completing the induction. With $s_m = 0$ for all $m \geq 0$, it follows from 11.24 that $\sum_{n=0}^{\infty} c_n z^n$ represents a rational function.

§ 3 Isolated Singularities on the Circle of Convergence

Here we look at two beautiful classical results which enable one to decide when a point on the circle of convergence of a power series is the unique pole of the function and to find that pole in terms of the power series coefficients.

Example 11.26 *Let $z_0 \in \mathbb{C}\backslash\{0\}$, U a neighborhood of $\bar{D}(0, |z_0|)$ and f a function holomorphic in $U\backslash\{z_0\}$, with a pole at z_0. Then*

$$\lim_{n \to \infty} \frac{(n+1)f^{(n)}(0)}{f^{(n+1)}(0)} \text{ exists and equals } z_0.$$

Proof: Let $g(z) = \sum_{k=1}^{m} b_k/(z - z_0)^k$ $(m \geq 1, b_m \neq 0)$ be the principal part of f at z_0. Thus $h = f - g$ is holomorphic in U and its power series $\sum_{n=0}^{\infty} a_n z^n$ converges at z_0 (in fact, in $D(0, |z_0| + \varepsilon)$ for some $\varepsilon > 0$). Therefore

(1) $\quad a_n z_0^n \to 0.$

We have

$$g^{(n)}(0) = \sum_{k=1}^{m} (-1)^k b_k (k+n-1)\cdots(k+1)k z_0^{-k-n}$$

and so

$$\begin{aligned}
&\frac{(n+1)f^{(n)}(0)}{f^{(n+1)}(0)} \\
&\quad= \frac{(n+1)h^{(n)}(0) + (n+1)g^{(n)}(0)}{h^{(n+1)}(0) + g^{(n+1)}(0)} \\
&\quad= \frac{(n+1)n!\, a_n + (n+1) \sum_{k=1}^{m} (-1)^k b_k (k+n-1)\cdots(k+1)k z_0^{-k-n}}{(n+1)!\, a_{n+1} + \sum_{k=1}^{m} (-1)^k b_k (k+n)(k+n-1)\cdots(k+1)k z_0^{-k-n-1}}
\end{aligned}$$

$$(2) \quad = \frac{\dfrac{(n+1)!}{(m+n)!} a_n z_0^n + \dfrac{n+1}{m+n} \displaystyle\sum_{k=1}^{m} (-1)^k b_k z_0^{-k} \frac{(k+n-1)\cdots(k+1)k}{(m+n-1)!}}{\dfrac{1}{z_0} \dfrac{(n+1)!}{(m+n)!} a_{n+1} z_0^{n+1} + \displaystyle\sum_{k=1}^{m} (-1)^k b_k z_0^{-k-1} \frac{(k+n)\cdots(k+1)k}{(m+n)!}}.$$

Now

$$0 \leq \frac{(n+1)!}{(m+n)!} \leq \frac{n+1}{n+m} \to 1,$$

while

$$\frac{(k+n)\cdots(k+1)k}{(m+n)!} \le \frac{1}{(m+n)(k-1)!} \to 0$$

if $k < m$. Therefore, recalling (1), we see that all the terms in both the numerator and the denominator of (2), save for $k = m$, converge to 0 as $n \to \infty$.

Exercise 11.27 *Let the power series $\sum_{n=0}^{\infty} c_n z^n$ have radius of convergence* 1 *and denote by f the holomorphic function it defines.*

(i) *Show that f has a holomorphic extension to a neighborhood of $\bar{D}(0, 1)\backslash\{1\}$ with a simple pole at* 1 *if and only if*

$$(1) \qquad \overline{\lim_{n\to\infty}} |c_{n+1} - c_n|^{1/n} < 1.$$

Hints: Note that

$$(2) \qquad (1-z)f(z) = (1-z)\sum_{n=0}^{\infty} c_n z^n = c_0 + \sum_{n=0}^{\infty} (c_{n+1} - c_n)z^{n+1} \quad \forall z \in D(0, 1).$$

Now suppose such an extension exists. Call it F. Since the pole which F has at 1 is supposed to be simple, the function $(1 - z)F(z)$ is holomorphic in a neighborhood of $\bar{D}(0, 1)$, say in $D(0, r)$, $r > 1$. The series on the right of (2) is the power series at 0 of the function $(1 - z)F(z)$ and so by 5.30 it converges throughout $D(0, r)$. By 3.3 this means that $\overline{\lim}_{n\to\infty}|c_{n+1} - c_n|^{1/n} \le 1/r < 1$.

Conversely suppose that (1) holds. Then by 3.3 the series on the right of (2) converges in a disk of radius greater than 1, say in $D(0, r)$. Let g denote the holomorphic function so defined. Then $g(z)/(1 - z)$ is the desired holomorphic extension of f to the neighborhood $D(0, r)\backslash\{1\}$ of $\bar{D}(0, 1)\backslash\{1\}$. The form of this function shows that 1 is either a first order pole (if $g(1) \neq 0$) or a removable singularity (if $g(1) = 0$). The latter would mean that f has a holomorphic extension to $D(0, r)$, contrary to the radius of convergence of $\sum c_n z^n$ being 1.

(ii) *f has a holomorphic extension to a neighborhood of $\bar{D}(0, 1)\backslash\{1\}$ with a simple pole at* 1 *if and only if*

$$(3) \qquad \overline{\lim_{n\to\infty}} \left|\frac{c_n}{c_{n+1}} - 1\right|^{1/n} < 1.$$

Hints: In view of (i) it suffices to establish the equivalence of (1) and (3). Suppose that (1) holds. Then, using the notation of the hints to the second part of (i), we have

$$0 \neq g(1) = c_0 + \sum_{n=0}^{\infty} (c_{n+1} - c_n) = \lim_{N\to\infty} c_N.$$

Consequently

$$(4) \qquad \lim_{n\to\infty} \frac{c_{n+1}}{c_n} = 1$$

and so from 3.9 $\lim_{n\to\infty}|c_n|^{1/n} = 1$. It follows that

$$(5)\qquad |c_{n+1}|^{1/n} = [|c_{n+1}|^{1/(n+1)}]^{1+1/n} \to 1, \qquad \text{as } n \to \infty.$$

Since

$$(6)\qquad |c_{n+1} - c_n|^{1/n} = |c_{n+1}|^{1/n}\left|1 - \frac{c_n}{c_{n+1}}\right|^{1/n},$$

we deduce (3) from (1) and (5).

Conversely from (3) follows easily (4), thence (5); and then (3), (5) and (6) yield (1).

Exercise 11.28 *Show that if* $\lim_{n\to\infty} c_n = 0$, *then* $\sum_{n=0}^{\infty} c_n z^n$ *converges in* $D(0, 1)$ *and no point of the unit circle is a pole for the function f so defined, i.e., f does not admit extension to a holomorphic function in* $D(z_0, r)\backslash\{z_0\}$ *with a pole at* z_0 *for any* $|z_0| = 1$, $r > 0$.

Hint: It suffices to show that 1 is not a pole. For this, show, by using the partial summation technique from 3.13(i), that $\lim_{r\uparrow 1}(1 - r)f(r) = 0$.

§ 4 The Residue Theorem and Some Applications

The main result here is the form which Cauchy's theorem takes when singularities are present:

Theorem 11.29 (Residue Theorem) *Let U be an open subset of* $\mathbb{C}$, *A a subset of U with no limit points in U,* γ *a piecewise smooth loop in* $U\backslash A$ *which is U-null-homologous, f holomorphic in* $U\backslash A$. *Then*

$$\frac{1}{2\pi i}\int_{\gamma} f = \sum_{a\in A} \operatorname{Res}(f, a)\operatorname{Ind}_{\gamma}(a)$$

and this sum contains only finitely many non-zero terms.

Proof: Let $B = \{a \in A\colon \operatorname{Ind}_\gamma(a) \neq 0\}$, C the unbounded component of $\mathbb{C}\backslash\gamma$. Thus $C \supset \mathbb{C}\backslash\bar{D}(0, r)$ for some $r > 0$. Each point b of B lies in a component C_b of $\mathbb{C}\backslash\gamma$ on which $\operatorname{Ind}_\gamma$ is constant and (therefore) non-zero; so C_b is disjoint from $\mathbb{C}\backslash U$ (since $\operatorname{Ind}_\gamma$ vanishes identically on $\mathbb{C}\backslash U$ by hypothesis). Thus $C_b \subset U$. Also $C_b \subset \mathbb{C}\backslash C \subset \bar{D}(0, r)$. Notice that the union K of the range of γ and all the components $C_b (b \in B)$ is a closed set. For $\mathbb{C}\backslash K$ is a union of components of $\mathbb{C}\backslash\gamma$ and each such is open by 1.30. Thus K is a closed and bounded, hence compact, subset of U. Since A has no limit point in U, $A \cap K$ is finite. Thus B (hence the number of non-zero terms in the sum in the statement of the theorem) is finite. Let, say, $a_1, a_2, \ldots, a_n$ be the distinct elements of B. If f_j is the principal part of f at a_j, then f_j is holomorphic in $\mathbb{C}\backslash\{a_j\}$ and $F = f - \sum_{j=1}^{n} f_j$ has removable singularities at $a_1, \ldots, a_n$ and so is (extendable to be) holomorphic in the set $U_0 = U\backslash(A\backslash B)$. This set is open, since A has no limit points in U. By definition of B and the fact that γ lies in $U\backslash A$ and is U-nullhomologous, we have that γ is U_0-nullhomologous and so by 7.19 $\int_\gamma F = 0$. Thus

$$(1) \quad \int_\gamma f = \sum_{j=1}^{n} \int_\gamma f_j.$$

By definition of principal part, there is an entire function $g_j(z) = \sum_{k=1}^\infty c_{jk} z^k$ such that $f_j(z) = g_j(1/(z - a_j))$. The series for g_j converges uniformly on the compact set $\{1/(\xi - a_j) : \xi \in \gamma\}$, so

$$(2) \quad \int_\gamma f_j = \int_\gamma \sum_{k=1}^{\infty} \frac{c_{jk}}{(\xi - a_j)^k}\, d\xi = \sum_{k=1}^{\infty} c_{jk} \int_\gamma \frac{d\xi}{(\xi - a_j)^k} = c_{j1} \int_\gamma \frac{d\xi}{\xi - a_j},$$

using 2.14. Since c_{j1} is $\operatorname{Res}(f, a_j)$, we have from (2)

$$\int_\gamma f_j = 2\pi i \operatorname{Res}(f, a_j) \operatorname{Ind}_\gamma(a_j)$$

and from this and (1) the assertion of the theorem follows.

Exercise 11.30 *Generalize* 11.29 *to the case of finitely many piecewise smooth loops* $\gamma_1, \ldots, \gamma_k$ *in* U *satisfying for some* $c_1, \ldots, c_k \in \mathbb{C}$, $\sum_{j=1}^k c_j \operatorname{Ind}_{\gamma_j}$ *identically* 0 *in* $\mathbb{C}\backslash U$.

Hint: The proof given above needs only the most superficial (to wit, notational) changes.

For the initial applications I have in mind it is convenient to dispose of certain computations in advance; this is done in the following multi-part exercise. It is also convenient to set some notation for the immediate future. The positive integer n is fixed and

$$S_n(z) = \sum_{k=0}^{n-1} \exp\left[\frac{2\pi i}{n}(z + k)^2\right] \quad \forall z \in \mathbb{C},$$

$$g(z) = e^{2\pi i z} - 1 \quad \forall z \in \mathbb{C},$$

$$f = \frac{S_n}{g} \quad \text{on } \mathbb{C}\backslash\mathbb{Z},$$

$$c = 1 + i,$$

$$I = \lim_{R \to \infty} \int_{-R}^{R} e^{-x^2} dx. \qquad \text{(See 8.26(i).)}$$

Exercise 11.31 (i) *For any* $a, r \in \mathbb{R}$ *with* $r > 2|a|$ *we have*

$$\left| \int_{-r}^{r} e^{2\pi i(a + tc)^2/n} dt - \int_{-r}^{r} e^{2\pi i(tc)^2/n} dt \right| \le 2|a| e^{-2\pi r^2/n}.$$

Hints: Let $\Gamma_r = [a - rc, -rc, rc, a + rc, a - rc]$. The function $h(z) = e^{2\pi i z^2/n}$ is entire so $\int_{\Gamma_r} h = 0$. That is,

$$-\int_{[a+rc,a-rc]} h - \int_{[-rc,rc]} h = \int_{[rc,a+rc]} h + \int_{[a-rc,-rc]} h.$$

$$-\int_0^1 h(a + rc - 2trc)(-2rc)dt - \int_0^1 h(-rc + 2trc)(2rc)dt$$

$$= \int_0^1 h(rc + ta)(a)dt + \int_0^1 h(a - rc - ta)(-a)dt.$$

Or, upon making the obvious changes of variable,

$$c\int_{-r}^{r} h(a + xc)dx - c\int_{-r}^{r} h(xc)dx = a\int_0^1 h(ta + rc)dt - a\int_0^1 h(ta - rc)dt.$$

Finally note that for $0 \le t \le 1$

$$|h(ta \pm rc)| = \exp\left[\frac{-2\pi}{n}\operatorname{Im}(ta \pm rc)^2\right] = \exp\left[\frac{-4\pi}{n}(r \pm ta)r\right]$$

$$< \exp\left[\frac{-4\pi}{n}\cdot\frac{r}{2}\cdot r\right] \quad \text{if } r > 2|a|.$$

(ii) $\lim_{r\to\infty}\int_{-1/2}^{1/2} f(t \pm rc)dt = 0.$

Hints: We have for all real t

$$|e^{2\pi i(t\pm rc)} - 1| = |e^{2\pi i(t\pm r)}e^{\pm 2\pi r} - 1|$$

and therefore

(1) $|e^{2\pi i(t+rc)} - 1| \ge 1 - e^{-2\pi r} > \frac{1}{2} \quad \forall r \ge \frac{1}{2\pi}$

(1)′ $|e^{2\pi i(t-rc)} - 1| \ge e^{2\pi r} - 1 > 1 \quad \forall r \ge \frac{1}{2\pi}.$

For any integer $0 \le k \le n - 1$ and all $-\frac{1}{2} \le t \le \frac{1}{2}$

$$\left|\exp\left[\frac{2\pi i}{n}(t \pm rc + k)^2\right]\right| = \exp\left[-\frac{4\pi}{n}\operatorname{Re}(t \pm rc + k)\operatorname{Im}(t \pm rc + k)\right]$$

$$= \exp\left[-\frac{4\pi}{n}(t \pm r + k)(\pm r)\right]$$

$$= \exp\left[-\frac{4\pi}{n}r^2\left(1 \pm \frac{t+k}{r}\right)\right]$$

(2)
$$\le \exp\left[-\frac{\pi}{n}r^2\right] \quad \forall r \ge 2(k + 1).$$

From (1), (1)′ and (2)

(3) $|f(t \pm rc)| \le 2ne^{-\pi r^2/n} \quad \forall r \ge 2n,\ t \in [-\frac{1}{2}, \frac{1}{2}].$

The assertion follows.

(iii) $f(z+1) - f(z) = e^{2\pi i z^2/n}[e^{2\pi i z} + 1] \quad \forall z \in \mathbb{C}\backslash\mathbb{Z}$.

Hints:

$$\begin{aligned}
&f(z+1) - f(z)\\
&= \frac{1}{e^{2\pi i z} - 1} \sum_{k=0}^{n-1} \left(\exp\left[\frac{2\pi i}{n}(z+1+k)^2\right] - \exp\left[\frac{2\pi i}{n}(z+k)^2\right] \right)\\
&= \frac{1}{e^{2\pi i z} - 1} \left(\exp\left[\frac{2\pi i}{n}(z+n)^2\right] - \exp\left[\frac{2\pi i}{n} z^2\right] \right)\\
&= e^{2\pi i z^2/n} \cdot \frac{\exp\left(\frac{2\pi i}{n}[(z+n)^2 - z^2]\right) - 1}{e^{2\pi i z} - 1}\\
&= e^{2\pi i z^2/n} \cdot \frac{\exp\left(\frac{2\pi i}{n}[2nz + n^2]\right) - 1}{e^{2\pi i z} - 1}\\
&= e^{2\pi i z^2/n} \cdot \frac{e^{4\pi i z} - 1}{e^{2\pi i z} - 1}.
\end{aligned}$$

(iv) *For any $r > n$*

$$\left| \int_{-r}^{r} f(\tfrac{1}{2} + tc) - f(-\tfrac{1}{2} + tc)]dt - [1 + (-i)^n] \int_{-r}^{r} e^{2\pi i (tc)^2/n} dt \right| \leq n e^{-2\pi r^2/n}.$$

Hints: Using (iii)

$$\begin{aligned}
&\int_{-r}^{r} [f(\tfrac{1}{2} + tc) - f(-\tfrac{1}{2} + tc)]dt\\
&= \int_{-r}^{r} \exp\left[\frac{2\pi i}{n}(-\tfrac{1}{2} + tc)^2\right][\exp 2\pi i(-\tfrac{1}{2} + tc) + 1]dt\\
&= \int_{-r}^{r} \exp\left[\frac{2\pi i}{n}(-\tfrac{1}{2} + tc)^2\right]dt\\
&\qquad + \int_{-r}^{r} \exp\left[\frac{2\pi i}{n}(-\tfrac{1}{2} + tc)^2 + 2\pi i(-\tfrac{1}{2} + tc)\right]dt\\
&= \int_{-r}^{r} \exp\left[\frac{2\pi i}{n}(-\tfrac{1}{2} + tc)^2\right]dt + \int_{-r}^{r} \exp\left(\frac{2\pi i}{n}\left[\left(\frac{n}{2} - \frac{1}{2} + tc\right)^2 - \frac{n^2}{4}\right]\right)dt\\
&= \int_{-r}^{r} \exp\left[\frac{2\pi i}{n}(-\tfrac{1}{2} + tc)^2\right]dt + e^{-n\pi i/2} \int_{-r}^{r} \exp\left[\frac{2\pi i}{n}\left(\frac{n}{2} - \frac{1}{2} + tc\right)^2\right]dt.
\end{aligned}$$

Note that $e^{-n\pi i/2} = [e^{-\pi i/2}]^n = (-i)^n$. Now make two citations of part (i).

Example 11.32 (i) $\lim_{R\to\infty} \int_{-R}^{R} e^{-x^2}dx$ *exists and equals* $\sqrt{\pi}$.

(ii) $\lim_{R\to\infty} \int_{-R}^{R} \cos(x^2)dx$, $\lim_{R\to\infty} \int_{-R}^{R} \sin(x^2)dx$ *each exist and each equals* $\sqrt{\pi/2}$.

(iii) $\sum_{k=0}^{n-1} e^{2\pi i k^2/n} = \sqrt{n}(1 + (-i)^n)/(1 - i)$ *for each positive integer n.*

Proof: We use the notation and results of the previous exercise. Notice that the poles of f are at the zeros of its denominator g, namely at the integers. Set

$$\gamma_r = [-rc - \tfrac{1}{2}, -rc + \tfrac{1}{2}, rc + \tfrac{1}{2}, rc - \tfrac{1}{2}, -rc - \tfrac{1}{2}], \quad r > 0.$$

Then a simple calculation reveals that $\gamma_r \cap \mathbb{R} = \{\pm\frac{1}{2}\}$. Therefore the unbounded connected sets $(-\infty, -1]$ and $[1, \infty)$ both lie in $\mathbb{C}\backslash\gamma_r$ and so by 4.3 $\operatorname{Ind}_{\gamma_r}$ is constantly 0 on each of these sets. In particular,

$$\operatorname{Ind}_{\gamma_r}(z) = 0 \quad \text{for all non-zero integers } z$$

and therefore

(1) $$\sum_{z\in\mathbb{Z}} \operatorname{Res}(f, z) \operatorname{Ind}_{\gamma_r}(z) = \operatorname{Res}(f, 0) \operatorname{Ind}_{\gamma_r}(0).$$

Now

$$g(z) = e^{2\pi i z} - 1 = \sum_{k=1}^{\infty} \frac{(2\pi i z)^k}{k!} = 2\pi i z h(z),$$

where $h(z) = \sum_{k=1}^{\infty} (2\pi i)^{k-1} z^{k-1}/k!$. Since $h(0) = 1$, S_n/h is holomorphic near 0, say

$$\frac{S_n(z)}{h(z)} = S_n(0) + \sum_{k=0}^{\infty} c_k z^{k+1}.$$

Then

$$f(z) = \frac{S_n(z)}{g(z)} = \frac{1}{2\pi i z}\frac{S_n(z)}{h(z)} = \frac{S_n(0)}{2\pi i z} + \sum_{k=0}^{\infty} \frac{c_k}{2\pi i} z^k.$$

Therefore

(2) $$\operatorname{Res}(f, 0) = \frac{S_n(0)}{2\pi i}.$$

Finally recall that $\operatorname{Ind}_{\gamma_r}(0) = 1$ by 4.17. Using this fact and (2) in (1),

(3) $$\sum_{z\in\mathbb{Z}} \operatorname{Res}(f, z) \operatorname{Ind}_{\gamma_r}(z) = \frac{S_n(0)}{2\pi i}.$$

According to the Residue Theorem (11.29 with $U = \mathbb{C}$) the left side of (3) equals $\frac{1}{2\pi i}\int_{\gamma_r} f$. Using 2.9 then, (3) becomes

$$S_n(0) = \int_{[-rc-1/2,\,-rc+1/2]} f + \int_{[-rc+1/2,\,rc+1/2]} f$$

$$+ \int_{[rc+1/2,\,rc-1/2]} f + \int_{[rc-1/2,\,-rc-1/2]} f$$

$$= \int_0^1 f(-rc - \tfrac{1}{2} + t)dt + \int_0^1 f(-rc + \tfrac{1}{2} + 2trc)(2rc)dt$$

$$- \int_0^1 f(rc + \tfrac{1}{2} - t)dt + \int_0^1 f(rc - \tfrac{1}{2} - 2trc)(-2rc)dt.$$

Whence, upon making the obvious changes of variable and combining the first with the third and the second with the fourth terms,

$$S_n(0) = \int_{-1/2}^{1/2} [f(-rc + t) - f(rc + t)]dt$$

$$+ c\int_{-r}^{r} [f(\tfrac{1}{2} + tc) - f(-\tfrac{1}{2} + tc)]dt.$$

Recalling 11.31(iv), we may then write, for some $\phi(r)$ with $\lim_{r\to\infty} \phi(r) = 0$,

$$S_n(0) = \int_{-1/2}^{1/2} [f(t - rc) - f(t + rc)]dt$$

$$+ c[1 + (-i)^n]\int_{-r}^{r} e^{2\pi i (tc)^2/n}dt + \phi(r).$$

Since $c^2 = 2i$, a change of variable here gives

$$(4) \quad S_n(0) = \int_{-1/2}^{1/2} [f(t - rc) - f(t + rc)]dt$$

$$+ \frac{c}{2}\sqrt{\frac{n}{\pi}}\,[1 + (-i)^n]\int_{-2r\sqrt{\pi/n}}^{2r\sqrt{\pi/n}} e^{-x^2}dx + \phi(r).$$

In (4) let $r \uparrow \infty$, taking account of 8.26(i) and 11.31(ii):

$$(5) \quad S_n(0) = \frac{c}{2}\sqrt{\frac{n}{\pi}}\,[1 + (-i)^n]I.$$

Since $S_1(0) = 1$, we get for I the value

$$(6) \quad I = \frac{2\sqrt{\pi}}{c[1 - i]} = \sqrt{\pi},$$

that is,

$$\lim_{R\to\infty}\int_{-R}^{R} e^{-x^2}dx = \sqrt{\pi},$$

establishing (i); and then substitution of (6) into (5) gives (iii).

For the proof of (ii) apply Cauchy's Theorem for Convex Regions to the function $h(z) = e^{-z^2}$ in $\mathbb{C}$ and the piecewise smooth loops Γ_r defined by

$$\Gamma_r(t) = \begin{cases} re^{\pi it/4} & 0 \le t \le 1 \\ (2-t)re^{\pi i/4} & 1 \le t \le 2 \\ (t-2)r & 2 \le t \le 3. \end{cases}$$

Thus $\int_{\Gamma_r} h = 0$, which after the appropriate changes of variable says that

$$(7)\qquad e^{\pi i/4}\int_0^r e^{-ix^2}dx - \int_0^r e^{-x^2}dx = ir\int_0^{\pi/4} \exp[-r^2e^{2ix}]e^{ix}dx.$$

For the right-hand side we have the bound

$$\left| ir\int_0^{\pi/4} \exp[-r^2e^{2ix}]e^{ix}dx\right| \le r\int_0^{\pi/4} |\exp[-r^2e^{2ix}]|dx$$

$$= r\int_0^{\pi/4} \exp[\mathrm{Re}(-r^2e^{2ix})]dx$$

$$(8)\qquad = r\int_0^{\pi/4} e^{-r^2\cos 2x}dx.$$

Put (8) into (7), and recall 3.31:

$$\left|\frac{1+i}{\sqrt{2}}\left[\int_0^r \cos(x^2)dx - i\int_0^r \sin(x^2)dx\right] - \int_0^r e^{-x^2}dx\right| \le r\frac{\pi}{4r^2}$$

$$(9)\qquad \left|\int_0^r \cos(x^2)dx + \int_0^r \sin(x^2)dx + i\left[\int_0^r \cos(x^2)dx - \int_0^r \sin(x^2)dx\right] - \sqrt{2}\int_0^r e^{-x^2}dx\right| \le \frac{\pi\sqrt{2}}{4r}.$$

Let $r\uparrow\infty$ in (9), keeping in mind the result (i). It follows that

$$(10)\qquad \lim_{r\to\infty}\left[\int_0^r \cos(x^2)dx - \int_0^r \sin(x^2)dx\right] = 0$$

and

$$(11)\qquad \lim_{r\to\infty}\left[\int_0^r \cos(x^2)dx + \int_0^r \sin(x^2)dx\right] = \sqrt{2}\cdot\frac{\sqrt{\pi}}{2} = \sqrt{\frac{\pi}{2}}.$$

Adding and subtracting (10) and (11), we learn that

$$\lim_{r\to\infty}\int_0^r \cos(x^2)dx,\ \lim_{r\to\infty}\int_0^r \sin(x^2)dx \text{ each exist and equal } \frac{1}{2}\sqrt{\frac{\pi}{2}},$$

which proves (ii).

Exercise 11.33 *Let $a > 0$, F the entire function $F(z) = e^{-\pi a z^2}$ and f the holomorphic function $f(z) = F(z)/(e^{2\pi i z} - 1)$, $z \in \mathbb{C}\setminus\mathbb{Z}$.*

(i) *Show that for each real y*

(1) $$\lim_{r\to\infty}\int_{-r}^{r} e^{-\pi a(x+iy)^2}dx = \lim_{r\to\infty}\int_{-r}^{r} e^{-\pi a x^2}dx. \quad \textit{Call this number } I(a).$$

Show that, in fact, uniformly in $y \in \mathbb{R}$ we have

(2) $$\lim_{r\to\infty} e^{-\pi a y^2}\left| I(a) - \int_{-r}^{r} e^{-\pi a(x+iy)^2}dx\right| = 0.$$

Hints: If R and t are real

$$|F(R+it)| = |e^{-\pi a(R^2+2iRt-t^2)}| = e^{-\pi a R^2}e^{\pi a t^2}, \quad \text{and so}$$

(3) $$\left|\int_{[R,R+iy]} F\right| \le \int_0^{|y|} |F(R \pm it)|dt \le e^{-\pi a R^2}\int_0^{|y|} e^{\pi a t^2}dt.$$

By Cauchy's theorem for rectangles

$$\int_{[-r+iy,r+iy]} F = \int_{[r,r+iy]} F - \int_{[-r,-r+iy]} F + \int_{[-r,r]} F.$$

By (3) the first two integrals on the right converge to 0 as $r \to \infty$. This gives (1). With (1) thus secured, look at $R > r > 0$. Since

$$|e^{-\pi a(x+iy)^2}| = e^{\pi a y^2}e^{-\pi a x^2},$$

we have

$$e^{-\pi a y^2}\left|\int_{-R}^{R} e^{-\pi a(x+iy)^2}dx - \int_{-r}^{r} e^{-\pi a(x+iy)^2}dx\right|$$

$$\le \int_{-R}^{-r} e^{-\pi a x^2}dx + \int_{r}^{R} e^{-\pi a x^2}dx$$

$$= \int_{-R}^{R} e^{-\pi a x^2}dx - \int_{-r}^{r} e^{-\pi a x^2}dx.$$

Let $R \to \infty$ and use (1) to get

$$e^{-\pi a y^2}\left| I(a) - \int_{-r}^{r} e^{-\pi a(x+iy)^2}dx\right| \le I(a) - \int_{-r}^{r} e^{-\pi a x^2}dx.$$

This proves (2), since by (1) the right side goes to 0 with $1/r$.

(ii) *Show that*

$$(4) \qquad \lim_{r\to\infty} \int_{[-r+i,r+i]} f = -I(a) \sum_{n=0}^{\infty} e^{-\pi n^2/a}.$$

Hints: If $z \in i + \mathbb{R}$, then $|e^{2\pi i z}| = e^{-2\pi} < 1$ and so the series $1/(e^{2\pi i z} - 1) = -\sum_{n=0}^{\infty} (e^{2\pi i z})^n$ converges uniformly for such z. Therefore

$$f(z) = -\sum_{n=0}^{\infty} e^{-\pi a z^2}(e^{2\pi i z})^n = -\sum_{n=0}^{\infty} e^{-\pi n^2/a} e^{-\pi a (z - in/a)^2}$$

with convergence uniform for z in any compact subset of $i + \mathbb{R}$. Consequently

$$\int_{[-r+i,r+i]} f = -\sum_{n=0}^{\infty} e^{-\pi n^2/a} \int_{[-r+i,r+i]} e^{-\pi a(z - in/a)^2} dz$$

$$(5) \qquad = -e^{\pi a} \sum_{n=0}^{\infty} \left[e^{-\pi a(1-n/a)^2} \int_{-r}^{r} e^{-\pi a(x+i-in/a)^2} dx \right] e^{-2\pi n}.$$

The geometric series $\sum (e^{-2\pi})^n$ is convergent and the square bracketed term above converges to $I(a)e^{-\pi a(1-n/a)^2}$ as $r \to \infty$, uniformly in n, according to (2). It follows easily from this and (5) that

$$\lim_{r\to\infty} \int_{[-r+i,r+i]} f = -e^{\pi a} \sum_{n=0}^{\infty} [I(a)e^{-\pi a(1-n/a)^2}] e^{-2\pi n},$$

which is (4).

(iii) *Show that*

$$\lim_{r\to\infty} \int_{[-r-i,r-i]} f = I(a) \sum_{n=1}^{\infty} e^{-\pi n^2/a}.$$

Hints: If $z \in -i + \mathbb{R}$, then $|e^{-2\pi i z}| = e^{-2\pi} < 1$ and so the series

$$\frac{1}{e^{2\pi i z} - 1} = \frac{-e^{-2\pi i z}}{e^{-2\pi i z} - 1} = \sum_{n=0}^{\infty} (e^{-2\pi i z})^{n+1} = \sum_{n=1}^{\infty} e^{-2\pi i n z}$$

converges uniformly for such z. The rest is the same as (ii).

(iv) *Prove that*

$$\sum_{n=-\infty}^{\infty} e^{-\pi a n^2} = I(a) \sum_{n=-\infty}^{\infty} e^{-\pi n^2/a}.$$

Hints: Integrate f over the rectangle $[-r - i, r - i, r + i, -r + i, -r - i]$, r of the form $N + \frac{1}{2}$, N an integer. Apply the Residue Theorem, parts (ii) and (iii) and show as in (i) that

$$(6) \qquad \lim_{|N|\to\infty} \int_{[N+1/2-i,N+1/2+i]} f = 0.$$

To prove (6), note that for $z \in N + \frac{1}{2} + i\mathbb{R}$

$$1 - e^{2\pi i z} = 1 + e^{-2\pi \operatorname{Im} z} \geq 1$$

and so $|f(z)| \leq |F(z)|$ for such z. Therefore (6) follows from the second inequality in (3).

(v) *Derive anew the identity* 11.32(i) *and show that for all* $z \in \mathbb{C}$ *with* $\operatorname{Re} z > 0$

$$(*) \qquad \sum_{n=-\infty}^{\infty} e^{-\pi z n^2} = \frac{1}{\sqrt{z}} \sum_{n=-\infty}^{\infty} e^{-\pi n^2/z},$$

where $\sqrt{z}$ *denotes the (holomorphic) square root in the right half-plane with positive real part.*

Hints: Take $a = 1$ in (iv) and get $I(1) = 1$. Then make a change of variable in the integral defining $I(a)$ to get $I(a) = (1/\sqrt{a})I(1) = 1/\sqrt{a}$. Taking $a = 1/\pi$ gives 11.32(i). Next prove that both sides of (*) converge locally uniformly in the open right half-plane and so define holomorphic functions there. By (iv) and the fact $I(a) = 1/\sqrt{a}$, the two functions agree at all positive real numbers.

Exercise 11.34 *Prove that* $\lim_{r\to\infty} \int_0^r e^{-x^2} \cos(x^2)dx$ *and* $\lim_{r\to\infty} \int_0^r e^{-x^2} \sin(x^2)dx$ *exist and equal*

$$\frac{\sqrt{\pi}\sqrt{\sqrt{2}+1}}{4} \quad \textit{and} \quad \frac{\sqrt{\pi}\sqrt{\sqrt{2}-1}}{4}$$

respectively.

Hint: Proceed as in the proof of 11.32(ii), using however the loop

$$\Gamma_r(t) = \begin{cases} re^{\pi i t/8} & 0 \leq t \leq 1 \\ (2-t)re^{\pi i/8} & 1 \leq t \leq 2 \\ (t-2)r & 2 \leq t \leq 3. \end{cases}$$

Exercise 11.35 $\left|\int_{-r}^{r} (\sin x/x)^3 dx - 3\pi/4\right| < 1/r^2$ *for all* $r > 0$.

Hint: $\sin^3 x = \operatorname{Im}[\frac{1}{4}(1 - e^{3ix}) - \frac{3}{4}(1 - e^{ix})]$.

Exercise 11.36 *Prove that* $\sum_{n=0}^{\infty} (-1)^n/(2n+1)^3 = \pi^3/32$.

Hints: Let

$$f(z) = \frac{1}{(2z+1)^3}, \quad g(z) = \frac{\pi}{\sin(\pi z)}, \quad \gamma_N = C(0, N + \tfrac{1}{2})$$

for each positive integer N. Use the Residue Theorem to write

$$\frac{1}{2\pi i}\int_{\gamma_N} fg = \sum_{n=-N}^{N} (-1)^n f(n) + \operatorname{Res}(fg, -\tfrac{1}{2}),$$

then show that $\lim_{N\to\infty}\int_{\gamma_N} fg = 0$. To this end, notice that $\sin(\pi z)$ is bounded away from 0 in $\mathbb{C}\setminus\bigcup_{n\in\mathbb{Z}} D(n, \frac{1}{2})$. For consideration of the cases $\operatorname{Im} z \geq 0$, $\operatorname{Im} z \leq 0$ separately, leads at once to the inequalities

$$\begin{aligned} 2|\sin z| &= |e^{iz} - e^{-iz}| \geq e^{|\operatorname{Im} z|} - e^{-|\operatorname{Im} z|} \\ &> \tfrac{1}{2}e^{|\operatorname{Im} z|} \geq \tfrac{1}{2}e \quad \text{whenever } |\operatorname{Im} z| \geq 1. \end{aligned}$$

On the other hand, $|\sin(\pi z)|$ has no zeros in the compact set

$$[0, 1] \times [-1, 1]\setminus[D(0, \tfrac{1}{2}) \cup D(1, \tfrac{1}{2})]$$

and so is bounded away from 0 there. Consequently there is a positive number M such that

$$|\sin(\pi z)| > M \quad \forall z \in [0, 1] \times \mathbb{R}\setminus[D(0, \tfrac{1}{2}) \cup D(1, \tfrac{1}{2})].$$

By periodicity then this inequality holds throughout $\mathbb{C}\setminus\bigcup_{n\in\mathbb{Z}} D(n, \frac{1}{2})$.

Exercise 11.37 *Let f be entire and satisfy for some $M > 0$*

(*) $\quad |f(z)| \leq Me^{|\operatorname{Im} z|} \quad \forall z \in \mathbb{C}.$

Define a holomorphic function F by

$$F(z) = \frac{f(z)}{z^2 \cos z}$$

for all z not equal to 0 or an odd multiple of $\pi/2$. For each positive integer n define $V_n = n\pi(1 + i)$ and

$$\Gamma_n = [V_n, -\bar{V}_n, -V_n, \bar{V}_n, V_n].$$

(i) *Show that*

$$\frac{1}{2\pi i}\int_{\Gamma_n} F = f'(0) - \sum_{k=-n}^{n-1} (-1)^k \frac{f((k + \frac{1}{2})\pi)}{(k + \frac{1}{2})^2\pi^2}.$$

(ii) *Show that $|\cos z| \geq \frac{1}{4}e^{|\operatorname{Im} z|}$ for all $z \in \Gamma_n$.*

Hint: If $\operatorname{Re} z = n\pi$, then $2|\cos z| = e^{-\operatorname{Im} z} + e^{\operatorname{Im} z}$. If $\operatorname{Im} z = n\pi$, then $2|\cos z| \geq e^{|n\pi|} - e^{-|n\pi|}$.

(iii) *Show that for all $z \in \Gamma_n$*

$$|F(z)| \leq \frac{4M}{|z|^2}.$$

(iv) *From* (i) *and* (iii) *deduce that*

$$f'(0) = \sum_{k=-\infty}^{\infty} (-1)^k \frac{f((k + \frac{1}{2})\pi)}{(k + \frac{1}{2})^2\pi^2}.$$

(v) *Apply* (iv) *to the function* $f = \sin$ *to deduce that*

$$\sum_{k=-\infty}^{\infty} \frac{1}{(2k+1)^2} = \frac{\pi^2}{4}. \quad \text{(Cf. 3.38.)}$$

(vi) *From* (iv) *and* (v) *deduce that in general for any* f *which satisfies* (*) *we have*

$|f'(0)| \leq \sup|f(\mathbb{R})|$.

(vii) (Bernstein) *Let* g *be an entire function and for some* $c > 0$, $M > 0$ *satisfy*

(*) $|g(z)| \leq Me^{c|\mathrm{Im}\, z|} \quad \forall z \in \mathbb{C}$.

Show that $|g'(x)| \leq c \sup|g(\mathbb{R})|$ *for all* $x \in \mathbb{R}$.

Hint: Given $x \in \mathbb{R}$, the function $f(z) = g(x + z/c)$ satisfies the hypotheses of (i)–(vi) above. Cite the conclusion (vi).

Remark: The same conclusion follows if it is only hypothesized that $|g(z)|e^{-c|z|}$ is bounded in $\mathbb{C}$ but g is bounded on $\mathbb{R}$. Using this conclusion together with 5.37, one sees that g' satisfies all the hypotheses which g does. Inductively one concludes that $g^{(n)}$ is bounded on $\mathbb{R}$ by $c^n \sup|g(\mathbb{R})|$. It follows easily that $|g(z)|e^{-c|\mathrm{Im}\, z|}$ is bounded in $\mathbb{C}$ after all (by its bound on $\mathbb{R}$, in fact). See Chapter XIV for the details.

Exercise 11.38 *Let*

$$f(z) = \frac{\pi^2}{\sin^2 \pi z}, \qquad g(z) = \sum_{n=-\infty}^{\infty} \frac{1}{(z-n)^2} \quad \textit{for } z \in \mathbb{C}\backslash\mathbb{Z}.$$

(i) *Show that the series converges locally uniformly in* $\mathbb{C}\backslash\mathbb{Z}$. *In fact this series converges uniformly in* $\mathbb{C}\backslash\bigcup_{n=-\infty}^{\infty} D(n, r)$ *for each* $r > 0$.

(ii) *Show that the principal part of* f *at* n *is* $(z-n)^{-2}$ *for each* $n \in \mathbb{Z}$.

Hint: It suffices to consider $n = 0$. By 3.27

$$\sin \pi z = \pi z\left[1 + \sum_{n=1}^{\infty} \frac{(-1)^n \pi^{2n} z^{2n}}{(2n+1)!}\right] \stackrel{\text{def.}}{=} \pi z[1 - z^2\psi(z)].$$

For all sufficiently small z (i.e., small enough that $|z^2\psi(z)| < 1$) we then have

$$\frac{\pi z}{\sin \pi z} = \frac{1}{1 - z^2\psi(z)} = \sum_{n=0}^{\infty} [z^2\psi(z)]^n = 1 + \sum_{n=1}^{\infty} z^{2n}\psi^n(z),$$

$$\frac{\pi}{\sin \pi z} = \frac{1}{z} + z\sum_{n=1}^{\infty} z^{2(n-1)}\psi^n(z) \stackrel{\text{def.}}{=} \frac{1}{z} + z\phi(z),$$

$$\frac{\pi^2}{\sin^2 \pi z} = \frac{1}{z^2} + 2\phi(z) + z^2\phi^2(z).$$

(iii) *Conclude from* (i) *and* (ii) *that the function* $h = f - g$, *which is holomorphic in* $\mathbb{C}\backslash\mathbb{Z}$, *actually extends to be entire.*

Hint: For each n there are functions f_n, g_n holomorphic near n such that $f(z) = (z - n)^{-2} + f_n(z)$, $g(z) = (z - n)^{-2} + g_n(z)$. Thus $h = f_n - g_n$ near n.

(iv) *Let h be extended as in* (iii). *Show that*

$$h\left(\frac{z}{2}\right) + h\left(\frac{z+1}{2}\right) = 4h(z) \quad \forall z \in \mathbb{C}.$$

Hint: Calculate. At one point the "double angle" formula $\sin(\pi z/2)\cos(\pi z/2) = \frac{1}{2}\sin(\pi z)$ will be needed.

(v) *Deduce from* (iv) *that $h = 0$, that is,*

$$\frac{\pi^2}{\sin^2 \pi z} = \sum_{n=-\infty}^{\infty} \frac{1}{(z-n)^2} \quad \forall z \in \mathbb{C}\backslash\mathbb{Z}.$$

Hint: The extended h is continuous on $\bar{D}(0, 1)$, so $M = \sup|h(D(0, 1))| < \infty$. By (iv) and convexity of $D(0, 1)$, we have

$$|h(z)| \leq \frac{1}{4}\left|h\left(\frac{z}{2}\right)\right| + \frac{1}{4}\left|h\left(\frac{z+1}{2}\right)\right| \leq \frac{1}{2}M \quad \forall z \in D(0, 1).$$

Thus $M \leq \frac{1}{2}M$, so $M = 0$. By 5.62, $h = 0$ throughout $\mathbb{C}$.

(vi) *Deduce from* (v) *that*

$$\pi \cot \pi z = \frac{1}{z} + \sum_{k=1}^{\infty} \frac{2z}{z^2 - k^2} = \lim_{N\to\infty} \sum_{n=-N}^{N} \frac{1}{z-n} \quad \forall z \in \mathbb{C}\backslash\mathbb{Z}.$$

Hints: For $z \in \mathbb{C}\backslash\mathbb{Z}$ define

(1) $\quad F(z) = \pi \cot \pi z - \dfrac{1}{z},$

(2) $\quad F_N(z) = \displaystyle\sum_{\substack{n=-N\\ n\neq 0}}^{N} \frac{1}{z-n}, \quad N = 1, 2, 3, \ldots.$

Combine the terms $n = k$ and $n = -k$ in the sum for F_N and get

(3) $\quad F_N(z) = \displaystyle\sum_{k=1}^{N} \frac{2z}{z^2 - k^2}.$

We have

$$F'(z) = \frac{1}{z^2} - \frac{\pi^2}{\sin^2 \pi z} \quad \text{and by (2)} \quad F_N'(z) = \sum_{\substack{n=-N\\ n\neq 0}}^{N} \frac{-1}{(z-n)^2}.$$

Therefore by (v)

(4) $\quad F_N' \to F'$ locally uniformly in $\mathbb{C}\backslash\mathbb{Z}$.

It is clear from (3) that $\{F_N\}$ converges locally uniformly in $\mathbb{C}\backslash\mathbb{Z}$, say to G. Then by 5.44(ii), $F_N' \to G'$ and we have from (4) that $F' = G'$ in $\mathbb{C}\backslash\mathbb{Z}$. As this set is connected (1.24), we infer from 2.10(iv) that $F - G$ is constant there. By (1) and (3), F and each F_N is an odd function. The constant function $F - G = F - \lim F_N$ is therefore odd, hence equal to 0.

(vii) *Deduce from* (vi) *that*

$$\frac{\pi}{\sin \pi z} = \frac{1}{z} + \sum_{k=1}^{\infty} \frac{(-1)^k 2z}{z^2 - k^2} = \lim_{N \to \infty} \sum_{n=-N}^{N} \frac{(-1)^n}{z - n} \quad \forall z \in \mathbb{C} \setminus \mathbb{Z}.$$

Hints: The series evidently converges; call it $H(z)$. For $N = 2K + 1$ separate even and odd indices into groups and get

$$\frac{1}{z} + \sum_{k=1}^{N} \frac{(-1)^k 2z}{z^2 - k^2} = \sum_{n=-N}^{N} \frac{(-1)^n}{z - n} = \sum_{n=-K}^{K} \frac{1}{z - 2n} - \sum_{n=-K-1}^{K} \frac{1}{z - 1 - 2n}.$$

From (vi) then $H(z) = \frac{1}{2}\pi \cot(\pi z/2) - \frac{1}{2}\pi \cot(\pi(z - 1)/2) = \pi/\sin(\pi z)$.

(viii) *By taking* $z = i$ *in* (vi) *compute* $\sum_{k=-\infty}^{\infty} (k^2 + 1)^{-1}$.

§ 5 Specifying Principal Parts—Mittag-Leffler's Theorem

Here we prove an analog for principal parts of Weierstrass' theorem 7.32 on prescribing the zeros of a holomorphic function and give some of the more prominent applications of this famous theorem. The relation to Weierstrass is more than mere analogy, as 11.40 illustrates. Moreover, a grand synthesis theorem allowing both poles and zeros to be specified can easily be fabricated from the two results (11.41).

Theorem 11.39 (Mittag-Leffler [1884a]) *Suppose that U is an open subset of $\mathbb{C}$, A a subset of U with no limit point in U and for each $a \in A$, f_a is an entire function which is zero at* 0. *Then there exists a holomorphic function f in $U \setminus A$ whose principal part at each $a \in A$ is the function $g_a(z) = f_a(1/(z - a))$.*

Proof: Let $K_1, K_2, \ldots$ be compact subsets of U as per 1.31. Thus

(1) $\quad K_n \subset \mathring{K}_{n+1}, \qquad U = \bigcup_{n=1}^{\infty} K_n.$

(2) $\quad$ Each bounded component of $\mathbb{C} \setminus K_n$ contains a component of $\mathbb{C} \setminus U$.

Put

(3) $\quad A_1 = A \cap K_1, \qquad A_n = A \cap (\mathring{K}_n \setminus K_{n-1}), \quad n = 2, 3, \ldots.$

Since A has no limit point in U, it meets each compact subset of U in only a finite set. Thus in particular each A_n is a finite set. Define

(4) $\quad F_n(z) = \sum_{a \in A_n} f_a\left(\frac{1}{z - a}\right), \quad z \in \mathbb{C} \setminus A_n.$

Thus

(5) $\quad F_n$ is holomorphic in $\mathbb{C} \setminus A_n$.

Since $\mathbb{C}\backslash A_n$ is an open neighborhood of $\mathbb{C}\backslash(\mathring{K}_n\backslash K_{n-1}) \supset K_{n-1}$, it follows from (2) and Rational Runge I (8.11, with $\mathbb{C}\backslash U$ in the role of the set A there) that there are rational functions R_n such that

(6) R_n is holomorphic in U,

and

(7) $|F_n(z) - R_n(z)| < 2^{-n} \quad \forall z \in K_{n-1}$.

We will show that the following series converges throughout $U\backslash A$ and defines a function f with the desired properties:

$$(8) \qquad f(z) = F_1(z) + \sum_{n=2}^{\infty} [F_n(z) - R_n(z)].$$

Fix $N \geq 2$. Then by (5) and (6), $F_n - R_n$ is holomorphic in $\mathring{K}_N$ for every $n \geq N + 1$ and on K_N we have $|F_n - R_n| < 2^{-n}$ for such n by (7). Therefore the series $\sum_{n=N+1}^{\infty} [F_n - R_n]$ converges uniformly on K_N, to a function f_N which is holomorphic in $\mathring{K}_N$. We have

$$f(z) = F_1(z) + \sum_{n=2}^{N} [F_n(z) - R_n(z)] + f_N(z), \quad z \in \mathring{K}_N\backslash A$$

$$(9) \qquad f(z) - F_1(z) - \cdots - F_N(z) = f_N(z) - R_2(z) - \cdots - R_N(z), \quad z \in \mathring{K}_N\backslash A.$$

Since $f_N - R_2 - \cdots - R_N$ is holomorphic in $\mathring{K}_N$, it follows from (9) that $f - F_1 - \cdots - F_N$ is also. Therefore, recalling (5), f is holomorphic in $\mathring{K}_N\backslash A_1 \cup \cdots \cup A_N = \mathring{K}_N\backslash A$ and f has precisely the principal part $f_a(1/(z - a))$ at each $a \in \mathring{K}_N$ (11.3). Since by (1) $U = \bigcup_{N=2}^{\infty} \mathring{K}_N$, the proof is complete.

Exercise 11.40 *Use Mittag-Leffler's theorem* 11.39 *and the Residue Theorem* 11.29 *to deduce the simply-connected version of the theorem of Weierstrass* 7.32 *on prescribing the zeros of a holomorphic function.*

Hints: Let U, A, $n(a)$ be as in 7.32 but with U simply-connected. It suffices (4.66) to treat the case U connected. In that case $U\backslash A$ is also connected and open by 1.24 and so by 1.28, if we fix a point $z_0 \in U\backslash A$, for each $z \in U\backslash A$ there is a polygon γ_z in $U\backslash A$ joining z_0 to z. 11.39 provides an $h \in H(U\backslash A)$ with principal part $n(a)/(z - a)$ at each $a \in A$. Form $H(z) = \frac{1}{2\pi i}\int_{\gamma_z} h$ and

(1) $F = e^{2\pi i H}$.

We will show that F is holomorphic in $U\backslash A$ with a removable singularity at each $a \in A$ and that each a is a zero of the extension of order $n(a)$.

Let $z_1 \in U\backslash A$ be given; choose $r > 0$ so that $D(z_1, r) \subset U\backslash A$ and consider only $z \in D(z_1, r)$. Then

$$(2) \qquad H(z) - H(z_1) - \frac{1}{2\pi i}\int_{[z_1, z]} h \quad \text{is an integer}$$

because by the Residue Theorem this sum, which is the integral of $\frac{1}{2\pi i}h$ around a piecewise smooth loop in $U\backslash A$, is a finite sum of residues of h times index values; the latter are integers and the former are the numbers $n(a)$, integers too. Therefore we have from (1) and (2)

$$(3)\qquad F(z) = e^{\phi(z)+2\pi i H(z_1)} \quad \forall z \in D(z_1, r),$$

where ϕ is the function

$$(4)\qquad \phi(z) = \int_{[z_1,z]} h, \quad z \in D(z_1, r).$$

This function is holomorphic and so therefore is its exponential. Thus F is holomorphic in $D(z_1, r)$, hence throughout $U\backslash A$.

Now to investigate F near an $a \in A$, choose $R > 0$ so small that $D(a, R) \subset U\backslash[A\backslash\{a\}]$. By definition of h there exists a function h_a holomorphic in $D(a, R)$ such that

$$h(z) = \frac{n(a)}{z - a} + h_a(z) \quad \forall z \in D(a, R)\backslash\{a\}.$$

For the z_1 above take any point on $C(a, R/2)$ and for the r above take $R/2$. Equation (4) now reads

$$(4)'\qquad \phi(z) = \int_{[z_1,z]} \frac{n(a)}{\xi - a}\,d\xi + \int_{[z_1,z]} h_a(\xi)d\xi \quad \forall z \in D(z_1, r).$$

We define

$$H_a(z) = \int_{[z_1,z]} h_a \qquad z \in D(a, R),$$

$$g_a(z) = (z - a)^{n(a)} \qquad z \in D(z_1, r),$$

$$G_a(z) = \int_{[z_1,z]} \frac{g_a'}{g_a} = \int_{[z_1,z]} \frac{n(a)}{\xi - a}\,d\xi, \quad z \in D(z_1, r).$$

These functions are holomorphic and in terms of them (4)′ reads

$$\phi(z) = G_a(z) + H_a(z), \quad z \in D(z_1, r),$$

which inserted into (3) gives

$$(5)\qquad F(z) = e^{G_a(z)}e^{H_a(z)+2\pi i H(z_1)}, \quad z \in D(z_1, r).$$

Since $G_a' = g_a'/g_a$, we have $[g_a e^{-G_a}]' = 0$ and so (as in the proof of 5.34) there is a constant c_a such that $g_a = e^{G_a + c_a}$ in $D(z_1, r)$. Therefore (5) yields finally

$$\begin{aligned} F(z) &= g_a(z)e^{H_a(z)+2\pi i H(z_1)-c_a} \\ &= (z - a)^{n(a)}e^{H_a(z)+2\pi i H(z_1)-c_a}, \quad z \in D(z_1, r). \end{aligned}$$

Both sides of this equation being holomorphic functions in $D(a, R)\backslash\{a\}$, the uniqueness theorem insures that this equality persists throughout $D(a, R)\backslash\{a\}$, thus proving the assertions about F at a.

Theorem 11.41 (Mittag-Leffler's "Anschmiegungssatz") *Let $U \subset \mathbb{C}$ be open, $A \subset U$ have no limit points in U. For each $a \in A$ let g_a be an entire function with $g_a(0) = 0$, P_a a polynomial of degree n_a. Then there exists $F \in H(U\backslash A)$ such that at each $a \in A$, $F(z) - g_a(1/(z-a)) - P_a(z-a)$ has a removable singularity and a zero of order at least $n_a + 1$.*

Proof: 7.32 provides an $f \in H(U)$ which at each $a \in A$ has a zero of order $n_a + 1$ and no others. In particular, f vanishes identically in no component of U. Let then h_a be the principal part at a of the function

$$\frac{1}{f(z)}\left[g_a\left(\frac{1}{z-a}\right) + P_a(z-a)\right].$$

Use 11.39 to come up with a function $h \in H(U\backslash A)$ which for each $a \in A$ has principal part h_a at a. Then set $F = fh$. By definition, about each $a \in A$ there is an open disk D_a in U (and disjoint from $A\backslash\{a\}$) and $f_a, F_a, H_a \in H(D_a)$ such that

$$f(z) = (z-a)^{n_a+1}f_a(z), \frac{1}{f(z)}\left[g_a\left(\frac{1}{z-a}\right) + P_a(z-a)\right] = h_a(z) + F_a(z)$$

and $h(z) = h_a(z) + H_a(z)$ for all $z \in D_a\backslash\{a\}$. For these z then

$$F(z) = g_a\left(\frac{1}{z-a}\right) + P_a(z-a) + (z-a)^{n_a+1}[f_aH_a - f_aF_a](z).$$

Since the square bracketed term is holomorphic in the neighborhood D_a of a, it is therefore clear that F has the desired behavior near a.

Corollary 11.42 *If U is an open subset of $\mathbb{C}$ and $f_1, \ldots, f_n$ are holomorphic in U and have no common zeros, then there exist $g_1, \ldots, g_n$ holomorphic in U such that $f_1g_1 + \cdots + f_ng_n \equiv 1$.*

Proof: Passing to components we may assume without loss of generality that U is connected. We go by induction on n and so suppose that no f_j is identically 0. The case $n = 1$ being trivial, we look at $n = 2$. If a is a zero of f_2 of order $n(a)$, then $f_1(a) \neq 0$, and so $1/f_1$ is holomorphic near a. Now $f_2^{-1}(0)$ has no limit point in U, since U is connected and f_2 is not identically 0. Therefore 11.41 provides a function g_1 holomorphic in U such that $g_1 - 1/f_1$ has a zero at a of order not less than $n(a)$, for each zero a of f_2. If

$$g_1(z) - \frac{1}{f_1(z)} = (z-a)^{n(a)}h_a(z), \quad z \text{ near } a,$$

then

$$1 - f_1(z)g_1(z) = -(z - a)^{n(a)}h_a(z)f_1(z), \quad z \text{ near } a,$$

so $(1 - f_1g_1)/f_2$ has a removable singularity at a. Thus $(1 - f_1g_1)/f_2$ extends to a holomorphic function g_2 in U. We have evidently $1 - f_1g_1 \equiv f_2g_2$.

Now suppose the result true for $n = N$ and consider $n = N + 1$. Let $A = f_1^{-1}(0) \cap \cdots \cap f_N^{-1}(0)$ and for each $a \in A$ let $m(a)$ be the minimum of the orders of a as a zero of $f_1, \ldots, f_N$. Weierstrass provides an $h \in H(U)$ with a zero of order $m(a)$ at each $a \in A$ and no others. Evidently then $F_j = f_j/h$ is holomorphic in U for each $1 \leq j \leq N$ and $F_1, \ldots, F_N$ have no common zero. By the induction hypothesis there exist $G_1, \ldots, G_N$ in $H(U)$ such that

$$F_1G_1 + \cdots + F_NG_N = 1$$

$$(*) \qquad f_1G_1 + \cdots + f_NG_N = h.$$

We have $h^{-1}(0) \cap f_{N+1}^{-1}(0) = A \cap f_{N+1}^{-1}(0) = f_1^{-1}(0) \cap \cdots \cap f_N^{-1}(0) \cap f_{N+1}^{-1}(0) = \varnothing$ and so by the case $n = 2$ considered above there exist $g, g_{N+1} \in H(U)$ such that

$$hg + f_{N+1}g_{N+1} \equiv 1,$$

that is, recalling (*) and setting $g_j = gG_j$ $(j = 1, 2, \ldots, N)$,

$$(f_1g_1 + \cdots + f_Ng_N) + f_{N+1}g_{N+1} \equiv 1.$$

Exercise 11.43 *Deduce from* 11.42 *that every finitely generated ideal in* $H(U)$ *is principal.*

Hints: The problem reduces at once to connected U. Given $F_1, \ldots, F_n \in H(U)$, none identically 0, the set $A = F_1^{-1}(0) \cap \cdots \cap F_n^{-1}(0)$ has no limit point in U (5.62) and 7.32 provides an $F \in H(U)$ such that at each $a \in A$, F has a zero of order $\min\{n(a, F_j): 1 \leq j \leq n\}$ and no other zeros in U. Then the functions $f_j = F_j/F \in H(U)$ and 11.42 applies, providing $g_j \in H(U)$ such that $\sum_{j=1}^n f_jg_j = 1$. Then $\sum_{j=1}^n F_jg_j = F$, so F belongs to the ideal generated by $F_1, \ldots, F_n$.

Exercise 11.44 *Let U be an open subset of $\mathbb{C}$, I an ideal in $H(U)$ which is closed with respect to uniform-on-compacta convergence.*

(i) *Show that if the functions in I have no common zero, then $I = H(U)$.*

Hints: There is no loss of generality in taking U connected. (Why?) We have $I \neq \{0\}$, so there exists $0 \not\equiv f \in I$. Since U is connected, the zero set A of f is therefore isolated (5.62), hence countable, say $A = \{a_1, a_2, \ldots\}$. Let n_j be the multiplicity of a_j. For each a_j there is, by the hypothesis of (i), an $f_j \in I$ such that $f_j(a_j) \neq 0$. Let $\{K_n\}$ be a compact exhaustion of U as in 1.31. Multiplying f_j by an appropriate positive constant, we may assume that

$$(1) \qquad \sup\left\{\left|\frac{f(z)f_j(z)}{(z - a_j)^{n_j}}\right| : z \in K_j\right\} \leq 2^{-j}, \quad j = 1, 2, \ldots$$

Then the series

$$(2) \qquad g(z) = \sum_{j=1}^{\infty} \frac{f(z)f_j(z)}{(z - a_j)^{n_j}}$$

converges for each $z \in U$ and uniformly on each compact subset of U, so $g \in H(U)$. By definition of n_j there is a holomorphic function g_j in U such that $f(z) = (z - a_j)^{n_j} g_j(z)$ and $g_j(a_j) \neq 0$. Thus the jth summand in g is the product $g_j f_j$. This function belongs to the ideal I, since f_j does; therefore $g \in I$, as the latter is closed. We see also that for each j, $g(a_j) = g_j(a_j) f_j(a_j) \neq 0$. [Recall that $f(a_j) = 0$, so if $k \neq j$, the kth summand in g is 0 at a_j.] Since the only zeros of f are the a_j, it follows that f and g have no common zeros and consequently there exist by 11.42 $h_1, h_2 \in H(U)$ such that $1 = fh_1 + gh_2 \in I$.

(ii) *Show that I is principal, that is, $I = h \cdot H(U)$, for some $h \in H(U)$.*

Hints: The result is trivial if $I = \{0\}$, so assume that $I \neq \{0\}$. Again assume without loss of generality that U is connected. Then the set A of common zeros of the functions in I is an isolated set in U. For each $a \in A$ let $n(a)$ be the smallest order of a as a zero of any function in I and cite 7.32 to produce an $h \in H(U)$ which has a zero of order $n(a)$ at each $a \in A$ and no other zeros. Then $f/h \subset H(U)$ for each $f \in I$ and the set of these quotients is an ideal J which has no common zeros. It is also easy to see that J is closed with respect to uniform-on-compacta convergence. Therefore by part (i), $J = H(U)$.

(iii) *Prove the converse of* (ii): *every principal ideal in $H(U)$ is closed with respect to uniform-on-compacta convergence.*

Hint: $h \cdot H(U)$ is the set of all $f \in H(U)$ which vanish at each zero of h to at least the order of that zero in h.

§ 6 Meromorphic Functions

Definition 11.45 If U is an open subset of $\mathbb{C}$, A a subset with no limit points in U and f is holomorphic in $U \backslash A$ with each point of A a pole or a removable singularity, we say f is *meromorphic* in U. The rationale for this terminology (compare its Greek origins) is

Theorem 11.46 *A function f is meromorphic in an open set U if and only if $f = g/h$ for some holomorphic functions g, h in U, where h does not vanish identically in any component of U.*

Proof: ($\Leftarrow$) The set $A = h^{-1}(0)$ has no limit points in U since h is not identically zero in any component of U. Then $f = g/h$ is holomorphic in $U \backslash A$ and every point of A is clearly either a removable singularity or a pole of f. Indeed if

$$h(z) = (z - a)^n H(z)$$

where H is holomorphic in U and $H(a) \neq 0$, then near a

$$f(z) = \frac{1}{(z-a)^n}\frac{g(z)}{H(z)} = \sum_{k=0}^{\infty} c_k(z-a)^{k-n}$$

where $\sum_{k=0}^{\infty} c_k(z-a)^k$ is the power series for g/H near a.

($\Rightarrow$) If f is meromorphic in U and A is its set of (isolated) singularities, we may suppose each $a \in A$ is in fact a pole and we let $n(a)$ denote its order. Weierstrass' theorem (7.32) provides an $h \in H(U)$ with a zero of order $n(a)$ at each $a \in A$ and no others. Consider $g = fh$ in $U\backslash A$. Evidently each point of A is a removable singularity for g, so g extends to be holomorphic in U. Of course $f = g/h$ in $U\backslash A$.

Exercise 11.47 *If f is meromorphic in the region Ω and $f^{-1}(0)$ has a limit point in Ω, then $f \equiv 0$.*

Hint: Let Ω_0 be the set of limit points in Ω of $f^{-1}(0)$. We will show that $z_0 \in \Omega_0$ and $\bar{D}(z_0, r) \subset \Omega$ imply $D(z_0, r) \subset \Omega_0$. The singularities of f are isolated in Ω so only finitely many, say $b_1, \ldots, b_n$ (counted according to multiplicity), occur in the compact subset $\bar{D}(z_0, r)$. The function $F(z) = f(z)\prod_{k=1}^{n}(z - b_k)$ is then holomorphic in $D(z_0, r)$ and z_0, being a limit point of zeros of f, is a limit point of zeros of F. By 5.62, $F = 0$ in $D(z_0, r)$, whence $f = 0$ in $D(z_0, r)\backslash\{b_1, \ldots, b_n\}$. Thus every point of $D(z_0, r)$ is a limit point of zeros of f, i.e., $D(z_0, r) \subset \Omega_0$. By hypothesis $\Omega_0 \neq \varnothing$, so by the Basic Connectedness Lemma, $\Omega_0 = \Omega$. It follows that f has no poles in Ω, for by 11.7(ii) a pole is surely not a limit point of zeros of f. Thus f is holomorphic in Ω and the result follows from 5.62.

Note that alternative short derivations from 5.62 can be based on 1.24 or on 11.46.

Exercise 11.48 *Does the conclusion of* 11.14 *hold for all f which are meromorphic in $\mathbb{C}$ and satisfy $|f(z)| = 1$ whenever $|z| = 1$?*

Exercise 11.49 *Does the conclusion of* 11.15 *hold for all f which are meromorphic in $\mathbb{C}$ and satisfy $f(z) > 0$ whenever $|z| = 1$? Similar question for* 11.18.

Just as the Residue Theorem is the final generalization of Cauchy's Theorem, there is a meromorphic function generalization of the various zero-counting integral theorems (e.g., the Argument Principle) also due to Cauchy:

Theorem 11.50 *Let U be an open subset of $\mathbb{C}$, F a function meromorphic in U and not identically zero in any component of U. Let P be the pole set and Z the zero set of F in U, $n(z, F)$ and $n(p, F)$ the multiplicities of z and p as zeros and poles of F respectively. Let $\gamma_1, \ldots, \gamma_k$ be piecewise smooth loops in $U\backslash(P \cup Z)$ such that $\sum_{j=1}^{k} \operatorname{Ind}_{\gamma_j}(w) = 0$ for all $w \in \mathbb{C}\backslash U$. Then*

$$\sum_{j=1}^{k} \frac{1}{2\pi i}\int_{\gamma_j} \frac{F'}{F} = \sum_{j=1}^{k}\left[\sum_{z\in Z} n(z, F)\operatorname{Ind}_{\gamma_j}(z) - \sum_{p\in P} n(p, F)\operatorname{Ind}_{\gamma_j}(p)\right],$$

both sums within the square brackets having only a finite number of non-zero terms.

Proof: We note that $P \cap Z = \varnothing$ and the set $A = P \cup Z$ has no limit point in U. Let $f = F'/F$ in $U\backslash A$. This function is holomorphic and we will show that each point of A is a pole such that

$$(*) \qquad \operatorname{Res}\left(\frac{F'}{F}, a\right) = \begin{cases} n(z, F) & \text{if } a = z \in Z \\ -n(p, F) & \text{if } a = p \in P. \end{cases}$$

The assertion of the theorem then follows from the Residue Theorem (in the form of 11.30). If $a \in A$, there is a non-zero integer n such that

$$(z - a)^n F(z) = H_a(z)$$

has a removable singularity and no zero at a: $|n| = n$ is the multiplicity of a as a pole if $a \in P$, $|n| = -n$ is the multiplicity of a as a zero if $a \in Z$. Then

$$n(z - a)^{n-1}F(z) + (z - a)^n F'(z) = H_a'(z)$$

$$\frac{F'(z)}{F(z)} = \frac{-n}{z - a} + \frac{H_a'(z)}{H_a(z)},$$

which confirms the assertion (*), since H_a'/H_a is holomorphic near a.

Exercise 11.51 *Formulate a version of Rouché's theorem involving the values the functions assume "inside" a loop and prove it from* 11.50.

Exercise 11.52 *Let* $0 < r < R < \infty$, $A = \{z \in \mathbb{C}: r < |z| < R\}$, $a \in A$. *Let* f *be holomorphic in* $A\backslash\{a\}$ *with a first order pole at* a, *be continuous on* $\bar{A}\backslash\{a\}$ *and satisfy* $|f| \leq 1$ on ∂A. *Then* f *assumes every value in* $\mathbb{C}\backslash\bar{D}(0, 1)$ *exactly once in* A.

Hints: Let $\varepsilon > 0$. Since f is continuous on $\bar{A}\backslash\{a\}$ and $1 + \varepsilon > 1 \geq |f(\partial A)|$, we have

$$1 + \varepsilon > |f(C(0, r_\varepsilon))| \quad \text{and} \quad 1 + \varepsilon > |f(C(0, R_\varepsilon))|$$

whenever $r_\varepsilon - r > 0$ and $R - R_\varepsilon > 0$ are sufficiently small. Choose and fix such r_ε, R_ε so that $a \in A_\varepsilon = \{z \in \mathbb{C}: r_\varepsilon < |z| < R_\varepsilon\}$. If $|w| \geq 1 + \varepsilon$, an application of 11.50 to $F = f - w$ shows that

$$N(w) = \frac{1}{2\pi i}\int_{C(0,R_\varepsilon)} \frac{f'}{f - w} - \frac{1}{2\pi i}\int_{C(0,r_\varepsilon)} \frac{f'}{f - w}$$

is the number of zeros of $f - w$ in A_ε minus the number of its poles. But N is evidently a continuous function of $w \in \mathbb{C}\backslash f(C(0, R_\varepsilon) \cup C(0, r_\varepsilon)) \supset \mathbb{C}\backslash D(0, 1 + \varepsilon)$ and is arbitrarily small when $|w|$ is large. Therefore $N(w) = 0$ for all w in the connected set $\mathbb{C}\backslash D(0, 1 + \varepsilon)$. Since by hypothesis $f - w$ has exactly one pole in A_ε and that of order one, it follows that $f - w$ has exactly one zero in A_ε and that of order one. This is the case for every $w \in \mathbb{C}\backslash D(0, 1 + \varepsilon)$. And $\varepsilon > 0$ is arbitrary.

Exercise 11.53 *In addition to the hypothesis of the last exercise suppose the pole a is on the negative real axis. Show that $|f(x)| \leq 1$ for all $r < x < R$ and equality occurs for such an x only if f is constant.*

Hints: Let $F(z) = \frac{1}{2}f(z) + \frac{1}{2}\bar{f}(\bar{z})$. Then F is meromorphic in A, its only possible pole being at a, F is real on the real axis and $|F(\partial A)| \leq 1$. If the singularity at a is removable, then $|F(z)| \leq 1$ for all $z \in A$ with equality for some such z only if F is constant, by the Maximum Modulus Principle. Since $F = \operatorname{Re} f$ on (r, R), it follows that $|\operatorname{Re} f(x)| \leq 1$ for all $x \in (r, R)$ with equality for some such x only if f is constant. On the other hand, suppose that F has a pole at a. Write $f(z) = c/(z - a) + g(z)$, for some $c \in \mathbb{C}$ and some g holomorphic in A. Then $F(z) = (\operatorname{Re} c)/(z - a) + \frac{1}{2}g(z) + \frac{1}{2}\bar{g}(\bar{z})$ and $\operatorname{Re} c \neq 0$. Now $\operatorname{Re} g(x)$ is bounded for x near a. Therefore, taking $x \in (-R, -r)$ near a, we see that $F(x)$ assumes arbitrarily large positive and arbitrarily large negative values. Since $|F(-r)| \leq 1$ and $|F(-R)| \leq 1$, it follows from connectedness considerations that $F(-R, a) \cup F(a, -r) \supset (-\infty, -1) \cup (1, \infty)$. But then by the last exercise, none of the values in $(-\infty, -1) \cup (1, \infty)$ is assumed by F outside $(-R, -r)$. Since F is not constant, by the Open Map Theorem $F(A\backslash(-R, -r))$ is an open set. It is disjoint from $(-\infty, -1) \cup (1, \infty)$ and therefore disjoint from $(-\infty, -1] \cup [1, \infty)$ too. In particular, $|\operatorname{Re} f(x)| = |F(x)| < 1$ for all $x \in (r, R)$.

In general, given $x_0 \in (r, R)$, multiply f by $e^{i\theta_0}$ for appropriate real θ_0 so that $|f(x_0)| = \operatorname{Re}(e^{i\theta_0}f)(x_0)$ and apply the foregoing considerations to $e^{i\theta_0}f$ in the role of f to conclude that $|f(x_0)| = |\operatorname{Re}(e^{i\theta_0}f)(x_0)| \leq 1$, with equality only if $e^{i\theta_0}f$ is constant, i.e., only if f is constant.

§ 7 Poisson's Formula in an Annulus and Isolated Singularities of Harmonic Functions

Exercise 11.54 (Poisson Formula for the Annulus) *Let* $0 < r < 1$.

(i) *Show that for all non-zero complex numbers z, w which satisfy*

(1) $$r^2 < \left|\frac{z}{w}\right| < r^{-2}$$

the series

(2) $$H(w, z) = 2\sum_{k=1}^{\infty} r^{2k}\left(\frac{w}{z - r^{2k}w} - \frac{z}{w - r^{2k}z}\right)$$

converges absolutely and satisfies

(3) $$H(w, z) = 2\sum_{k=1}^{\infty} \frac{r^{2k}}{1 - r^{2k}}\left[\left(\frac{w}{z}\right)^k - \left(\frac{z}{w}\right)^k\right].$$

Hint:

$$\left|\frac{w}{z - r^{2k}w}\right| \leq \frac{|w|}{|z| - r^{2k}|w|} < \frac{|w|}{|z| - r^2|w|} = \frac{1}{c - r^2} \quad (c = |z/w|).$$

For (3), develop into power series in z/w the rational functions in (2) and invert orders of summation. (See 3.12.)

(4) *If* $|z| = |w| \neq 0$, *then* $\operatorname{Re} H(w, z) = 0$.

Hint:

$$\left(\frac{w}{z}\right)^k - \left(\frac{z}{w}\right)^k = \frac{(w\bar{z})^k}{|z|^{2k}} - \frac{(z\bar{w})^k}{|w|^{2k}} = \frac{1}{|z|^{2k}} 2i \operatorname{Im}(w\bar{z})^k.$$

(5) *If* $|z| = 1$ *and* $|w| = r$, *then* $\operatorname{Re}\left[H(w, z) - \frac{w+z}{w-z} - 1\right] = 0$.

Hint:

$$\frac{w+z}{w-z} + 1 = -\frac{2w/z}{1 - w/z} = -2\sum_{k=1}^{\infty}\left(\frac{w}{z}\right)^k.$$

Consequently

$$\begin{aligned} H(w, z) &- \frac{w+z}{w-z} - 1 \\ &= 2\sum_{k=1}^{\infty} \frac{r^{2k}}{1 - r^{2k}}\left[\left(\frac{w}{z}\right)^k - \left(\frac{z}{w}\right)^k + \frac{1 - r^{2k}}{r^{2k}}\left(\frac{w}{z}\right)^k\right] \\ &= 2\sum_{k=1}^{\infty} \frac{r^k}{1 - r^{2k}}\left[\left(\frac{w}{rz}\right)^k - \left(\frac{rz}{w}\right)^k\right]. \end{aligned}$$

Now apply to w and rz the equality in the hint to (4).

(6) *If* $|z| = r$ *and* $|w| = 1$, *then* $\operatorname{Re}\left[H(w, z) - \frac{w+z}{w-z} + 1\right] = 0$.

(ii) *Let* $A = \{z \in \mathbb{C}: r < |z| < 1\}$, f *a continuous real-valued function on* ∂A *which satisfies*

(7) $$\int_{-\pi}^{\pi} f(re^{i\theta})d\theta = \int_{-\pi}^{\pi} f(e^{i\theta})d\theta = 2\pi c, \quad say.$$

Define

(8) $$\begin{aligned} F(z) = {} & \frac{1}{2\pi}\int_{-\pi}^{\pi}\left[\frac{e^{i\theta} + z}{e^{i\theta} - z} - H(e^{i\theta}, z)\right] f(e^{i\theta})d\theta \\ & - \frac{1}{2\pi}\int_{-\pi}^{\pi}\left[\frac{re^{i\theta} + z}{re^{i\theta} - z} - H(re^{i\theta}, z)\right] f(re^{i\theta})d\theta - c. \end{aligned}$$

Show that F *is holomorphic in* A.

Hint: H is continuous on the compact set $\partial A \times \bar{A}$. Cite 7.18.

(iii) $\lim_{z \to z_0, z \in A} \operatorname{Re} F(z) = f(z_0)$ *for each* $z_0 \in \partial A$.

Hints: From (4), (5) and the continuity of H on $\partial A \times \bar{A}$

$$\begin{aligned}
\lim_{\substack{z\to e^{i\phi}\\ z\in A}} \operatorname{Re} F(z) &= \lim_{\substack{z\to e^{i\phi}\\ z\in A}} \frac{1}{2\pi}\int_{-\pi}^{\pi} \operatorname{Re}\left[\frac{e^{i\theta}+z}{e^{i\theta}-z}\right] f(e^{i\theta})d\theta + \frac{1}{2\pi}\int_{-\pi}^{\pi} f(re^{i\theta})d\theta - c\\
&= \lim_{\substack{z\to e^{i\phi}\\ z\in A}} \frac{1}{2\pi}\int_{-\pi}^{\pi} \operatorname{Re}\left[\frac{e^{i\theta}+z}{e^{i\theta}-z}\right] f(e^{i\theta})d\theta \quad \text{by (7)}\\
&= f(e^{i\theta}) \quad \text{by 5.20.}
\end{aligned}$$

Whereas (4) and (6) yield

$$\begin{aligned}
&\lim_{\substack{z\to re^{i\phi}\\ z\in A}} \operatorname{Re} F(z)\\
&= \frac{1}{2\pi}\int_{-\pi}^{\pi} f(e^{i\theta})d\theta + \lim_{\substack{z\to re^{i\phi}\\ z\in A}} \frac{1}{2\pi}\int_{-\pi}^{\pi} \operatorname{Re}\left[-\frac{re^{i\theta}+z}{re^{i\theta}-z}\right] f(re^{i\theta})d\theta - c\\
&= \lim_{\substack{z\to re^{i\phi}\\ z\in A}} \frac{1}{2\pi}\int_{-\pi}^{\pi} \operatorname{Re}\left[\frac{e^{-i\theta}+r/z}{e^{-i\theta}-r/z}\right] f(re^{i\theta})d\theta \quad \text{by (7)}\\
&= \lim_{\substack{w\to e^{-i\phi}\\ w\in A}} \frac{1}{2\pi}\int_{-\pi}^{\pi} \operatorname{Re}\left[\frac{e^{it}+w}{e^{it}-w}\right] f(re^{-it})dt \quad (w = r/z,\ t = -\theta)\\
&= f(re^{i\phi}) \quad \text{by 5.20 applied to the function } f(re^{-it}).
\end{aligned}$$

Exercise 11.55 (Dirichlet Problem for the Annulus) *Let* $0 < r < R < \infty$ *and* f *be a continuous real-valued function on* $C(0, r) \cup C(0, R)$. *Define*

$$a_0 = \frac{1}{2\pi}\int_0^{2\pi} f(re^{i\theta})d\theta, \qquad A_0 = \frac{1}{2\pi}\int_0^{2\pi} f(Re^{i\theta})d\theta, \quad b = \frac{A_0 - a_0}{\log R - \log r}.$$

(i) *Show that there is a holomorphic function* F *in*

$$A(r, R) = \{z \in \mathbb{C} : r < |z| < R\}$$

such that $\operatorname{Re} F$ *is (extendable to be) continuous on* $\bar{A}(r, R)$ *and satisfies*

$$f(z) = b\log|z| + \operatorname{Re} F(z) \quad \forall z \in C(0, r) \cup C(0, R).$$

Hints: Define f_0 on $\partial A(r, R)$ by

$$f_0(z) = f(z) - b\log|z|.$$

Then the definition of b is just such as to insure that

$$\int_{-\pi}^{\pi} f_0(re^{i\theta})d\theta = \int_{-\pi}^{\pi} f_0(Re^{i\theta})d\theta.$$

Apply 11.54, after a change of scale, to come up with a function F which is holomorphic in $A(r, R)$ and such that $\operatorname{Re} F$ is continuous on $\bar{A}(r, R)$ with

$$\operatorname{Re} F = f_0 \quad \text{on } \partial A(r, R).$$

(ii) *Express the Laurent coefficients of F directly in terms of f. (F is only determined by the data up to an imaginary constant, so for the* 0th *Laurent coefficient only the real part is expressible in terms of f.)*

Hints: Let c_n be the nth Laurent coefficient of F:

$$F(z) = \sum_{n=-\infty}^{\infty} c_n z^n, \quad z \in A(r, R).$$

Then for $r < \rho < R$

$$\begin{aligned}\operatorname{Re} F(\rho e^{i\theta}) &= \frac{1}{2}\sum_{n=-\infty}^{\infty} c_n\rho^n e^{in\theta} + \frac{1}{2}\sum_{n=-\infty}^{\infty} \bar{c}_n\rho^n e^{-in\theta} \\ &= \sum_{n=-\infty}^{\infty} \frac{1}{2}(c_n\rho^n + \bar{c}_{-n}\rho^{-n})e^{in\theta},\end{aligned}$$

whence

$$\frac{1}{2\pi}\int_0^{2\pi} \operatorname{Re} F(\rho e^{i\theta})e^{-in\theta}d\theta = \tfrac{1}{2}(c_n\rho^n + \bar{c}_{-n}\rho^{-n}), \quad n \in \mathbb{Z}, r < \rho < R.$$

By continuity of $\operatorname{Re} F$ in $\bar{A}(r, R)$, these equations continue to hold for $\rho = r$ and $\rho = R$ and lead to

$$(*) \quad \frac{1}{2\pi}\int_0^{2\pi} [f(\rho e^{i\theta}) - b\log\rho]e^{-in\theta}d\theta = \tfrac{1}{2}(c_n\rho^n + \bar{c}_{-n}\rho^{-n}), \quad n \in \mathbb{Z}, \rho \in \{r, R\}.$$

Since $\int_0^{2\pi} e^{-in\theta}d\theta = 0$ if $n \neq 0$, this gives

$$\left.\begin{aligned} A_n &\overset{\text{def.}}{=} \frac{1}{2\pi}\int_0^{2\pi} f(Re^{i\theta})e^{-in\theta}d\theta = \tfrac{1}{2}(c_nR^n + \bar{c}_{-n}R^{-n}) \\ a_n &\overset{\text{def.}}{=} \frac{1}{2\pi}\int_0^{2\pi} f(re^{i\theta})e^{-in\theta}d\theta = \tfrac{1}{2}(c_nr^n + \bar{c}_{-n}r^{-n}) \end{aligned}\right\} \quad n = \pm 1, \pm 2, \ldots.$$

Therefore

$$c_n = \frac{2A_nR^n - 2a_nr^n}{R^{2n} - r^{2n}}, \quad n = \pm 1, \pm 2, \ldots.$$

Take $n = 0$ and $\rho = r$ in (*) to get in addition

$$\operatorname{Re} c_0 = \frac{1}{2\pi}\int_0^{2\pi} f(re^{i\theta})d\theta - b\log r = \frac{a_0\log R - A_0\log r}{\log R - \log r}.$$

(iii) *Use* (ii) *to explain the origin of the Poisson formula in* 11.54.

Hint: Take $R = 1$ for simplicity, as before. Assume that on each of $C(0, 1)$ and $C(0, r)$, f is given by an absolutely convergent Fourier series. Then term-by-term integration reveals that the coefficients of these series are the A_n and a_n

above. This justifies all the steps necessary to derive the Poisson formula for F from its Laurent series by substituting the integral formulas for A_n, a_n into the expression for c_n and interchanging the sum in the Laurent series with the resulting integrals.

Exercise 11.56 *Let* $0 \le a < b < \infty$, $h: A(a, b) \to \mathbb{R}$ *harmonic. Show that there exists a constant* B *and a holomorphic function* F *in* $A(a, b)$ *such that*

$$(*) \qquad h(z) = \operatorname{Re} F(z) + B \log|z| \quad \forall z \in A(a, b).$$

Hints: For each pair r, R with $a < r < (a + b)/2 < R < b$ apply 11.55(i) to find a holomorphic function $F_{r,R}$ such that

$$(1) \qquad h(z) = \operatorname{Re} F_{r,R}(z) + b(r, R) \log|z| \quad \forall z \in C(0, r) \cup C(0, R)$$

where

$$(2) \qquad b(r, R) = \frac{A(R) - A(r)}{\log R - \log r}, \qquad A(\rho) = \frac{1}{2\pi}\int_0^{2\pi} h(\rho e^{i\theta})d\theta, \quad a < \rho < b.$$

By the Maximum Principle (5.11) equality (1) holds throughout $A(r, R)$. Let $c(r, R)$ denote the 0th Laurent coefficient of $F_{r,R}$ and integrate (1) to get

$$(3) \qquad A(\rho) = \operatorname{Re} c(r, R) + b(r, R) \log \rho, \quad r < \rho < R.$$

If $r' < r$ and $R' > R$ are another eligible pair and we subtract the two versions of (3) we see that

$$\operatorname{Re} c(r, R) - \operatorname{Re} c(r', R') = [b(r', R') - b(r, R)] \log \rho \quad \forall r < \rho < R.$$

It follows that

$$b(r', R') - b(r, R) = 0.$$

Thus b is a constant function. Call its value B. Then (1) shows that

$$(4) \qquad \operatorname{Re} F_{r,R}(z) = h(z) - B \log|z| \quad \forall z \in A(r, R).$$

Add an appropriate imaginary constant to $F_{r,R}$ to secure

$$(5) \qquad \operatorname{Re} F_{r,R}\left(\frac{a + b}{2}\right) = F_{r,R}\left(\frac{a + b}{2}\right).$$

From (4), (5) and the Uniqueness Theorem it follows that a holomorphic function F is well-defined in $A(a, b)$ by the decree

$$F = F_{r,R} \quad \text{in } A(r, R).$$

By (4) this function satisfies (*).

Exercise 11.57 *Let $b > 0$ and $h\colon A(0, b) \to \mathbb{R}$ be harmonic.*

(i) *Suppose h satisfies the (weaker than boundedness) condition*

$$(**) \quad \lim_{r \downarrow 0} \frac{\int_0^{2\pi} |h(re^{i\theta})| d\theta}{-\log r} = 0.$$

Show that h has a harmonic extension to $D(0, b)$.

Hints: According to 11.56 there is a holomorphic function F in $A(0, b)$ such that

$$(1) \quad h(z) = \operatorname{Re} F(z) + B \log|z| \quad \forall z \in A(0, b),$$

where

$$(2) \quad B = \frac{A(R) - A(r)}{\log R - \log r}$$

for any pair $0 < r < R < b$. Here

$$(3) \quad A(r) = \frac{1}{2\pi} \int_0^{2\pi} h(re^{i\theta}) d\theta, \quad 0 < r < b.$$

(For the formula (2) one can integrate and subtract two versions of 11.56 (*) or, alternatively, examine our derivation thereof.) If c_n is the nth Laurent coefficient of F, the uniform convergence of the Laurent series to F on $C(0, r)$ permits an interchange of summation and integration which leads to

$$(4) \quad \frac{1}{2\pi} \int_0^{2\pi} 2 \operatorname{Re} F(re^{i\theta}) e^{-in\theta} d\theta = c_n r^n + \bar{c}_{-n} r^{-n}, \quad n = \pm 1, \pm 2, \ldots.$$

Use (**) to select $b > R > r_1 > r_2 > \cdots$ convergent to 0 such that

$$(5) \quad \lim_{k \to \infty} \frac{\int_0^{2\pi} |h(r_k e^{i\theta})| d\theta}{-\log r_k} = 0.$$

From (3) and (5) we evidently have

$$\lim_{k \to \infty} \frac{A(r_k)}{\log R - \log r_k} = 0,$$

from which it follows that $B = 0$. Thus $\operatorname{Re} F = h$, so (4) yields

$$(6) \quad |c_n r^n + \bar{c}_{-n} r^{-n}| \le \frac{1}{\pi} \int_0^{2\pi} |h(re^{i\theta})| d\theta.$$

Now $r^N/(-\log r)$ converges to 0 as $r \downarrow 0$ if $N > 0$, but converges to ∞ if $N < 0$ (3.18). Therefore if we consider $n > 0$, divide by $-\log r$, take $r = r_k$ in (6) and let $k \to \infty$, we are forced to conclude via (5) that $c_{-n} = 0$.

(ii) *Suppose $h \ge 0$. Show that in this case there is a constant B and a holomorphic function F in $D(0, b)$ such that*

$$h(z) = \operatorname{Re} F(z) + B \log|z| \quad \forall z \in A(0, b).$$

Hints: In the above reasoning (4) yields ($n \neq 0$)

$$\begin{aligned}\pi|c_n r^n + \bar{c}_{-n} r^{-n}| &= \left|\int_0^{2\pi} \operatorname{Re} F(re^{i\theta})e^{-in\theta}d\theta\right| \\ &= \left|\int_0^{2\pi} [h(re^{i\theta}) - B\log r]e^{-in\theta}d\theta\right| \\ &= \left|\int_0^{2\pi} h(re^{i\theta})e^{-in\theta}d\theta\right|, \quad \text{since } n \neq 0, \\ &\leq \int_0^{2\pi} |h(re^{i\theta})e^{-in\theta}|d\theta \\ &= \int_0^{2\pi} h(re^{i\theta})d\theta, \quad \text{since } h \geq 0, \\ &= \int_0^{2\pi} [\operatorname{Re} F(re^{i\theta}) + B\log r]d\theta \\ &= 2\pi \operatorname{Re} c_0 + 2\pi B \log r.\end{aligned}$$

It follows as before that $c_{-n} = 0$ for all positive n.

(iii) *Draw the same conclusion as in* (ii) *under the* (*weaker than non-negativity*) *hypothesis that*

$$\overline{\lim_{r\downarrow 0}} \frac{\min h(C(0, r))}{-\log r} > -\infty.$$

Hints: Choose a real c less than this limit superior and then $r_1 > r_2 \cdots \to 0$ such that

$$\frac{\min h(C(0, r_k))}{-\log r_k} > c, \quad k = 1, 2, \ldots.$$

Then $h(z) + c\log|z| > 0$ on $\bigcup_{k=1}^{\infty} C(0, r_k)$. Apply the Maximum Principle (5.11) to the harmonic function $H(z) = -h(z) - c\log|z|$ in each annulus $A(r_{k+1}, r_k)$ to conclude that it is negative throughout $A(0, r_1)$. Thus (ii) may be applied to $-H$.

Remark 11.58 Another hypothesis under which the conclusions of 11.57(ii) and (iii) hold is that $\iint_{A(0,r)} |h(x, y)|^2 dxdy < \infty$ for some $0 < r < b$. See, e.g., pp. 15–20 of Hervé [1970].

The next exercise outlines an alternative, short proof of the removable singularity theorem for harmonic functions which is independent of the representation theorem 11.56. Cf. Zaremba [1909].

Exercise 11.59 (i) *Let u be subharmonic in $\Omega = D(0, r)\backslash\{0\}$ and satisfy $\overline{\lim}_{z\to\zeta} u(z) \leq 0$ for every $\zeta \in C(0, r)$. Show that if in addition*

$$\overline{\lim_{z\to 0}} \frac{u(z)}{-\log|z|} \leq 0,$$

then $u \leq 0$ throughout Ω.

Hint: Examine the proof of 5.46, i.e., apply 5.12 to $u_\varepsilon(z) = u(z) + \varepsilon \log|z|$ for each $\varepsilon > 0$ and let $\varepsilon \downarrow 0$ in the end. [A proof from the subharmonic three circles theorem, mentioned in the Chapter V notes, is also possible when u is bounded above in Ω.]

(ii) *From* (i) *derive anew* (cf. 11.57(i)) *the fact that an isolated singularity of a bounded harmonic function is removable. Show in fact that if h is harmonic in $D(0, R)\backslash\{0\}$ and $\lim_{z\to 0} h(z)/\log|z| = 0$, then h extends to a harmonic function in $D(0, R)$.*

Hint: Pick $0 < r < R$ and let f be the function which is harmonic in $D(0, r)$, continuous on $\bar{D}(0, r)$ and coincides with h on $C(0, r)$. Apply (i) to each of the functions $u = \pm(h - f)$.

(iii) *Show that the Dirichlet problem is not solvable for a punctured disk, even if only subharmonic extensions of the boundary data are sought.*

Hint: Let $D = D(0, 1)$, $\Omega = D\backslash\{0\}$, $f: \partial\Omega \to \mathbb{R}$ the (continuous) function which is 1 at 0 and 0 on ∂D. If $h: \bar{\Omega} \to \mathbb{R}$ is continuous, subharmonic in Ω and equal to f on $\partial D \subsetneq \partial\Omega$, then by (i), $h(0) \leq 0$, so $h = f$ on $\partial\Omega$ is impossible.

Exercise 11.60 (i) *Show that if h is harmonic in $A(a, b)$ and*

$$H(r) = \int_0^{2\pi} h(re^{i\theta})d\theta \quad \textit{for } a < r < b,$$

then there are two real constants α, β such that $H(r) \equiv \alpha + \beta \log r$.

Hints: 11.56 and 5.5.

(ii) (F. Riesz [1922/23]) *Show that if u is subharmonic in $A(a, b)$ and $U(r) = \int_0^{2\pi} u(re^{i\theta})d\theta$ for $a < r < b$, then $U(r)$ is a convex function of* $\log r$.

Hints: Given $a < r_1 < r_2 < b$, 11.55 provides a function h which is harmonic in $A(r_1, r_2)$, continuous in $\bar{A}(r_1, r_2)$ and coincides on the boundary of $A(r_1, r_2)$ with u. By the harmonic majorant property of u we have $u \leq h$ in $A(r_1, r_2)$, hence $U \leq H$ in (r_1, r_2). Therefore for each $0 \leq \lambda \leq 1$

$$\begin{aligned} U(e^{\lambda \log r_1 + (1-\lambda)\log r_2}) &\leq H(e^{\lambda \log r_1 + (1-\lambda)\log r_2}) \\ &\overset{(i)}{=} \lambda H(r_1) + (1 - \lambda)H(r_2) \\ &= \lambda U(r_1) + (1 - \lambda)U(r_2). \end{aligned}$$

The last equality is because $u = h$ on $C(0, r_j)$ entails $U(r_j) = H(r_j)$, $j = 1, 2$.

Remarks: If $a = 0$ in 11.60(ii), it follows from the most elementary properties of convex functions (see, e.g., STROMBERG [1980]) that $U(r)$ and $U(r)/\log r$ are monotonic for small r and consequently have limits in $[-\infty, \infty]$ as $r \downarrow 0$. In fact, one can even assert that $\lim_{r\downarrow 0} u(re^{i\theta})/\log r$ exists and equals $\lim_{r\downarrow 0} U(r)/2\pi \log r$ for most θ. (See ARSOVE and HUBER [1973].)

Notes to Chapter XI

The bilateral series expansion in 11.2 was known to WEIERSTRASS in 1841 but not published then. (See his *Werke* I, pp. 51–66.) It was announced in a short note by P. A. LAURENT [1843], in which he remarked only that it followed easily from Cauchy's earlier work. (A proof of his appeared posthumously in 1863.) On pp. 938–940 of the same journal Cauchy and Liouville comment on this communiqué and then CAUCHY [1843] gives a detailed proof. (See pp. 34–35 of MITTAG-LEFFLER [1923].) The treatment of 11.1 and 11.2 in the text is Cauchy's. For other treatments of 11.2 see pp. 1028–1030 of BURKHARDT [1914], MITTAG-LEFFLER [1884b], SCHEEFFER [1884], and BIERMANN [1908]. (The treatment in PRINGSHEIM [1896], [1925] is also interesting; as remarked in the Chapter V notes, he replaces integrals with limits of special Riemann sums, a device which Cauchy himself had on occasion used. See also BEESACK [1972].) 11.6 (CASORATI [1868], WEIERSTRASS [1876], § 8) was independently discovered by the Russian mathematician J. W. Sochozki. See COLLINGWOOD and LOHWATER [1966], pp. 4–5 for the history, HÖLDER [1882] for another treatment. On 11.10–11.12 see § 11, chapter III of BEHNKE and SOMMER [1965]. For an even shorter proof of 11.12, proceeding independently of 5.37, see the development in BOAS [1964] (which is only sixteen lines long!).

On 11.15 see KAKEYA [1923]. 11.16, which is due to Fejér and F. Riesz (FEJÉR [1915], pp. 55–59; SZÁSZ [1919], pp. 166–167), is an important ingredient in the classical proof of the Spectral Theorem for unitary operators in a Hilbert space. See p. 282 of RIESZ and SZ.-NAGY [1955]. There is a nice generalization of 11.16 to certain classes of entire functions of exponential type. See, e.g., ACHIEZER [1956], p. 154, pp. 437–443 of B. LEVIN [1964] and BRICKMAN [1964]. The compactness argument in 11.16 is rather heavy-handed and the reader may simply prefer to give an *ab initio* proof modelled on 11.15. (See, e.g., PÓLYA and SZEGÖ [1976], exercise VI.40.) Alternatively, he can extend 11.15 to cover the case where zeros (and poles) are permitted on the unit circle. [The proof of this is essentially the same as that of 11.15 but the pairing off of poles and zeros is slightly more complicated.] There is also an obvious version of 11.16 involving polynomials which are non-negative on $\mathbb{R}$ (see BRICKMAN and STEINBERG [1962]), and both this and 11.16 itself have operator analogs. E.g., if P_j are bounded linear operators on a Hilbert space H and $P(x) = \sum_{j=0}^{2n} x^j P_j$ is a positive operator for each real x, then there exist bounded linear operators Q_j on H such that $P(x) = Q^*(x)Q(x)$ for all real x, where $Q(x) = \sum_{j=0}^{n} x^j Q_j$. See ROSENBLUM and ROVNYAK [1971], pp. 312–313, and MIAMEE and SALEHI [1976].

For more on 11.14 and 11.48 see GRONWALL [1912–13] and on this general theme, PÓLYA and SZEGÖ [1976], exercises VI. 44–VI. 49. For material related to 11.19 see STYER and MINDA [1974] and R. K. WILLIAMS [1976] for a variety of other proofs. E.g., 11.19 is a corollary of 12.28. Another of Williams' proofs occurs in Chapter XVIII. An important feature of 11.19 was the deduction of surjectivity from injectivity. For more on that interesting theme see BORSUK [1933b] and BANACH and MAZUR [1935] and the references given in the latter, as well as BROWDER [1954], GORDON [1972], PLASTOCK [1974] and HO [1975]. 11.20 occurs in CAUCHY [1844] (perhaps earlier too) and MÉRAY [1855].

11.23 is stated in FATOU [1906], p. 369. The suggested proof is from PÓLYA [1916] who attributes it to Hurwitz.

The reader may search for 11.26 in DARBOUX [1878]; he will find it explicit in J. KÖNIG [1884] and LECORNU [1887]. HADAMARD [1892b], p. 109 gave the following example to show that the converse of 11.26 fails:

$$\frac{1}{1-z} + \log(1+z) = 1 + \sum_{n=1}^{\infty}\left[1 + \frac{(-1)^{n+1}}{n}\right]z^n.$$

Here $c_n/c_{n+1} \to 1$ yet there are two singular points (1 and -1) on the circle of convergence. Moreover, other examples show that even if there is only one singularity z_0 on the circle of convergence, but it is not a pole, then the ratio c_n/c_{n+1} need not converge to z_0. 11.27(i) is due to PRINGSHEIM [1929] pp. 117–124 and 11.27(ii) to HADAMARD [1892b]. For the case where the (unique) pole is not simple see MILLER [1935]. Notice that by a change of variable 11.27(ii) reads as follows: a necessary and sufficient condition for f to have a holomorphic extension to a neighborhood of $\bar{D}(0, 1)\backslash\{z_0\}$ with a simple pole at z_0, for some unimodular z_0, is that

$$\overline{\lim_{n\to\infty}}\left|\frac{c_n}{c_{n+1}} - z_0\right|^{1/n} < 1.$$

HADAMARD [1892b] also provides the following related criterion which does not explicitly involve the unique pole z_0: $\overline{\lim}_{n\to\infty}|c_{n-1}c_{n+1} - c_n^2|^{1/n} < 1$. For a perspicuous proof see OSTROWSKI [1926b]. There is a related (deeper) result due to FABRY [1896a] (On this attribution see, however, pp. 97 and 298 of PRINGSHEIM [1929].): If c_n/c_{n+1} converges to some z_0, then the series has radius of convergence $|z_0|$ (by 3.9 and 3.3) and f does not have a holomorphic extension to any neighborhood of z_0. In § 131 of PRINGSHEIM [1932] the reader will find a proof of the following stronger form of this theorem: If r is the radius of convergence of $\sum c_n z^n$ and $(c_n/c_{n+1})|c_{n+1}/c_n|$ has the (unimodular) limit u_0, then $z_0 = ru_0$ is a singular point in the above sense. For other proofs of (various stronger forms of) this Fabry result, see pp. 307–315 of BIEBERBACH [1931], pp. 64 ff. of BIEBERBACH [1955], pp. 76–86 of LANDAU [1929a], pp. 377 ff. of DIENES [1931], or S. K. SINGH [1964/65]. See also AGMON [1948], BROGGI [1935], V. BERNSTEIN [1935], and DUFRESNOY and PISOT [1951] (p. 122). (SINGH's proof seems incomplete to me—and to the *Zentralblatt* reviewer as well.) For the more elementary version of Fabry's theorem when the series represents a rational function see BROGGI [1931]. 11.28 is from PÓLYA and SZEGÖ [1972], exercise III.246 and is also contained, along with generalizations, in TSCHAKALOFF [1948]. For a good survey of results and history on questions of singularities on the circle of convergence see § 45 of BIEBERBACH's encyclopedia article. The matter recurs in Chapters XVI and XVII below.

I have followed RUDIN [1974] in the proof of 11.29 (for its history see § 35 of BURKHARDT [1914]) and HEINS [1968] in the development of 11.31, 11.32. The latter treatment is due to MORDELL [1933], pp. 351–352. For a somewhat different one see pp. 285–286 of DIEUDONNÉ [1971]. The integral in 11.32(i) seems to have been first computed by Euler; see VAN VEEN [1943]. The most popular derivation of 11.32(i) (see almost any text) involves an improper iterated integral, its conversion to a double integral and the polar coordinate change of variable in this integral. The shortest proof is perhaps the nice one-page one ANONYMOUS [1889], which uses (an elementary version of) 7.18. The integrals in 11.32(ii) are known in the literature as Fresnel integrals and are important in optics. For other treatments see FABRY [1896b], JAMET [1896], [1903], BLOCH [1919] and OLDS [1968]. The sum in (iii) first occurred in an 1811 paper of C. F. GAUSS and is called a Gauss sum, the related expression $\sum_{k=0}^{n-1} e^{2\pi i h k^2/n}$ is called a generalized Gauss sum. For the contributions of Gauss, Cauchy and Dirichlet to their evaluation see the notes in KRONECKER [1889a]; the evaluation of the Gauss sums via residues first occurs here; see also p. 243 of LANDSBERG [1893]. For a discussion of Kronecker's proof as well as proofs due to SCHUR [1921] and MERTENS [1896] and the one due to Dirichlet (based on Fourier analysis) see pp. 197–218 of LANDAU [1958]. See also LERCH [1903a] (who corrects errors in Kronecker and Bachmann), SALVADORI [1904], § 38 of LINDELÖF [1905] and pp. 3–11 vol. IV of LEVEQUE [1974] for recent literature on Gauss sums. For an algebraic treatment, close to Gauss' original, which also evaluates the generalized Gauss sums (as does SCHUR *op. cit.*), see chapter 11 of RADEMACHER [1964] and § 7 of BACHMANN [1894]. (Another algebraic treatment occurs in SHANKS [1958].) These sums are related to the important law of quadratic reciprocity and there is a reciprocity relationship for the generalized Gauss sums too. This is also elucidated in RADEMACHER's and BACHMANN's books, in KRONECKER [1890], in LANDSBERG [1893], in KRAZER [1903], pp. 184–186 and in BELLMAN [1961], pp. 38–39. In turn there are important relationships between the Gauss sums and the formula 11.33(v). For these see KRAZER [1903], LANDSBERG [1893] and pp. 222–226 of DYM and MCKEAN [1972]; for more on the connection of this formula with the important special functions θ, Γ, ζ see also HAMBURGER [1922]. The proof in the text is from Landsberg. There is another standard proof based on Fourier series (i.e., on 11.10 and 11.11); this is beautifully set out in SANSONE and GERRETSEN [1960], pp. 109–111. A version of this proof will be offered in Chapter XIII;

cf. DYM and MCKEAN [1972], pp. 52–54. A different proof is offered in PÓLYA [1927] (reproduced on pp. 40–41 of BELLMAN *loc. cit.*). See also DYM and MCKEAN [1972], pp. 60–66 for the relation of this identity to the heat equation of physics.

For help with 11.35 see exercise 3.7.7 of KRZYŻ [1971] or MITRINOVIĆ [1966], pp. 32–34. See the latter also for many other interesting series and integrals evaluated via residues. The reader intrigued with contour integrals can be sated by looking into GARNIR and GOBERT [1965].

11.37 is from S. N. BERNSTEIN [1923]; the equivalence in 3.35(ii) provides the nexus. (For a Banach algebra proof see Chapter XIX.) The proof in the text is attributed by DONOGHUE [1969] to Lars Gårding; another proof occurs in Chapter XIV. For (finite) exponential polynomials see O'HARA [1973]. See also BOAS [1969]. For a converse see p. 139 of ACHIEZER [1956]. There are many treatments of 11.38 in the literature. The device in parts (iv) and (v) is a modification of one used by LERCH [1903b] and MOHR [1953b]. For a different treatment see pp. 110–115 of TITCHMARSH [1939] or § 7, chapter III of BEHNKE and SOMMER [1965]. See also Chapter XIV where (vi) and (vii) will be derived in a different way and (vi) will be parlayed (by integration and exponentiation) into the product formula $\sin \pi z = \lim_{N\to\infty} \pi z \prod_{n=1}^{N} (1 - z^2/n^2)$. Cf. PRINGSHEIM [1925], pp. 485 ff., NEVILLE [1951], VENKATACHALIENGAR [1962] and § 23 of the encyclopedia article of PRINGSHEIM, FABER and MOLK [1911].

I have followed RUDIN [1974] in the proof of 11.39. A more constructive proof is offered in § 7 chapter III of BEHNKE and SOMMER [1965]; it is very similar to that in MITTAG–LEFFLER [1884a]. The latter work also contains 11.41 (as theorems B, p. 43; D, p. 53 and A′, p. 72). Earlier Weierstrass (7.32) had shown how to construct a function with poles of specified orders and locations a_n where $|a_n| \to \infty$. (Take the reciprocal of the function provided by 7.32.) MITTAG–LEFFLER's work originated in papers written in Swedish and in the French Academy notes of [1882]. The paper [1884a] represents the grand synthesis and generalization. On page 20 and at various other points throughout this paper he records this evolution and the contribution of others (who had read his Swedish papers), notably SCHERING [1880], WEIERSTRASS [1880b], DINI [1881], HERMITE [1881], CAZZANIGA [1882], CASORATI [1882] and GUICHARD [1884]. See also STÄCKEL [1890], FABER [1905], JOURDAIN [1905a], PÓLYA [1941] and GERMAY [1946], [1948]. GELFOND [1967] proves the following: If Ω is open and simply-connected, $A \subset \Omega$, then there exists a non-polynomial $f \in H(\Omega)$ which together with all its derivatives takes only integer values in A if and only if A has no limit point in Ω. See also MAHLER [1971] and FRANKLIN [1926].

The interested reader should try to modify the ideas in the hints to 11.40 to cover the non-simply-connected case. (See, e.g., BEHNKE and SOMMER [1965], § 8 chapter III.) Mittag–Leffler himself proved Weierstrass' theorem in this way. The key lies in showing that $\int_\gamma h$ is an integer multiple of $2\pi i$ whenever γ is a piecewise smooth loop in U (γ, U and h as in the hints to 11.40). For this the more detailed knowledge of the approximating rational functions provided in Behnke and Sommer's proof is needed.

11.42 and 11.43 are due to HELMER [1940] and 11.44 to SCHILLING [1946]. See also KAKUTANI [1955], COHEN [1961], chapters 6 and 10 of HOFFMAN [1962], AGUILÓ FUSTER [1965], ALLING [1968] (esp. pp. 25–28 for history), RUBEL [1970], the nice treatment in § 2 of SHAPIRO [1970] and VON RENTELN [1976]. Problem 10, p. 341 of RUDIN [1974] is relevant to 11.43.

11.46 for $U = \mathbb{C}$ is due to WEIERSTRASS [1876], for general U to MITTAG-LEFFLER [1884a]. 11.52 and 11.53 are from ROBINSON [1943].

On the Dirichlet problem for an annulus see DINI [1913] (and his earlier paper referenced there), VILLAT [1912], [1921], GRONWALL [1912/13] (the sources of 11.54), KOEBE [1913b], PICONE [1926a], JULIA [1930a], DEMTCHENKO [1931], CISOTTI [1931], KOMATU [1945b], CONSIGLIO [1950] and § 14 of LICHTENSTEIN [1919]. When the boundary data in 11.55 are trigonometric polynomials the problem reduces to an easy algebraic one. The general problem can then be solved via 13.1, the Maximum Principle, and a limit argument. For the details of this approach (which is very similar to but less explicit than that of Gronwall and Picone), see pp. 63–64 of HEINS [1962] or pp. 300–302 of HEINS [1968]. See also pp. 155–158 of GAIER [1964] and the references there. The whole theory in § 7 can be derived *ab initio* by Fourier methods, as hinted in 11.55(iii). In fact, by differentiating under the integral sign (in the definition of a_n in 11.55(ii)) and using the polar form of Laplace's Equation, differential equations for these coefficients (as functions of r) are found which can be solved by inspection, leading to the appropriate form of a_n ($n \neq 0$) and a_0 (see 11.60(i)). For the details see pp. 56–58 of HEINS [1962]. For the material on removable singularities (11.57) see PICARD [1923], [1924], LEBESGUE [1923], STOŻEK [1926], RAYNOR [1926], PICONE [1926b], [1929], KELLOGG [1926a], NOAILLON [1927], ASCOLI [1928a]. A version of 11.57(ii) appears in BÔCHER [1902/03]. Much of the material of § 7 can be found on pp. 204–210, vol. II of H. A. SCHWARZ' *Gesammelte Mathematische Abhandlungen*. (Isolated singularities of subharmonic functions are treated in BRELOT [1934] and chapter VII of RADÓ [1937].)

We have by now seen Poisson formulas for disks, annuli and half-planes. There are also simple and useful ones for strips and half-disks. See WIDDER [1961] and DINGHAS [1954] (or p. 61 of [1961]), respectively. (These are produced by transferring from the disk via a conformal map.) For some other regions see KOEBE [1913b].

Chapter XII
Omitted Values and Normal Families

The center-piece of this chapter is the following remarkable result due to Friedrich Schottky: if f is holomorphic in $D(0, 1)$ and assumes neither of the values 0 or 1, then f is bounded in $D(0, r)$ by a bound depending on r and $|f(0)|$ only, for each $r < 1$. [In fact we prove a generalization in which f does not assume 0 and $f^{(n)}$ (for some $n \geq 0$) does not assume 1.] The principal application is the almost immediate fact that a family of holomorphic functions on a common domain, none of which has 0 or 1 in its range, if bounded at a point, is uniformly bounded in a neighborhood of that point. The compactness theorems of Chapter VII then come into play with astounding consequences. Of course "0" and "1" here are convenient normalizations: any two distinct complex numbers would serve as well.

To scale these peaks we start with some modest preparation of gear.

§1 Logarithmic Means and Jensen's Inequality

Lemma 12.1 *The function* $\log^+$ *defined on* $[0, \infty)$ *by*

$$\log^+(x) = \log(\max\{1, x\})$$

has the following properties:

(i) $\log^+$ *is continuous, non-decreasing and non-negative*

(ii) $\log^+(x_1 \cdots x_n) \leq \log^+(x_1) + \cdots + \log^+(x_n) \quad \forall x_1, \ldots, x_n \in [0, \infty)$

(iii) $\log^+(x_1 + \cdots + x_n) \leq \log^+(x_1) + \cdots + \log^+(x_n) + \log(n)$ $\forall x_1, \ldots, x_n \in [0, \infty)$

(iv) $\log^+(x) \leq \log(1 + x) \leq \sqrt{x}$ *for all* $x \geq 0$

(v) $|\log x| = \log^+(x) + \log^+(1/x)$ *and* $\log x = \log^+(x) - \log^+(1/x)$ *for all* $x > 0$.

Proof: As the composite of the continuous non-decreasing function $x \to \max\{1, x\}$ of $[0, \infty)$ onto $[1, \infty)$ and the continuous non-decreasing function $\log: [1, \infty) \to [0, \infty)$, the function $\log^+$ is continuous, non-decreasing and non-negative.

Now (ii) is trivial if its left side is 0. Otherwise, some $x_j > 1$ and we may suppose the notation chosen so that

$$0 \leq x_1 \leq x_2 \leq \cdots \leq x_{j-1} \leq 1 \leq x_j \leq \cdots \leq x_n.$$

Then

$$x_1 \cdots x_n \leq x_j \cdots x_n$$

and so

$$\log^+(x_1\cdots x_n) \le \log^+(x_j\cdots x_n) = \log(x_j\cdots x_n) = \log(x_j) + \cdots + \log(x_n)$$
$$\le \log^+(x_1) + \cdots + \log^+(x_j) + \cdots + \log^+(x_n).$$

Similarly, (iii) is trivial if each $x_j < 1/n$. Otherwise, let the points be monotone in their indices as before and have $x_n \ge 1/n$ and

$$\log^+(x_1 + \cdots + x_n) \le \log^+(nx_n) = \log(nx_n) = \log(n) + \log(x_n)$$
$$\le \log(n) + \log^+(x_1) + \cdots + \log^+(x_n).$$

Since $\log^+$ is non-decreasing, we have

$$\log^+(x) \le \log^+(1 + x) = \log(1 + x).$$

Also the function $f(x) = \sqrt{x} - \log(1 + x)$ has derivative

$$f'(x) = \frac{(1 - \sqrt{x})^2}{2\sqrt{x}(1 + x)} \ge 0$$

on $(0, \infty)$ and so f is non-decreasing. Since $f(0) = 0$, we have $f(x) \ge 0$ for all $x \ge 0$, completing the proof of (iv).

For (v) consider the two cases $x \ge 1$ or $x < 1$.

Definition 12.2 If $r \ge 0$ and f is a continuous complex-valued function on $C(0, r)$, define the means

$$M(r,f) = \sup_{\theta\in\mathbb{R}}|f(re^{i\theta})|$$

$$L(r,f) = \frac{1}{2\pi}\int_0^{2\pi} \log^+|f(re^{i\theta})|d\theta.$$

If, moreover, f is never zero, define

$$V(r,f) = \frac{1}{2\pi}\int_0^{2\pi} |\log|f(re^{i\theta})|\,|d\theta.$$

Then by 12.1(v) we see that

$$V(r,f) = L(r,f) + L\left(r,\frac{1}{f}\right).$$

Theorem 12.3 *If f is holomorphic in $D(0, r)$, $0 < R < r$ and f has no zeros in $\bar{D}(0, R)$, then*

$$\log|f(z)| = \frac{1}{2\pi}\int_0^{2\pi} \log|f(Re^{i\theta})| \operatorname{Re}\left[\frac{Re^{i\theta} + z}{Re^{i\theta} - z}\right]d\theta \quad \forall|z| < R.$$

Proof: Let F be a holomorphic logarithm for f in some neighborhood of $\bar{D}(0, R)$, apply the Cauchy–Schwarz formula (5.18 appropriately scaled to the disk $\bar{D}(0, R)$) to F and take real parts, remembering that $\operatorname{Re} F = \log|f|$.

Theorem 12.4 *If F is holomorphic in $D(0, 1)$ and $0 < R < 1$, then*

(i) $$\log^+|F(z)| \le \frac{1}{2\pi}\int_0^{2\pi} \log^+|F(Re^{i\theta})| \operatorname{Re}\left[\frac{Re^{i\theta}+z}{Re^{i\theta}-z}\right] d\theta \quad \forall z \in D(0, R).$$

If in addition F is not identically 0, then

(ii) $$\log M(r, F) \le \frac{R+r}{R-r} L(R, F) \quad \forall 0 < r < R.$$

If in addition F has no zeros on $C(0, R)$, then

(iii) $$\log|F(z)| \le \frac{R+|z|}{R-|z|} L(R, F) - \frac{R-|z|}{R+|z|} L\left(R, \frac{1}{F}\right) \quad \forall z \in D(0, R)\backslash F^{-1}(0).$$

Proof: First consider the case where F has no zeros on $C(0, R)$. Then F is not identically zero in $\bar{D}(0, R)$, hence has only finitely many zeros there (5.62). Let these zeros counted according to multiplicity be $z_1, \ldots, z_n$. We have $|z_j| < R$ by assumption, so each of the functions

$$\phi_j(z) = \frac{R(z - z_j)}{R^2 - \bar{z}_j z}$$

is holomorphic in a neighborhood of $\bar{D}(0, R)$. Thus $f = F/\phi_1 \cdots \phi_n$ is holomorphic in a neighborhood of $\bar{D}(0, R)$ and has no zeros in $\bar{D}(0, R)$. We apply the last theorem to this function, noting that $|\phi_j(z)| = 1$ when $|z| = R$:

(1) $$\log\left|\frac{F(z)}{\phi_1(z)\cdots\phi_n(z)}\right| = \frac{1}{2\pi}\int_0^{2\pi} \log|F(Re^{i\theta})| \operatorname{Re}\left[\frac{Re^{i\theta}+z}{Re^{i\theta}-z}\right] d\theta, \quad |z| < R.$$

Since $|\phi_j(z)| < 1$ for all $z \in D(0, R)$, we have $\log|\phi_j(z)| < 0$ whenever $z \in D(0, R)\backslash F^{-1}(0)$, so for such z the left side of (1) dominates $\log|F(z)|$ and we get the so-called *Jensen Inequality*:

(2) $$\log|F(z)| \le \frac{1}{2\pi}\int_0^{2\pi} \log|F(Re^{i\theta})| \operatorname{Re}\left[\frac{Re^{i\theta}+z}{Re^{i\theta}-z}\right] d\theta, \quad z \in D(0, R)\backslash F^{-1}(0).$$

Recall that $\log(x) = \log^+(x) - \log^+(1/x)$ for all $x > 0$, to see from (2) that

(3) $$\log|F(z)| \le \frac{1}{2\pi}\int_0^{2\pi} \log^+|F(Re^{i\theta})| \operatorname{Re}\left[\frac{Re^{i\theta}+z}{Re^{i\theta}-z}\right] d\theta - \frac{1}{2\pi}\int_0^{2\pi} \log^+\left|\frac{1}{F(Re^{i\theta})}\right| \operatorname{Re}\left[\frac{Re^{i\theta}+z}{Re^{i\theta}-z}\right] d\theta.$$

Now (equation (2) in the proof of 5.20)

$$\operatorname{Re}\left[\frac{Re^{i\theta}+z}{Re^{i\theta}-z}\right] = \frac{R^2 - |z|^2}{|Re^{i\theta} - z|^2}$$

and therefore

(4) $$\frac{R+|z|}{R-|z|} \ge \operatorname{Re}\left[\frac{Re^{i\theta}+z}{Re^{i\theta}-z}\right] \ge \frac{R-|z|}{R+|z|} > 0 \quad \text{for } |z| < R.$$

Since $\log \le \log^+$, it follows from (4) that

$$(5)\quad \frac{1}{2\pi}\int_0^{2\pi} \log|F(Re^{i\theta})| \operatorname{Re}\left[\frac{Re^{i\theta}+z}{Re^{i\theta}-z}\right]d\theta$$
$$\le \frac{1}{2\pi}\int_0^{2\pi} \log^+|F(Re^{i\theta})| \operatorname{Re}\left[\frac{Re^{i\theta}+z}{Re^{i\theta}-z}\right]d\theta.$$

But also $\log^+ \ge 0$, so the right side of (5) is non-negative and it follows from (2) and (5) that

$$\log^+|F(z)| = \max\{0, \log|F(z)|\}$$
$$\le \frac{1}{2\pi}\int_0^{2\pi} \log^+|F(Re^{i\theta})| \operatorname{Re}\left[\frac{Re^{i\theta}+z}{Re^{i\theta}-z}\right]d\theta \quad \forall z \in D(0, R)\backslash F^{-1}(0),$$

giving (i). While (3) and the inequalities (4) lead to (iii).

Now consider the general case and a fixed $z \in D(0, 1)$. An easy uniform continuity argument shows that the right side of (i), call it $\Phi(R)$, is a continuous function of $R \in (|z|, 1)$. If F is identically 0, then (i) is trivial. Otherwise the zeros of F do not accumulate in $D(0, 1)$, hence there are only finitely many in each $\bar{D}(0, 1-1/n)$, hence only countably many in $D(0, 1)$. Thus for a dense set of $R \in [0, 1)$, F has no zeros on $C(0, R)$. Given $R \in (|z|, 1)$, pick R_n from this dense set with $|z| < R_n < R$ and $R_n \to R$. Then $\log^+|F(z)| \le \Phi(R_n)$ by the result of the first paragraph and $\Phi(R_n) \to \Phi(R)$ by continuity of Φ. Thus $\log^+|F(z)| \le \Phi(R)$, which is (i).

Similarly $L(R, F)$ is a continuous function of $R \in [0, 1)$ and (ii) is already established for the dense set of R for which F has no zeros on $C(0, R)$. Therefore (ii) follows for all $R \in [0, 1)$ by continuity.

Remark: Notice that (i) also follows from 5.22(i) and the subharmonicity of $\log^+|F|$ (5.26(ii)).

Theorem 12.5 *If f is holomorphic in $D(0, 1)$, then $L(r, f)$ is a non-decreasing function of $r \in [0, 1)$.*

Proof: 12.4 or 5.26(ii) shows that $\log^+|f|$ is a subharmonic function (5.6) and so the conclusion follows from 5.26(i). Here is an alternative, more prosaic proof. Let $0 < r < R < 1$. We may apply 12.4(i) to write

$$\log^+|f(re^{it})| \le \frac{1}{2\pi}\int_0^{2\pi} \log^+|f(Re^{i\theta})| \operatorname{Re}\left[\frac{Re^{i\theta}+re^{it}}{Re^{i\theta}-re^{it}}\right]d\theta, \quad t \in \mathbb{R}.$$

Integrate this:

$$L(r, f) = \frac{1}{2\pi}\int_0^{2\pi} \log^+|f(re^{it})|\,dt$$
$$\le \frac{1}{2\pi}\int_0^{2\pi} \frac{1}{2\pi}\int_0^{2\pi} \log^+|f(Re^{i\theta})| \operatorname{Re}\left[\frac{Re^{i\theta}+re^{it}}{Re^{i\theta}-re^{it}}\right]d\theta\,dt.$$

Since the integrand on the right is a continuous function of $(\theta, t) \in [0, 2\pi] \times [0, 2\pi]$, it is possible to invert the orders of integration in this iterated integral (see 8.26(vi)) to get

$$L(r,f) \leq \frac{1}{2\pi}\int_0^{2\pi} \log^+|f(Re^{i\theta})|\left(\frac{1}{2\pi}\int_0^{2\pi} \mathrm{Re}\left[\frac{Re^{i\theta}+re^{it}}{Re^{i\theta}-re^{it}}\right]dt\right)d\theta.$$

But (formula (4) in the proof of 5.20) this inner integral is 1 and so we get

$$L(r,f) \leq \frac{1}{2\pi}\int_0^{2\pi} \log^+|f(Re^{i\theta})|\,d\theta = L(R,f).$$

Theorem 12.6 *Let F be a holomorphic function in $D(0, 1)$ which is not identically 0 but satisfies*

$$(*)\qquad \sup_{r<1}\int_0^{2\pi} \log^+|F(re^{i\theta})|\,d\theta = M < \infty.$$

Then F has at most countably many zeros. Let them be, with due regard to multiplicity, $z_1, z_2, \ldots$. Then

$$\sum_{j=1}^{\infty}(1-|z_j|) < \infty.$$

Proof: Since for each $R < 1$, F has only finitely many zeros in $\bar{D}(0, R)$, we may suppose the notation chosen so that $|z_1| \leq |z_2| \leq \cdots$. Moreover, by dividing off an appropriate power of z, we may suppose without loss of generality that 0 is not a zero of F. Given n, choose R so that $|z_n| < R < 1$ and $R \neq |z_k|$ for each k. Let $z_1, \ldots, z_N$ $(N \geq n)$ be the zeros of F in $\bar{D}(0, R)$ and for each $1 \leq j \leq N$ define

$$\phi_j(z) = \frac{R(z-z_j)}{R^2-\bar{z}_j z},\qquad f = \frac{F}{\phi_1\cdots\phi_N}.$$

Then f is holomorphic and zero-free in a neighborhood of $\bar{D}(0, R)$. Apply 12.3 with $z = 0$, remembering that $|\phi_j| \equiv 1$ on $C(0, R)$, to see that

$$(**)\qquad \log|F(0)| + \sum_{j=1}^{N}\log\frac{R}{|z_j|} = \frac{1}{2\pi}\int_0^{2\pi}\log|F(Re^{i\theta})|\,d\theta.$$

Consequently, from (*)

$$\sum_{j=1}^{n}\log\frac{R}{|z_j|} \leq \sum_{j=1}^{N}\log\frac{R}{|z_j|} \leq \frac{M}{2\pi} - \log|F(0)|.$$

Let $R \uparrow 1$ through admissible values to have

$$\sum_{j=1}^{n}\log\frac{1}{|z_j|} \leq \frac{M}{2\pi} - \log|F(0)|.$$

Since for any $x > 1$ we have

$$\log x = \int_1^x \frac{1}{t}\,dt > \int_1^x \frac{1}{x}\,dt = \frac{x-1}{x} = 1 - \frac{1}{x},$$

it follows that

$$\sum_{j=1}^{n} (1 - |z_j|) \le \sum_{j=1}^{n} \log \frac{1}{|z_j|} \le \frac{M}{2\pi} - \log|F(0)|,$$

holding for *all* positive integers n.

Remark: The hypothesis (*) is certainly fulfilled if F is bounded, so this result generalizes the final conclusion in 6.9(iii). The equality (**) is known as the *Jacobi-Jensen* formula. See KOWALEWSKI [1918].

Exercise 12.7 (i) (PRIVALOV [1924a]) *Let F be holomorphic in $D(0, 1)$ and satisfy*

$$(*) \qquad \sup_{\rho<1} \int_0^{2\pi} \log^+|F(\rho e^{i\theta})|\,d\theta = K < \infty.$$

Show that F satisfies

$$|F(z)| \le e^{K/(\pi(1-|z|))} \quad \forall z \in D(0, 1).$$

Hints: We may assume that F is not identically 0, the conclusion being trivial otherwise. Then F has only countably many zeros. Given $r < 1$, there exists then an $R \in (r, 1)$ such that F is zero-free on $C(0, R)$. Apply 12.4(ii) to get

$$\log|F(z)| \le \frac{R + |z|}{R - |z|} L(R, F) \le \frac{2}{R - |z|} L(R, F) \quad \forall z \in D(0, R)\backslash F^{-1}(0).$$

Using hypothesis (*), remembering that $R > r$ and focusing only on $z \in D(0, r)$, we get

$$\log|F(z)| \le \frac{2}{R - |z|}\cdot\frac{K}{2\pi}$$

$$\le \frac{K}{\pi(r - |z|)}$$

$$|F(z)| \le e^{K/(\pi(r-|z|))} \qquad \forall z \in D(0, r)\backslash F^{-1}(0).$$

The restriction $z \notin F^{-1}(0)$ is now unnecessary, so we have

$$|F(z)| \le e^{K/(\pi(r-|z|))} \quad \forall z \in D(0, r),\ \forall r < 1.$$

Given $z \in D(0, 1)$, pick $r \in (|z|, 1)$ and let $r \uparrow 1$. From the last inequality follows

$$|F(z)| \le e^{K/(\pi(1-|z|))}.$$

(ii) *Use* (i) *to show that the boundedness hypothesis in* 7.4 *can be weakened to* $\sup_n \sup_{r<1} \int_0^{2\pi} \log^+|f_n(re^{i\theta})|\,d\theta < \infty$ *and the conclusion of* 7.4 *still follows.*

Hint: The finiteness of this supremum plus (i) insure that $\{f_n\}$ is locally uniformly bounded in $D(0, 1)$. (See PRIVALOV [1924c] for a related extension of 7.4.)

§ 2 Miranda's Theorem

This section is somewhat long and a bit arduous. It culminates in a generalization of Schottky's theorem mentioned in the introductory remarks. Here the function f omits 0 but one of its derivatives $f^{(n)}$ omits 1. The case $n = 0$ is Schottky's theorem and that is adequate for all the applications we make here, save one (12.18). Moreover, a self-contained proof of Schottky's theorem is offered in the last section of this chapter. Therefore the reader can skip this section and do § 7 instead. He can then cite Schottky instead of Miranda in the subsequent sections and get the case $n = 0$ of all subsequent theorems. As noted, this will secure for him all the applications except 12.18.

Lemma 12.8 *If f is holomorphic and zero-free in the open set U and $F = f'/f$, then for every $n \geq 1$ the quotient $f^{(n)}/f$ is a sum of $n!$ or fewer terms, each of which is a product of n or fewer factors of the form $F^{(j)}$ with $0 \leq j < n$.*

Proof: By induction on n. Trivial for $n = 1$. If true for $n = N$, consider $n = N + 1$. We have that $f^{(N)}$ is a sum of $N!$ or fewer terms of the form $f \cdot F^{(j_1)} \cdots F^{(j_k)}$ $(0 \leq j_1, \ldots, j_k < N, k \leq N)$. Since

$$\frac{[fF^{(j_1)} \cdots F^{(j_k)}]'}{f} = F^{(0)}F^{(j_1)} \cdots F^{(j_k)} + \sum_{l=1}^{k} F^{(j_1)} \cdots F^{(j_l+1)} \cdots F^{(j_k)},$$

a sum of $N + 1$ or fewer terms, each of which is a product of $N + 1$ or fewer factors of the form $F^{(j)}$ with $0 \leq j < N + 1$, the assertion for $n = N + 1$ follows.

Theorem 12.9 *For each integer $n \geq 1$ there exists a constant A_n such that for any f which is holomorphic and zero-free in $D(0, 1)$*

$$L\left(r, \frac{f^{(n)}}{f}\right) \leq A_n + n(n+2) \log \frac{1}{R-r} + n \log^+ V(R, f), \qquad 0 < r < R < 1.$$

Proof: Fix such an n and R, and let such an f be given. Let h be a holomorphic logarithm of f in $D(0, 1)$ and apply the Cauchy–Schwarz formula (5.18) to h in $D(0, R)$:

$$\begin{aligned} h(z) &= \frac{1}{2\pi} \int_0^{2\pi} \operatorname{Re} h(Re^{i\theta}) \left[\frac{Re^{i\theta} + z}{Re^{i\theta} - z}\right] d\theta + i \operatorname{Im} h(0) \\ &= \frac{1}{2\pi} \int_0^{2\pi} \frac{\operatorname{Re} h(Re^{i\theta}) Re^{i\theta}}{Re^{i\theta} - z} d\theta + \frac{z}{2\pi} \int_0^{2\pi} \frac{\operatorname{Re} h(Re^{i\theta})}{Re^{i\theta} - z} d\theta + i \operatorname{Im} h(0). \end{aligned}$$

Apply 2.14 to each of these latter integrals and amalgamate the results to conclude that

$$h'(z) = \frac{1}{2\pi}\int_0^{2\pi} \operatorname{Re} h(Re^{i\theta}) \frac{2Re^{i\theta}}{(Re^{i\theta} - z)^2}\, d\theta, \quad |z| < R.$$

Since $f = e^h$, $h' = f'/f$ and $\operatorname{Re} h = \log|f|$, so

$$(1) \quad \frac{f'(z)}{f(z)} = \frac{1}{2\pi}\int_0^{2\pi} \log|f(Re^{i\theta})| \frac{2Re^{i\theta}}{(Re^{i\theta} - z)^2}\, d\theta, \quad |z| < R.$$

Let $F = f'/f$. Repeated differentiations under the integral in (1) (i.e., appeals to 2.14) give

$$(2)\ F^{(j)}(z) = \frac{(j+1)!}{2\pi}\int_0^{2\pi} \log|f(Re^{i\theta})| \frac{2Re^{i\theta}}{(Re^{i\theta} - z)^{j+2}}\, d\theta, \quad |z| < R, j = 0, 1, \ldots.$$

It follows that

$$|F^{(j)}(z)| \leq \frac{(j+1)!}{2\pi}\int_0^{2\pi} |\log|f(Re^{i\theta})||\, \frac{2R}{(R - |z|)^{j+2}}\, d\theta, \quad |z| < R, j = 0, 1, \ldots$$

$$|F^{(j)}(re^{it})| \leq \frac{2R(j+1)!}{(R-r)^{j+2}} \frac{1}{2\pi}\int_0^{2\pi} |\log|f(Re^{i\theta})||\, d\theta$$

$$(3) \quad = \frac{2R(j+1)!}{(R-r)^{j+2}} V(R,f), \quad t \in \mathbb{R}, r < R, j = 0, 1, \ldots.$$

Thus, since $0 < R - r < R < 1$,

$$(4)\ |F^{(j)}(re^{it})| \leq 2(n+1)!\,(R-r)^{-n-2}V(R,f), \quad t \in \mathbb{R}, r < R, j = 0, 1, \ldots, n.$$

Use the fact that $\log^+$ is a non-decreasing function and

$$\log^+(n!\,x) \leq \log(n!) + \log^+(x) \quad (12.1(\text{ii}))$$

together with the last lemma and (4), to see that

$$\log^+\left|\frac{f^{(n)}(re^{it})}{f(re^{it})}\right| \leq \log(n!) + \log^+[2(n+1)!\,(R-r)^{-n-2}V(R,f)]^n,$$

$t \in \mathbb{R}$, $r < R$. Then use 12.1(ii) again to get

$$\log^+\left|\frac{f^{(n)}(re^{it})}{f(re^{it})}\right|$$

$$\le \log(n!) + n\left[\log(2(n+1)!) + (n+2)\log\frac{1}{R-r} + \log^+ V(R,f)\right]$$

$$(5) \qquad = \log(n!\,(2(n+1)!)^n) + n(n+2)\log\frac{1}{R-r} + n\log^+ V(R,f),$$

$t \in \mathbb{R}$, $r < R$. Christen this unpleasant constant A_n and integrate in (5) to get

$$L\left(r, \frac{f^{(n)}}{f}\right) \le A_n + n(n+2)\log\frac{1}{R-r} + n\log^+ V(R,f), \quad r < R,$$

as desired.

Lemma 12.10 (Bureau's Bootstrap Lemma) *Let $0 \le a < b \le 1$ and f be a non-negative, non-decreasing function on (a, b). Suppose that for some positive constants A and B, f satisfies*

$$(1) \qquad f(r) \le A\log^+ f(R) + A\log\frac{1}{R-r} + B \quad \forall a < r < R < b.$$

If $B \ge 2A(A+2)$, then f also satisfies

$$(2) \qquad f(r) \le 2A\log\frac{1}{b-r} + 2B \quad \forall a < r < b.$$

Hence, regardless of the size of B, f satisfies

$$(3) \qquad f(r) \le 2A\log\frac{1}{b-r} + 2B + 4A(A+2) \quad \forall a < r < b.$$

Proof: Suppose that

$$(4) \qquad B \ge 2A(A+2)$$

and that nevertheless (2) fails:

$$(5) \qquad f(r_0) > 2A\log\frac{1}{b-r_0} + 2B \quad \text{for some } a < r_0 < b.$$

Then (5) also holds with b replaced by $b' < b$, if b' is sufficiently close to b. Moreover, f is bounded on (a, b'), by $f(b')$. Therefore without loss of generality we can suppose, in trying to deduce a contradiction from (1), (4) and (5), that f is bounded on (a, b). Set

$$r_1 = \frac{r_0 + b}{2} = r_0 + \frac{b - r_0}{2}$$

and apply (1) with $r = r_0$, $R = r_1$ to get

$$f(r_0) \le A \log^+ f(r_1) + A \log \frac{2}{b - r_0} + B$$

$$\log^+ f(r_1) \ge \frac{1}{A} f(r_0) + \log \frac{b - r_0}{2} - \frac{B}{A},$$

whence, recalling 12.1(iv),

$$\sqrt{f(r_1)} \ge \frac{1}{A} f(r_0) + \log \frac{b - r_0}{2} - \frac{B}{A}$$

$$> 2 \log \frac{1}{b - r_0} + \frac{2B}{A} + \log \frac{b - r_0}{2} - \frac{B}{A} \quad \text{by (5)}$$

$$= \log \frac{2}{b - r_0} + \frac{B}{A} - \log 4 \tag{6}$$

$$> \frac{B}{A} - \log 4 \quad (\text{since } b - r_0 < b - a \le 1). \tag{6$'$}$$

Now from (4)

$$B \ge 2A(A + 2) = 2A^2 + 4A$$

$$B - 4A \ge 2A^2$$

$$\frac{B}{A} - \log 4 > \frac{B}{A} - 4 \ge 2A \tag{7}$$

and also from (4)

$$B \ge 2A(A + 2) \ge 2A(A + \log 4)$$

$$B^2 - 2AB(\log 4 + A) + A^2 \log^2 4 \ge 0$$

$$(B - A \log 4)^2 \ge 2A^2 B$$

$$\left(\frac{B}{A} - \log 4\right)^2 \ge 2B. \tag{8}$$

Now multiply inequality (6) by (6)′, keeping (7) and (8) in mind:

$$f(r_1) > \left(\frac{B}{A} - \log 4\right)\left(\log \frac{2}{b - r_0} + \frac{B}{A} - \log 4\right)$$

$$\ge 2A \log \frac{2}{b - r_0} + 2B$$

$$= 2A \log \frac{1}{b - r_1} + 2B. \tag{9}$$

From the (5) to (9) implication it follows that if we define

$$r_n = \frac{r_{n-1} + b}{2}, \quad n = 1, 2, \ldots,$$

then

$$f(r_n) > 2A \log \frac{1}{b - r_n} + 2B.$$

Since $b - r_n = (b - r_0)/2^n$, this gives

$$f(r_n) > 2A \log \frac{2^n}{b - r_0} + 2B$$

and this makes f unbounded, a contradiction.

Thus (2) is established. For (3), just replace B by $B + 2A(A + 2)$ in (1) and quote (2) for this new constant.

Lemma 12.11 *For each non-negative integer n there exists a constant B_n such that every f which is holomorphic and zero-free in $D(0, 1)$ and satisfies*

(i) $\quad |f^{(n)}(z)| \geq 1 \quad \forall z \in D(0, 1)$

also satisfies

(ii) $\quad \log M(r,f) \leq \dfrac{1}{1 - r}\left[8 \log^+|f(0)| + B_n \log \dfrac{2}{1 - r}\right] \quad \forall 0 \leq r < 1.$

Proof: We have from 12.2 and 12.1(v)

(1) $\quad V(r,f) = L(r,f) + L\left(r, \dfrac{1}{f}\right)$

and from 12.3 and 12.1(v)

$$\log|f(0)| = \frac{1}{2\pi}\int_0^{2\pi} \log|f(re^{i\theta})|\,d\theta = L(r,f) - L\left(r, \frac{1}{f}\right).$$

Therefore

(2) $\quad V(r,f) = 2L\left(r, \dfrac{1}{f}\right) + \log|f(0)|.$

By hypothesis

$$\left|\frac{1}{f}\right| \leq \left|\frac{f^{(n)}}{f}\right|$$

and therefore from 12.9

(3) $\quad L\left(r, \dfrac{1}{f}\right) \leq L\left(r, \dfrac{f^{(n)}}{f}\right) \leq A_n + n(n + 2) \log \dfrac{1}{R - r} + n \log^+ V(R,f)$

holds for $0 < r < R < 1$. Use (3) in (2) to get

$$(4)\qquad V(r,f) \le 2A_n + \log^+|f(0)| + 2n(n+2)\log\frac{1}{R-r} + 2n\log^+ V(R,f),$$

holding for $0 < r < R < 1$. Now by (1) and 12.5, $V(r,f)$ is a non-decreasing function of r, so we can apply Bureau's lemma 12.10 to conclude that

$$V(r,f) \le 4A_n + 2\log^+|f(0)| + 4(2n(n+2))(2n(n+2)+2) + 4n(n+2)\log\frac{1}{1-r}$$

$$(5)\qquad = 2\log^+|f(0)| + a_n + b_n\log\frac{1}{1-r}, \quad 0 < r < 1$$

(with the obvious definitions of a_n and b_n). Next apply 12.4(ii) with $R = (1+r)/2$ to see that

$$(6)\qquad \log M(r,f) \le \frac{4}{1-r}L\left(\frac{1+r}{2},f\right) \overset{(1)}{\le} \frac{4}{1-r}V\left(\frac{1+r}{2},f\right), \quad 0 < r < 1.$$

Apply (5) and (6) together to get

$$(7)\qquad \log M(r,f) \le \frac{1}{1-r}\left[8\log^+|f(0)| + 4a_n + 4b_n\log\frac{2}{1-r}\right], \quad 0 < r < 1.$$

Choose for B_n any number which satisfies

$$B_n \ge 8b_n \quad\text{and}\quad B_n \ge \frac{8a_n}{\log 2}.$$

Then

$$B_n\log\frac{2}{1-r} \ge \tfrac{1}{2}B_n\log 2 + \tfrac{1}{2}B_n\log\frac{2}{1-r}$$

$$\ge 4a_n + 4b_n\log\frac{2}{1-r}, \quad 0 \le r < 1$$

and (7) gives

$$(8)\qquad \log M(r,f) \le \frac{1}{1-r}\left[8\log^+|f(0)| + B_n\log\frac{2}{1-r}\right], \quad 0 < r < 1.$$

(By continuity the inequality (8) holds as well at $r = 0$.)

Lemma 12.12 *For each non-negative integer n there exists a constant C_n such that every f which is holomorphic and zero-free in $D(0, 1)$ and satisfies*

(i) $|f^{(n)}(z)| \le 1 \quad \forall z \in D(0, 1)$

also satisfies

(ii) $$\log M(r,f) \leq \frac{1}{1-r}\left[16 \log^{+}|f(0)| + C_n \log \frac{2}{1-r}\right] \quad \forall 0 \leq r < 1.$$

Proof: We go by induction on n. If $n = 0$, (i) says that $C_0 = 0$ will work in (ii). Suppose the result true for $0 \leq n < N$ and consider $n = N$. Let such a function and an $r \in [0, 1)$ be given and set $r' = (1 + r)/2$. Consider two cases:

Case I: There exists $z' \in D(0, r')$ such that

(1) $$|f^{(N-1)}(z')| \leq 1.$$

Then for all $z \in D(0, 1)$

$$\begin{aligned} |f^{(N-1)}(z)| &\leq |f^{(N-1)}(z')| + |f^{(N-1)}(z) - f^{(N-1)}(z')| \\ &\leq 1 + \left|\int_{[z',z]} f^{(N)}\right| \leq 1 + |z - z'| \quad \text{by (i)} \\ &< 3. \end{aligned}$$

Apply the induction hypothesis to the function $\frac{1}{3}f$:

$$\log M(r, \tfrac{1}{3}f) \leq \frac{1}{1-r}\left[16 \log^{+}|\tfrac{1}{3}f(0)| + C_{N-1} \log \frac{2}{1-r}\right],$$

whence, since $\log^{+}|\frac{1}{3}f(0)| \leq \log^{+}|f(0)|$,

$$\begin{aligned} \log M(r,f) &= \log M(r, \tfrac{1}{3}f) + \log 3 \\ &\leq \frac{1}{1-r}\left[16 \log^{+}|f(0)| + C_{N-1} \log \frac{2}{1-r}\right] + \log 3 \end{aligned}$$

(2) $$\leq \frac{1}{1-r}\left[16 \log^{+}|f(0)| + \left(C_{N-1} + \frac{\log 3}{\log 2}\right) \log \frac{2}{1-r}\right].$$

Case II:

(3) $$|f^{(N-1)}(z)| > 1 \quad \forall z \in D(0, r').$$

In this case the function $F(z) = f(r'z)/(r')^{N-1}$ ($|z| < 1$) satisfies the hypotheses of the last lemma with $n = N - 1$ and so we have

(4) $$\log M(R, F) \leq \frac{1}{1-R}\left[8 \log^{+}|F(0)| + B_{N-1} \log \frac{2}{1-R}\right], \quad 0 \leq R < 1.$$

We have for $0 \leq R < 1$

$$\sup_{|z| \leq R} |F(z)| = \frac{1}{(r')^{N-1}} \sup_{|z| \leq r'R} |f(z)|.$$

In particular then,

$$\log M\left(r, \frac{1}{(r')^{N-1}} f\right) = \log M(r/r', F)$$

and (4) with $R = r/r'$ gives

$$\begin{aligned}
&\log M(r,f)\\
&\le \log M\left(r, \frac{1}{(r')^{N-1}} f\right)\\
&\le \frac{1}{1 - r/r'}\left[8 \log^+\left|\frac{f(0)}{(r')^{N-1}}\right| + B_{N-1} \log \frac{2}{1 - r/r'}\right]\\
&\le \frac{1+r}{1-r}\left[8 \log^+|f(0)| + (N-1) 8 \log \frac{2}{1+r} + B_{N-1} \log \frac{2(1+r)}{1-r}\right],
\end{aligned}$$

by definition of r' and by 12.1(ii),

$$\begin{aligned}
&\le \frac{1}{1-r}\left[16 \log^+|f(0)| + 16(N-1) \log \frac{2}{1+r} + 2B_{N-1} \log \frac{4}{1-r}\right]\\
&\le \frac{1}{1-r}\left[16 \log^+|f(0)| + (16(N-1) + 2B_{N-1}) \log 2 \right.\\
&\qquad \left. + 2B_{N-1} \log \frac{2}{1-r}\right]
\end{aligned}$$

$$\le \frac{1}{1-r}\left[16 \log^+|f(0)| + (16(N-1) + 4B_{N-1}) \log \frac{2}{1-r}\right]. \tag{5}$$

From (2) and (5) we are lead to define

$$C_N = \max\left\{C_{N-1} + \frac{\log 3}{\log 2}, 16(N-1) + 4B_{N-1}\right\}$$

and thereby have (ii) valid for $n = N$.

Lemma 12.13 *For each non-negative integer n there exists a constant K_n such that every f which is holomorphic and zero-free in $D(0, 1)$ and satisfies*

(i) $f^{(n)} - 1$ *is zero-free and* $\left|\dfrac{f^{(n+1)}}{f^{(n)} - 1}\right| \le 1$ *in* $D(0, 1)$

also satisfies

(ii) $\log M(r,f) \le \dfrac{1}{1-r}\left[32 \log^+|f(0)| + K_n \log \dfrac{2}{1-r}\right] \quad \forall 0 \le r < 1.$

Proof: Let such a function and an $r \in [0, 1)$ be given and set $r' = (1 + r)/2$. Consider two cases as before:

Case I: There exists $z' \in D(0, r')$ such that

(1) $|f^{(n)}(z')| \leq 1.$

Let F be a holomorphic logarithm for $f^{(n)} - 1$ in $D(0, 1)$:

(2) $f^{(n)} - 1 = e^F$

(3) $\operatorname{Re} F(z') = \log|f^{(n)}(z') - 1| \overset{(1)}{\leq} \log 2.$

As in the last proof

$$\operatorname{Re} F(z) \leq \operatorname{Re} F(z') + \left|\operatorname{Re}\int_{[z',z]} F'\right| \leq \log 2 + \left|\operatorname{Re}\int_{[z',z]} \frac{f^{(n+1)}}{f^{(n)} - 1}\right|$$

$$\leq \log 2 + |z' - z| \quad \text{by (i)}$$

(4) $$< \log 2 + 2 \qquad \forall z \in D(0, 1).$$

It follows that

(5) $|f^{(n)}| \leq 1 + |f^{(n)} - 1| \overset{(2)}{=} 1 + e^{\operatorname{Re} F} \overset{(4)}{\leq} 1 + 2e^2.$

Let $c = 1 + 2e^2$ and form

(6) $\tilde{F}(z) = \dfrac{f(r'z)}{c(r')^n}, \quad z \in D(0, 1).$

Because of (5) we may apply the last lemma to $\tilde{F}$:

(7) $$\log M(R, \tilde{F}) \leq \frac{1}{1 - R}\left[16 \log^+|\tilde{F}(0)| + C_n \log \frac{2}{1 - R}\right] \quad \forall 0 \leq R < 1.$$

We have for $0 \leq R < 1$

$$\sup_{|z| \leq R} |\tilde{F}(z)| = \frac{1}{c(r')^n} \sup_{|z| \leq r'R} |f(z)|.$$

In particular then,

$$\log M\left(r, \frac{1}{c(r')^n} f\right) = \log M(r/r', \tilde{F})$$

and (7) with $R = r/r'$ gives

$$\log M(r,f) \leq \log c + \log M\left(r, \frac{1}{c(r')^n} f\right) = \log c + \log M(r/r', \tilde{F})$$

$$\leq \log c + \frac{1}{1 - r/r'}\left[16 \log^+\left|\frac{f(0)}{c(r')^n}\right| + C_n \log \frac{2}{1 - r/r'}\right]$$

$$\leq \log c + \frac{1}{1 - r/r'}\left[16 \log^+\left|\frac{f(0)}{(r')^n}\right| + C_n \log \frac{2}{1 - r/r'}\right],$$

since $c > 1$.

Appraisals like those in the proof of inequality (5) of the last lemma then lead us to

$$\log M(r,f) \le \log c + \frac{1}{1-r}\left[32\log^+|f(0)| + (32n + 4C_n)\log\frac{2}{1-r}\right]$$

$$\text{(8)} \qquad \le \frac{1}{1-r}\left[32\log^+|f(0)| + \left(32n + 4C_n + \frac{\log c}{\log 2}\right)\log\frac{2}{1-r}\right].$$

Case II:

(9) $|f^{(n)}(z)| > 1 \quad \forall z \in D(0, r')$.

This is exactly case II of the last proof and so, as was deduced there,

$$\text{(10)} \quad \log M(r,f) \le \frac{1}{1-r}\left[16\log^+|f(0)| + (16n + 4B_n)\log\frac{2}{1-r}\right].$$

From (8) and (10) it appears that the choice

$$K_n = \max\left\{16n + 4B_n, 32n + 4C_n + \frac{\log c}{\log 2}\right\}$$

will validate (ii).

Theorem 12.14 (Miranda [1935]) *For each non-negative integer n there exists a constant M_n such that every f which is holomorphic in $D(0, 1)$ and for which f and $f^{(n)} - 1$ are zero-free satisfies*

$$\log M(r,f) \le \frac{1}{1-r}\left[64\log^+|f(0)| + \frac{M_n}{(1-r)^4}\right] \quad \forall 0 \le r < 1.$$

Proof: Let such a function f and an $r \in [0, 1)$ be given and set $r' = (1 + r)/2$. Three cases arise, of which the first two are trivial reductions to previous lemmas:

Case I:

(1) $|f^{(n)}(z)| \le 2 \quad \forall z \in D(0, r')$.

Then apply 12.12 to the function $F(z) = f(r'z)/2(r')^n$ and, via the by-now-familiar routine, translate the resulting inequality about F into one about f. There results

$$\log M(r,f)$$

$$< \log 2 + \frac{1+r}{1-r}\left[16\log^+|f(0)| + 16n\log\left(\frac{2}{1+r}\right) + C_n\log\frac{2(1+r)}{1-r}\right]$$

$$< \log 2 + \frac{1}{1-r}\left[32\log^+|f(0)| + 32n\log\frac{2}{1+r} + 2C_n\log\frac{4}{1-r}\right]$$

$$\le \frac{1}{1-r}\left[32\log^+|f(0)| + (32n + 2C_n + 1)\log 2 + 2C_n\log\frac{2}{1-r}\right]$$

$$\text{(2)} \quad \le \frac{1}{1-r}\left[32\log^+|f(0)| + (32n + 4C_n + 1)\log\frac{2}{1-r}\right].$$

Case II:

(3) $$\left|\frac{f^{(n+1)}(z)}{f^{(n)}(z)-1}\right| \le 1 \quad \forall z \in D(0,r').$$

Then for the function $F(z) = f(r'z)/(r')^n$ we have $F^{(n)}(z) = f^{(n)}(r'z)$, whence

$$\left|\frac{F^{(n+1)}(z)}{F^{(n)}(z)-1}\right| = r'\left|\frac{f^{(n+1)}(r'z)}{f^{(n)}(r'z)-1}\right| \le r' < 1 \quad \forall z \in D(0,1)$$

and so 12.13 is applicable to F. After the canonical maneuvers there results

$$\log M(r,f) \le \frac{1}{1-r}\left[64\log^+|f(0)| + 64n\log\frac{2}{1+r} + 2K_n\log\frac{4}{1-r}\right]$$

(4) $$< \frac{1}{1-r}\left[64\log^+|f(0)| + (64n + 4K_n)\log\frac{2}{1-r}\right].$$

Case III:

(5) $\quad |f^{(n)}(z')| > 2 \quad$ for some $z' \in D(0,r')$

and

(6) $$\left|\frac{f^{(n+1)}(z'')}{f^{(n)}(z'')-1}\right| > 1 \quad \text{for some } z'' \in D(0,r').$$

From (6) we have in particular that $f^{(n+1)}$ is not identically 0 in $D(0, 1)$ and so, as noted before (penultimate paragraph of the proof of 12.4), $f^{(n+1)}$ has a zero on $C(0, \rho)$ for at most countably many $\rho \in [0, 1)$. In what follows consider

(7) $$\rho > \frac{1+r'}{2} \quad \text{such that } 0 \notin f^{(n+1)}(C(0,\rho)).$$

Then on $C(0, \rho)$ we may write

$$\frac{1}{f} = \frac{f^{(n)}}{f} - \frac{f^{(n)}-1}{f^{(n+1)}}\cdot\frac{f^{(n+1)}}{f}$$

and apply 12.1(ii) and (iii) to see that

$$L\left(\rho,\frac{1}{f}\right) \le \log 2 + L\left(\rho,\frac{f^{(n)}}{f}\right) + L\left(\rho,\frac{f^{(n+1)}}{f}\right) + L\left(\rho,\frac{f^{(n)}-1}{f^{(n+1)}}\right).$$

As noted in the opening lines of the proof of 12.11,

$$V(\rho,f) = 2L\left(\rho,\frac{1}{f}\right) + \log|f(0)|,$$

and so

(8) $$V(\rho,f) \le \log|f(0)| + \log 4 + 2L\left(\rho,\frac{f^{(n)}}{f}\right)$$

$$+ 2L\left(\rho,\frac{f^{(n+1)}}{f}\right) + 2L\left(\rho,\frac{f^{(n)}-1}{f^{(n+1)}}\right).$$

Apply 12.4(iii) to $F = f^{(n+1)}/(f^{(n)} - 1)$, $z = z''$ and $R = \rho$:

$$0 \overset{(6)}{<} \log\left|\frac{f^{(n+1)}(z'')}{f^{(n)}(z'') - 1}\right| \le \frac{\rho + |z''|}{\rho - |z''|} L\left(\rho, \frac{f^{(n+1)}}{f^{(n)} - 1}\right) - \frac{\rho - |z''|}{\rho + |z''|} L\left(\rho, \frac{f^{(n)} - 1}{f^{(n+1)}}\right),$$

$$\begin{aligned} L\left(\rho, \frac{f^{(n)} - 1}{f^{(n+1)}}\right) &\le \left(\frac{\rho + |z''|}{\rho - |z''|}\right)^2 L\left(\rho, \frac{f^{(n+1)}}{f^{(n)} - 1}\right) \\ &\le \frac{4}{(\rho - r')^2} L\left(\rho, \frac{f^{(n+1)}}{f^{(n)} - 1}\right) \\ &\le \frac{64}{(1 - r)^2} L\left(\rho, \frac{f^{(n+1)}}{f^{(n)} - 1}\right). \end{aligned} \tag{9}$$

Apply 12.9 with the n there equal 1 and the f there equal $f^{(n)} - 1$:

$$L\left(\rho, \frac{f^{(n+1)}}{f^{(n)} - 1}\right) \le A_1 + 3 \log \frac{1}{R' - \rho} + \log^+ V(R', f^{(n)} - 1), \quad \rho < R' < 1. \tag{10}$$

Similarly appraise the terms $L(\rho, f^{(n)}/f)$ and $L(\rho, f^{(n+1)}/f)$ in (8) via 12.9 and use (9) and (10) to get

$$\begin{aligned} V(\rho, f) \le \log|f(0)| &+ (\log 4 + 2A_n + 2A_{n+1}) \\ &+ 2n(n + 2) \log \frac{1}{R - \rho} + 2(n + 1)(n + 3) \log \frac{1}{R - \rho} \\ &+ 2n \log^+ V(R, f) + 2(n + 1) \log^+ V(R, f) \\ &+ \frac{128}{(1 - r)^2}\left[A_1 + 3 \log \frac{1}{R' - \rho} + \log^+ V(R', f^{(n)} - 1)\right], \end{aligned}$$

holding for all $\rho < R' < R < 1$. Take $R' = \frac{1}{2}(\rho + R)$ and get

$$\begin{aligned} V(\rho, f) \le \log|f(0)| &+ (\log 4 + 2A_n + 2A_{n+1}) \\ &+ 2(n(n + 2) + (n + 1)(n + 3)) \log \frac{1}{R - \rho} \\ &+ 2(2n + 1) \log^+ V(R, f) + \frac{128A_1}{(1 - r)^2} + \frac{384}{(1 - r)^2} \log \frac{2}{R - \rho} \\ &+ \frac{128}{(1 - r)^2} \log^+ V(R', f^{(n)} - 1), \end{aligned}$$

holding for all $\rho < R < 1$. Set

$$\begin{aligned} H_n = \log 4 &+ 2A_n + 2A_{n+1} + 2n(n + 2) \\ &+ 2(n + 1)(n + 3) + 128A_1 + 384 \end{aligned} \tag{11}$$

and have then *a fortiori*

$$(12)\quad V(\rho,f) \le \log|f(0)| + H_n + \frac{H_n}{(1-r)^2}\log\frac{2}{R-\rho}$$

$$+ \frac{H_n}{(1-r)^2}[\log^+ V(R,f) + \log^+ V(R',f^{(n)}-1)],$$

holding for all $\rho < R < 1$, $R' = \frac{1}{2}(\rho + R)$. As in the derivation of (9), use the fact $|f^{(n)}(z') - 1| > 1$ (from (5)) and 12.4(iii) to deduce that

$$L\left(R', \frac{1}{f^{(n)}-1}\right) \le \left(\frac{R'+|z'|}{R'-|z'|}\right)^2 L(R', f^{(n)}-1)$$

$$\le \frac{4}{(R'-\rho)^2} L(R', f^{(n)}-1)$$

$$= \frac{16}{(R-\rho)^2} L(R', f^{(n)}-1).$$

Therefore

$$V(R', f^{(n)}-1)$$

$$= L(R', f^{(n)}-1) + L\left(R', \frac{1}{f^{(n)}-1}\right)$$

$$\le L(R', f^{(n)}-1) + \frac{16}{(R-\rho)^2} L(R', f^{(n)}-1) \le \frac{17}{(R-\rho)^2} L(R', f^{(n)}-1)$$

$$\le \frac{17}{(R-\rho)^2}[L(R', f^{(n)}) + \log 2], \quad \text{by 12.1(iii)}$$

$$\le \frac{17}{(R-\rho)^2}\left[L\left(R', \frac{f^{(n)}}{f}\right) + L(R',f) + \log 2\right], \quad \text{by 12.1(ii)}$$

$$(13)\quad \le \frac{17}{(R-\rho)^2}\left[L\left(R', \frac{f^{(n)}}{f}\right) + V(R,f) + \log 2\right], \quad \text{since } R' < R.$$

Now apply $\log^+$ to (13), again making use of 12.1:

$$\log^+ V(R', f^{(n)}-1)$$

$$\le \log\frac{17}{(R-\rho)^2} + \log^+ L\left(R', \frac{f^{(n)}}{f}\right) + \log^+ V(R,f) + \log^+ \log 2 + \log 3$$

$$\le \log\frac{17}{(R-\rho)^2} + L\left(R', \frac{f^{(n)}}{f}\right) + \log^+ V(R,f) + \log 3,$$

using the obvious facts $\log^+ x \le x$ and $\log 2 < 1$.

An application of 12.9 then gives

$$\log^+ V(R', f^{(n)} - 1) \le \log \frac{17}{(R-\rho)^2} + A_n + n(n+2)\log\frac{1}{R-R'}$$
$$+ n\log^+ V(R,f) + \log^+ V(R,f) + \log 3$$

(14)
$$= A_n + \log\frac{51}{4} + 2\log\frac{2}{R-\rho} + n(n+2)\log\frac{2}{R-\rho}$$
$$+ (n+1)\log^+ V(R,f).$$

We set

(15) $H_n' = A_n + \log \frac{51}{4} + (n+1)(n+2)$

and have *a fortiori* from (14)

(16) $\log^+ V(R', f^{(n)} - 1) \le H_n' + H_n' \log\frac{2}{R-\rho} + H_n' \log^+ V(R,f).$

If we insert (16) into (12),

$$V(\rho,f) \le \log|f(0)| + H_n + \frac{H_n H_n'}{(1-r)^2} + \frac{H_n(H_n'+1)}{(1-r)^2}\log\frac{2}{R-\rho}$$
$$+ \frac{H_n(H_n'+1)}{(1-r)^2}\log^+ V(R,f)$$
$$\le \log^+|f(0)| + H_n + \frac{H_n(2H_n'+1)}{(1-r)^2} + \frac{H_n(H_n'+1)}{(1-r)^2}\log\frac{1}{R-\rho}$$
$$+ \frac{H_n(H_n'+1)}{(1-r)^2}\log^+ V(R,f)$$

(17)
$$\le \log^+|f(0)| + \frac{2H_n(H_n'+1)}{(1-r)^2} + \frac{H_n(H_n'+1)}{(1-r)^2}\log\frac{1}{R-\rho}$$
$$+ \frac{H_n(H_n'+1)}{(1-r)^2}\log^+ V(R,f).$$

This inequality is valid for a dense set of $\rho \in ((1+r')/2, 1)$ and for each such ρ it is valid for all $\rho < R < 1$. But an easy uniform continuity argument on the integrand defining $V(\rho,f)$ shows that it is a continuous function of ρ and so it follows that (17) is valid for all $\rho \in ((1+r')/2, 1)$ and all $\rho < R < 1$. Since (12.5), $V(\rho,f)$ is a non-decreasing function of ρ, we are in a position to apply Bureau's lemma 12.10. We get

(18) $V(\rho,f) \le 2\log^+|f(0)| + \dfrac{4H_n(H_n'+1)}{(1-r)^2}$

$$+ \frac{4H_n(H_n'+1)}{(1-r)^2}\left[\frac{H_n(H_n'+1)}{(1-r)^2} + 2\right]$$
$$+ \frac{2H_n(H_n'+1)}{(1-r)^2}\log\frac{1}{1-\rho}, \qquad \frac{1+r'}{2} < \rho < 1.$$

Let

(19) $L_n = 4H_n(H_n' + 1)$

and have *a fortiori* from (18)

$$(20)\quad V(\rho,f) \le 2\log^+|f(0)| + \frac{L_n}{(1-r)^2} + \frac{L_n}{(1-r)^2}\left[\frac{L_n}{(1-r)^2} + 2\right]$$

$$+ \frac{L_n}{(1-r)^2}\log\frac{1}{1-\rho}, \qquad \frac{1+r'}{2} < \rho < 1.$$

An eligible ρ is

$$(21)\quad \rho = \frac{3+r'}{4} = \frac{7+r}{8}.$$

We have $\rho - r = \frac{7}{8}(1-r) > (1-r)/2$ so we may apply 12.4(ii) to see that

$$\log M(r,f) \le \frac{\rho + r}{\rho - r} L(\rho,f) \le \frac{\rho + r}{\rho - r} V(\rho,f)$$

$$(22)\quad \le \frac{2}{\rho - r} V(\rho,f) \le \frac{4}{1-r} V(\rho,f).$$

Use (21) in (20) and insert the result into (22):

$$\log M(r,f) \le \frac{8}{1-r}\log^+|f(0)| + \frac{4L_n}{(1-r)^3} + \frac{4L_n}{(1-r)^3}\left[\frac{L_n}{(1-r)^2} + 2\right]$$

$$+ \frac{4L_n}{(1-r)^3}\log\frac{8}{1-r}$$

$$\le \frac{8}{1-r}\log^+|f(0)| + \frac{12L_n}{(1-r)^3} + \frac{4L_n^2}{(1-r)^5} + \frac{32L_n}{(1-r)^4}$$

(since $\log x \le x$ for all $x > 0$)

$$(23)\quad \le \frac{8}{1-r}\log^+|f(0)| + \frac{48L_n^2}{(1-r)^5} \quad (\text{since } L_n \ge 1).$$

From (2), (4) and (23) we see that we may take for M_n the number

$$M_n = \max\{64n + 8C_n + 2, 128n + 8K_n, 48L_n^2\}.$$

Exercise 12.15 (Schottky) *When $n = 0$, a stronger conclusion is actually valid in Miranda's theorem, namely*

$$\log M(r,f) \le \frac{1}{1-r}\left[12\log^+|f(0)| + M_0'\log\frac{2}{1-r}\right], \quad 0 \le r < 1$$

for some constant M_0'. Prove this by refining the argument for case III above, notably by citing 12.4(iii) (*with* $F = f'/(f-1)$) *to assert that*

$$(9)' \quad L\left(r, \frac{f-1}{f'}\right) \le L\left(r, \frac{f'}{f-1}\right) + \log\left|\frac{f(0)-1}{f'(0)}\right|$$

if $f'(0) \neq 0$, *and using* (9)′ *instead of* (9) *in the rest of the proof. Similarly, get a simple estimate in* (13) *free of the factor* $1/(R-\rho)^2$.

§ 3 Immediate Applications of Miranda

Corollary 12.16 *If F is a zero-free entire function and for some non-negative integer n the range of $F^{(n)}$ does not contain all non-zero complex numbers, then F is a (non-zero) constant.*

Proof: If $\mathbb{C}\backslash\{0\} \not\subset F^{(n)}(\mathbb{C})$, then by scaling we may suppose that $1 \notin F^{(n)}(\mathbb{C})$. Consider $R \ge \frac{1}{2}$. Set $f(z) = F(2Rz)/(2R)^n$. Then f is holomorphic and zero-free in $D(0, 1)$ and $f^{(n)}(z) = F^{(n)}(2Rz)$, so $f^{(n)} - 1$ is also zero-free. Take $r = \frac{1}{2}$ in Miranda's theorem to conclude that

$$\log M(\tfrac{1}{2}, f) \le 2[64 \log^+|f(0)| + 16M_n] = 128 \log^+\left|\frac{F(0)}{(2R)^n}\right| + 32M_n$$

$$(1) \qquad \le 128 \log^+|F(0)| + 32M_n,$$

since $\log^+$ is non-decreasing and $2R \ge 1$. Let B denote the number on the right side of (1). Then (1) says that $\sup_{|z| \le 1/2} |f(z)| \le e^B$, that is,

$$\sup_{|z| \le R} \frac{|F(z)|}{(2R)^n} \le e^B$$

$$(2) \qquad \sup_{|z| \le R} |F(z)| \le (2^n e^B) R^n,$$

holding for all sufficiently large R. It follows from this and 6.33 that F is a polynomial of degree not greater than n. Since F has no zeros, it must in fact be constant (4.49).

Corollary 12.17 (Picard's Little Theorem) *At most one complex number is absent from the range of a non-constant entire function.* (Cf. 12.28.) *At most two complex numbers are absent from the range of a non-constant function which is meromorphic in the whole plane.*

Proof: If f is entire and $a, b \notin f(\mathbb{C})$ where $a \neq b$, then the entire function $F = f - a$ is zero-free and $0 \neq b - a \notin F(\mathbb{C})$, so by 12.16 F is constant. If G is meromorphic in $\mathbb{C}$ and a, b, c are distinct complex numbers not in the range of G, then $1/(G - a)$ extends to an entire function which omits $1/(b - a)$ and $1/(c - a)$.

Remark: The non-constant functions e^z and $1/(1 - e^z)$ are examples where one or two values, respectively, are omitted.

Corollary 12.18 (Pólya–Saxer) *If f is an entire function and $f \cdot f' \cdot f''$ is zero-free, then $f(z) \equiv e^{az+b}$ for some constants a and b.*

Proof: f never vanishes, hence is of the form $f = e^g$ for some entire g and $g' = f' \cdot e^{-g}$ never vanishes, hence has the form $g' = e^h$ for some entire h. Then $f' = g'e^g = e^{g+h}$, so $g' + h' = f'' \cdot e^{-(g+h)}$ never vanishes, that is, $e^h + h'$ never vanishes. Therefore $e^h(1 + h'e^{-h}) = e^h(1 - (e^{-h})')$ never vanishes. It follows that $(e^{-h})'$ does not take the value 1. 12.16 then implies that e^{-h} must be constant. Since $g' = e^h$, it follows that g is linear and since $f = e^g$, the desired conclusion about the form of f is obtained.

Corollary 12.19 *If f is entire and not of the form $f(z) = z + b$, then $f \circ f$ has a fixed point.*

Proof: Suppose $f \circ f$ has no fixed point. Then of course f has none either and therefore we may form

$$g(z) = \frac{f(f(z)) - z}{f(z) - z},$$

an entire function. We have g is zero-free since $f \circ f$ has no fixed points. Also $g(z) \neq 1$, else $f(f(z)) - z = f(z) - z$ and $f(z)$ is a fixed point for f. By Picard's Little Theorem g is then constant, say c, where $c \neq 0, 1$.

(1) $\quad f(f(z)) - z = c(f(z) - z) \quad \forall z \in \mathbb{C}.$

Differentiate (1) to get

(2) $\quad f'(z)[f'(f(z)) - c] = 1 - c \quad \forall z \in \mathbb{C}.$

It follows that $f'(f(z)) \neq c$ for any z and that $f'(z) \neq 0$ for any z. *A fortiori* from the latter, $f'(f(z)) \neq 0$ for any z. One more application of Picard's Little Theorem shows that $f' \circ f$ is constant. Then (2) shows that f' is constant, say a. Setting $b = f(0)$, we thus have $f(z) \equiv az + b$. This function has a fixed point (namely $-b/(a - 1)$) unless $a = 1$.

Theorem 12.20 *Let f and g be entire functions and for some positive integer n satisfy the identity*

$$f^n + g^n = 1.$$

(i) *If $n = 2$, then there is an entire function h such that $f = \cos \circ h$, $g = \sin \circ h$.*
(ii) *If $n > 2$, then f and g are each constant.*

Proof: (i) When $n = 2$ the identity is

(1) $(f + ig)(f - ig) = f^2 + g^2 = 1.$

Thus the entire function $f + ig$ has no zeros and consequently is of the form e^{ih} for some entire h. It follows then from (1) that

$$f - ig = \frac{1}{f + ig} = e^{-ih}.$$

Therefore

$$f = \tfrac{1}{2}(f + ig) + \tfrac{1}{2}(f - ig) = \frac{e^{ih} + e^{-ih}}{2} = \cos \circ h$$

$$g = \frac{1}{2i}(f + ig) - \frac{1}{2i}(f - ig) = \frac{e^{ih} - e^{-ih}}{2i} = \sin \circ h.$$

(ii) Let u be a primitive nth root of -1, say $u = e^{i\pi/n}$. Then $u, u^3, u^5, \ldots, u^{2n-1}$ are all the nth roots of -1 and so

(2) $z^n + 1 = (z - u)(z - u^3)(z - u^5)\cdots(z - u^{2n-1}), \quad \forall z \in \mathbb{C}.$

If $v, w \in \mathbb{C}$ and $w \neq 0$, put $z = v/w$ and multiply by w^n to get

$$v^n + w^n = (v - uw)(v - u^3w)(v - u^5w)\cdots(v - u^{2n-1}w).$$

Of course this last equality is valid for $w = 0$ also and so

$$1 = f^n + g^n = (f - ug)(f - u^3g)(f - u^5g)\cdots(f - u^{2n-1}g).$$

Thus each factor on the right is an entire function without zeros and there are at least three such factors since $n \geq 3$. Consequently there exist entire functions $h_0, \ldots, h_{n-1}$ such that

(3.k) $f - u^{2k-1}g = e^{h_{k-1}}, \quad k = 1, 2, 3, \ldots, n.$

Subtract equation (3.2) from equation (3.1) and equation (3.3) from equation (3.2) and get

(4) $(u^3 - u)g = e^{h_0} - e^{h_1}$

(5) $(u^5 - u^3)g = e^{h_1} - e^{h_2}.$

Now $u \neq 0$ and $u^2 = e^{2\pi i/n} \neq \pm 1$ (since $n > 2$) and so (4), and (5) give

$$u^2(e^{h_0} - e^{h_1}) = e^{h_1} - e^{h_2}$$

$$u^2e^{h_0} + e^{h_2} = (u^2 + 1)e^{h_1}$$

$$\frac{u^2}{u^2 + 1}\, e^{h_0 - h_1} + \frac{1}{u^2 + 1}\, e^{h_2 - h_1} = 1$$

(6) $$\left(\frac{u}{v}\, e^{(h_0 - h_1)/2}\right)^2 + \left(\frac{1}{v}\, e^{(h_2 - h_1)/2}\right)^2 = 1,$$

where v is any square root of the complex number $u^2 + 1$. Applying (i) to equation (6) provides an entire function h such that

$$\frac{u}{v} e^{(h_0 - h_1)/2} = \cos \circ h, \tag{7}$$

$$\frac{1}{v} e^{(h_2 - h_1)/2} = \sin \circ h. \tag{8}$$

It follows from (7) that h is constant. For if not, then by Picard's Little Theorem there exists a $z \in \mathbb{C}$ such that either $h(z) = \pi/2$ or $h(z) = -\pi/2$. In either case $\cos h(z) = 0$ and equation (7) is violated, for an exponential is never 0. From the constancy of h follows that of $e^{h_0 - h_1}$, by (7). Let, say,

$$e^{h_0 - h_1} = a. \tag{9}$$

Then from (4)

$$g = \frac{e^{h_0} - e^{h_1}}{u(u^2 - 1)} = e^{h_1} \frac{e^{h_0 - h_1} - 1}{u(u^2 - 1)} = e^{h_1} \frac{a - 1}{u(u^2 - 1)}$$

$$= Ae^{h_1}, \quad \text{where } A = \frac{a - 1}{u(u^2 - 1)}, \tag{10}$$

while from equation (3.2)

$$f = u^3 g + e^{h_1} = u^3 A e^{h_1} + e^{h_1}$$

$$= Be^{h_1}, \quad \text{where } B = u^3 A + 1. \tag{11}$$

From (10), (11) and the basic identity

$$1 = f^n + g^n = (A^n + B^n)e^{nh_1},$$

it follows that $A^n + B^n \neq 0$ and then that e^{nh_1} is constant. Thus by (10), (11), f^n and g^n are each constant. The range of the holomorphic function f thus lies in the finite set composed of nth roots of the constant f^n. By the Open Map Theorem (5.77) f is therefore constant. Similarly for g.

Exercise 12.21 (Cf. 7.25) *Let $\mathscr{S}_0$ denote the set of all holomorphic functions f in $D = D(0, 1)$ with $f(0) = 0, f'(0) = 1$ and $0 \notin f(D\backslash\{0\})$. Show that there exists an $\alpha > 0$ such that $D(0, \alpha) \subset f(D)$ for every $f \in \mathscr{S}_0$.*

Hints: Suppose contrariwise that there exist $\alpha_n \in \mathbb{C}\backslash\{0\}$ with $\alpha_n \to 0$ and $f_n \in \mathscr{S}_0$ such that $\alpha_n \notin f_n(D)$. Then $1 - f_n/\alpha_n$ is zero-free and so there exists for it a holomorphic logarithm L_n in D, which may be normalized by $L_n(0) = 0$. The numbers $\pm 2\pi i$ are then absent from the range of each L_n and so by 12.14 $\{L_n\}$ is uniformly bounded in $D(0, \frac{1}{2})$. From this and 5.32 follows the boundedness of the set $\{(1 - f_n/\alpha_n)'(0)\} = \{-1/\alpha_n\}$.

§ 4 Normal Families and Julia's Extension of Picard's Great Theorem

Definition 12.22 A set $\mathscr{F}$ of functions holomorphic in an open subset V of $\mathbb{C}$ is called *normal* (in V) if every sequence in $\mathscr{F}$ contains a subsequence which either converges locally uniformly in V or diverges to ∞ locally uniformly in V.

Lemma 12.23 (MONTEL [1916], p. 227) *Let Ω be a region, $\mathscr{F}$ a family of functions holomorphic in Ω with the property that about each point $z \in \Omega$ there is an open disk D_z in which $\mathscr{F}$ is normal. Then $\mathscr{F}$ is normal in Ω.*

Proof: Let $\{f_n\}$ be a sequence in $\mathscr{F}$. Since Ω is a countable union of compacta (e.g., the sets $\{z \in \mathbb{C}: |z| \leq n \text{ and } d(z, \mathbb{C}\backslash\Omega) \geq 1/n\}$), it is easy to see that countably many of the disks D_z cover Ω. Since $\mathscr{F}$ is normal in each of these, a diagonal argument produces a single subsequence $\{f_{n_j}\}$ of $\{f_n\}$ with the property that every point z of Ω lies in some open disk on which $\{f_{n_j}\}$ either converges uniformly or diverges uniformly to ∞. If the two classes of points z are denoted by C and D respectively, then by their very definitions each is open, $C \cap D = \varnothing$ and $C \cup D = \Omega$. It follows from the connectedness of Ω that either $C = \Omega$ or $D = \Omega$. A simple covering compactness argument shows that in the first case $\{f_{n_j}\}$ converges uniformly on each compact subset of Ω and in the second case $\{f_{n_j}\}$ diverges uniformly to ∞ on each compact subset of Ω.

Theorem 12.24 (MONTEL [1911]) *If U is an open subset of $\mathbb{C}$, $\mathscr{F} \subset H(U)$ is a normal family and for some $a \in U$ the set $\mathscr{F}(a) = \{f(a): f \in \mathscr{F}\}$ is bounded, then $\mathscr{F}$ is locally uniformly bounded on U.*

Proof: Suppose contrariwise that $\mathscr{F}$ is not uniformly bounded on some compact $K \subset U$. So there exist $f_n \in \mathscr{F}$ with

$$(*) \qquad \sup|f_n(K)| > n.$$

A subsequence $\{f_{n_j}\}$ of $\{f_n\}$ either converges locally uniformly on U or diverges locally uniformly to ∞ on U. Since $\mathscr{F}(a)$ is bounded, the sequence $\{f_{n_j}(a)\}$ is bounded. Hence the second part of the alternative is untenable, the first part prevails and (*) is contradicted.

Theorem 12.25 *Let Ω be a region, N a non-negative integer, $a, b \in \mathbb{C}$, $b \neq 0$; and $a \neq b$ if $N = 0$. Let $\mathscr{F}$ be a family of holomorphic functions in Ω such that for each $f \in \mathscr{F}$, $a \notin f(\Omega)$ and $b \notin f^{(N)}(\Omega)$. Then $\mathscr{F}$ is a normal family.*

Proof: For each $f \in \mathscr{F}$ define

$$\tilde{f} = \begin{cases} \dfrac{f-a}{b-a} & \text{if } N = 0 \\[2ex] \dfrac{f-a}{b} & \text{if } N \neq 0 \end{cases}$$

and let $\tilde{\mathscr{F}} = \{\tilde{f}: f \in \mathscr{F}\}$. Manifestly $\mathscr{F}$ is normal if and only if $\tilde{\mathscr{F}}$ is. Moreover for each $\tilde{f} \in \tilde{\mathscr{F}}$ we have $0 \notin \tilde{f}(\Omega)$ and $1 \notin \tilde{f}^{(N)}(\Omega)$. We are therefore reduced to considering the case $a = 0$, $b = 1$.

We first establish that, for each $R > 0$ there is a constant $M_N(R)$ such that

(1) $D(w, 2R) \subset \Omega \Rightarrow \log|f(z)| \leq 128 \log^+|f(w)| + M_N(R)\ \forall\, z \in D(w, R), f \in \mathscr{F}$.

Indeed, given that $D(w, 2R) \subset \Omega$, it is possible to define for each $f \in \mathscr{F}$ a function F on $D(0, 1)$ by $F(z) = f(2Rz + w)/(2R)^N$. This function meets the hypotheses of Miranda's theorem and so, taking $r = \frac{1}{2}$ in that theorem,

$$\sup_{|z| \leq 1/2} \log|F(z)| \leq 2[64 \log^+|F(0)| + 16M_N],$$

that is,

$$\log \frac{|f(z)|}{(2R)^N} \leq 2\left[64 \log^+\left|\frac{f(w)}{(2R)^N}\right| + 16M_N\right]$$

$$\log|f(z)| \leq 128 \log^+|f(w)| + 32M_N + 127|\log(2R)^N| \quad \forall z \in D(w, R).$$

We set $M_N(R) = 32M_N + 127N|\log(2R)|$ and have (1). It follows at once from (1) that

(2) $$D(w, 2R) \subset \Omega \Rightarrow \sup_{z \in D(w,R)} \log|f(z)| \leq 128 \log^+|f(w)| + M_N(R) \quad \forall f \in \mathscr{F}$$

and also that

(3) $$D(z, 3R) \subset \Omega \Rightarrow \log|f(z)| \leq 128 \inf_{w \in D(z,R)} \log^+|f(w)| + M_N(R) \quad \forall f \in \mathscr{F},$$

since $D(z, 3R) \subset \Omega$ implies $D(w, 2R) \subset \Omega$ for every $w \in D(z, R)$.

Given any point $u \in \Omega$, choose $R = R(u) > 0$ so that $D(u, 3R) \subset \Omega$. We will show that $\mathscr{F}$ is normal in $D(u, R)$. Then we can cite 12.23 to assert that $\mathscr{F}$ is normal in Ω, as desired. To prove that $\mathscr{F}$ is normal in $D(u, R)$, let a sequence $\{f_n\}$ in $\mathscr{F}$ be given. We want to show that either there is a subsequence of $\{f_n\}$ locally uniformly convergent on $D(u, R)$ or there is a subsequence of $\{f_n\}$ divergent to ∞ uniformly on $D(u, R)$. In case $\{f_n(u)\}$ is bounded, take $w = u$ in (2) and learn that $\{f_n\}$ is uniformly bounded in $D(u, R)$. Then cite 7.7 to come up with a subsequence of $\{f_n\}$ locally uniformly convergent on $D(u, R)$. In case $\{f_n(u)\}$ is unbounded, there exist $1 \leq n_1 < n_2 < \cdots$ such that $\lim_{k \to \infty}|f_{n_k}(u)| = \infty$. Then (3) with $z = u$ shows that $\lim_{k \to \infty}|f_{n_k}(w)| = \infty$ uniformly for $w \in D(u, R)$.

Lemma 12.26 *Let $U = \{z \in \mathbb{C}: 0 < |z| < 1\}$ and let f be holomorphic in U and have an essential singularity at* 0. *Define $f_n(z) = f(2^{-n}z)$ $(z \in U)$. Then $\{f_n\}$ is not a normal family.*

Proof: (MONTEL [1927], pp. 79, 80) According to Casorati–Weierstrass (11.6) there are points z arbitrarily near 0 on which f takes arbitrarily small values, that is, there exist z_k such that

(1) $1/4 > |z_1| > |z_2| > \cdots \to 0$

and

(2) $\lim_{k\to\infty} f(z_k) = 0.$

Choose integers $n_k \geq 1$ such that

$$2^{-n_k-2} \leq |z_k| < 2^{-n_k-1}$$

for each k. Then $\{n_k\}$ is a non-decreasing and unbounded sequence. We have

(3) $w_k = 2^{n_k} z_k \in \bar{A}(\tfrac{1}{4}, \tfrac{1}{2}), \quad k = 1, 2, \ldots$

and

(4) $f_{n_k}(w_k) = f(z_k) \to 0 \quad \text{as } k \to \infty.$

Now let us suppose that $\{f_n\}$ is a normal family and derive a contradiction. The sequence $\{f_{n_k}\}$ in the normal family $\{f_n\}$ contains a subsequence $\{f_{n_{k_j}}\}$ which is either uniformly convergent on each compact subset of U or is uniformly divergent to ∞ on each compact subset of U. On the compact subset $\bar{A}(\tfrac{1}{4}, \tfrac{1}{2})$, however, $\{f_{n_{k_j}}\}$ cannot diverge *uniformly* to ∞, as (3) and (4) show. So the first alternative holds. In particular, $\{f_{n_{k_j}}\}$ is uniformly bounded on the compact subset $\bar{A}(\tfrac{1}{4}, \tfrac{1}{2})$, say by M. Given $z \in D(0, 2^{-n_{k_1}-1})\backslash\{0\}$, there exists $j \geq 1$ such that $2^{-n_{k_j}-2} \leq |z| < 2^{-n_{k_j}-1}$. Then $2^{n_{k_j}} z \in \bar{A}(\tfrac{1}{4}, \tfrac{1}{2})$ and so

$$|f(z)| = |f_{n_{k_j}}(2^{n_{k_j}} z)| \leq M.$$

This shows that f is bounded by M in $D(0, 2^{-n_{k_1}-1})\backslash\{0\}$, making 0 a removable singularity (by 5.41), contrary to hypothesis.

Theorem 12.27 (Julia) *If the holomorphic function f has an essential singularity at a, then there exist real number θ_0 and complex number b such that for every (sufficiently small) $\varepsilon > 0$*

(1) $\mathbb{C}\backslash\{b\} \subset f\{a + re^{i\theta}: |\theta - \theta_0| < \varepsilon, \ 0 < r < \varepsilon\}.$

Remark: Any such $e^{i\theta_0}$ is called a *Julia direction* and the ray $\{a + re^{i\theta_0}: r \in \mathbb{R}^+\}$ is called a *Julia line* (for the function f).

Proof: Without loss of generality we may suppose that $a = 0$ and that f is holomorphic in $U = \{z \in \mathbb{C}: 0 < |z| < 1\}$. Let $r_n = 2^{-n}$ and f_n be as defined in

12.26. By 12.23 then there is a point $z_0 \in U$ such that $\{f_n\}$ is not normal in any neighborhood of z_0. For every pair of positive integers n and m set

$$\text{(2)} \qquad \begin{cases} D_m = \left\{z \in \mathbb{C}: |z - z_0| < \dfrac{1}{m}|z_0|\right\} \\ D_{n,m} = \left\{z \in \mathbb{C}: |z - r_n z_0| < \dfrac{1}{m} r_n|z_0|\right\} = r_n D_m. \end{cases}$$

We will show that for any two distinct complex numbers b and c and any m, there exist infinitely many n for which either $b \in f(D_{n,m})$ or $c \in f(D_{n,m})$. [For a set S not known to be in the domain of f we write simply $f(S)$ for the f image of those points in S which are in the domain of f. If there are none, $f(S)$ means the void set.] For $f(D_{n,m}) = f_n(D_m)$, so if the assertion is false there is an m and an N such that all the functions f_n for $n \geq N$ omit the distinct values b and c on D_m. But then by 12.25, $\{f_n\}_{n=N}^{\infty}$ (and so obviously also $\{f_n\}_{n=1}^{\infty}$) is normal in the neighborhood D_m of z_0, contrary to the definition of z_0.

It follows that for each m there is a complex number b_m such that

$$\text{(3)} \qquad \mathbb{C}\backslash\{b_m\} \subset \bigcap_{k=1}^{\infty} \bigcup_{n=k}^{\infty} f(D_{n,m}).$$

Pick θ_0 so that $z_0 = |z_0|e^{i\theta_0}$ and let $\varepsilon > 0$ be given and consider the sector

$$\text{(4)} \qquad S = S(\varepsilon) = \{re^{i\theta}: |\theta - \theta_0| < \varepsilon, \quad 0 < r < \varepsilon\}.$$

It is open and if k is such that $r_k < \varepsilon/|z_0|$, then $r_n z_0 = r_n|z_0|e^{i\theta_0}$ belongs to S for all $n \geq k$ (because for such n, $r_n|z_0| \leq r_k|z_0| < \varepsilon$). We may then choose $m = m(\varepsilon)$ large enough that

$$D_{k,m} = \left\{z \in \mathbb{C}: |z - r_k z_0| < \frac{1}{m} r_k|z_0|\right\} \subset S.$$

Then for any $n \geq k$

$$D_{n,m} = \frac{r_n}{r_k} D_{k,m} \subset S,$$

since S is clearly closed under multiplication by positive scalars less than 1. Thus

$$\text{(5)} \qquad \bigcup_{n=k}^{\infty} D_{n,m} \subset S.$$

From (3) and (5) follows

$$\text{(6)} \qquad \mathbb{C}\backslash\{b_m\} \subset f(S).$$

The assertion of the theorem follows from this. For if there are two *different* complex numbers b' and b'' such that $b' \notin f(S(\varepsilon'))$ for some $\varepsilon' > 0$ and $b'' \notin f(S(\varepsilon''))$

for some $\varepsilon'' > 0$, then for $\varepsilon = \min\{\varepsilon', \varepsilon''\}$, *both* the complex numbers b' and b'' would lie outside $f(S(\varepsilon))$, contradicting (6).

Exercise 12.28 *Deduce from* 12.27 *the following stronger form of Picard's Little Theorem: a non-polynomial entire function assumes infinitely often every complex value with at most one exception*—Picard [1880].

Hints: If $a \neq b$ and a, b are each assumed only finitely many times by the entire function F, then neither is assumed outside $D(0, 1/r)$ for sufficiently small positive r. Consequently the function $f(z) = F(1/z)$, $0 < |z| < r$, does not have an essential singularity at 0, i.e., its Laurent series contains only finitely many negative powers of z. Evidently then, F is a polynomial.

Exercise 12.29 *Let Ω be a region, $\{f_n\}$ a sequence of univalent holomorphic functions in Ω. The object of* (i)–(iv) *below is to show that there exists a point $a \in \Omega$ and a subsequence $\{h_k\}$ of $\{f_n\}$ which is normal in $\Omega\backslash\{a\}$;* (v) *and* (vi) *are applications of this fact.*

(i) *There is a point $a \in \Omega$ and a subsequence $\{f_{n_j}\}$ of $\{f_n\}$ such that in each compact $K \subset \Omega\backslash\{a\}$, for only finitely many j does the function f_{n_j} have a zero.*

Hints: If the set of subsequential limits of $\{f_n^{-1}(0)\}$ meets Ω, let a be any point in the intersection. Otherwise, let a be any point whatsoever in Ω. Let $\{K_j\}$ be a compact exhaustion of $\Omega\backslash\{a\}$ as furnished by 1.31. For each j let N_j be a neighborhood of a in $D(a, 1/j) \cap \Omega$ which is disjoint from K_j. In the first case there exists a subsequence $\{f_{n_j}\}$ such that $f_{n_j}^{-1}(0) \in N_j$ for each j. In the second case no set $\{f_n^{-1}(0)\}_{n=k}^{\infty}$ $(k = 1, 2, 3, \ldots)$ lies wholly in any K_j and we can choose the n_j successively such that $1 \le n_1 < n_2 < \cdots$ and $f_{n_j}^{-1}(0) \notin K_j$. Thus in either case $K_j \cap \{f_{n_i}^{-1}(0)\}_{i=j}^{\infty} = \varnothing$, for each j. Since any compact subset K of $\Omega\backslash\{a\}$ lies in some K_j, the claim (i) is established.

(ii) *There is a point $b \in \Omega\backslash\{a\}$ and a subsequence $\{h_k\}$ of $\{f_{n_j}\}$ such that in any compact $K \subset \Omega\backslash\{a, b\}$, for only finitely many k does the function h_k take either of the values* 0 *or* 1.

Hints: Apply the argument of part (i) to the sequence $\{1 - f_{n_j}\}$ of univalent functions in the open set $\Omega\backslash\{a\}$.

(iii) *The sequence $\{h_k\}$ of* (ii) *is normal in $\Omega\backslash\{a, b\}$.*

Hints: By (ii) and 12.25 $\{h_k\}$ is normal in every open disk D such that $\bar{D} \subset \Omega\backslash\{a, b\}$. Since the latter is connected (1.24), it follows from 12.23 that $\{h_k\}$ is normal therein.

(iv) *The sequence $\{h_k\}$ of* (iii) *is normal in $\Omega\backslash\{a\}$.*

Hints: Since $\{h_k\}$ is normal in $\Omega\backslash\{a, b\}$, any subsequence of $\{h_k\}$ contains a further subsequence $\{g_l\}$ such that either (1): $\{g_l\}$ converges locally uniformly in

$\Omega\backslash\{a, b\}$ or (2): $\{|g_l|\}$ diverges to ∞ locally uniformly in $\Omega\backslash\{a, b\}$. If (1) prevails, then $\{g_l\}$ converges uniformly on the boundaries of two disjoint closed disks D_a, D_b centered at a and b and lying in Ω (since such circles are compact subsets of $\Omega\backslash\{a, b\}$). Then by the Maximum Modulus Principle $\{g_l\}$ converges uniformly in D_a and D_b. Thus (1) implies (1)′: $\{g_l\}$ converges locally uniformly throughout Ω. If (2) prevails, select $r > 0$ so that $\bar{D}(b, r) \subset \Omega\backslash\{a\}$. Then $|g_l| \to \infty$ uniformly on $C(b, r)$. Bearing in mind the definition of the f_{n_j} from (i) and the fact that $\{g_l\}$ is a subsequence of $\{f_{n_j}\}$, we know that for only finitely many l does g_l have a zero in $\bar{D}(b, r)$, so $\{1/g_l\}$ is a sequence of holomorphic functions in $D(b, r)$. From $|1/g_l| \to 0$ uniformly on $C(b, r)$ and the Maximum Modulus Principle it follows that $|1/g_l| \to 0$ uniformly in $D(b, r)$, so $|g_l| \to \infty$ uniformly in $D(b, r)$. Thus (2) implies (2)′: $\{|g_l|\}$ diverges to ∞ locally uniformly throughout $\Omega\backslash\{a\}$.

(v) *(Extension of* 7.29(i)) *Show that if*

(*) $\{f_n(c), f_n(d): n \in \mathbb{N}\}$ *is bounded for two distinct points* $c, d \in \Omega$,

then $\{f_n\}$ *is locally uniformly bounded in* Ω.

Hints: Consider any subsequence of $\{f_n\}$. Apply (iv) to this subsequence to come up with a point $a \in \Omega$ and a further subsequence $\{g_l\}$ of the given subsequence, such that either (1): $\{g_l\}$ converges locally uniformly in $\Omega\backslash\{a\}$ or (2): $\{|g_l|\}$ diverges to ∞ in $\Omega\backslash\{a\}$. Since one of the points c, d must be in $\Omega\backslash\{a\}$, alternative (2) cannot prevail because of (*). Now argue as in (iv) that $\{g_l\}$ converges uniformly in some disk centered at a, hence locally uniformly throughout Ω. We have shown that $\{f_n\}$ is normal in Ω. Now conclude by citing 12.24.

(vi) *Prove the following extension of* 7.11 *(due to* Montel [1925], p. 253): *If* $\lim_{n\to\infty} f_n(c)$, $\lim_{n\to\infty} f_n(d)$ *exist in* $\mathbb{C}$ *and coincide, equal* L, *say, for two distinct points* $c, d \in \Omega$, *then* $\{f_n\}$ *converges to* L *locally uniformly in* Ω.

Hints: Use (v) above, 7.11 and 7.8.

§ 5 Sectorial Limit Theorems

Montel [1912] used normal families in a simple but ingenious way to investigate boundary behavior of holomorphic functions in angular domains. (Later he used these ideas to treat the boundary problem for conformal maps which we analyzed in Chapter IX.) We have already explored this theme, on the boundary itself in 5.16 and internally in 5.56. Most of the theorems of this section look at internal approach. The bounded version of 12.30 is, of course, contained in 5.56. It is interesting to note that, in spite of the diversity of proof techniques involved, the bounded 12.30 plus 5.16 is equivalent to 12.31.

Theorem 12.30 *Let* $\alpha < \gamma < \beta$, $S = \{re^{i\theta}: r > 0, \alpha < \theta < \beta\}$ *and let* f *be holomorphic in* S *and for some complex* c *satisfy* $\lim_{r\to\infty} f(re^{i\gamma}) = c$. *Suppose* *that there are two distinct complex numbers absent from the range of f. (For example, f might be bounded.) Then* $\lim_{r\to\infty} f(re^{i\theta}) = c$ *for every* $\theta \in (\alpha, \beta)$ *and uniformly in any compact subinterval.*

Proof: Note that S is an open, convex (hence connected) subset of $\mathbb{C}$. Define for each positive integer n functions f_n in S by

(1) $\quad f_n(z) = f(nz) \quad z \in S.$

Notice that

(2) $\quad \lim_{n\to\infty} f_n(re^{i\gamma}) = c \quad \forall r > 0.$

In particular, the set $\{f_n(e^{i\gamma})\}$ is bounded. Since also the two distinct numbers absent from the range of f are both absent from the range of each f_n, it follows from 12.24 that $\{f_n\}$ is uniformly bounded on each compact subset of S. But then from (2) and the Vitali–Porter theorem 7.6 we know that f_n converges to a holomorphic function f_0 uniformly on each compact subset of S. On the ray $(0, \infty)e^{i\gamma}$ we have $f_0 = c$ so by the fundamental uniqueness theorem, $f_0 \equiv c$. If I is a compact subinterval of (α, β), let $K(I) = \{re^{i\theta}: 1 \le r \le 2, \theta \in I\}$, a compact subset of S. Given $\varepsilon > 0$, there exists then $N = N(I, \varepsilon)$ such that

(3) $\quad |f_n(z) - c| < \varepsilon \quad \forall n \ge N, \forall z \in K(I).$

In particular,

(4) $\quad |f(re^{i\theta}) - c| < \varepsilon \quad \forall r \ge N, \forall \theta \in I.$

Indeed, given such θ and r, let n be the integer such that $n \le r < n + 1$. Then the number $z = (r/n)e^{i\theta}$ belongs to $K(I)$ and $n \ge N$. Since $f(re^{i\theta}) = f_n(z)$, the inequality in (4) follows from (3) and the proof is finished.

Exercise 12.31 (LINDELÖF [1915]) *Let* $0 < \alpha \le \pi$, $S = \{re^{i\theta}: r > 0, -\alpha < \theta < \alpha\}$, $\tilde{S} = \{re^{i\theta}: r > 0, -\alpha \le \theta < \alpha\}$ *and suppose that* $f: \tilde{S} \to \mathbb{C}$ *is bounded, continuous, holomorphic in S and satisfies*

(1) $\quad \lim_{r\to\infty} f(re^{-i\alpha}) = 0.$

Show that

(2) $\quad \lim_{r\to\infty} f(re^{i\theta}) = 0$ *uniformly for* θ *in any compact subset of* $[-\alpha, \alpha)$.

Hints: Introduce

$$S_0 = \{re^{i\theta}: r > 0, -\alpha/2 < \theta < \alpha/2\}$$

and form

$$F(z) = f(ze^{-i\alpha/2})\bar{f}(\bar{z}e^{-i\alpha/2}), \quad z \in \bar{S}_0.$$

This function is continuous and bounded in $\bar{S}_0$, holomorphic in S_0. Since $F(re^{-i\alpha/2})$ equals $f(re^{-i\alpha})$ times a bounded factor and $F(re^{i\alpha/2})$ equals $\bar{f}(re^{-i\alpha})$ times a bounded factor, we see from (1) that

$$\lim_{r\to\infty} F(re^{\pm i\alpha/2}) = 0.$$

Therefore by 5.16, $\lim_{r\to\infty} F(re^{i\theta}) = 0$ for each $\theta \in [-\alpha/2, \alpha/2]$. In particular,

$$0 = \lim_{r\to\infty} F(r) = \lim_{r\to\infty} |f(re^{-i\alpha/2})|^2$$

and so

(3) $$\lim_{r\to\infty} f(re^{-i\alpha/2}) = 0.$$

From (1), (3) and 5.16 it follows that

(4) $$\lim_{r\to\infty} f(re^{i\theta}) = 0 \quad \text{uniformly for } \theta \in [-\alpha, -\alpha/2],$$

while from (3) and 12.30 it follows that

(5) $$\lim_{r\to\infty} f(re^{i\theta}) = 0 \quad \text{uniformly for } \theta \text{ in any compact subset of } (-\alpha, \alpha).$$

From (4) and (5) the claim (2) follows.

Exercise 12.32 *Give an* ab initio *proof of* 12.31 *using the harmonic majorant technique.*

Hints: It is convenient to put the problem in the following form: Let $H = \{z \in \mathbb{C}: \operatorname{Im} z > 0\}$, $f: H \cup (0, \infty) \to \mathbb{C}$ bounded, continuous and holomorphic in H. Suppose that $f(x) \to 0$ as real $x \to +\infty$. We are to show that

(*) $$\lim_{\substack{|z|\to\infty \\ z\in H_\varepsilon}} f(z) = 0$$

for each set $H_\varepsilon = \{re^{i\theta}: r > 0, 0 \le \theta < (1 - \varepsilon)\pi\}$, $\varepsilon > 0$. We may assume that $|f| \le 1$. Let $\varepsilon > 0$ and $0 < \delta < 1$ be given. Choose $x_\delta > 0$ so large that

(1) $$|f(x)| \le \delta \quad \forall x \ge x_\delta.$$

Let L denote the holomorphic logarithm in $\mathbb{C}\backslash\{iy: y \le 0\}$ which satisfies

(2) $$L(re^{i\theta}) = \log r + i\theta \quad \text{for } r > 0, \frac{-\pi}{2} < \theta < \frac{3\pi}{2}.$$

(Cf. 3.43.) Look at the function

(3) $$h(z) = 1 - \frac{1}{\pi} \operatorname{Im} L(z - x_\delta), \quad z \in \bar{H}\backslash\{x_\delta\}.$$

It is harmonic and bounded in H, equals 0 on $(-\infty, x_\delta)$ and equals 1 on (x_δ, ∞). Consequently the function

$$\log(\max\{|f|, \delta\}) - (\log \delta)h = \log \delta + \log^+\left|\frac{f}{\delta}\right| - (\log \delta)h$$

is bounded and subharmonic in H (by 5.26(ii)), and has non-positive limit superior at each point of $\partial H\backslash\{x_\delta\} = (-\infty, x_\delta) \cup (x_\delta, \infty)$. Consequently, by 7.15 this function is non-positive throughout H:

(4) $\quad \log(\max\{|f|, \delta\}) \leq (\log \delta)h \quad$ on H.

Since $\operatorname{Im} L(re^{i\theta} - x_\delta) = \operatorname{Im} L(e^{i\theta} - x_\delta/r)$ [for all $r > x_\delta$, $0 \leq \theta \leq \pi$] and since there exists [by uniform continuity of L on the compact set $\bar{H} \cap C(0, 1)$] an $x'_\delta > x_\delta$ such that

$$\begin{aligned}\operatorname{Im} L(e^{i\theta} - x_\delta/r) &\leq \operatorname{Im} L(e^{i\theta}) + \frac{\varepsilon\pi}{2}\\ &= \theta + \frac{\varepsilon\pi}{2} \quad \forall r \geq x'_\delta, 0 \leq \theta \leq \pi,\end{aligned}$$

it follows that

$$\operatorname{Im} L(z - x_\delta) \leq (1 - \varepsilon)\pi + \frac{\varepsilon\pi}{2} = \left(1 - \frac{\varepsilon}{2}\right)\pi,$$

(5) $\quad h(z) \geq \frac{\varepsilon}{2} \quad \forall z \in H_\varepsilon$ with $|z| \geq x'_\delta$.

From (4) and (5), remembering that $\log \delta < 0$, we get

$$\log(\max\{|f(z)|, \delta\}) \leq (\log \delta)\frac{\varepsilon}{2} \quad \forall z \in H_\varepsilon \text{ with } |z| \geq x'_\delta.$$

Exponentiating gives us finally

$$|f(z)| \leq \delta^{\varepsilon/2} \quad \forall z \in H_\varepsilon \text{ with } |z| \geq x'_\delta.$$

Since $\lim_{\delta \downarrow 0} \delta^{\varepsilon/2} = 0$ for each fixed $\varepsilon > 0$, the claim (*) is established.

Exercise 12.33 *Deduce both* 12.30 *and* 5.16 *from* 12.31.

Exercise 12.34 *Try to deduce* 5.56 *from* 9.20 *and* 12.30 *by a reflection argument like that used in the proof of Milloux's theorem* 5.50. (*See* p. 26 *of* CARTWRIGHT [1935].)

If we restrict to "non-tangential" approach, then the method of 12.30 can be further exploited to yield (cf. 5.29(ii))

Exercise 12.35 *Let* $\alpha < \beta$, $S = \{re^{i\theta} : r > 0, \alpha < \theta < \beta\}$, f *a holomorphic function in* S *for which some pair of complex numbers is absent from* $f(S)$. *Suppose that there exist* $\alpha < \alpha' < \beta' < \beta$ *such that*

$$(*) \qquad \sup_{\rho > r} \inf\{|f(\rho e^{i\theta})| : \alpha' < \theta < \beta'\} \to 0 \quad \text{as } r \to \infty.$$

Show that $\lim_{r\to\infty} f(re^{i\theta}) = 0$ *uniformly for* θ *in any compact subset of* (α, β).

Hints: (MONTEL [1917], p. 20) Define $f_n(z) = f(2^n z)$, $z \in S$, $n = 1, 2, \cdots$ and $K = \{re^{i\theta} : 1 \le r \le 2, \alpha' \le \theta \le \beta'\}$. For each $r \in [1, 2]$ select $\theta_n(r) \in [\alpha', \beta']$ such that

$$\begin{aligned} |f_n(re^{i\theta_n(r)})| &= |f(2^n re^{i\theta_n(r)})| = \inf\{|f(2^n re^{i\theta})| : \alpha' < \theta < \beta'\} \\ (**) \qquad &\to 0 \quad \text{as } n \to \infty, \text{ by } (*). \end{aligned}$$

This shows that no subsequence of $\{|f_n|\}$ can diverge to ∞ *uniformly* in the compact set K. However, by 12.25 $\{f_n\}$ is a normal family in S. We infer that $\{f_n\}$ is locally uniformly bounded in S (see 12.24). Using the Cauchy integral formula in small disks, we further infer that $\{f_n\}$ is locally uniformly equicontinuous in S. In particular, $\{f_n\}$ is uniformly equicontinuous on K and so from (**) it follows easily that if g is the limit of a convergent subsequence $\{f_{n_j}\}$ of $\{f_n\}$, then $g(z) = 0$ for each cluster point z of $\{re^{i\theta_n(r)} : n = n_1, n_2, \ldots\}$. We have this for each $r \in [1, 2]$, so such z constitute an infinite subset of K and by 5.62, g is the 0 function. By 7.8 $\{f_n\}$ therefore converges to 0 locally uniformly in S. The proof is concluded from this point just as was 12.30.

Exercise 12.36 *Use the method of* 12.30 *to prove this result of* LINDELÖF [1909]: *if f is holomorphic in an open sector S, never* 0 *nor* 1, *and is bounded on some half-line in S, then f is bounded in every closed subsector of S.*

Next we are going to investigate the analog of 12.30 when $|f(re^{i\theta})|$ converges as $r \to \infty$, rather than $f(re^{i\theta})$ itself. (An easy result along these lines is 7.31.) Also we will look at a harmonic function analog of 12.30. Here it is more convenient to work with strips and half-strips rather than sectors, so we begin with

Exercise 12.37 (i) *Check that* 12.30 *remains valid if the sector S there is replaced with a truncated sector* $S \backslash \bar{D}(0, R)$ *for any* $R > 0$. [*No modifications whatsoever are necessary in the proof if one first re-scales to get* $R < 1$.]

(ii) *If* $a, b, c \in \mathbb{R}$ *with* $a < b$, *then there is an exponential which maps the semi-strip* $(a, b) \times (c, \infty)$ *conformally onto a truncated sector* $S \backslash \bar{D}(0, R)$. *Using it, formulate a semi-strip version of* 12.30.

Exercise 12.38 *Let* $\alpha, \beta, a, b, c \in \mathbb{R}$ *with* $\alpha < a < b < \beta$ *and form* $S = (\alpha, \beta) \times (c, \infty)$. *Let* $f: S \to \mathbb{C}$ *be holomorphic and bounded, by* K, *say.*

(i) *Suppose that*

$$(1) \qquad \lim_{y\to\infty} |f(a + iy)| = \lim_{y\to\infty} |f(b + iy)| = 1,$$

(2) $b - a < \frac{1}{2}(\beta - \alpha)$,

(3) $a - \alpha \le b - a$ *and* $\beta - b \le b - a$.

Show that for any $\delta > 0$ *there exists* $c(\delta) > c$ *such that*

(4) $|f(x + iy)| \ge \dfrac{1}{2K}$ *whenever* $\alpha + \delta \le x \le \beta - \delta$, $y \ge c(\delta)$.

Hints: For $z \in S_a = (\alpha, 2a - \alpha) \times (c, \infty)$ define $g(z) = \overline{f(\overline{2a - z})}$. Note that $\overline{2a - z} = 2a - \bar{z} \in S_a \overset{(3)}{\subset} S$, so g is well-defined and holomorphic. Then $F = fg$ is holomorphic and bounded in S_a and satisfies

$$F(a + iy) = |f(a + iy)|^2 \to 1 \quad \text{as } y \to \infty \text{, by (1).}$$

Since $\alpha < a < 2a - \alpha$, we may cite 12.37(ii) to conclude that $\lim_{y\to\infty} F(x + iy) = 1$ uniformly for x in any compact subset of $(\alpha, 2a - \alpha)$. In particular, there exists $c_1(\delta) > c$ such that

(5) $|F(x + iy)| > \frac{1}{2} \quad \forall (x, y) \in (\alpha + \delta, 2a - \alpha - \delta) \times (c_1(\delta), \infty)$.

Because $|f| \le K$, (5) and the definition of F show that

(6) $|f(x + iy)| > \dfrac{1}{2K} \quad \forall (x, y) \in (\alpha + \delta, 2a - \alpha - \delta) \times (c_1(\delta), \infty)$.

Argue similarly to produce $c_2(\delta) > c$ such that

(7) $|f(x + iy)| > \dfrac{1}{2K} \quad \forall (x, y) \in (2b - \beta + \delta, \beta - \delta) \times (c_2(\delta), \infty)$.

Now by (2), if δ is sufficiently small, the strips (6) and (7) overlap and so for $c(\delta) = \max\{c_1(\delta), c_2(\delta)\}$ the inequalities (6) and (7) yield (4).

(ii) *Under hypotheses* (1), (2), (3) *show that*

(8) $\lim\limits_{y\to\infty} |f(x + iy)| = 1$ *uniformly for* x *in any compact subset of* (α, β).

Hints: Let $\varepsilon > 0$ be given. There is then by (1) a $y(\varepsilon) > c$ such that

(9) $|f(a + iy)| \le 1 + \varepsilon, |f(b + iy)| \le 1 + \varepsilon \quad \forall y \ge y(\varepsilon)$.

The function

$$h_\varepsilon(z) = \frac{z - (a + iy(\varepsilon))}{z - (a + iy(\varepsilon)) + K(b - a)}, \quad z \in [a, b] \times [y(\varepsilon), \infty)$$

is bounded by 1 and on $[a, b] \times \{y(\varepsilon)\}$ it is bounded by $1/K$. Consequently, because of (9), the holomorphic function fh_ε is bounded in this set by K and bounded by $1 + \varepsilon$ on its boundary. From 5.13 it follows then that fh_ε is bounded by $1 + \varepsilon$ throughout this set. Since $\lim_{y\to\infty} h_\varepsilon(x + iy) = 1$ uniformly for

$x \in [a, b]$, we may select $y_1(\varepsilon) > y(\varepsilon)$ sufficiently large that $|h_\varepsilon(x + iy)| > (1 + \varepsilon)/(1 + 2\varepsilon)$ for all $y \geq y_1(\varepsilon)$ and all $x \in [a, b]$. We shall then have

$$|f(x + iy)| \leq 1 + 2\varepsilon \quad \forall (x, y) \in [a, b] \times [y_1(\varepsilon), \infty).$$

Because of (4), the same argument may be applied to $1/f$. Conclude that

$$\text{(10)} \qquad \lim_{y \to \infty} |f(x + iy)| = 1 \quad \text{uniformly for } x \in [a, b].$$

Now consider $\delta > 0$. In (i) we saw that $\lim_{y \to \infty} F(x + iy) = 1$ uniformly for $x \in [\alpha + \delta, a]$. Thus, by definition of F,

$$\text{(11)} \qquad \lim_{y \to \infty} |f(x + iy)|\,|f(2a - x + iy)| = 1 \quad \text{uniformly for } x \in [\alpha + \delta, a].$$

Since $a \leq 2a - x \leq 2a - \alpha \overset{(3)}{\leq} b$ for $x \in [\alpha + \delta, a]$, it follows that for these x the behavior of the second factor on the left side of (11) is governed by (10). It follows therefore from (11) that

$$\lim_{y \to \infty} |f(x + iy)| = 1 \quad \text{uniformly for } x \in [\alpha + \delta, a],$$

and similarly for $x \in [b, \beta - \delta]$. As $\delta > 0$ is arbitrary, (8) is established.

(iii) *Prove that* (8) *follows from* (1) *and* (2) *alone.*

Hints: Choose $\alpha_0 \in [\alpha, a)$, $\beta_0 \in (b, \beta]$ so that all the hypotheses of (ii) hold in $S_0 = (\alpha_0, \beta_0) \times (c, \infty)$. Set $d_0 = \frac{1}{2}(\beta_0 - \alpha_0)$, $\alpha_1 = \max\{\alpha, \alpha_0 - d_0\}$, $\beta_1 = \min\{\beta, \beta_0 + d_0\}$. Then if points a_1, b_1 with $a_1 < b_1$ are appropriately selected in (α_0, β_0), part (ii) will be applicable to $S_1 = (\alpha_1, \beta_1) \times (c, \infty)$. Construct successively strips $S_0 \subset S_1 \subset \cdots \subset S$ in which (ii) applies and observe that $S_n = S$ must occur after some finite number n of steps.

Exercise 12.39 *Let $\alpha, \beta, a, b, c \in \mathbb{R}$ with $\alpha < a < b < \beta$, $F\colon (\alpha, \beta) \times (c, \infty) \to \mathbb{C}$ a bounded holomorphic function. Suppose that*

$$\text{(1)} \qquad \lim_{y \to \infty} |F(a + iy)| = A > 0, \qquad \lim_{y \to \infty} |F(b + iy)| = B > 0,$$

$$\text{(2)} \qquad b - a < \tfrac{1}{2}(\beta - \alpha).$$

Show that $\lim_{y \to \infty} |F(x + iy)|$ exists for each $x \in (\alpha, \beta)$, uniformly in compact subsets of such x. Find the value of this limit.

Hint: Show that the function $f(z) = A^{-(b-z)/(b-a)} B^{-(z-a)/(b-a)} F(z)$ fulfills the hypotheses of 12.38(iii).

Exercise 12.40 *Let $\alpha, \beta, c \in \mathbb{R}$ with $\alpha < \beta$, $F\colon (\alpha, \beta) \times (c, \infty) \to \mathbb{C}$ a bounded holomorphic function. Suppose that there are three distinct numbers $x_1, x_2, x_3 \in (\alpha, \beta)$ such that $\lim_{y \to \infty} |F(x_k + iy)|$ exists for $k = 1, 2, 3$. Prove that then $\lim_{y \to \infty} |F(x + iy)|$ exists for every $x \in (\alpha, \beta)$ and has the form $e^{ax + b}$ for appropriate real constants a and b.*

Exercise 12.41 (i) *Suppose $\alpha < a < b < \beta$ and $h: S = (\alpha, \beta) \times (c, \infty) \to \mathbb{R}$ a bounded harmonic function satisfying*

$$\lim_{y\to\infty} h(a + iy) = A, \qquad \lim_{y\to\infty} h(b + iy) = B.$$

Show that $\lim_{y\to\infty} h(x + iy) = B(x - a)/(b - a) + A(b - x)/(b - a)$ *for each* $x \in (\alpha, \beta)$.

Hints: Replace $h(z)$ by $h(z) - B(\operatorname{Re} z - a)/(b - a) - A(b - \operatorname{Re} z)/(b - a)$ to achieve $A = B = 0$. Select (10.2) $F \in H(S)$ with $h = \operatorname{Re} F$ and form $f = e^F$. Since *both* f and $1/f$ are then bounded, the first part of the proof of 12.38(ii) shows, without appeal to 12.38(i), that $\lim_{y\to\infty}|f(x + iy)| = 1$ for all $x \in [a, b]$. Now cite 12.40. For an alternative short and elegant *Beweisschluss* at this point see Hardy [1926].

(ii) (Loomis [1943]) *Let h be a positive harmonic function in the open right half-plane. Suppose that $-\pi/2 < \theta_1 < \theta_2 < \pi/2$ and* $\lim_{r\downarrow 0} h(re^{i\theta_1}) = A$, $\lim_{r\downarrow 0} h(re^{i\theta_2}) = B$. *Show that* $\lim_{r\downarrow 0} h(re^{i\theta}) = B(\theta - \theta_1)/(\theta_2 - \theta_1) + A(\theta_2 - \theta)/(\theta_2 - \theta_1)$ *for every $\theta \in (-\pi/2, \pi/2)$ and uniformly for each compact set of such θ.*

Hints: According to 5.29, h is bounded on each set $S_\varepsilon = \{re^{i\theta}: 0 < r < 1, |\theta| < \pi/2 - \varepsilon\}$ $(\varepsilon > 0)$. The exponential function $E(z) = e^{iz}$ maps $(-\pi/2, \pi/2) \times (0, \infty)$ conformally onto S_0 and each vertical half-line $\{\theta\} \times (0, \infty)$ onto the segment $\{re^{i\theta}: 0 < r < 1\}$. Therefore $h \circ E$ fulfills the hypotheses of (i).

Exercise 12.42 *Let f be entire and suppose that for each real θ we have*

$$(*) \qquad \begin{cases} \text{either } \lim\limits_{r\to\infty} |f(re^{i\theta}) - c(\theta)| = 0 & \text{for some } c(\theta) \in \mathbb{C} \\ \text{or } \lim\limits_{r\to\infty} |f(re^{i\theta})| = \infty. \end{cases}$$

(Examples of such functions will be produced in Chapter XV.)

(i) *Show that in any non-degenerate interval I there is a non-degenerate subinterval J such that*

either (1.J) $\quad |f(re^{i\theta})| \geq 1$ *for all $\theta \in J$ and all sufficiently large r*

or (2.J) $\quad |f(re^{i\theta})| \leq 2$ *for all $\theta \in J$ and all sufficiently large r.*

Hints: Suppose that no such J exists. Construct non-degenerate closed intervals $J_n \subset I$ and $r_n > n$ inductively as follows: given J_{2n-1}, the failure of $(1.J_{2n-1})$ means $\theta_{2n} \in J_{2n-1}$ and $r_{2n} > 2n$ exist such that $|f(r_{2n}e^{i\theta_{2n}})| < 1$. By continuity there is then a whole (non-degenerate) subinterval J_{2n} of J_{2n-1} such that

(1.n) $\quad |f(r_{2n}e^{i\theta})| < 1 \quad \forall\theta \in J_{2n}$.

Similarly, given J_{2n}, the failure of $(2.J_{2n})$ means $\theta_{2n+1} \in J_{2n}$ and $r_{2n+1} > 2n + 1$ exist such that $|f(r_{2n+1}e^{i\theta_{2n+1}})| > 2$. By continuity there is then a whole (non-degenerate) subinterval J_{2n+1} of J_{2n} such that

(2.n) $\quad |f(r_{2n+1}e^{i\theta})| > 2 \quad \forall\theta \in J_{2n+1}$.

For any $\theta \in \bigcap_{n=1}^{\infty} J_n$ (and there is at least one, by compactness) (1.n) and (2.n) each hold for all n. It follows that $\lim_{n\to\infty}|f(r_n e^{i\theta})|$ does not exist in $[0, \infty]$. Since $r_n \to \infty$, this contradicts (*).

(ii) *Call a non-degenerate interval I a convergence sector (for f) if*

$$(**) \quad \begin{cases} \textit{either } \lim\limits_{r\to\infty} \sup\limits_{\substack{\rho > r \\ \theta\in I}} |f(\rho e^{i\theta}) - c| = 0 \quad \textit{for some } c \in \mathbb{C} \\ \textit{or } \lim\limits_{r\to\infty} \inf\limits_{\substack{\rho > r \\ \theta\in I}} |f(\rho e^{i\theta})| = \infty. \end{cases}$$

Show that every non-degenerate interval contains a convergence sector.

Hint: Because of (i) and (*), 12.37(i) is applicable to one of f or $1/f$.

Exercise 12.43 *Let $a < b$, $c < d$ be real numbers, $R = (a, b) \times (c, d)$ and f continuous on $\bar{R}$, holomorphic in R. Prove that if*

$$\min_{c\le y\le d} [\operatorname{Re} f(b + iy) - \operatorname{Re} f(a + iy)] \ge b - a,$$

then

$$\max_{a\le x\le b} [\operatorname{Im} f(x + id) - \operatorname{Im} f(x + ic)] \ge d - c.$$

Hints: For each sufficiently small $\varepsilon > 0$ form

$$R_\varepsilon = (a + \varepsilon, b - \varepsilon) \times (c + \varepsilon, d - \varepsilon).$$

By Cauchy's theorem $\int_{\partial R_\varepsilon} f = 0$. Let $\varepsilon \downarrow 0$ and appeal to the (uniform) continuity of f on $\bar{R}$ to conclude that $\int_{\partial R} f = 0$. Writing out this integral and equating its imaginary part to 0 yields.

$$\int_c^d [\operatorname{Re} f(b + iy) - \operatorname{Re} f(a + iy)]dy = \int_a^b [\operatorname{Im} f(x + id) - \operatorname{Im} f(x + ic)]dx,$$

whence

$$(d - c) \min_{c\le y\le d} [\operatorname{Re} f(b + iy) - \operatorname{Re} f(a + iy)]$$
$$\le (b - a) \max_{a\le x\le b} [\operatorname{Im} f(x + id) - \operatorname{Im} f(x + ic)].$$

Exercise 12.44 *Let f be holomorphic in $S = (a, b) \times (0, \infty)$, continuous on $\bar{S}$ and suppose that $\overline{\lim}_{y\to\infty} \operatorname{Re} f(a + iy) < \underline{\lim}_{y\to\infty} \operatorname{Re} f(b + iy)$. Then f is unbounded in S.*

Hints: (Pólya [1933]) By translating and scaling we may suppose that

$$\overline{\lim_{y\to\infty}} \operatorname{Re} f(a + iy) < a < b < \underline{\lim_{y\to\infty}} \operatorname{Re} f(b + iy).$$

There exists then $c > 0$ such that

(*) $\quad \operatorname{Re} f(a + iy) < a$ and $\operatorname{Re} f(b + iy) > b \quad \forall y \ge c$.

Set $m = \max_{a \le x \le b}|f(x + ic)|$ and let $M > 0$ be given. Consider $d = c + m + M$. By (*) we have

$$\min_{c \le y \le d} [\operatorname{Re} f(b + iy) - \operatorname{Re} f(a + iy)] \ge b - a$$

and therefore from the last exercise

$$\max_{a \le x \le b} [\operatorname{Im} f(x + id) - \operatorname{Im} f(x + ic)] \ge d - c.$$

It follows then from the definition of m that

$$\max_{a \le x \le b} |f(x + id)| \ge d - c - m = M.$$

§ 6 Applications to Iteration Theory

Exercise 12.45 *Prove that in* 7.37–7.41 *the hypothesis that the region* Ω *be bounded can be weakened to the simple assumption that* $\mathbb{C}\backslash\Omega$ *contain at least two points.*

Hints: Examine the proofs of these results to see that boundedness of Ω is only used to ensure that the iterates of the self-map f constitute a normal family. Since our assumption is that there are at least two distinct points a, b in $\mathbb{C}\backslash\Omega$, we have $a, b \notin f^{[n]}(\Omega) \subset \Omega$ for all n, so the desired normality is a consequence of 12.25. One subtle point occurs in the proof of 7.37. We use normality to choose the m_{j_k} so that $\{f^{[m_{j_k}]}\}$ either diverges to ∞ locally uniformly on Ω or else converges locally uniformly on Ω to a holomorphic limit Φ. We have to show that it is the second alternative which occurs. Since ϕ is assumed non-constant, $\phi(\Omega) \subset \Omega$. Indeed, if $\phi(z_0) = w_0 \notin \Omega$ for some z_0, then $f^{[n_j]} - w_0$ are never 0 in Ω but converge locally uniformly on Ω to $\phi - w_0$ which is 0 at the point z_0. By 7.11 it would follow that ϕ is constant. Now we may fix a $z \in \Omega$ and then $\{\phi(z)\} \cup \{f^{[n_j]}(z) : j = 1, 2, \ldots\}$ is a compact subset K of Ω. The identity

$$f^{[m_{j_k}]}(f^{[n_{j_k}]}(z)) = f^{[n_{j_k+1}]}(z)$$

then shows that $\{f^{[m_{j_k}]}\}$ does not diverge *uniformly* to ∞ on K. Hence the first alternative for $\{f^{[m_{j_k}]}\}$ does not occur and the proof proceeds as in Chapter VII. To get the needed local boundedness of the sequence of iterates in 7.40, cite 12.24.

Here is the ultimate generalization of the above:

Exercise 12.46 *Let* Ω *be a region such that* $\mathbb{C}\backslash\Omega$ *contains at least two points. Let* $f \in H(\Omega)$ *and* $f(\Omega) \subset \Omega$. *Suppose that for some* $z_0 \in \Omega$ *the sequence* $\{f^{[n]}(z_0)\}$ *has a cluster point* w_0 in Ω. *Show that then either* f *is one-to-one and onto or else* $f(w_0) = w_0$ *and* $\{f^{[n]}\}$ *converges locally uniformly on* Ω *to* w_0.

Hints: If some iterate $f^{[m]}$ has a fixed point α in Ω, then 12.45 may be applied to the function $f^{[m]}$ to conclude that either $|f^{[m]\prime}(\alpha)| < 1$ or else $f^{[m]}$ is one-to-one and onto. The latter clearly implies that f itself is one-to-one and onto. The former implies by 7.35 and normality (see the Hints to 12.45) that the iterates of $f^{[m]}$ converge to α throughout Ω. In particular,

$$f(\alpha) = \lim_{k\to\infty} f(f^{[km]}(\alpha)) = \lim_{k\to\infty} f^{[km]}(f(\alpha)) = \alpha.$$

Then by a simple computation with the Chain Rule, $f^{[m]\prime}(\alpha) = [f'(\alpha)]^m$, so the assumption $|f^{[m]\prime}(\alpha)| < 1$ means that $|f'(\alpha)| < 1$. Another appeal to 7.35 and normality leads to the conclusion that $\{f^{[n]}\}$ converges in Ω to α. But then $\alpha = \lim f^{[n]}(z_0)$ must coincide with the cluster point w_0.

Therefore the proof is reduced to the case where

(1) no iterate of f has a fixed point in Ω.

Now mimic the proof of 7.37. Choose $n_1 < n_2 < \cdots$ so that $f^{[n_j]}(z_0) \to w_0$ and set $m_j = n_{j+1} - n_j$. Since $\{f^{[n]}\}$ is a normal family, $\{m_j\}$ has a subsequence $\{m_{j_k}\}$ such that either $\{f^{[m_{j_k}]}\}$ diverges to ∞ uniformly on compact subsets of Ω or converges to a holomorphic function Φ. Argue as in the hints to 12.45 that the first alternative cannot obtain: Since

(2) $\quad f^{[m_{j_k}]}(f^{[n_{j_k}]}(z_0)) = f^{[n_{j_k}+1]}(z_0),$

the sequence $\{f^{[m_{j_k}]}\}$ cannot diverge uniformly to ∞ on the compact subset $K = \{w_0\} \cup \{f^{[n_j]}(z_0): j = 1, 2, \ldots\}$. Next, go to the limit in (2) uniformly on K to see that

(3) $\quad \Phi(w_0) = w_0.$

The assumption (1) means that $f^{[n]} - I$ has no zero in Ω for any n, where I denotes the identity function. Since $f^{[m_{j_k}]} - I$ converges locally uniformly on Ω to the function $\Phi - I$ which does have a zero by (3), we learn from 7.11 that $\Phi - I \equiv 0$. It follows then from 7.36 that f is one-to-one and onto.

§ 7 Ostrowski's Proof of Schottky's Theorem

Exercise 12.47 *Let f be holomorphic in $D(0, r), f(0) = 1$ and $0 < |f| \le M$. Then for all $0 < \rho < r$*

$$|f(z)| \ge M^{-2\rho/(r-\rho)} \quad \forall z \in D(0, \rho).$$

Hint: Let F be a holomorphic logarithm for M/f and apply the upper Harnack inequality (5.28).

Lemma 12.48 *Let f be holomorphic in a neighborhood of $\bar{D}(0, 1)$, $f\cdot(f-1)$ zero-free and for all $0 < r < 1$ define*

$$M(r) = \max_{|z| \le r}|f(z)|, \qquad m(r) = \max_{|z| \le r}\frac{1}{|f(z)|}$$

$$\tilde{M}(r) = \max\{e, M(r)\}, \qquad \tilde{m}(r) = \max\{e, m(r)\}$$

$$c = \max\{1, |\log|f(0)||\}.$$

Then for any $0 < \rho < r < 1$

(*) $\quad \tilde{m}(\rho) \le e^{10c/(r-\rho)}[\log \tilde{M}(r)]^{2/(r-\rho)}$.

Proof: Fix such an r and ρ. Notice that $\tilde{M}(r) \ge e$ and $c \ge 1$, so $10c/(r-\rho) \ge 10/(r-\rho) > 10$. Consequently the right side of (*) exceeds e and the proposed inequality is true if $\tilde{m}(\rho) \le e$, that is, if $m(\rho) \le e$. So we need only address ourselves to the case where

(1) $\quad \tilde{m}(\rho) = m(\rho) > e.$

Select z_ρ with $|z_\rho| = \rho$ such that

(2) $\quad \tilde{m}(\rho) = \dfrac{1}{|f(z_\rho)|}.$

Let n be the positive integer specified by

(3) $\quad n \le \log \tilde{M}(r) < n + 1.$

Now f is holomorphic and $1 - f$ is zero-free in $V = D(0, R)$ for some $R > 1$. Therefore by 5.35 there is a holomorphic function F such that

(4) $\quad F^n = 1 - f \quad \text{in } V.$

We set $f^* = n(1 - F)$. Let s denote the holomorphic nth root in $D(1, 1)$ which is positive on $(0, 2)$, that is, $s = e^{L/n}$ where L is the Principal Branch of the Logarithm in $\mathbb{C}\setminus(-\infty, 0]$ (3.43). We will show that F can be chosen so that

(5) $\quad F = s \circ (1 - f) \quad \text{near } z_\rho.$

Indeed we have

(6) $\quad |f(z_\rho)| = \dfrac{1}{\tilde{m}(\rho)} \le \dfrac{1}{e} < \dfrac{1}{2},$

so for some connected neighborhood U of z_ρ in V we have

$$|f(z)| < 1 \quad \forall z \in U.$$

For these z then $F(z)$ and $s(1 - f(z))$ are each (non-zero) nth roots of $1 - f(z)$ and so their quotient is an nth root of 1. As $s \circ (1 - f)/F$ is continuous on the connected set U, it is therefore constantly equal to some one nth root of 1, say $e^{2\pi ij/n}$. We may replace F by $e^{2\pi ij/n}F$ without prejudice to (4) and upon so doing we shall have (5) valid throughout U. Computing $s^{(k)}$ as $[s|(0, 2)]^{(k)}$ and remembering that on (0, 2), s is just the unique positive nth root function, we see

$$s^{(k)}(1) = \frac{1}{n}\left(\frac{1}{n} - 1\right)\cdots\left(\frac{1}{n} - k + 1\right)$$

$$s(w) = \sum_{k=0}^{\infty} \frac{1}{n}\left(\frac{1}{n} - 1\right)\cdots\left(\frac{1}{n} - k + 1\right)\frac{1}{k!}(w - 1)^k \quad \forall w \in D(1, 1).$$

It follows that $|s^{(k)}(1)| \le (k - 1)!/n$ and so

$$|1 - s(w)| \le \sum_{k=1}^{\infty} \frac{1}{n}|w - 1|^k = \frac{1}{n}\frac{|w - 1|}{1 - |w - 1|}$$

(7) $$< \frac{2}{n}|w - 1| \quad \forall w \in D(1, \tfrac{1}{2}).$$

Thus for z near z_ρ we have from (5), (6) and (7)

$$|f^*(z)| < 2|f(z)|.$$

In particular,

$$|f^*(z_\rho)| < 2|f(z_\rho)| = \frac{2}{\tilde{m}(\rho)} \quad \text{by (2)}$$

and so

$$\tilde{m}(\rho) < \frac{2}{|f^*(z_n)|} \le 2 \operatorname*{Max}_{|z| \le \rho} \frac{1}{|f^*(z)|} \stackrel{\text{def.}}{=} 2m^*(\rho),$$

(8) $$\frac{\tilde{m}(\rho)}{2} < m^*(\rho).$$

Moreover, from the definitions of F and f^*

$$\begin{aligned} M^*(r) &\le n[1 + (1 + \tilde{M}(r))^{1/n}] \\ &< [\log \tilde{M}(r)][1 + (1 + e^{n+1})^{1/n}] \quad \text{by definition (3) of } n \\ &= [1 + e(e^{-n} + e)^{1/n}] \log \tilde{M}(r) \\ &< [1 + e(e^{-n} + e)] \log \tilde{M}(r) \\ &\le (e^2 + 2) \log \tilde{M}(r) \end{aligned}$$

(9) $$< 10 \log \tilde{M}(r).$$

In the last step we used the fact that $e^2 < 8$ (exercise). Next note that $|F(0)| \leq (1 + |f(0)|)^{1/n}$ by (4), so by (4) and the definition of f^*

$$\frac{1}{|f^*(0)|} = \frac{1}{n|f(0)|}\left|\frac{1 - F^n(0)}{1 - F(0)}\right|$$

$$\leq \frac{1}{n|f(0)|}(1 + |F(0)| + \cdots + |F(0)|^{n-1}) < \frac{n(1 + |f(0)|)}{n|f(0)|}$$

$$= 1 + \frac{1}{|f(0)|} \leq 1 + e^c, \quad \text{since } c \geq -\log|f(0)| \text{ by definition,}$$

(10) $$< e^{2c}.$$

Now 12.47 applied to $f^*/f^*(0)$ [note that f^* is zero-free] says that

$$|f^*(0)|m^*(\rho) = \sup_{|z| \leq \rho} \frac{1}{\left|\dfrac{f^*(z)}{f^*(0)}\right|} \leq \left[\sup_{|z| \leq r}\left|\frac{f^*(z)}{f^*(0)}\right|\right]^{2\rho/(r-\rho)} = \left[\frac{M^*(r)}{|f^*(0)|}\right]^{2\rho/(r-\rho)},$$

$$m^*(\rho) \leq \left[\frac{1}{|f^*(0)|}\right]^{(r+\rho)/(r-\rho)} M^*(r)^{2\rho/(r-\rho)}.$$

Using this and (8), (9), (10)

$$\frac{\tilde{m}(\rho)}{2} \leq [e^{2c}]^{2/(r-\rho)}[10 \log \tilde{M}(r)]^{2/(r-\rho)} = [10e^{2c}]^{2/(r-\rho)}[\log \tilde{M}(r)]^{2/(r-\rho)},$$

(11) $$\tilde{m}(\rho) \leq 2(10)^{2/(r-\rho)}e^{4c/(r-\rho)}[\log \tilde{M}(r)]^{2/(r-\rho)}.$$

But $0 < r - \rho < 1$, $c \geq 1$ and $e^6 > 200$ (exercise), so (11) yields

$$\tilde{m}(\rho) \leq e^{10c/(r-\rho)}[\log \tilde{M}(r)]^{2/(r-\rho)},$$

as claimed.

Lemma 12.49 *Let the notation and hypotheses be those of the previous lemma. Then*

(*) $$M(1 - \theta) \leq e^{168c(1/\theta)\log(e/\theta)} \quad \forall 0 < \theta < 1.$$

Proof: First consider only $\theta \in (0, 1/e]$ and suppose, with a view toward ultimately reaching a contradiction, that for some $\theta \in (0, 1/e]$

(1) $$M(1 - \theta) > e^{56c(1/\theta)\log(1/\theta)}.$$

Now apply the last lemma to $1/f$ in the role of f. The conclusion is then

(2) $$\tilde{M}(\rho) \leq e^{10c/(r-\rho)}[\log \tilde{m}(r)]^{2/(r-\rho)}.$$

Here let us take $\rho = 1 - \theta$ and $r = 1 - \theta/2$, $(r - \rho)/2 = \theta/4$. We get

$$\log \tilde{m}(r) \geq e^{-5c}\tilde{M}(\rho)^{(r-\rho)/2} = e^{-5c}\tilde{M}(1 - \theta)^{\theta/4} \quad \text{by (2)}$$

$$> e^{-5c}e^{-14c \log \theta} \quad \text{by (1)}$$

$$= \left(\frac{1}{\theta}\right)^{14c}\left(\frac{1}{e}\right)^{5c} = \left(\frac{1}{\theta}\right)^{2c}\left(\frac{1}{\theta e}\right)^{5c}\left(\frac{1}{\theta}\right)^{7c}$$

$$\geq \left(\frac{1}{\theta}\right)^{2} e^{7c} \quad (\text{since } c \geq 1 \text{ and } \theta \leq 1/e)$$

$$> \frac{1}{\theta^2}(e^{7c} - 1) = \frac{7c}{\theta^2}\cdot\frac{e^{7c} - 1}{7c} > \frac{7c}{\theta^2}\cdot\frac{e^7 - 1}{7}$$

(since $(e^x - 1)/x$ is an increasing function of $x \in [1, \infty)$)

$$= \left(\frac{2}{\theta}\right)^2 c\,\frac{e^7 - 1}{4} > \left(\frac{2}{\theta}\right)^2 c\cdot 100$$

(from $e^6 > 200$, above, follows $e^7 > 401$)

$$> 100c\,\frac{2}{\theta}\log\frac{2}{\theta} \quad (\text{since } x > \log x).$$

Thus

$$\tilde{m}(1 - \theta/2) > e^{100c(2/\theta)\log(2/\theta)}.$$

The right side exceeds e so the definition of $\tilde{m}$ shows that $\tilde{m}(1 - \theta/2) = m(1 - \theta/2)$ and so our inequality says that

$$m(1 - \theta/2) > e^{100c(2/\theta)\log(2/\theta)}$$

$$> e^{56c(2/\theta)\log(2/\theta)}. \tag{3}$$

Now just as (3) was deduced from (2) via 12.48 applied to $1/f$, we can deduce from (3) via 12.48 applied to f that

$$M(1 - \theta/4) > e^{56c(4/\theta)\log(4/\theta)}. \tag{4}$$

The whole implication "(1) $\Rightarrow$ (4)" can now simply be iterated n times (for any positive integer n) and there results the inequality

$$M(1 - \theta/4^n) > e^{56c(4^n/\theta)\log(4^n/\theta)}, \quad n = 1, 2, \ldots \tag{5}$$

Now $(4^n/\theta)\log(4^n/\theta) \to \infty$ as $n \to \infty$ (see 3.18) and so for large values of n, (5) contradicts the boundedness of f in $D(0, 1)$. (Remember, f is holomorphic in a neighborhood of $\bar{D}(0, 1)$.) Thus (1) is untenable and we have proved that

$$M(1 - \theta) \leq e^{56c(1/\theta)\log(1/\theta)}, \quad 0 < \theta \leq 1/e. \tag{6}$$

Now $(1/\theta)\log(e/\theta)$ is clearly a decreasing function of $\theta \in (0, 1]$, so its smallest value on this interval occurs at $\theta = 1$ and is 1. It follows that for $\theta \in [1/e, 1)$

$$M(1 - \theta) \leq M(1 - 1/e) \overset{(6)}{\leq} e^{56ce} < e^{168c} \leq e^{168c(1/\theta)\log(e/\theta)}. \tag{7}$$

Since the right hand of (6) is not greater than the right hand of (*), the conjunction of (6) and (7) yields (*).

Corollary 12.50 (Schottky) *Let f be holomorphic in $D(0, 1)$ and omit* 0 *and* 1 *from its range. Set $d = \max\{1, \log|f(0)|\}$. Then*

$$(*) \qquad \sup_{|z| \le 1-\theta} |f(z)| \le e^{169d(1/\theta)\log(e/\theta)} \quad \forall 0 < \theta < 1.$$

Proof: For each $r \in (0, 1)$ the function $f_r(z) = f(rz)$ is holomorphic in the neighborhood $D(0, 1/r)$ of $\bar{D}(0, 1)$ and so the last lemma applies to it and affirms that

$$|f(rz)| \le e^{168c(1/\theta)\log(e/\theta)} \quad \forall z \in D(0, 1 - \theta).$$

This holding for all $r < 1$, we let $r \uparrow 1$ and conclude that

$$(**) \quad |f(z)| \le e^{168c(1/\theta)\log(e/\theta)} \quad \forall z \in D(0, 1 - \theta).$$

Here $c = \max\{1, |\log|f(0)||\}$ and of course $d \le c$. If $d = c$, we have in (**) (a bit more than) the desired inequality (*).

If $d < c$, then $d = 1$ and $c > 1$. Thus $\log|f(0)| \ne |\log|f(0)||$, so $\log|f(0)| = -|\log|f(0)|| = -c < -1$, $|f(0)| < 1/e < \frac{1}{2}$. Then certainly $e > |1 - f(0)| > 1/e$. Consequently the appropriate c-constant for the function $1 - f$ is 1 and (**) for this function reads

$$|1 - f(z)| \le e^{168(1/\theta)\log(e/\theta)} \quad \forall z \in D(0, 1 - \theta),$$

whence

$$|f(z)| \le 1 + e^{168(1/\theta)\log(e/\theta)} < e^{169(1/\theta)\log(e/\theta)} \quad \forall z \in D(0, 1 - \theta)$$

(since $(1/\theta)\log(e/\theta) > 1$, as noted in the last proof). This confirms (*) in the complementary case $d < c$ and finishes the proof.

Remarks: This is perhaps the simplest of the elementary derivations of Schottky's theorem which nevertheless give quantitative estimates for the majorant, free of unspecified constants. It is from OSTROWSKI [1925a]. He also presents the same proof in the appendix of his monograph [1931] but with minor modifications which lead to the better inequality

$$\sup_{|z| \le 1-\theta} |f(z)| \le e^{20d(1/\theta)\log(e/\theta) - 16}.$$

Notes to Chapter XII

The trek to the *Hauptsatz* 12.14 (MIRANDA [1935]) is laborious but, I think, worth the trouble. For one thing, these techniques have a quite broad applicability; they represent the elementary phase of a very fruitful and deep chapter of function theory inaugurated in this century by R. NEVANLINNA. See his book [1953], [1970] and HAYMAN [1964]. My account follows VALIRON [1937]. I have, however, tried to watch over the constants more carefully and to keep before the reader at all times their independence from the particular functions being analyzed. The inequality we obtain is not the best. It can be improved, as regards its dependence on $1/(1 - r)$, for example, by using techniques from MILLOUX [1940]. HIONG [1958a] proves that for each non-negative integer n there are constants H_n, K_n such that

$$(*)\qquad \log M(r,f) \leq \frac{1}{1-r}\left[H_n|\log|f(0)|| + K_n \log\frac{2}{1-r}\right], \quad 0 \leq r < 1$$

for all f as in 12.14. In its dependence on $1/(1 - r)$ this is a marked improvement over the inequality in the text. Contrary to that author's claim (p. 987, *op. cit.*), I found that when the proof details and supporting material had all been supplied, the proof of (*) was longer than that in the text. However, a more serious disadvantage of (*) is the fact that the bound on f which it provides gets coarser as $|f(0)| < 1$ gets smaller, a circumstance somewhat at variance with intuition, and because of this I was unable, despite the author's claim (p. 994, *op. cit.*), to deduce from (*) the important normality criterion 12.25. [For the history and priorities on the latter see p. 987, *op. cit.*] So I settled for the version of Miranda's Theorem in the text. The reader may wish to investigate this matter on his own. He will find in HIONG [1968] an English translation of (a Chinese translation of) HIONG [1958a] and interesting related matter in HIONG and HO [1961], SHIEH [1962] and YANG and CHANG [1965].

The special case of the theorem in 12.15 and 12.50 is due to SCHOTTKY [1904], [1906], [1917] (and BOUTROUX [1905], [1906] independently, though his derivation contains a non-trivial lacuna), and is already a very deep and striking result. His original theorem asserted a bound on $\log M(r,f)$ dependent only on $f(0)$ and $(1 - r)^{-4}$. LANDAU [1906] improved this to $(1 - r)^{-2}$ and finally LÉVY [1912] showed that the dependence on r was of the order $(1 - r)^{-1}$. For Schottky's theorem there exist very polished direct proofs, the products of many hands: LANDAU [1926], BLOCH [1924], [1925], VALIRON [1926], [1930], H. KÖNIG [1957], ESTERMANN [1971]. See, for example, the account in SAKS and ZYGMUND [1971], pp. 341–350 (cf. MACKI [1968]) or CONWAY [1973] or LANDAU [1940] or [1929a]. (See especially the latter, pp. 18–19 for history.) For a somewhat different proof see TITCHMARSH [1939], CARTWRIGHT [1956] and HOLLAND [1973]. For an exhaustive account of Schottky's theorem exploring many of its applications see the treatise of OSTROWSKI [1931], chapters 1 and 4. For extensions of a different kind from Miranda's see BIEBERBACH [1922], LANDAU [1922b], FEKETE [1922], SAXER [1934] and KRAJKIEWICZ [1977].

In our version neither the constants nor the order of dependence on $1/(1 - r)$ are best possible. In HAYMAN [1947] occurs one of the sharpest forms of the inequality:

$$\log M(r,f) \leq [\pi + \log^+|f(0)|]\frac{1+r}{1-r}.$$

JENKINS [1955] gives a short proof of this and some improvements which depend on the size of $|f(0)|$. Earlier AHLFORS [1938] had proved a corresponding inequality with the (larger) constant $4 + \log 10$ instead of π. See also ROBINSON [1939]. In Chapter XVIII we will prove a sharp Schottky-type inequality for *univalent* functions. OSTROWSKI [1933] finds asymptotic expressions for $M(r,f)$ as $r \uparrow 1$.

The amazing lemma 12.10 is from BUREAU [1932]; the ideas in it go back to BOREL [1896], [1896–97] and BLUMENTHAL [1910]. A version of it features also in Ostrowski's proof. (See the proof of 12.49.)

12.16 is due to BUREAU [1931] and 12.17 to PICARD [1879a]. The beautiful corollary 12.18 is asserted in PÓLYA [1921] and [1922] and proved in SAXER [1923], VAROPOULOS [1928] and CSILLAG [1928], [1935]. This result has a remarkable generalization due to TUMURA [1937] (whose proof contained gaps) and CLUNIE [1962]: if f is entire and for some $k \geq 2$, $f \cdot f^{(k)}$ is zero-free, then f has the form $f(z) \equiv e^{az+b}$ for some constants a and b. See pp. 67 ff. of HAYMAN [1964]. For a short, but not elementary, proof when $k = 2$, (first proved by Hayman) see C.-C. YANG [1970]. See also S. S. MACINTYRE [1949] and EDREI [1955]. Finally, FRANK [1976] has proved a conjecture of Hayman that if meromorphic f are admitted, the only new members of the club are $f(z) = (az + b)^{-n}$, n a positive integer.

For 12.19 see ROSENBLOOM [1948] and [1952]. It is pointed out in the review of the first paper that (part of) the result is actually contained in an earlier one of FATOU [1926] (p. 346). For an interesting application of 12.19 see the *American Mathematical Monthly* 73 (1966), 404–405 and for a modest extension of 12.18, the same journal, 66 (1959), 155.

12.20 is from chapter IV of MONTEL [1916]. (See the last paragraph of the review of this paper.) A little more is true: if p, q are integers and $1/p + 1/q < 1$, then there are no non-constant entire functions F, G such that $F^p + G^q = 1$. See VALIRON [1929a], JATEGAONKAR [1965] (where the proof follows the pattern of 12.20) and GROSS [1966]. Actually 12.20 is a special case of a result of Picard affirming that if F and G are meromorphic in $D(0, 1)\backslash\{0\}$ and satisfy an algebraic relation of genus greater than 1, then 0 is a removable singularity of F and of G. [One takes $F(z) = f(1/z)$, $G(z) = g(1/z)$.]

12.21 is *Satz* VIII in HURWITZ [1904], with $\alpha \geq 1/58$. LANDAU [1906], p. 285 gave a proof of this also. 12.21 was rediscovered by BOCHNER [1926] and FEKETE [1927a]. CARATHÉODORY [1907] and Bochner each confirmed Hurwitz' conjecture that 1/16 is the best value for α. (See also NEHARI [1952], p. 323 or SANSONE and GERRETSEN [1969], p. 502.) The proofs use a modular function; see below. For a related result see LANDAU [1922a], BOHR [1923] and MONTEL [1929].

The important special case of 12.25 where $N = 0$ is due to MONTEL [1912] (but is adumbrated in *Satz* VI of CARATHÉODORY and LANDAU [1911]; see also the footnotes there.) For another proof in that case see DE LA VALLÉE POUSSIN [1915–16]. This theorem is the principal ingredient in our development of the Julia theorem (12.27); in turn its principal ingredient is Schottky's theorem. If we let our functions take their values on the Riemann sphere $\mathbb{C}_\infty$ (see Chapter I notes), the idea of normal family extends naturally to meromorphic functions and becomes synonymous with the modern topological idea of pre-compact set (in the appropriately topologized space of functions). This was first pointed out by OSTROWSKI [1926b], § 3. In terms of the so-called chordal metric in $\mathbb{C}_\infty$ a very useful normality criterion was given by F. Marty. (See HAYMAN [1964], pp. 158–160 for references and a statement and proof of the result.) This can be made an alternative starting point, from which Montel's criterion, thence Julia's theorem *and* (a qualitative version of) Schottky's theorem can be deduced. For the elegant details see ZALCMAN [1975]. The important paper of CARATHÉODORY [1929b] should also be mentioned in this context.

MANDELBROJT [1929] shows that for a family $\mathscr{F}$ of zero-free holomorphic functions in a region Ω the local uniform boundedness in $\Omega \times \Omega$ of the family of auxiliary functions $\tilde{f}(z, w) = f(z)/f(w)$, $f \in \mathscr{F}$, is necessary and sufficient for the normality of $\mathscr{F}$. Cf. 7.28.

For an elementary proof of a slightly weakened version of 12.28 see ZALCMAN [1978].

Before 12.27, PICARD [1879b] had proved that there exists $b \in \mathbb{C}$ such that $\mathbb{C}\backslash\{b\} \subset f(D(a, R))$ for all (sufficiently small) $R > 0$, i.e., in every neighborhood of a, f assumes all complex values with at most one exception. This is the so-called Big (or Great) Picard Theorem [in contrast to the little (or lesser) one in 12.17.] See FRANKLIN [1925a] for a modest extension and CARATHÉODORY [1912b] (also OSTROWSKI [1929b], BIEBERBACH [1929], LANDAU [1929c] and FENCHEL [1931]) for a quantitative extension: there exists a constant $r \in (0, 1)$ such that each f which is holomorphic in $A(0, 1)$ and omits 0 and 1 from its range satisfies either $f(A(0, r)) \subset D(0, 2)$ or $f(A(0, r)) \subset \mathbb{C}\backslash D(0, \frac{1}{2})$. For Julia's extension of Picard's theorem see p. 102 of JULIA [1924]. Interesting extensions of Julia's theorem may be found in OSTROWSKI [1931]. An extension to harmonic functions appears in MONTEL [1932c].

There is an entirely different, rather geometric, development of Schottky's theorem and many of its corollaries like the Picard theorems, based on the Monodromy Theorem (Chapter X notes) and a modular function. The latter is a holomorphic map of the upper half-plane onto $\mathbb{C}\backslash\{0, 1\}$ which is invariant under a certain group of conformal self-maps of the upper half-plane. For a careful treatment of it see pp. 353–356 of RUDIN [1974] or chapter 23 of ESTERMANN [1962]. This approach was the historically first one, used for example by Picard himself, and it is via a modular function that the sharpest inequalities have been obtained. See LANDAU [1906] and the works referenced there, CARATHÉODORY [1960], DINGHAS [1961], §§ 71, 72; HILLE [1962], § 14.5; VEECH [1967], pp. 126–136 and LEHNER [1969]. The proof given by PERRON [1929] is a more elementary version of the modular function proofs; he manufactures on the spot a substitute for the modular function. For a more detailed exposition of Perron's work see § 140 of PRINGSHEIM [1932]. Cf. also DE LA VALLÉE POUSSIN [1915–16].

After Picard proved his results with the modular function a search began for an elementary proof (meaning one which did not use the modular function or any analytic results of comparable depth). In [1896] BOREL achieved this for the little theorem (see also the appendix to his book [1921]) and in [1904], building on Borel's technique, SCHOTTKY did it for the great theorem. For a nice treatment of their work (including contributions of Carathéodory and Landau) see the congress lecture of LINDELÖF [1910]. The next phase occurred with the arrival of Bloch's theorem 7.26(iii) (see Chapter VII notes), already presaged in the work of Landau, Hurwitz and Koebe. From this follows a relatively easy proof of Schottky's theorem, thence the Picard theorems. This is the approach followed in almost all textbooks which want to avoid the modular function.

For a detailed history of the Schottky and Picard theorems up to 1920, see pp. 409–417 of BIEBERBACH [1921]; for related history see VALIRON [1932].

In 12.29 (taken from pp. 66–70 of MONTEL [1927]) the reader has a brief introduction to Montel's theory of quasi-normal families. See his book just cited and VALIRON [1929b] for more on this topic.

12.30 is due to MONTEL [1912]. Nice generalizations in which we only hypothesize that $f(z_n) \to c$ for a certain sequence of points $z_n \in S$ may be found in EGGLESTON [1951], A. EVANS [1953] and BOWEN and MACINTYRE [1954] (cf. 7.31). For a good summary statement of the classical results, containing 5.16 $\cup$ 5.56 $\cup$ 12.36 $\cup$ 12.31, plus references to important work of W. Gross and F. Iversen on the subject, see p. 420 of the encyclopedia article of BIEBERBACH [1921]. For a unified treatment of all these results via harmonic majorants see pp. 301–308 of TSUJI [1959]. For other proofs of 12.31 see § 21 of MONTEL [1917], exercise III.277 of PÓLYA and SZEGÖ [1972] and pp. 12–13 of F. and R. NEVANLINNA [1922]. For 12.36 see p. 518 of MONTEL [1912] and OU [1957].

12.38–12.40 are from HARDY, INGHAM and PÓLYA [1928]. These authors show that equality can be tolerated in hypothesis (2) of 12.38 and 12.39 and provide a counterexample to the conclusion when $b - a > \frac{1}{2}(\beta - \alpha)$. But they show how 12.41(i) can be used to eliminate all hypotheses on a, b (beyond $a \neq b$, of course) when the function has no zeros. See CARTWRIGHT [1962] and HAYMAN [1962] for extensions. 12.41(i) is from HARDY [1926]; for a generalization see TSUJI [1930b], SHIMIZU [1931a], BONSALL [1949] and BEARDON [1971] (subharmonic functions); see also pp. 392 ff. of OSTROWSKI [1937]. On 12.41(ii) see YANAGIHARA [1965] and pp. 42–43 of DONOGHUE [1974].

I took 12.42 from ROTH [1938], though it is adumbrated in BOHR [1930]. Compare also LAURITZEN [1950]. Bohr proves an analog of 12.42 for entire functions f which are bounded on every ray: given any non-degenerate closed interval $J \subset [0, 2\pi]$, there is a non-degenerate closed interval $I \subset J$ such that $\{f(re^{i\theta}): \theta \in I, r \geq 0\}$ is bounded. He also shows that if $\{I_n\}$ are disjoint open intervals in $[0, 2\pi]$ whose union is dense in $[0, 2\pi]$, then there exists a non-constant entire f which is bounded on every ray and uniformly bounded in each set $\{re^{i\theta}: \theta \in I_n, r \geq 0\}$. Without the uniformity feature 12.42(ii) is essentially contained in BOHR [1927]. Non-constant entire functions bounded on every ray through 0 are not too hard to construct: see D. J. NEWMAN [1976]. As to non-constant entire functions which have limits at infinity along every ray, these too exist; in Chapter XV we will construct a few. [The easiest example I know of is the function $f(z)f(iz)$, where f is the function NEWMAN constructs, *op. cit.*]

Both 12.43 and 12.44 are taken from PÓLYA [1933]. See also AHLFORS [1933], MILLOUX [1935], KERÉKJÁRTÓ [1934/35], A. J. MACINTYRE [1939] and JENKINS [1953]. An even briefer proof of a related result is offered in LANDAU and OSSERMAN [1959]. They use the Cauchy–Riemann equations and a double integral instead of Cauchy's theorem.

12.45 may fail if only one point is omitted. Specifically, for $\Omega = \mathbb{C}\backslash\{-2\}$ and $f(z) = 2(e^z - 1)$, we have $f(0) = 0$ and $f(\Omega) = \Omega$. Since, however, $f'(0) = 2$, the conclusion of 7.40(i) does not hold here.

In 12.46 I have followed HEINS [1941b].

Regarding 12.6 and its converse 7.33, OSTROWSKI [1925b] proved the following definitive result: Let $\mathscr{L} = \{f \in H(D): \sup_{r<1} \int_0^{2\pi} \log|f(re^{i\theta})| d\theta < \infty\}$. Here $D = D(0, 1)$ and the integral is in Lebesgue's sense. A necessary and sufficient condition on points $z_n \in D$ that there exist $f \in \mathscr{L}$ whose zero set is $\{z_1, z_2, \ldots\}$ is that $\sum_{n=1}^{\infty} (1 - |z_n|) < \infty$. For a related result see DENJOY [1929] (cf. also the Chapter VII notes). Proofs of 12.6 and 12.7(ii) may also be found on pp. 26–28 of F. and R. NEVANLINNA [1922]. They show that the hypothesis (*) of 12.6 is equivalent to F being a quotient of two bounded holomorphic functions in the disk (this follows easily from 7.14(ii)), so the generality of 12.6 over 6.9(iii) is illusory. Ostrowski's result above (resp., 6.9(iii)) together with 7.33 enable one to factor any function of class $\mathscr{L}$ (resp., $H^{\infty}(D)$) into the product of a (bounded) zero-free holomorphic function in D, i.e., the exponential of a holomorphic function in D, and a Blaschke product. This representation is extremely useful. See the elementary discussion on pp. 40–45 of MONTEL [1927]. For a comprehensive treatment see chapter VII of R. NEVANLINNA [1970].

If hypothesis (*) in 12.6 is weakened to $\sup_{r<1}\int_0^r \int_0^{2\pi} \log^+|F(\rho e^{i\theta})| d\theta d\rho < \infty$, then the weaker conclusion $\sum_{j=1}^{\infty} (1 - |z_j|)^2 < \infty$ can be inferred; see KABAILA [1971].

Bibliography

I repeat the prefatory disclaimer that this bibliography is rather selective. E.g., hardly any papers on Riemann surfaces, elliptic functions, Dirichlet series, the order and type theory of entire functions or Nevanlinna theory appear because these topics are not discussed in the text. In general too the references and notes do not attempt to lead the reader into the differentiable manifold, global analysis or several complex variable extensions of the material presented in the text.

Most of the entries carry a reference to one or two of the review journals: **MR** = *Mathematical Reviews* (published by the American Mathematical Society, Providence), **Zbl** = *Zentralblatt für Mathematik und ihre Grenzgebiete* (published by Springer-Verlag, Berlin), **FM** = *Jahrbuch über die Fortschritte der Mathematik* (published by Verlag Georg Reimer, Berlin, prior to 1913 and thereafter by Walter de Gruyter and Co., Berlin). The first covers 1940 to the present, the second covers 1931 to the present, and the third covers 1868 to 1942. Every paper which was reviewed is equipped with its review coordinates. In addition, many of the books and monographs here have been accorded longer, more critical reviews in the *Jahresbericht der Deutschen Mathematiker Vereinigung* or in **BAMS** = *Bulletin of the American Mathematical Society*. For the reader's further convenience, books reviewed in the latter are also supplied with the volume and page numbers where that review can be found. The reader should always consult the several reviews of each paper (some being more detailed than others) because the reviews point out contact with related results in the literature and frequently discuss questions of priority or indicate errors. I hope, too, that reference to the review will compensate for any bibliographic inadequacies or inaccuracies in the individual books and papers listed here.*

If any journal title abbreviation in this bibliography confounds the reader, he should consult the review. There the title will also be abbreviated (differently perhaps) and he will have to consult the code of abbreviations for that reviewing organ. This is a table of full-name-equivalents for all abbreviations used; each of **MR**, **Zbl** and **FM** publishes such tables periodically with its reviews.

With books I try to list (only) the latest edition and rely on the review to mention all previous editions, reprints and translations and the coordinates of their reviews; almost always the review supplies this information. The only disadvantage of this is a chronological distortion when, for example, one cites "the great classic SAKS and ZYGMUND [1971]." Some journals (like the *Bulletin of the American Mathematical Society* and the *Comptes Rendus de l'Académie des Sciences Paris*) carry research announcements with no proof details. Often (but not always!) when an author announces results like this and follows with a later detailed paper elsewhere, my bibliography only lists the latter paper.

If a recent book or paper on a certain result supersedes several older ones and mentions them in its bibliography, I sometimes list only the former here and refer my readers to it (and, implicitly, to its bibliography).

* Since its founding in 1953 *Referativnyĭ Žurnal Matematika* has also provided comprehensive reviews (in Russian) of the world's mathematical literature and the reader might wish to check there works published after 1953.

I hope the reader will not be scandalized by the paucity of entries under the names of the founding fathers Cauchy and Gauss (and the complete absence of any under that of the great Euler). The author has no pretensions to being a historian of mathematics and must leave the interesting task of exegesis and evaluation of the early work to others. Most of it is pretty well documented (for example, in the encyclopedia articles of Burkhardt, Lichtenstein, Osgood and Pringsheim *et al.*) and I have been content to cite such secondary sources, for virtually everything before 1868. As noted, that was the first year of the *Jahrbuch* and consequently the first year that a moderately complete search of the literature was feasible.

For some of the authors cited here collected (or selected) works have been published and it was my original intention to list these and give references to them as well as to the original works, but limitations of space have precluded that. Also I had hoped to include whenever available an obituary notice (from one of the journals). These usually provide a short biography, an overview and assessment of the person's work, a complete list of publications and sometimes a photograph. This project also proved unfeasible and must therefore be commended to the initiative of the interested reader. I mention in passing only the 30-year index published in 1913 by *Acta Mathematica*; this contains very short *curricula vitae* and photographs (almost 200) of early contributors to that journal, many of whom appear in this bibliography.

ABIAN, A.

"A proof and extension of Brouwer's fixed point theorem for the closed 2-cell," *Boll. della Un. Mat. Ital.* (3) 16 (1961), 281–284. **Zbl** 147, p. 419. **MR** 25 #1546.

"The identity theorem for analytic functions," *Jour. Math. Anal. and Applic.* 45 (1974), 682–683. **Zbl** 275 #30001. **MR** 49 #9168.

"Hurwitz' theorem implies Rouché's theorem," *Jour. Math. Anal. and Applic.* 61 (1977), 113–115. **MR** 57 #12823.

ABIAN, A. and BROWN, A. B.

"A fixed point theorem for the closed 2-cell," *Portugaliae Math.* 21 (1962), 93–98. **Zbl** 107, p. 166. **MR** 26 #3032.

ACHIEZER, N. I.

Theory of Approximation (translated from Russian by C. J. Hyman), Frederick Ungar Publishing Co. (1956), New York. **MR** 36 #5567. **Zbl** 152, p. 253. **BAMS** 63, p. 163.

ADEL'SON-VEL'SKIĬ, G. M. and KRONROD, A. S.

"Cauchy's theorem without integration," *Dokl. Akad. Nauk SSSR* (*N.S.*) 50 (1945), 7–9 (Russian). **MR** 14, p. 546.

AGMON, S.

"Sur deux théorèmes de Fabry," *C.R. Acad. Sci. Paris* 226 (1948), 1673–1674. **MR** 9, p. 576. **Zbl** 30, p. 153.

AGOSTINI, A.

"Il teorema fondamentale dell'algebra," *Periodico di Mat.* (4) 4 (1924), 307–327. **FM** 50, p. 8.

AGOSTON, M. K.

Algebraic Topology, A First Course, Pure and Applied Mathematics, vol. 32. Marcel Dekker, Inc. (1976), New York. **Zbl** 337 #55001. **MR** 56 #3825.

AGUILÓ FUSTER, R.

"Untersuchung der Ideale des Ringes der ganzen Funktionen," *Collect. Math.* 17 (1965), 105–134. Correction, *ibid.*, 297 (Spanish). **Zbl** 163, p. 35. **MR** 34 #2608 and 35 #4204.

AHLFORS, L. V.

"Untersuchungen zur Theorie der konformen Abbildung und der ganzen Funktionen," *Acta Soc. Sci. Fennicae* (2) A, 1, No. 9 (1930). **FM** 56, p. 984.

(a) "Sur une généralisation du théorème de Picard," *C.R. Acad. Sci. Paris* 194 (1932), 245–247. **Zbl** 3, p. 407. **FM** 58, p. 343.

(b) "Ein Satz über die charakteristische Funktion und den Maximalmodul einer meromorpher Funktion," *Soc. Sci. Fennicae Comment. phys.-math.* (= *Finska Vetenskaps-Societeten*) 6, No. 9 (1932). **FM** 58, p. 333. **Zbl** 5, p. 300.

"Sur les domaines dans lesquels une fonction méromorphe prend des valeurs appartenant à une région donnée," *Acta Soc. Sci. Fennicae* (2) A2, No. 2 (1933). **FM** 59, p. 1033. **Zbl** 8, p. 262.

"On Phragmén-Lindelöf's principle," *Trans. Amer. Math. Soc.* 41 (1937), 1–8. **FM** 63, p. 286. **Zbl** 6, p. 32.

"An extension of Schwarz' Lemma," *Trans. Amer. Math. Soc.* 43 (1938), 359–364. **Zbl** 18, p. 410. **FM** 64, p. 315.

(a) *Complex Analysis*, McGraw-Hill Book Co. (2nd ed., 1966), New York. **MR** 14, p. 857 and 32 #5844. **Zbl** 154, p. 319. **BAMS** 59, p. 464.

(b) "Remarks on Carleman's formula for functions in a half-plane," *SIAM Jour. on Num. Anal.* 3 (1966), 183–187. **Zbl** 145, p. 150. **MR** 34 #2874.

Conformal Invariants: Topics in Geometric Function Theory, McGraw-Hill Book Co. (1973), New York. **Zbl** 272 #30012. **MR** 50 #10211.

AHLFORS, L. V. and HEINS, M.

"Questions of regularity connected with the Phragmén–Lindelöf Principle," *Annals of Math.* (2) 50 (1949), 341–346. **Zbl** 36, p. 47. **MR** 10, p. 522.

AHLFORS, L. V. and SARIO, L.

Riemann Surfaces, Princeton Mathematical Series, No. 26. Princeton University Press (1960), Princeton. **MR** 22 #5729. **Zbl** 196, p. 338.

ÅKERBERG, B.

"Proof of Poisson's formula," *Proc. Camb. Phil. Soc.* 57 (1961), 186. **Zbl** 91, p. 97. **MR** 22 #8234.

ALBRECHT, R.

"Iterationsverfahren zur konformen Abbildung eines Ringgebietes auf einen konzentrischen Kreisring," *Sitzungsber. Bayer. Akad. Wiss. München* (1954), 169–178. **MR** 17, p. 26. **Zbl** 64, p. 76.

ALEXANDER, J. W., II

"A proof and extension of the Jordan-Brouwer separation theorem," *Trans. Amer. Math. Soc.* 23 (1922), 333–349. **FM** 49, p. 403.

ALLING, N.

"The valuation theory of meromorphic function fields," pp. 8–29 of *Entire Functions and Related Parts of Analysis*, Proceedings of Symposia in Pure Mathematics, vol. 11, American Mathematical Society (1968), Providence. **Zbl** 182, p. 63. **MR** 38 #4700.

ANGER, G.

"Die Entwicklung der Potentialtheorie im Hinblick auf ihre grundlegenden Existenzsätze," *Jahresber. Deutsch. Math. Verein.* 64 (1961/62), 1te Abteilung, 51–78 and 100–134. **Zbl** 106, p. 77. **MR** 26 #5170a,b.

ANONYME, U.

"Sur l'intégrale $\int_0^\infty e^{-x^2}\,dx$," *Bull. Sci. Math.* (2) 13 (1889), 84. **FM** 21, p. 277.

ANTOINE, L.

"Sur l'homéomorphie de deux figures et de leurs voisinages," *Jour. de Math. Pures et Appliq.* (8) 4 (1921), 221–325. **FM** 48, p. 650.

"Sur les voisinages de deux figures homéomorphes," *Fund. Math.* 5 (1924), 264–287. **FM** 50, p. 369.

APOSTOL, T. M.

"Term-wise differentiation of power series," *Amer. Math. Monthly* 59 (1952), 323–326. (See also **MR** 40 #270.)

Introduction to Analytic Number Theory, Springer-Verlag (1976), New York. **Zbl** 335 #10001. **MR** 55 #7892.

ARAKÉLIAN, N. U.

"Uniform approximation on closed sets by entire functions," *Izv. Akad. Nauk SSSR Ser. Math.* 28 (1964), 1187–1206 (Russian). **MR** 30 #258. **Zbl** 143, p. 296.

"Certain questions of approximation theory and the theory of entire functions," *Math. Notes* 9 (1971), 267–271. **MR** 44 #440. (See also **Zbl** 234 #30029 and **MR** 54 #10609.)

ARSOVE, M. G.

"On the definition of an analytic function," *Amer. Math. Monthly* 62 (1955), 22–25. **MR** 16, p. 683. **Zbl** 64, p. 66.

"Intrinsic characterization of regions bounded by closed curves," *Duke Math. Jour.* 34 (1967), 425–429. **MR** 36 #367. **Zbl** 165, p. 97.

(a) "The Osgood–Taylor–Carathéodory theorem," *Proc. Amer. Math. Soc.* 19 (1968), 38–44. **MR** 36 #3966. **Zbl** 164, p. 97.

(b) "Some boundary properties of the Riemann mapping function," *Proc. Amer. Math. Soc.* 19 (1968), 560–568. **MR** 37 #1572. **Zbl** 162, p. 103.

"A correction to 'Some boundary properties of the Riemann mapping function'," *Proc. Amer. Math. Soc.* 22 (1969), 711–712. **MR** 39 #4368. **Zbl** 186, p. 400.

ARSOVE, M. G. and HUBER, A.

"Local behavior of subharmonic functions," *Indiana Univ. Math. Jour.* 22 (1973), 1191–1199. **MR** 48 #11536. **Zbl** 264 #31001.

ARTIN, E. and SCHREIER, O.

"Algebraische Konstruktion reeller Körper," *Abhand. Math. Sem. Univ. Hamburg* 5 (1926), 85–99. **FM** 52, p. 120.

ASCOLI, G.

(a) "Sulle singolarità isolate delle funzioni armoniche," *Boll. della Un. Mat. Ital.* (2) 7 (1928), 230–237. **FM** 54, p. 510.

(b) "Sulla unicità della soluzione nel problema di Dirichlet," *Rend. Accad. d. Lincei Roma* (6) 8 (1928), 348–351. **FM** 54, p. 509.

AUMANN, G. and CARATHÉODORY, C.

"Ein Satz über konforme Abbildung mehrfach zusammenhängender ebener Gebiete," *Math. Annalen* 109 (1934), 756–763. **Zbl** 9, p. 26. **FM** 60, p. 285.

AYOUB, R.

"Euler and the Zeta function," *Amer. Math. Monthly* 81 (1974), 1067–1086. **MR** 50 #12566. **Zbl** 293 #10001. (See also **MR** 51 #7778.)

BACHMANN, P.

Zahlentheorie: Zweiter Teil: Die Analytische Zahlentheorie, B. G. Teubner (1894), Leipzig. **FM** 25, p. 249. Reprinted by Johnson Reprint Corp. (1968), New York and London.

BAGEMIHL, F.
"Some identity and uniqueness theorems for normal meromorphic functions," *Ann. Acad. Scient. Fennicae Ser. AI*, No. 299 (1961). **Zbl** 100, p. 71. **MR** 23 #A302.

"A boundary condition for a holomorphic function in a Jordan region to be schlicht," *Proc. Nat. Acad. Sci. U.S.A.* 58 (1967), 1102–1103. **Zbl** 155, p. 400. **MR** 36 #1633.

BAGEMIHL, F. and SEIDEL, W. P.
"Some boundary properties of analytic functions," *Math. Zeit.* 61 (1954), 186–199. **MR** 16, p. 460. **Zbl** 58, p. 61.

"Sequential and continuous limits of meromorphic functions," *Ann. Acad. Scient. Fennicae Ser. AI*, No. 280 (1960). **Zbl** 95, p. 58. **MR** 22 #12226.

"Koebe arcs and Fatou points of normal functions," *Comm. Math. Helv.* 36 (1961), 9–18. **Zbl** 125, p. 317. **MR** 25 #5183.

BAKER, A.
Transcendental Number Theory, Cambridge University Press (1975), Cambridge. **Zbl** 297 #10013. **MR** 54 #10163. **BAMS** 84, p. 1370.

BAKER, I. N.
"Zusammensetzungen ganzer Funktionen," *Math. Zeit.* 69 (1958), 121–163. **MR** 20 #4000. **Zbl** 178, p. 75.

"The existence of fix-points of entire functions," *Math. Zeit.* 73 (1960), 280–284. **MR** 22 #4838b. **Zbl** 129, p. 291.

BALK, M. B.
"A theorem on entire functions," *Moskov. Oblast. Pedagog. Inst. Uč. Zap.* 57 (1957), 51–53 (Russian). **MR** 20 #5283. **Zbl** 90, p. 291.

BANACH, S.
"Über die Baire'sche Kategorie gewisser Funktionenmengen," *Studia Math.* 3 (1931), 174–179. **FM** 57, p. 305. **Zbl** 3, p. 297.

BANACH, S. and MAZUR, S.
"Über mehrdeutige stetige Abbildungen," *Studia Math.* 5 (1935), 174–178. **FM** 60, p. 1227. **Zbl** 13, p. 82.

BANK, S. B. and ORLAND, G. H.
"A note on Rouché's theorem," *Fund. Math.* 63 (1968), 137–141. **MR** 39 #1656. **Zbl** 165, p. 401.

BARROW, D. F.
"Infinite exponentials," *Amer. Math. Monthly* 43 (1936), 150–160. **Zbl** 13, p. 254. **FM** 62, p. 210.

BARTH, K. F. and SCHNEIDER, W. J.
"On a problem of Erdös concerning the zeros of the derivatives of an entire function," *Proc. Amer. Math. Soc.* 32 (1972), 229–232. **MR** 45 #2168. **Zbl** 236 #30032.

BAŠMAKOVA, I. G.
"On the proof of the fundamental theorem of algebra," *Istoriko-mat. Issledovanija* 10 (1957), 257–304 (Russian). **Zbl** 101, p. 246.

"Le théorème fondamental de l'algèbre et la construction des corps algébriques," *Arch. Internat. Histoirie Sci.* 13 (1961), 211–222. **Zbl** 126, p. 258.

BASYE, R. E.
"Simply connected sets," *Trans. Amer. Math. Soc.* 38 (1935), 341–356. **FM** 61, p. 636. **Zbl** 12, pp. 250 and 1353.

BEAR, H. S.
"A new look at the three circles theorem," *Amer. Math. Monthly* 81 (1974), 487–490. **Zbl** 288 #30019. **MR** 49 #9186.

BEARDON, A. F.
"Montel's theorem for subharmonic functions and solutions of partial differential equations," *Proc. Camb. Phil. Soc.* 69 (1971), 123–150. **Zbl** 207, p. 110. **MR** 42 #4756.

BECK, A.
"On rings on rings," *Proc. Amer. Math. Soc.* 15 (1964), 350–353. **MR** 29 #1347. **Zbl** 135, p. 289.

BECKENBACH, E. F.
"The stronger form of Cauchy's integral theorem," *Bull. Amer. Math. Soc.* 49 (1943), 615–618. **MR** 5, p. 35.
"Concerning the definition of harmonic functions," *Bull. Amer. Math. Soc.* 51 (1945), 240–245. **MR** 6, p. 227.

BEESACK, P. R.
"The Laurent expansion without Cauchy's theorem," *Canad. Math. Bull.* 15 (1972), 473–480. **MR** 46 #9301. **Zbl** 266 #30040.

BEHAN, D. F.
"Commuting analytic functions without fixed points," *Proc. Amer. Math. Soc.* 37 (1973), 114–120. **MR** 46 #7492. **Zbl** 251 #30009.

BEHNKE, H. and SOMMER, F.
Theorie der analytischen Funktionen einer komplexen Veränderlichen, Die Grundlehren der mathematischen Wissenschaften in Einzeldarstellungen, Bd. 77. Springer-Verlag (3rd ed., 1965), Berlin. **MR** 26 #5137. **Zbl** 273 #30001.

BELLMAN, R.
A Brief Introduction to Theta Functions, Holt, Rinehart and Winston, Inc. (1961), New York. **MR** 23 #A2556. **Zbl** 98, p. 283.

BENDIXSON, J.
"Sur une extension à infini de la formule d'interpolation de Gauss," *Acta Math.* 9 (1886/87), 1–34. **FM** 18, p. 208.

BERENŠTEĬN, M. N.
"Certain sufficient conditions for the analyticity of functions," *Isv. Vysš. Učebn. Zaved. Matematika* No. 11 (114) (1971), 11–18 (Russian). **Zbl** 234 #30002. **MR** 45 #5322.

BERENSTEIN, C. A. and TAYLOR, B. A.
"The 'three squares' theorem for continuous functions," *Arch. Rational Mech. and Anal.* 63 (1977), 253–259. **Zbl** 353 #46027. **MR** 56 #3335.

BERGMAN, S.
The Kernel Function and Conformal Mapping, Mathematical Surveys, Vol. V. American Mathematical Society (2nd ed. 1970), Providence. **MR** 12, p. 402. **Zbl** 40, p. 190 and 208, p. 343. **BAMS** 58, p. 76.

BERNARDI, S. D.
Bibliography of Schlicht Functions, Courant Institute of Mathematical Sciences (1966), New York. **MR** 34 #2849. *Part II (1966–1975)*, ibid. (1977).

BERNDT, B. C.
"On Gaussian sums and other exponential sums with periodic coefficients," *Duke Math. Jour.* 40 (1973), 145–156. **MR** 47 #1757. **Zbl** 255 #10042.
"Elementary evaluation of $\zeta(2n)$," *Math. Mag.* 48 (1975), 148–154. **Zbl** 303 #10038. **MR** 51 #3078.

Bernstein, F.

"Über einen Schönflies'schen Satz der Theorie der stetigen Funktionen zweier reeller Veränderlichen," *Göttingen Nachr.* (1900), 98–102. **FM** 31, p. 478.

Bernstein, S. N.

"Sur le principe de Dirichlet et le développement des fonctions harmoniques en séries de polynômes," *C.R. Acad. Sci. Paris* 148 (1909), 1306–1308. **FM** 40, p. 450.

"Sur une propriété des fonctions entières," *C.R. Acad. Sci. Paris* 176 (1923), 1603–1605. **FM** 49, p. 215.

Leçons sur les propriétés extrémales et la meilleure approximation des fonctions analytiques d'une variable réelle, Gauthier-Villars (1926), Paris. Reprinted by Chelsea Publishing Co. (1970), New York. **MR** 40 #2511. **FM** 52, p. 256. **Zbl** 237 #01043. **BAMS** 33, p. 369.

Bernstein, V.

"Alcune osservazioni sopra un teorema di Fabry," *Rend. Accad. d. Lincei Roma* (6) 21 (1935), 780–785. **Zbl.** 12, p. 261. **FM** 61, p. 312.

Bers, L.

"On rings of analytic functions," *Bull. Amer. Math. Soc.* 54 (1948), 311–315. **MR** 9, p. 575. **Zbl** 32, p. 203.

Besicovitch, A. S.

"Quelques théorèmes générales sur l'interversion des intégrations et l'intégration des séries," *Permĭ, Jour. Soc. Phys. et Math.* 1 (1918–1919), 99–139. **FM** 50, p. 632.

"On sufficient conditions for a function to be analytic, and on behaviour of analytic functions in the neighbourhood of non-isolated singular points," *Proc. Lon. Math. Soc.* (2) 32 (1930), 1–9. **FM** 56, p. 272.

Beurling, A.

"Études sur un problème de majoration," (thesis), Almqvist & Wiksell (1933), Upsala. **FM** 59, p. 1042. **Zbl** 8, p. 318.

Bieberbach, L.

(a) "Über einen Satz des Herrn Carathéodory," *Göttingen Nachr.* (1913), 552–560. **FM** 44, p. 760.

(b) "Über den Jordanschen Kurvensatz, die Schönfliesschen Sätze von Erreichbarkeit und Unbewalltheit und den Satz von der Invarianz des ebenen Gebietes," *Jahresber. Deutsch. Math. Verein.* 22 (1913), 144–153. **FM** 44, p. 559.

"Zur Theorie und Praxis der konformen Abbildung," *Rend. Circ. Mat. Palermo* 38 (1914), 98–112. **FM** 45, p. 670.

"Über einige Extremalprobleme im Gebiete der konformen Abbildung," *Math. Annalen* 77 (1916), 153–172. **FM** 46, p. 549.

"Über eine Vertiefung des Picardschen Satzes bei ganzen Funktionen endlicher Ordnung," *Math. Zeit.* 3 (1919), 175–190. **FM** 47, p. 299.

"Neuere Untersuchungen über Funktionen von komplexen Variablen," *Encyklopädie der Mathematischen Wissenschaften* Bd. II, 3rd Part, 1st Half, 379–532. B. G. Teubner (1921), Leipzig. **FM**. 48, p. 313.

"Über die Verteilung der Null- und Einsstellen analytischer Funktionen," *Math. Annalen* 85 (1922), 141–148. **FM** 48, p. 324.

"Über schlichte Abbildungen des Einheitskreises durch meromorphe Funktionen," *Sitzungsber. Preuss. Akad. Wiss. Berlin, Phys.-math. Kl.* (1929), 620–623. **FM** 55, p. 210.

Lehrbuch der Funktionentheorie I (4th ed., 1934), II (2nd ed., 1931), B. G. Teubner, Leipzig. Reprinted by Chelsea Publishing Co. (1945), New York and by Johnson Reprint Corp. (1968), New York. **MR** 6, p. 261 and 40 #2823a,b. **Zbl** 1, p. 211. **FM** 56, p. 258 and 57, p. 340. **BAMS** 28, p. 467; 34, p. 244; 38, pp. 19, 790.

"Über einen Satz Pólyascher Art," *Arch. der Math.* 4 (1953), 23–27. **MR** 14, p. 1074. **Zbl** 51, p. 57.

Analytische Fortsetzung, Ergebnisse der Mathematik und ihrer Grenzgebiete, Neue Folge, Heft 3. Springer-Verlag (1955), Berlin. **MR** 35 #5585. **Zbl** 174, p. 368. **BAMS** 62, p. 184.

Einführung in die konforme Abbildung, Sammlung Göschen Band 768/768a, Walter de Gruyter and Co. (6th ed., 1967), Berlin. **MR** 38 #4661. **Zbl** 158, p. 323. **FM** 63, p. 294. **BAMS** 34, p. 24. English translation of 4th ed. by F. Steinhardt, Chelsea Publishing Co. (1953), New York. **Zbl** 50, p. 84. **MR** 14, p. 462.

BIEHLER, CH.

"Sur la limite de $(1 + x/m)^m$ quand m augmente indéfiniment," *Nouvelles Annales de Math.* (3) 6 (1887), 60–67. **FM** 19, p. 235.

BIERMANN, O.

"Eine formale Ableitung der Laurentschen Reihe einer Funktion aus einer ihrer rationalgebrochenen Näherungsfunktionen," *Jour. für. Reine und Angew. Math.* 135 (1908), 142–145. **FM** 39, p. 466.

BIRKHOFF, G.

A Source Book in Classical Analysis, Harvard University Press (1973), Cambridge, Massachusetts. **Zbl** 275 #01009. **MR** 57 #9395.

BISHOP, E.

Foundations of Constructive Analysis, McGraw-Hill Book Co. (1967), New York. **Zbl** 183, p. 15. **MR** 36 #4930. **BAMS** 76, pp. 301–323.

BLASCHKE, W.

"Eine Erweiterung des Satzes von Vitali über Folgen analytischer Funktionen," *Ber. Verhandl. Kön. Sächs. Gesell. Wiss. Leipzig* 67 (1915), 194–200. **FM** 45, p. 638.

"Ein Mittelwertsatz und eine kennzeichnende Eigenschaft des logarithmischen Potentials," *Ber. Verhandl. Kön. Sächs. Gesell. Wiss. Leipzig* 68 (1916), 3–7. **FM** 46, p. 742.

"Mittelwertsätze der Potentialtheorie," *Jahresber. Deutsch. Math. Verein.* 27 (1918), 157–160. **FM** 46, p. 742.

BLOCH, A.

"Sur les intégrales de Fresnel," *Bull. Sci. Math.* (2) 43 (1919), 179–180 and 46 (1922), 34–35. **FM** 47, p. 226 and 48, p. 259.

(a) "Démonstration directe de théorèmes de M. Picard," *C.R. Acad. Sci. Paris* 178 (1924), 1593–1595. **FM** 50, p. 217.

(b) "Les théorèmes de M. Valiron sur les fonctions entières, et la théorie de l'uniformisation," *C.R. Acad. Sci. Paris* 178 (1924), 2051–2053. **FM** 50, p. 217.

"Les théorèmes de M. Valiron sur les fonctions entières, et la théorie de l'uniformisation," *Annales de la Faculté des Sciences de Toulouse* (3) 17 (1925), 1–22. **FM** 52, p. 324.

Les fonctions holomorphes et méromorphes dans le cercle-unité, Mémorial des Sciences Mathématiques, fasc. XX, Gauthier-Villars (1926), Paris. **FM** 52, p. 324.

BLUMENTHAL, O.
Principes de la théorie des fonctions entières d'ordre infini, Gauthier-Villars (1910), Paris. **FM** 41, p. 462. **BAMS** 21, p. 36.

BLUTEL, E.
"Sur une méthode d'approximation," *Bull. Soc. Math. France* 39 (1911), 155–159. **FM** 42, p. 113.

BOAS, R. P., Jr.
Entire Functions, Pure and Applied Mathematics, vol. V, Academic Press (1954), New York. **MR** 16, p. 914. **Zbl** 58, p. 302. **BAMS** 62, p. 57.
"Periodic entire functions," *Amer. Math. Monthly* 71 (1964), 782. **Zbl** 128, p. 72.
"Inequalities for the derivatives of polynomials," *Math. Mag.* 42 (1969), 165–174. **MR** 41 #4067. **Zbl** 179, p. 378.

BOBOC, N.
Complex Functions (Rumanian), Editura Didactică şi Pedagogică (1969), Bucureşti. **Zbl** 212, p. 97 and 220, p. 433.

BÔCHER, M.
"Singular points of functions which satisfy partial differential equations of elliptic type," *Bull. Amer. Math. Soc.* 9 (1902/03), 455–465. **FM** 34, p. 386.
"On harmonic functions in two dimensions," *Proc. Amer. Acad. Arts & Sci.* 41 (1906), 577–583. **FM** 37, p. 481.
"On semi-analytic functions," *Annals of Math.* (2) 12 (1910–11), 18–26. **FM** 41, p. 485.

BOCHNER, S.
"Remarks on the theorems of Picard–Landau and Picard–Schottky," *Jour. Lon. Math. Soc.* 1 (1926), 100–103. **FM** 52, p. 325.

BOGGIO, T.
"Sulle funzioni di variabile complessa in un'area circolare," *Atti della R. Accad. d. Scienze di Torino* 47 (1911–12), 22–37. **FM** 43, p. 487.

BOHR, H.
"Über streckentreue und konforme Abbildung," *Math. Zeit.* 1 (1918), 403–420. **FM** 46, p. 558.
"Über einen Satz von Edmund Landau," *Scripta Univ. atque Biblio. Hierosolymitanarum* 1, No. 2 (1923). **FM** 49, p. 711.
"On the limit values of analytic functions," *Jour. Lon. Math. Soc.* 2 (1927), 180–181. **FM** 53, p. 284.
"Über ganze transzendente Funktionen, die auf jeder durch den Nullpunkt gehenden Geraden beschränkt sind," pp. 39–46 of *Opuscula Mathematica A. Wiman Dedicata*, Lundequist (1930). Upsala.
"Zum Picardschen Satz," *Mat. Tidsskr. B* (1940), 1–6. **FM** 66, p. 361. **Zbl** 23, p. 139. **MR** 2, p. 183.

BONSALL, F. F.
"Note on a theorem of Hardy and Rogosinski," *Quart. Jour. of Math.* (*Oxford Series*) 20 (1949), 254–256. **Zbl** 34, p. 363. **MR** 11, p. 358.

BOREL, É.
"Démonstration élémentaire d'un théorème de M. Picard sur les fonctions entières," *C.R. Acad. Sci. Paris* 122 (1896), 1045–1048. **FM** 27, p. 321.
"Sur les zéros des fonctions entières," *Acta Math.* 20 (1896–97), 357–396. **FM** 28, p. 360.

Leçons sur les fonctions entières, Gauthier-Villars (2nd ed. 1921), Paris. **FM** 48, p. 316.

Leçons sur les fonctions de variables réelles et les développements en séries de polynômes, Gauthier-Villars (2nd ed. 1928), Paris. **FM** 54, p. 268. **BAMS** 12, p. 399.

BOREL, É. and ROSENTHAL, A.
"Neuere Untersuchungen über Funktionen reeller Veränderlichen," *Encyklopädie der Mathematischen Wissenschaften* Bd. II, 3rd Part, 2nd Half, 851–1187. B. G. Teubner (1924), Leipzig. **FM** 50, p. 176. **BAMS** 31, p. 453.

BORSUK, K.
"Über Schnitte der *n*-dimensionalen Euklidischen Räume," *Math. Annalen* 106 (1932), 239–248. **FM** 58, p. 628. **Zbl** 4, p. 73.

(a) "Drei Sätze über die *n*-dimensionale euklidische Sphäre," *Fund. Math.* 20 (1933), 177–190. **Zbl** 6, p. 424. **FM** 59, p. 560.

(b) "Über stetige Abbildungen der euklidischen Räume," *Fund. Math.* 21 (1933), 236–243. **FM** 59, p. 1254. **Zbl** 8, p. 133.

"Sur les prolongements des transformations continues," *Fund. Math.* 28 (1937), 99–110. **FM** 63, p. 1168. **Zbl** 15, p. 321.

"Set theoretical approach to the disconnection theory of the Euclidean space," *Fund. Math.* 37 (1950), 217–241. **MR** 13, p. 150. **Zbl** 40, p. 102.

BOULIGAND, G.
"Démonstration élémentaire d'un théorème de détermination à un facteur constant près d'une fonction harmonique," *Mathematica (Cluj)* 6 (1932), 80–85. **FM** 58, p. 503. **Zbl** 4, p. 395.

BOURBAKI, N.
Elements of Mathematics, General Topology. Part 2, Addison-Wesley Publishing Co. (1966), Reading. **MR** 34 #5044b, 40, p. 1704 and 41 #984. **Zbl** 145, p. 193; 197, p. 304; 301 #54002 and 337 #54001.

BOUTROUX, P.
"Propriétés d'une fonction holomorphe dans un cercle où elle ne prend pas les valeurs 0 et 1," *C.R. Acad. Sci. Paris* 141 (1905), 305–307 and *Bull. Soc. Math. France* 34 (1906), 30–39. **FM** 36, p. 466 and 37, p. 421.

BOWEN, N. A.
(a) "On the limit of the modulus of a bounded regular function," *Proc. Edin. Math. Soc.* (2) 14 (1964–65), 21–24. **MR** 29 #1309. **Zbl** 145, p. 84.

(b) "On mean value limits for bounded regular functions," *Proc. Edin. Math. Soc.* (2) 14 (1964–65), 103–108. **MR** 31 #5981. **Zbl** 146, p. 99.

BOWEN, N. A. and MACINTYRE, A. J.
"Interpolatory methods for theorems of Vitali and Montel type," *Proc. Royal Soc. Edin. Sect. A* 64 (1954), 71–79. **MR** 16, p. 232. **Zbl** 56, p. 72.

BRAWN, F. T.
"Positive harmonic majorization of subharmonic functions in strips," *Proc. Lon. Math. Soc.* (3) 27 (1973), 261–289. **MR** 48 #8819. **Zbl** 265 #31008.

BRELOT, M.
"Einige neuere Untersuchungen über das Dirichletsche Problem," *Jahresber. Deutsch. Math. Verein.* 42 (1933), 111–129. **FM** 58, p. 508. **Zbl** 6, p. 15. French translation: **FM** 59, p. 481. **Zbl** 6, p. 349.

Étude des fonctions sousharmoniques au voisinage d'un point, Actualités Scientifiques et Industrielles no. 139, Hermann et Cie. (1934), Paris. **FM** 60, p. 1134. **Zbl** 9, p. 19. **BAMS** 43, p. 458.

"Familles de Perron et problème de Dirichlet," *Acta Litt. ac Scient. Univ. Hung. (Szeged)* 9 (1938), 133–153. **FM** 65, p. 418. **Zbl** 23, p. 233. **MR** 1, p. 121.

"Quelques applications aux fonctions holomorphes de la théorie moderne du potentiel et du problème de Dirichlet," *Bull. Soc. Roy. Sci. Liége* 8 (1939), 385–391. **FM** 65, p. 419. **Zbl** 21, p. 319. **MR** 1, p. 114.

"Sur l'allure à la frontière des fonctions harmoniques, sousharmoniques ou holomorphes," *Bull. Soc. Roy. Sci. Liége* 8 (1939), 468–477. **FM** 65, p. 419. **Zbl** 23, p. 234. **MR** 1, p. 122.

"La théorie moderne du potentiel," *Ann. Inst. Fourier (Grenoble)* 4 (1952), 113–140. **Zbl** 55, p. 89. **MR** 15, p. 527. See also **Zbl** 201, p. 139 and **MR** 42 #6260.

Éléments de la théorie classique du potentiel, Centre de Documentation Universitaire (4th ed. 1969), Paris. **MR** 31 #2412 and #1387. **Zbl** 84, p. 309 and 116, p. 75.

"Les étapes et les aspects multiples de la théorie du potentiel," *L'Enseignement Mathématique* (2) 18 (1972), 1–36. **Zbl** 235 #31002. **MR** 51 #8436 and 53, p. 2382.

BRICKMAN, L.

"Non-negative entire and meromorphic functions," *Jour. Lon. Math. Soc.* 39 (1964), 15–18. **MR** 28 #4112. **Zbl** 123, p. 268.

BRICKMAN, L. and STEINBERG, L.

"On non-negative polynomials," *Amer. Math. Monthly* 69 (1962), 218–221.

BRILL, A. and NOETHER, M.

"Die Entwickelung der Theorie der algebraischen Functionen in älterer und neuerer Zeit," *Jahresber. Deutsch. Math. Verein.* 3 (1894), 107–566. FM 25, p. 70.

BRILLINGER, D. R.

"The analyticity of the roots of a polynomial as functions of the coefficients," *Math. Mag.* 39 (1966), 145–147. **MR** 34 #4455. **Zbl** 164, p. 375.

BRISSE, CH.

"Démonstration du théorème de d'Alembert," *Jour. École Polyt.* 56 (1886), 163–169. **FM** 18, p. 60.

BRODOVICH, M. T.

"Monogeneity conditions for discontinuous mappings," *Teor. Funkcii Funkcional. Anal. i Priložen. Vyp.* 12 (1970), 94–103 (Russian). **Zbl** 218, p. 169. **MR** 45 #5323.

"A certain sufficient condition for the conformality of an arbitrary one-to-one mapping," *Dopovidi Akad. Nauk Ukraïn. RSR Ser. A* (1974), 489–492, 572 (Ukrainian). **MR** 51 #8388. **Zbl** 283 #30008. See also **MR** 55 #3269.

BROGGI, U.

"Un teorema sulle serie di potenze che rappresentano funzioni razionali," *Rend. del R. Istituto Lombardo di Scienze e Lettere* (2) 64 (1931), 238–243. **FM** 57, p. 416. **Zbl** 4, p. 10.

"Su di un teorema concernente le serie di potenze," *Rend. del R. Istituto Lombardo di Scienze e Lettere* (2) 68 (1935), 859–862. **Zbl** 14, p. 23. **FM** 61, p. 1134.

BROMWICH, T. J. I'A.

An Introduction to the Theory of Infinite Series, Cambridge University Press (2nd ed. revised, 1926), London. **FM** 42, p. 208. **BAMS** 34, p. 244. **Zbl** 133, p. 8.

BROUWER, L. E. J.

"Zur Analysis Situs," *Math. Annalen* 68 (1910), 422–433. **FM** 41, p. 543.

"Beweis der Invarianz der Dimensionszahl," *Math. Annalen* 70 (1911), 161–165. **FM** 42, p. 416.

(a) "Beweis der Invarianz des *n*-dimensionalen Gebiets," *Math. Annalen* 71 (1912), 305–313 and 72 (1912), 55–56. Correction *ibid.* 82 (1921), 286. **FM** 42, p. 418, 43, p. 479.

(b) "Über Abbildung von Mannigfaltigkeiten," *Math. Annalen* 71 (1912), 97–115. Correction *ibid.* 71 (1912), 598 and 82 (1921), 286. **FM** 42, p. 417.

(c) "Beweis der Invarianz der geschlossenen Kurve," *Math. Annalen* 72 (1912), 422–425. **FM** 43, p. 569.

Collected Works, vol. II: Geometry, Analysis, Topology and Mechanics, H. Freudenthal ed., North-Holland Publishing Co. (1976), Amsterdam. **Zbl** 328 #01020.

BROWDER, F.

"Covering spaces, fibre spaces, and local homeomorphisms," *Duke Math. Jour.* 21 (1954), 329–336. **MR** 15, p. 978. **Zbl** 56, p. 166.

"On the proof of Mergelyan's approximation theorem," *Amer. Math. Monthly* 67 (1960), 442–444. **MR** 22 #6894. **Zbl** 106, p. 45.

BROWN, A. B.

"Extensions of the Brouwer fixed point theorem," *Amer. Math. Monthly* 69 (1962), 643. **Zbl** 134, p. 189.

BROWN, L., GAUTHIER, P. M. and SEIDEL, W.

"Possibility of complex asymptotic approximation on closed sets," *Math. Annalen* 218 (1975), 1–8. **Zbl** 304 #30026. **MR** 54 #10610.

BROWN, L., SCHREIBER, B. and TAYLOR, B. A.

"Spectral synthesis and the Pompeiu problem," *Ann. Inst. Fourier* (*Grenoble*) 23 (1973), Fasc. 3, 125–154. **MR** 50 #4979. **Zbl** 265 #46044.

BROWN, R. F.

"Elementary consequences of the noncontractibility of the circle," *Amer. Math. Monthly* 81 (1974), 247–252. **Zbl** 284 #55009. **MR** 48 #9629.

BRUCKNER, A. M., LOHWATER, A. J. and RYAN, F.

"Some non-negativity theorems for harmonic functions," *Ann. Acad. Scient. Fennicae Ser. AI*, No. 452 (1969). **Zbl** 187, p. 358. **MR** 42 #529. (See also **MR** 46 #3795.)

BRUNÉ, C.

"Sulla rappresentazione conforme dei campi piani semplici pluriconnessi," *Atti del Reale Istituto Veneto di Scienze, Lettere ed Arti* 91 (1932), 359–380. **FM** 58, p. 1092.

BUHL, A.

Séries analytiques. Sommabilité, Mémorial des Sciences Mathématiques fasc. 7, Gauthier-Villars (1925), Paris. **FM** 51, p. 242. **BAMS** 33, p. 120.

BURDICK, D. and LESLEY, F. D.

"Some uniqueness theorems for analytic functions," *Amer. Math. Monthly* 82 (1975), 152–155. **MR** 50 #10234. **Zbl** 306 #30004.

BUREAU, F.

"Sur les fonctions holomorphes dans un cercle de rayon fini et sur les fonctions entières," *C.R. Acad. Sci. Paris* 192 (1931), 1629–1630. **Zbl** 2, p. 38. **FM** 57, p. 1418.

"Sur quelques propriétés des fonctions uniformes au voisinage d'un point singulier essentiel isolé," *C.R. Acad. Sci. Paris* 192 (1931), 1350–1352. **Zbl** 1, p. 398. **FM** 57, p. 1418.

"Mémoire sur les fonctions uniformes à point singulier essentiel isolé," *Mém. Soc. Roy. Sci. Liége* (3) 17 No. 3 (1932). **Zbl** 6, p. 408. **FM** 58, p. 1085.

BURGERS, W. A. M.
Radiale randlimieten van holomorfe functies, Centrale Drukkerij (1929), Nijmegen. **FM** 55, p. 770.

BURKHARDT, H.
"Trigonometrische Reihen und Integrale," *Encyklopädie der Mathematischen Wissenschaften* Bd. II, 1st Part, 2nd Half, Heft 7, 819–1354. B. G. Teubner (1914), Leipzig. **FM** 45, p. 374.

BUTLER, R.
"On the evaluation of $\int_0^\infty (\sin^m t)/t^m\, dt$ by the trapezoidal rule," *Amer. Math. Monthly* 67 (1960), 566–569. **MR** 22 #4841. **Zbl** 100, p. 56.

CAIRNS, S. S.
"An elementary proof of the Jordan–Schoenflies theorem," *Proc. Amer. Math. Soc.* 2 (1951), 860–867. **Zbl** 44, p. 198. **MR** 13, p. 764.

CANTOR, D. G. and PHELPS, R.
"An elementary interpolation theorem," *Proc. Amer. Math. Soc.* 16 (1965), 523–525. **MR** 31 #358. **Zbl** 146, p. 96.

CARATHÉODORY, C.
"Sur quelques applications du théorème de Landau–Picard," *C.R. Acad. Sci. Paris* 144 (1907), 1203–1206. **FM** 38, p. 432.

"Über den Variabilitätsbereich der Fourier'schen Konstanten von positiven harmonischen Funktionen," *Rend. Circ. Mat. Palermo* 32 (1911), 193–217. **FM** 42, p. 429.

(a) "Untersuchungen über die konformen Abbildungen von festen und veränderlichen Gebieten," *Math. Annalen* 72 (1912), 107–144. **FM** 43, p. 524.

(b) "Sur le théorème de M. Picard," *C.R. Acad. Sci. Paris* 154 (1912), 1690–1693. **FM** 43, p. 506.

(a) "Sur la représentation conforme des polygones convexes," *Annales de la Soc. Sci. de Bruxelles* 37 (1913), 2e partie, 1–10. **FM** 44, p. 756.

(b) "Über die gegenseitige Beziehung der Ränder bei der konformen Abbildung des Inneren einer Jordanschen Kurve auf einen Kreis," *Math. Annalen* 73 (1913), 305–320. **FM** 44, p. 757.

(c) "Zur Ränderzuordnung bei konformer Abbildung," *Göttingen Nachr.* (1913), 509–518. **FM** 44, p. 758.

"Elementarer Beweis für den Fundamentalsatz der konformen Abbildungen," pp. 19–41 of *Mathematische Abhandlungen H.A. Schwarz Gewidmet*, Verlag Julius Springer (1914), Berlin. **FM** 45, p. 667. Reprinted by Chelsea Publishing Co. (1974), New York.

"Bemerkungen zu den Existenztheoremen der konformen Abbildung," *Bull. Calcutta Math. Soc.* 20 (1928), 125–134. **FM** 56, p. 295.

(a) "Über die Winkelderivierten von beschränkten analytischen Funktionen," *Sitzungsber. Preuss. Akad. Wiss. Berlin, Phys.-math. Kl.* (1929), 39–54. **FM** 55, p. 209.

(b) "Stetige Konvergenz und normale Familien von Funktionen," *Math. Annalen* 101 (1929), 515–533. **FM** 55, p. 198.

"Über die Abbildungen, die durch Systeme von analytischen Funktionen von mehreren Veränderlichen erzeugt werden," *Math. Zeit.* 34 (1932), 758–792. **Zbl** 3, p. 407. **FM** 58, p. 349.

(a) "On Dirichlet's problem," *Amer. Jour. of Math.* 59 (1937), 709–731. **FM** 63, p. 454. **Zbl** 17, p. 260.

(b) "The most general transformations of plane regions which transform circles into circles," *Bull. Amer. Math. Soc.* 43 (1937), 573–579. **FM** 63, p. 294. **Zbl** 17, p. 229.

"A proof of the first principal theorem on conformal representation," pp. 75–83 of *Studies and Essays Presented to R. Courant on his 60th Birthday*, Interscience Publishers, Inc. (1948), New York. **Zbl** 38, p. 52. **MR** 9, p. 232.

Conformal Representation, Cambridge Tracts in Mathematics and Physics, No. 28. Cambridge University Press (2nd ed. 1952), Cambridge. **FM** 58, p. 354. **Zbl** 47, p. 79. **MR** 13, p. 734. **BAMS** 60, p. 281.

Theory of Functions of a Complex Variable I, (2nd ed. 1958), II (2nd ed. 1960), (translated from German by F. Steinhardt) Chelsea Publishing Co., New York. **MR** 15, p. 612; 16, p. 346 and 24 #A209. **Zbl** 96, p. 49. **BAMS** 57, p. 190.

CARATHÉODORY, C. and FEJÉR, L.

"Remarques sur le théorème de M. Jensen," *C.R. Acad. Sci. Paris* 145 (1907), 163–165. **FM** 38, p. 427.

CARATHÉODORY, C. and LANDAU, E.

"Beiträge zur Konvergenz von Funktionenfolgen," *Sitzungsber. Kön. Preuss. Akad. Wiss. Berlin* (1911), 587–613. **FM** 42, p. 275.

CARATHÉODORY, C. and RADEMACHER, H.

"Über die Eineindeutigkeit im Kleinen und im Grossen stetiger Abbildungen von Gebieten," *Arch. der Math. und Physik* (3) 26 (1917), 1–9. **FM** 46, p. 834.

CARLEMAN, T.

"Über die Approximation analytischer Funktionen durch lineare Aggregate von vorgegebenen Potenzen," *Arkiv för Mat., Astro. och Fysik* 17, No. 9 (1922). **FM** 49, p. 708.

"Sur un théorème de Weierstrass," *Arkiv för Mat., Astro. och Fysik* 20B, No. 4 (1927). **FM** 53, p. 237.

CARLESON, L.

"Mergelyan's theorem on uniform approximation," *Math. Scand.* 15 (1965), 167–175. **MR** 33 #6368. **Zbl** 163, p. 86.

Selected Problems on Exceptional Sets, Van Nostrand Mathematical Studies #13, D. Van Nostrand Co., Inc. (1967), Princeton. **MR** 50 #4932. **Zbl** 224 #31001.

CARLSON, F.

"Sur le module maximum d'une fonction analytique uniforme. I," *Arkiv för Mat., Astro. och Fysik* 26A, No. 9 (1938). **FM** 64, p. 1071. **Zbl** 19, p. 221.

CARTWRIGHT, M. L.

"Some generalizations of Montel's theorem," *Proc. Camb. Phil. Soc.* 31 (1935), 26–30. **Zbl** 11, p. 29. **FM** 61, p. 307.

Integral Functions, Cambridge Tracts in Mathematics and Physics, No. 44. Cambridge University Press (1956), Cambridge. **MR** 17, p. 1067. **Zbl** 75, p. 59. **BAMS** 67, p. 454.

"A generalization of Montel's theorem," *Jour. Lon. Math. Soc.* 37 (1962), 179–184. **MR** 25 #4106. **Zbl** 137, p. 50.

CASORATI, F.

"Un teorema fondamentale nella teoria delle discontinuità delle funzioni," *Rend. del R. Istituto Lombardo di Scienze e Lettere* (2) 1 (1868), 123–125. **FM** 1, p. 128.

Teorica delle funzioni di variabili complesse, Vol. I, Tipografi fratelli Fusi (1868), Pavia. **FM** 1, p. 128.

"Aggiunte a recenti lavori dei signori Weierstrass e Mittag-Leffler sulle funzioni di una variabile complessa," *Annali Mat. Pura Appl.* (2) 10 (1882), 261–278. **FM** 14, p. 327. (See also **FM** 25, p. 707.)

CAUCHY, A.-L.

Cours d'analyse de l'École Royale Polytechnique. Première partie: analyse algébrique (1821), Paris. (See **FM** 28, p. 12.) Reprinted by Wissenschaftliche Buchgesellschaft (1968), Darmstadt. German translation by C. Itzigsohn, Verlag Julius Springer (1885), Berlin. **FM** 17, p. 200.

"Note sur le développement des fonctions en séries convergentes ordonnées suivant les puissances entières des variables," *C.R. Acad. Sci. Paris* 17 (1843), 940–942.

"Mémoire sur quelques propositions fondamentales du calcul des résidus, et sur la théorie des intégrales singulières," *C.R. Acad. Sci. Paris* 19 (1844), 1337–1344.

"Mémoire sur les fonctions complémentaires," *C.R. Acad. Sci. Paris* 19 (1844), 1377–1384.

"Sur les intégrals qui s'étendent à tous les points d'une courbe fermée," *C.R. Acad. Sci. Paris* 23 (1846), 251–255 and 689–704.

"Abhandlung über bestimmte Integrale zwischen imaginären Grenzen," translation (from French) and commentary by P. Stäckel in *Ostwald's Klassiker der exakten Wissenschaften*, No. 112. Verlag von W. Engelmann (1900), Leipzig. **FM** 31, p. 314. (See also **FM** 7, p. 155 and **Zbl** 274 #01026 for reprintings of original French and IACOBACCI [1965] for an English translation.)

CAUER, W.

"The Poisson integral for functions with positive real part," *Bull. Amer. Math. Soc.* 38 (1932), 713–717. **FM** 58, p. 1087. **Zbl** 5, p. 361.

CAZZANIGA, P.

"Espressione di funzioni intere che in posti dati arbitrariamente prendono valori prestabiliti," *Annali Mat. Pura Appl.* (2) 10 (1882), 279–290. **FM** 14, p. 324.

ČECH, E.

"Une démonstration du théorème de Cauchy et de la formule de Gauss," *Rend. Accad. d. Lincei Roma* (6) 11 (1930), 884–887. **FM** 56, p. 958.

"Encore sur le théorème de Cauchy," *Rend. Accad. d. Lincei Roma* (6) 12 (1930), 286–289. **FM** 56, p. 958.

Point Sets (translated from Czech by A. Pultr), Academic Press (1969), New York. **MR** 41 #1000. **Zbl** 191, p. 532.

ČERNÝ, I.

Foundations of Analysis in the Complex Plane (Czech), Academia-Nakladatelství Čezkoslovenské akademie věd (1967), Prague. **Zbl** 168, p. 34. **MR** 36 #5309.

"A simple proof of Cauchy theorem," *Časopis pro pěstování matematiky* 101 (1976), 366–369. **Zbl** 353 #30001. **MR** 56 #5900.

CHESSIN, A. S.

"On the singularities of single-valued and generally analytic functions," *Annals of Math.* 11 (1895–96), 52–56. **FM** 27, p. 301.

CHOQUET, G.

"Sur un type de transformation analytique généralisant la représentation conforme et définie au moyen de fonctions harmoniques," *Bull. Sci. Math.* (2) 69 (1945), 156–165. **MR** 8, p. 92.

CHOU, HSIN-TI
"On the uniqueness theorem for analytic functions and its applications," *Chinese Math.* 9 (1967), 433–438. **MR** 21 #5717. **Zbl** 172, p. 373.

CIMMINO, G.
"Formole di maggiorazione nel problema di Dirichlet per le funzioni armoniche," *Rend. Seminario Mat. Padova* 3 (1932), 46–66. **FM** 58, p. 508. **Zbl** 5, p. 16.

CIORANESCO, N.
"Le problème de Dirichlet harmonique et les médiations successives," *Bull. Sci. Math.* (2) 54 (1930), 191–200. **FM** 56, p. 1068.

CISOTTI, U.
"Determinazione della funzione di variabile complessa regolare in una corona circolare note: La parte reale sulla circonferenza esterna e la parte immaginaria sull'interna," *Rend. Accad. d. Lincei Roma* (6) 13 (1931), 395–399. **Zbl** 2, p. 33. **FM** 57, p. 348.

CLUNIE, J.
"On integral and meromorphic functions," *Jour. Lon. Math. Soc.* 37 (1962), 17–27. **MR** 26 #1456. **Zbl** 104, p. 295.

COHEN, P. J.
"A note on constructive methods in Banach algebras," *Proc. Amer. Math. Soc.* 12 (1961), 159–163. **MR** 23 #A1827. **Zbl** 97, p. 106.

COLLINGWOOD, E. F.
"On Three Circles Theorems (I)," *Jour. Lon. Math. Soc.* 7 (1932), 162–166. **Zbl** 5, p. 17. **FM** 58, p. 331.

COLLINGWOOD, E. F. and LOHWATER, A. J.
The Theory of Cluster Sets, Cambridge Tracts in Mathematics and Mathematical Physics, No. 56. Cambridge University Press (1966), Cambridge. **MR** 50 #13528. **Zbl** 212, p. 423. (See also the Russian supplement: **MR** 53 #3311. **Zbl** 283 #30032.)

COLLINGWOOD, E. F. and PIRANIAN, G.
"The mapping theorems of Carathéodory and Lindelöf," *Jour. de Math. Pures et Appliq.* (9) 43 (1964), 187–199. **MR** 29 #6022. **Zbl** 133, p. 327.

CONNELL, E. H.
"A classical theorem in complex variable," *Amer. Math. Monthly* 72 (1965), 729–732. **Zbl** 152, p. 60. **MR** 32 #1360.

CONSIGLIO, A.
"Su la risoluzione in termini finiti dei problemi di Villat e Dirichlet relativi alla corona circolare," *Atti Accad. Gioenia Sci. natur. Catania* (6) 6, Mem. 7 (1950). **Zbl** 39, p. 324. **MR** 12, p. 88.

CONWAY, J. B.
Functions of One Complex Variable, Graduate Texts in Mathematics, vol. 11. Springer-Verlag (1973), New York. **Zbl** 277 #30001. **MR** 56 #5843.

COROMINAS, F. and SUNYER BALAGUER, F.
"Conditions for an infinitely differentiable function to be a polynomial," *Revista Mat. Hisp.-Amer.* (4) 14 (1954), 26–43 (Spanish). **MR** 15, p. 942. **Zbl** 55, p. 288.

CORPUT, J. G., VAN DER
"Le théorème fondamental de l'algèbre," *Actual. Math. Centrum Amsterdam*, Scriptum 2, 1. Teil (1950). **Zbl** 39, p. 11. **MR** 12, p. 475.

CORRÁDI, K.
"Über Konvergenz-Eigenschaften von Potenzreihen," *Mat. Lapok* 10 (1959), 136–141 (Hungarian). **MR** 22 #8108. **Zbl** 89, p. 48.

COURANT, R.

(a) "Über eine Eigenschaft der Abbildungsfunktionen bei konformer Abbildung," *Göttingen Nachr.* (1914), 101–109 and (1922), 69–70. **FM** 45, p. 668 and 48, p. 1235.

(b) "Über die Existenztheoreme der Potential- und Funktionentheorie," *Jour. für Reine und Angew. Math.* 144 (1914), 190–211. **FM** 45, p. 668.

Dirichlet's Principle, Conformal Mapping, and Minimal Surfaces, Interscience Publishers, Inc. (1950), New York. Reprinted by Springer-Verlag (1977), New York. **MR** 12, p. 90. **Zbl** 354 #30012. **BAMS** 58, p. 95.

COWELL, P.

"Julia points of functions meromorphic in a disk," *Bull. Lon. Math. Soc.* 4 (1972), 327–329. **MR** 47 #5259. **Zbl** 262 #30033.

COWLING, V. F.

"On analytic functions having a positive real part in the unit circle," *Amer. Math. Monthly* 63 (1956), 329–330. **MR** 17, p. 1070. **Zbl** 70, p. 72.

CRAIG, C., Jr. and MACINTYRE, A. J.

"Inequalities for functions regular and bounded in a circle," *Pac. Jour. of Math.* 20 (1967), 449–454. **MR** 34 #7806. **Zbl** 182, p. 408.

CRAVEN, B. D.

"A note on Green's theorem," *Jour. Austral. Math. Soc.* 4 (1964), 289–292. **Zbl** 135, p. 114. **MR** 32 #5807.

CREMER, H.

"Über die Iteration rationaler Funktionen," *Jahresber. Deutsch. Math. Verein.* 33 (1925), 185–210. **FM** 51, p. 262.

"Ein Existenzbeweis der Kreisringabbildung zweifach zusammenhängender schlichter Bereiche," *Ber. Verhandl. Sächs. Akad. Wiss. Leipzig* 82 (1930), 190–192. **FM** 56, p. 296.

CSÁSZÁR, Á.

"Sur les auxiliaires topologiques des éléments de la théorie des fonctions analytiques," *Mat. Lapok* 13 (1962), 73–94 (Hungarian). **Zbl** 122, p. 75. **MR** 26 #3918.

CSILLAG, P.

"Untersuchungen über die Borelschen Verallgemeinerungen des Picardschen Satzes," *Math. Annalen* 100 (1928), 367–383. **FM** 54, p. 347.

"Über ganze Funktionen, welche drei nicht verschwindende Ableitungen besitzen," *Math. Annalen* 110 (1935), 745–752. **Zbl** 10, p. 265. **FM** 61, p. 341.

CUNNINGHAM, F., Jr.

"Taking limits under the integral sign," *Math. Mag.* 40 (1967), 179–186 and *Amer. Math. Monthly* 75 (1968), 827. **Zbl** 156, p. 62.

CURTISS, D. R.

"Note on the sufficient conditions for an analytic function," *Bull. Amer. Math. Soc.* 8 (1901/02), 329–331. **FM** 33, p. 395.

"A proof of the theorem concerning artificial singularities," *Annals of Math.* (2) 7 (1905–06), 161–162. **FM** 37, p. 416.

"On certain theorems of mean value for analytic functions of a complex variable," *Annals of Math.* (2) 8 (1906–07), 118–126. **FM** 38, p. 337.

CURTISS, J. H.

"Faber polynomials and Faber series," *Amer. Math. Monthly* 78 (1971), 577–596. Correction, *ibid.* 79 (1972), 363. **MR** 45 #2183, #8893. **Zbl** 215, p. 415; 237 #30001.

DARBOUX, J. G.
"Mémoire sur les fonctions discontinues," *Ann. Sci. École Norm. Sup.* (2) 4 (1875), 57–112. **FM** 7, p. 243 and 11, p. 274.
"Mémoire sur l'approximation des fonctions de très-grands nombres, et sur une classe étendue de développements en série," *Jour. de Math. Pures et Appliq.* (3) 4 (1878), 5–56 and 377–416. **FM** 10, p. 279 and 8, p. 304.
"Un peu de géométrie à propos de l'intégral de Poisson," *Bull. Sci. Math.* (2) 34 (1910), 287–300. **FM** 41, p. 319.

DARST, R.
"Most infinitely differentiable functions are nowhere analytic," *Canad. Math. Bull.* 16 (1973), 597–598. **MR** 49 #10838. **Zbl** 285 #26015. (See also **MR** 52 #6685.)

DAUBEN, J. W.
"The invariance of dimension: Problems in the early development of set theory and topology," *Historia Math.* 2 (1975), 273–288. **Zbl** 313 #01007. **MR** 57 #15888.

DEAUX, R. and DELCOURTE, M.
"Calcul des intégrales $(m, n) = \int_0^\infty (\sin^m x/x^n)\,dx$, m et n entiers positifs, $m \geq n$," *Mathesis* 66 (1957), 16–22. **Zbl** 85, p. 43. **MR** 19, p. 29.

DEBRANGES, L. and ROVNYAK, J.
Square Summable Power Series, Holt, Rinehart and Winston, Inc. (1966), New York. **MR** 35 #5909. **Zbl** 153, p. 396.

DEEDS, J. B.
"The Caley–Hamilton Theorem via complex integration," *Amer. Math. Monthly* 78 (1971), 1003–1004. **Zbl** 229 #15007.

DEKOK, F.
"Sur quelques propriétés d'une fonction à partie réelle positive," *C.R. Acad. Sci. Paris* 197 (1933), 476–478. **Zbl** 7, p. 214. **FM** 59, p. 324.

DELANGE, H.
"Sur la forme forte du théorème de Cauchy relatif à l'intégrale d'une fonction de variable complexe," *Bull. Sci. Math.* (2) 80 (1956), 156–160. **MR** 19, p. 23. **Zbl** 72, p. 290.

DELSARTE, J.
"Note sur une propriété nouvelle des fonctions harmoniques," *C.R. Acad. Sci. Paris* 246 (1958), 1358–1360. **MR** 20 #2548. **Zbl** 84, p. 94.
Lectures on Topics in Mean Periodic Functions and the Two Radius Theorem, Notes by K. B. Vedak, Tata Institute of Fundamental Research (1961), Bombay.

DEMIN, E.
"Remarque sur une formule de Laurent," *Bull. Soc. Roy. Sci. Liége* 15 (1946), 84–86. **Zbl** 61, p. 148. **MR** 8, p. 507.

DEMTCHENKO, B.
"Sur la formule de M. H. Villat résolvant le problème de Dirichlet dans un anneau circulaire," *Jour. de Math. Pures et Appliq.* (9) 10 (1931), 201–211. **FM** 57, p. 1422.

DENJOY, A.
"Sur les fonctions définies par des séries de fractions rationnelles," *C.R. Acad. Sci. Paris* 174 (1922), 95–98. **FM** 48, p. 320.
"Sur l'itération des fonctions analytiques," *C.R. Acad. Sci. Paris* 182 (1926), 255–257. **FM** 52, p. 309.

"Sur une classe de fonctions analytiques," *C.R. Acad. Sci. Paris* 188 (1929), 140–142 and 1084–1086. **FM** 55, pp. 180, 181.

"Sur les polygones d'approximation d'une courbe rectifiable," *C.R. Acad. Sci. Paris* 196 (1933), 29–33. **FM** 59, p. 313. **Zbl** 6, p. 62.

"Représentation conforme des aires limitées par des continus cycliques," *C.R. Acad. Sci. Paris* 213 (1941), 975–977. **FM** 67, p. 285. **Zbl** 26, p. 218. **MR** 5, p. 115.

"Les continus cycliques et la représentation conforme," *Bull. Soc. Math. France* 70 (1942), 97–124. **Zbl** 28, p. 403. **MR** 6, p. 207.

"Le théorème de Cauchy–Goursat," *C.R. Acad. Sci. Paris* 240 (1955), 386–389 and 473–476. **MR** 16, p. 683. **Zbl** 64, p. 67.

de Possel, R.

"Zum Parallelschlitztheorem unendlich-vielfach zusammenhängender Gebiete," *Göttingen Nachr.* (1931), 199–202. **FM** 57, p. 400. **Zbl** 3, p. 314.

"Quelques problèmes de représentation conforme," *Jour. École Polyt.* (2) 30 (1932), 1–98. **FM** 58, p. 1093. **Zbl** 6, p. 263.

"Sur la représentation conforme d'un domaine à connexion infinie sur un domaine à fentes parallèles," *Jour. de Math. Pures et Appliq.* (9) 18 (1939), 285–290. **FM** 65, p. 1242. **Zbl** 23, p. 55. **MR** 1, p. 111.

Derrick, W. R.

"A condition under which a mapping is a homeomorphism," *Amer. Math. Monthly* 80 (1973), 554–555. **MR** 48 #2977. **Zbl** 285 #57001.

Deuring, M.

"Eine Bemerkung zum Cauchyschen Integralsatz," *Arch. der Math.* 1 (1949), 321–322. **Zbl** 31, p. 352. **MR** 11, p. 91.

Dienes, P.

The Taylor series: An introduction to the theory of functions of a complex variable, Clarendon Press (1931), Oxford. Reprinted by Dover Publications, Inc. (1957), New York. **FM** 57, p. 339. **Zbl** 78, p. 59. **MR** 19, p. 735. **BAMS** 38, p. 792.

Dieudonné, J.

Foundations of Modern Analysis I, Pure and Applied Mathematics, vol. 10-I. Academic Press (1969), New York. **MR** 50 #1782. **Zbl** 176, p. 5 and 264 #26001. **BAMS** 67, p. 246.

Infinitesimal Calculus, Houghton Mifflin Co. (1971), Boston; Kershaw Publishing Co., Ltd. (1973), London. **MR** 50 #1780. **Zbl** 155, p. 100.

Dinghas, A.

"Über das Phragmén-Lindelöfsche Prinzip und den Julia-Carathéodoryschen Satz," *Sitzungsber. Preuss. Akad. Wiss. Berlin, Phys.-math. Kl.* (1938), 32–48. **Zbl** 18, p. 262. **FM** 64, p. 301.

"A simple proof of a formula in the theory of functions," *Math. Student* 22 (1954), 101–102. **Zbl** 57, p. 307. **MR** 16, p. 231.

Vorlesungen über Funktionentheorie, Die Grundlehren der mathematischen Wissenschaften in Einzeldarstellungen, Bd. 110. Springer-Verlag (1961), Berlin. **MR** 31 #3577. **Zbl** 102, p. 293. **BAMS** 68, p. 311.

Dini, U.

"Alcuni teoremi sulle funzioni di una variabile complessa," pp. 258–276 of *In memoriam Dominici Chelini. Collectanea mathematica* (L. Cremona and E. Beltrami, editors), Hopli (1881), Milan. **FM** 13, p. 310.

"Il problema di Dirichlet in un' area anulare, e nello spazio compreso fra due sfere concentriche," *Rend. Circ. Mat. Palermo* 36 (1913), 1–28.

DIXON, J. D.
"A brief proof of Cauchy's Integral Theorem," *Proc. Amer. Math. Soc.* 29 (1971), 625–626. **MR** 43 #3432. **Zbl** 205, p. 378.

DOETSCH, G.
"Über die obere Grenze des absoluten Betrages einer analytischen Funktion auf Geraden," *Math. Zeit.* 8 (1920), 237–240. **FM** 47, p. 274.

DONOGHUE, W. F., Jr.
Distributions and Fourier Transforms, Pure and Applied Mathematics, vol. 32. Academic Press (1969), New York. **Zbl** 188, p. 181.

Monotone Matrix Functions and Analytic Continuation, Die Grundlehren der mathematischen Wissenschaften in Einzeldarstellungen, Bd. 207. Springer-Verlag (1974), New York. **Zbl** 278 #30004. **BAMS** 81, p. 847.

DÖRGE, K.
"Über den Fundamentalsatz der Algebra," *Sitzungsber. Preuss. Akad. Wiss. Berlin, Phys.-math. Kl.* (1928), 87–89. **FM** 54, p. 117.

DOUGLAS, J.
"Solution of the problem of Plateau," *Trans. Amer. Math. Soc.* 33 (1931), 263–321. **FM** 57, pp. 661 and 1542. **Zbl** 1, p. 141.

DUFFIN, R. J.
"A note on Poisson's integral," *Quarterly of Applied Math.* 15 (1957), 109–111. **Zbl** 79, p. 122. **MR** 19, p. 261.

DUFRESNOY, J. and PISOT, C.
"Prolongement analytique de la série de Taylor," *Ann. Sci. École Norm. Sup.* (3) 68 (1951), 105–124. **MR** 13, p. 221. **Zbl** 43, p. 294.

DULST, D., VAN
"A functional analytic proof of Rouché's Theorem," *Amer. Math. Monthly* 78 (1971), 770–771. **MR** 45 #2504. **Zbl** 224 #30004.

DVORETSKY, A. and ERDÖS, P.
"On power series diverging everywhere on the circle of convergence," *Mich. Math. Jour.* 3 (1955), 31–35. **MR** 17, p. 138. **Zbl** 73, p. 62.

DYM, H. and MCKEAN, H. P.
Fourier Series and Integrals, Probability and Mathematical Statistics, vol. 14. Academic Press (1972), New York. **Zbl** 242 #42001. **MR** 56 #945. **BAMS** 79, p. 641.

DZEWAS, J.
"Konstruktion der Winkelfunktionen mittels der Theorie der topologischen Gruppen," *Math.-Phys. Semesterber.* (new series) 15 (1968), 163–171. **MR** 38 #3473. **Zbl** 164, p. 80.

EATON, J. E.
"The fundamental theorem of algebra," *Amer. Math. Monthly* 67 (1960), 578–579.

EBERLEIN, W. F.
"The elementary transcendental functions," *Amer. Math. Monthly* 61 (1954), 386–392. **MR** 15, p. 791. **Zbl** 56, p. 281. (See also **MR** 40 #270.)

"The circular function(s)," *Math. Mag.* 39 (1960), 197–201. **Zbl** 147, p. 60.

"The Gauss–Green and Cauchy integral theorems," *Amer. Math. Monthly* 82 (1975), 625–629. **MR** 51 #3465. **Zbl** 311 #30036.

ECHOLS, W. H.
"On the expansion of functions in infinite series," *Amer. Jour. of Math.* 15 (1893), 223–228. **FM** 25, p. 676.

EDREI, A.
"On the zeros of successive derivatives," *Proc. Amer. Math. Soc.* 6 (1955), 386–391. **MR** 16, p. 914. **Zbl** 64, p. 70.

EGGLESTON, H. G.
"A Tauberian lemma," *Proc. Lon. Math. Soc.* (3) 1 (1951), 28–45. **MR** 13, p. 22. **Zbl** 45, p. 335.

EILENBERG, S.
"Sur les transformations d'espaces métriques en circonférence," *Fund. Math.* 24 (1935), 160–176. **FM** 61, p. 621. **Zbl** 10, p. 277.

"Transformations continues en circonférence et la topologie du plan," *Fund. Math.* 26 (1936), 61–112. **FM** 62, p. 680. **Zbl** 13, p. 420.

"An invariance theorem for subsets of S^n," *Bull. Amer. Math. Soc.* 47 (1941), 73–75. **FM** 67, p. 738. **MR** 2, p. 179. **Zbl** 25, p. 234.

Lectures on Topology, mimeographed lecture notes (1948), New York University.

EISENACK, G. and FENSKE, C.
Fixpunkttheorie, Bibliographisches Institut (1978), Mannheim. **Zbl** 369 #47001.

ELJOSEPH, N.
"On the iteration of linear fractional transformations," *Amer. Math. Monthly* 75 (1968), 362–366. **Zbl** 159, p. 367.

ELKINS, J. M.
"A Borel–Carathéodory inequality and approximation of entire functions by polynomials with restricted zeros," *Jour. Approx. Theory* 4 (1971), 274–278. **MR** 44 #4191. **Zbl** 234 #30030.

ELY, G. S.
"Bibliography of Bernoulli's Numbers," *Amer. Jour. of Math.* 5 (1882), 228–235. **FM** 15, p. 21.

EPSTEIN, B.
Orthogonal Families of Analytic Functions, The Macmillan Company (1965), New York. **MR** 31 #4916. **Zbl** 145, p. 89.

ERDÖS, P., HERZOG, F., and PIRANIAN, G.
"Schlicht Taylor series whose convergence on the unit circle is uniform but not absolute," *Pac. Jour. of Math.* 1 (1951), 75–82. **MR** 13, p. 335. **Zbl** 43, p. 80.

ESSÉN, M.
The cos $\pi\lambda$ *Theorem*, Lecture Notes in Mathematics Vol. 467, Springer-Verlag (1975), New York. **Zbl** 335 #31001. **MR** 57 #6464.

ESTERMANN, T.
"Über die totale Variation einer stetigen Funktion und den Cauchyschen Integralsatz," *Math. Zeit.* 37 (1933), 556–560 and 38 (1934), 641. **FM** 59, p. 313 and 60, p. 243. **Zbl** 7, p. 213 and 9, p. 24.

"On the fundamental theorem of algebra," *Jour. Lon. Math. Soc.* 31 (1956), 238–240. **Zbl** 71, p. 249. **MR** 18, p. 4.

Complex Numbers and Functions, Athlone Press (1962), London. **Zbl** 115, p. 60.

"Notes on Landau's proof of Picard's 'Great' Theorem," pp. 101–106 of *Studies in Pure Mathematics*, L. Mirsky editor, Academic Press (1971), New York. **MR** 44 #1809. **Zbl** 215, p. 420.

ETTLINGER, H. J.
"Cauchy's paper of 1814 on definite integrals," *Annals of Math.* (2) 23 (1921–22), 255–270. **FM** 49, p. 209. (See also **FM** 16, p. 25.)

"On the inversion of the order of integration of a two fold iterated integral," *Annals of Math.* (2) 28 (1926–27), 65–68. **FM** 52, p. 245.

EUSTICE, D. J.
"Holomorphic idempotents and common fixed points on the 2-disk," *Mich. Math. Jour.* 19 (1972), 347–352. **MR** 47 #2040. **Zbl** 254 #32008.

EVANS, A.
"The application of complex variable methods to Tauberian theorems," *Jour. Lon. Math. Soc.* 28 (1953), 94–102. **MR** 14, p. 551. **Zbl** 50, p. 67.

EVANS, G. C.
"Fundamental points of potential theory," pp. 252–329 of *The Rice Institute Pamphlet* 7, No. 4 (1920), Houston. **FM** 48, p. 1268.

"Problems of potential theory," *Proc. Nat. Acad. Sci. U.S.A.* 7 (1921), 89–98. **FM** 48, p. 568.

"Sur l'intégral de Poisson," *C.R. Acad. Sci. Paris* 177 (1923), 241–242. **FM** 49, p. 339.

"Note on a theorem of Bôcher," *Amer. Jour. of Math.* 50 (1928), 123–126, **FM** 54, p. 508.

EVERITT, W. N.
"An elementary inequality in function theory," *Glasgow Math. Jour.* 10 (1969), 162–168. **Zbl** 192, p. 432. **MR** 40 #2824.

EVGRAFOV, M. A., *et al.*
Recueil de problèmes sur la théorie des fonctions analytiques, Éditions Mir (1974), Moscow. **MR** 51 #3406. **Zbl** 286 #30001.

FABER, G.
"Über polynomische Entwicklungen," *Math. Annalen* 57 (1903), 389–408 and 64 (1907), 116–135. **FM** 34, p. 430 and 38, p. 439.

"Über analytische Funktionen mit vorgeschriebenen Singularitäten," *Math. Annalen* 60 (1905), 379–397. **FM** 36, p. 457.

"Neuer Beweis eines Koebe-Bieberbachschen Satzes über konforme Abbildung," *Sitzungsber. Kön. Bayer. Akad. Wiss. München* (1916), 39–42. **FM** 46, p. 550.

"Über den Hauptsatz aus der Theorie der konformen Abbildung," *Sitzungsber. Bayer. Akad. Wiss. München* (1922), 91–100. **FM** 48, p. 1231.

FABRY, E.
(a) "Sur les points singuliers d'une fonction donnée par son développement en série et l'impossibilité du prolongement analytique dans des cas très généraux," *Ann. Sci. École Norm. Sup.* (3) 13 (1896), 367–399. **FM** 27, p. 303.

(b) "Sur les intégrales de Fresnel," *Nouvelles Annales de Math.* (3) 15 (1896), 504–505. **FM** 27, p. 235.

"Sur les séries de Taylor qui ont une infinité de points singuliers," *Acta Math.* 22 (1898–99), 65–87. **FM** 29, p. 209.

"Sur les séries les plus générales," *C.R. Acad. Sci. Paris* 211 (1940), 245–247. **FM** 66, p. 333. **Zbl** 24, p. 418. **MR** 3, p. 76.

FALLIN, H. I., Jr. and GOULD, H. W.
"Chronological bibliography of the Cauchy integral theorem," *Mathematica Monongaliae*, No. 10 (1965), Department of Mathematics, West Virginia University, Morgantown, West Virginia.

FARKAS, J.
"Sur les fonctions itératives," *Jour. de Math. Pures et Appliq.* (3) 10 (1884), 101–108. **FM** 16, p. 375.

FARRELL, O. J.
"On approximation to a mapping function by polynomials," *Amer. Jour. of Math.* 54 (1932), 571–578. **FM** 58, p. 356. **Zbl** 4, p. 403.

FATOU, P.
"Sur les séries entières à coéfficients entiers," *C.R. Acad. Sci. Paris* 138 (1904), 342–344. **FM** 35, p. 257.

"Séries trigonométriques et séries de Taylor," *Acta Math.* 30 (1906), 335–400. **FM** 37, p. 283.

"Sur les équations fonctionnelles," *Bull. Soc. Math. France* 47 (1919), 161–271 and 48 (1920), 33–94 and 208–314. **FM** 47, p. 921.

"Sur les fonctions holomorphes et bornées à l'intérieur d'un cercle," *Bull. Soc. Math. France* 51 (1923), 191–202. **FM** 49, p. 221.

"Sur l'itération des fonctions transcendantes entières," *Acta Math.* 47 (1926), 337–370. **FM** 52, p. 309.

FEIGL, G.
"Über einige Eigenschaften der einfachen stetigen Kurven, Teil I," *Math. Zeit.* 27 (1928), 161–186. **FM** 53, p. 567.

FEJÉR, L.
"Untersuchungen über Fouriersche Reihen," *Math. Annalen* 58 (1904), 51–69. **FM** 34, p. 287. (See also **MR** 23 #A789.)

"Über gewisse Potenzreihen an der Konvergenzgrenze," *Sitzungsber. Kön. Bayer. Akad. Wiss. München* (1910), 3. Abhandlung. **FM** 41, p. 284.

"La convergence sur son cercle de convergence d'une série de puissance effectuant une représentation conforme du cercle sur le plan simple," *C.R. Acad. Sci. Paris* 156 (1913), 46–49. **FM** 44, p. 290.

"Über die Konvergenz der Potenzreihe an der Konvergenzgrenze in Fällen der konformen Abbildung auf die schlichte Ebene," pp. 42–53 of *Mathematische Abhandlungen H.A. Schwarz Gewidmet*, Verlag Julius Springer (1914), Berlin. **FM** 45, p. 670. Reprinted by Chelsea Publishing Co. (1974), New York.

"Über trigonometrische Polynome," *Jour. für Reine und Angew. Math.* 146 (1915), 53–82. **FM** 45, p. 406.

"Über Potenzreihen, deren Summe im abgeschlossenen Konvergenzkreise überall stetig ist," *Sitzungsber. Kön. Bayer. Akad. Wiss. München* (1917), 33–50. FM 46, p. 480.

"Interpolation und konforme Abbildung," *Göttingen Nachr.* (1918), 319–331. **FM** 46, p. 517.

FEJÉR, L. and RIESZ, F.
"Über einige funktionentheoretische Ungleichungen," *Math. Zeit.* 11 (1921), 305–314. **FM** 48, p. 327.

FEKETE, M.
"Eine Bemerkung zu der Arbeit des Herrn Bieberbach 'Über die Verteilung der Null- und Einsstellen analytischer Funktionen,'" *Math. Annalen* 88 (1922), 166–168. **FM** 48, p. 325.

"Zum Koebeschen Verzerrungssatz," *Göttingen Nachr.* (1925), 142–150. **FM** 51, p. 272.

(a) "Über die Wurzelverteilung analytischer Funktionen, deren Wert an zwei Stellen gegeben ist," *Jahresber. Deutsch. Math. Verein.* 36 (1927), 216–222. **FM** 53, p. 279.

(b) "Zur Theorie der konformen Abbildung," *Acta Litt. ac Scient. Univ. Hung.* (*Szeged*) 3 (1927), 25–31. **FM** 53, p. 322.

FEKETE, M. and SZEGÖ, G.
"Eine Bemerkung über ungerade schlichte Funktionen," *Jour. Lon. Math. Soc.* 8 (1933), 85–89. **Zbl** 6, p. 353. **FM** 59, p. 347.

FENCHEL, W.
"Bemerkungen über die im Einheitskreis meromorphen schlichten Funktionen," *Sitzungsber. Preuss. Akad. Wiss. Berlin, Phys.-math. Kl.* (1931), 431–436. **FM** 57, p. 404. **Zbl** 2, p. 269.

FENTON, P. C.
"Functions having the restricted mean value property," *Jour. Lon. Math. Soc.* (2) 14 (1976), 451–458. **Zbl** 344 #31002. **MR** 55 #10703.

FERRAR, W. L.
"Note on a paper of L. Neder," *Math. Zeit.* 31 (1930), 519–520. **FM** 56, p. 261.

FEYEL, D. and PRADELLE, A. DE LA
Exercices sur les fonctions analytiques. Maîtrise de mathématiques. Série "Mathématiques," Librairie Armand Colin (1973), Paris. **Zbl** 351 #30001.

FICHTENHOLZ, G.
"Un théorème sur l'intégration sous le signe intégrale," *Rend. Circ. Mat. Palermo* 36 (1913), 111–114. **FM** 44, p. 346.

FINKELSTEIN, M. and SCHEINBERG, S.
"Kernels for solving problems of Dirichlet type in a half-plane," *Adv. in Math.* 18 (1975), 108–113. **Zbl** 309 #31001. **MR** 52 #3559.

FISHER, S.
"The convex hull of the finite Blaschke products," *Bull. Amer. Math. Soc.* 74 (1968), 1128–1129. **MR** 38 #2316. **Zbl** 164, p. 83.

FLATTO, L. and SHISHA, O.
"A proof of Cauchy's Integral Theorem," *Jour. Approx. Theory* 7 (1973), 386–390. **Zbl** 258 #30033. **MR** 49 #7456. See also **Zbl** 317 #30038 and **MR** 49 #9166.

FLETT, T. M.
"A note on conformal mapping," *Jour. Lon. Math. Soc.* 29 (1954), 118–121. **Zbl** 55, p. 72. **MR** 15, p. 303.

FLOYD, E. E.
"On the extension of homeomorphisms on the interior of a two cell," *Bull. Amer. Math. Soc.* 52 (1946), 654–658. **Zbl** 60, p. 401. **MR** 8, p. 50.

FORT, M. K., Jr.
"Some properties of continuous functions," *Amer. Math. Monthly* 59 (1952), 372–375. **MR** 13, p. 925. **Zbl** 49, p. 338.

"The complements of bounded, open, connected subsets of euclidean space," *Bull. Acad. Polon. Sci.* 9 (1961), 457–460. **MR** 24 #A2953. **Zbl** 103, p. 394.

"Continuous square roots of mappings," pp. 169–180 of *Lectures on Calculus*, K. O. May, editor. Holden-Day, Inc. (1967), San Francisco. **Zbl** 204, p. 77.

FOUËT, ÉD. A.
Leçons sur la théorie des fonctions analytiques. Seconde partie. Théorèmes d'existence—Étude des fonctions analytiques au point de vue de Cauchy, de Weierstrass,

de Riemann. Gauthier-Villars (1904), Paris. **FM** 35, p. 375. See also **FM** 38, p. 415; 40, p. 435 and **BAMS** 18, p. 30.

FRANK, G.
"Eine Vermutung von Hayman über Nullstellen meromorpher Funktionen," *Math. Zeit.* 149 (1976), 29–36. **Zbl** 312 #30032. **MR** 54 #10601.

FRANKLIN, P.
"A qualitative definition of the potential functions," *Bull. Amer. Math. Soc.* 30 (1924), 41–50. **FM** 50, p. 332.

(a) "Functions with an essential singularity," *Bull. Amer. Math. Soc.* 31 (1925), 157–162. **FM** 51, p. 261.

(b) "The Weierstrass approximation theorem," *Jour. of Math. & Physics of Massachusetts Institute of Technology* 4 (1925), 148–152. **FM** 51, pp. 210 and 215.

"Functions of a complex variable with assigned derivatives at an infinite number of points, and an analogue of Mittag-Leffler's theorem," *Acta Math.* 47 (1926), 371–385. **FM** 52, p. 303.

"Analytic functions with assigned values," *Bull. Amer. Math. Soc.* 33 (1927), 461–466. **FM** 53, p. 279.

"A qualitative definition of the sub- and superharmonic functions," *Jour. of Math. & Physics of Massachusetts Institute of Technology* 7 (1928), 86–92. **FM** 54, p. 518.

FREUDENTHAL, H.
"The cradle of modern topology, according to Brouwer's inedita," *Historia Math.* 2 (1975), 495–502. **Zbl** 334 #01027. (See also **Zbl** 275 #01011.)

FUCHS, W. H. J.
"A uniqueness theorem for mean values of analytic functions," *Proc. Lon. Math. Soc.* (2) 48 (1943), 35–47. **MR** 5, p. 36. **Zbl** 28, p. 402.

Topics in the Theory of Functions of One Complex Variable, Van Nostrand Mathematical Studies #12, D. Van Nostrand Co., Inc. (1967), Princeton. **MR** 36 #3954. **Zbl** 155, p. 115.

Théorie de l'approximation des fonctions d'une variable complexe, Séminaire de Mathématiques Supérieures, No. 26. The University of Montreal Press (1968). **MR** 41 #5630. **Zbl** 199, p. 397.

GAIER, D.
"Schlichte Potenzreihen, die auf $|z| = 1$ gleichmässig, aber nicht absolut konvergieren," *Math. Zeit.* 57 (1953), 349–350. **MR** 14, p. 737. **Zbl** 50, p. 78.

"Schlichte Potenzreihen an der Konvergenzgrenze," *Math. Zeit.* 58 (1953), 456–458. **MR** 15, p. 113. **Zbl** 52, p. 80.

"Über ein Iterationsverfahren von Komatu zur konformen Abbildung von Ringgebieten," *Jour. Math. and Mech.* 6 (1957), 865–883. **MR** 20 #105. **Zbl** 79, p. 298.

Konstruktive Methoden der konformen Abbildung, Springer Tracts in Natural Philosophy, Vol. 3. Springer-Verlag (1964), Berlin. **MR** 33 #7507. **Zbl** 132, p. 367.

"Konforme Abbildung mehrfach zusammenhängender Gebiete," *Jahresber. Deutsch. Math. Verein.* 81 (1978), 25–44.

GAIER, D. and POMMERENKE, CH.
"On the boundary behavior of conformal maps," *Mich. Math. Jour.* 14 (1967), 79–82. **Zbl** 182, p. 102. **MR** 34 #4470.

GAMELIN, T. W.
Uniform Algebras, Prentice-Hall, Inc. (1969), Englewood Cliffs. **Zbl** 213, p. 404. **MR** 53 #14137. **BAMS** 76, p. 1226.

GARABEDIAN, P. R.
"Univalent functions and the Riemann mapping theorem," *Proc. Amer. Math. Soc.* 61 (1976), 242–244. **Zbl** 346 #30008. **MR** 54 #13052.

GARNETT, J.
Analytic Capacity and Measure, Lecture Notes in Mathematics Vol. 297, Springer-Verlag (1972), New York. **Zbl** 253 #30014. **MR** 56 #12257.

GARNIR, H. G. and GOBERT, J.
Fonctions d'une Variable Complexe, Dunod (1965), Paris. **Zbl** 146, p. 298.

GASAPINA, U.
"Il teorema fondamentale dell' algebra," *Periodico di Mat.* (4) 35 (1957), 149–163. **Zbl** 79, p. 242. **MR** 19, p. 1034.

GATTEGNO, C. and OSTROWSKI, A.
Représentation conforme à la frontière: domaines généraux, Mémorial des Sciences Mathématiques fasc. 109, Gauthier-Villars (1949), Paris. **MR** 11, p. 425. **Zbl** 37, p. 180.

Représentation conforme à la frontière: domaines particuliers, Mémorial des Sciences Mathématiques fasc. 110, Gauthier-Villars (1949), Paris. **MR** 11, p. 426. **Zbl** 40, p. 331.

GAUSS, C. F.
"Die vier Beweise für die Zerlegung ganzer algebraischer Functionen in reelle Factoren ersten oder zweiten Grades (1799–1849)," translation from Latin and commentary by E. Netto in *Ostwald's Klassiker der exacten Wissenschaften*, No. 14. Verlag von W. Engelmann (1890), Leipzig. **FM** 22, p. 105. (3rd ed. 1913). **FM** 44, p. 128.

GAUTHIER, P.
"*Cercles de remplissage* and asymptotic behaviour," *Canad. Jour. Math.* 21 (1969), 447–455. **MR** 38 #6074. **Zbl** 181, p. 355.

GAUTHIER, P. and HENGARTNER, W.
"Complex approximation and simultaneous interpolation on closed sets," *Canad. Jour. Math.* 29 (1977), 701–706. **Zbl** 352 #30028.

GAVRILOV, V. I.
"On holomorphic functions bounded on point sequences," *Sibirsk. Mat. Žurnal* 6 (1965), 1227–1233 (Russian). **Zbl** 163, p. 313. **MR** 33 #285.

GEHMAN, H. M.
"On extending a continuous (1–1) correspondence of two plane continuous curves to a correspondence of their planes," *Trans. Amer. Math. Soc.* 28 (1926), 252–265. **FM** 52, p. 604.

GEHRING, F. W. and LOHWATER, A. J.
"On the Lindelöf theorem," *Math. Nachr.* 19 (1958), 165–170. **MR** 21 #4246. **Zbl** 89, p. 53.

GELFOND, A. O.
"Sur quelques questions concernant les fonctions entières à valeurs entières," *C.R. Acad. Sci. Paris* 264 (1967), A932–A934. **MR** 37 #414. **Zbl** 183, p. 72.

GERBER, L.
"Complex iterated radicals," *Proc. Amer. Math. Soc.* 41 (1973), 205–210. **MR** 47 #7267. **Zbl** 271 #40002.

GERMAY, R. H. J.
(a) "Sur une application des théorèmes de Weierstrass et de Mittag-Leffler de la théorie générale des fonctions," *Annales de la Soc. Sci. de Bruxelles* (1) 60 (1946), 190–195. **MR** 8, p. 324.
(b) "Extension du théorème d'E. Picard sur la décomposition en facteurs primaires des fonctions uniformes ayant une ligne de points singuliers essentiels," *Bull. Soc. Roy. Sci. Liége* 15 (1946), 9–13. **MR** 8, p. 507.
(a) "Sur une proposition de la théorie générale des fonctions analytiques," *Bull. Soc. Roy. Sci. Liége* 17 (1948), 62–65. **MR** 11, p. 21. **Zbl** 34, p. 47.
(b) "Sur une application d'un théorème de E. Picard relatif aux produits indéfinis de facteurs primaires," *Bull. Soc. Roy. Sci. Liége* 17 (1948), 138–143. **Zbl** 34, p. 48. **MR** 11, p. 21.
(c) "Extension d'un théorème d'E. Picard sur les produits indéfinis de facteurs primaires," *Bull. Soc. Roy. Sci. Liége* 17 (1948), 180–185. **Zbl** 34, p. 48. **MR** 11, p. 21.

GILLESPIE, D. C.
"Repeated integrals," *Annals of Math.* (2) 20 (1918–19), 224–228. **FM** 46, p. 396.

GIRONZA SOLANAS, J. A.
"Representación conforme de recintos doblemente conexos sobre la corona circular," *Universidad* (*Zaragoza*) (1932), 11–30 and 41–69. **FM** 58, p. 1093.

GIUDICE, F.
"Esistenza, calcolo e differenze di radici d'equazioni numeriche," *Rend. Circ. Mat. Palermo* 16 (1902), 180–184. **FM** 33, p. 109.
"Sulla dimostrazione di Clifford del teorema fondamentale d'algebra," *Giornale di Mat. di Battaglini* 45 [(2), 14] (1907), 7–15. **FM** 38, p. 117.

GLICKSBERG, I.
"A remark on Rouché's theorem," *Amer. Math. Monthly* 83 (1976), 186–187. **Zbl** 318 #30002. **MR** 53 #779.

GOLDBERG, J. L.
"Functions with positive real part in a half-plane," *Duke Math. Jour.* 29 (1962), 333–339. **MR** 29 #1340. **Zbl** 101, p. 297.
"Bounds on the derivatives of positive functions," *SIAM Review* 8 (1966), 343–345. **MR** 35 #381. **Zbl** 147, p. 69.

GOLUZIN, G. M.
Geometric Theory of Functions of a Complex Variable, Translation of Mathematical Monographs, vol. 26. American Mathematical Society (1969), Providence. **MR** 40 #308. **Zbl** 148, p. 306.

GOLUZIN, G. M. and KRYLOW, W. J.
"Verallgemeinerung einer Formel von Carleman und ihre Anwendung zur analytischen Fortsetzung," *Recueil Math. Moscou* (= *Mat. Sbornik*) 40 (1933), 144–149 (Russian). **FM** 59, p. 1021. **Zbl** 7, p. 416.

GONÇALVES, J. V.
"Sur l'intégrale prise sur un contour variable," *Portugaliae Math.* 1 (1940), 343–345. **FM** 66, p. 328. **Zbl** 24, p. 25. **MR** 2, p. 78.
"Contours de Jordan et intégrale de Cauchy," *Portugaliae Math.* 2 (1941), 166–172 and 3 (1942), 253–257. **FM** 67, p. 256. **Zbl** 25, p. 151 and 27, p. 396. **MR** 3, p. 76 and 4, p. 241.

GORDON, W. B.
"On the diffeomorphisms of euclidean space," *Amer. Math. Monthly* 79 (1972), 755–759 and 80 (1973), 674–675. **MR** 46 #4548, 47 #7758. **Zbl** 263 #57015–6.

GOULD, H. W.
"Explicit formulas for Bernoulli numbers," *Amer. Math. Monthly* 79 (1972), 44–51. **Zbl** 227 #10010. **MR** 46 #5229.

GOURSAT, E.
"Sur la définition générale des fonctions analytiques d'après Cauchy," *Trans. Amer. Math. Soc.* 1 (1900), 14–16. **FM** 31, p. 398.
"Sur un théorème de M. Jensen," *Bull. Sci. Math.* (2) 26 (1902), 298–302. **FM** 33, p. 401.

GRABINER, S.
"A short proof of Runge's theorem," *Amer. Math. Monthly* 83 (1976), 807–808. **MR** 55 #670.

GRAESER, E.
"Über konforme Abbildung des allgemeinen zweifach zusammenhängenden schlichten Bereichs auf die Fläche eines Kreisrings," Dissertation (1930), Leipzig. **FM** 56, p. 987.

GRANAS, A.
"The theory of compact vector fields and some of its applications to topology of functional spaces (I)," *Rozprawy Matematyczne* (= *Dissertationes Mathematicae*) 30 (1960). **MR** 26 #6743. **Zbl** 111, p. 110.

GRAY, J. D. and MORRIS, S. A.
"When is a function which satisfies the Cauchy-Riemann equations analytic?", *Amer. Math. Monthly* 85 (1978), 246–256. **MR** 57 #594 and #9940.

GREEN, J. W.
"Mean values of harmonic functions on homothetic curves," *Pac. Jour. of Math.* 6 (1956), 279–282. **Zbl** 71, p. 320. **MR** 18, p. 295.
"Functions that are harmonic or zero," *Amer. Jour. of Math.* 82 (1960), 867–872. **MR** 22 #11232. **Zbl** 97, p. 282.

GRONWALL, T. H.
"On analytic functions of constant modulus on a given contour," *Annals of Math.* (2) 14 (1912–13), 72–80. **FM** 43, p. 491.
"Some special boundary problems in the theory of harmonic functions," *Bull. Amer. Math. Soc.* 19 (1912/13), 227–233. **FM** 44, p. 433.
"On the maximum modulus of an analytic function," *Annals of Math.* (2) 16 (1914–15), 77–81. **FM** 45, p. 641.

GROSS, F.
"On the functional equation $f^n + g^n = h^n$," *Amer. Math. Monthly* 73 (1966), 1093–1096. **MR** 34 #4494. **Zbl** 154, p. 401.
Factorization of Meromorphic Functions, Mathematics Research Center, Naval Research Laboratory (1972), Washington, D.C. **Zbl** 266 #30006. **MR** 53 #11030.

GRUNSKY, H.
"Zur Theorie der beschränkten Funktionen," *Deutsche Math.* 3 (1938), 679–683. **FM** 64, p. 303. **Zbl** 19, p. 420.
"Der Hauptsatz der komplexen Analysis," *Jahresber. Deutsch. Math. Verein.* 68 (1966), 1–6. **Zbl** 131, p. 304. **MR** 33 #2800.
Lectures on Theory of Functions in Multiply Connected Domains, Studia Mathematika. Skript 4, Vandenhoeck & Ruprecht (1978), Göttingen. **MR** 57 #3365.

GUGGENHEIMER, H.
"The Jordan Curve Theorem and an unpublished manuscript by Max Dehn," *Arch. History Exact Sci.* 17 (1977), 193–200. **Zbl** 357 #01023.

GUICHARD, C.
"Sur les fonctions entières," *Ann. Sci. École Norm. Sup.* (3) 1 (1884), 427–432. **FM** 16, p. 367.

GUICHARDET, A.

Analyse harmonique commutative, Monographies universitaires de mathématiques, No. 26. Dunod (1968), Paris. **MR** 39 #1901. **Zbl** 159, p. 186.

GUTZMER, A.

"Ein Satz über Potenzreihen," *Math. Annalen* 32 (1888), 596–600. **FM** 20, p. 248.

"Sur certaines moyennes arithmétiques des fonctions d'une variable complexe," *Teixeira Jour.* 8 (1887), 147–156 and *Nouvelles Annales de Math.* (3) 8 (1889), 101–111. **FM** 19, p. 370 and 21, p. 233.

HADAMARD, J.

"Sur le rayon de convergence des séries ordonnées suivant les puissances d'une variable," *C.R. Acad. Sci. Paris* 106 (1888), 259–262. **FM** 20, p. 247.

(a) "Sur les fonctions entières de la forme $e^{G(x)}$," *C.R. Acad. Sci. Paris* 114 (1892), 1053–1055. **FM** 24, p. 400.

(b) "Essai sur l'étude des fonctions données par leur développement de Taylor," *Jour. de Math. Pures et Appliq.* (4) 8 (1892), 101–186. **FM** 24, p. 359 and 33, p. 448. (See also **FM** 62, p. 1039 and **Zbl** 13, p. 9.)

"Étude sur les propriétés des fonctions entières et en particulier d'une fonction considérée par Riemann," *Jour. de Math. Pures et Appliq.* (4) 9 (1893), 171–215. **FM** 25, p. 698 and 33, p. 448. (See also **FM** 62, p. 1039 and **Zbl** 13, p. 9.)

"Sur les fonctions entières," *Bull. Soc. Math. France* 24 (1896), 186–187. **FM** 27, p. 321. (See also **FM** 62, p. 1039 and **Zbl** 13, p. 9.)

HADWIGER, H.

"Elementare Begründung ausgewählter stetigkeitsgeometrischer Sätze für Kreis und Kugelfläche," *Elemente der Math.* 14 (1959), 49–60. **Zbl** 90, p. 393. **MR** 21 #5188.

"Ein Satz über stetige Funktionen auf der Kugelfläche," *Arch. der Math.* 11 (1960), 65–68. **Zbl** 117, p. 170. **MR** 22 #4050.

"Elementare Kombinatorik und Topologie," *Elemente der Math.* 15 (1960), 49–60. **Zbl** 96, p. 173. **MR** 22 #7132.

HALL, T.

"Sur la mesure harmonique de certains ensembles," *Arkiv för Mat., Astro. och Fysik* 25A, No. 28 (1937). **FM** 63, p. 287. **Zbl** 16, p. 216.

HALMOS, P.

Naive Set Theory, Van Nostrand Reinhold Co. (1960), New York. Reprinted by Springer-Verlag (1975), New York. **MR** 22 #5575. **Zbl** 321 #04001.

HAMBURGER, H.

"Über einige Beziehungen, die mit der Funktionalgleichung der Riemannschen ζ-Funktion äquivalent sind," *Math. Annalen* 85 (1922), 129–140. **FM** 48, p. 1214.

HAMEL, G.

"Über die Umkehrung einer Potenzreihe," *Deutsche Math.* 5 (1940), 338–339. **FM** 66, p. 329. **Zbl** 24, p. 214. **MR** 7, p. 193.

HAMILTON, H. J.

"The partial fraction decomposition of a rational function," *Math. Mag.* 45 (1972), 117–119. **Zbl** 241 #12103.

HARDY, G. H.

"Notes on some points in the integral calculus XXXVIII. On the definition of an analytic function by means of a definite integral," *Messenger of Math.* 43 (1914), 29–33. **FM** 44, p. 348.

"The mean value of the modulus of an analytic function," *Proc. Lon. Math. Soc.* (2) 14 (1915), 269–277. **FM** 45, p. 1331.

"A theorem concerning harmonic functions," *Jour. Lon. Math. Soc.* 1 (1926), 130–131. **FM** 52, p. 497.

HARDY, G. H. and CARLEMAN, T.

"Fourier's series and analytic functions," *Proc. Royal Soc. Lon.* (A) 101 (1922), 124–133. **FM** 48, p. 300.

HARDY, G. H., INGHAM, A. E. and PÓLYA, G.

"Theorems concerning mean values of analytic functions," *Proc. Royal Soc. Lon.* (A) 113 (1927), 542–569. **FM** 53, p. 304.

"Notes on moduli and mean values," *Proc. Lon. Math. Soc.* (2) 27 (1928), 401–409. **FM** 54, p. 331.

HARDY, G. H. and LITTLEWOOD, J. E.

"Abel's theorem and its converse," *Proc. Lon. Math. Soc.* (2) 18 (1920), 205–235. **FM** 47, p. 279.

HARDY, G. H. and ROGOSINSKI, W.

"Theorems concerning functions subharmonic in a strip," *Proc. Royal Soc. Lon.* (A) 185 (1946), 1–14. **Zbl** 61, p. 232. **MR** 7, p. 448.

HARNACK, A.

"Existenzbeweise zur Theorie des Potentiales in der Ebene und im Raume," *Ber. Verhandl. Kön. Sächs. Gesell. Wiss. Leipzig* (1886), 144–169; reproduced in *Math. Annalen* 35 (1890), 19–40. **FM** 18, p. 919 and 21, p. 994.

HARTOGS, F.

"Über die Grenzfunktionen beschränkter Folgen von analytischen Funktionen," *Math. Annalen* 98 (1928), 164–178. **FM** 53, p. 281.

HARTOGS, F. and ROSENTHAL, A.

"Über Folgen analytischer Funktionen," *Math. Annalen* 100 (1928), 212–263. **FM** 54, p. 358.

"Über Folgen analytischer Funktionen," *Math. Annalen* 104 (1931), 606–610. **Zbl** 1, p. 213. **FM** 57, p. 371.

HARUKI, H.

"On a functional equation for the exponential function of a complex variable," *Glasgow Math. Jour.* 12 (1971), 31–34. **MR** 45 #5359. **Zbl** 224 #30016.

"A proof of the principle of circle-transformation by use of a theorem on univalent functions," *L'Enseignement Mathématique* (2) 18 (1972), 145–146. **MR** 48 #4312. **Zbl** 241 #30017. (See also **MR** 43 #492 and **Zbl** 223 #30003.)

"On the equivalence of Hille's and Robinson's functional equations," *Annales Polonici Mathematici* 28 (1973), 261–264. **Zbl** 268 #30002. **MR** 49 #7430.

HASELEN, A. VAN

"Sur la représentation conforme," *Kon. Akad. Wetensch. Amsterdam, Proc.* 35 (1932), 867–869. **Zbl** 5, p. 169. **FM** 58, p. 360.

HAUSDORFF, F.

Grundzüge der Mengenlehre, Veit & Company (1914), Leipzig. **FM** 45, p. 123. Reprinted by Chelsea Publishing Co. (1955), New York. **MR** 11, p. 88. **BAMS** 27, p. 116.

"Über halbstetige Funktionen und deren Verallgemeinerung," *Math. Zeit.* 5 (1919), 292–309. **FM** 47, p. 240.

Set Theory (translated from German by J. Aumann, *et al.*), Chelsea Publishing Co. (1957), New York. **FM** 61, p. 60. **Zbl** 12, p. 203. **MR** 25 #4999. **BAMS** 33, p. 778.

HAYASHI, T.

"Relation between the zeros of a rational integral function and its derivate," *Annals of Math.* (2) 15 (1914–15), 112–113. **FM** 45, p. 167.

"On the integral $\int_0^\infty (\sin^n x/x^m)\,dx$," *Nieuw Archief voor Wiskunde* (2) 14 (1922), 13–18. **FM** 48, p. 257.

"On the integral $\int_0^\infty x^{-m} \sin^n dx$, with an appendix to an application to a theory of approximation of a function," *Tôhoku Math. Jour.* 22 (1923), 165–170. **FM** 48, p. 258.

HAYMAN, W. K.

"Some remarks on Schottky's theorem," *Proc. Camb. Phil. Soc.* 43 (1947), 442–454. **MR** 9, p. 84. **Zbl** 29, p. 123.

"On the limits of moduli of analytic functions," *Annales Polonici Mathematici* 12 (1962), 143–150. **MR** 26 #324 and p. 1544. **Zbl** 107, p. 283.

Meromorphic Functions, Clarendon Press (1964, reprinted with appendix, 1975), Oxford. **MR** 35 #5617. **Zbl** 149, p. 30.

HAYMAN, W. K. and KENNEDY, P. B.

Subharmonic Functions, vol. I, London Mathematical Society Monographs No. 9, Academic Press (1976), New York. **MR** 57 #665. **BAMS** (2) 1, p. 376.

HEFFTER, L.

Begründung der Funktionentheorie auf alten und neuen Wegen, Springer-Verlag (2nd ed. 1960), Berlin. **MR** 22 #9587. **Zbl** 89, p. 47. **BAMS** 62, p. 271.

HEILBRONN, H.

"Zu dem Integralsatz von Cauchy," *Math. Zeit.* 37 (1933), 37–38. **FM** 59, p. 312. **Zbl** 6, p. 260.

"On the representation of homotopic classes by regular functions," *Bull. Acad. Polon. Sci.* 6 (1958), 181–184. **MR** 20 #5481. **Zbl** 80, p. 160.

HEINHOLD, J.

"Zur konformen Abbildung schlichter Gebiete," *Math. Zeit.* 67 (1957), 133–138. **MR** 19, p. 401. **Zbl** 79, p. 298.

HEINS, M.

(a) "A generalization of the Aumann–Carathéodory 'Starrheitssatz'," *Duke Math. Jour.* 8 (1941), 312–316. **MR** 3, p. 81. **Zbl** 25, p. 173. **FM** 67, p. 283.

(b) "On the iteration of functions which are analytic and single-valued in a given multiply-connected region," *Amer. Jour. of Math.* 63 (1941), 461–480. **MR** 2, p. 275. **Zbl** 25, p. 46. **FM** 67, p. 283.

(c) "A note on a theorem of Radó concerning the (1, m) conformal maps of a multiply-connected region into itself," *Bull. Amer. Math. Soc.* 47 (1941), 128–130. **FM** 67, p. 283. **Zbl** 25, p. 173. **MR** 2, p. 186.

"The problem of Milloux for functions analytic throughout the interior of the unit circle," *Amer. Jour. of Math.* 67 (1945), 212–234. **Zbl** 60, p. 217. **MR** 6, p. 262.

(a) "On the Phragmén–Lindelöf Principle," *Trans. Amer. Math. Soc.* 60 (1946), 238–244. **MR** 8, p. 371. **Zbl** 60, p. 221.

(b) "On the number of 1–1 directly conformal maps which a multiply-connected plane region of finite connectivity p (> 2) admits onto itself," *Bull. Amer. Math. Soc.* 52 (1946), 454–457. **MR** 8, p. 21.

"Entire functions with bounded minimum modulus; subharmonic function analogues," *Annals of Math.* (2) 49 (1948), 200–213. **Zbl** 29, p. 298. **MR** 9, p. 341.

Selected topics in the Classical Theory of Functions of a Complex Variable, Holt, Rinehart and Winston, Inc. (1962), New York. **MR** 29 #217.

Complex Function Theory, Pure and Applied Mathematics, vol. 28. Academic Press (1968), New York. **MR** 39 #413. **Zbl** 155, p. 115. **BAMS** 76, p. 968.

HEINZ, E.

"An elementary analytic theory of the degree of mapping in n-dimensional space," *Jour. of Math. and Mech.* 8 (1959), 231–247. **MR** 21 #1370. **Zbl** 85, p. 171.

HEJHAL, D. A.

"On the loci $|f(z)| = R$, $f(z)$ entire," *Glasgow Math. Jour.* 10 (1969), 94–102. **MR** 41 #470. **Zbl** 189, p. 362.

HELMER, O.

"Divisibility properties of integral functions," *Duke Math. Jour.* 6 (1940), 345–356. **FM** 66, p. 105. **Zbl** 23, p. 239. **MR** 1, p. 307.

HERGLOTZ, G.

"Über Potenzreihen mit positivem reellen Teil im Einheitskreis," *Ber. Verhandl. Kön. Sächs. Gesell. Wiss. Leipzig* 63 (1911), 501–511. **FM** 42, p. 438.

HERMITE, C.

"Sur la formule d'interpolation de Lagrange. Extrait d'une lettre à M. Borchardt," *Jour. für Reine und Angew. Math.* 84 (1878), 70–79. **FM** 9, p. 312.

"Sur quelques points de la théorie des fonctions. Extrait d'une lettre à M. Mittag-Leffler," *Jour. für Reine und Angew. Math.* 91 (1881), 54–78. **FM** 13, p. 307.

HERVÉ, M.

"Quelques propriétés des transformations intérieures d'un domaine borné," *Ann. Sci. École Norm. Sup.* (3) 68 (1951), 125–168. **Zbl** 44, p. 303. **MR** 13, p. 734.

Compléments sur les fonctions d'une variable complexe (lecture notes), Faculté des Sciences de Paris. Mathématiques approfondies 1965/66. Secrétariat mathématique (1970), Paris. **MR** 42 #1980. **Zbl** 195, p. 81.

HERZIG, A.

"Die Winkelderivierte und das Poisson-Stieltjes-Integral," *Math. Zeit.* 46 (1940), 129–156. **MR** 1, p. 213. **Zbl** 22, p. 238. **FM** 66, p. 362.

HERZOG, F.

"A note on power series which diverge everywhere on the unit circle," *Mich. Math. Jour.* 2 (1953–54), 175–177. **MR** 16, p. 578. **Zbl** 58, p. 60.

HERZOG, F. and PIRANIAN, G.

"Sets of convergence of Taylor series, I," *Duke Math. Jour.* 16 (1949), 529–534. **MR** 11, p. 91. **Zbl** 34, p. 48.

"Sets of convergence of Taylor series, II," *Duke Math. Jour.* 20 (1953), 41–54. **MR** 14, p. 738. **Zbl** 50, p. 78. (See also **MR** 28 #4091.)

HESSENBERG, G.

Transzendenz von e und π. Ein Beitrag zur höheren Mathematik vom elementaren Standpunkt aus, Biblioteca Mathematica Teubneriana, Bd. 2. B. G. Teubner (1912), Leipzig & Berlin. **FM** 43, p. 530. **BAMS** 20, p. 421. Reprinted by Johnson Reprint Corp. (1965), New York and London. **MR** 32 #1169.

HESTENES, M. R.

"An analogue of Green's theorem in the calculus of variations," *Duke Math. Jour.* 8 (1941), 300–311. **Zbl** 25, p. 181. **MR** 3, p. 53. **FM** 67, p. 366.

HEUSER, P.

"Über Entwicklungen analytischer Funktionen nach gebietsabhängigen Polynomen," *Math. Annalen* 110 (1934), 1–11. **FM** 60, p. 248. **Zbl** 9, p. 118.

HILBERT, D.

"Über die Entwicklung einer beliebigen analytischen Function einer Variabeln in eine unendliche nach ganzen rationalen Functionen fortschreitende Reihe," *Göttingen Nachr.* (1897), 63–70. **FM** 28, p. 359.

"Zur Theorie der konformen Abbildung," *Göttingen Nachr.* (1909), 314–323. **FM** 40, p. 732.

HILLE, E.

"Über die Variation der Bogenlänge bei konformer Abbildung von Kreisbereichen," *Arkiv för Mat., Astro. och Fysik* 11, No. 27 (1916–17). **FM** 46, p. 537.

"A Pythagorean functional equation," *Annals of Math.* (2) 24 (1922–23), 175–180. **FM** 49, p. 327.

"Essai d'une bibliographie de la représentation analytique d'une fonction monogène," *Acta Math.* 52 (1928), 1–80. **FM** 55, p. 172.

Analytic Function Theory, I, II, Ginn and Co. (1959, 1962), Boston; 2nd ed. Chelsea Publishing Co. (1974), New York. **MR** 21 #6415, **MR** 34 #1490. **Zbl** 88, p. 52 and 273 #30002. **Zbl** 102, p. 294. **BAMS** 66, p. 253.

HIONG, KING-LAI

"Sur l'impossibilité de quelques relations identiques entre des fonctions entières," *C.R. Acad. Sci. Paris* 243 (1956), 222–225. **MR** 18, p. 27. **Zbl** 70, p. 297.

Sur les fonctions méromorphes et les fonctions algébroïdes. Extensions d'un théorème de M. R. Nevanlinna, Mémorial des Sciences Mathématiques, fasc. 139. Gauthier-Villars (1957), Paris. **MR** 19, p. 950. **Zbl** 79, p. 296.

(a) "Sur le cycle de Montel-Miranda dans la théorie des familles normales," *Scientia Sinica* 7 (1958), 987–1000. **MR** 21 #4243a,b. **Zbl** 86, p. 62.

(b) "Sur un problème de M. Montel concernant la théorie des familles normales des fonctions," *Sci. Record* (*N.S.*) 2 (1958), 189–192. **MR** 21 #3562a. **Zbl** 82, p. 291.

"On the Montel–Miranda cycle in the theory of normal families," *Chinese Math.* 9 (1967), 390–399 (1968). **Zbl** 173, p. 320.

HIONG, KING-LAI and HO, YU-TSAIM

"Sur les valeurs multiples des fonctions méromorphes et de leurs dérivées," *Scientia Sinica* 10 (1961), 267–285 and *Chinese Math.* 3 (1963), 156–168. **MR** 27 #5910 and #2629. **Zbl** 104, p. 296 and 163, p. 92.

HO, C.-W.

"A note on proper maps," *Proc. Amer. Math. Soc.* 51 (1975), 237–241. **MR** 51 #6698. **Zbl** 303 #54002.

HOFFMAN, K.

Banach Spaces of Analytic Functions, Prentice-Hall, Inc. (1962), Englewood Cliffs. **MR** 24 #A2844. **Zbl** 117, p. 340. **BAMS** 69, p. 687.

Analysis in Euclidean Space, Prentice-Hall, Inc. (1975), Englewood Cliffs. **Zbl** 314 #26001. **MR** 56 #12186.

HOISCHEN, L.

"A note on the approximation of continuous functions by integral functions," *Jour. Lon. Math. Soc.* 42 (1967), 351–354. **MR** 35 #2038. **Zbl** 146, p. 300.

"Eine Verschärfung eines Approximationssatzes von Carleman," *Jour. Approx. Theory* 9 (1973), 272–277. **Zbl** 271 #30035. **MR** 51 #3459.

"Approximation und Interpolation durch ganze Funktionen," *Jour. Approx. Theory* 15 (1975), 116–123. **Zbl** 318 #30034. **MR** 52 #14309.

HÖLDER, O.
"Beweis des Satzes, dass eine eindeutige analytische Function in unendlicher Nähe einer wesentlich singulären Stelle jedem Werth beliebig nahe kommt," *Math. Annalen* 20 (1882), 138–143. **FM** 14, p. 337.

HOLLAND, A. S. B.
Introduction to the Theory of Entire Functions, Pure and Applied Mathematics, vol. 56. Academic Press (1973), New York. **Zbl** 278 #30001. **MR** 56 #5882.

HOPF, E.
"Bemerkungen zur Aufgabe 49," *Jahresber. Deutsch. Math. Verein.* 39 (1930), 2te Abteilung, 4–6. **FM** 56, p. 1071.

HOROWITZ, L.
"A proof of the 'fundamental theorem of algebra' by means of Galois theory and 2-Sylowgroups," *Nieuw Archief voor Wiskunde* (3) 14 (1966), 95–96. **MR** 33 #5617. **Zbl** 143, p. 58.

HU, K.
"On Cauchy's integral theorem," *Sci. Reports of Nat. Tsing Hua Univ. Peking* A3 (1935), 179–183. **FM** 61, p. 1132. **Zbl** 12, p. 261.

HUBER, H.
"Ein Mittelwertsatz für Funktionen einer komplexen Veränderlichen," *Comment. Math. Helv.* 21 (1948), 58–66. **Zbl** 29, p. 393. **MR** 9, p. 506.

"Über analytische Abbildungen von Ringgebieten in Ringgebiete," *Compositio Math.* 9 (1951), 161–168. **MR** 13, p. 337. **Zbl** 43, p. 302.

HUREWICZ, W.
"Über ein topologisches Theorem," *Math. Annalen* 101 (1929), 210–218. **FM** 55, p. 313.

HURWITZ, A.
"Über die Nullstellen der Bessel'schen Functionen," *Math. Annalen* 33 (1889), 245–255. **FM** 20, p. 502.

"Über die Anwendung der elliptischen Modulfunktionen auf einen Satz der allgemeinen Funktionentheorie," *Vierteljahrschr. Naturforsch. Gesell. Zürich* 49 (1904), 242–253. **FM** 35, p. 401.

IACOBACCI, R. F.
"Augustin–Louis Cauchy and the development of mathematical analysis," thesis (1965), New York University.

IGLISCH, R.
"Über den Fundamentalsatz der Algebra," *Deutsche Math.* 5 (1940), 339–340. **FM** 66, p. 58. **Zbl** 23, p. 400. **MR** 8, p. 127.

INOUE, M.
"Sur un procédé pour construire la solution du problème de Dirichlet," *Proc. Imper. Acad. Tokyo* 14 (1938), 368–372. **FM** 64, p. 1161. **Zbl** 20, p. 130.

IVERSEN, F.
"Recherches sur les fonctions inverses des fonctions méromorphes," thesis (1914), Helsingfors. **FM** 45, p. 656.

IZUMI, S.
"Analytic function with assigned values at the given point set," *Tôhoku Math. Jour.* 30 (1929), 182–184. **FM** 55, p. 178.

JACOBSTHAL, E.
"Bemerkungen zum Vitalischen Satze," *Jahresber. Deutsch. Math. Verein.* 36 (1927), 361–363. **FM** 53, p. 279.

JAMET, V.

"Sur le théorème de d'Alembert," *Mathesis* (2) 4 (1894), 5–14. **FM** 25, p. 142.

"Sur les intégrales de Fresnel," *Nouvelles Annales de Math.* (3) 15 (1896), 372–376 and (4) 3 (1903), 357–359. **FM** 27, p. 235 and 34, p. 335.

JANISZEWSKI, Z.

"Über die durch Kontinua bewirkte Zerschneidung der Ebene," *Prace Mat.-Fiz.* 26(1915), 11–63. (Polish, French résumé). **FM** 45, p. 132. (See also **MR** 25 #3803.)

JATEGAONKAR, A. V.

"Elementary proof of a theorem of P. Montel on entire functions," *Jour. Lon. Math. Soc.* 40 (1965), 166–170. **MR** 30 #248. **Zbl** 134, p. 54.

JENKINS, J.

"Some results related to extremal length," *Annals of Math. Studies*, No. 30 (1953), 87–94. **MR** 15, p. 115. **Zbl** 52, p. 81.

"On explicit bounds for Schottky's theorem," *Canad. Jour. Math.* 7 (1955), 76–82. **MR** 16, p. 579. **Zbl** 64, p. 73.

JENSEN, J. L. W. V.

"Sur un nouvel et important théorème de la théorie des fonctions," *Acta Math.* 22 (1898–99), 359–364. **FM** 30, p. 364.

"Sur les fonctions convexes et les inégalités entre les valeurs moyennes," *Acta Math.* 30 (1906), 175–193. **FM** 37, p. 422.

"Investigation of a class of fundamental inequalities in the theory of analytic functions," Authorized translation from the Danish by T. H. Gronwall, *Annals of Math.* (2) 21 (1919–20), 1–29. **FM** 47, p. 271.

JENTZSCH, R.

"Untersuchungen zur Theorie der Folgen analytischen Funktionen," *Acta Math.* 41 (1918), 219–251. **FM** 46, p. 516.

JØRGENSEN, V.

"Über die Randableitung einer beschränkten analytischen Funktion nebst einer Anwendung auf den Picardschen Satz," *Math. Annalen* 116 (1939), 658–663. **Zbl** 21, p. 418. **FM** 65, p. 336.

JOURDAIN, P. E. B.

(a) "On the general theory of functions," *Jour. für Reine und Angew. Math.* 128 (1905), 169–210. **FM** 36, p. 443.

(b) "The theory of functions with Cauchy and Gauss," *Biblio. Math.* (3) 6 (1905), 190–207. **FM** 36, p. 59.

JULIA, G.

"Mémoire sur l'itération des fractions rationnelles," *Jour. de Math. Pures et Appliq.* (8) 1 (1918), 47–245. **FM** 46, p. 520 and 47, p. 47.

(a) "Une propriété générale des fonctions entières liée au théorème de M. Picard," *C.R. Acad. Sci. Paris* 168 (1919), 502–504 and 598–600. **FM** 47, p. 312.

(b) "Sur quelques problèmes relatifs à l'itération des fractions rationnelles," *C.R. Acad. Sci. Paris* 168 (1919), 147–149. **FM** 47, p. 311.

"Sur quelques propriétés nouvelles des fonctions entières ou méromorphes," *Ann. Sci. École Norm. Sup.* (3) 36 (1919), 93–125; 37 (1920), 165–218; 38 (1921), 165–181. **FM** 47, p. 312 and 48, p. 363.

"Extension nouvelle d'un lemme de Schwarz," *Acta Math.* 42 (1920), 349–355. **FM** 47, p. 272.

Leçons sur les fonctions uniformes à point singulier essentiel isolé, Gauthier-Villars (1924), Paris. **FM** 50, p. 254. **BAMS** 31, p. 59.

(a) "Développement en série de polynômes ou de fonction rationnelle de la fonction que fournit la représentation conforme d'une aire simplement connexe sur un cercle," *Ann. Sci. École Norm. Sup.* (3) 44 (1927), 289–316. **FM** 53, p. 323.

(b) "Sur la représentation conforme des aires simplement connexes," *C.R. Acad. Sci. Paris* 184 (1927), 1107–1109. **FM** 53, p. 323.

(c) "Sur une classe de polynômes," *C.R. Acad. Sci. Paris* 184 (1927), 1227–1228. **FM** 53, p. 323.

(d) "Sur les moyennes des modules de fonctions analytiques," *Bull. Sci. Math.* (2) 51 (1927), 198–214. **FM** 53, p. 306.

"Sur une suite double de polynômes liée à la représentation conforme des aires planes simplement connexes," *Jour. de Math. Pures et Appliq.* (9) 7 (1928), 381–407. **FM** 54, p. 377.

(a) "Remarques sur les problèmes de 'meilleure approximation' et sur le problème de Dirichlet," pp. 95–101 of vol 3 of *Atti del Congresso Internazionale dei Matematici Bologna, 3–10 September 1928*. N. Zanichelli (1930), Bologna. Reprinted by Kraus Reprint, Ltd. (1967), Nendel/Liechtenstein. **FM** 56, p. 423.

(b) *Principes géométriques d'analyse*, Part I (1930), Part II (1932), Cahiers scientifiques, fasc. 6 and 11. Gauthier-Villars, Paris. **FM** 56, p. 294 and 58, p. 355. **Zbl** 4, p. 10. **BAMS** 36, p. 789 and 39, p. 15.

"Sur quelques majorantes utiles dans la théorie des fonctions analytiques ou harmoniques," *Ann. Sci. École Norm. Sup.* (3) 48 (1931), 15–64. **FM** 57, p. 349. **Zbl** 1, p. 279.

"Essai sur le développement de la théorie des fonctions de variables complexes," pp. 102–125 of Vol. I of *Verhandlungen des Internationalen Mathematiker-Kongresses Zürich 1932*. Orell Füssli, Verlag (1932), Zürich & Leipzig. Reprinted by Kraus Reprint, Ltd. (1967), Nendeln/Liechtenstein. **FM** 58, p. 298. **Zbl** 6, p. 260. **MR** 53 #7713. Reprinted with a Foreword by Gauthier-Villars et Cie. (1933), Paris. **FM** 59, p. 311. **Zbl** 6, p. 211. **BAMS** 40, p. 521.

"La représentation conforme des aires multiplement connexes," *L'Enseignement Mathématique* 33 (1935), 137–168. **Zbl** 11, p. 169. **FM** 61, p. 363.

Leçons sur la représentation conforme des aires simplement connexes, Cahiers scientifiques, fasc. 8. Gauthier-Villars (2nd ed. 1950), Paris. **Zbl** 41, p. 50. **FM** 57, p. 397.

Leçons sur la représentation conforme des aires multiplement connexes, Cahiers scientifiques, fasc. 14. Gauthier-Villars (2nd ed. 1955), Paris. **Zbl** 11, p. 312. **FM** 60, p. 285.

Exercices d'analyse. II. Fonctions analytiques. Développements en série. Résidus. Transformations analytiques. Représentation conforme, Gauthier-Villars (2nd ed. 1958, reprinted 1969), Paris. **FM** 59, p. 1018. **Zbl** 169, p. 90. **MR** 39 #414. **BAMS** 40, p. 23.

JÜRGENS, E.
Allgemeine Sätze über Systeme von zwei eindeutigen und stetigen reellen Funktionen, B. G. Teubner (1879), Leipzig. **FM** 11, p. 269.

KABAILA, V. P.
"An analogue of the Poisson–Jensen formula with a double integral," *Litovsk. Mat. Sb.* 11 (1971), 241–253 (Russian). **Zbl** 224 #30055. **MR** 45 #3718.

KAKEYA, S.
"On the mean modulus of polynomial," *Proc. Phys.-Math. Soc. Japan* (3) 5 (1923), 77–89.

KAKUTANI, S.
"Rings of analytic functions," pp. 71–83 of *Lectures on Functions of a Complex Variable*, W. Kaplan, editor; University of Michigan Press (1955), Ann Arbor. **Zbl** 67, p. 54. **MR** 16, p. 1125.

KAMIYA, H.
"An elementary proof of a theorem of L. Moore's on continuous curves," *Japan. Jour. of Math.* 7 (1931), 301–304. **FM** 57, p. 742. **Zbl** 1, p. 407.

KAMOWITZ, H.
"The spectra of endomorphisms of algebras of analytic functions," *Pac. Jour. of Math.* 66 (1976), 433–442. **Zbl** 358 #46038. **MR** 58 #2426.

KAMPKE, E.
"Verallgemeinerung des Jordanschen Kurvensatzes und stetige Winkelfunktionen," *Sitzungsber. Preuss. Akad. Wiss. Berlin, Phys.-math. Kl.* (1928), 283–299. **FM** 54, p. 625.

"Zu dem Integralsatz von Cauchy," *Math. Zeit.* 35 (1932), 539–543. FM 58, p. 300. **Zbl** 4, p. 247.

KAPLAN, W.
"Approximation by entire functions," *Mich. Math. Jour.* 3 (1955–56), 43–52. **MR** 17, p. 31. **Zbl** 70, p. 62.

KAWAKAMI, Y.
"On Montel's theorem," *Nagoya Math. Jour.* 10 (1956), 125–127. **Zbl** 71, p. 71. **MR** 18, p. 292.

KELDYCH, M. and LAVRENTIEFF, M.
"Sur un problème de M. Carleman," *C.R. Acad. Sci. URSS* (2) 23 (1939), 746–748. **Zbl** 21, p. 335. **FM** 65, p. 1227. **MR** 2, p. 82.

KELLOGG, O. D.
(a) "On some theorems of Bôcher concerning isolated singular points of harmonic functions," *Bull. Amer. Math. Soc.* 32 (1926), 664–668. **FM** 52, p. 496.

(b) "Recent progress with the Dirichlet problem," *Bull. Amer. Math. Soc.* 32 (1926), 601–625 and 34 (1928), 154. **FM** 52, p. 491 and 54, p. 507.

"Les moyennes arithmétiques dans la théorie du potentiel," *L'Enseignement Mathématique* 27 (1928), 14–26. **FM** 54, p. 502.

Foundations of Potential Theory, Die Grundlehren der mathematischen Wissenschaften in Einzeldarstellungen, Bd. 31, Verlag Julius Springer (1929), Berlin. Reprinted by Dover Publications, Inc. (1953), New York and by Springer-Verlag (1967), New York. **FM** 55, p. 282. **Zbl** 53, p. 73 and 152, p. 313. **MR** 36 #5369. **BAMS** 37, p. 141.

"Converses of Gauss' theorem on the arithmetic mean," *Trans. Amer. Math. Soc.* 36 (1934), 227–242. **FM** 60, p. 430. **Zbl** 9, p. 112.

KERÉKJÁRTÓ, B. v.
Vorlesungen über Topologie I, Die Grundlehren der mathematischen Wissenschaften in Einzeldarstellungen, Bd. 8. Verlag Julius Springer (1923), Berlin. **FM** 49, p. 396.

"Sur l'indice des transformations analytiques," *Acta Litt. ac Scient. Univ. Hung.* (*Szeged*) 7 (1934/35), 163–172. **FM** 61, pp. 626, 1353. **Zbl** 11, p. 407.

KHAJALIA, G.
"Sur la représentation conforme des aires doublement connexes sur l'anneau circulaire," *Trav. Inst. Math. Tbilissi* [= *Trudy Tbiliss Mat. Inst.*] 1 (1937), 89–107 (Russian). **Zbl** 16, p. 407. **FM** 63, p. 982.

"Sur la représentation conforme des domaines doublement connexes," *Trav. Inst. Math. Tbilissi* [= *Trudy Tbiliss Mat. Inst.*] 4 (1938), 123–133 (Russian). **Zbl** 19, p. 126. **FM** 64, p. 1074.

KHINTCHINE, A. I.
"Sur les suites de fonctions analytiques bornées dans leur ensemble," *Fund. Math.* 4 (1923), 72–75. **FM** 49, p. 212.
"On sequences of analytic functions," *Recueil Math. Moscou* (= *Mat. Sbornik*) 31 (1922–24), 147–151 (Russian). **FM** 48, p. 374.

KLEINER, W.
"Démonstration du théorème de Carathéodory par la méthode des points extrémaux," *Annales Polonici Mathematici* 2 (1955), 67–72 and 11 (1962), 217–224. **MR** 17, p. 473 and 25 #202. **Zbl** 65, p. 310 and 111, p. 78.

KLINE, J. R.
"A new proof of a theorem due to Schönflies," *Proc. Nat. Acad. Sci. U.S.A.* 6 (1920), 529–531. **FM** 47, p. 519.
"What is the Jordan curve theorem?", *Amer. Math. Monthly* 49 (1942), 281–286. **MR** 3, p. 318. **Zbl** 61, p. 402.

KLOOSTERMAN, H. D.
"Über die Pole auf dem Rande des Konvergenzgebiets gewisser Potenzreihen," *Math. Zeit.* 26 (1927), 733–743. **FM** 53, p. 288.

KNASTER, B. and KURATOWSKI, K.
"Sur les continus non-bornés. (Application de la méthode d'inversion)," *Fund. Math.* 5 (1924), 23–58. **FM** 50, p. 139.

KNASTER, B., KURATOWSKI, K. and MAZURKIEWICZ, S.
"Ein Beweis des Fixpunktsatzes für *n*-dimensionale Simplexe," *Fund. Math.* 14 (1929), 132–137. **FM** 55, p. 972. (See also **MR** 55 #5357 and **Zbl** 352 #01010.)

KNESER, A.
"Arithmetische Begründung einiger algebraischer Fundamentalsätze," *Jour. für Reine und Angew. Math.* 102 (1887), 20–55. **FM** 19, p. 66.
"Die elementare Theorie der analytischen Funktionen und die komplexe Integration," *Sitzungsber. Bayer. Akad. Wiss. München* 49 (1920), 65–81. **FM** 47, p. 267.

KNESER, H.
"Lösung der Aufgabe 41," *Jahresber. Deutsch. Math. Verein.* 35 (1926), 2te Abteilung, 123–124. **FM** 52, p. 498.
"Laplace, Gauss und der Fundamentalsatz der Algebra," *Deutsche Math.* 4 (1939), 318–322. **FM** 65, p. 49. **Zbl** 21, p. 196.
"Der Fundamentalsatz der Algebra und der Intuitionismus," *Math. Zeit.* 46 (1940), 287–302. **FM** 66, p. 58. **Zbl** 22, p. 301. **MR** 1, p. 322.
"Über den Beweis des Cauchyschen Integralsatzes bei streckbarer Randkurve," *Arch. der Math.* 1 (1949), 318–321. **Zbl** 31, p. 352. **MR** 11, p. 91.
Funktionentheorie, Studia Mathematica Bd. 13, Vandenhoeck & Ruprecht (1958), Göttingen. **MR** 26 #6361. **Zbl** 80, p. 280. **BAMS** 65, p. 337.

KNESER, H. and ULLRICH, E.
"Funktionentheorie," chapter 12 of Part I of Pure Mathematics section of *FIAT Review of German Science 1939–1946*, (W. Süss, editor). Dieterich'sche Verlagsbuchhandlung (1948), Wiesbaden. **MR** 11, pp. 91 and 148. **Zbl** 30, p. 152.

KNOPP, K.

"Einheitliche Erzeugung und Darstellung der Kurven von Peano, Osgood und von Koch," *Archiv der Math. und Physik* (3) 26 (1917), 103–115. **FM** 46, p. 400.

Theory and Application of Infinite Series, Blackie & Son, Ltd. (2nd ed. 1951), Glasgow (translated from German by R. C. H. Young). **FM** 54, p. 222. **Zbl** 1, p. 392; 42, p. 292; 124, p. 283. **MR** 17, p. 1074; 32 #1473. **BAMS** 36, p. 614; 37, p. 810.

Problem Book in the Theory of Functions, Vol. I (translated from German by L. Bers, 1948), Vol. II (translated from German by F. Bagemihl, 1953), Dover Publications, Inc., New York. **FM** 57, p. 341 and 68, p. 159. **Zbl** 3, p. 62; 32, p. 65 and 27, p. 305. **MR** 10, p. 288 and 21 #4225. **BAMS** 31, p. 183; 35, p. 415; 38, p. 471.

KOCH, H. v.

"Sur une classe remarquable de fonctions entières et transcendantes," *Arkiv för Mat., Astro. och Fysik* 1 (1903–04), 205–208. **FM** 34, p. 456 and 35, p. 417.

KOEBE, P.

(a) "Über konforme Abbildung mehrfach zusammenhängender ebener Bereiche, insbesondere solcher Bereiche, deren Begrenzung von Kreisen gebildet wird," *Jahresber. Deutsch. Math. Verein.* 15 (1906), 142–153. **FM** 37, p. 684.

(b) "Herleitung der partiellen Differentialgleichung der Potentialfunktion aus der Integraleigenschaft," *Sitzungsber. Berlin Math. Gesell.* 5 (1906), 39–42. **FM** 37, p. 384.

"Über die Uniformisierung beliebiger analytischen Kurven," *Göttingen Nachr.* (1907), 191–210. Zweite Mitteilung, *ibid.* (1907), 633–669. Dritte Mitteilung, *ibid.* (1908), 337–358. Vierte Mitteilung, *ibid.* (1909), 324–361. **FM** 38, pp. 454 and 455; 40, pp. 467 and 468.

(a) "Über die Uniformisierung der algebraischen Kurven durch automorphe Funktionen mit imaginärer Substitutionsgruppe," *Göttingen Nachr.* (1909), 68–76. **FM** 40, p. 468.

(b) "Über die Uniformisierung der algebraischen Kurven. I," *Math. Annalen* 67 (1909), 145–224. **FM** 40, p. 470. II, *ibid.* 69 (1910), 1–81. **FM** 41, p. 480.

"Über eine neue Methode der konformen Abbildung und Uniformisierung (Voranzeige)," *Göttingen Nachr.* (1912), 844–848. **FM** 43, p. 520.

(a) "Ränderzuordnung bei konformer Abbildung," *Göttingen Nachr.* (1913), 286–288. **FM** 44, p. 760.

(b) "Lösung der Randwertaufgabe der Potentialtheorie für Kreisring, Ellipse und Rechteck mittels des Poissonschen Integrals," *Ber. Verhandl. Kön. Sächs. Gesell. Wiss. Leipzig* 65 (1913), 210–213. **FM** 44, p. 759.

"Abhandlungen zur Theorie der konformen Abbildung I. Die Kreisabbildung des allgemeinsten einfach und zweifach zusammenhängenden schlichten Bereichs und die Ränderzuordnung bei konformer Abbildung," *Jour. für Reine und Angew. Math.* 145 (1915), 177–223. **FM** 45, p. 667.

"Über das Schwarzsche Lemma und einige damit zusammenhängende Ungleichheitsbeziehungen der Potentialtheorie und Funktionentheorie," *Math. Zeit.* 6 (1920), 52–84. **FM** 47, p. 271.

KOENIGS, G.

"Recherches sur les substitutions uniformes," *Bull. Sci. Math.* (2) 7 (1883), 340–357. **FM** 15, p. 114.

KOMATU, Y.

(a) "Ein alternierendes Approximationsverfahren für konforme Abbildung von einem Ringgebiete auf einen Kreisring," *Proc. Japan Acad.* 21 (1945), 146–155. **MR** 11, p. 341. **Zbl** 60, p. 236.

(b) "Sur la représentation de Villat pour les fonctions analytiques définies dans un anneau circulaire concentrique," *Proc. Imper. Acad. Tokyo* 21 (1945), 94–96. **MR** 7, p. 286. **Zbl** 60, p. 236.

"Existence theorem of conformal mapping of doubly-connected domains," *Kōdai Math. Sem. Reports*, No. 5–6 (1949), 3–4. **MR** 11, p. 590. **Zbl** 45, p. 41.

"On angular derivative," *Kōdai Math. Sem. Reports* 13 (1961), 167–179. **MR** 25 #2210. **Zbl** 119, p. 62.

KOMMERELL, K.

"Der Fundamentalsatz der Algebra," *Archiv der Math. und Physik* (3) 13 (1908), 99–102. **FM** 39, p. 119.

KÖNIG, H.

"Über die Landausche Verschärfung des Schottkyschen Satzes," *Arch. der Math.* 8 (1957), 112–114. **MR** 19, p. 1045. **Zbl** 77, p. 285.

KÖNIG, J.

"Die Factorenzerlegung ganzer Functionen und damit zusammenhängende Eliminationsprobleme," *Math. Annalen* 15 (1879), 161–173. **FM** 11, p. 62.

"Über eine Eigenschaft der Potenzreihen," *Math. Annalen* 23 (1884), 447–450. **FM** 16, p. 202.

KOVANKO, A.

"Sur les suites de fonctions à une variable complexe," *C.R. Acad. Sci. Paris* 179 (1924), 873–875. **FM** 50, p. 245.

KOWALEWSKI, G.

"Ein funktionentheoretischer Satz Jacobis," *Jahresber. Deutsch. Math. Verein.* 27 (1918), 53–55 and 160. **FM** 46, p. 537.

KRAFFT, M.

"Der Satz von der Gebietstreue," *Jour. für Reine und Angew. Math.* 167 (1932), 388–389. **FM** 58, p. 356. **Zbl** 3, p. 210.

"Elementare Ermittlung des Wertes des Integrals $\int_0^\infty e^{-x^2}\,dx$," *Jahresber. Deutsch. Math. Verein.* 57 (1955), 31–33. **MR** 15, p. 609. **Zbl** 57, p. 292.

"Elementare Bestimmung des Wertes des Wahrscheinlichkeitsintegrals," *Math.-Phys. Semesterber.* 5 (1956), 120–122. **MR** 18, p. 368. **Zbl** 74, p. 290.

KRAJKIEWICZ, P.

"A new proof of a theorem of Saxer," *Annales Polonici Mathematici* 33 (1977), 293–309. **Zbl** 348 #30011. **MR** 55 #8344.

KRAZER, A.

Lehrbuch der Thetafunktionen, B. G. Teubner (1903), Leipzig. **FM** 34, p. 492. Reprinted by Chelsea Publishing Co. (1970), New York. **Zbl** 212, p. 429. **BAMS** 11, p. 375.

KRONECKER, L.

"Zwei Sätze über Gleichungen mit ganzzahligen Coefficienten," *Jour. für Reine und Angew. Math.* 53 (1857), 173–175.

"Zur Theorie der Elimination einer Variabeln aus zwei algebraischen Gleichungen," *Monatsber. Kön. Preuss. Akad. Wiss. Berlin* (1881), 535–600. **FM** 13, p. 114.

(a) "Summirung der Gauss'schen Reihen $\sum_{h=1}^{h=n-1} e^{2h^2\pi i/n}$," *Jour. für Reine und Angew. Math.* 105 (1889), 267–268. **FM** 21, p. 251.

(b) "Bemerkung über die Darstellung von Reihen durch Integrale," *Jour. für Reine und Angew. Math.* 105 (1889), 157–159 and 345–354. **FM** 21, p. 281.

"Über die Dirichlet'sche Methode der Werthbestimmung der Gauss'schen Reihen," *Festschrift der Math. Gesell. Hamburg* (1890), 32–36. **FM** 22, p. 205.

Krzyż, J. G.

Problems in Complex Variable Theory, Modern Analytic and Computational Methods in Science and Mathematics, No. 36. American Elsevier Publishing Co. (1971), New York. **Zbl** 152, p. 59 and 239 #30001. **MR** 56 #5844.

Kunugui, K.

"Sur l'allure d'une fonction analytique uniforme au voisinage d'un point frontière de son domaine de définition," *Japan. Jour. of Math.* 18 (1942), 1–39. **Zbl** 60, p. 229. **MR** 8, p. 24.

Kuran, Ü.

(a) "Classes of subharmonic functions in $R^n \times (0, \infty)$," *Proc. Lon. Math. Soc.* (3) 16 (1966), 473–492. **Zbl** 142, p. 87. **MR** 34 #2917.

(b) "Generalizations of a theorem on harmonic functions," *Jour. Lon. Math. Soc.* 41 (1966), 145–152. **Zbl** 138, p. 364. **MR** 33 #298.

"Harmonic majorization in half-balls and half-spaces," *Proc. Lon. Math. Soc.* (3) 21 (1970), 614–636. **MR** 47 #3697. **Zbl** 207, p. 416.

"On the mean-value property of harmonic functions," *Bull. Lon. Math. Soc.* 4 (1972), 311–312. **MR** 47 #8887. **Zbl** 257 #31007. (See also **MR** 52 #8464.)

Kuratowski, K.

"Sur les continus de Jordan et le théorème de M. Brouwer," *Fund. Math.* 8 (1926), 137–150. **FM** 52, p. 599.

"Théorèmes sur l'homotopie des fonctions continues de variable complexe et leurs rapports à la théorie des fonctions analytiques," *Fund. Math.* 33 (1945), 316–367. **MR** 8, p. 50. **Zbl** 60, p. 415.

"Remark on an invariance theorem," *Fund. Math.* 37 (1950), 251–252. **MR** 13, p. 150. **Zbl** 39, p. 394.

"Sur quelques invariants topologiques dans l'espace euclidien," *Jour. de Math. Pures et Appliq.* (9) 36 (1957), 191–200. **Zbl** 78, p. 365. **MR** 19, p. 970.

"Sur l'extension de la notion de fonction rationnelle à l'espace euclidien n-dimensionnel," *Bull. Acad. Polon. Sci.* 6 (1958), 281–287. **Zbl** 89, p. 389. **MR** 20 #3532.

"Un critère de coupure de l'espace euclidien par un sous-ensemble arbitraire," *Math. Zeit.* 72 (1959), 88–94. **Zbl** 88, p. 383. **MR** 21 #6574.

Topology, Vol. II (translated from French by A. Kirkor), Academic Press (1968), New York. **MR** 41 #4467. **Zbl** 102, p. 376. **BAMS** 58, p. 265.

Introduction to Set Theory and Topology (translated from Polish by L. F. Boron) Pergamon Press (2nd edition, 1972), New York. **MR** 49 #11449. **Zbl** 247 #54001 and 267 #54002.

Labelle, G.

"On the theorems of Gauss-Lucas and Grace," *Ann. Univ. Mariae Curie–Skłodowska*, Sect. A 20 (1966), 5–21 (1971). **MR** 46 #7487. **Zbl** 266 #30002.

Lagrange, J.

"Calcul des intégrales $I_m = \int_0^\infty (\sin^m x/x^a)\, dx$, $J_m = \int_0^\infty (\cos^m x/x^a)\, dx$, m entier positif, a quelconque," *Mathesis* 66 (1957), 363–369. **Zbl** 85, p. 43. **MR** 20 #3240.

Lai, W.-t.

"Über den Satz von Schottky," *Sci. Record* (*N.S.*) 3 (1959), 381–384. **MR** 23 #A1802. **Zbl** 102, p. 296.

LAMMEL, E.

"Eine Bemerkung zum Satz von Vitali über Konvergenz von Funktionenfolgen," *Math. Nachr.* 18 (1958), 309–312. **MR** 20 #2423. **Zbl** 81, p. 297.

LANDAU, E.

"Über eine Verallgemeinerung des Picardschen Satzes," *Sitzungsber. Kön. Preuss. Akad. Wiss. Berlin* (1904), 1118–1133. **FM** 35, p. 401.

"On a familiar theorem of the theory of functions," *Bull. Amer. Math. Soc.* 12 (1905/06), 155–156. **FM** 37, p. 416.

"Über den Picardschen Satz," *Vierteljahrschr. Naturforsch. Gesell. Zürich* 51 (1906), 252–318. **FM** 37, p. 418.

(a) "Über die Approximation einer stetiger Funktion durch eine ganze rationale Funktion," *Rend. Circ. Mat. Palermo* 25 (1908), 337–346. **FM** 39, p. 472.

(b) "Beiträge zur analytischen Zahlentheorie," *Rend. Circ. Mat. Palermo* 26 (1908), 169–302. **FM** 39, p. 267.

"Neuer Beweis eines Hardyschen Satzes," *Arch. der Math. und Physik* (3) 25 (1916), 173–178. **FM** 46, p. 536.

"Über die Blaschkesche Verallgemeinerung des Vitalischen Satzes," *Ber. Verhandl. Sächs. Gesell. Wiss. Leipzig* 70 (1918), 156–159. **FM** 46, p. 517 and 47, p. 311.

(a) "Zum Koebeschen Verzerrungssatz," *Rend. Circ. Mat. Palermo* 46 (1922), 347–348. **FM** 48, p. 406.

(b) "Bemerkung zu der Arbeit des Herrn Bieberbach 'Über die Verteilung der Null- und Einsstellen analytischer Funktionen'," *Math. Annalen* 86 (1922), 158–160. **FM** 48, p. 324.

"Der Picard–Schottkysche Satz und die Blochsche Konstante," *Sitzungsber. Preuss. Akad. Wiss. Berlin, Phys.-math. Kl.* (1926), 467–474. **FM** 52, p. 324.

(a) *Darstellung und Begründung einiger neuerer Ergebnisse der Funktionentheorie*, Verlag Julius Springer (2nd ed. 1929), Berlin. Reprinted by Chelsea Publishing Co. (1946 and 1960), New York. **FM** 55, p. 171. **Zbl** 93, p. 1. **MR** 22 #10886. **BAMS** 37, p. 656.

(b) "Über die Blochsche Konstante und zwei verwandte Weltkonstanten," *Math. Zeit.* 30 (1929), 608–634. **FM** 55, p. 770.

(c) "Über die Carathéodorysche Verschärfung des grossen Picardschen Satzes," *Math. Zeit.* 30 (1929), 208–210 and 796. **FM** 55, p. 193.

(d) "Bestimmung der genauen Konstanten in einem Koebeschen Hilfssatz," *Jour. für Reine und Angew. Math.* 161 (1929), 135–136. **FM** 55, p. 211.

"Über den Millouxschen Satz," *Göttingen Nachr.* (1930), 1–9. **FM** 56, p. 270.

"Ausgewählte Kapitel der Funktionentheorie," *Trav. Inst. Math. Tbilissi* [= *Trudy Tbiliss Mat. Inst.*] 8 (1940), 23–68. **MR** 3, p. 78. **Zbl** 24, p. 420. **FM** 66, p. 350.

Differential and Integral Calculus (translated from German by M. Hausner and M. Davis), Chelsea Publishing Co. (1950), New York. **MR** 12, p. 397. **Zbl** 8, p. 303 and 42, p. 55. **FM** 60, p. 167. **BAMS** 41, p. 317.

Foundations of Analysis (translated from German by F. Steinhardt), Chelsea Publishing Co. (1951), New York. **FM** 56, p. 191. **Zbl** 42, p. 278 and 198, p. 385. **MR** 22 #661. (See also **Zbl** 319 #00001; 349 #04002 and **MR** 54 #9987.)

Elementary Number Theory (translated from German by J. E. Goodman), Chelsea Publishing Co. (1958), New York. **MR** 19, p. 1159. **Zbl** 79, p. 62. **FM** 53, p. 123.

LANDAU, E. and VALIRON, G.
"A deduction from Schwarz's Lemma," *Jour. Lon. Math. Soc.* 4 (1929), 162–163. **FM** 55, p. 769.

LANDAU, H. J. and OSSERMAN, R.
"Some distortion theorems for multivalent mappings," *Proc. Amer. Math. Soc.* 10 (1959), 87–91. **MR** 21 #3553. **Zbl** 99, p. 59.

"On analytic mappings of Riemann surfaces," *Jour. d'Analyse Math.* 7 (1959–60), 249–279. **Zbl** 101, p. 55. **MR** 23 #A311.

LANDSBERG, G.
"Zur Theorie der Gauss'schen Summen und der linearen Transformation der Thetafunctionen," *Jour. für Reine und Angew. Math.* 111 (1893), 234–253. **FM** 25, p. 283.

LAURENT, H.
"Sur les séries de polynômes," *Jour. de Math. Pures et Appliq.* (5) 8 (1902), 309–328. **FM** 33, p. 410.

LAURENT, P. A.
"Extension du théorème de M. Cauchy relatif à la convergence du développement d'une fonction suivant les puissances ascendantes de la variable," *C.R. Acad. Sci. Paris* 17 (1843), 348–349.

LAURITZEN, SVEND
"On entire transcendental functions which approach a definite limit along every ray from the origin," *Mat. Tidsskr. B* (1950), 42–48 (Danish). **MR** 12, p. 326. **Zbl** 40, p. 35.

LAVRENTIEFF, M. A.
Sur les fonctions d'une variable complexe représentables par des séries de polynômes, Actualités Scientifiques et Industrielles, no. 441. Hermann et Cie. (1936), Paris. **Zbl** 17, p. 206. **FM** 62, p. 1205. **BAMS** 44, p. 22.

LAVRENTIEFF, M. A. and KWASSELAVA, D.
"Über einen Ostrowskischen Satz," *Mitt. Georg. Abt. Akad. Wiss. USSR* 1 (1940), 171–174 (Russian). **FM** 66, p. 367. **Zbl** 23, p. 139. **MR** 2, p. 83.

LAVRENTIEFF, M. M.
"Numerical estimates in interior theorems of uniqueness," *Dokl. Akad. Nauk SSSR* (*N.S.*) 110 (1956), 731–734 (Russian). **Zbl** 72, p. 70. **MR** 18, p. 728. (See also **Zbl** 81, p. 69 and **MR** 20 #1759.)

LEAU, L.
"Nouvelle démonstration du théorème de d'Alembert," *L'Enseignement Mathématique* 11 (1909), 110–118. **FM** 40, p. 119.

"Cas extrême d'une intégrale de Cauchy et limite de certaines intégrales," *Jour. de Math. Pures et Appliq.* (9) 5 (1926), 211–218. **FM** 52, p. 289.

Les suites de fonctions en général (*domaine complexe*), Mémorial des Sciences Mathématiques, fasc. 59, Gauthier-Villars (1932), Paris. **Zbl** 6, p. 352. **FM** 58, p. 301.

LEBESGUE, H.
"Sur l'approximation des fonctions," *Bull. Sci. Math.* (2) 22 (1898), 278–287. **FM** 29, p. 352.

"Sur le problème de Dirichlet," *Rend. Circ. Mat. Palermo* 24 (1907), 371–402. **FM** 38, p. 392.

"Sur la représentation approchée des fonctions," *Rend. Circ. Mat. Palermo* 26 (1908), 325–328. **FM** 39, p. 473.

"Sur les singularités des fonctions harmoniques," *C.R. Acad. Sci. Paris* 176 (1923), 1097–1099 and 1270–1271. **FM** 49, p. 338.

LECORNU, L.
"Sur les séries entières," *C.R. Acad. Sci. Paris* 104 (1887), 349–352. **FM** 19, p. 228.

LEHNER, J.
"The Picard theorems," *Amer. Math. Monthly* 76 (1969), 1005–1012. **MR** 40 #2866. **Zbl** 186, p. 394.

LEHTO, O.
"Anwendung orthogonaler Systeme auf gewisse funktionentheoretische Extremal- und Abbildungsprobleme," *Ann. Acad. Scient. Fennicae*, Ser. A I, No. 59 (1949). **MR** 11, p. 170. **Zbl** 35, p. 56.

LEJA, F.
"Construction de la fonction analytique effectuant la représentation conforme d'un domaine plan quelconque sur le cercle," *Math. Annalen* 111 (1935), 501–504. **Zbl** 12, p. 214. **FM** 61, p. 356.

"Sur une suite de polynômes et la représentation conforme d'un domaine plan quelconque sur le cercle," *Ann. Soc. Polon. Math.* 14 (1936), 116–134. **Zbl** 18, p. 260. **FM** 62, p. 1216.

"Une méthode élémentaire de résolution du problème de Dirichlet dans le plan," *Ann. Soc. Polon. Math.* 23 (1950), 230–245. **Zbl** 39, p. 325. **MR** 12, p. 703.

"Polynômes extrémaux et la représentation conforme des domaines doublement connexes," *Annales Polonici Mathematici* 1 (1954), 13–28. **MR** 16, p. 348. **Zbl** 56, p. 74.

"Construction of the function mapping conformally an arbitrary simply connected domain upon a circle," *Zastosowania Mat.* 2 (1955), 117–122 (Polish). **MR** 16, p. 917. **Zbl** 64, p. 325.

LELAND, K. O.
"Topological analysis of analytic functions," *Fund. Math.* 56 (1964–65), 157–182. **Zbl** 145, p. 80. **MR** 30 #2157.

"A polynomial approach to topological analysis, III," *Jour. Approx. Theory* 4 (1971), 433–441. **Zbl** 235 #30047. **MR** 55 #8375.

LELONG-FERRAND, J.
Représentation conforme et transformations à intégrale de Dirichlet bornée, Cahiers scientifiques, Fasc. XXII. Gauthier-Villars (1955), Paris. **MR** 16, p. 1096. **Zbl** 64, p. 322.

LERCH, M.
(a) "Zur Theorie der Gauss'schen Summen," *Math. Annalen* 57 (1903), 554–567. **FM** 34, p. 228.

(b) "Démonstration élémentaire de la formule $\pi^2/\sin^2 x\pi = \sum_{\nu=-\infty}^{\infty} 1/(x+\nu)^2$," *L'Enseignement Mathématique* 5 (1903), 450–453. **FM** 34, p. 311.

"Sur quelques applications des sommes de Gauss," *Annali Mat. Pura Appl.* (3) 11 (1904), 79–91. **FM** 35, p. 209.

LEVAVASSEUR, R.
"Sur la représentation conforme de deux aires planes à connexion multiple, d'après M. Schottky," *Annales de la Faculté des Sciences de Toulouse* (2) 4 (1902), 45–100. **FM** 33, p. 707.

LEVEQUE, W. J.
Reviews in Number Theory, 6 volumes, American Mathematical Society (1974), Providence. **Zbl** 287 #10001. **MR** 50 #2040–2045.

LEVI, E. E.

"Sopra une proprietà caratteristica delle funzioni armoniche," *Rend. Accad. d. Lincei Roma* (5) 18 (1909), 10–15. **FM** 40, p. 452.

LEVIN, B.

Distribution of Zeros of Entire Functions, Translation of Mathematical Monographs, vol. 5, American Mathematical Society (1964), Providence. **MR** 28 #217. **Zbl** 152, p. 67.

LEVIN, V. I.

"Ein Beitrag zu dem Milloux–Landauschen Satz," *Jahresber. Deutsch. Math. Verein.* 44 (1934), 262–265. **FM** 60, p. 255. **Zbl** 10, p. 308.

LÉVY, P.

"Sur une généralisation des théorèmes de MM. Picard, Landau et Schottky," *C.R. Acad. Sci. Paris* 153 (1911), 658–660. **FM** 42, p. 427.

"Remarques sur le théorème de M. Picard," *Bull. Soc. Math. France* 40 (1912), 25–39. **FM** 43, p. 506.

LEWIS, C. J.

"The problem of Milloux for functions analytic in an open annulus," *Duke Math. Jour.* 25 (1958), 591–600. **Zbl** 116, p. 60. **MR** 20 #1783.

LICHTENSTEIN, L.

"Über die Integration eines bestimmten Integrals in Bezug auf einen Parameter," *Göttingen Nachr.* (1910), 468–475. **FM** 41, p. 331.

"Über die zweimalige Integration von Funktionen zweier reellen Veränderlichen," *Sitzungsber. Berlin Math. Gesell.* 10 (1911), 55–69. **FM** 42, p. 318.

"Über die erste Randwertaufgabe der Potentialtheorie," *Sitzungsber. Berlin Math. Gesell.* 15 (1916), 92–96. **FM** 46, p. 738.

"Neuere Entwicklung der Potentialtheorie. Konforme Abbildung," *Encyklopädie der Mathematischen Wissenschaften* Bd. II, 3rd part, 1st half, 181–377. B. G. Teubner (1919), Leipzig. **FM** 47, p. 450.

LIEBECK, H.

"The convergence of sequences with linear fractional recurrence relation," *Amer. Math. Monthly* 68 (1961), 353–355. **Zbl** 115, p. 277. **MR** 23 #A1181.

"A proof of the equality of column and row rank of a matrix," *Amer. Math. Monthly* 73 (1966), 1114. **Zbl** 143, p. 50.

LINDELÖF, E.

"Sur la détermination de la croissance des fonctions entières définies par un développement de Taylor," *Bull. Sci. Math.* (2) 27 (1903), 213–226. FM 34, p. 440.

Le calcul des résidus et ses applications à la théorie des fonctions, Gauthier-Villars (1905), Paris. **FM** 36, p. 468. **BAMS** 12, p. 134. Reprinted by Chelsea Publishing Co. (1947), New York.

"Mémoire sur certaines inégalités dans la théorie des fonctions monogènes et sur quelques propriétés nouvelles de ces fonctions dans le voisinage d'un point singulier essentiel," *Acta Soc. Sci. Fennicae* 35, No. 7 (1909). **FM** 40, p. 439.

"Sur le théorème de M. Picard dans la théorie des fonctions monogènes," pp. 112–136 of *Compte rendu du congrès des mathématiciens tenu à Stockholm 22–25 septembre 1909*. B. G. Teubner (1910), Berlin and Leipzig. **FM** 41, p. 465.

"Démonstration nouvelle d'un théorème fondamental sur les suites de fonctions monogènes," *Bull. Soc. Math. France* 41 (1913), 171–178. **FM** 44, p. 489.

"Sur la représentation conforme," *C.R. Acad. Sci. Paris* 158 (1914), 245–247. **FM** 45, p. 665.

"Sur un principe général de l'analyse et ses applications à la théorie de la représentation conforme," *Acta Soc. Sci. Fennicae* 46, No. 4 (1915). **FM** 45, p. 665.

"Sur la représentation conforme d'une aire simplement connexe sur l'aire d'un cercle," pp. 59–90 of *Quatrième congrès des mathématiciens scandinaves à Stockholm 1916*, Almqvist & Wicksells (1920), Upsala. **FM** 47, p. 322.

LINDWART, E. and PÓLYA, G.

"Über einen Zusammenhang zwischen der Konvergenz von Polynomfolgen und der Verteilung ihrer Wurzeln," *Rend. Circ. Mat. Palermo* 37 (1914), 297–304. **FM** 45, p. 650.

LIPKA, S.

"Eine Verallgemeinerung des Rouchéschen Satzes," *Jour. für Reine und Angew. Math.* 160 (1929), 143–150. **FM** 55, p. 175.

LITTLEWOOD, J. E.

"On inequalities in the theory of functions," *Proc. Lon. Math. Soc.* (2) 23 (1925), V–IX and 481–519. **FM** 51, p. 247.

"Mathematical notes (6): On the definition of subharmonic function," *Jour. Lon. Math. Soc.* 2 (1927), 189–192. **FM** 53, p. 467.

"Mathematical notes (14): 'Every polynomial has a root'," *Jour. Lon. Math. Soc.* 16 (1941), 95–98. **Zbl** 28, p. 198. **FM** 67, p. 976. **MR** 3, p. 110.

Some Problems in Real and Complex Analysis, D. C. Heath & Co. (1968), Lexington, Massachusetts. **Zbl** 185, p. 115. **MR** 39 #5777.

LLOYD, N. G.

Degree Theory, Cambridge Tracts in Mathematics 73, Cambridge University Press (1978), Cambridge. **Zbl** 367 #47001.

LOHWATER, A. J.

(a) "A uniqueness theorem for a class of harmonic functions," *Proc. Amer. Math. Soc.* 3 (1952), 278–279. **Zbl** 46, p. 326. **MR** 13, p. 743.

(b) "The boundary values of a class of meromorphic functions," *Duke Math. Jour.* 19 (1952), 243–252. **Zbl** 46, p. 300. **MR** 14, p. 34.

LOOMAN, H.

"Über die Cauchy–Riemannschen Differentialgleichungen," *Göttingen Nachr.* (1923), 97–108. **FM** 49, p. 709.

"Bemerkung zur Definition der harmonischen Funktionen," *Christiaan Huygens* 3 (1923–24), 41–42. **FM** 50, p. 646.

"Über eine Erweiterung des Cauchy–Goursatschen Integralsatzes," *Nieuw Archief voor Wiskunde* (2) 14 (1925), 234–239. **FM** 50, p. 250.

LOOMIS, L. H.

"The converse of Fatou's theorem for positive harmonic functions," *Trans. Amer. Math. Soc.* 53 (1943), 239–250. **MR** 4, p. 199. **Zbl** 61, p. 233.

"An elementary proof of the strong form of the Cauchy theorem," *Bull. Amer. Math. Soc.* 50 (1944), 831–833. **MR** 6, p. 122.

LOOMIS, L. H. and WIDDER, D. V.

"The Poisson integral representation of functions which are positive and harmonic in a half-plane," *Duke Math. Jour.* 9 (1942), 643–645. **MR** 4, p. 101. **Zbl** 61, p. 233.

LORIA, G.

"Il teorema fondamentale della teoria delle equazioni algebriche," *Rivista di Mat.* 1 (1891), 185–248 and 3 (1893), 105–108. **FM** 23, p. 34 and 25, p. 61.

"Esame di alcune ricerche concernenti l'esistenza di radici nelle equazioni algebriche," *Biblio. Math.* (2) 5 (1891). **FM** 23, p. 35.

LÖWNER, K.
"Untersuchungen über die Verzerrung bei konformen Abbildungen des Einheitskreises $|z| < 1$, die durch Funktionen mit nicht verschwindender Ableitung geliefert werden," *Ber. Verhandl. Kön. Sächs. Gesell. Wiss. Leipzig* 69 (1917), 89–106. **FM** 46, p. 556.

LÖWNER, K. and RADÓ, T.
"Bemerkung zu einem Blaschkeschen Konvergenzsatze," *Jahresber. Deutsch. Math. Verein.* 32 (1923), 198–200. **FM** 49, p. 231.

MACGREGOR, T. H.
"Geometric problems in complex analysis," *Amer. Math. Monthly* 79 (1972), 447–468. **MR** 42 #7034. **Zbl** 242 #30016.

MACINTYRE, A. J.
"On Bloch's theorem," *Math. Zeit.* 44 (1938), 536–540. **Zbl** 19, p. 419. **FM** 64, p. 1075.

"On a theorem concerning functions regular in an annulus," *Recueil Math. Moscou* (= *Mat. Sbornik*) (2) 5 (1939), 307–308 (Russian). **FM** 65, p. 337. **Zbl** 22, p. 238. **MR** 1, p. 308.

"Euler's limit for e^x and the exponential series," *Edin. Math. Notes* no. 37 (1949), 26–28. **MR** 10, p. 446. **Zbl** 36, p. 35.

"Convergence of $i^{i^{\cdots}}$," *Proc. Amer. Math. Soc.* 17 (1966), 67. **MR** 32 #5855. **Zbl** 133, p. 324.

MACINTYRE, A. J. and ROGOSINSKI, W. W.
"Some elementary inequalities in function theory," *Edin. Math. Notes* no. 35 (1945), 1–3. **MR** 7, p. 150. **Zbl** 60, p. 217.

MACINTYRE, S. S.
"On the zeros of successive derivatives of integral functions," *Trans. Amer. Math. Soc.* 67 (1949), 241–251. **MR** 11, p. 340. **Zbl** 35, p. 49.

MACKI, J. W.
"On Julia's corollary to Picard's great theorem," *Amer. Math. Monthly* 75 (1968), 655–656. **MR** 38 #324. **Zbl** 157, p. 398.

MAHLER, K.
"An elementary existence theorem for entire functions," *Bull. Austral. Math. Soc.* 5 (1971), 415–419. **MR** 45 #515. **Zbl** 219 #30014.

Lectures on Transcendental Numbers, Lecture Notes in Mathematics vol. 546, Springer-Verlag (1976), New York. **Zbl** 332 #10019. **BAMS** 84, p. 1370.

MAITLAND, B. J.
"A note on functions regular and bounded in the unit circle and small at a set of points near the circumference of the circle," *Proc. Camb. Phil. Soc.* 35 (1939), 382–388. **FM** 65, p. 333. **Zbl** 21, p. 240. **MR** 1, p. 112.

MANDELBROJT, S.
"Sur les suites de fonctions holomorphes. Les suites correspondantes des fonctions dérivées. Fonctions entières," *Jour. de Math. Pures et Appliq.* (9) 8 (1929), 173–195. **FM** 55, p. 762.

Les singularités des fonctions analytiques représentées par une série de Taylor, Mémorial des Sciences Mathématiques, fasc. 54, Gauthier-Villars (1932), Paris. **Zbl** 4, p. 300. **FM** 58, p. 1079.

MANNING, K. R.
"The emergence of the Weierstrassian approach to complex analysis," *Arch. History Exact Sci.* 14 (1975), 297–383. **Zbl** 338 #01009. **MR** 56 #2746.

MANSION, P.
"Toute équation algébrique a une racine," *Annales de la Soc. Sci. de Bruxelles* 4 (B) (1880), 99–124. Correction, 17(A) (1893), 65–66. **FM** 12, p. 63 and 25, p. 143.

MARDEN, A., RICHARDS, I. and RODIN, B.
"Analytic self-mappings of Riemann surfaces," *Jour. d'Analyse Math.* 18 (1967), 197–225. **Zbl** 152, p. 274. **MR** 35 #3057.

MARDEN, M.
Geometry of Polynomials, Mathematical Surveys No. 3, American Mathematical Society (2nd ed. 1966), Providence. **MR** 37 #1562. **Zbl** 162, p. 371. **BAMS** 56, p. 78.
"On the zeros of the derivative of an entire function," *Amer. Math. Monthly* 75 (1968), 829–839. **MR** 38 #3436. **Zbl** 165, p. 401.
"Much ado about nothing," *Amer. Math. Monthly* 83 (1976), 788–798. **MR** 54 #7762. **Zbl** 352 #30001.

MARKUSCHEWITSCH, A. I.
Skizzen zur Geschichte der Analytischen Funktionen (translated from Russian by W. Ficker), VEB Deutscher Verlag der Wissenschaften (1955), Berlin. **Zbl** 64, p. 310. **MR** 17, p. 445.
Theory of Functions of a Complex Variable (translated from Russian and edited by R. A. Silverman), I (1965); II (1965); III (1967), Prentice-Hall, Inc. Englewood Cliffs. Reprinted by Chelsea Publishing Co. (1977), New York. **Zbl** 357 #30002. **MR** 30 #2125; 31 #5965; 35 #6799. (See also **MR** 35 #5581 and 38 #304.)
Selected chapters in the theory of analytic functions (Russian), Izdat. "Nauka" (1976), Moscow. **MR** 55 #8324.

MARSHALL, D. E.
"An elementary proof of the Pick–Nevanlinna interpolation theorem," *Mich. Math. Jour.* 21 (1974), 219–223. **Zbl** 281 #30032. **MR** 51 #903.
"Blaschke products generate H^∞," *Bull. Amer. Math. Soc.* 82 (1976), 494–496. **Zbl** 327 #30029. **MR** 53 #5877.

MAZURKIEWICZ, S.
"Sur les lignes de Jordan," *Fund. Math.* 1 (1920), 166–209. **FM** 47, p. 521.
(a) "Extension du théorème du Phragmén–Brouwer aux ensembles non bornés," *Fund. Math.* 3 (1922), 20–25. **FM** 48, p. 213.
(b) "Sur les séries de puissances," *Fund. Math.* 3 (1922), 52–58. **FM** 48, p. 337.
"Sur les fonctions non-dérivables," *Studia Math.* 3 (1931), 92–93. **FM** 57, p. 305. **Zbl** 3, p. 297.
"Sur le théorème de Rouché," *Sprawozdania Z Posiedzeń Towarzystwa Naukowego Warszawskiego* (= *C.R. Soc. Sci. Lett. Varsovie*) 28 (1936), 78–79. **Zbl** 13, p. 311. **FM** 62, p. 1199.

MCSHANE, E. J.
"On the Osgood–Carathéodory theorem," *Amer. Math. Monthly* 44 (1937), 288–291. **FM** 63, p. 299. **Zbl** 16, p. 266.

MEIER, K. E.
"Über die Randwerte meromorpher Funktionen und hinreichende Bedingungen für Regularität von Funktionen einer komplexen Variablen," *Comm. Math. Helv.* 24 (1950), 238–259. **MR** 12, p. 490. **Zbl** 41, pp. 52 and 661.

"Zum Satz von Looman–Menchoff," *Comm. Math. Helv.* 25 (1951), 181–195. **MR** 14, p. 150. **Zbl** 44, p. 78.

"Eine hinreichende Bedingung für die Regularität einer komplexen Funktion," *Comm. Math. Helv.* 34 (1960), 67–70. **MR** 22 #5714. **Zbl** 90, p. 47.

"Hinreichende Bedingungen für die Regularität einer komplexen Funktion," *Comm. Math. Helv.* 45 (1970), 256–264. **MR** 43 #6406. **Zbl** 194, p. 375.

Meisters, G. and Olech, C.

"Locally one-to-one mappings and a classical theorem on schlicht functions," *Duke Math. Jour.* 30 (1963), 63–80. **Zbl** 112, p. 377. **MR** 26 #1471.

Menchoff, D.

"Sur les fonctions monogènes," *Bull. Soc. Math. France* 59 (1931), 141–182. **Zbl** 4, p. 117. **FM** 57, p. 344.

"Sur la généralisation des conditions de Cauchy–Riemann," *Fund. Math.* 25 (1935), 59–97. **Zbl** 12, p. 82. **FM** 61, p. 305.

Les conditions de monogénéité, Actualités Scientifiques et Industrielles no. 329, Hermann et Cie. (1936), Paris. **Zbl** 14, p. 167. **FM** 62, p. 318. **BAMS** 44, p. 18.

"Sur une généralisation d'un théorème de M. H. Bohr," *Recueil Math. Moscou* (= *Mat. Sbornik*) (2) 2 (1937), 339–356. **FM** 63, p. 965. **Zbl** 17, p. 215.

Méray, Ch.

"Mémoire sur les fonctions doublement périodiques, monogènes et monodromes," *C.R. Acad. Sci. Paris* 40 (1855), 787–789.

"Démonstration analytique de l'existence et des propriétés des racines des équations binômes," *Ann. Sci. École Norm. Sup.* (3) 2 (1885), 337–356. **FM** 17, p. 60.

"Méthode directe fondée sur l'emploi des séries pour prouver l'existence des racines des équations entières à une inconnue par la simple exécution de leur calcul numérique," *Bull. Sci. Math.* (2) 15 (1891), 236–252. **FM** 23, p. 95.

Mergelyan, S. N.

"Uniform approximation to functions of a complex variable," *Uspehi Mat. Nauk* (*N.S.*) 7, No. 2 (48) (1952), 31–122. American Mathematical Society Translation No. 101 (1954, reprinted 1961), Providence. **MR** 15, p. 612 and 38 #1979. **Zbl** 49, p. 327.

Mertens, F.

"Der Fundamentalsatz der Algebra," *Monatshefte für Math. und Physik* 3 (1892), 293–308 and *Sitzungsber. Kaiserl. Akad. Wiss. Wien* 101 (1892), Zweite Abteilung, 415–424. **FM** 24, p. 87.

"Über die Gaussischen Summen," *Sitzungsber. Kön. Preuss. Akad. Wiss. Berlin* (1896), 217–219. **FM** 27, p. 144.

"Über die Zerfällung einer ganzen Funktion einer Veränderlichen in zwei Faktoren," *Sitzungsber. Kaiserl. Akad. Wiss. Wien* 120 (1911), Zweite Abteilung, 1485–1502. **FM** 42, p. 114.

Miamee, A. and Salehi, H.

"Positive operators on a Banach space and the Fejér–Riesz theorem," *Proc. Amer. Math. Soc.* 56 (1976), 189–192. **Zbl** 324 #30044. **MR** 53 #3788.

Mignosi, G.

"Sul teorema fondamentale dell'algebra nell'algebra classica e nell'algebra moderna," *Esercitazioni Mat. Catania* (2) 13 (1941), 28–47. **FM** 67, p. 61. **Zbl** 24, p. 145.

MILLER, E. W.
"On singularities of an analytic function," *Bull. Amer. Math. Soc.* 41 (1935), 561–565. **FM** 61, p. 310. **Zbl** 12, p. 170.

MILLOUX, H.
"Le théorème de M. Picard, suites de fonctions holomorphes, fonctions méromorphes et fonctions entières," *Jour. de Math. Pures et Appliq.* (9) 3 (1924), 345–401. **FM** 50, p. 211.

"Sur le théorème de Picard," *Bull. Soc. Math. France* 53 (1925), 181–207 and *C.R. des Séances Soc. Math. France* 54 (1926), 25–29. **FM** 52, p. 320.

"Les cercles de remplissage des fonctions méromorphes ou entières et le théorème de Picard–Borel," *Acta Math.* 52 (1928), 189–255. **FM** 55, p. 193.

"Sur certaines fonctions holomorphes et bornées dans un cercle," *Mathematica (Cluj)* 4 (1930), 182–185. **FM** 56, p. 971.

"Sur une inégalité de la théorie des fonctions et ses applications," *C.R. Acad. Sci. Paris* 194 (1932), 587–589. **FM** 58, p. 332. **Zbl** 3, p. 406.

"Sur les valeurs asymptotiques des fonctions entières d'ordre infini," *Compositio Math.* 1 (1935), 305–313. **FM** 60, p. 1019. **Zbl** 10, p. 364.

Les fonctions méromorphes et leurs dérivées. Extensions d'un théorème de M. R. Nevanlinna. Applications. Actualités Scientifiques et Industrielles no. 888, Hermann et Cie. (1940), Paris. **MR** 7, p. 427. **Zbl** 26, p. 316. **FM** 66, p. 1249.

(a) "Une application de la théorie des familles normales," *Bull. Sci. Math.* (2) 72 (1948), 12–16. **Zbl** 33, p. 116. **MR** 10, p. 289.

(b) "Le problème de la distribution des valeurs d'une fonction uniforme," *Mathematica, Timişoara* 23 (1948), 76–84. **MR** 10, p. 28. **Zbl** 31, p. 22.

Principes. Méthodes générales. Fasc. I. Avec la collaboration de Charles Pisot. (*Traité de théorie des fonctions. Tome I.*) Gauthier-Villars (1953), Paris. **MR** 15, p. 300. **Zbl** 53, p. 43.

MINAMI, U.
"Sur un théorème de Phragmén–Lindelöf," *Tôhoku Math. Jour.* 44 (1937), 85–93. **FM** 63, p. 978. **Zbl** 18, p. 141.

"An extension of Phragmén–Lindelöf's theorem," *Proc. Imper. Acad. Tokyo* 13 (1937), 241–243. **FM** 63, p. 978. **Zbl** 19, p. 419.

"On the Cauchy's integral theorem," *Proc. Imper. Acad. Tokyo* 18 (1942), 440–445. **Zbl** 60, p. 200. **MR** 7, p. 284.

MINDA, C. D.
"Analytic functions on nonopen sets," *Math. Mag.* 46 (1973), 223–224. **MR** 48 #2349.

MINSKER, S.
"A familiar combinatorial identity proved by complex analysis," *Amer. Math. Monthly* 80 (1973), 1051. **Zbl** 277 #05005.

MIODUSZEWSKI, J.
"On two-to-one continuous functions," *Rozprawy Matematyczne* (= *Dissertationes Mathematicae*) 24 (1961). **Zbl** 104, p. 173. **MR** 26 #3021.

MIRANDA, C.
"Sur un nouveau critère de normalité pour les familles de fonctions holomorphes," *Bull. Soc. Math. France* 63 (1935), 185–196. **Zbl** 13, p. 272. **FM** 61, p. 1155.

MITCHELMORE, M. C.
"A matter of definition," *Amer. Math. Monthly* 81 (1974), 643–647.

MITRINOVIĆ, D. S.

Calculus of Residues, Tutorial Text No. 4, (translated from Serbian by J. H. Michael) P. Noordhoff, Ltd. (1966), Groningen. **MR** 34 #323. **Zbl** 136, p. 241.

MITTAG-LEFFLER, G.

"Sur la théorie des fonctions uniformes d'une variable. Extrait d'une lettre adressée à M. Hermite," *C.R. Acad. Sci. Paris* 94 (1882), 414–416, 511–514, 713–715, 781–783, 938–941, 1040–1042, 1105–1108, 1163–1166; 95 (1882), 335–336. **FM** 14, p. 325.

(a) "Sur la représentation analytique des fonctions monogènes uniformes d'une variable indépendante," *Acta Math.* 4 (1884), 1–79. **FM** 16, p. 351.

(b) "Démonstration nouvelle du théorème de Laurent," *Acta Math.* 4 (1884), 80–88 and *Mém. Soc. Roy. Sci. Liége* (2) 11 (1885). **FM** 16, pp. 218, 350 and 17, p. 383.

"Sur la représentation analytique des fonctions d'une variable réelle," *Rend. Circ. Mat. Palermo* 14 (1900), 217–224. **FM** 31, p. 409.

"Sur le théorème de M. Jensen," *Bull. Soc. Math. France* 32 (1904), 1–4. **FM** 35, p. 393.

(a) "Sur une classe de fonctions entières," pp. 258–264 of *Verhandlungen des Dritten Internationalen Mathematiker-Kongresses Heidelberg 1904*. B. G. Teubner (1905), Leipzig. **FM** 36, p. 469. Reprinted by Kraus Reprint, Ltd. (1967), Nendeln/Liechtenstein.

(b) "Sur la représentation analytique d'une branche uniforme d'une fonction monogène. (Cinquième note.)," *Acta Math.* 29 (1905), 101–181. **FM** 36, p. 469.

"Sur la représentation analytique d'une branche uniforme d'une fonction monogène. (Sixième note.)," *Acta Math.* 42 (1920), 285–308. **FM** 47, p. 269.

"Die ersten 40 Jahre des Lebens von Weierstrass," *Acta Math.* 39 (1923), 1–57. **FM** 49, p. 8.

MOHR, E.

"Beweis des sogenannten Fundamentalsatzes der Algebra im reellen Gebiete," *Jour. für Reine und Angew. Math.* 184 (1942), 175–177. **FM** 68, p. 34. **Zbl** 27, p. 7. **MR** 5, p. 169. Correction, *ibid.* 189 (1952), 250–252. **Zbl** 46, p. 245. **MR** 12, p. 938.

"Der sogenannte Fundamentalsatz der Algebra als Satz der reellen Analysis," *Math. Nachr.* 6 (1951/52), 65–69. **Zbl** 44, p. 8. **MR** 12, p. 938. Correction, *ibid.* 6 (1951/52), 385–386. **Zbl** 46, p. 14. **MR** 14, p. 465.

(a) "Ein elementarer Beweis für den Fundamentalsatz der Algebra in Reellen," *Annali Mat. Pura Appl.* (4) 34 (1953), 407–410. **Zbl** 51, p. 10. **MR** 15, p. 418.

(b) "Elementarer Beweis für die Partialbruchzerlegung des Cotangens," *Zeit. für Angew. Math. und Mech.* 33 (1953), 247–248. **Zbl** 50, p. 288. **MR** 14, p. 1080.

"Elementarer Beweis für die Produktentwicklung des Sinus und die Partialbruchzerlegung des Cotangens," *Zeit. für Angew. Math. und Mech.* 39 (1959), 78–80. **Zbl** 90, p. 46. **MR** 21 #1404.

MONNA, A. F.

Dirichlet's principle: a mathematical comedy of errors and its influence on the development of analysis, Oosthoek, Scheltema & Holkema (1975), Utrecht. **Zbl** 312 #31001.

MONTEL, P.

"Sur les suites infinies de fonctions," *Ann. Sci. École Norm. Sup.* (3) 24 (1907), 233–334. **FM** 38, p. 440.

Leçons sur les séries de polynômes à une variable complexe, Gauthier-Villars (1910), Paris. **FM** 41, p. 277.

"Sur l'indétermination d'une fonction uniforme dans les voisinages de ses points essentiels," *C.R. Acad. Sci. Paris* 153 (1911), 1455–1456. **FM** 42, p. 426.

"Sur les familles de fonctions analytiques, qui admettent des valeurs exceptionnelles dans un domaine," *Ann. Sci. École Norm. Sup.* (3) 29 (1912), 487–535. **FM** 43, p. 509.

"Sur le théorème de d'Alembert et la continuité des fonctions algébriques," *Nouvelles Annales de Math.* (4) 13 (1913), 481–492. **FM** 44, p. 116.

"Sur les familles normales de fonctions analytiques," *Ann. Sci. École Norm. Sup.* (3) 33 (1916), 223–302. **FM** 46, p. 519.

"Sur la représentation conforme," *Jour. de Math. Pures et Appliq.* (7) 3 (1917), 1–54. **FM** 46, p. 548.

"Sur les suites de fonctions analytiques qui ont pour limite une constante," *Bull. Soc. Math. France* 53 (1925), 246–257. **FM** 52, p. 306.

Leçons sur les familles normales de fonctions analytiques et leurs applications, Gauthier-Villars (1927), Paris. **FM** 53, p. 303. Reprinted by Chelsea Publishing Co. (1974), New York.

"Sur les fonctions convexes et les fonctions sousharmoniques," *Jour. de Math. Pures et Appliq.* (9) 7 (1928), 29–60. **FM** 54, p. 517.

"Sur les domaines formés par les points représentant les valeurs d'une fonction analytique," *Ann. Sci. École Norm. Sup.* (3) 46 (1929), 1–23. **FM** 55, p. 198.

(a) "Sur les séries de fractions rationnelles," *Publications Math. Univ. Belgrade* 1 (1932), 157–169. **Zbl** 6, p. 211. **FM** 58, p. 302.

(b) "Sur un théorème de Rouché," *C.R. Acad. Sci. Paris* 195 (1932), 1214–1216. **Zbl** 6, p. 62. **FM** 58, p. 300.

(c) "Sur les fonctions harmoniques qui admettent des valeurs exceptionnelles," *C.R. Acad. Sci. Paris* 194 (1932), 40–41. **FM** 58, p. 503. **Zbl** 3, p. 263.

(d) "Sur une formule de Darboux et les polynomes d'interpolation," *Ann. Scuola Norm. Sup. Pisa* (2) 1 (1932), 371–384. **FM** 58, p. 299. **Zbl** 5, p. 290. (See also **MR** 5, p. 92 and **Zbl** 60, p. 200.)

(a) "Sur un théorème de M. Pompeiu," *Bull. Math. Soc. Roum. Sci.* 35 (1933), 179–181. **Zbl** 8, p. 317. **FM** 59, p. 1023.

(b) *Leçons sur les fonctions univalentes ou multivalentes*, Gauthier-Villars (1933), Paris. **Zbl** 6, p. 351. **FM** 59, p. 346. **BAMS** 42, p. 463.

"Le rôle des familles normales," *L'Enseignement Mathématique* 33 (1934), 5–21. **FM** 60, p. 1007. **Zbl** 10, p. 264.

"Sur les relations de Cauchy," *Bull. Math. Soc. Roum. Sci.* 38 (1936), 97–99. **FM** 62, p. 1222. **Zbl** 16, p. 264.

"Sur les fonctions localement univalentes ou multivalentes," *Ann. Sci. École Norm. Sup.* (3) 54 (1937), 39–54. **FM** 63, p. 290. **Zbl** 17, p. 24.

"Harmonische und subharmonische Funktionen," *Publ. Inst. Mat. Rosario* 2 (1940), 1–23 (Spanish). **FM** 66, p. 448. **Zbl** 24, p. 116. **MR** 2, p. 76.

"Sur le rôle des familles de fonctions dans l'analyse moderne," *Bull. Soc. Roy. Sci. Liége* 15 (1946), 262–267. **Zbl** 61, p. 149. **MR** 8, p. 507. (Italian translation: **MR** 12, p. 490.)

Leçons sur les récurrences et leurs applications, Gauthier-Villars (1957), Paris. **MR** 19, p. 427. **Zbl** 77, p. 66.

MOORE, E. H.

"A simple proof of the fundamental Cauchy–Goursat theorem," *Trans. Amer. Math. Soc.* 1 (1900), 499–506. **FM** 31, p. 398.

MOORE, R. L.

"A theorem concerning continuous curves," *Bull. Amer. Math. Soc.* 23 (1916/17), 233–236. **FM** 46, pp. 308, 829.

"A characterization of Jordan regions by properties having no reference to their boundaries," *Proc. Nat. Acad. Sci. U.S.A.* 4 (1918), 364–370. **FM** 46, p. 1451.

"Conditions under which one of two given closed linear point sets may be thrown into the other one by a continuous transformation of a plane into itself," *Amer. Jour. of Math.* 48 (1926), 67–72. **FM** 52, p. 604.

MORDELL, L. J.

"On a simple summation of the series $\sum_{s=0}^{n-1} e^{2s^2\pi i/n}$," *Messenger of Math.* 48 (1918), 54–56. **FM** 46, p. 538.

"The definite integral $\int_{-\infty}^{\infty} (e^{ax^2+bx})/(e^{cx}+d)\,dx$ and the analytic theory of numbers," *Acta Math.* 61 (1933), 323–360. **Zbl** 8, p. 55. **FM** 59, p. 287.

MORERA, G.

"Un teorema fondamentale nella teorica delle funzioni di una variabile complessa," *Rend. del R. Istituto Lombardo di Scienze e Lettere* (2) 19 (1886), 304–307. **FM** 18, p. 338.

"Sulla definizione di funzione di una variabile complessa," *Atti della R. Accad. d. Scienze di Torino* 37 (1902), 99–102. **FM** 33, p. 396.

MORGENSTERN, D.

"Unendlich oft differenzierbare nicht-analytische Funktionen," *Math. Nachr.* 12 (1954), 74. **Zbl** 58, p. 289. **MR** 16, p. 342.

MÜLLER, C.

"Über die Umkehrung des Cauchyschen Integralsatzes," *Arch. der Math.* 6 (1954), 47–51. **Zbl** 59, p. 65. **MR** 16, p. 346.

MÜLLER, M.

"Zur konformen Abbildung angenähert kreisförmiger Gebiete," *Math. Zeit.* 43 (1938), 628–636. **FM** 64, p. 1073. **Zbl** 17, p. 408.

MYLLER-LÉBÉDEFF, V.

"Sur une application du lemme de M. Carleman," *Ann. Sci. Univ. Jassy. I: Math.* 24 (1938), 1–4. **FM** 64, p. 313. **Zbl** 18, p. 141.

NABETANI, K.

"On Study's theorem in the theory of conformal representation," *Tôhoku Math. Jour.* 41 (1935–36), 406–410. **Zbl** 14, p. 24. **FM** 62, p. 374.

NAGASAWA, M.

"Isomorphisms between commutative Banach algebras with an application to rings of analytic functions," *Kōdai Math. Sem. Reports* 11 (1959), 182–188. **MR** 22 #12379. **Zbl** 166, p. 400.

NAKAI, M.

"On rings of analytic functions on Riemann surfaces," *Proc. Japan Acad.* 39 (1963), 79–84. **MR** 27 #295. **Zbl** 113, p. 58.

"Divisors on meromorphic function fields," *Proc. Japan Acad.* 51 (1975), 507–509. **MR** 52 #5956. **Zbl** 368 #46030.

NEDER, L.

"Konvergenzdefekte der Potenzreihen stetiger Funktionen auf dem Rande des Konvergenzkreises," *Math. Zeit.* 6 (1920), 262–269. **FM** 47, p. 276.

NEHARI, Z.

Conformal Mapping, McGraw-Hill Book Co. (1952), New York. **MR** 13, p. 640. **Zbl** 48, p. 315. **BAMS** 58, p. 515. Reprinted by Dover Publications, Inc. (1975), New York. **MR** 51 #13206.

NERSESJAN, A. A.

"Carleman sets," *Izv. Akad. Nauk Armjan. SSR Ser. Mat.* 6, no. 6 (1971), 465–471 (Russian). **MR** 46 #367. **Zbl** 235 #30041. (See also **MR** 51 #3460, #3461 and **Zbl** 251 #30040, 292 #30037.)

NETTO, E.

"Beweis der Wurzelexistenz algebraischer Gleichungen," *Jour. für Reine und Angew. Math.* 88 (1880), 16–21. **FM** 11, p. 61.

NETTO, E. and LE VAVASSEUR, R.

"Les fonctions rationnelles," *Encyclopédie des sciences mathématiques pures et appliquées*, t. I, v. 2, fasc. 1, 1–232. Gauthier-Villars (1907), Paris. **FM** 30, p. 95.

NEUMANN, C.

"Revision einiger allgemeinen Sätze aus der Theorie des logarithmischen Potentials," *Math. Annalen* 3 (1871), 325–349. **FM** 3, p. 491.

NEVANLINNA, F. and NEVANLINNA, R.

"Über die Eigenschaften analytischer Funktionen in der Umgebung einer singulären Stelle oder Linie," *Acta Soc. Sci. Fennicae* 50, No. 5 (1922). **FM** 48, p. 358.

NEVANLINNA, F. and NIEMINEN, T.

"Das Poisson-Stieltjes'sche Integral und seine Anwendung in der Spektraltheorie des Hilbert'schen Raumes," *Ann. Acad. Scient. Fennicae Ser. AI*, No. 207 (1955). **Zbl** 67, p. 90. **MR** 17, p. 648.

NEVANLINNA, R.

"Über beschränkte Funktionen, die in gegebenen Punkten vorgeschriebene Werte annehmen," *Ann. Acad. Scient. Fennicae Ser.* (*A*) 13, No. 1 (1920). **FM** 46, p. 1466 and 47, p. 271.

(a) "Asymptotische Entwicklungen beschränkter Funktionen und das Stieltjessche Momentenproblem," *Ann. Acad. Scient. Fennicae Ser.* (*A*) 18, No. 5 (1922). **FM** 48, p. 1226.

(b) "Kriterien für die Randwerte beschränkter Funktionen," *Math. Zeit.* 13 (1922), 1–9. **FM** 48, p. 322.

(a) "Untersuchungen über den Picardschen Satz," *Acta Soc. Sci. Fennicae* 50, No. 6 (1924). **FM** 50, p. 219.

(b) "Beweis des Picard–Landauschen Satzes," *Göttingen Nachr.* (1924), 151–154. **FM** 50, p. 219.

"Über eine Erweiterung des Poissonschen Integrals," *Ann. Acad. Scient. Fennicae Ser.* (*A*) 24, No. 4 (1925). **FM** 51, p. 360.

(a) "Remarques sur le lemme de Schwarz," *C.R. Acad. Sci. Paris* 188 (1929), 1027–1029. **FM** 55, p. 768.

(b) *Le théorème de Picard–Borel et la théorie des fonctions méromorphes*, Gauthier-Villars (1929), Paris. Reprinted by Chelsea Publishing Co. (1974), New York. **FM** 55, p. 773. **MR** 54 #5468. **Zbl** 357 #30019.

(c) "Über beschränkte analytische Funktionen," *Ann. Acad. Scient. Fennicae Ser.* (*A*) 32 (Lindelöf-Festschrift), No. 7 (1929). **FM** 55, p. 768.

(d) "Sur un problème d'interpolation," *C.R. Acad. Sci. Paris* 188 (1929), 1224–1226. **FM** 55, p. 768.

"Über eine Minimumaufgabe in der Theorie der konformen Abbildung," *Göttingen Nachr.* (1933), 103–115. **FM** 59, p. 348. **Zbl** 6, p. 408.

"Das harmonische Mass von Punktmengen und seine Anwendung in der Funktionentheorie," pp. 116–133 of *Comptes rendus du huitième congrès des mathématiciens scandinaves tenu à Stockholm 14–18 août 1934*. Håkon Ohlssons Boktryckeri (1935), Lund. **FM** 61, p. 306. **Zbl** 12, p. 78.

Eindeutige Analytische Funktionen, Die Grundlehren der mathematischen Wissenschaften in Einzeldarstellungen, Bd. 46. Springer-Verlag (2nd ed. 1953), Berlin. **FM** 62, p. 315. **Zbl** 278 #30002. **MR** 15, p. 208 and 49 #9165. **BAMS** 45, p. 52.

Analytic Functions (translated from German by P. Emig), Die Grundlehren der mathematischen Wissenschaften in Einzeldarstellungen, Bd. 162. Springer-Verlag (1970), Berlin. **MR** 43 #5003. **Zbl** 199, p. 125.

NEVILLE, E. H.

"A trigonometrical identity," *Proc. Camb. Phil. Soc.* 47 (1951), 629–632. **MR** 13, p. 19. **Zbl** 43, p. 285.

NEWMAN, D. J.

"An entire function bounded in every direction," *Amer. Math. Monthly* 83 (1976), 192–193. **Zbl** 321 #30033. **MR** 52 #8433.

NEWMAN, M. H. A.

Elements of the Topology of Plane Sets of Points, Cambridge University Press (2nd ed. 1951), Cambridge. **FM** 65, p. 873. **Zbl** 123, p. 343. **MR** 24 #A2374. **BAMS** 58, p. 101.

NICOLESCO, M.

"Sur un critère d'harmonicité de Volterra et Vitali," *C.R. Acad. Sci. Roum.* 7 (1945), 16–19. **MR** 9, p. 142.

NIELSEN, N.

Traité élémentaire des nombres de Bernoulli, Gauthier-Villars (1923), Paris. **FM** 49, p. 99 and 50, p. 170.

NITSCHE, J. C. C.

"On the module of doubly connected regions under harmonic mappings," *Amer. Math. Monthly* 69 (1962), 781–782. **Zbl** 109, p. 305.

NOAILLON, P.

"Point singulier isolé non critique d'une fonction harmonique," *C.R. Acad. Sci. Paris* 184 (1927), 360–362. **FM** 53, p. 464.

NÖBELING, G.

"Eine allgemeine Fassung des Hauptsatzes der Funktionentheorie von Cauchy," *Math. Annalen* 121 (1949), 54–66. **Zbl** 33, p. 360. **MR** 10, p. 691.

"Ein gemeinsamer Beweis für den Jordanschen Kurvensatz und zwei damit zusammenhängende Sätze," *Jour. für Reine und Angew. Math.* 188 (1950), 22–39. **Zbl** 38, p. 362. **MR** 14, p. 192.

NOSHIRO, K.

"Some theorems on a cluster-set of an analytic function," *Proc. Imper. Acad. Tokyo* 13 (1937), 27–29. **FM** 63, p. 283. **Zbl** 17, p. 407.

NOVINGER, W. P.

(a) "An elementary approach to the problem of extending conformal maps to the boundary," *Amer. Math. Monthly* 82 (1975), 279–282. **MR** 51 #8390. **Zbl** 315 #30007.

(b) "Some theorems from geometric function theory: applications," *Amer. Math. Monthly* 82 (1975), 507–510. **Zbl** 315 #30008. **MR** 51 #858.

NOWINSKI, K.
"A generalization of the Borsuk–Ulam antipodal theorem," pp. 513–515 of *Topics in Topology* (Proceedings of a Colloquium at Keszthely, 1972), Coll. Math. Soc. János Bolyai, 8. North-Holland Publishing Co. (1974), Amsterdam. **MR** 50 #11199. **Zbl** 349 #47045.

O'HARA, P. J.
"Another proof of Bernstein's theorem," *Amer. Math. Monthly* 80 (1973), 673–674. **MR** 48 #507. **Zbl** 271 #30037.

OLDS, C. D.
"The Fresnel integrals," *Amer. Math. Monthly* 75 (1968), 285–286. **Zbl** 161, p. 248.

ONICESCU, O.
"Sur le théorème fondamental de l'algèbre," *Mathematica, Timişoara* 22 (1946), 208–214. **Zbl** 60, p. 46. **MR** 8, p. 127.

OSGOOD, W. F.
"Some points in the elements of the theory of functions," *Bull. Amer. Math. Soc.* (2) 2 (1895/96), 296–302. **FM** 27, p. 301.

"Über einen Satz des Herrn Schönflies aus der Theorie der Functionen zweier reeller Veränderlichen," *Göttingen Nachr.* (1900), 94–97. **FM** 31, p. 478.

"Allgemeine Theorie der analytischen Funktionen a) einer und b) mehrerer komplexen Grössen," *Encyklopädie der Mathematischen Wissenschaften* Bd. II, 2nd part, Heft 1, 1–114. B. G. Teubner (1901), Leipzig. **FM** 33, p. 389.

"Note on the functions defined by infinite series whose terms are analytic functions of a complex variable; with corresponding theorems for definite integrals," *Annals of Math.* (2) 3 (1901–02), 25–34. **FM** 32, p. 399.

"On Neumann's existence proof," *Annals of Math.* (2) 25 (1923–24), 238–240. **FM** 50, p. 648.

Lehrbuch der Funktionentheorie I, B. G. Teubner (5th ed. 1928), Leipzig & Berlin. **FM** 54, p. 326. **BAMS** 37, p. 330. Reprinted by Chelsea Publishing Co. (1965), New York. **MR** 33 #4233a. **Zbl** 184, p. 297.

OSGOOD, W. F., BOUTROUX, P. and CHAZY, J.
"Fonctions analytiques," *Encyclopédie des sciences mathématiques pures et appliquées*, t. II, v. 2, fasc. 1, 94–190. Gauthier-Villars (1911), Paris. **FM** 42, p. 458.

OSGOOD, W. F. and TAYLOR, E. H.
"Conformal transformations on the boundaries of their regions of definition," *Trans. Amer. Math. Soc.* 14 (1913), 277–298. **FM** 44, p. 758.

OSTROWSKI, A.
"Über vollständige Gebiete gleichmässiger Konvergenz von Folgen analytischer Funktionen," *Abhand. Math. Sem. Univ. Hamburg* 1 (1922), 327–350. **FM** 48, p. 372.

"Über die Bedeutung der Jensenschen Formel für einige Fragen der komplexen Funktionentheorie," *Acta Litt. ac Scient. Univ. Hung.* (*Szeged*) 1 (1922/23), 80–87. **FM** 49, p. 713.

(a) "Über Potenzreihen die überkonvergente Abschnittsfolgen besitzen," *Sitzungsber. Preuss. Akad. Wiss. Berlin, Phys.-math. Kl.* (1923), 185–192. **FM** 49, p. 230.

(b) "Über allgemeine Konvergenzsätze der komplexen Funktionentheorie," *Jahresber. Deutsch. Math. Verein.* 32 (1923), 185–194. **FM** 49, p. 231.

(a) "Über den Schottkyschen Satz und die Borelschen Ungleichungen," *Sitzungsber. Preuss. Akad. Wiss. Berlin, Phys.-math. Kl.* (1925), 471–484. **FM** 51, p. 262.

(b) "Mathematische Miszellen III: Über Nullstellen gewisser im Einheitskreis regulärer Funktionen und einige Sätze zur Konvergenz unendlicher Reihen," *Jahresber. Deutsch. Math. Verein.* 34 (1925), 161–171. **FM** 51, p. 246.

(a) "Mathematische Miszellen VI: Über einen Satz von Herrn Hadamard," *Jahresber. Deutsch. Math. Verein.* 35 (1926), 179–182. **FM** 52, p. 292.

(b) "Über Folgen analytischer Funktionen und einige Verschärfungen des Picardschen Satzes," *Math. Zeit.* 24 (1926), 215–258 and 38 (1934), 642. **FM** 51, p. 260 and 60, p. 243.

(a) "Mathematische Miszellen, XV: Zur konformen Abbildung einfach zusammenhängender Gebiete," *Jahresber. Deutsch. Math. Verein.* 38 (1929), 168–182. **FM** 55, p. 788.

(b) "Über Schwankungen analytischer Funktionen, die gegebene Werte nicht annehmen," *Sitzungsber. Preuss. Akad. Wiss. Berlin, Phys.-math. Kl.* (1929), 276–289. **FM** 55, p. 194.

"Über konforme Abbildungen annähernd kreisförmiger Gebiete," *Jahresber. Deutsch. Math. Verein.* 39 (1930), 78–81. **FM** 56, p. 297.

Studien über den Schottkyschen Satz, B. Wepf et Cie. (1931), Basel. **Zbl** 3, p. 211. **FM** 57, p. 350.

"Asymptotische Abschätzung des absoluten Betrages einer Funktion, die die Werte 0 und 1 nicht annimmt," *Comm. Math. Helv.* 5 (1933), 55–87. **FM** 59, p. 327. **Zbl** 5, p. 251.

"Zur Randverzerrung bei konformer Abbildung," *Prace Mat. Fiz.* 44 (1937), 371–471. **FM** 62, p. 381. **Zbl** 20, p. 238.

(a) "Sur l'inverse d'une transformation continue et biunivoque," *C.R. Acad. Sci. Paris* 223 (1946), 229–230. **Zbl** 60, p. 496. **MR** 8, p. 49.

(b) "Nouvelle démonstration du théorème de Schoenflies pour les espaces à *n* dimensions," *C.R. Acad. Sci. Paris* 223 (1946), 530–531. **Zbl** 60, p. 406. **MR** 8, p. 164.

Ou, Šo-Mo

"Some properties of analytic function omitting two values," *Rev. Roum. Math. Pures et Appliq.* 2 (1957), 145–150. **Zbl** 88, p. 286. **MR** 20 #3284.

Painlevé, P.

"Sur le développement en série de polynômes d'une fonction holomorphe dans une aire quelconque," *C.R. Acad. Sci. Paris* 102 (1886), 672–675. **FM** 18, p. 346.

"Sur les lignes singulières des fonctions analytiques," *Annales de la Faculté des Sciences de Toulouse* (1) 2 (1888), 1–130. **FM** 20, p. 404.

"Sur la représentation des fonctions analytiques uniformes," *C.R. Acad. Sci. Paris* 126 (1898), 200–202. **FM** 29, p. 361.

"Sur le développement des fonctions uniformes ou holomorphes dans un domaine quelconque," *C.R. Acad. Sci. Paris* 126 (1898), 318–321. **FM** 29, p. 361.

Pál, J.

"Zur Topologie der Ebene," *Acta Litt. ac Scient. Univ. Hung.* (*Szeged*) 1 (1922/23), 226–239. **FM** 49, p. 409.

Patil, D. J.

"Representation of H^p-functions," *Bull. Amer. Math. Soc.* 78 (1972), 617–620. **MR** 45 #7069. **Zbl** 255 #30010. (See also **MR** 52 #11010 and **Zbl** 317 #60027.)

Perkins, F. W.

"An intrinsic treatment of Poisson's integral," *Amer. Jour. of Math.* 50 (1928), 389–414. **FM** 54, p. 512.

PERRON, O.

"Eine neue Behandlung der ersten Randwertaufgabe für $\Delta u = 0$," *Math. Zeit.* 18 (1923), 42–54. **FM** 49, p. 340.

"Über die Picard–Landauschen Sätze," *Göttingen Nachr.* (1929), 65–72. **FM** 55, p. 192.

PERRY, A. D. and YOUNGS, J. W. T.

"Remarks on analyticity and integration," *Amer. Math. Monthly* 54 (1947), 313–318. **Zbl** 29, p. 31. **MR** 9, p. 138.

PESCHL, E.

Funktionentheorie. I, B. I. Hochschultaschenbücher Bd. 131/131a. Bibliographisches Institut (1967, revised printing 1968), Mannheim. **MR** 37 #4240. **Zbl** 159, p. 361.

PETERSEN, J.

"Quelques remarques sur les fonctions entières," *Acta Math.* 23 (1899), 85–90. **FM** 30, p. 378.

PETROVA, S. S.

"Über den ersten Beweis des Fundamentalsatzes der Algebra," *Istor. Metodolog. estestv. Nauk* 11 (1971), 123–127 (Russian). **Zbl** 261 #01013. (See also **Zbl** 263 #01022.)

"Aus der Geschichte der analytischen Beweise des Fundamentalsatzes der Algebra," *Istor. Metodolog. estestv. Nauk* 14 (1973), 167–172 (Russian). **Zbl** 285 #01013.

"Sur l'histoire des démonstrations analytiques du théorème fondamental de l'algèbre," *Historia Math.* 1 (1974), 255–261. **Zbl** 281 #01008. **MR** 57 #5609.

PHRAGMÉN, E.

"Ein elementarer Beweis des Fundamentalsatzes der Algebra," *Öfversigt af Kongl. Svenska Vetenskaps-Akademiens Förhandlingar* 48 (1891), 113–129. **FM** 23, p. 86.

"Sur la représentation analytique des fonctions réelles, données sur un ensemble quelconque de points," *Rend. Circ. Mat. Palermo* 14 (1900), 256–261. **FM** 31, p. 410.

"Sur une extension d'un théorème classique de la théorie des fonctions," *Acta Math.* 28 (1904), 351–368. **FM** 35, p. 404.

PHRAGMÉN, E. and LINDELÖF, E.

"Sur une extension d'un principe classique de l'analyse et sur quelques propriétés des fonctions monogènes dans le voisinage d'un point singulier," *Acta Math.* 31 (1908), 381–406. **FM** 39, p. 465.

PICARD, É.

(a) "Sur une propriété des fonctions entières," *C.R. Acad. Sci. Paris* 88 (1879), 1024–1027. **FM** 11, p. 267.

(b) "Sur les fonctions analytiques uniformes dans le voisinage d'un point singulier essentiel," *C.R. Acad. Sci. Paris* 89 (1879), 745–747. **FM** 11, p. 267.

"Mémoire sur les fonctions entières," *Ann. Sci. École Norm. Sup.* (2) 9 (1880), 145–166. **FM** 12, p. 327.

"Deux théorèmes élémentaires sur les singularités de fonctions harmoniques," *C.R. Acad. Sci. Paris* (1923), 933–935. **FM** 48, p. 338.

"Quelques théorèmes élémentaires sur les fonctions harmoniques," *Bull. Soc. Math. France* 52 (1924), 162–166. **FM** 50, p. 333.

PICK, G.
"Über die Beschränkungen analytischer Funktionen, welche durch vorgegebene Funktionswerte bewirkt werden," *Math. Annalen* 77 (1915), 7–23. **FM** 45, p. 642.
"Über den Koebeschen Verzerrungssatz," *Ber. Verhandl. Kön. Sächs. Gesell. Wiss. Leipzig* 68 (1916), 58–64. **FM** 46, p. 550.
"Über beschränkte Funktionen mit vorgeschriebenen Wertzuordnungen," *Ann. Acad. Scient. Fennicae Ser. (A)* 15, No. 3 (1920).

PICONE, M.
(a) "Sul problema di Dirichlet per la corona circolare," *Boll. della Un. Mat. Ital.* 5 (1926), 114–118. **FM** 52, p. 493.
(b) "Sulle singolarità delle funzioni armoniche," *Rend. Accad. d. Lincei Roma* (6) 3 (1926), 655–660. **FM** 52, p. 496.
"Sulle singolarità isolate delle funzioni armoniche," *Rend. Accad. d. Lincei Roma* (6) 9 (1929), 1067–1073. **FM** 55, p. 889.
"Sul calcolo delle funzioni olomorphe di una variabile complessa," pp. 118–126 of *Studies in Mathematics and Mechanics presented to Richard von Mises*, Academic Press, Inc. (1954), New York. **MR** 16, p. 459. **Zbl** 58, p. 301.

PLASTOCK, R.
"Homeomorphisms between Banach spaces," *Trans. Amer. Math. Soc.* 200 (1974), 169–183. **MR** 50 #8593. **Zbl** 291 #54009.

PLEMELJ, J.
"Die Grenzkreis-Uniformisierung analytischer Gebilde," *Monatshefte für Math. und Physik* 23 (1912), 297–304. **FM** 43, p. 519.
"Ein Abschätzungssatz der Potentialtheorie," *Publications Math. Univ. Belgrade* 2 (1933), 150–153. **FM** 59, p. 1133. **Zbl** 8, p. 357.

PLESSNER, A.
"Über das Verhalten analytischer Funktionen am Rande ihres Definitionsbereichs," *Jour. für Reine und Angew. Math.* 158 (1927), 219–227. **FM** 53, p. 284. **Zbl** 177, p. 106. **MR** 34 #4505.

POINCARÉ, H.
"Sur les fonctions à espaces lacunaires," *Amer. Jour. of Math.* 14 (1892), 201–221. **FM** 24, p. 388.
"Sur l'uniformisation des fonctions analytiques," *Acta Math.* 31 (1907), 1–63. **FM** 38, p. 452.

POLLARD, S.
"On the conditions for Cauchy's theorem," *Proc. Lon. Math. Soc.* (2) 21 (1923), 456–482. **FM** 49, p. 209.

PÓLYA, G.
"Über ganzwertige ganze Funktionen," *Rend. Circ. Mat. Palermo* 40 (1915), 1–16. **FM** 45, p. 655.
"Über Potenzreihen mit ganzzahligen Koeffizienten," *Math. Annalen* 77 (1916), 497–513. **FM** 46, p. 481.
"Bestimmung einer ganzen Funktion endlichen Geschlechts durch viererlei Stellen," *Mat. Tidsskr. B* (1921), 16–21. **FM** 48, p. 354.
"Über die Nullstellen sukzessiver Derivierten," *Math. Zeit.* 12 (1922), 36–60. **FM** 48, p. 370.
"Elementarer Beweis einer Thetaformel," *Sitzungsber. Preuss. Akad. Wiss. Berlin, Phys.-math. Kl.* (1927), 158–161. **FM** 53, p. 344.

"Untersuchungen über Lücken und Singularitäten von Potenzreihen," *Math. Zeit.* 29 (1929), 549–640; French résumés in *Boll. Un. Mat. Ital.* 8 (1929), 211–214 and *Atti del Congresso Internazionale, Bologna* (1928), Vol. III, 243–247. Part II, *Annals of Math.* (2) 34 (1933), 731–777. **Zbl** 8, p. 62. **FM** 55, p. 186; 56, p. 271; 59, p. 319.

"Über analytische Deformationen eines Rechtecks," *Annals of Math.* (2) 34 (1933), 617–620. **FM** 59, p. 348. **Zbl** 7, p. 169.

"Sur l'existence de fonctions entières satisfaisant à certaines conditions linéaires," *Trans. Amer. Math. Soc.* 50 (1941), 129–139. **FM** 67, p. 269. **Zbl** 25, p. 166. **MR** 2, p. 356.

PÓLYA, G. and SZEGÖ, G.

Aufgaben und Lehrsätze aus der Analysis, I and II. Die Grundlehren der mathematischen Wissenschaften in Einzeldarstellungen, Bd. 19, 20. Springer-Verlag (3rd ed. 1964), Berlin. **FM** 51, p. 173. **Zbl** 201, p. 381. **MR** 42 #6160 and 49 #8781. **BAMS** 34, p. 233.

Problems and Theorems in Analysis, vol. I (translated from German by D. Aeppli), Die Grundlehren der mathematischen Wissenschaften in Einzeldarstellungen, Bd. 193 (1972). Vol. II (translated from German by C. E. Billingheimer), *ibid.*, Bd. 216 (1976). Springer-Verlag, New York. **Zbl** 338 #00001, 311 #00002 and 359 #00003. **MR** 49 #8782 and 53 #2. **BAMS** 84, p. 53.

POMEY, É.

"Nouvelle démonstration du théorème de Dalembert," *Nouvelles Annales de Math.* (4) 5 (1905), 388–394. **FM** 36, p. 121.

POMMERENKE, Ch.

"Polynome und konforme Abbildung," *Monatshefte für Math.* 69 (1965), 58–61. **Zbl** 151, p. 93. **MR** 30 #4919.

Univalent Functions, Studia Mathematica Bd. 25, Vandenhoeck & Ruprecht (1975), Göttingen. **Zbl** 298 #30014.

POMPEIU, D.

(a) "Sur la continuité des fonctions de variables complexes," *Annales de la Faculté des Sciences de Toulouse* (2) 7 (1905), 264–315. **FM** 36, p. 454.

(b) "Sur l'extension du théorème des accroissements finis aux fonctions analytiques d'une variable complexe," *Rend. Circ. Mat. Palermo* 19 (1905), 309–313. **FM** 36, p. 454.

"Einige Sätze über monogene Funktionen," *Sitzungsber. Kaiserl. Akad. Wiss. Wien* 120 (1911), Zweite Abteilung, 1249–1250. **FM** 42, p. 422. (See also **FM** 57, p. 346 and **Zbl** 3, p. 300.)

"Sur une propriété des fonctions holomorphes," *Ann. Scuola Norm. Sup. Pisa* (2) 2 (1933), 227–230. **FM** 59, p. 1023. **Zbl** 6, p. 316.

PONOMAREV, S. P.

"On a condition of analyticity," *Siberian Math. Jour.* 11 (1970), 360–363. **MR** 41 #8636. **Zbl** 203, p. 72.

PORTER, M. B.

(a) "On functions defined by an infinite series of analytic functions of a complex variable," *Annals of Math.* (2) 6 (1904–1905), 45–48. **FM** 35, p. 397.

(b) "Concerning series of analytic functions," *Annals of Math.* (2) 6 (1904–1905), 190–192. **FM** 36, p. 459.

"On the polynomial convergents of a power series," *Annals of Math.* (2) 8 (1906–1907), 189–192. **FM** 38, p. 302.

PRACHAR, K.

"Über einige einfache Folgen und Reihen im Schulunterricht," *Elemente der Math.* 30 (1975), 36–39. **MR** 51 #1188.

PRINGSHEIM, A.

"Über Functionen, welche in gewissen Punkten endliche Differentialquotienten jeder endlichen Ordnung, aber keine Taylor'sche Reihenentwickelung besitzen," *Math. Annalen* 44 (1894), 41–56. **FM** 25, p. 389.

(a) "Zur Theorie der synektischen Functionen," *Sitzungsber. Kön. Bayer. Akad. Wiss. München* 26 (1896), 167–182. **FM** 27, p. 300.

(b) "Über Vereinfachungen in der elementaren Theorie der analytischen Functionen," *Math. Annalen* 47 (1896), 121–154. **FM** 27, p. 300.

"Zur Geschichte des Taylor'schen Lehrsatzes," *Biblio. Math.* (3) 1 (1900), 433–479. **FM** 31, p. 42.

"Über den Goursat'schen Beweis des Cauchy'schen Integralsatzes," *Trans. Amer. Math. Soc.* 2 (1901), 413–421. **FM** 32, p. 308.

"Zur Theorie der ganzen transcendenten Functionen," *Sitzungsber. Kön. Bayer. Akad. Wiss. München* 32 (1902), 163–192 and 295–304. **FM** 33, p. 416.

"Der Cauchy–Goursatsche Integralsatz und seine Übertragung auf reelle Kurven-Integrale," *Sitzungsber. Kön. Bayer. Akad. Wiss. München* 33 (1903), 673–682. **FM** 34, p. 339.

"Elementare Theorie der ganzen transcendenten Funktionen von endlicher Ordnung," *Math. Annalen* 58 (1904), 257–342. **FM** 35, p. 405.

"Elementare Funktionentheorie und komplexe Integration," *Sitzungsber. Bayer. Akad. Wiss. München* 49 (1920), 145–182 and 50 (1921), 255–258. **FM** 47, p. 267 and 48, p. 317.

"Kritisch-historische Bemerkungen zur Funktionentheorie," *Sitzungsber. Bayer. Akad. Wiss. München* 59 (1929), 95–112, 113–124, 281–306. **FM** 55, p. 760.

Vorlesungen über Zahlen- und Funktionenlehre Vol. II: *Funktionenlehre* (Part 1: *Grundlagen der Theorie der analytischen Funktionen einer komplexen Veränderlichen.* Part 2: *Eindeutige analytische Funktionen.*) Bibliotheca Mathematica Teubneriana, Bd. 29. B. G. Teubner (1925), (1932), Leipzig and Berlin. **FM** 51, p. 237 and 58, p. 296. **Zbl** 5, p. 199. **BAMS** 39, p. 487. Reprinted by Johnson Reprint Corp. (1968), New York. **MR** 40 #5407b.

PRINGSHEIM, A., FABER, G. and MOLK, J.

"Analyse algébrique," *Encyclopédie des sciences mathématiques pures et appliquées*, t. II, v. 2, fasc. 1, 1–93. Gauthier-Villars (1911), Paris. **FM** 40, p. 300 and 42, p. 458.

PRIVALOV, I. I.

(a) "Eine Erweiterung des Satzes von Vitali über Folgen analytischer Funktionen," *Math. Annalen* 93 (1924), 149–152. **FM** 50, p. 247.

(b) "Über einen Mittelwertsatz in der Theorie der analytischen Funktionen," *Recueil Math. Moscou* (= *Mat. Sbornik*) 32 (1924), 50–53 (Russian). **FM** 50, p. 251.

(c) "Über die Folgen von analytischen Funktionen," *Recueil Math. Moscou* (= *Mat. Sbornik*) 32 (1924), 45–49 (Russian). **FM** 50, p. 247.

(d) "Sur les suites de fonctions analytiques," *C.R. Acad. Sci. Paris* 178 (1924), 178–180. **FM** 50, p. 246.

"Sur certaines questions de la théorie des fonctions subharmoniques et des fonctions analytiques," *Recueil Math. Moscou* (= *Mat. Sbornik*) 41 (1935), 527–550 (Russian). **FM** 60, p. 1135. **Zbl** 11, p. 314.

Randeigenschaften Analytischer Funktionen (translated from Russian by J. Auth), VEB Deutscher Verlag der Wissenschaften (1956), Berlin. **MR** 18, p. 727. **Zbl** 73, p. 65.

PUCCIANO, G.

"Sulle condizioni di validità del teorema di Cauchy," *Giornale di Mat. di Battaglini* 47 [(2), 16] (1909), 55–64. **FM** 40, p. 440.

PUCCIO, L.

"Su di una dimostrazione puramente algebrica ed elementare del teorema fondamentale dell'algebra," *Bollettino di Mat.* (2) 12 (1933), 110–118. **FM** 59, p. 121.

RADEMACHER, H.

(a) "Über streckentreue und winkeltreue Abbildung," *Math. Zeit.* 4 (1919), 131–138. **FM** 47, p. 268.

(b) "Bemerkungen zu den Cauchy–Riemannschen Differentialgleichungen und zum Moreraschen Satz," *Math. Zeit.* 4 (1919), 177–185. **FM** 47, p. 268.

Lectures on Elementary Number Theory, Blaisdell Publishing Co. (1964), New York. Reprinted by Robert E. Krieger Publishing Company, Inc. (1977), Huntington. **MR** 30 #1079. **Zbl** 119, p. 278.

RADO, R.

"Über stetige Fortsetzung reeller Funktionen," *Sitzungsber. Bayer. Akad. Wiss. München* (1931), 81–84. **FM** 57, p. 300. **Zbl** 3, p. 154.

RADÓ, T.

(a) "Bemerkung zu einem Unitätssatze der konformen Abbildung," *Acta Litt. ac Scient. Univ. Hung. (Szeged)* 1 (1922/23), 101–103. **FM** 49, p. 247.

(b) "Über die Fundamentalabbildungen schlichter Gebiete," *Acta Litt. ac Scient. Univ. Hung. (Szeged)* 1 (1922/23), 240–251. **FM** 49, p. 712.

(c) "Zur Theorie der mehrdeutigen konformen Abbildungen," *Acta Litt. ac Scient. Univ. Hung. (Szeged)* 1 (1922/23), 55–64. **FM** 48, p. 1235.

(d) "Sur la représentation conforme de domaines variables," *Acta Litt. ac Scient. Univ. Hung. (Szeged)* 1 (1922/23), 180–186. **FM** 49, p. 247.

"Über die konformen Abbildungen schlichter Gebiete," *Acta Litt. ac Scient. Univ. Hung. (Szeged)* 2 (1924–26), 47–60. **FM** 50, pp. 255 and 640.

"Über eine nicht fortsetzbare Riemannsche Mannigfaltigkeit," *Math. Zeit.* 20 (1924), 1–6. **FM** 50, p. 255.

"Bemerkung über die konformen Abbildungen konvexer Gebiete," *Math. Annalen* 102 (1930), 428–429. **FM** 55, p. 208 and 57, p. 1425.

"On continuous transformations in the plane," *Fund. Math.* 27 (1936), 201–211. **FM** 62, p. 809. **Zbl** 15, p. 418.

Subharmonic Functions, Ergebnisse der Mathematik und ihrer Grenzgebiete, Bd. 5. Verlag Julius Springer (1937), Berlin. Reprinted by Chelsea Publishing Co. (1949), New York. **FM** 63, p. 458. **Zbl** 16, p. 249. **BAMS** 43, p. 758.

RADÓ, T. and RIESZ, F.

"Über die erste Randwertaufgabe für $\Delta u = 0$," *Math. Zeit.* 22 (1925), 41–44. **FM** 51, p. 361.

RÅDSTRÖM, H.

"On the iteration of analytic functions," *Math. Scand.* 1 (1953), 85–92. **MR** 15, p. 115. **Zbl** 50, p. 300.

RAJAGOPAL, C. T.
"On inequalities for analytic functions," *Amer. Math. Monthly* 60 (1953), 693–695. **Zbl** 53, p. 46. **MR** 15, p. 412.

RAYNOR, G. E.
"Isolated singular points of harmonic functions," *Bull. Amer. Math. Soc.* 32 (1926), 537–544. **FM** 52, p. 495.
"Note on the expansion of harmonic functions in the neighborhood of isolated singular points," *Annals of Math.* (2) 31 (1930), 35–42. **FM** 56, p. 1066.

READE, M. O.
"Some remarks on subharmonic functions," *Duke Math. Jour.* 10 (1943), 531–536. **MR** 5, p. 7.

REDHEFFER, R.
"The fundamental theorem of algebra," *Amer. Math. Monthly* 64 (1957), 582–585. **MR** 19, p. 537. **Zbl** 77, p. 25.
"What! Another note just on the fundamental theorem of algebra?", *Amer. Math. Monthly* 71 (1964), 180–185. **Zbl** 117, p. 38.
"The homotopy theorems of function theory," *Amer. Math. Monthly* 76 (1969), 778–787. **MR** 41 #7126. **Zbl** 194, p. 375.

REICH, E.
"Elementary proof of a theorem on conformal rigidity," *Proc. Amer. Math. Soc.* 17 (1966), 644–645. **MR** 33 #4247. **Zbl** 153, p. 100.

REICH, E. and WARSCHAWSKI, S. E.
"On canonical conformal maps of regions of arbitrary connectivity," *Pac. Jour. of Math.* 10 (1960), 965–985. **MR** 22 #8120. **Zbl** 91, p. 255.
"Canonical conformal maps onto a circular slit annulus," *Scripta Math.* 25 (1960), 137–146. **MR** 22 #12228. **Zbl** 99, p. 60.

REID, W. T.
"Green's lemma and related results," *Amer. Jour. of Math.* 63 (1941), 563–574. **FM** 67, p. 185. **Zbl** 25, p. 151. **MR** 3, p. 75.

REIJNIERSE, J. M.
"Sur la limite d'une suite de polynômes," *Nieuw Archief voor Wiskunde* (2) 19 (1938), 241–248. **FM** 64, p. 1056. **Zbl** 18, p. 207.

REMAK, R.
"Über winkeltreue und streckentreue Abbildung an einem Punkte und in der Ebene," *Rend. Circ. Mat. Palermo* 38 (1914), 193–246. **FM** 45, p. 674.
"Über potentialkonvexe Funktionen," *Math. Zeit.* 20 (1924), 126–130. **FM** 50, p. 333.
"Über die erste Randwertaufgabe der Potentialtheorie," *Jour. für Reine und Angew. Math.* 156 (1926), 227–230. **FM** 53, p. 460.

RENGEL, E.
"Über einige Schlitztheoreme der konformen Abbildung," *Schriften Math. Seminars und Inst. Angew. Math. Univ. Berlin* 1 (1933), 141–162. **FM** 59, p. 351. **Zbl** 7, p. 21.
"Existenzbeweise für schlichte Abbildungen mehrfach zusammenhängender Bereiche auf gewisse Normalbereiche," *Jahresber. Deutsch. Math. Verein.* 44 (1934), 51–55. **FM** 60, p. 286. **Zbl** 9, p. 173.

RENNIE, B. C.
"Repeated Riemann integrals," *Proc. Camb. Phil. Soc.* 76 (1974), 187–189. **Zbl** 283 #26005. **MR** 50 #4857.

RENTELN, M. VON
"Divisibility structure and finitely generated ideals in the disc algebra," *Monatshefte für Math.* 82 (1976), 51–56. **Zbl** 339 #30035. **MR** 54 #13576.

RICHARDS, I.
"More on rings on rings," *Bull. Amer. Math. Soc.* 74 (1968), 677. **MR** 37 #1980. **Zbl** 169, p. 409.

"Axioms for analytic functions," *Adv. in Math.* 5 (1970), 311–338. **Zbl** 205, p. 136. **MR** 55 #10701. (See also **Zbl** 298 #46052 and **MR** 51 #8413.)

RIDDER, J.
(a) "Über den Cauchyschen Integralsatz für reelle und komplexe Funktionen," *Math. Annalen* 102 (1930), 132–156. **FM** 55, p. 761.

(b) "Ein Satz über iterierte Integrale (*R*)," *Nieuw Archief voor Wiskunde* (2) 16 (1930), 43–45. **FM** 56, p. 914.

(c) "Ein Satz über iterierte Integrale und seine Anwendung zur Untersuchung der Analytizität von komplexen Funktionen," *Math. Zeit.* 31 (1930), 141–148. **FM** 55, p. 139.

"Harmonische, subharmonische und analytische Funktionen," *Ann. Scuola Norm. Sup. Pisa* (2) 9 (1940), 277–287. **FM** 66, p. 448. **Zbl** 24, p. 206. **MR** 3, p. 125.

"Über den Greenschen Satz in der Ebene," *Nieuw Archief voor Wiskunde* (2) 21 (1941), 28–32. **FM** 67, p. 185. **Zbl** 26, p. 207. **MR** 7, p. 376.

"Über harmonische Funktionen," *Nieuw Archief voor Wiskunde* (2) 22 (1946), 162–170. **Zbl** 61, p. 233. **MR** 8, p. 461.

"Ein einfacher Eindeutigkeitssatz für analytische Funktionen," *Kon. Nederl. Akad. Wetensch., Proc. Ser. A* 70 = *Indag. Math.* 29 (1967), 373–374. **Zbl** 184, p. 147. **MR** 36 #3992.

RIEMANN, B.
Gesammelte Mathematische Werke, edited by H. Weber with supplement (1902) by M. Noether and W. Wirtinger, B. G. Teubner (2nd ed. 1892), Berlin & Leipzig. **FM** 24, p. 21 and 33, p. 25. Reprinted, with English introduction by H. Lewy, by Dover Publications, Inc. (1953), New York. **Zbl** 53, p. 194. **MR** 14, p. 610. French translation by L. Laugel with a preface by Hermite and an address by F. Klein, Gauthier-Villars et Fils (1898), Paris. **FM** 29, p. 9. Reprinted by Librairie Scientifique et Technique Albert Blanchard (1968), Paris. **Zbl** 188, p. 301. **MR** 36 #4952.

RIESZ, F.
(a) "Sur les suites de fonctions analytiques," *Acta Litt. ac Scient. Univ. Hung.* (*Szeged*) 1 (1922/23), 88–97. **FM** 49, p. 713.

(b) "Sur les valeurs moyennes du module des fonctions harmoniques et des fonctions analytiques," *Acta Litt. ac Scient. Univ. Hung.* (*Szeged*) 1 (1922/23), 27–32. **FM** 48, p. 1268.

"Über subharmonische Funktionen und ihre Rolle in der Funktionentheorie und der Potentialtheorie," *Acta Litt. ac Scient. Univ. Hung.* (*Szeged*) 2 (1924–26), 87–100. **FM** 51, p. 363.

"Sur une inégalité de M. Littlewood dans la théorie des fonctions," *Proc. Lon. Math. Soc.* (2) 23 (1925), XXXVI–XXXIX. **FM** 51, p. 247.

RIESZ, F., and SZ.-NAGY, B.
Functional Analysis (translated from French by Leo F. Boron), Frederick Ungar

Publishing Co. (1955), New York. **MR** 17, p. 175. **Zbl** 64, p. 354 and 70, p. 109. **BAMS** 62, p. 423.

RIESZ, M.
"Sur certaines inégalités dans la théorie des fonctions avec quelques remarques sur les géométries non-euclidiennes," *Kungl. Fysiografiska Sällskapets i Lund Förhandlingar* Bd. 1, No. 4 (1931). **FM** 56, p. 983 and 57, p. 347. **Zbl** 3, p. 259.

RITT, J. F.
"On the conformal mapping of a region into a part of itself," *Annals of Math.* (2) 22 (1920–21), 157–160. **FM** 48, p. 363.

ROBERTS, J. H.
"Two-to-one transformations," *Duke Math. Jour.* 6 (1940), 256–262. FM 66, p. 982. **Zbl** 27, p. 96. **MR** 1, p. 319.

ROBINSON, R. M.
"On numerical bounds in Schottky's theorem," *Bull. Amer. Math. Soc.* 45 (1939), 907–910. **Zbl** 23, p. 51. **FM** 65, p. 332. **MR** 1, p. 112.

"Analytic functions in circular rings," *Duke Math. Jour.* 10 (1943), 341–354. **MR** 4, p. 241. **Zbl** 60, p. 218.

"Univalent majorants," *Trans. Amer. Math. Soc.* 61 (1947), 1–35. **MR** 8, p. 370. **Zbl** 32, p. 156.

"A curious trigonometric identity," *Amer. Math. Monthly* 64 (1957), 83–85. **MR** 18, p. 568. **Zbl** 81, p. 340.

ROBISON, G. R.
"A new approach to circular functions, π and lim $(\sin x)/x$" *Math. Mag.* 41 (1968), 66–70. **Zbl** 165, p. 400.

ROGOSINSKI, W. W.
"On subordinate functions," *Proc. Camb. Phil. Soc.* 35 (1939), 1–26. **Zbl** 20, p. 140. **FM** 65, p. 332.

"On the coefficients of subordinate functions," *Proc. Lon. Math. Soc.* (2) 48 (1943), 48–82. **MR** 5, p. 36. **Zbl** 28, p. 355.

"On the order of the derivatives of a function analytic in an angle," *Jour. Lon. Math. Soc.* 20 (1945), 100–109. **MR** 8, p. 324.

ROHDE, H.-W.
"Complex iterated radicals," *Amer. Math. Monthly* 81 (1974), 14–21. **Zbl** 282 #30008. **MR** 49 #3098.

ROSENBLATT, A.
"On the modulus of functions analytic in the unit circle," *Actas Acad. Ci. Lima* 8 (1945), 27–44 (Spanish). **MR** 8, p. 19.

ROSENBLOOM, P. C.
"An elementary constructive proof of the fundamental theorem of algebra," *Amer. Math. Monthly* 52 (1945), 562–570. **MR** 7, p. 295. **Zbl** 60, p. 47.

"L'itération des fonctions entières," *C.R. Acad. Sci. Paris* 227 (1948), 382–383. **MR** 10, p. 187. **Zbl** 30, p. 251.

"The fix-points of entire functions," *Medd. Lunds Univ. Mat. Sem.* (1952), tome supplémentaire, 186–192. **MR** 14, p. 546. **Zbl** 47, p. 316.

ROSENBLUM, M. and ROVNYAK, J.
"The factorization problem for non-negative operator valued functions," *Bull. Amer. Math. Soc.* 77 (1971), 287–318. **MR** 42 #8315. **Zbl** 214, p. 121.

ROTH, A.

"Approximationseigenschaften und Strahlengrenzwerte meromorpher und ganzer Funktionen," *Comm. Math. Helv.* 11 (1938), 77–125. **Zbl** 20, p. 235. **FM** 64, p. 1067.

"Meromorphe Approximationen," *Comm. Math. Helv.* 48 (1973), 151–176. **Zbl** 275 #30035. **MR** 57 #16615.

"Uniform and tangential approximation by meromorphic functions on closed sets," *Canad. Jour. Math.* 28 (1976), 104–110. **Zbl** 322 #30035. **MR** 57 #9978.

ROUCHÉ, E.

"Mémoire sur la série de Lagrange," *Jour. École Polyt.* 22 (1862), 193–224.

ROUX, D.

"Una dimostrazione del teorema fondamentale dell'algebra," *Boll. della Un. Mat. Ital.* (3) 14 (1959), 563–567. **Zbl** 93, p. 18. **MR** 22 #5721.

ROYDEN, H. L.

"A modification of the Neumann–Poincaré method for multiply connected regions," *Pac. Jour. of Math.* 2 (1952), 385–394. **MR** 14, p. 182. **Zbl** 47, p. 79.

"A generalization of Morera's theorem," *Annales Polonici Mathematici* 12 (1962), 199–202. **MR** 25 #5163. **Zbl** 105, p. 281.

RUBEL, L. A.

"Some applications of the Gauss–Lucas Theorem," *L'Enseignement Mathématique* (2) 12 (1966), 33–39. Errata, *ibid.* (2) 16 (1970), 113. **MR** 34 #4457. **Zbl** 144, p. 331.

"Lectures on vector spaces of analytic functions," pp. 191–269 of *Symposia on Theoretical Physics and Mathematics*, Vol. 9 (Symposium, Madras, 1968, A. Ramakrishnan, editor). Plenum Press (1969), New York. **Zbl** 179, p. 176. **MR** 56 #3315.

"Nonmaximal prime ideals in the ring of holomorphic functions," pp. 341–342 of *Mathematical Essays Dedicated to A. J. MacIntyre*, Ohio University Press (1970), Athens, Ohio. **Zbl** 209, p. 152. **MR** 42 #7919.

"How to use Runge's theorem," *L'Enseignement Mathématique* (2) 22 (1976), 185–190. Errata, *ibid.* (2) 23 (1977), 149. **Zbl** 347 #30033 and 357 #30027. **MR** 55 #3223.

RUBEL, L. A. and VENKATESWARAN, S.

"Simultaneous approximation and interpolation by entire functions," *Arch. der Math.* 27 (1976), 526–529. **Zbl** 348 #30029. **MR** 55 #5869.

RUBINSTEIN, Z.

"On analytic functions satisfying the mean value theorem and a conjecture of W. G. Dotson," *Math. Mag.* 42 (1969), 256–259. **MR** 41 #435. **Zbl** 184, p. 297.

RUDIN, W.

"A theorem on subharmonic functions," *Proc. Amer. Math. Soc.* 2 (1951), 209–212. **Zbl** 43, p. 104. **MR** 12, p. 825.

(a) "Some theorems on bounded analytic functions," *Trans. Amer. Math. Soc.* 78 (1955), 333–342. **MR** 16, p. 685. **Zbl** 64, p. 312.

(b) "Multiplicative groups of analytic functions," *Proc. Amer. Math. Soc.* 6 (1955), 83–87. **MR** 16, p. 578. **Zbl** 64, p. 312.

Principles of Mathematical Analysis, McGraw-Hill Book Co. (3rd ed. 1976), New York. **MR** 52 #5893. **Zbl** 346 #26002. **BAMS** 59, p. 572.

Real and Complex Analysis, McGraw-Hill Book Co. (1966), New York. **MR** 35 #1420. **Zbl** 142, p. 17. **BAMS** 74, p. 79. (2nd ed. 1974). **MR** 52 #10292. **Zbl** 333 #28001.

RUNG, D. C.

"Behavior of holomorphic functions in the unit disk on arcs of positive hyperbolic diameter," *Jour. Math. Kyoto Univ.* 8 (1968), 417–464. Corrigendum, *ibid.* 9 (1969), 337. **MR** 39 #4403 and 40 #2875. **Zbl** 176, p. 29 and 184, p. 305.

RUNGE, C.

"Zur Theorie der eindeutigen analytischen Functionen," *Acta Math.* 6 (1885), 229–244. **FM** 17, p. 379.

SAALSCHÜTZ, L.

Vorlesungen über die Bernoullischen Zahlen, ihren Zusammenhang mit den Secanten-Coefficienten und ihre wichtigeren Anwendungen, Verlag Julius Springer (1893), Berlin. **FM** 24, p. 236.

SAKASHITA, H.

"On the conformal mapping of nearly circular domains," *Bull. Kyoto Gakugei Univ. Ser. B*, No. 8 (1956), 10–14. **MR** 20 #5278.

SAKS, S.

"Sur une inégalité de la théorie des fonctions," *Acta Litt. ac Scient. Univ. Hung.* (*Szeged*) 4 (1928/29), 51–55. **FM** 54, p. 334.

"On subharmonic functions," *Acta Litt. ac. Scient. Univ. Hung.* (*Szeged*) 5 (1930–32), 187–193. **FM** 58, p. 518. **Zbl** 4, p. 116.

"Note on defining properties of harmonic functions," *Bull. Amer. Math. Soc.* 38 (1932), 380–382. **FM** 58, p. 1141. **Zbl** 5, p. 17.

Theory of the Integral (translated from Polish by L. C. Young), G. E. Stechert (2nd ed. 1937), New York. **Zbl** 17, p. 300. **FM** 63, p. 183. **BAMS** 44, p. 615. Reprinted by Dover Publications, Inc. (1964), New York. **MR** 29 #4850.

SAKS, S. and ZYGMUND, A.

Analytic Functions (translated from Polish by E. J. Scott), American Elsevier Publishing Co. (3rd ed. 1971), New York. **MR** 50 #2456. **Zbl** 199, p. 126. **FM** 64, p. 1048. **BAMS** 60, p. 495.

SALVADORI, M.

Esposizione della teoria delle somme di Gauss, Tipografia fratelli Nistri (1904), Pisa. **FM** 35, pp. 210 and 977.

SALZMANN, H. and ZELLER, K.

"Singularitäten unendlich oft differenzierbarer Funktionen," *Math. Zeit.* 62 (1955), 354–367. **MR** 17, p. 134. **Zbl** 64, p. 299.

SAMUELSSON, Å.

"A local mean value theorem for analytic functions," *Amer. Math. Monthly* 80 (1973), 45–46. **MR** 47 #8822. **Zbl** 261 #30002.

SANSONE, G. and GERRETSEN, J.

Lectures on the Theory of Functions of a Complex Variable I (1960), II (1969), P. Noordhoof Ltd., Groningen. **MR** 22 #4819. **Zbl** 93, p. 268. **MR** 41 #3714. **Zbl** 188, p. 381. **BAMS** 69, p. 39.

SAXER, W.

"Über die Picardschen Ausnahmewerte sukzessiver Derivierten," *Math. Zeit.* 17 (1923), 206–227. **FM** 49, p. 219.

"Sur les valeurs exceptionnelles des dérivées successives des fonctions méromorphes," *C.R. Acad. Sci. Paris* 182 (1926), 831–833. **FM** 52, p. 323.

"Über eine Verallgemeinerung des Satzes von Schottky," *Compositio Math.* 1 (1934), 207–216. **FM** 60, p. 255. **Zbl** 9, p. 216.

SCARF, H.
"Group invariant integration and the fundamental theorem of algebra," *Proc. Nat. Acad. Sci. U.S.A.* 38 (1952), 439–440. **Zbl** 46, p. 244. **MR** 14, p. 126.

SCHEEFFER, L.
"Beweis des Laurent'schen Satzes," *Acta Math.* 4 (1884), 375–380. **FM** 16, p. 350.

SCHEINBERG, S.
"Uniform approximation by entire functions," *Jour. d'Analyse Math.* 29 (1976), 16–18. **Zbl** 343 #41022.

SCHERING, E.
"Das Anschliessen einer Function an algebraische Functionen in unendlich vielen Stellen," *Göttingen Abhand.* 27 (1880). **FM** 12, p. 315.

SCHIFFER, M. M.
"On the modulus of doubly-connected domains," *Quart. Jour. of Math. (Oxford Series)* 17 (1946), 197–213. **MR** 8, p. 325. **Zbl** 60, p. 237.

SCHILLING, O. F. G.
"Ideal theory on open Riemann surfaces," *Bull. Amer. Math. Soc.* 52 (1946), 945–963. **MR** 8, p. 454.

SCHMIDT, H.
"Zur Faktorenzerlegung reeller Polynome einer Veränderlichen," *Jahresber. Deutsch. Math. Verein.* 46 (1936), 2te Abteilung, 64–67. **FM** 62, p. 68. **Zbl** 19, p. 101.

SCHÖNFLIES, A.
"Über einen Satz der Analysis Situs," *Göttingen Nachr.* (1899), 282–290. **FM** 30, p. 434.

"Beiträge zur Theorie der Punktmengen, I," *Math. Annalen* 58 (1904), 195–234. II, *Math. Annalen* 59 (1904), 129–160. III, *Math. Annalen* 62 (1906), 286–328. **FM** 34, p. 74; 35, p. 88; 37, p. 73.

"Die Entwicklung der Lehre von den Punktmannigfaltigkeiten, II," *Jahresber. Deutsch. Math. Verein., Der Ergänzungsbände* II. Bd. (1908), 1–331. **FM** 39, p. 95.

"Bemerkung zu dem vorstehenden Aufsatz des Herrn L. E. J. Brouwer," *Math. Annalen* 68 (1910), 435–444. **FM** 41, p. 543.

SCHÖNHAGE, A.
"Zur Ränderzuordnung beim Riemannschen Abbildungssatz," *Math. Annalen* 179 (1969), 97–100. **MR** 38 #1662. **Zbl** 167, p. 65.

SCHOTTKY, F.
"Über die conforme Abbildung mehrfach zusammenhängender ebener Flächen," *Jour. für Reine und Angew. Math.* 83 (1877), 300–351. **FM** 9, p. 584.

"Über den Picard'schen Satz und die Borel'schen Ungleichungen," *Sitzungsber. Kön. Preuss. Akad. Wiss. Berlin* (1904), 1244–1262 and (1906), 32–36. **FM** 35, p. 401 and 37, p. 417.

"Über das Cauchysche Integral," *Jour. für Reine und Angew. Math.* 146 (1916), 234–244. **FM** 46, p. 470.

"Problematische Punkte und die elementaren Sätze, die zum Beweis des Picardschen Theorems dienen," *Jour. für Reine und Angew. Math.* 147 (1917), 161–173. **FM** 46, p. 515.

SCHOU, E.
"Sur la théorie des fonctions entières," *C.R. Acad. Sci. Paris* 125 (1897), 763–764. **FM** 28, p. 361.

SCHRÖDER, E.
"Über unendlich viele Algorithmen zur Auflösung der Gleichungen," *Math. Annalen* 2 (1870), 317–363. **FM** 2, p. 42.

SCHUR, I.
"Über Potenzreihen, die im Innern des Einheitskreises beschränkt sind. I, II," *Jour. für Reine und Angew. Math.* 147 (1917), 205–232; 148 (1918), 122–145. **FM** 46, p. 475.

"Über die Gaussschen Summen," *Göttingen Nachr.* (1921), 147–153. **FM** 48, p. 130.

SCHUSKE, G. and THRON, W.
"On periodic infinite radicals," *Ann. Acad. Scient. Fennicae, Ser. A I*, No. 307 (1962). **MR** 25 #1385. **Zbl** 103, p. 44.

SCHWARZ, H. A.
Gesammelte Mathematische Abhandlungen, (2 vols.), Verlag Julius Springer (1890), Berlin. Reprinted by Chelsea Publishing Co. (1972), New York. **FM** 22, p. 31. **MR** 52 #13287. See *Sitzungsber. Berlin Math. Gesell.* 21 (1922), 47–52 (**FM** 48, p. 17) for works not included here.

SEIDEL, W.
"Über die Ränderzuordnung bei konformen Abbildungen," *Math. Annalen* 104 (1931), 182–243. **Zbl** 1, p. 19. **FM** 57, p. 398.

"Bibliography of numerical methods in conformal mapping," pp. 269–280 of *Construction and Applications of Conformal Maps*, National Bureau of Standards Applied Math. Series No. 18, E. F. Beckenbach editor, (1952), Washington, D.C. **MR** 14, p. 589. **Zbl** 49, p. 335.

SERRIN, J.
"A note on harmonic functions defined in a half plane," *Duke Math. Jour.* 23 (1956), 523–526. **Zbl** 73, p. 82. **MR** 18, p. 120.

SEVERINI, G.
"Alcune ricerche sulla teoria della funzioni analitiche," *Rend. del R. Istituto Lombardo di Scienze e Lettere* (2) 34 (1901), 891–904. **FM** 32, p. 395.

"Sopra una proprietà caratteristica delle funzioni armoniche," *Boll. Accad. Gioenia* (2) fasc. 16 (1911), 2–4. **FM** 42, p. 495.

SHAH, S. M.
"On the singularities of a class of functions on the unit circle," *Bull. Amer. Math. Soc.* 52 (1946), 1053–1056. **MR** 8, p. 322. **Zbl** 61, p. 146.

SHANKS, D.
"Two theorems of Gauss," *Pac. Jour. of Math.* 8 (1958), 609–612. **Zbl** 84, p. 60. **MR** 20 #5994.

SHAPIRO, H. S.
"Some function-theoretic problems motivated by the study of Banach algebras," pp. 95–113 of *Proceedings of the NRL Conference on Classical Function Theory* (1970), Washington, D.C., F. Gross, editor. **MR** 48 #2775. **Zbl** 286 #30030.

SHASHKEVICH, M.
"Remark on Féjer's inequality which is used in the Weierstrass factorization theorem," *L'Enseignement Mathématique* (2) 8 (1962), 279–280. **MR** 26 #6375. **Zbl** 111, p. 273. (See also **FM** 34, p. 477.)

SHEFFER, I. M.
"A proof of the fundamental theorem of algebra," *Bull. Amer. Math. Soc.* 35 (1929), 227–230. **FM** 55, p. 69.

SHELL, D. L.
"On the convergence of infinite exponentials," *Proc. Amer. Math. Soc.* 13 (1962), 678–681. **MR** 25 #5307. **Zbl** 109, p. 299.

SHIEH, HUI-CHUN
"On functions holomorphic in the unit circle with two exceptional values *B*," *Chinese Math.* 1 (1962), 226–234. **MR** 28 #223. **Zbl** 152, p. 68.

SHIELDS, A. L.
"On fixed points of commuting analytic functions," *Proc. Amer. Math. Soc.* 15 (1964), 703–706. **MR** 29 #2790. **Zbl** 129, p. 291.

SHIMIZU, T.
(a) "On the domain of indetermination of a regular function," *Japan. Jour. of Math.* 7 (1931), 275–300. **FM** 57, p. 358. **Zbl** 1, p. 397.

(b) "On equi-modular functions and their applications," *Proc. Phys.-Math. Soc. Japan* (3) 13 (1931), 79–92. **FM** 57, p. 369. **Zbl** 1, p. 344.

SHISHA, O.
"On sequences of power series with restricted coefficients," *Amer. Math. Monthly* 72 (1965), 533–537. **MR** 36 #5316. **Zbl** 158, p. 66.

SHOLANDER, M.
"On defining the sine and cosine," *Math. Mag.* 43 (1970), 72–75. **Zbl** 191, p. 349.

SIDON, S.
"Über einen Satz von Herrn Bohr," *Math. Zeit.* 26 (1927), 731–732. **FM** 53, p. 281.

SIERPINSKI, W.
"Sur une série potentielle qui, étant convergente en tout point de son cercle de convergence, représente sur ce cercle une fonction discontinue," *Rend. Circ. Mat. Palermo* 41 (1916), 187–190. **FM** 46, p. 1466.

SIMONART, F.
"Sur les transformations ponctuelles et leurs applications géométriques. Première partie: les transformations ponctuelles," *Annales de la Soc. Sci. de Bruxelles* 49A (1929), 121–144 and 50A (1930), 35–51. "Deuxième partie: la représentation conforme," *ibid.* 50A (1930), 81–104 and 51A (1931), 49–72. **FM** 55, p. 994; 56, pp. 1145, 982; 57, p. 1425. **Zbl** 2, p. 195.

"Limitations du module d'une fonction holomorphe et de sa dérivée sur un cercle quelconque," *Annales de la Soc. Sci. de Bruxelles* 50A (1930), 128–131. **FM** 56, p. 974.

SINCLAIR, A.
"A general solution for a class of approximation problems," *Pac. Jour. of Math.* 8 (1958), 857–866. **MR** 21 #2746. **Zbl** 93, p. 65.

"$|\epsilon(z)|$-closeness of approximation," *Pac. Jour. of Math.* 15 (1965), 1405–1413. **MR** 32 #7759. **Zbl** 142, p. 38.

SINGH, A. N.
"On the evaluation of a class of definite integrals," *Bull. Calcutta Math. Soc.* 15 (1925), 139–158. **FM** 51, p. 196.

SINGH, S. K.
"A note on a theorem of Fabry," *Jour. Karnatak Univ. Sci.* 9–10 (1964/65), 163–164. **MR** 32 #2563. **Zbl** 192, p. 169.

SMIRNOV, V. I. and LEBEDEV, N. A.
Functions of a Complex Variable, Constructive Theory (translated from Russian

by Scripta Technica, Ltd.), M.I.T. Press (1968), Cambridge, Massachusetts. **MR** 37 #5369. **Zbl** 164, p. 375.

SMITH, J. D.
"Determination of polynomials and entire functions," *Amer. Math. Monthly* 82 (1975), 822–825. **Zbl** 317 #30002. **MR** 52 #5966.

SPECHT, W.
"Algebraische Gleichungen mit reellen oder komplexen Koeffizienten," *Enzyklopädie der Mathematischen Wissenschaften* Bd. I, 1, Heft 3, Teil II, 1–76. B. G. Teubner (2nd ed. 1958), Stuttgart. **Zbl** 82, p. 245. **MR** 21 #5008. **BAMS** 66, p. 63.

"Eine Bemerkung zum Satze von Gauss–Lucas," *Jahresber. Deutsch. Math. Verein.* 62 (1959), 85–92. **Zbl** 87, p. 20. **MR** 22 #4821.

SPERNER, E.
"Neuer Beweis für die Invarianz der Dimensionszahl und des Gebietes," *Abhand. Math. Sem. Univ. Hamburg* 6 (1928), 265–272. **FM** 54, p. 614.

SPIVAK, M.
Calculus, W. A. Benjamin, Inc. (1967), New York. **Zbl** 159, p. 343.

SPRINGER, G.
"On Morera's theorem," *Amer. Math. Monthly* 64 (1957), 323–331. **MR** 19, p. 22. **Zbl** 77, p. 75.

SPRINGER, T. A.
"Der Satz von Cauchy," *Simon Stevin* 32 (1958), 68–79 (Dutch). **MR** 20 #4269. **Zbl** 87, p. 285.

STÄCKEL, P.
"Zur Theorie der eindeutigen Functionen," *Jour. für Reine und Angew. Math.* 106 (1890), 189–192. **FM** 22, p. 396.

"Integration durch imaginäres Gebiet. Ein Beitrag zur Geschichte der Functionentheorie," *Biblio. Math.* (3) 1 (1900), 109–128. **FM** 31, p. 43.

STARK, E. L.
"A new method of evaluating the sums of $\sum (-1)^{k+1}k^{-2p}$, $p = 1, 2, 3, \ldots$ and related series," *Elemente der Math.* 27 (1972), 32–34. **MR** 45 #3996. **Zbl** 225 #40002.

"The series $\sum_{k=1}^{\infty} k^{-s}$, $s = 2, 3, 4, \ldots$ once more," *Math. Mag.* 47 (1974), 197–202. **Zbl** 291 #40004. **MR** 50 #5261.

STEEN, L. and SEEBACH, J., Jr.
Counterexamples in Topology, Holt, Rinehart and Winston, Inc. (1970), New York. 2nd edition, Springer-Verlag (1978), New York. **MR** 42 #1040. **Zbl** 211, p. 544.

STEINITZ, E.
"Bedingt konvergente Reihen und konvexe Systeme," *Jour. für Reine und Angew. Math.* 143 (1913), 128–175. **FM** 44, p. 287.

STIELTJES, T. J., Jr.
"Recherches sur les fractions continues," *Annales de la Faculté des Sciences de Toulouse* 8 (1894), 1–122. **FM** 25, p. 326 and 35, p. 978.

STOÏLOW, S.
"Sur les transformations continues des espaces topologiques," *Bull. Math. Soc. Roum. Sci.* 35 (1934), 229–235. **FM** 59, p. 1254. **Zbl** 8, p. 326.

Leçons sur les principes topologiques de la théorie des fonctions analytiques, Gauthier-Villars (2nd ed. 1956), Paris. **FM** 64, p. 309. **Zbl** 121, p. 61. **MR** 32 #5899. **BAMS** 44, p. 758.

STONE, M. H.
"Hilbert space methods in conformal mapping," pp. 409–425 of *Proceedings of the International Symposium on Linear Spaces*, Jerusalem Academic Press, Jerusalem and Pergamon Press (1960), New York. **MR** 25 #178. **Zbl** 118, p. 299.
"Topological aspects of conformal mapping theory," pp. 343–346 of *General Topology and its Relations to Modern Analysis and Algebra*, Academic Press, New York and Publishing House of the Czechoslovakian Academy of Science (1962), Prague. **MR** 31 #323. **Zbl** 109, p. 303.

STOUT, E. L.
The Theory of Uniform Algebras, Bogden and Quigley Inc., Publishers (1971), Tarrytown-on-Hudson. **Zbl** 286 #46049. **MR** 54 #11066.

STOŻEK, W.
"Sur l'allure de fonctions harmoniques dans le voisinage d'un point exceptionnel," *C.R. Acad. Sci. Paris* 180 (1925), 727–728. **FM** 51, p. 362.
"Sur l'allure de fonctions harmoniques dans le voisinage d'un point exceptionnel," *Ann. Soc. Polon. Math.* 4 (1926), 52–58. **FM** 52, p. 497.

STROMBERG, K.
An Introduction to Classical Real Analysis, Prindle, Weber & Schmidt, Inc. (1980), Boston.

STUDY, E.
Vorlesungen über ausgewählte Gegenstände der Geometrie, Heft 2, "Konforme Abbildung einfach-zusammenhängender Bereiche," B. G. Teubner (1913), Leipzig & Berlin. **FM** 44, p. 755. **BAMS** 20, p. 443.

STYER, D. and MINDA, C. D.
"The use of the Monodromy Theorem and entire functions with nonvanishing derivative," *Amer. Math. Monthly* 81 (1974), 639–642. **Zbl** 288 #30003. **MR** 49 #5348.

SU, LI PI
"Rings of analytic functions on any subset of the complex plane," *Pac. Jour. of Math.* 42 (1972), 535–538. **MR** 47 #2076. **Zbl** 245 #30038.

SUETIN, P. K.
"The basic properties of Faber polynomials," *Uspehi Mat. Nauk* 19 (1964), No. 4 (118), 125–154. English translation in *Russian Mathematical Surveys* 19 (1964), No. 4, 121–149 (published by the London Mathematical Society). **MR** 29 #6029. **Zbl** 138, p. 294. (See also **MR** 56 #15896 and **Zbl** 313 #42018.)

ŠURA-BURA, M. R.
"Zur Theorie der bikompakten Räume," *Recueil Math. Moscou* (= *Mat. Sbornik*) new series 9 (1941), 385–388 (Russian). **MR** 3, p. 137. **Zbl** 25, p. 95. **FM** 67, p. 756.

ŚWIĄTKOWSKI, T.
"On the holomorphism of the integral with respect to a complex parameter," *Colloq. Math.* 16 (1967), 61–65. **MR** 35 #1765. **Zbl** 191, p. 369.

SZÁSZ, O.
"Über Potenzreihen und Bilinearformen," *Math. Zeit.* 4 (1919), 163–176. **FM** 47, p. 273.

SZEGÖ, G.
"Über orthogonale Polynome, die zu einer gegebenen Kurve der komplexen Ebene gehören," *Math. Zeit.* 9 (1921), 218–270. **FM** 48, p. 374.
"Conformal mapping of the interior of an ellipse onto a circle," *Amer. Math. Monthly* 57 (1950), 474–478. **MR** 12, p. 401. **Zbl** 41, p. 413.

TAKAHASHI, S.
"Bemerkung zu einer Arbeit von Herrn Radó," *Japan. Jour. of Math.* 7 (1930), 161–162. **FM** 56, p. 986.

TAKENAKA S.
"On the approximation of a function of two variables by polynomials," *Tôhoku Math. Jour.* 16 (1919), 16–25. **FM** 47, p. 251.

TAMMI, O.
"Note on Gutzmer's coefficient theorem," *Rev. Fac. Sci. Univ. Istanbul* A 22 (1957), 9–12. **MR** 20 #5293. **Zbl.** 83, p. 67.

TANAKA, C.
"On E. Lindelöf's theorem on the meromorphic function of bounded characteristic in the unit circle," *Mem. School Sci. Engineering Waseda Univ.* No. 28 (1964), 57–61. **Zbl** 284 #30022. **MR** 42 #6242.

"On the asymptotic values for regular functions with bounded characteristic," *Kōdai Math. Sem. Reports* 27 (1976), 94–115. **MR** 53 #3313. **Zbl** 329 #30026.

TAR, M. M.
"On a criterion for the analyticity of a function of a complex variable," *Ukrain. Math. Jour.* 21 (1969), 354–357 (1970). **MR** 40 #310. **Zbl** 201, p. 87.

"Über gewisse hinreichende Bedingungen der Analytizität von Funktionen einer komplexen Veränderlichen," *Dopovidī Akad. Nauk Ukraīn. RSR, Ser. A* (1971), 212–214, 284 (Ukrainian). **MR** 45 #510. **Zbl** 234 #30003.

"Analyticity of functions of a complex variable," *Ukrain. Math. Jour.* 25 (1973), 59–65. **Zbl** 261 #30001. **MR** 47 #3642.

TAYLOR, A. E.
(a) "A note on the Poisson kernel," *Amer. Math. Monthly* 57 (1950), 478–479. **MR** 12, p. 411. **Zbl** 37, p. 347.

(b) "New proofs of some theorems of Hardy by Banach space methods," *Math. Mag.* 23 (1950), 115–124. **MR** 11, p. 507. **Zbl** 35, p. 172.

TAYLOR, E. H.
"An extension of a theorem of Painlevé," *Bull. Amer. Math. Soc.* 19 (1913), 403–406. **FM** 44, p. 470.

TEISSIER DU CROS, F.
"Sur la convergence d'une série entière dont le terme général a sa partie réelle bornée en deux points de la circonférence-unité," *C.R. Acad. Sci. Paris* 219 (1944), 44–45. **Zbl** 60, p. 214. **MR** 7, p. 200.

TERASAKA, H.
"Über eine Verschärfung des Rouché–Lipkaschen Satzes," *Proc. Phys.-Math. Soc. Tokyo* (3) 11 (1929), 90–94. **FM** 55, p. 175.

THOMAS, J. M.
"The resolvents of a polynomial," *Amer. Math. Monthly* 47 (1940), 686–694. **MR** 2, p. 242. **Zbl** 61, p. 19. **FM** 66, p. 1199.

THRON, W. J.
"Convergence of infinite exponentials with complex elements," *Proc. Amer. Math. Soc.* 8 (1957), 1040–1043. **Zbl** 81, p. 57. **MR** 20 #2552.

TIDEMAN, M.
"Elementary proof of a uniqueness theorem for positive harmonic functions," *Nordisk Mat. Tidskrift* 2 (1954), 95–96. **Zbl** 55, p. 332. **MR** 16, p. 129.

TIETZE, H.
"Über einfach zusammenhängende Flächen und ihre Deformationen in sich,"

Sitzungsber. Kaiserl. Akad. Wiss. Wien 122 (1913), Zweite Abteilung, 1653–1658. **FM** 44, p. 562.

"Über Funktionen, die auf einer abgeschlossenen Menge stetig sind," *Jour. für Reine und Angew. Math.* 145 (1914), 9–14. **FM** 45, p. 628.

"Über stetige Kurven, Jordansche Kurvenbögen und geschlossene Jordansche Kurven," *Math. Zeit.* 5 (1919), 284–288. **FM** 47, p. 520.

TITCHMARSH, E.

The Theory of Functions, Clarendon Press (2nd ed. 1939), Oxford. **FM** 65, p. 302. **Zbl** 336 #30001. **MR** 33 #5850. **BAMS** 39, p. 650.

TOLSTOV, G. P.

"Über beschränkte Funktionen, die der Laplaceschen Differentialgleichung genügen," *Mat. Sbornik* new series (71) 29 (1951), 559–564 (Russian). **Zbl** 44, p. 100. **MR** 13, p. 943.

TOMIĆ, M.

"Sur un théorème de H. Bohr," *Math. Scand.* 11 (1962), 103–106. **MR** 31 #316. **Zbl** 109, p. 302.

TONELLI, L.

"Sopra une proprietà caratteristica delle funzioni armoniche," *Rend. Accad. d. Lincei Roma* (5) 18 (1909), 577–582. **FM** 40, p. 452.

"Sulla rappresentazione analitica delle funzioni di più variabili reali," *Rend. Circ. Mat. Palermo* 29 (1910), 1–36. **FM** 41, p. 490.

TROKHIMCHUK, Y. Y.

"Applications continues et fonctions analytiques," pp. 7–29 of *Fonctions d'une variable complexe. Problèmes contemporains*, A. I. Markouchevitch ed. (translated from Russian by L. Nicolas), Gauthier-Villars (1962), Paris. **MR** 22 #11099. **Zbl** 116, pp. 288 and 281.

Continuous Mappings and Conditions of Monogeneity (translated from Russian by R. Mandl), Israel Program for Scientific Translations, Ltd. (1964), Jerusalem. **MR** 33 #2801. **Zbl** 133, p. 38.

TSCHAKALOFF, L.

"Sur les singularités polaires des séries entières," *C.R. Acad. Bulgare Sci. Mat. Nat.* 1 (1948), 9–12. **MR** 10, p. 691. **Zbl** 36, p. 331. (See also **MR** 15, p. 514 and **Zbl** 53, p. 375.)

TSCHEBOTAREFF, N.

"Über die Realität von Nullstellen ganzer transzendenter Funktionen," *Math. Annalen* 99 (1928), 660–686. **FM** 54, p. 351.

TSUJI, M.

"On Blaschke's theorem," *Japan. Jour. of Math.* 3 (1926), 65–68. **FM** 52, p. 312.

(a) "On the theorems of Carathéodory and Lindelöf in the theory of conformal representation," *Japan. Jour. of Math.* 7 (1930), 91–99. **FM** 56, p. 983.

(b) "Theorems concerning Poisson integrals," *Japan. Jour. of Math.* 7 (1930), 227–253. **FM** 56, p. 1065.

"On a positive harmonic function in a half-plane," *Japan. Jour. of Math.* 15 (1939), 277–285. **FM** 65, p. 1239. **Zbl** 33, p. 54.

"A remark on Schottky's Theorem," *Jour. Math. Soc. of Japan* 1 (1949), 266–269. **MR** 11, p. 718. **Zbl** 38, p. 53.

"On a positive harmonic function in a half-plane," *Jour. Math. Soc. of Japan* 7 (1955), 76–78. **MR** 16, p. 819. **Zbl** 66, p. 87.

"On a non-negative subharmonic function in a half-plane," *Kōdai Math. Sem. Reports* 8 (1956), 134–141. **Zbl** 72, p. 315. **MR** 18, p. 885.

Potential Theory in Modern Function Theory, Maruzen Co., Ltd. (1959), Tokyo. Reprinted by Chelsea Publishing Co. (1975), New York. **MR** 54 #2990. **Zbl** 322 #30001.

"Huber's theorem on analytical mappings of a ring domain in a ring domain," *Comm. Math. Univ. Sancti Pauli* 8 (1960), 41–43. **MR** 22 #5727. **Zbl** 91, p. 255.

TUCKER, A. W.

"Some topological properties of disk and sphere," pp. 285–309 of *Proceedings of the First Canadian Mathematical Congress*, Montreal, 1945. University of Toronto Press (1946), Toronto. **MR** 8, p. 525. **Zbl** 61, p. 403.

TUMURA, Y.

"On extensions of Borel's theorem and Saxer–Csillag's theorem," *Proc. Phys.-Math. Soc. Japan* (3) 19 (1937), 29–35. **Zbl** 16, p. 217. **FM** 63, p. 275.

TUTSCHKE, W.

"Über eine spezielle Form des Maximumprinzips für holomorphe Funktionen in Zusammenhang mit einem Satz von Rossberg," *Annales Polonici Mathematici* 22 (1970), 341–343. **Zbl** 191, p. 373. **MR** 41 #3759.

UHERKA, D. J. and SERGOTT, A. M.

"On the continuous dependence of the roots of a polynomial on its coefficients," *Amer. Math. Monthly* 84 (1977), 368–370. **MR** 55 #8394.

UNKELBACH, H.

"Über die Randverzerrung bei konformer Abbildung," *Math. Zeit.* 43 (1938), 739–742. **FM** 64, p. 313. **Zbl** 18, p. 224.

"Über die Randverzerrung bei schlichter konformer Abbildung," *Math. Zeit.* 46 (1940), 329–336. **FM** 66, p. 364. **Zbl** 23, p. 50. **MR** 2, p. 83.

VAHLEN, K. TH.

"Der Fundamentalsatz der Algebra und die Auflösung der Gleichungen durch Quadratwurzeln," *Acta Math.* 21 (1897), 287–299. **FM** 28, p. 91.

VALIRON, G.

(a) "Remarque sur un théorème de M. Julia," *Bull. Sci. Math.* (2) 49 (1925), 68–73 and 270–275. **FM** 51, p. 261.

(b) *Fonctions entières et fonctions méromorphes d'une variable*, Mémorial des Sciences Mathématiques fasc. 2, Gauthier-Villars (1925), Paris. **FM** 51, p. 249.

(c) "Sur la formule d'interpolation de Lagrange," *Bull. Sci. Math.* (2) 49 (1925), 181–192, 203–224. **FM** 51, p. 250.

"Sur les théorèmes de MM. Bloch, Landau, Montel et Schottky," *C.R. Acad. Sci. Paris* 183 (1926), 728–730. **FM** 52, p. 324.

(a) "Sur un théorème de MM. Koebe et Landau," *Bull. Sci. Math.* (2) 51 (1927), 34–42. **FM** 53, p. 303.

(b) "Compléments au théorème de Picard–Julia," *Bull. Sci. Math.* (2) 51 (1927), 167–183. **FM** 53, p. 301.

"Le théorème de M. Picard et le complément de M. Julia," *Jour. de Math. Pures et Appliq.* (9) 7 (1928), 113–126. **FM** 54, p. 347.

(a) "Sur le théorème de M. Picard," *L'Enseignement Mathématique* 28 (1929), 55–59. **FM** 55, p. 193.

(b) *Familles normales et quasi-normales de fonctions méromorphes*, Mémorial des Sciences Mathématiques fasc. 38, Gauthier-Villars (1929), Paris. **FM** 55, p. 762.

(c) "Sur un théorème de M. Julia étendant le lemme de Schwarz," *Bull. Sci. Math.* (2) 53 (1929), 70–76. **FM** 55, p. 769.

"Sur le théorème de Bloch," *Rend. Circ. Mat. Palermo* 54 (1930), 76–82. **FM** 56, p. 269.

"Sur l'itération des fonctions holomorphes dans un demi-plan," *Bull. Sci. Math.* (2) 55 (1931), 105–128. **Zbl** 1, p. 281. **FM** 57, p. 381.

"Le théorème de Borel–Julia dans la théorie des fonctions méromorphes," pp. 270–279 of Vol. I of *Verhandlungen des Internationalen Mathematiker-Kongresses Zürich 1932*. Orell Füssli, Verlag (1932), Zürich & Leipzig. Reprinted by Kraus Reprint, Ltd. (1967), Nendeln/Liechtenstein. **FM** 58, p. 335. **Zbl** 6, p. 409. **MR** 53 #7713.

Sur les valeurs exceptionnelles des fonctions méromorphes et de leurs dérivées, Actualités Scientifiques et Industrielles, no. 570, Hermann et Cie. (1937), Paris. **Zbl** 19, p. 419. **FM** 63, p. 977. **BAMS** 47, p. 7.

"Sur l'approximation des nombres réels et un théorème de M. Teissier du Cros," *C.R. Acad. Sci. Paris* 219 (1944), 45–47. **Zbl** 60, p. 214. **MR** 7, p. 200.

"Directions de Julia et directions de Picard des fonctions entières," *Revista Unión Math. Argentina* 12 (1946), 49–54. **MR** 9, p. 84. **Zbl** 61, p. 150.

Fonctions analytiques, Presses Universitaires de France (1954), Paris. **MR** 15, p. 861. **Zbl** 55, p. 67. **BAMS** 61, p. 85.

Théorie des fonctions, Cours d'analyse mathématique. I. Masson et Cie. (2nd ed. 1955), Paris. **MR** 7, p. 283. **Zbl** 28, p. 208. **FM** 68, p. 99.

Vallée Poussin, C. de la

"Sur les applications de la notion de convergence uniforme," *Annales de la Soc. Sci. de Bruxelles* 17 (B) (1893), 323–330. **FM** 25, p. 463.

"Démonstration simplifiée du théorème fondamental de M. Montel sur les familles normales de fonctions," *Annals of Math.* (2) 17 (1915–16), 5–11. **FM** 45, p. 637.

"Application de l'intégrale de Lebesgue au problème de la représentation d'une aire simplement connexe sur un cercle," *Annales de la Soc. Sci. de Bruxelles* 50 (A) (1930), 23–34. **FM** 56, p. 983.

"Utilisation de la méthode du balayage dans la théorie de la représentation conforme," *Bull. Acad. Bruxelles* (5) 18 (1932), 385–400. **FM** 58, p. 1091. **Zbl** 4, p. 355.

Le potentiel logarithmique. Balayage et représentation conforme, Louvain Librairie Universitaire, Gauthier-Villars (1949), Paris. **Zbl** 37, p. 346.

Valson, C.-A.

La vie et les travaux du Baron Cauchy. Tome I: Partie historique, Tome II: Partie scientifique. Réimpression augmentée d'une introduction par René Taton. Librairie Scientifique et Technique Albert Blanchard (1970), Paris. **FM** 1, p. 15. **Zbl** 225 #01006. (See also **FM** 2, pp. 19–20.)

Van Yzeren, J.

"A rehabilitation of $(1 + z/n)^n$," *Amer. Math. Monthly* 77 (1970), 995–998. **Zbl** 204, p. 77.

Varopoulos, Th.

"Sur les valeurs exceptionnelles des fonctions dérivées et le théorème de M. Saxer," *Acta Math.* 51 (1928), 23–29. **FM** 53, p. 296.

Vasilesco, F.

"Sur les singularités des fonctions harmoniques," *Jour. de Math. Pures et Appliq.* (9) 9 (1930), 81–111. **FM** 56, p. 1067.

VAUGHN, H. E.
"On two theorems of plane topology," *Amer. Math. Monthly* 60 (1953), 462–468. **MR** 15, p. 146. **Zbl** 51, p. 146.

VEECH, W.
A Second Course in Complex Analysis, W. A. Benjamin, Inc. (1967), New York. **MR** 36 #3955. **Zbl** 145, p. 299.

VEEN, S. C., VAN
"Historische Besonderheiten über $\int_0^\infty e^{-x^2}\,dx$," *Mathematica, Zutphen.* B. (= *Simon Stevin*) 12 (1943), 1–4 (Dutch). **Zbl** 28, p. 194. **MR** 7, p. 354.

VENKATACHALIENGAR, K.
"Elementary proofs of the infinite product for sin z and allied formulae," *Amer. Math. Monthly* 69 (1962), 541–545. **Zbl** 100, p. 61.

VERBLUNSKY, S.
"On positive harmonic functions in a half-plane," *Proc. Camb. Phil. Soc.* 31 (1935), 482–507. **FM** 61, p. 523. **Zbl** 13, p. 158.

"A theorem on positive harmonic functions," *Proc. Camb. Phil. Soc.* 45 (1949), 207–212. **Zbl** 33, p. 63, **MR** 10, p. 533.

"On a fundamental formula of potential theory," *Jour. Lon. Math. Soc.* 26 (1951), 25–30. **Zbl** 42, p. 106. **MR** 12, p. 703.

VIJAYARAGHAVAN, T.
"A power series that converges and diverges at everywhere dense sets of points on its circle of convergence," *Jour. Indian Math. Soc.* (*N.S.*) 11 (1947), 69–72. **MR** 10, p. 25. **Zbl** 30, p. 153.

VILLAT, H.
"Sur le problème de Dirichlet relatif au cercle," *Bull. Soc. Math. France* 39 (1911), 443–456.

"Le problème de Dirichlet dans une aire annulaire," *Rend. Circ. Mat. Palermo* 33 (1912), 134–174. **FM** 43, p. 490.

"Sur la représentation conforme des aires doublement connexes," *Ann. Sci. École Norm. Sup.* (3) 38 (1921), 183–227. **FM** 48, p. 406.

VISSER, C.
"Sur les limites des fonctions bornées à la frontière de leur domaine d'existence," *Nieuw Archief voor Wiskunde* (2) 17 (1932), 147–150. **Zbl** 3, p. 402. **FM** 58, p. 329.

"Über beschränkte analytische Funktionen und die Randverhältnisse bei konformen Abbildungen," *Math. Annalen* 107 (1933), 28–39. **Zbl** 4, p. 405. **FM** 58, p. 361.

VITALI, G.
"Sopra le serie di funzioni analitiche," *Rend. del R. Istituto Lombardo di Scienze e Lettere* (2) 36 (1903), 772–774. **FM** 34, p. 419.

"Sopra le serie di funzioni analitiche," *Annali Mat. Pura Appl.* (3) 10 (1904), 65–82. **FM** 35, p. 397.

"Sul teorema fondamentale dell'algebra," *Periodico di Mat.* (4) 8 (1928), 102–105. **FM** 54, p. 117.

VITO, L. DE
"Sulla connessione dei campi piani," *Rend. Mat. e Appl.* (5) 16 (1957), 297–314. **Zbl** 105, p. 167. **MR** 19, p. 1186.

VIVANTI, G.

"Sul *valor medio* di Pringsheim e sulla sua applicazione alla teoria delle funzioni analitiche," *Math. Annalen* 58 (1904), 457–468. **FM** 35, p. 393.

VIZZINI, G.

"Su di un importante teorema sulle funzioni continue," *Bollettino di Mat.* (2) 11 (1932), 121–126. **FM** 58, p. 245.

VOLKOVYSKII, L., LUNDS, G. and ARAMANOVICH, I.

Collection of Problems on Complex Analysis (translated from Russian by J. Berry), Addison-Wesley Publishing Co. (1965), Reading. **MR** 49 #9164; 56 #3259. **Zbl** 357 #30001; 272 #30002.

VOLTERRA, V.

"Alcune osservazioni sopra proprietà atte ad individuare una funzione," *Rend. Accad. d. Lincei Roma* (5) 18 (1909), 263–266. **FM** 40, p. 453.

WALECKI, F.

"Démonstration du théorème de d'Alembert," *Nouvelles Annales de Math.* (3) 2 (1883), 241–248. **FM** 15, p. 45.

WALL, C. T. C.

A Geometric Introduction to Topology, Addison-Wesley Publishing Co. (1972), Reading. **Zbl** 261 #55001. **MR** 57 #17617.

WALSH, J. L.

"Über die Entwicklung einer analytischen Funktion nach Polynomen," *Math. Annalen* 96 (1927), 430–436. **FM** 52, p. 299.

"On the expansion of analytic functions in series of polynomials and in series of other analytic functions," *Trans. Amer. Math. Soc.* 30 (1928), 307–332. **FM** 54, p. 360.

"On approximation by rational functions to an arbitrary function of a complex variable," *Trans. Amer. Math. Soc.* 31 (1929), 477–502. **FM** 55, p. 762.

"On the overconvergence of sequences of polynomials of best approximation," *Trans. Amer. Math. Soc.* 32 (1930), 794–816. **FM** 56, p. 263.

"The Cauchy–Goursat theorem for rectifiable Jordan curves," *Proc. Nat. Acad. Sci. U.S.A.* 19 (1933), 540–541. **FM** 59, p. 312.

Approximation by polynomials in the complex domain, Mémorial des Sciences Mathématiques Fasc. 73, Gauthier-Villars (1935), Paris. **Zbl** 11, p. 298. **FM** 61, p. 319.

Interpolation and Approximation by Rational Functions in the Complex Domain, American Mathematical Society Colloquium Publications, Vol. 20 (5th ed. 1969), Providence. **MR** 36 #1672b. **Zbl** 146, p. 299. **FM** 61, p. 315. **BAMS** 42, p. 604.

"History of the Riemann Mapping Theorem," *Amer. Math. Monthly* 80 (1973), 270–276. **MR** 48 #2348. **Zbl** 273 #30003.

WALTHER, A.

"Maximum und Minimum einer harmonischen Funktion auf Kreisen," *Math. Zeit.* 11 (1921), 157–160. **FM** 48, p. 557.

WARSCHAWSKI, S. E.

"On differentiability at the boundary in conformal mapping," *Proc. Amer. Math. Soc.* 12 (1961), 614–620. **MR** 24 #A1374. **Zbl** 100, p. 288.

WATSON, G. N.

Complex Integration and Cauchy's Theorem, Cambridge Tracts in Mathematics and Physics, No. 15. Cambridge University Press (1914), London. Reprinted by Hafner Publishing Co. (1960), New York. **FM** 45, p. 664. **MR** 22 #12210.

WAVRE, R.

"Sur la réduction des domaines par une substitution à *m* variables complexes et l'existence d'un seul point invariant," *L'Enseignement Mathématique* 25 (1926), 218–234. **FM** 53, p. 554.

WAVRE, R. and BRUTTIN, A.

"Sur une transformation continue et l'existence d'un point invariant," *C.R. Acad. Sci. Paris* 183 (1926), 843–845. **FM** 53, p. 574.

WEIERSTRASS, K. T. W.

"Zur Theorie der eindeutigen analytischen Functionen," *Abhandl. Kön. Preuss. Akad. Wiss. Berlin* (1876), 11–60. **FM** 10, p. 282. French translation by É. Picard in *Ann. Sci. École Norm. Sup.* (2) 8 (1879), 111–150. **FM** 11, p. 264.

(a) "Zur Functionenlehre," *Monatsber. Kön. Preuss. Akad. Wiss. Berlin* (1880), 719–743. **FM** 12, p. 310. French translation in *Bull. Sci. Math.* (2) 5 (1881), 157–183. **FM** 13, p. 306.

(b) "Über einen functionstheoretischen Satz des Herrn G. Mittag-Leffler," *Monatsber. Kön. Preuss. Akad. Wiss. Berlin* (1880), 707–717. **FM** 12, p. 311. French translation in *Bull. Sci. Math.* (2) 5 (1881), 113–124. **FM** 13, p. 307.

"Über die analytische Darstellbarkeit sogennanter willkürlicher Functionen einer reellen Veränderlichen," *Sitzungsber. Kön. Preuss. Akad. Wiss. Berlin* (1885), 633–640, 789–806. **FM** 17, p. 384. French translation by L. Laugel in *Jour. de Math. Pures et Appliq.* (4) 2 (1886), 105–136. **FM** 18, p. 344.

"Neuer Beweis des Satzes, dass jede ganze rationale Function einer Veränderlichen dargestellt werden kann als ein Product aus linearen Functionen derselben Veränderlichen," *Sitzungsber. Kön. Preuss. Akad. Wiss. Berlin* (1891), 1085–1101. **FM** 23, p. 83. French translation by C. Bourlet in *Ann. Sci. École Norm. Sup.* (3) 12 (1895), 317–335. **FM** 26, p. 114.

Mathematische Werke I (1894); II (1895); III (1903), Meyer & Müller, Berlin. **FM** 25, p. 49; 26, p. 41; 34, p. 23. Reprinted by Georg Olms Verlagsbuchhandlung, Hildesheim and Johnson Reprint Corp., New York (1967). **MR** 36 #1279–1281.

"Zur Funktionentheorie," *Acta Math.* 45 (1925), 1–10. **FM** 50, p. 210. (This extract from an 1884 seminar is not in the *Werke*.)

WEIL, A.

Elliptic Functions according to Eisenstein and Kronecker, Ergebnisse der Mathematik und ihrer Grenzgebiete, Bd. 88. Springer-Verlag (1976), New York. **Zbl** 318 #33004.

WHITNEY, H.

"Note on Perron's solution of the Dirichlet problem," *Proc. Nat. Acad. Sci. U.S.A.* 18 (1932), 68–70. **FM** 58, p. 509. **Zbl** 3, p. 349.

WHITTAKER, J. M.

Interpolatory Function Theory, Cambridge Tracts in Mathematics and Physics No. 33, Cambridge University Press (1935), London. Reprinted by Stechert-Hafner Service Agency, Inc. (1972), New York. **Zbl** 12, p. 155. **FM** 61, p. 331. **MR** 32 #2798. **BAMS** 42, p. 305.

(a) "An inequality involving derivatives," *Bull. Lon. Math. Soc.* 2 (1970), 297–300. **MR** 43 #488. **Zbl** 206, p. 87.

(b) "The basis of Vitali's theorem," pp. 353–357 of *Mathematical Essays Dedicated to A. J. Macintyre*, Ohio University Press (1970), Athens, Ohio. **MR** 43 #2200. **Zbl** 209, p. 100.

WHYBURN, G. T.

"On the construction of simple arcs," *Amer. Jour. of Math.* 54 (1932), 518–524. **FM** 32, p. 640. **Zbl** 5, p. 27.

"An open mapping approach to Hurwitz's theorem," *Trans. Amer. Math. Soc.* 71 (1951), 113–119. **Zbl** 43, p. 178. **MR** 13, p. 149.

"Developments in topological analysis," *Fund. Math.* 50 (1961–62), 305–318. **MR** 29 #3651. **Zbl** 100, p. 67.

Topological Analysis, Princeton Mathematical Series, No. 23. Princeton University Press (2nd ed. 1964), Princeton. **MR** 29 #2758. **Zbl** 186, p. 559. **BAMS** 66, p. 255.

"What is a curve?", pp. 23–38 of *Studies in Modern Topology*, MAA Studies in Math. vol. 5, (P. J. Hilton, editor). Mathematical Association of America and Prentice-Hall (1968), Englewood Cliffs. **Zbl** 187, p. 448. **MR** 37 #895.

WHYBURN, W. M.

"A connectedness theorem in abstract sets," *Bull. Amer. Math. Soc.* 41 (1935), 365–366. **FM** 61, p. 630. **Zbl** 12, p. 57.

WIDDER, D. V.

"Functions harmonic in a strip," *Proc. Amer. Math. Soc.* 12 (1961), 67–72. **MR** 24 #A2674. **Zbl** 96, p. 77.

WIGERT, S.

"Sur un théorème concernant les fonctions entières," *Arkiv för Mat., Astro. och Fysik* 11, No. 21 (1916). **FM** 46, p. 505.

"Sur le théorème fondamental de l'algèbre," *Arkiv för Mat., Astro. och Fysik* 25 B, No. 17 (1936). **FM** 63, p. 856. **Zbl** 15, p. 97.

WIGNER, E. P.

"Simplified derivation of the properties of elementary transcendentals," *Amer. Math. Monthly* 59 (1952), 669–683. **MR** 14, p. 460. **Zbl** 48, p. 53.

WILKOSZ, W.

"Sur un point fondamental de la théorie du potentiel," *C.R. Acad. Sci. Paris* 174 (1922), 435–437. **FM** 48, p. 556.

Les propriétés topologiques du plan euclidien, Mémorial des Sciences Mathématiques Fasc. 45, Gauthier-Villars (1931), Paris. **FM** 56, p. 1142. **Zbl** 1, p. 171.

"Sur le théorème intégral de Cauchy," *Ann. Soc. Polon. Math.* 11 (1933), 19–27 and 56–57. **Zbl** 7, p. 105. **FM** 59, p. 1020.

WILLE, F.

"Verallgemeinerung der Sätze von Borsuk–Ulam und Lusternik–Schnirelmann–Borsuk auf nichtantipodische Punktpaare," *Monatshefte für Math.* 74 (1970), 351–370. **MR** 43 #4004. **Zbl** 218 #52007.

WILLIAMS, K. S.

"On $\sum_{n=1}^{\infty}(1/n^{2k})$," *Math. Mag.* 44 (1971), 273–276. **MR** 45 #3997. **Zbl** 224 #40008.

WILLIAMS, R. K.

"A note on conformality," *Amer. Math. Monthly* 80 (1973), 299–300. **Zbl** 273 #30004. **MR** 48 #506.

"On the linearity of one-to-one entire functions," *Delta* (*Waukesha*) 6 (1976), 72–77. **Zbl** 347 #30011. **MR** 55 #3256.

WINTERNITZ, A.

"Beweis für die Invarianz des ebenen Gebiets," *Math. Zeit.* 26 (1927), 165–169. **FM** 53, p. 564.

WOLF, F.
"Ein Eindeutigkeitssatz für analytische Funktionen," *Math. Annalen* 117 (1940), 383. **MR** 2, p. 81. **Zbl** 23, p. 49. **FM** 66, p. 363.
"On majorants of subharmonic and analytic functions," *Bull. Amer. Math. Soc.* 48 (1942), 925–932. **Zbl** 61, p. 231. **MR** 4, p. 144.
"Extension of analytic functions," *Duke Math. Jour.* 14 (1947), 877–887. **Zbl** 29, p. 123. **MR** 9, p. 420.

WOLFENSTEIN, S.
"Proof of the fundamental theorem of algebra," *Amer. Math. Monthly* 74 (1967), 853–854. **Zbl** 153, p. 96.

WOLFF, J.
"Sur les séries $\sum A_k/(z - a_k)$," *C.R. Acad. Sci. Paris* 173 (1921), 1327–1328. **FM** 48, p. 320.
"Über die Loomansche Erweiterung eines Satzes von Pompéiu," *Nieuw Archief voor Wiskunde* (2) 14 (1924), 337–339 and pp. 457–459 of vol. 1 of *Proceedings of the International Mathematical Congress, Toronto* (1924), University of Toronto Press. Reprinted by Kraus Reprint, Ltd. (1967), Nendeln/Liechtenstein. **FM** 50, p. 250 and 54, p. 328.
(a) "Sur une généralisation d'un théorème de Schwarz," *C.R. Acad. Sci. Paris* 182 (1926), 918–920 and 183 (1926), 500–502. **FM** 52, p. 309.
(b) "Sur l'itération des fonctions bornées," *C.R. Acad. Sci. Paris* 182 (1926), 42–43 and 200–201. **FM** 52, p. 309.
"Über die Iteration derjenigen in einem Gebiete regulären Funktionen, deren Werte dem Gebiete angehören," *Math. Zeit.* 26 (1927), 125–127. **FM** 53, p. 304.
"Sur les limites radiales d'une fonction holomorphe dans un cercle," *Bull. Soc. Math. France* 56 (1928), 167–173. **FM** 54, p. 346.
"Sur l'itération des fonctions holomorphes dans un demi-plan," *Bull. Soc. Math. France* 57 (1929), 195–203. **FM** 55, p. 769.
(a) "Sur la représentation conforme," *Kon. Akad. Wetensch. Amsterdam, Proc.* 33 (1930), 96–97. **FM** 56, p. 983.
(b) "Quelques propriétés des fonctions holomorphes dans un demi-plan dont toutes les valeurs sont dans ce demi-plan," *Kon. Akad. Wetensch. Amsterdam, Proc.* 33 (1930), 1185–1188. **FM** 56, p. 983.

WOLFF, J. and DE KOK, F.
"Les fonctions holomorphes à partie réelle positive et l'intégral de Stieltjes," *Bull. Soc. Math. France* 60 (1932), 221–227. **Zbl** 6, p. 171. **FM** 59, p. 324.

WÖLFFING, E.
Mathematischer Bücherschatz. I. Teil: Reine Mathematik, B. G. Teubner (1903), Leipzig. **FM** 34, p. 46. **BAMS** 10, p. 261.
"Über die bibliographischen Hilfsmittel der Mathematik," *Jahresber. Deutsch. Math. Verein.* 12 (1903), 408–426. **FM** 34, p. 47.

YANAGIHARA, N.
"On the difference of positive harmonic functions," *Jour. College Arts Sci. Chiba Univ.* 4 (1965), 235–239. **MR** 36 #3993.

YANG, CHUNG-CHUN
"On the zeros of an entire function and its second derivative," *Rend. Accad. d. Lincei Roma* (8) 49 (1970), 27–29. **MR** 45 #3705. **Zbl** 218, p. 181.

YANG LO and CHANG KUAN-HEO
"Recherches sur la normalité des familles de fonctions analytiques à des valeurs multiples. I. Un nouveau critère et quelques applications," *Scientia Sinica* 14 (1965), 1258–1271. **MR** 33 #7550. **Zbl** 166, p. 326. "II. Généralisations," *ibid.* 15 (1966), 433–453. **MR** 34 #2843. **Zbl** 173, p. 320.

YOUNG, W. H.

"On parametric integration," *Monatshefte für Math. und Physik* 21 (1910), 125–149. **FM** 41, p. 325.

"On the fundamental theorem in the theory of functions of a complex variable," *Proc. Lon. Math. Soc.* (2) 10 (1912), 1–6. **FM** 42, p. 421.

YÛJÔBÔ, Z.

"A theorem concerning subharmonic functions defined in a strip domain," *Comm. Math. Univ. Sancti Pauli* 1 (1952), 1–4. **Zbl** 48, p. 341. **MR** 15, p. 526.

YUŠKEVIČ, A. P.

"On the origins of Cauchy's definition of the integral," *Akad. Nauk SSSR. Trudy Inst. Istorii Estestvoznaniya* 1 (1947), 373–411 (Russian). **MR** 11, p. 572.

ZALCMAN, L.

(a) *Analytic Capacity and Rational Approximation*, Lecture Notes in Mathematics Vol. 50, Springer-Verlag (1968), New York. **MR** 37 #3018. **Zbl** 171, p. 37.

(b) "Null sets for a class of analytic functions," *Amer. Math. Monthly* 75 (1968), 462–470. **MR** 37 #6448. **Zbl** 181, p. 351.

"Analyticity and the Pompeiu problem," *Arch. Rational Mech. and Anal.* 47 (1972), 237–254. **Zbl** 251 #30047. **MR** 50 #582.

"Real proofs of complex theorems (and vice-versa)," *Amer. Math. Monthly* 81 (1974), 115–137 and 83 (1976), 84–85. **Zbl** 279 #30001. **MR** 48 #6370.

"A heuristic principle in complex function theory," *Amer. Math. Monthly* 82 (1975), 813–817. **Zbl** 315 #30036. **MR** 52 #757.

"Picard's theorem without tears," *Amer. Math. Monthly* 85 (1978), 265–268.

ZARANKIEWICZ, K.

"Sur la représentation conforme d'un domaine doublement connexe sur un anneau circulaire," *C.R. Acad. Sci. Paris* 198 (1934), 1347–1349. **Zbl** 9, p. 26. **FM** 60, p. 286.

"Über ein numerisches Verfahren zur konformen Abbildung zweifach zusammenhängender Gebiete," *Zeit. für Angew. Math. und Mech.* 14 (1934), 97–104. **Zbl** 9, p. 26. **FM** 60, p. 1207.

ZAREMBA, S.

"Contributions à la théorie d'une équation fonctionelle de la physique," *Rend. Circ. Mat. Palermo* 19 (1905), 140–150. **FM** 36, p. 822.

"Sur l'unicité de la solution du problème de Dirichlet," *Krakau Anz.* 1 (1909), 561–564. **FM** 40, p. 452.

"Sur une propriété générale des fonctions harmoniques," pp. 171–176 of *Conférences de la Réunion Internationale des Mathématiciens, Paris, 1937*. Gauthier-Villars (1939), Paris. **FM** 64, p. 469.

ZASSENHAUS, H.

"On the fundamental theorem of algebra," *Amer. Math. Monthly* 74 (1967), 485–497 and 75 (1969), 827. **MR** 36 #2605. **Zbl** 145, p. 279.

ZIN, G.

"Esistenza e rappresentazione di funzioni analitiche, le quali, su una curva di Jordan, si reducono a una funzione assegnata," *Annali Mat. Pura Appl.* (4) 34 (1953), 365–405. **MR** 14, p. 1073. **Zbl** 51, p. 309.

ZINTERHOF, P.

"Konstruktion von schlichten Funktionen mit unendlich vielen Fixpunkten," *Rend. Ist. Mat. Univ. Trieste* 3 (1971), 125–134 (1972). **MR** 46 #9331. **Zbl** 225 #30018. (See also **MR** 52 #8405 and **Zbl** 353 #30015.)

ZORETTI, L.

Leçons sur le prolongement analytique, Gauthier-Villars (1911), Paris. **FM** 42, p. 414. **BAMS** 20, p. 321.

Name Index

Abel, N. H. 59
Abian, A. 118, 189, 190, 463
Achiezer, N. I. 406, 409, 463
Adel'son-Vel'skiĭ, G. M. 183, 463
Aeppli, D. 521
Agmon, S. 407, 463
Agostini, A. 118, 463
Agoston, M. K. 115, 463
Aguiló Fuster, R. 410, 463
Ahlfors, L. V. 41, 159, 187, 189, 216, 292, 338, 356, 357, 458, 461, 464
Åkerberg, B. 185, 464
Albrecht, R. 357, 464
Alexander, J. W., II 464
Alling, N. 356, 410, 464
Anger, G. 343, 464
Anonyme, U. 408, 464
Antoine, L. 104, 465
Apostol, T. M. 81, 465
Arakélian, N. U. 291, 465
Aramovich, I. 539
Arsove, M. G. 51, 338, 340, 341, 406, 465
Artin, E. 117, 465
Arzelà, C. 252
Ascoli, G. 252, 253, 410, 465
Aumann, G. 246, 254, 465
Aumann, J. 491
Auth, J. 523
Ayoub, R. 82, 465

Bachmann, P. 408, 465
Bagemihl, F. 339, 341, 466, 500
Baire, R. 10, 18, 166, 227
Baker, A. 82, 466
Baker, I. N. 217, 255, 466
Balk, M. B. 291, 466
Banach, S. 186, 407, 466
Bank, S. B. 291, 466
Barrow, D. F. 255, 466
Barth, K. F. 254, 466
Bašmakova, I. G. 118, 466
Basye, R. E. 356, 466
Bear, H. S. 187, 467
Beardon, A. F. 254, 460, 467
Beck, A. 356, 467
Beckenbach, E. F. 341, 342, 467, 530
Beesack, P. R. 406, 467
Behan, D. F. 255, 467
Behnke, H. 82, 190, 253, 406, 409, 410, 467
Bellman, R. 408, 409, 467
Beltrami, E. 480
Bendixson, J. 189, 467
Berenstein, C. A. 188, 467
Berenšteĭn, M. N. 51, 467
Bergman, S. 338, 358, 467
Bernardi, S. D. 186, 215, 216, 217, 340, 357, 467
Berndt, B. C. 82, 467
Bernhart, J. 11
Bernoulli, J. 75, 76, 82
Bernstein, F. 118, 468
Bernstein, S. N. 185, 253, 388, 409, 468
Bernstein, V. 407, 468
Berry, J. 539
Bers, L. 356, 468, 500
Besicovitch, A. S. 291, 468
Beurling, A. 159, 468
Bieberbach, L. 10, 11, 50, 118, 215, 232, 251, 252, 254, 289, 290, 291, 337, 338, 341, 357, 407, 408, 457, 459, 460, 463, 468–469
Biehler, Ch. 67, 469
Biermann, O. 406, 469
Billingheimer, C. E. 521
Birkhoff, G. 469
Bishop, E. 298, 303, 469
Blaschke, W. 215, 219, 251, 254, 342, 469
Bloch, A. 232, 253, 408, 457, 460, 469
Blumenthal, O. 458, 470
Blutel, E. 117, 470
Boas, R. P. Jr. 292, 406, 409, 470
Boboc, N. 470
Bôcher, M. 185, 253, 342, 410, 470
Bochner, S. 458, 470
Boggio, T. 185, 470
Bohr, H. 10, 37, 50, 340, 458, 461, 470
Bonsall, F. F. 187, 460, 470
Borel, É. 116, 210, 213, 291, 458, 460, 470, 471
Boron, L. F. 502, 525
Borsuk, K. 9, 38, 95, 110, 116, 119, 407, 471
Bouligand, G. 188, 291, 471

Bourbaki, N. 60, 471
Bourlet, C. 540
Boutroux, P. 184, 457, 471, 517
Bowen, N. A. 251, 254, 460, 471
Brawn, F. T. 253, 471
Brelot, M. 184, 189, 190, 343, 410, 471–472
Brickman, L. 406, 472
Brill, A. 183, 472
Brillinger, D. R. 472
Brisse, Ch. 117, 472
Brodovich, M. T. 51, 472
Broggi, U. 407, 408, 472
Bromwich, T. J. I'A. 53, 81, 255, 472
Brouwer, L. E. J. 39, 109, 115, 116, 118, 119, 472–473
Browder, F. E. 190, 290, 407, 473
Brown, A. B. 118, 463, 473
Brown, L. 188, 291, 473
Brown, R. F. 115, 473
Bruckner, A. M. 253, 292, 473
Bruné, C. 246, 473
Bruttin, A. 108, 540
Buckner, G. 12
Buhl, A. 290, 473
Burdick, D. 189, 473
Bureau, F. 419, 458, 473–474
Burgers, W. A. M. 190, 474
Burkhardt, H. 81, 82, 183, 185, 186, 406, 408, 463, 474
Butler, R. 184, 474

Cairns, S. S. 341, 474
Caldwell, L. 11
Cantor, D. G. 215, 474
Cantor, G. 118
Carathéodory, C. 10, 11, 40, 159, 162, 186, 189, 190, 192, 193, 194, 198, 210, 215, 216, 217, 227, 246, 251, 252, 254, 284, 293, 298, 303, 304, 309, 311, 315, 317, 337, 338, 339, 343, 458, 459, 460, 465, 474, 475
Carleman, T. 10, 158, 188, 273, 276, 291, 339, 356, 475, 491
Carleson, L. 290, 341, 475
Carlson, F. 10, 187, 475
Cartwright, M. L. 163, 254, 444, 457, 460, 475
Casorati, F. 361, 406, 409, 438, 475–476
Cauchy, A.-L. 51, 52, 57, 71, 77, 80, 82, 120, 135, 168, 170, 183, 186, 396, 406, 407, 408, 463, 476
Cauer, W. 292, 476
Cazzaniga, P. 409, 476
Čech, E. 9, 115, 116, 341, 476
Černý, I. 115, 264, 476
Chang, Kuan-heo 457, 542
Chazy, J. 517
Chessin, A. S. 188, 476
Choquet, G. 341, 476
Chou, Hsin-ti 339, 477
Christoffel, E. B. 338
Cimmino, G. 477
Cioranesco, N. 477
Cisotti, U. 410, 477
Clowes, W. 12
Clunie, J. 458, 477
Cohen, P. J. 410, 477
Collingwood, E. F. 163, 187, 190, 338, 339, 340, 406, 477
Connell, E. H. 183, 477
Consiglio, A. 410, 477
Conway, J. B. 39, 253, 290, 333, 342, 356, 457, 477
Corominas, F. 166, 477
Corput, J. G. van der 117, 477
Corrádi, K. 477
Courant, R. 338, 339, 342, 475, 478
Cowell, P. 478
Cowling, V. F. 217, 478
Craig, C. Jr. 255, 478
Craven, B. D. 341, 478
Cremer, H. 255, 357, 478
Cremona, L. 480
Császár, Á. 115, 478
Csillag, P. 10, 458, 478
Cunningham, F. Jr. 253, 478
Curtiss, D. R. 50, 188, 190, 478
Curtiss, J. H. 340, 478

d'Alembert, J. 118
Darboux, J. G. 185, 188, 310, 407, 479
Darst, R. 186, 479
Dauben, J. W. 118, 479
Davis, Marie 12
Davis, M. 503
Deaux, R. 184, 479
de Branges, L. 479
Dedekind, R. 14
Deeds, J. B. 479
Dehn, M. 489
de Kok, F. 216, 292, 479, 542
Delange, H. 341, 479
Delcourte, J. 184, 479
Delsarte, J. 188, 479
Demin, E. 189, 479

Demtchenko, B. 410, 479
Denjoy, A. 249, 252, 289, 339, 341, 461, 479–480
de Possel, R. 10, 357, 480
Derrick, W. R. 341, 480
Deuring, M. 341, 480
Dienes, P. 252, 407, 480
Dieudonné, J. 40, 408, 480
Dinghas, A. 216, 292, 337, 340, 410, 459, 480
Dini, U. 409, 410, 480
Dirichlet, P. G. L. 82, 408
Dixon, J. D. 229, 481
Doetsch, G. 187, 481
Donoghue, W. F. Jr. 166, 409, 460, 481
Dörge, K. 117, 481
Douglas, J. 338, 481
Douglas, L. 12
Duffin, R. J. 10, 185, 481
Dufresnoy, J. 407, 481
Dulst, D. van 190, 481
Dvoretsky, A. 81, 481
Dym, H. 408, 409, 481
Dzewas, J. 60, 481

Eaton, J. E. 117, 481
Eberlein, W. F. 81, 253, 481
Echols, W. H. 189, 481
Edrei, A. 458, 482
Eggleston, H. G. 133, 460, 482
Eilenberg, S. 9, 99, 100, 115, 116, 118, 482
Eisenack, G. 115, 119, 482
Eljoseph, N. 255, 482
Elkins, J. M. 217, 482
Ely, G. S. 82, 482
Emig, P. 516
Epstein, B. 338, 482
Erdös, P. 81, 481, 482
Essén, M. 292, 482
Estermann, T. 117, 291, 341, 457, 459, 482
Ettlinger, H. J. 51, 183, 291, 482–483
Euler, L. 15, 64, 82, 408, 463
Eustice, D. J. 255, 483
Evans, A. 460, 483
Evans, G. C. 253, 342, 483
Everitt, W. N. 483
Evgrafov, M. A. 11, 483

Faber, G. 80, 232, 289, 290, 338, 339, 409, 483
Fabry, E. 10, 407, 408, 483
Fallin, H. I. 183, 341, 483
Farkas, J. 254, 484
Farrell, O. J. 341, 484
Fatou, P. 189, 197, 198, 251, 254, 255, 338, 372, 407, 458, 484
Feigl, G. 104, 484
Fejér, L. 81, 185, 194, 289, 290, 293, 303, 315, 357, 406, 475, 484
Fekete, M. 254, 290, 457, 458, 484–485
Fenchel, W. 291, 459, 485
Fenske, C. 115, 119, 482
Fenton, P. C. 185, 485
Ferrar, W. L. 188, 485
Feyel, D. 11, 485
Fichtenholz, G. 291, 485
Ficker, W. 509
Finkelstein, M. 291, 485
Fisher, S. 269, 485
Flatto, L. 253, 485
Flett, T. M. 339, 485
Floyd, E. E. 338, 341, 485
Fort, M. K. Jr. 115, 116, 485
Fouët, Éd. A. 52, 485–486
Frank, G. 458, 486
Franklin, P. 189, 291, 409, 459, 486
Fresnel, A.-J. 408
Freudenthal, H. 39, 119, 473, 486
Frost, R. 186
Fubini, G. 10, 264, 273
Fuchs, W. H. J. 163, 189, 291, 486

Gaier, D. 81, 340, 357, 358, 410, 486
Gamelin, T. W. 290, 487
Garabedian, P. R. 338, 357, 487
Gårding, L. 409
Garnett, J. 290, 487
Garnir, H. G. 409, 487
Gasapina, U. 118, 487
Gattegno, C. 339, 487
Gauss, C. F. 51, 117, 118, 183, 408, 463, 487
Gauthier, P. 291, 473, 487
Gavrilov, V. I. 339, 487
Gehman, H. M. 341, 487
Gehring, F. W. 163, 186, 487
Gelfond, A. O. 409, 487
Gerber, L. 255, 487
Germay, R. H. J. 409, 488
Gerretsen, J. 338, 408, 458, 528
Gillespie, D. C. 291, 488
Gironza Solanas, J. A. 357, 488
Giudice, F. 117, 488
Glicksberg, I. 190, 488

Gobert, J. 409, 487
Goldberg, J. L. 186, 216, 488
Goluzin, G. M. 190, 246, 252, 339, 356, 357, 358, 488
Gonçalves, J. V. 341, 488
Goodman, J. E. 503
Gordon, W. B. 190, 215, 407, 488
Gould, H. W. 82, 183, 341, 483, 489
Goursat, E. 120, 149, 183, 489
Grabiner, S. 284, 489
Graeser, E. 357, 489
Granas, A. 118, 119, 489
Grandjot, K. 291
Gray, J. D. 51, 489
Green, G. 336, 337
Green, J. W. 185, 489
Gronwall, T. H. 185, 215, 407, 410, 489, 496
Gross, F. 255, 458, 460, 489, 530
Grötzsch, H. 10, 357
Grunsky, H. 216, 341, 358, 489
Guggenheimer, H. 104, 116, 489
Guichard, C. 409, 489
Guichardet, A. 81, 490
Gutzmer, A. 81, 490

Hadamard, J. 80, 82, 147, 187, 210, 211, 407, 490
Hadwiger, H. 119, 490
Hall, T. 163, 490
Halmos, P. 13, 490
Hamburger, H. 408, 490
Hamel, G. 190, 490
Hamilton, H. J. 367, 490
Hardy, G. H. 10, 81, 185, 187, 253, 292, 448, 460, 490, 491
Harnack, A. 81, 143, 253, 491
Hartogs, F. 184, 252, 253, 290, 491
Haruki, H. 193, 491
Haselen, A. van 491
Hausdorff, F. 39, 356, 491
Hausner, M. 503
Hayashi, T. 117, 184, 492
Hayman, W. K. 184, 187, 456, 457, 458, 459, 460, 492
Heffter, L. 183, 492
Heilbronn, H. 119, 341, 492
Heine, E. 18
Heinhold, J. 357, 492
Heins, M. 40, 159, 215, 237, 246, 253, 254, 255, 290, 292, 339, 356, 408, 410, 461, 464, 492–493
Heinz, E. 115, 493
Hejhal, D. A. 215, 493
Helmer, O. 410, 493
Helly, E. 292
Hengartner, W. 291, 487
Herglotz, G. 292, 493
Hermite, C. 170, 409, 493, 525
Hervé, M. 187, 254, 356, 404, 493
Herzig, A. 206, 216, 217, 493
Herzog, F. 81, 482, 493
Hessenberg, G. 82, 493
Hestenes, M. R. 341, 493
Heuser, P. 340, 493
Hewitt, E. 12
Hilbert, D. 289, 293, 322, 357, 494
Hille, E. 11, 81, 185, 189, 190, 216, 252, 290, 292, 337, 338, 459, 494
Hilton, P. J. 541
Hiong, King-lai 457, 494
Ho, C.-W. 190, 407, 494
Ho, Yu-tsaim 457, 494
Hoffman, K. 11, 49, 187, 277, 410, 494
Hoischen, L. 273, 291, 494
Hölder, O. 188, 406, 495
Holland, A. S. B. 217, 457, 495
Hopf, E. 342, 495
Hopf, H. 119
Horowitz, L. 117, 495
Hu, K. 341, 495
Huber, A. 406, 465
Huber, H. 190, 216, 356, 495
Huebsch, W. 341
Hurewicz, W. 119, 495
Hurwitz, A. 221, 253, 301, 407, 458, 460, 495
Hyman, C. J. 463

Iacobacci, R. F. 476, 495
Iglisch, R. 117, 495
Ingham, A. E. 10, 292, 460, 491
Inoue, M. 342, 495
Itzigsohn, C. 476
Iversen, F. 184, 460, 495
Izumi, S. 189, 495

Jacobi, C. 416
Jacobson, N. 117
Jacobsthal, E. 251, 495
Jamet, V. 117, 408, 496
Janiszewski, Z. 100, 105, 106, 496
Jategaonkar, A. V. 458, 496
Jenkins, J. 458, 461, 496
Jensen, J. L. W. V. 186, 194, 215, 217, 251, 413, 416, 496

Jentzsch, R. 251, 496
Jordan, C. 116
Jørgensen, V. 164, 496
Jourdain, P. E. B. 183, 189, 409, 496
Julia, G. 10, 185, 215, 216, 254, 255, 284, 337, 338, 341, 343, 356, 358, 362, 410, 496–497
Jürgens, E. 118, 497

Kabaila, V. P. 461, 497
Kakeya, S. 406, 497
Kakutani, S. 356, 410, 498
Kamiya, H. 40, 498
Kamowitz, H. 254, 498
Kampke, E. 116, 341, 498
Kaplan, W. 291, 498
Karamata, J. 291, 498
Kawakami, Y. 163, 498
Keldych, M. 291, 498
Kellogg, O. D. 10, 185, 343, 410, 498
Kennedy, P. B. 184, 187, 492
Kerékjártó, B. v. 115, 340, 461, 498
Khajalia, G. 216, 357, 498–499
Khintchine, A. I. 252, 499
Kirkor, A. 502
Klein, F. 525
Kleiner, W. 338, 499
Kline, J. R. 104, 116, 499
Kloosterman, H. D. 118, 499
Knaster, B. 118, 357, 499
Kneser, A, 117, 499
Kneser, H. 117, 340, 341, 357, 499
Knopp, K. 11, 53, 81, 116, 500
Koch, H. von 189, 500
Koebe, P. 10, 185, 192, 216, 232, 235, 246, 252, 253, 254, 293, 298, 304, 337, 338, 339, 342, 356, 357, 410, 460, 500
Koenigs, G. 254, 500
Komatu, Y. 216, 357, 410, 500–501
Kommerell, K. 501
König, H. 12, 457, 501
König, J. 117, 407, 501
Kovanko, A. 252, 501
Kowalewski, G. 416, 501
Krafft, M. 56, 270, 501
Krajkiewicz, P. 457, 501
Krazer, A. 408, 501
Krein, M. 291
Kronecker, L. 118, 373, 408, 501–502
Kronrod, A. S. 183, 463
Krylow, W. J. 190, 488
Krzyż, J. G. 11, 311, 409, 502
Kunugui, K. 252, 502
Kuran, Ü. 185, 292, 502
Kuratowski, K. 115, 116, 118, 119, 357, 499, 502
Kwasselava, D. 337, 504

Labelle, G. 117, 502
Lagrange, J. 184, 502
Lai, W.-t. 502
Lammel, E. 251, 503
Landau, E. 9, 11, 13, 14, 18, 81, 82, 158, 159, 184, 185, 187, 188, 203, 216, 217, 251, 252, 253, 254, 291, 407, 408, 457, 458, 459, 460, 475, 503–504
Landau, H. J. 198, 216, 356, 461, 504
Landsberg, G. 408, 504
Laplace, P.-S. de 41, 322, 323, 342, 410
Laugel, L. 525, 540
Laurent, H. 189, 504
Laurent, P. A. 406, 504
Lauritzen, S. 461, 504
Lavrentieff, M. A. 290, 291, 337, 356, 498, 504
Lavrentieff, M. M. 194, 504
Leau, L. 341, 504
Lebedev, N. A. 290, 531
Lebesgue, H. 37, 115, 118, 291, 322, 342, 410, 504–505
Lecornu, L. 407, 505
Lehner, J. 460, 505
Lehto, O. 337, 338, 358, 505
Leibniz, G. W. 145
Leja, F. 290, 337, 341, 342, 357, 505
Leland, K. O. 183, 505
Lelong-Ferrand, J. 338, 339, 505
Lerch, M. 408, 409, 505
Lesley, F. D. 189, 473
LeVavasseur, R. 118, 357, 367, 505, 515
LeVeque, W. J. 82, 408, 505
Levi, E. E. 185, 506
Levin, B. 278, 291, 406, 506
Levin, V. I. 158, 251, 506
Lévy, P. 457, 506
Lewis, C. J. 159, 506
Lewy, H. 525
Lichtenstein, L. 192, 253, 291, 337, 338, 342, 343, 410, 463, 506
Liebeck, H. 255, 371, 506
Lindelöf, E. 10, 11, 82, 130, 132, 161, 162, 184, 186, 187, 188, 193, 216, 217, 224, 251, 252, 338, 339, 408, 442, 445, 460, 506–507, 519
Lindwart, E. 251, 507
Liouville, J. 71, 406

Lipka, S. 340, 507
Littlewood, J. E. 10, 117, 140, 186, 491, 507
Lloyd, N. G. 115, 116, 507
Logan, M. 12
Lohwater, A. J. 108, 163, 190, 215, 253, 292, 339, 340, 406, 473, 477, 487, 507
Looman, H. 51, 342, 507
Loomis, L. H. 282, 291, 292, 341, 448, 507
Loria, G. 118, 507–508
Löwner, K. 194, 216, 219, 254, 508
Lucas, F. 117
Lunds, G. 539
Lüroth, J. 118
Lusin, N. N. 81, 189

MacGregor, T. H. 216, 508
Macintyre, A. J. 67, 203, 217, 251, 254, 255, 460, 461, 471, 478, 508
Macintyre, S. S. 458, 508
Macki, J. W. 457, 508
Mahler, K. 82, 409, 508
Maitland, B. J. 159, 508
Mandelbrojt, S. 147, 188, 254, 459, 508
Mandl, R. 535
Manning, K. R. 52, 509
Mansion, P. 117, 509
Marden, A. 202, 246, 356, 509
Marden, M. 117, 509
Markuschewitsch, A. I. 51, 356, 509, 535
Marshall, D. E. 215, 291, 509
Marty, F. 459
May, K. O. 485
Mazur, S. 186, 407, 466
Mazurkiewicz, S. 40, 81, 118, 186, 356, 499, 509
McKean, H. P. 408, 409, 481
McShane, E. J. 338, 509
Meier, K. E. 51, 509–510
Meisters, G. 340, 510
Menchoff, D. 51, 510
Méray, Ch. 52, 117, 407, 510
Mergelyan, S. N. 290, 291, 320, 341, 510
Mertens, F. 117, 408, 510
Miamee, A. 406, 510
Michael, J. H. 512
Mignosi, G. 117, 510
Miller, E. W. 407, 511
Milloux, H. 158, 188, 254, 444, 457, 461, 511
Minami, U. 188, 341, 511
Minda, C. D. 51, 358, 407, 511
Minsker, S. 511
Mioduszewski, J. 216, 511
Miranda, C. 10, 417, 426, 456, 457, 511
Mirsky, L. 482
Mitchelmore, M. C. 255, 511
Mitrinović, D. S. 409, 512
Mittag-Leffler, G. 10, 183, 237, 251, 254, 289, 291, 390, 393, 406, 409, 410, 512
Mohr, E. 117, 409, 512
Molk, J. 80, 409, 522
Monna, A. F. 293, 512
Montel, P. 10, 51, 82, 117, 153, 184, 185, 186, 187, 220, 235, 251, 252, 253, 254, 255, 289, 311, 338, 339, 340, 356, 438, 441, 445, 458, 459, 460, 461, 512–513
Moore, E. H. 183, 514
Moore, R. L. 40, 340, 341, 514
Mordell, L. J. 408, 514
Morera, G. 149, 188, 514
Morgenstern, D. 186, 514
Morris, S. A. 51, 489
Morse, M. 341
Müller, C. 188, 514
Müller, M. 337, 514
Müntz, C. H. 11
Myller-Lébédeff, V. 339, 514

Nabetani, K. 216, 514
Nagasawa, M. 356, 514
Nakai, M. 356, 514
Neder, L. 81, 514
Nehari, Z. 357, 458, 515
Nersesjan, A. A. 291, 515
Netto, E. 117, 118, 367, 487, 515
Neumann, C. 184, 322, 515
Nevanlinna, F. 187, 188, 189, 253, 285, 292, 460, 461, 515
Nevanlinna, R. 10, 11, 158, 159, 187, 188, 189, 215, 216, 252, 253, 285, 291, 292, 456, 460, 461, 515–516
Neville, E. H. 409, 516
Newman, D. J. 461, 516
Newman, M. H. A. 104, 116, 341, 516
Nicolas, L. 535
Nicolesco, M. 185, 516
Nielsen, N. 82, 516
Nieminen, T. 292, 515
Nitsche, J. C. C. 356, 516
Noaillon, P. 410, 516
Nöbeling, G. 104, 118, 341, 516
Noether, M. 183, 472, 525
Noshiro, K. 216, 516

Novinger, W. P. 81, 339, 516
Nowinski, K. 119, 517

Ockham, W. 41
O'Hara, P. J. 409, 517
Olds, C. D. 408, 517
Olech, C. 340, 510
Onicescu, O. 517
Orland, G. H. 291, 466
Osgood, W. F. 39, 116, 118, 185, 188, 190, 227, 253, 254, 303, 304, 309, 322, 338, 340, 342, 463, 517
Osserman, R. 198, 216, 356, 461, 504
Ostrowski, A. 10, 11, 12, 108, 118, 147, 151, 187, 188, 189, 251, 252, 298, 337, 338, 339, 407, 456, 457, 458, 459, 460, 461, 487, 517–518
Ou, Šo-Mo 186, 518

Painlevé, P. 133, 184, 289, 321, 518
Pál, J. 115, 118, 340, 518
Parseval, M. A. 71, 81
Patil, D. J. 190, 518
Perkins, F. W. 185, 518
Perron, O. 10, 342, 460, 519
Perry, A. D. 45, 188, 519
Peschl, E. 357, 519
Petersen, J. 194, 215, 519
Petrova, S. S. 118, 519
Phelps, R. 215, 474
Phragmén, E. 40, 184, 187, 224, 519
Picard, É. 10, 82, 188, 310, 362, 410, 432, 440, 458, 459, 460, 519, 540
Pick, G. 215, 254, 520
Pickel, P. 12
Picone, M. 185, 190, 410, 520
Piranian, G. 338, 477, 482, 493
Pisot, C. 407, 481, 511
Plastock, R. 190, 407, 520
Plemelj, J. 186, 253, 254, 520
Plessner, A. 520
Poincaré, H. 192, 273, 303, 322, 342, 520
Poisson, S.-D. 51, 134
Pollard, S. 341, 520
Pólya, G. 10, 11, 158, 184, 186, 188, 189, 196, 216, 217, 251, 255, 292, 337, 340, 356, 406, 407, 409, 433, 449, 458, 460, 461, 491, 507, 520–521
Pomey, É. 117, 521
Pommerenke, Ch. 337, 339, 486, 521
Pompeiu, D. 51, 175, 188, 190, 341, 521
Ponomarev, S. P. 51, 521
Porter, M. B. 219, 251, 521
Prachar, K. 53, 522
Pradelle, A. de la 11, 485
Pringsheim, A. 52, 80, 81, 82, 85, 117, 118, 183, 186, 188, 406, 407, 409, 460, 463, 522
Privalov, I. I. 10, 186, 189, 253, 254, 416, 417, 522–523
Pucciano, G. 341, 523
Puccio, L. 117, 523
Pultr, A. 476

Quitsch, E. 12

Rademacher, H. 50, 188, 190, 198, 311, 408, 475, 523
Rado, R. 40, 523
Radó, T. 171, 184, 194, 197, 198, 216, 219, 246, 254, 317, 338, 340, 342, 356, 410, 508, 523
Rådström, H. 255, 523
Rajagopal, C. T. 217, 524
Ramakrishnan, A. 527
Raynor, G. E. 410, 524
Reade, M. O. 185, 524
Redheffer, R. 45, 117, 253, 524
Reich, E. 202, 358, 524
Reid, W. T. 341, 524
Reijnierse, J. M. 356, 524
Remak, R. 50, 342, 524
Rengel, E. 357, 524
Rennie, B. C. 291, 524
Renteln, M. von 410, 525
Richards, I. 188, 202, 246, 356, 525
Ridder, J. 51, 188, 189, 291, 341, 342, 525
Riemann, B. 52, 82, 165, 184, 188, 293, 322, 337, 525
Riesz, F. 40, 184, 185, 186, 252, 293, 303, 342, 357, 406, 484, 523, 525
Riesz, M. 216, 526
Ritt, J. F. 243, 254, 356, 526
Roberts, J. H. 216, 526
Robertson, M. S. 213
Robinson, R. M. 187, 216, 410, 458, 526
Robison, G. R. 81, 526
Rodin, B. 202, 246, 356, 509
Rogosinski, W. W. 133, 187, 213, 216, 217, 491, 508, 526
Rohde, H.-W. 255, 526
Rosenblatt, A. 189, 526
Rosenbloom, P. C. 117, 458, 526
Rosenblum, M. 406, 526
Rosenthal, A. 116, 253, 290, 291, 471, 491
Rosten, L. 9

Roth, A. 291, 461, 527
Rouché, E. 118, 179, 190, 256, 527
Roux, D. 527
Rovnyak, J. 406, 479, 526
Royden, H. L. 188, 357, 358, 527
Rubel, L. A. 117, 291, 410, 527
Rubinstein, Z. 190, 527
Rudin, W. 9, 11, 13, 14, 17, 18, 21, 44, 53, 81, 119, 184, 185, 187, 188, 289, 290, 339, 356, 358, 408, 409, 410, 459, 527
Rung, D. C. 339, 528
Runge, C. 10, 263, 265, 289, 528
Ryan, F. 253, 292, 473

Saalschütz, L. 82, 528
Saeki, S. 189
Sakashita, H. 337, 528
Saks, S. 11, 40, 41, 51, 140, 184, 189, 190, 254, 289, 290, 291, 338, 340, 342, 356, 457, 462, 528
Salehi, H. 406, 510
Salvadori, M. 408, 528
Salzmann, H. 186, 528
Samuelsson, Å. 190, 528
Sansone, G. 338, 408, 458, 528
Sario, L. 356, 357, 464
Saxer, W. 10, 433, 457, 458, 528
Scarf, H. 117, 529
Schaeffer, A. C. 10
Scheeffer, L. 406, 529
Scheinberg, S. 273, 291, 485, 529
Schering, E. 409, 529
Schiffer, M. M. 216, 356, 529
Schilling, O. F. G. 410, 529
Schmidt, E. 192
Schmidt, H. 117, 529
Schneider, W. J. 254, 466
Schönflies, A. 9, 10, 104, 116, 118, 314, 341, 320, 529
Schönhage, A. 339, 529
Schottky, F. 10, 187, 188, 356, 411, 431, 456, 457, 460, 529
Schou, E. 215, 529
Schreiber, B. 188, 473
Schreier, O. 117, 465
Schröder, E. 254, 530
Schur, I. 215, 254, 408, 530
Schuske, G. 255, 530
Schwarz, H. A. 135, 184, 185, 186, 189, 191, 216, 292, 322, 410, 470, 484, 530
Scott, E. J. 528
Seebach, J. Jr. 40, 532
Seidel, W. 216, 291, 339, 358, 466, 473, 530
Sergott, A. M. 536
Serrin, J. 216, 292, 530
Severini, G. 253, 342, 530
Shah, S. M. 530
Shanks, D. 408, 530
Shapiro, H. S. 410, 530
Shashkevich, M. 530
Sheffer, I. M. 530
Shell, D. L. 255, 531
Shieh, Hui-chun 457, 531
Shields, A. L. 249, 255, 531
Shimizu, T. 215, 460, 531
Shisha, O. 251, 253, 485, 531
Sholander, M. 531
Sidon, S. 82, 531
Sierpinski, W. 81, 531
Silverman, R. A. 509
Simonart, F. 337, 531
Sinclair, A. 265, 291, 531
Singh, A. N. 184, 531
Singh, S. K. 407, 531
Smirnov, V. I. 290, 531
Smith, J. D. 189, 532
Sochozki, J. W. 406
Sommer, F. 82, 190, 253, 406, 409, 410, 467
Specht, W. 117, 118, 532
Sperner, E. 119, 532
Spivak, M. 117, 270, 532
Springer, G. 188, 532
Springer, T. A. 532
Stäckel, P. 51, 183, 407, 476, 532
Stark, E. L. 82, 532
Steen, L. 40, 532
Steinberg, L. 406, 472
Steinhardt, F. 469, 475, 503
Steinhaus, H. 119
Steinitz, E. 40, 532
Stieltjes, T. J. Jr. 148, 251, 532
Stoïlow, S. 216, 532
Stokes, G. G. 337
Stone, M. H. 338, 358, 533
Stout, E. L. 290, 533
Stożek, W. 410, 533
Stromberg, K. 13, 255, 406, 533
Study, E. 216, 338, 339, 533
Styer, D. 358, 407, 533
Su, Li Pi 356, 533
Suetin, P. K. 340, 533
Sunyer Balaguer, F. 166, 477
Šura-Bura, M. R. 32, 533

Süss, W. 499
Świątkowski, T. 253, 533
Szász, O. 11, 406, 533
Szegö, G. 10, 11, 184, 186, 188, 189, 216, 217, 255, 337, 340, 356, 406, 407, 408, 460, 485, 521, 533
Sz.-Nagy, B. 406, 525

Takahashi, S. 216, 534
Takenaka, S. 291, 534
Tammi, O, 82, 534
Tanaka, C. 162, 534
Tar, M. M. 51, 534
Taylor, A. E. 186, 534
Taylor, B. A. 188, 467, 473
Taylor, E. 12
Taylor, E. H. 171, 304, 309, 338, 517, 534
Teissier du Cros, F. 81, 534
Terasaka, H. 340, 534
Thomas, J. M. 117, 534
Thron, W. J. 255, 530, 534
Tideman, M. 291, 534
Tietze, H. 40, 104, 534–535
Titchmarsh, E. 409, 457, 535
Toburen, J. 12
Tolstov, G. P. 342, 535
Tomić, M. 82, 535
Tonelli, L. 185, 291, 535
Trokhimchuk, Y. Y. 51, 535
Tschakaloff, L. 408, 535
Tschebotareff, N. 291, 535
Tsuji, M. 143, 215, 216, 292, 338, 339, 356, 460, 535–536
Tucker, A. W. 115, 536
Tumura, Y. 458, 536
Tutschke, W. 536

Uherka, D. J. 536
Ulam, S. 119
Ullrich, E. 499
Unkelbach, H. 206, 536

Vahlen, K. Th. 117, 536
Valiron, G. 10, 81, 189, 203, 216, 249, 254, 255, 290, 456, 457, 458, 460, 504, 536–537
Vallée Poussin, C. de la 188, 253, 338, 459, 460, 537
Valson, C.-A. 183, 537
Van Yzeren, J. 67, 537
Varopoulos, Th. 458, 537
Vasilesco, F. 341, 537
Vaughn, H. E. 538
Vedak, K. B. 479
Veech, W. 82, 232, 246, 337, 356, 459, 538
Veen, S. C. van 408, 538
Venkatachaliengar, K. 409, 538
Venkateswaran, S. 291, 527
Verblunsky, S. 216, 292, 337, 538
Vijayaraghavan, T. 81, 538
Villat, H. 185, 410, 538
Visser, C. 216, 538
Vitali, G. 117, 219, 251, 538
Vito, L. de 40, 538
Vivanti, G. 190, 539
Vizzini, G. 40, 539
Volkovyskii, L. 11, 539
Volterra, V. 185, 539

Waerden, B. L. van der 117
Walecki, F. 117, 539
Wall, C. T. C. 115, 539
Walsh, J. L. 189, 215, 289, 290, 320, 321, 337, 341, 343, 539
Walther, A. 187, 539
Warschawski, S. E. 358, 524, 539
Watson, G. N. 341, 539
Wavre, R. 108, 254, 540
Weierstrass, K. T. W. 10, 52, 81, 117, 186, 188, 190, 237, 254, 269, 291, 361, 390, 406, 409, 410, 540
Weil, A. 81, 540
Wermer, J. 8
Whitney, H. 342, 540
Whittaker, J. M. 189, 251, 252, 540
Whyburn, G. T. 40, 183, 187, 341, 541
Whyburn, W. M. 23, 541
Widder, D. V. 282, 291, 292, 410, 507, 541
Wiener, N. 10, 343
Wigert, S. 117, 215, 541
Wigner, E. P. 81, 541
Wilkosz, W. 51, 342, 541
Wille, F. 119, 541
Williams, K. S. 82, 541
Williams, R. K. 50, 407, 541
Wiman, A. 470
Winternitz, A. 118, 541
Wirtinger, W. 525
Wolf, F. 189, 292, 542
Wolfenstein, S. 117, 542
Wolff, J. 188, 190, 216, 249, 255, 284, 289, 292, 338, 542
Wölffing, E. 542

Yanagihara, N. 460, 542
Yang, Chung-chun 458, 542
Yang Lo 457, 542
Young, L. C. 528
Young, R. C. H. 500
Young, W. H. 187, 253, 291, 543
Youngs, J. W. T. 45, 188, 519
Yûjôbô, Z. 292, 543
Yuškevič, A. P. 51, 543

Zalcman, L. 185, 188, 190, 217, 290, 340, 341, 358, 459, 543
Zarankiewicz, K. 357, 543
Zaremba, S. 188, 341, 342, 404, 543
Zassenhaus, H. 117, 543
Zeller, K. 186, 528
Zeus, 51
Zin, G. 190, 543
Zinterhof, P. 215, 543
Zoretti, L. 341, 543
Zygmund, A. 11, 40, 41, 190, 254, 289, 290, 291, 338, 340, 356, 457, 462, 528

Subject Index

Abel's partial summation formula 59, 377
abscissa 15
accessible point 25, 116
– –, density of 25
– –, in the mapping theorem proof 308, 399
accumulation point 17
Acta Mathematica 463
adherence 16
algebra 14
algebraic determination of the radius ratio 350, 351, 356
algebraic relations between entire functions 433–435, 458
algebraic topology 9, 115
almost uniform convergence 251
American Mathematical Monthly 11, 291, 458
analytic continuation (see also: removable singularity)
– – along curves 246, 358
– – and simple-connectivity 358
– – by overconvergence 290
– – by summability 289
– – over an arc 321
analytic function 41
angular derivative 205
annulus 16
–, bilateral series in 359–361
–, Cauchy formula in 359
–, conformal automorphisms of 348, 349, 356
–, connectedness of 27
–, Dirichlet problem in 400 ff., 405
–, harmonic functions in 398 ff.
–, holomorphic functions in 359 ff.
–, holomorphic logarithms in 348
–, Huber's theorem in 201
–, meromorphic functions in 397–398
–, Poisson formula in 398–401
–, prototype of double-connectivity 352
–, radius ratio invariance 349, 351, 356
–, – – algebraically determined 350–351, 356
Anschmiegungssatz 393
antiderivative (see: primitive)
Antipodensatz 110, 119
approximate identity 138, 269
approximation and analytic continuation 289, 290
– and interpolation 265, 290, 291
– in solving the Dirichlet problem 185, 327, 343, 410
– pointwise 356
–, role in the *Hauptsatz* 344, 345
approximation theorem of Carleman 273, 276, 291
– – of Fejér 185
– – of Fisher 269
– – of Lavrentieff 290
– – of Marshall 291
– – of Mergelyan 290, 291, 341
– – of Montel 356
– – of Sinclair 265
– – of Walsh 320, 341
– – of Weierstrass 272, 291
area method 10, 232, 315, 338, 356, 357
areal mean value property 185, 342
arc 24, 40
–, simple 24
arc lemma of Flett 339
– – of Koebe 304, 339
– – of Lindelöf 130, 171, 319
– – of Radó 171
arcsin series 176
arcwise connectedness of curves 24, 40, 163
Argument Principle 179, 190, 291, 310
– –, applications of 311 ff., 340, 397
– – for meromorphic functions 396
– –, topological version of 340

Baire Category Theorem 10, 18, 166, 227
balayage 34
ball, closed 16
–, open 16
Banach algebra 11, 409
barrier 325
– in simply-connected regions 330–332
– in solution of Dirichlet problem 327, 332, 333
Basic Connectedness Lemma 28

basic construction 325–326, 333
Basic Integral Representation 256, 260
– – –, in approximation theory 262, 289
– – –, in Carleman's theorem 273
– – –, in Cauchy's theorem 264
– – –, in Rouché's theorem 265
– – –, in production of loops 267
Bernoulli functions 75
– polynomials 75
– numbers 76, 82
Bernstein inequality 388, 409
– polynomials 253
bicontinuous 20, 87
bijective 13
bilateral sequence 17
– (power) series (see: Laurent series)
Blaschke product, finite 194, 268
– –, infinite 241, 461
– convergence theorem 219, 416
– theorem on zeros 195, 241
– uniqueness theorem 195, 415
bootstrap lemma (see: Bureau's Lemma)
Borsuk–Ulam *Antipodensatz* 110, 119
boundary 17
– behavior of Riemann map 303 ff., 338–340
– of simply-connected sets 345
boundary-value problem 136, 342
bounded above 18
–, locally uniformly 218
–, – – and normal families 436
–, – – and omitted values 436
–, – – and pointwise convergence 265
bounded set 16
bounds, absolute from relative 419
– for derivatives 145, 147, 388
– for $|f|$ in terms of its log mean 416
– for $|f|$ in terms of $\operatorname{Re} f$ 208
– for $f(z)$ in terms of $f(w)$ 146, 148, 195, 215, 459
– for $f'(z)/f'(w)$ 235
– for functions omitting two values 426, 431, 456
– for $\operatorname{Im} f$ 142
– for $\operatorname{Re} f$ 209
– for normal families 436
– in class $\mathscr{S}$ 231 ff.
– on circles 147
– on lines 132, 147, 441, 442
Brouwer Fixed Point Theorem 106
Bulletin of the American Mathematical Society 462
Bureau's lemma 419, 422, 430, 458

canonical conformal type 357–358
Carleman's integral formula 190
– approximation theorem 273, 276, 291
Casorati–Weierstrass theorem 361, 370, 438
– – – and J. W. Sochozki 406
– – –, Picard's and Julia's improvements 438
Cauchy estimates 71, 81
Cauchy–Goursat theorem 120
– – –, converse of 147 (see also: Morera's theorem)
Cauchy–Hadamard formula 80
Cauchy Integral Formula for an annulus 359
– – – for an arc 190
– – – for homologically-connected regions 229
– – – for starlike regions 145
– – – using continuity of f' 77
Cauchy's Integral Theorem, catalog of proofs 341
– – – for general regions 229, 264, 290, 341
– – – homology versions 229, 264, 377, 378
– – – homotopy versions 230
– – – in presence of singularities 377, 378
– – – on the boundary 321, 341
– – – role in *Hauptsatz* 344, 345
– – – for simply-connected regions 263, 321, 344
– – – for starlike sets 145
– – – via approximation 264, 290, 321
– – – via Fubini 264
Cauchy product of series 144–145
Cauchy–Riemann equation 49
– – –, weak sufficient conditions for 51
Cauchy–Schwarz inequality 16, 21, 296
– – formula 135
Cauchy sequence 17
– –, locally uniform 218
Chain Rule 21, 42
chord of a Jordan-curve 311
chordal metric 459
circle 16, 64
–, characterization by harmonic functions 185
circle of convergence, behavior on 81
– – –, poles on 375–377, 407

circular slit region 357
circulation index (see: index of a point)
circumferential mean 127
– – and subordination 186
– – for $\log|F|$, $\log^+|F|$ 412 ff.
– – log convexity of 405
– – monotonicity of 142, 414
circumferential mean value property 141, 185
clopen 18
closed-and-open 18
closed curve 23
closed ideals of holomorphic functions 394–395
closed set 16
closure 16
cluster point 17
coefficients and convergence 220
– estimates 71, 212, 213
– and growth of series 71–72
– and radius of convergence 54, 80
– and poles on circle of convergence 375–377, 407
– in terms of derivatives 144
– square summability 71
– via differential equations 410
cohomology 9
column rank 14
common fixed point 249, 255
common zeros 393, 394
compact 17, 18
compact convergence 251
compact-open topology 251
compactness in euclidean space 18
– in function spaces 220, 231, 252–253, 459
–, relation to normality 459
complete metric 17, 218
complex conjugate 15
complex numbers 14
component 29
– of simply-connected set 99, 114, 345
– of double-connected set 355
Comptes Rendus de l'Académie des Sciences Paris 462
conditionally compact 18
conformal 41
– at the boundary 339
conformal automorphisms of annuli 349 ff., 356
– – of $\mathbb{C}$ 305, 370
– – of disks 43, 192
– – – – preserve subdisks 193
– – fixing a point 244, 246, 254, 346
– – of half-planes 80, 199
– – – – –, role in mapping doubly-connected regions 353 ff.
conformal mapping via Hilbert space 338
conjugate bilinear 16
conjugate, harmonic (see: harmonic conjugate)
connected 18
–, polygonally 24
– *im kleinen* 340
connectedness and isolated points 27
– of annuli 27
– of curves 26
– of starlike sets 27
– of the complement of an arc 27, 103
– under continuous functions 22
connectivity 33
– and continuous logarithms 96–99, 111, 112, 348
–, positional invariance 98, 99, 114, 116
–, internal characterizations 98–99, 111–113, 344–345, 355
continuity of the derivative 78, 122, 183, 186, 341
– of the Riemann map at the boundary 303 ff.
continuous at a point 20
– on a set 20
–, uniformly 20
continuous and nowhere differentiable 186
contour integral 44, 51, 409
convergence, compact 251
– on a dense open set 227
– induced 188, 219
– of inverses 227, 301–303, 319
– of Jordan regions 315–320
–, locally uniform 218
– of Riemann maps in variable regions 316, 320
– of univalent functions 221, 235, 236, 315–320, 441
convergence of iterates to a constant 243, 450
– – – in the disk 247–251
– – – of the exponential 247, 255
– – – to a fixed point 244, 247–251, 450
– – – of rational functions 247, 248
convergence sector 449
convergence theorem of Blaschke 219
– – – – generalized 416

– – of Carathéodory 227, 320
– – of Denjoy and Wolff 249 ff.
– – of Farkas and Ritt 243
– – of Fejér 315
– – for harmonic functions 151, 223
– – of Harnack 221
– – of Hurwitz 221, 301, 316
– – of Koebe 304
– – of Montel 153, 220, 235, 441
– – of Montel and Privalov 236
– – of Schröder 242
– – of Shell 247
– – of Stieltjes 147
– – from uniqueness theorems 252
– – of Vitali and Porter 219, 251
– – – – – –, Carathéodory and Landau extension 252
– – – – – –, Nevanlinna extension 252
– – of Weierstrass 151, 188
convergent sequence 17
– –, locally uniformly 218
convex (set) 25
convex function 10, 208
– combination 25
– function 186, 406
– hull 25
– –, diameter of 26
– –, compactness of 26
countable 13
cover 17
covering space 115, 356–357
covering theorem of Bloch 232–233, 254, 460
– – of Hurwitz 253, 435, 458
– – of Koebe *et al.* 232, 253
cross-cuts 25, 29
– – and analytic continuation 321
– – in Jordan regions 105
– – in proof of the mapping theorem 308–309, 338
curve 33
–, closed 23
–, piecewise smooth 24
curvilinear integral (see: contour integral)
cycle 14, 260

Darboux–Picard univalence theorem 310–311
– – – –, topological extension 340–341
decreasing function 20
Dedekind cut 14
degree of a map 115
Denjoy uniqueness theorem 252
dense set 18
derivative 41
–, angular 205
–, continuity of 78, 122, 183, 186, 341
–, directional 336
–, normal 336, 342
– of sums, products, composites 42
determinants 14
–, in rationality criteria 374–375
diameter 19
– of convex set 26
difference of positive harmonic functions 224
differentiability on the boundary 205, 339
– of harmonic functions 187
– of holomorphic functions 122, 144
–, weak sufficient conditions 49, 51
differentiable 41
differential characterization of exponential 69
– – of sin, cos 168
differentiation of inverse functions 173, 174, 180–181
– of power series 55
– of uniform limits 151
–, continuity as an operator 218
– under the integral 48, 57, 228, 410, 418
dimension 14, 118
Dirichlet problem 136
– – for annulus 185, 400
– – via approximation of boundary data 185, 327, 343, 410
– –, constructive solutions 291, 336–337
– – for disk 137, 185
– – for punctured disk 405
– – with discontinuous data 185, 278
– –, Fejér's solution 185
– – for general regions 327, 332–333
– – solved with Green's function 336
– – for half-plane 276, 291, 327–329
– –, insolubility of 405
– – and mapping multiply-connected regions 357
– – and one point components 332–333, 405
– –, relation to Riemann map theorem 334, 336
– – for simply-connected regions 332–333, 342
– – via subharmonic functions 325 ff., 342–343

Dirichlet series 82, 462
disconnection 18
disk 16
–, conformal automorphisms of 43, 192
–, connectedness of 27
–, characterization by harmonic functions 185
–, punctured (see: annulus)
– and simple-connectivity 32, 344
distortion theorem of Koebe 235, 357
– – of Pólya 449–450, 461
– – for zero-free functions 146
divergent sequence 17
double angle formula 156, 209
double-connectivity and components 355
– – and compact complement in $\mathbb{C}$ 351
– –, three canonical types 352, 354
– – and covering spaces 357
– –, internal characterizations 112, 355
– –, other treatments 357
– –, positional invariance 114, 355
doubly-connected 33

Eilenberg separation theorem 100
Elemente der Mathematik 11
end-cut 25
endpoint 15
l'Enseignement Mathématique 11
entire function 41
– –, bounded 71
– –, bounded on rays 461
– –, convergent along rays 448–449, 461
– –, growth and coefficients 71, 72
– –, unimodular on a circle 197
– –, univalent 370, 407
equicontinuity 20
– from boundedness 252, 445
– in the mapping theorems 317–320
essential singularity 361
– – and Casorati–Weierstrass 361
– – and Julia–Picard 438, 459
– – and normal families 437
euclidean algorithm 108
– metric 17
– norm 16
explosion point 40
exponential function 64, 67, 81
– – and Bernoulli functions 73 ff.
– –, differential characterization 69
– –, Euler's formula for 67
– –, homomorphic characterization 60, 69, 81
– –, in definition of index 84
– –, in treatment of double-connectivity 351–352
– –, inverse (see: logarithms)
– –, local inverse 65
exponential sums 365–366, 369, 381, 385, 408
extended plane (= Riemann sphere) 39, 357
extension of harmonic functions 403 ff., 410
– of holomorphic functions 149, 187, 341
– of homeomorphisms 38, 104
– of the Riemann map (see: Osgood–Taylor–Carathéodory theorem)
– – – – –, continuous but not univalent 339
– – – – –, differentiable on the boundary 339
extension theorem of Borsuk 38, 95
– – of Lebesgue and Tietze 37, 115
– – of Osgood–Taylor–Carathéodory 303 ff., 314, 315, 338–340
– – of Schönflies 9, 10, 104, 116, 314, 320, 341
extension theorems and components of the complement 97
– – and continuous logarithms 95, 96, 99

Faber polynomials 290, 340
factorization by Blaschke products 194, 461
– in class $\mathscr{L}$ 461
– of polynomials 107
– of positive operators 406
– of positive trig. polynomials 369
– of rational functions 367–370
– to remove zeros 167, 194
Fatou uniqueness theorem 189, 251
– radial limit theorem 251, 338
field 14
field axioms 14
finite cover 17
finite intersection property 17
finitely generated group 98, 112
– – ideal in $H(U)$ 394
first boundary value problem 136
first category 18
fixed-endpoint homotopy 230, 358
fixed point, attractive 242
– – and conformal automorphisms 244, 248, 346, 450–451

– – on the boundary 199, 248, 250
– – for a commuting family 249, 255
– – and convergence of iterates 242–244, 247–251, 450–451
fixed point theorem for continuous functions 108–109
– – – for entire functions 433
– – – for holomorphic functions 193, 215, 244, 249
Fourier expansion 365–366, 401
– –, heuristics in Poisson formula 401
Fourier theory 187, 340
– – in Gauss sums 408
– – in treating harmonic functions in annulus 410
– – proof of theta formula 408
Fresnel integrals 381, 408
frontier 17
Fubini theorem 10, 264, 273, 288, 415
– – for Riemann integrals 291
– – for series 59
function 13
fundamental sequence 17
Fundamental Theorem of Algebra 106, 117–118, 132
– – of Calculus 46
– – of Cauchy theory 344–345
– – on Double-Connectivity 352, 357
– – on Simple-Connectivity 344
– – on Symmetric Polynomials 118

gap theorems 8, 10
Gauss–Lucas theorem 107, 117
Gauss mean value theorem 185
Gauss sums 381, 408
greatest lower bound 19
Green's function 336
Green's–Stokes' formula 337
Green's theorem 185
group 14
– of exponentials 94, 98, 133, 349
– of zero-free functions 94, 98, 113

Hadamard's extension of Liouville 211
– radius of convergence formula 80
half-plane, conformal automorphisms of 199, 200
Hardy spaces 11
Hardy's theorem on circumferential means 142
– – – – – generalized 186
– sectorial limit theorem 448, 460 (see also 447)
harmonic 41
harmonic conjugate 337
– – in disks 139
– – and holomorphic logarithms 334
– – produced via property (*P*) 346
– – and Riemann mapping theorem 334–336
– – and simple-connectivity 344–345
harmonic functions in annulus 400 ff.
– – in half-plane 276 ff., 291–292
– –, higher derivations of 187, 341
– – and Laplace's equation 322–323, 342
– – and mean value 141, 185
– – and multiplication 324
– – and normal derivatives 336, 342
– – which are positive 143, 154, 160, 163, 165, 224, 278 ff., 292, 403, 405, 448
harmonic majorant property 139
harmonic majorization 151 ff., 188–189, 223, 285, 443–444
harmonic measure 189
Harnack inequalities 143, 211, 451
– convergence theorem 221
Hauptsätze of Cauchy Theory 344–345, 355
Heine–Borel theorem 18
Helly's selection theorem 292
hermiticity 11
Hilbert space 11, 338, 406
Hiong's version of Miranda's theorem 457
holes and connectivity 33
holomorphic function 41
– – depending on a parameter 150, 182–183, 228
– –, *n*th roots 93, 166, 346, 349–350
holomorphic logarithms 65, 111–113, 348
– – and analytic continuation 358
– – in annuli 348
– – in disks 65
– –, role in *Hauptsatz* 344–345
– – in starlike sets 146
holomorphic square roots 111, 113, 166, 346
– – – and existence of logarithms 93
– – – in the *Hauptsatz* 344–345
– – – in the Riemann map theorem 293, 298
homeomorphism 20

– and connectivity (see: positional invariance)
–, extension of 38–39, 104 (see also: Schönflies theorem and Osgood–Taylor–Carathéodory theorem)
homologically-connected 88, 113
– – implies simply-connected 268
– –, role in *Hauptsatz* 344–345
homologous in a set 88
– implies loophomotopic 90–91
homology basis 347–348
homomorphism of function rings 350–351, 356
– of ℝ into 𝕋 60, 69, 81
homotopic in a set 90
homotopically-connected 89
– –, role in *Hauptsatz* 344–345
homotopy 90
– with fixed endpoints 230, 358
– and logarithms 92–93, 95
– extension theorem 38
– form of Cauchy's theorem 230
– – – – – via Bernstein polynomials 253
Huber's ring theorem 201, 348
Hurwitz' convergence theorem 221, 301, 316
– covering theorem 253, 435, 458

ideal 14
– in $H(U)$ 394–395, 410
image 13
imaginary part 15
– unit 15
Implicit Function Theorem 181
increasing function 19
index of a point 84
– – – –, dependence on the curve 87
– – – –, dependence on the point 84
– – – –, constant under homotopy 90
– – – –, integral representation 85
– – – –, for Jordan-curves 87, 103
– – – –, and multiplicity of a zero 179, 310
– – – –, in a rectangle 85–87
– – – –, for simple loops 87, 103
induced convergence 188, 219 (see also: convergence theorems)
inequality of Borel on coefficients 213
– of Cauchy–Schwarz 16, 21, 296
– of Hadamard–Borel–Carathéodory 210, 214, 217, 222
– of Harnack 143, 211, 451
–, triangle 15
infimum 19
infinitely differentiable, non-analytic function 186
initial point 23
injective 13, 41
– entire functions 370, 407
– holomorphic functions (see: conformal)
inside of a Jordan-curve 104
– – – – –, role in *Hauptsatz* 344–345
integrability of partial derivatives 150, 182, 183, 228
integral along a curve 44
integral characterization of harmonicity 141
integral domain 14
integral function 41
integral representation (see: Basic Integral Representation, Cauchy Integral Formula, Cauchy–Schwarz formula, Poisson formula)
integration-by-parts 21, 69
integration in foundation of function theory 183
interior (point) 16, 290
Intermediate Value Theorem 21
interpolation 189, 265, 290, 291, 390–393, 409
– and approximation 265
– on a convergent sequence 169–171, 189
interval 15, 24
–, closed 14
–, open 14
–, half-open 15
–, connectedness of 23
invariance of connectivity (see: positional invariance)
– of radius ratio 349, 351, 356
Invariance of Dimension 118
– of Domain 9, 110, 118
inverse function 13
– –, continuity of 20, 110, 172
– –, differentiability of 173, 180
– –, integral formula for 180
– –, power series for 181
– –, for sin 176
inverse image 13
inversion technique 240–241, 314, 331–333, 351, 357
inverting order of integrations (see: Fubini's theorem)
isolated point 17

– – and connectedness 27
isolated singularity of holomorphic function 359
– – – – –, three types 361
– – on circle of convergence 375–377, 407
– – of harmonic function 403 ff., 410
iterate 242
iteration of exponentials 247, 255
– of rational functions 247
– of self-maps of disks 249–251
– of self-maps of regions 242 ff.
– of square roots 255

Jacobian 10
Jacobi–Jensen formula 416
Jahrbuch über die Fortschritte der Mathematik 51, 190, 255, 462, 463
Jahresbericht der Deutschen Mathematiker Vereinigung 254, 462
Jensen inequality 186, 413
Jordan-curve 24 (see also: Jordan Curve Theorem)
– – of positive area 116
– – representable by power series 315
Jordan Curve Theorem 9, 102–104, 116, 161–162, 341, 356
Jordan region 104–106, 303 ff.
– –, convergence of sequences of 315 ff., 337
– –, strong Cauchy theorem in 321
– –, and extendability of Riemann map 303 ff.
– –, polynomial approximation in 321
– –, solubility of Dirichlet problem in 332, 336–337
– –, topological characterization 340
Julia direction 438
– line 438–440
Julia's extension of Picard's theorem 438–439, 459

Koebe arc lemma 304, 307, 339
– $\frac{1}{4}$-(covering) theorem 232, 253
– map 293–296, 298, 299, 301, 303, 307
– radius 293–296, 299
– region 293–296, 298, 303
Kronecker's rationality criterion 373

Landau–Milloux theorem 158
Laplace's equation 41, 322
– – and continuity of partials 341, 342
– – and harmonicity 322, 323, 342
– – in physics 342
– – in polar coordinates 410
Laplacian 322
– and superharmonicity 322, 323
Laurent coefficients 361
– –, in terms of boundary data 401
– –, and nature of singularity 361
Laurent series expansion 359, 361, 406
– – – for composite function 363
– – – for Dirichlet problem in annulus 401
– – – for inverse function 365
law of exponents 64
least upper bound 18
– – – axiom 19
Lebesgue extension theorem 37, 115
Lebesgue integration 21, 185, 337, 338, 461
Lebesgue measure 51, 189, 251, 338
left-hand continuity 21
Leibniz formula 145
length (function) 16
length of a curve 44
– of a polygon 45
limit at a point 20
limit inferior 19
limit point 17
limit superior 19
linearization of conformal maps 199, 200
linear transformation 14
Liouville's theorem 71, 117, 230, 370
– –, generalized 211
– –, for subharmonic functions 187
Littlewood's theorem on subharmonic functions 140
– subordination inequality 186
local behavior of holomorphic functions 177 ff.
local maximum principle 347
locally uniformly 218
locally univalent 41
local univalence of holomorphic functions 56, 174, 180
– – of power series 56
logarithm function 64, 78
– principal branches 80
– in starlike sets 93
logarithms and components of the complement 96–100, 109, 111–112, 348
–, role in the *Hauptsatz* 344–345
logarithmic convexity of circumferential means 405
– – of maximum modulus 187

logarithmic means and the Blaschke condition 415, 461
– –, non-decreasing in r 414
– – and factorization 461
– – and uniform bounds 416
Loomis' sectorial limit theorem 448
loop 24
loophomotopic 90
loophomotopically-connected 90, 115
– –, and analytic continuation 358
– –, role in *Hauptsatz* 344–345
loophomotopy in characterization of double-connectivity 355
– and continuous logarithms 92, 93
– and index 90, 91
loopnullhomotopic 90
loops around holes 267, 347
lower limits 19
lower semicontinuity of zero-counting function 267
Lusin and Privalov uniqueness theorem 189

mapping theorem for double-connected regions (see: doubly-connected region)
– – for multiply-connected regions 357–358
– – for simply-connected regions (see: Riemann Mapping Theorem)
Marty's normality criterion 459
Mathematical Reviews 45, 51, 185, 216, 255, 462
Mathematics Magazine 11
Mathematische Annalen 118
Maximum Modulus Principle 128–129
– – – and Fundamental Theorem of Algebra 132
– – – for sums 168–169
Maximum Principles 127–128, 154, 184, 224
– –, local 347
mean values over $\mathbb{R}$ 276 ff., 292
Mean Value Theorem (of differential calculus) 21, 50, 70, 77, 167, 188
mean value theorem for harmonic functions (see: circumferential mean and areal mean)
– – – for complex valued functions 190
– – –, converse of 175, 311
meromorphic function 395
– –, argument principle for 396
– –, as quotient of holomorphic functions 395
– –, uniqueness theorem for 396
metric 15
– in $\mathbb{C}$, $\mathbb{R}^n$ 16
–, complete 17
– in $H(U)$ 218
–, locally uniform 218
–, supremum 21
–, translation invariant 218
metric space 15
Milloux's theorem 158–159, 444
minimum modulus, limiting behavior 285–286, 292
Miranda's theorem 426, 456–457
– –, relation to normality 436–437, 457
– –, sharper forms 457
Mittag-Leffler's theorem 10, 237, 390, 393, 409
– – –, relation to Weierstrass' 391–393
modular function 459
– –, role in Picard–Schottky theory 459–460
Monodromy theorem (continuous) 92, 100, 115
– – (holomorphic) 358
– – –, carelessness with 358
monogenic 41
monotonic 20
Montel's convergence theorems 153, 235
– compactness theorem 220, 293
– – – implies Vitali–Porter 252
Montel's sectorial limit theorems 441, 445, 460
– – – –, role in Lindelöf's 443
Morera's theorem 149, 188
– –, extensions of 188
m-to-1 map 197–198, 216, 343
multiplicity of a zero 167
multiply-connected region (see: connectivity, double-connectivity)
– – –, mapping theorems for 357–358

natural boundary 10
natural numbers 14
negatively oriented 104
neighborhood 16
Nevanlinna's extension of Vitali–Porter 252
Nevanlinna theory 456, 462
n-gon 24
non-decreasing function 19
non-increasing function 20
non-tangential convergence 144, 161, 189, 203–206, 255, 444–445

norm (function) 16
normal derivative 336, 342
normal family 436, 459
– –, boundedness properties 436
– –, and essential singularities 437
– –, relation to compactness 459
– –, role in iteration theory 450–451
– –, role in Julia–Picard theorem 439
– –, local implies global 436
– –, of meromorphic functions 459
– –, from omitted values 436, 458–459
– –, from univalence 440–441
nowhere dense 18
*n*th root, continuous 93
– –, holomorphic 166, 346
nullhomologous in a set 88
nullhomotopic in a set 90

omitted values implying universal bounds 426, 431, 456
– – implying constancy 432
– – implying normality 436
one-to-one 13, 41
– – – entire functions 370, 407
– – – holomorphic functions (see: conformal)
onto 13
open-and-closed 18
Open Map Theorem for continuous functions 110
– – – for holomorphic functions 172, 180
open set 16
order completeness 18
order of a point (see: index of a point)
– of a pole 361
– of a zero 167
order preserving 19
ordinate 15
orientation of Jordan-curve 104
oscillation 47
Osgood–Taylor–Carathéodory theorem 303 ff.
– – – – in Argument Principle 310
– – – –, role in Dirichlet Problem 336
– – – –, and characterization of Jordan regions 340
– – – –, other proofs 339
– – – – in Schönflies theorem 314
– – – – in sectorial limit theorem 315
Ostrowski's proof of Schottky 451 ff.
outside of Jordan-curve 104
overconvergence 10, 290

Painlevé's analytic continuation theorem 321–322
– rotation trick 131, 171, 184, 306
parallel slit region 357
Parseval's formula 71
Partial fraction decomposition 367
– summation 59, 377
periodic function (see: Fourier expansion)
permutation 14
Perron–Wiener–Kellogg method 325 ff., 342–343
Phragmén–Lindelöf technique 154, 224ff., 253
Picard's Great Theorem 459
– – –, elementary proofs 460
Picard's Little Theorem 82, 432
– – –, applications 433, 435
– – –, elementary proofs 460
– – –, extended form 440
– – –, and modular function 459
Poincaré's uniqueness theorem 303
pointwise limit of polynomials 264, 356
Poisson formula for annulus 398–399
– – from Cauchy formula 185
– – for disk 134
– – – –, other proofs 185
– – from Green's theorem 185, 337
– – involving Green's function 336
– – for half-plane 277 ff., 291–292
– – for strips and half-disks 410
Poisson–Stieltjes formula 217
polar singularity (see: pole)
pole of meromorphic function 361
–, characterization of 361–362
– on circle of convergence 375–377, 407
–, order of 361
– of rational function 260
–, determined by Taylor coefficients 375–377, 407
Pólya's *Rechtecksatz* 449–450, 461
– uniqueness theorem 196
Pólya–Saxer theorem 433, 458
polygon 24
polygonal connectedness of regions 29
polygonally connected 24
polynomial approximation (see: approximation)
polynomial Runge theorems 263, 289, 290 (see also: Runge approximation theorems)
positional invariance of connectivity 98, 115, 116
– – of simple-connectivity 98, 99, 114

– – of double-connectivity 355
positive harmonic function (see: harmonic function)
positive operator 406
positively oriented 104, 336
potential theory 136, 322, 337, 342
power series 53
– – for arcsin 176
– – on circle of convergence 81
– – expansion of holomorphic function 122, 136, 144
– – – – – – without integration 187
– – – – – – via Fourier theory 187
– – expansion of integrals with kernel 56
– – of inverse function 181
– – of logarithm 56–66
– –, multiplication of 81, 144
– – of tan, cot 77
pre-compact 18
prescribing principal parts 390, 393, 409
– zeros 237, 241, 254, 391, 393
prime end 339
Prime Number Theorem 82
primitive (functions) 120, 344
– – yield harmonic conjugates 346
– – yield holomorphic logarithms 146
– –, role in *Hauptsatz* 344–345
– – produced via integrals 46, 122, 149, 346, 391–392
– – in starlike regions 122
Principal Branch of the Logarithm 80
principal ideals in $H(U)$ 394–395
principal part 361, 390
Proceedings of the Saxon Academy 357
product, in $\mathbb{C}$ 14
–, dot, inner, scalar 16
– of disjoint cycles 14, 260
– of harmonic functions 324
– of holomorphic functions 42
– of power series 144
product representation of sin 409
property (X) (see the Symbol Index)

quadratic reciprocity 408
quasi-normal family 460
quotient group $\mathcal{G}/\mathcal{E}$ 94, 98, 113
– of holomorphic functions 395
– of polynomials 260

radial slit region 357
radius of convergence 54, 80
radius ratio (see: annulus)
Radó's uniqueness theorem 171
range 24
rank (of a group) 98
rational function 52, 260, 366 ff.
– –, characterized among meromorphic functions 370
– – of constant modulus on $\mathbb{R}$ 370
– – of constant modulus on $\mathbb{T}$ 367, 407
– – with integer Taylor coefficients 372
– –, positive on a circle 368, 406
rationality criterion of Kronecker 373
rational numbers 14
rational Runge theorems 263, 289, 290 (see also: Runge approximation theorems)
real numbers 14
real part 15
reciprocity relation for Gauss sums 408
Referativnyĭ Žurnal Matematika 462
reflection principle 281, 291, 356
region 28
regular function 41
relatively closed 17
relatively compact 18
relatively open 17
removable arc 321, 341
removable singularity 361
– – for harmonic functions 403, 405, 410
– – theorem 149, 187, 362
residual 18
residue 361
residue theorem 377–378
– –, application to integrals and sums 378 ff.
– – in proof of argument principle 397
restriction of a function 13
Riemann mapping theorem 52, 246, 251, 293, 298, 337 ff.
– – –, constructive proofs 293, 298 ff., 337–339
– – –, de Possel's proof 357
– – – via Dirichlet problem 322, 334–336, 339
– – –, extension to boundary 303 ff.
– – –, explicit formulas 338
– – –, Fejér–Riesz proof 293, 303, 357
– – –, gap in original proof 293, 337
– – – via Hilbert space methods 338
– – – for multiply-connected regions 357–358
– – – for polygonal regions 338

– – –, survey of proof methods 338–339
– – – and variable regions 315 ff., 337
Riemann sphere (= extended plane) 39, 357, 459
– uniqueness theorem 165
– sums 183, 228, 256, 262, 406
– Zeta function 82
Riemann–Stieltjes integral 21, 292
right-hand continuity 21
ring (algebraic) 14
ring (geometric) (see: annulus)
ring of holomorphic functions, ideals in 394–395, 410
– – – –, relation to conformal structure 349–351, 356
rotation trick 131, 171, 184, 306
Roth's sectorial limit theorem 448–449, 461
Rouché's theorem from argument principle 190, 340, 397
– – from Basic Integral Representation 265, 291
– – for disks 179, 190
– – and Fundamental Theorem of Algebra 118
– –, general form 265
– –, generalization and other proofs 190, 291, 340
row rank 14
Runge approximation theorems 263, 273
– – –, in proof of Cauchy's theorem 264, 290
– – – with interpolation 265
– – – in proof of Mittag-Leffler's theorem 391
– – – in proof of Walsh's theorem 321

$\mathcal{S}$ 230, 270
–, local boundedness 231, 234
–, compactness 231
–, covering properties 232
schlicht 41
Schmiegungsverfahren 10, 337
Schönflies converse of Jordan Curve Theorem 116
– extension theorem 9, 10, 104, 116, 314, 320, 341
Schottky's theorem 10, 411, 417, 431, 456–459
– – via Bloch's 460
– –, constants in 456, 457–458
– –, dependence on 1–*r* 457–458
– – via modular function 459
– –, Ostrowski's proof 451 ff.
– –, other proofs 457, 459, 460
Schwarz–Christoffel formulas 338
Schwarz' Lemma 191, 215
– –, generalizations 192–193
– –, geometric version 208
– –, half-plane version 198
– –, strip version 200
sectorial limits on the boundary 132, 315, 442, 460
– –, internally 144, 161–163, 441–442, 445, 448, 460
sectorial limit theorems for $|f|$ 445–447, 448–449
– – – for harmonic functions 448
– – – via harmonic majorization 443–444
– – – of Hall 163
– – – of Lindelöf 132–133, 161–162, 442–444, 460
– – – of Loomis 448
– – – of Montel 441–442, 445, 460
– – – of Roth 448–449, 461
– – – of Tsuji 144
semigroup 14, 269
separation of points by a set 100
sequence 17
simple-connectivity 33, 40
– – and analytic continuation 358
– – characterized by boundary properties 345
– – characterized by conformal automorphisms 246, 346
– – internal, external characterizations 345
– – topological, analytic characterizations 345, 358
– – and components 99, 114
– – of convergence set 253
– –, equivalent properties for regions 344
– –, equivalent properties for open sets 113, 268, 345
– – and logarithms 99
– –, positional invariance 99, 114
simple loop 24, 87
simple pole (= pole of order 1) 361, 376–377, 397–398
simple zero (= zero of order 1) 165
simply-connected 33 (see: simple connectivity)
singularity (see: isolated singularity)
slit-plane 79

slit region 357
smooth curve 24
smoothness in Dirichlet problem 336–337
snuggling up 10, 337, 393
spectral theorem 406
spectrum of a function 350
square roots in connected sets 23
– –, holomorphic (see: holomorphic square root)
– – of odd functions 93
starlike (set) 25
starlike function 10, 208
stereographic projection 39
streckentreue 50
strictly increasing function 19
strip limit theorems 445 ff., 460
– – – for $|f|$ 445–447, 460
– – – for harmonic functions 448–449, 460
Study's theorem 208
subcover 17
subharmonic 127 (see: subharmonicity)
subharmonic Liouville theorem 187
subharmonicity, conformal invariance 140
– of $\log^+|f|$ and $\log\circ f$ 142, 184
– of harmonic functions 139
– and existence of harmonic majorants 223
–, equivalence to (HM) 139
– and Jensen inequality 186
– and the Laplacian 322–323
–, weaker sufficient conditions 140
subordination, principle of 208 ff.
– inequality of Littlewood 186
subsequence 17
subspace 18
successive approximation 174, 269
successor function 14
summability methods 289
superharmonic 127
superharmonic extension 141, 333
superharmonicity in solving Dirichlet problem 325 ff., 333, 342–343
supremum 19
supremum norm 21
surjective 13
surjectivity from injectivity 407

Taylor series 186 (see: power series)
term-by-term differentiation 55
terminal point 23
théorème de d'Alembert 118
theta formula 386, 408
Three Circles Theorem 147, 187, 405
Three Lines Theorem 147, 187 (see also 447)
Tietze extension theorem 40
torsion-free 94, 112
totally disconnected 40
transcendence of e, π 82
triangle inequality 15
triangulation 120–121, 183
trigonometric functions 70
– polynomials 366, 406, 410
– –, positive 369, 406
– series 365–366, 386, 408
Tsuji's sectorial limit theorem 144
two-constants theorem 152, 189

uniform continuity 20
– – of Riemann map 338
uniform limit of harmonic functions 151 (see: convergence)
– – of holomorphic functions 151 (see: convergence)
uniform metric 21
uniqueness theorem of Blaschke 195
– – – –, generalized 415
– –, constructive proof 169–170
– –, relation to convergence theorem 252
– – of Denjoy 252
– – of Fatou 189
– – of Flett 339
– – of Kabaila 461
– – of Koebe 304, 339
– – of Lusin and Privalov 189
– – for meromorphic functions 396
– – of Poincaré 303
– – of Pólya 196
– – of Radó 171
– – for rational functions 370
– – of Riemann and Weierstrass 165
unitary operator 406
unit function 268–269
univalence and local boundedness 236, 441
– and induced convergence 236, 441
– and non-vanishing of f' 56, 174, 180, 190
– and fixed points 193, 215, 244
– of limit functions 221, 301–303
univalence theorem of Darboux and Picard 310–311

univalent 41
– entire functions 370, 407
– holomorphic functions (see: conformal)
upper limit 19

variable regions 337
vector space 14, 16
Verzerrungssatz of Koebe 235
Vorbereitungssatz 190

Weierstrass' convergence theorem 151, 188 (see also: convergence theorems)
– function 186
– polynomial approximation theorem 269, 272–273, 290–291
– – – –, role in Dirichlet problem 327, 333
– *Vorbereitungssatz* 190
– theorem on zero-sets 237, 254
– – – – –, proved from Mittag-Leffler 391
– – – – –, role in characterizing meromorphic functions 396
– – – – –, role in ideal theory of $H(U)$ 394, 395
– – – – –, role in producing nth roots 346

winding number 84 (see: index of a point)
winkeltreue 50

Zentralblatt für Mathematik und ihre Grenzgebiete 51, 202, 291, 341, 408, 462
zero-counting integral 179, 396
– – –, lower semi-continuity of 267
– – via argument principle 311 ff.
zeros of f' and univalence of f 56, 174, 180, 190
zeros of holomorphic functions and Blaschke products 241, 461
– – – –, arbitrarily prescribed 237, 241, 254, 391
– – – –, bounds on size and number 194
– – – –, common to an ideal 393–394
– – – –, convergence properties 195, 461
– – – –, isolation 165, 396
– – – –, order 165
zeros of polynomials (see: Fundamental Theorem of Algebra and Gauss–Lucas Theorem)
– – – via Argument Principle 311–314
Zeta function 82, 408

Symbol Index

$\in, \forall, \exists, \cup, \cap, \subset, \supset, \setminus, \varnothing$ 13
$\cup \mathscr{S}, \cap \mathscr{S}$ 14
$\mathbb{N}, \mathbb{Z}, \mathbb{Z}^+, \mathbb{Q}, \mathbb{Q}^+, \mathbb{R}, \mathbb{R}^+, \mathbb{C}$ 14
$\mathbb{C}^n, \mathbb{R}^n$ 16
$\mathbb{T}$ 16
$\overline{\mathbb{R}}$ 19
$\mathbb{C}_\infty$ 39, 357, 459
i 15
e 64
π 60, 62, 68, 82
$-\infty, \infty, +\infty$ 15, 19
$a < b, a \leq b, a > b, a \geq b$ 14
$(a, b), (a, b], [a, b), [a, b]$ 14, 15
$[a, \infty), (a, \infty), (-\infty, b], (-\infty, b)$ 15
$(a, b)(c, d)$ 14
$(a_0, a_1, \ldots, a_n), (a_0, a_1, \ldots, a_n],$
$[a_0, a_1, \ldots, a_n), [a_0, a_1, \ldots, a_n]$ 24
$0 \cdot \gamma, (-1) \cdot \gamma, n \cdot \gamma$ 24, 91, 268, 355
$\| \ \|_\infty, \| \ \|_{C(X)}, d_\infty(\ , \ )$ 21, 218
$\bar{z}, |z|, \|z\|$ 15, 16
$\| \ \|$ 349
$\| \ \|_K, d_{\mathscr{K}}(\ , \ )$ 218
$f: S \to T, S \xrightarrow{f} T$ 13
$f(A), f^{-1}(A)$ 13
$f|A, f|_A$ 13
f^{-1} 13
$f \circ g$ 13
$f^{[n]}$ 242
F_w 260
f_a 43, 192, 294
ω_f 47
∇^2 322
Ind_γ 84
sin, cos, tan, cot 70
exp, log, e^z 64, 82
E 60–64, 351–354
$\mathscr{S}$ 230, 370
$\mathscr{G}, \mathscr{E}, \mathscr{G}(U), \mathscr{E}(U), \mathscr{G}(K), \mathscr{E}(K)$ 94, 98, 113, 349
$\mathscr{G}_a$ 246, 346
$H(U), H^\infty(U)$ 41
$C(X), C_{\mathbb{R}}(X)$ 21
$\mathscr{R}$ 108, 373
$\mathscr{U}$ 349
$\mathscr{L}$ 461
$\mathscr{I}(J), \mathscr{O}(J)$ 104
$\text{Res}(f, a), n(a, f)$ 361, 167
$B(x, r), D(z, r), C(z, r), A(c, r, R),$
$A(r, R)$ 16
$M(r, f), L(r, f), V(r, f)$ 412
$A_n, B_n, C_n, K_n, M_n, M_0'$ 417, 421, 422, 424, 426, 432
$\bar{S}, \dot{S}$ 16
$\partial S, \partial \bar{S}, bdry(S)$ 17, 24
l.u.b.(S), sup(S), g.l.b.(S), inf(S),
diam(S) 19
$d(x, S)$ 19
co S 25
lim sup, $\overline{\lim}$, lim inf, $\underline{\lim}$ 19, 20
$x \uparrow a, x \to a^-, x \downarrow a, x \to a^+,$
$\lim_{x \to a, x \in S}$ 21
$(CL), (HL), (CS), (HS), (HS_1)$ 113
(HM) 139
(CM) 141
$(C), (P), (AP)$ 344
(H) 333
(C.R.) 49
FM, MR, Zbl, BAMS 462
$l(\gamma), \int_\gamma f, \int_\gamma f(\xi) d\xi$ 44
$\int_\Gamma d|\xi|$ 336
$\sigma(f)$ 350
$D_k, d/dx$ 21
$'$ 21, 41
D_N 336

Series Summed

$\sum_{k=1}^{\infty} k^{-2n}$ 76, 82

$\sum_{k=1}^{\infty} (2k-1)^{-2n}$ 76, 388

$\sum_{k=1}^{\infty} (-1)^n k^{-2n}$ 76

$\sum_{k=-\infty}^{\infty} (k^2+1)^{-1}$ 390

$\sum_{n=0}^{\infty} (-1)^n (2n+1)^{-3}$ 386

$\sum_{k=1}^{\infty} \frac{1\cdot 3\cdots(2k-1)}{2\cdot 4\cdots(2k)}\cdot\frac{1}{2k+1}$ 176

$\sum_{n=-\infty}^{\infty} (z-n)^{-2}$ 389

$\sum_{k=1}^{\infty} 2z/(z^2-k^2)$ 389

$\sum_{k=1}^{\infty} (-1)^k 2z/(z^2-k^2)$ 390

$\sum_{k=1}^{N} k^n$ 76, 82

$\sum_{k=0}^{n-1} e^{2\pi i k^2/n}$ 381, 408

tan, cot 77, 389

$1/\sin z$ 390

$1/\sin^2 z$ 389

arcsin 176

theta formula 386

Integrals Evaluated

$\int_{-\infty}^{\infty} \frac{1}{1+x^2}\,dx$ 68

$\int_{-\infty}^{\infty} \frac{\sin x}{x}\,dx$ 123

$\int_{-\infty}^{\infty} \left(\frac{\sin x}{x}\right)^2 dx$ 123

$\int_{-\infty}^{\infty} \left(\frac{\sin x}{x}\right)^3 dx$ 386

$\int_{-\infty}^{\infty} \left(\frac{\sin x}{x}\right)^4 dx$ 124

$\int_{-\infty}^{\infty} \frac{\sin^4 x}{x^2}\,dx$ 124

$\int_{-\infty}^{\infty} \frac{\sin(x^2)}{x}\,dx$ 125

$\int_{-\infty}^{\infty} e^{-x^2} dx$ 381, 386

$\int_{-\infty}^{\infty} \sin(x^2)dx, \int_{-\infty}^{\infty} \cos(x^2)dx$ 381, 408

$\int_{0}^{\infty} e^{-x^2} \sin(x^2)dx, \int_{0}^{\infty} e^{-x^2} \cos(x^2)dx$ 386

Errata and Addenda to

R. B. Burckel:

An Introduction to Classical Complex Analysis I

p. 101 lines –4 and –1.	"+" to "–"
p. 133 line –4.	Change closed interval to open interval.
p. 161 line –12.	"(iii)" to "(ii)"
p. 163 lines –4 and –3.	"γ" to "($\gamma \cap \bar{S}_r$)"
p. 183 line 5.	Replace this remark with the following: It is easy to show, using 5.33, that continuity of F and existence of D_2F already imply the continuity of the latter needed in 5.93.
p. 186 line 17.	5.27 was earlier noted by LINDELÖF [1909], p. 10.
p. 187 line 8ff.	Fourier methods of deducing Cauchy's theorem were used much earlier. See, for example, A. HARNACK *Math. Annalen* 21 (1883), 305–326. **FM** 15, p. 185.
p. 212 line 13.	Theorem 6.34 is due to K. LÖWNER [1917].
p. 249 line 23.	WOLFF [1926a] is a very elegant proof of the theorem in full generality.
p. 255 line 4.	A nice treatment may also be found in J. J. ANDREWS and R. C. LACHER *Aequationes Math.* 16 (1977), 137–147. **MR** 57 #6331.
p. 264 line 19.	The proof based on Fubini occurs also in the 3rd edition (1978) of the book AHLFORS [1966a].
p. 265 line 10.	This result occurs already in WALSH [1928].
p. 278 line 18.	A direct proof of (ix), independent of (vii) and (viii), follows from 7.15.
p. 281 line –1.	For an alternative elegant proof of 8.33 see E. PHRAGMÉN *Acta Math.* 14 (1890–91), 225–232. **FM** 23, p. 890.
p. 289 line 18.	An important precursor of Runge was P. APPELL *Acta Math.* 1 (1882–83), 145–152 and *Math. Annalen* 21 (1883), 118–124. **FM** 15, pp. 323, 324.
p. 289 line –14.	Add: See also JULIA [1927a].
p. 292 line 1.	Delete "149–151"
p. 292 line 19.	The integral representation is due independently to F. RIESZ *Ann. Sci. Ecole Norm. Sup.* (3) 28 (1911), 33–62. **FM** 42, p. 374.
p. 339 line 5.	Add MAITLAND [1939].
p. 340 line –13.	"simply-connected" to "bounded, simply-connected"
p. 341 line –5.	Add BESICOVITCH [1930].
p. 361 line –6.	For detailed history of this theorem see E. NEUENSCHWANDER *Historia Math.* 5 (1978), 139–166. **Zbl** 374 #01010.
p. 408 line 17ff.	Different treatments of the Fresnel integrals occur in FEYEL and PRADELLE [1973], problem [6.25] and in M. GODEFROY *Nouvelles Annales de Math.* (3) 17 (1898), 205–206. **FM** 29, p. 245.
p. 449 line –12.	Delete period.
p. 457 line –15.	The $(1 - r)^{-1}$ dependence was shown earlier by H. BOHR and E. LANDAU *Göttingen Nachr.* (1910), 303–330. **FM** 41, p. 290.
p. 466 line –1.	Delete "and 1353."
p. 495 line 28.	"245–255" to "246–266"
p. 511 line 8.	After "and" add: *C. R. des Séances Soc. Math. France*
p. 519 line 14.	"(1899)" to "(1900)"
p. 519 line –3.	Before "(1923)" insert volume numbers "176"
p. 522 line –7.	Delete "(Russian)."
p. 526 line 6.	After "(1931)" add page numbers "18–38."
p. 569 line 3.	"$(-1)^n$" to "$(-1)^k$"

Errata and Addenda to

p. 509 line 3 … "(−1)" to "(−1)"

MIX
Papier aus verantwortungsvollen Quellen
Paper from responsible sources
FSC® C105338

FSC
www.fsc.org

If you have any concerns about our products,
you can contact us on
ProductSafety@springernature.com

In case Publisher is established outside the EU,
the EU authorized representative is:
Springer Nature Customer Service Center GmbH
Europaplatz 3, 69115 Heidelberg, Germany

Printed by Libri Plureos GmbH
in Hamburg, Germany